Computational Formalisms in Euclidean Geometry

Volume I: A Trigonometric, Vectorial and Complex Numbers Approach

Other World Scientific Titles by the Author

An Informal Introduction to Stochastic Calculus with Applications
ISBN: 978-981-4678-93-3
ISBN: 978-981-4689-91-5 (pbk)

An Informal Introduction to Stochastic Calculus with Applications
Second Edition
ISBN: 978-981-12-4709-5
ISBN: 978-981-12-4756-9 (pbk)

Deterministic and Stochastic Topics in Computational Finance
ISBN: 978-981-320-307-5
ISBN: 978-981-320-308-2 (pbk)

Stochastic Geometric Analysis with Applications
ISBN: 978-981-12-8327-7

Computational Formalisms in Euclidean Geometry

Volume I: A Trigonometric, Vectorial and Complex Numbers Approach

Ovidiu Calin

Eastern Michigan University, USA

World Scientific

NEW JERSEY • LONDON • SINGAPORE • BEIJING • SHANGHAI • TAIPEI • CHENNAI

Published by

World Scientific Publishing Co. Pte. Ltd.
5 Toh Tuck Link, Singapore 596224
USA office: 27 Warren Street, Suite 401-402, Hackensack, NJ 07601
UK office: 57 Shelton Street, Covent Garden, London WC2H 9HE

British Library Cataloguing-in-Publication Data
A catalogue record for this book is available from the British Library.

COMPUTATIONAL FORMALISMS IN EUCLIDEAN GEOMETRY
Volume I: A Trigonometric, Vectorial and Complex Numbers Approach

ISBN 978-981-98-0616-4 (hardcover)
ISBN 978-981-98-0713-0 (paperback)
ISBN 978-981-98-0617-1 (ebook for institutions)
ISBN 978-981-98-0618-8 (ebook for individuals)

For any available supplementary material, please visit
https://www.worldscientific.com/worldscibooks/10.1142/14127#t=suppl

Preface

This book is based on lectures given by the author at Eastern Michigan University over the past 23 years. Originally intended as a textbook for the College Geometry course, its initial content was designed to align with the standards of various national mathematical organizations at the university level, such as the Mathematical Association of America (MAA), the Committee on the Undergraduate Program in Mathematics (CUPM), the Conference Board of the Mathematical Sciences (CBMS), and the Mathematics Education Trust (MET). As additional material was introduced, the textbook became more comprehensive, while still maintaining the author's primary goal of cultivating mathematical thinking and communication skills among students by helping them develop the ability to produce written proofs.

The present version of this book gradually evolved over the years from several early stages. At each stage, the author employed a different methodology for approaching geometry, all of which are now included in the book as distinct component parts. These parts work together to provide an understanding of geometry from various perspectives, such as trigonometric, complex numbers, and vectorial approaches. Each approach requires different basic axioms and techniques, and in the end, they all solve the same geometry problems, each with its own advantages and disadvantages.

Book overview

This volume is organized into three main parts, each centered on a specific geometric approach. The first section focuses on trigonometric computations, particularly in the context of triangle and quadrilateral geometry. The second section explores geometry through the use of complex numbers. The third section employs vectors to demonstrate geometric concepts. Additionally, a final section offers solutions and hints to the problems presented.

Part I

This part presents an approach to Euclidean geometry from a metric perspective. It is based on elementary applications of trigonometric calculus to geometry. The goal is to prove geometric properties by calculating distances between points and angles. In addition to being a robust method, it is also the oldest approach in geometry.

Chapter 1 introduces an axiomatic system based on several groups of axioms (such as incidence, metric, separation, congruence, and parallel postulate). It then presents some elementary properties of angles and circles. At the end of the chapter, some well-known algebraic inequalities that are often used in geometry are mentioned.

Chapter 2 is dedicated to the geometry of the triangle. It contains elementary metric properties of triangles (Pythagorean relation, law of sines, law of cosines) as well as more advanced metric tools, such as Stewart's relation, Steiner's theorem and the power of a point with respect to a circle. It also provides a group of theorems concerning the intersection of lines and collinearity of points in a triangle, such as Menelaus' theorem, Ceva's theorem, and Oppel's relation.

An important space is dedicated to some remarkable centers in a triangle, such as circumcenter, incenter, centroid, orthocenter, and symmedian point. The aforementioned metric tools are used to compute some of the distances between these remarkable points. The chapter contains several formulas for the area of a triangle, some of them bearing the names of their discoverers, such as Heron's formula, or Tziteica's formula.

The concept of area has proved itself to be a very good metric relator among different elements of a triangle and can be used successfully as a metric tool. Furthermore, several metric properties can be proved using the concept of area, such as Menelaus' or Ceva's theorems.

A distinguished section deals with some relations between geometry and some branches of physics, such as mechanics, optics, or electricity. Some geometric minimization properties can be interpreted as the Fermat's minimum principle from Physics. In relation to this, we have also discussed the metric properties of the orthic triangle and its minima properties.

The quadrilateral geometry is emphasized in **Chapter 3**. A distinguished topic presented here deals with cyclic quadrilaterals. Their properties were captured in classical theorems such as Ptolemy's and Pappus' and contains properties such as Simson's line, Miquel's point, or Aubert's line. The properties of the centroid and anticenter of a cyclic quadrilateral are also included.

Other topics of interest included in this chapter are the cocyclic quadrilaterals and orthodiagonal quadrilaterals. The chapter includes important metric relations in a quadrilateral such as Archimedes' area formula, or Bretschneider's relation, and Carlitz' identity for bicentric quadrilaterals.

Part II

This part models Euclidean geometry from the perspective of complex numbers. Each geometric property (such as collinearity, concurrence, concyclicity, segment quotients, congruence) can be written in the language of complex numbers. Each figure is represented on an Argand diagram, which is similar to a Cartesian system of coordinates. This way, to prove geometric properties reduces to some computation with complex numbers. Complex numbers are particulary useful when dealing with centroids and rotations. The complex numbers formalism, including results such as Euler's exponential representation, and De' Moivre's formula, provides a powerful tool of computation in Euclidean geometry. These results are contained in **Chapter 4**.

Part III

This section focuses on a vector-based approach to Euclidean geometry. The primary tools used are linear algebra techniques, which allow students to explore a wide range of applications in both plane and solid geometry. Several classical theorems, as well as some new results, are derived using vector calculus methods.

Chapter 5 contains an introduction to basic set theory notions, which leads to the vector concept by means of equivalence relations. It introduces operations with vectors (triangle and parallelogram rules), and the linear independence of vectors.

Chapter 6 presents some elementary applications of vectors to the proof of Thales' theorem, as well as to describing medians and bimedians.

Chapter 7 deals with applications of the inner product to geometry. This is used to tackle problems regarding segment lengths and perpendicularity of lines. A distinguished result of this chapter is Leibnitz' formula and its applications to computing distances between remarkable centers of a triangle.

Chapter 8 introduces the cross product and its properties. Its mainly geometric applications regard problems involving areas and collinearity.

Chapter 9 presents the mixed product of three vectors and its algebraic properties, including Gibbs' formula.

Chapter 10 contains applications of the concepts introduced in previous chapters to classical geometry problems, such as Napoleon's problem, Menelaus' theorem, Euler's line, Neuberg's theorem, Desargues' theorem, Stewart's relation and Euler's circle.

Chapter 11 defines and uses baricentric coordinates, and their vectorial properties in the triangle geometry.

Chapter 12 defines momenta of inertia with respect to a point, and to an axis. It solves the problem of finding the axis with the smallest momenta for a triangle.

Chapter 13 is devoted to solid geometry, and mostly deals with the properties of tetrahedrons.

Target Audience

The students who have taken this course over the years were primarily mathematics majors, especially those studying secondary education. Other students, such as computer science or physics majors needing a minor in mathematics, also enrolled in the class. Although originally conceived as an undergraduate textbook for junior students at an American university, it can also benefit gifted high school students and their mathematics teachers. This book is also useful for math educators who are interested in diverse proofs of geometric problems and need inspiration for their classwork.

Acknowledgements

The work on this book began in 2002 when I was assigned to teach the College Geometry course at Eastern Michigan University as a new assistant professor. Several early versions of the chapters were used as teaching materials for many years. I would like to thank Walter Parry, who spent long hours proofreading the first manuscript in 2004. Heartfelt thanks go to Joanne Caniglia for assisting me in understanding the principles and standards for school mathematics in the U.S. I am also indebted to the Department of Mathematics and Statistics at Eastern Michigan University for providing excellent working conditions. Many thanks go to my many generations of students for their invaluable assistance during the book's preparation.

The final stages of the book were prepared while the author was supported by a Fulbright scholarship at the University of Bucharest.

Finally, I would like to express my gratitude to World Scientific and its editors for making this endeavor a reality.

The Author

Ypsilanti, Michigan, 2024

Interdependence chart of the chapters

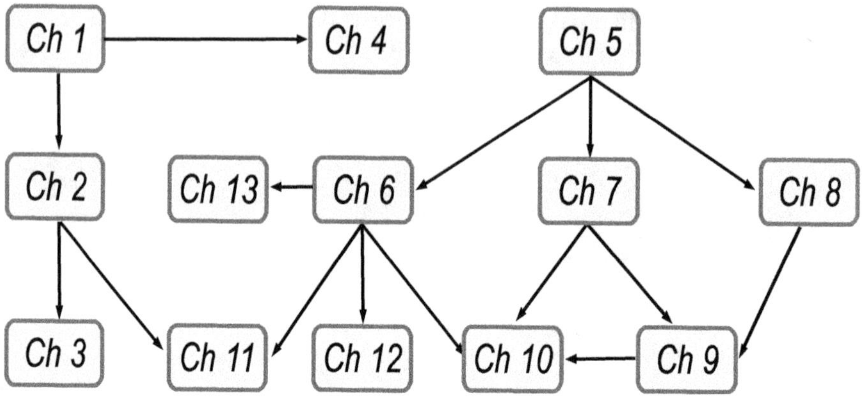

This chart illustrates the relationship between each chapter and its preceding chapters.

List of Notations

A, B, C	points
ℓ, m	lines
$\mathcal{P}, \alpha, \beta$	planes
$\widehat{A}$	angle with vertex at A
$\widehat{ABC}$	angle with center at B
$\lvert AB \rvert$	length of segment AB
AB	line containing A and B
ΔABC	triangle ABC
$ABCD$	quadrilateral
$\mathcal{C}(O, R)$	circle of center O and radius R
$\Vert$	parallel
$\equiv$	congruent
$\perp$	perpendicular
$\cup$	union
$\cap$	intersection
$\{\dots\}$	set
$\Longleftrightarrow$	equivalent
$\longrightarrow$	logic implication
I	incenter
I_a, I_b, I_c	centers of excircles
O	circumcenter
H	orthocenter
G	centroid

a, b, c	sides lengths of a triangle
p	semiperimeter
$\mathcal{A}_{ABC}$	area of triangle ABC
r	inradius
r_a, r_b, r_c	excircles radii
R	circumradius
r_0	orthic triangle inradius
R_0	orthic triangle circumradius
a_0, b_0, c_0	sides lengths of the orthic triangle
$d(A, B)$	Euclidean distance between A and B
z_A	the affix of the point A
$(A, B; C)$	the quotient of three points
$(A, B; C, D)$	the biquotient of four points
$\omega_0, \ldots, \omega_{n-1}$	roots of unity of order n
$\overline{AB}$	the oriented segment AB
$\vec{u}, \vec{v}, \vec{w}$	vectors
$\overrightarrow{AB}$	the vector AB

Contents

Introduction

Many individuals develop an aversion to mathematics, often from a young age. This aversion may stem from the fact that mathematics is not inherently intuitive for most people. One possible explanation lies in the evolutionary development of the brain's structure. The frontal lobe, responsible for logical thinking, is one of the most recently evolved parts of the brain. In contrast, the occipital lobe, housing the visual cortex, was among the first to develop. As a result, visual concepts are typically more easily grasped by the brain than logical or abstract concepts.

In contrast to other branches of mathematics, which often rely on abstract analytical reasoning, Euclidean geometry stands out for its reliance on visualization. This distinctive feature sets it apart, as it engages the brain in a unique manner. Thus, geometry, as a discipline centered on the logical analysis of shapes, establishes a bridge between the frontal cortex and the occipital lobe. This may explain why geometry emerged as one of the earliest branches of mathematics, predating other branches of science.

Engaging in geometric thinking fosters unique neural connections between the visual cortex and the frontal lobe, facilitating robust and abstract cognitive processes. According to the Hebbian theory, *neurons that fire together wire together*. Thus, practicing geometry forms connections between the frontal and occipital lobes, facilitating deeper understanding and cognitive development.

Unlike other analytically abstract branches of mathematics, Euclidean geometry is heavily based on visualization, which makes it unique in the way that it operates in the brain. In classical geometry, we depict a figure and utilize the capabilities of the visual cortex to deduce certain properties of the figure. For example, through drawing various distinct triangles, and their respective medians, we can deduce, through visual observation, that the medians of a triangle consistently intersect. Subsequently, we assert the yet unproven statement that medians in any triangle always intersect.

The visual cortex formulates conjectures, but the scientific proof of these statements follows a distinct process that requires logical reasoning from the frontal cortex. This validation process entails a logical deductive construction, beginning with foundational truths (axioms) and undefined concepts (point, line, plane, number), and culminating in the statement initially formulated intuitively by the visual cortex. The deductive process employs various techniques consistent with our logical principles inherent in the frontal lobe, including induction, modus ponens, modus tollens, and transitivity.

We observe that this method of validation is founded upon fundamental concepts that can be attributed to the visual cortex. For instance, notions like point, line, and plane find their significance through the activation of specific neurons. Within this neural framework, certain neurons specialize in perceiving spatial positions, while others are attuned to directional cues. Moreover, distinct groups of neurons exhibit heightened activity in response to particular geometric shapes; one group may activate in the presence of circles, while another fires in response to triangles. In short, there is a lot going on in the cortex trying to figure out the geometric properties of shapes. Recently, computer scientists have tried to simulate this process artificially as follows.

Geometry and AI

Given the increasing role that Artificial Intelligence (AI) plays today, we will discuss a parallel between the functions of the visual cortex and those of an AI system. In the subsequent stage, the inference provided by the AI (which is an analog of the visual cortex) necessitates validation through a logical framework congruent with the frontal lobe. We shall illustrate this with a classic example: in the 1700s, Johannes Kepler, upon examining centuries' worth of astronomical data, deduced three laws governing planetary motion, which now bear his name. This process is analogous to the AI stage wherein it proposes an inference or a pattern.

We observe that in every instance, we encounter the following two components:

(*i*) Data, from which an AI system derives inferences.

(*ii*) A logical framework employed to validate these inferences, originating from a defined set of axioms.

In certain domains like Euclidean geometry or Newtonian mechanics, these axioms align intuitively with our sensory perception and therefore, seem innate. However, in more abstract fields such as quantum mechanics or the general theory of relativity, certain axioms, like Schrödinger's equation, Minkowski's space-time manifold, or Einstein's field equations, are less congenial to our sensory cortex, which builds its connections from the daily routine experience.

Therefore, an important question arises: *Given certain results inferred from data, can one reconstruct a system of axioms capable of validating these findings?* These axioms should be limited in number, such that they are independent of each other. This issue, which has long been resolved in the case of Euclidean geometry, remains pertinent across various new domains of study.

We observe that the system of axioms must inherently possess sufficient information to facilitate the development or validation of inferences. The logical manipulations within the axiomatic system serve as processors of information, yet they cannot augment the existing information unless new hypotheses are introduced. Combining these systems yields an axiomatic foundation for formulating a novel theory meant to explain the data inferences.

The same procedure should be performed by an AI system. How this system of axioms would look like is a good question, since it might not be compatible with our visual cortex.

Several undefined concepts (point, line, plane, number, etc.), and numerous axioms (incidence, separation, etc.), serve as the foundation for constructing an entire axiomatic system for geometry. It is these undefined concepts and initial axioms that inherently contain the aforementioned information. By combining these elements, a multitude of propositions and results can be derived. These derivations can be proved through several formalisms.

Computational formalisms

Computation within a formalism refers to the process of systematically deriving conclusions or performing operations according to the rules and structures defined within a formal system. In this context, a formalism typically consists of a set of symbols, rules, and axioms that govern how these symbols can be manipulated and combined. Computation involves applying these rules and operations to generate new information or results within the framework of the formal system. This can include tasks such as theorem proving, problem-solving, or performing calculations within the constraints of the defined formalism.

In this book, we consider several computational formalisms. Part I uses a trigonometric-based computational formalism. This enables us to compute almost any distance between two points or angle measures. It can also allow for the computation of the area of any geometric shape in terms of the given information. This formalism assumes, as a prerequisite, a course in trigonometry, even if we have presented the bare minimum of trigonometry needed to follow the chapters of the book.

Part II is dedicated to a computational formalism using complex numbers. These numbers can be used to perform similar operations as those performed on real numbers (addition, subtraction, product, power, quotient and inversion). It is based on the association of each point in the plane with a complex number. This formalism transposes geometric properties of figures (collinearity, concurrency, etc.) into relations involving complex numbers, which can be solved by a symbolic manipulation of complex numbers.

Part III deals with vector formalism as applied to plane and solid geometry. This formalism harnesses the power of linear algebra, involving vectors and their operations, to solve geometric problems. Several geometric properties naturally lend themselves to description using vectors, including the perpendicularity of lines, the collinearity of points, and the concepts of distance, area, and volume.

There is no such thing as a perfect formalism. Each comes with its own advantages and disadvantages. For instance, the complex numbers formalism is useful for dealing with problems involving rotations in the plane but cannot be applied to solid geometry. The vector computation formalism is useful both in the plane and in space for computing volumes, areas, and distances, as well as for proving the perpendicularity of lines. While it is extremely useful for problems involving centroids, it can be more challenging to apply to problems involving concyclicity.

Part I

Trigonometric Approach

Chapter 1

Foundations of Geometry

Knowledge is two-fold, and consists not only in the affirmation of what is true, but in the negation of that which is false.
Charles Caleb Colton, 1825

1.1 Axiomatic systems

At the foundation of any science, or in particular, of any theory, there are some fundamental building blocks. There are truths which are taken for granted, and cannot be proved or disproved and are more or less "obvious" to the human mind. These are called *axioms*. They constitute the building blocks the entire deductive science is based on.

The first axiomatic system was developed by Euclid around 300 B.C. in his book called "Elements". Each science today is based on some axioms. Among these, Geometry is the oldest example. These axioms represent simple relations among points, lines and planes. For instance, given any two points, there is a line that contains them.

In Euclidean geometry, all axioms look obvious to the human intellect. However, this is not always a rule, and not all axioms are obvious in other branches of science. For instance, in Schrödinger's formalization of quantum mechanics, the fact that particles are described by a probability density function, which verifies Schrödinger's equation, is taken as an axiom. This fact was difficult to comprehend by most physicists in 1920s, including Schrödinger himself. The choice of axioms is done based on real world evidence, and as the case of quantum mechanics have shown, there are cases when writing down the axioms is a lot more difficult than in classical geometry.

The same happens with general relativity, where Einstein considered that the space-time has four dimensions and it has a metric which satisfies some tensorial structural equations. The equation is not obvious to any graduate student. However, it is then taken as an axiom. The whole general relativity is based on Einstein's intuition and it has value because it explains accurately the red-shift effect of stars and galaxies, the curvature of the light around large mass bodies like our sun and the advance of Mercury's perihelion with 42 seconds per century. These effects were observed before in the real world and couldn't be explained by any other theories, like Newton's theory.

Another example is Arithmetics. The number 1 is defined as an initial element together with a successor function, as an axiom. The successor of 1 is called 2, the successor of 6 is called 7, etc. We write the successor function additively as $1 + 1 = 2$, $6 + 1 = 7$, to denote the successors of 1 and 6. All the arithmetic calculations are based on Peano's axioms. These axioms have been laid out in 1800s, even if people have been used natural numbers for millennia! Something similar should have happened in the case of Euclidean geometry, namely, ancient mathematicians were using geometric concepts without having in mind the axiomatic construction of geometry. Crystallizing an axiomatic system might take centuries or even Millennia.

It is very interesting that modern AI systems are able to extract the important features from huge data and infer rules in a matter of hours or days, making the entire process of finding the building blocks of some body of information much faster.

1.2 Deduction rules

There are several ways to process the information while performing a mathematical proof. The essence of a proof is to model the information using certain transformations to show that the information required by the conclusion can be obtained by a modification of the information provided in the hypothesis, using certain processing tools. We shall discuss these tools next.

Modus ponens

This corresponds to the usual direct proof, where the information is processed directly from hypothesis to conclusion. Using the information provided by some axioms (or theorems previously proved), we use some deduction rules to imply other statements. The deduction scheme in this case is given in the following:

$$\frac{\begin{array}{c} A_1, A_2, \ldots, A_n \text{ axioms} \\ (A_1, A_2, \ldots, A_n) \to B \text{ (logical implication)} \end{array}}{\text{Then } B \text{ is a true statement.}}$$

Even if the direct proof is the most used type of proofs, it cannot cover all mathematical statements. Only a certain subset of mathematical theorems can be proved using a direct way.

Modus tollence

This type of deduction model is called an *indirect proof* or *proof by contradiction.* It is a very powerful method, which in most cases cannot be replaced by a direct proof. It consists of negating the conclusion and arriving at a statement which is in contradiction either with the hypothesis (or with something we know a priori that holds true).

The negation of a statement is the statement which consists of the contrary action. For instance, the sentence

A: I believe in science

has the negation

non A: I don't believe in science.

The modus tollence scheme can be represented as follows:

$$\frac{\begin{array}{c} A \text{ axiom} \\ non\, B \to non\, A \end{array}}{\text{Then } B \text{ is a true statement.}}$$

The sentence $non\, B \to non\, A$ is called the *contrapositive* of the sentence $A \to B$ and both sentences have the same truth value. Hence, the proof by contradiction is equivalent to proving that the contrapositive statement holds true.

To continue the example, let

B: I do research

be another sentence. The sentence

$$A \to B$$

is read as

If I believe in science, then I do research.

The contrapositive of this statement becomes

If I do not do research, then I do not believe in science.

Transitivity

This type of deduction rule assures that the implication is preserved along a chain of implications. In this case the scheme is:

$$\frac{\begin{array}{c} B_1 \rightarrow B_2 \\ B_2 \rightarrow B_3 \end{array}}{\text{Then } B_1 \rightarrow B_3 \text{ is a true statement.}}$$

Whenever we apply a proof involving an algebraic computation, we obtain a sequence of expressions which are related to each other by an equal sign, which state that they are computationally equivalent. Then the first and the last expressions in the chain are also equal due to the transitivity property of the equality relation. Most proofs involve a computation claiming to show that a certain expression involving certain geometric concepts holds. In this case, the transitivity is used tacitly most of the time. However, transitivity is usually stated explicitly when dealing with a sequence of inequalities.

1.3 Undecidability and completeness

That knowledge which stops at what it does not know is the highest knowledge.
Chuang Tzu, 4th c. B.C.

In order to be acceptable, a system of axioms has to satisfy certain properties. This makes the process of choosing the best axiomatic system a difficult task.

If starting from a given system of axioms it is not possible to prove both sentences A and *non* A, then the system is called *non-contradictory* or *consistent*. Euclidean geometry is an example of such a non-contradictory system. It says that the statements can only be true or only false but not both.

An axiomatic system is called *complete* if any statement can be proved or disproved using the axioms and theorems developed within the axiomatic system. An axiomatic system is called *incomplete* if there are some "undecidable" propositions, i.e., propositions which cannot be proved nor disproved by the approved procedures. In 1931, a brilliant Austrian logician Kurt Gödel produced an important theorem dealing with "undecidability":

Any non-contradictory system of axioms and procedures or rules is incomplete.

Hence "undecidability" is inherent to all axiomatic systems.

To exemplify this in the case of geometry, we invoke a famous example of a statement that people had tried to prove for more than two thousand years. It is the case of the parallel postulate, which appears to be the last axiom in Euclid's approach:

Given a point which does not belong to a line, there is an unique line parallel to the first one passing through the given point.

This statement looks obvious to any creature who lives on a small scale, like us. But at macro-scale, it seems that space may have a different behavior than the usual medium-scale we are accustomed to; here we can have more parallel lines through a given point to a given line. This happens in the hyperbolic geometry discovered by Gauss, Bolyai and Lobachevsky in the 1830's. We may also have no parallel lines, like in the case of elliptic geometries.

The parallel postulate is an example of a statement which can be formulated in the realm of absolute geometry and cannot be either proved or disproved using the other axioms. This is because this statement is independent of the axioms of absolute geometry and it is considered as an extra axiom, which is used to construct the Euclidean geometry.

In the so-called "non-Euclidean" worlds where the parallel axiom does not hold, the uniqueness or existence fails and we may end up having buildings with more ceilings parallel to a given floor. This reminds us about some drawings of Escher, who is well-known for his work at the intersection of mathematics and art.

The Dutch graphic artist Maurits Cornelis Escher (1898–1972) made lots of unusual lithographs like those that we just pointed out. One of them is named *Prententententoonstelling* and was made in 1956. The picture represents a young man standing in an exhibition gallery viewing a print of a seaport. As we follow the buildings from left to right and then down, we discover the same gallery he is standing in.

The previous discussion emphasised that absolute geometry is incomplete. By Gödel's theorem, even Euclidean geometry is incomplete, in the sense that we should be able to formulate statements that cannot be either proved or disproved using the accepted axioms. Fortunately, these statements are difficult to be formulated in a form that is appealing to the human intellect.

In the next sections, we shall present one axiomatic system of geometry based on the following groups of axioms: incidence axioms, metric axioms, angle measure axioms, plane separation axioms, triangle congruence axioms and the parallel postulate.

1.4 Incidence axioms

In the logical construction of geometry every concept involves other more primitive concepts. It is natural to express complicated concepts using simple ones. The construction of the classical geometry allows the concepts of *point*, *line* and *plane* to be undefined. Other concepts like the triangle or the circle can be defined using the above primitive concepts.

The points, lines and planes are geometric objects with certain relations among them. These relations are given by the following *incidence axioms*:

Axiom I1 *Each two distinct points determine a unique line.*

Axiom I2 *If two points lie in a plane, then any line containing those points lies in the plane.*

Axiom I3 *Three non-collinear points determine a unique plane.*

Axiom I4 *If two planes meet, then their intersection is a line.*

Axiom I5 *Each line contains at least two distinct points. Each plane contains at least three non-collinear points. Space has at least 4 non-coplanar points.*

Axiom I1 states the fact that a line is a geometric shape determined by 2 points. We shall see later that other more complex types of shapes are determined by a certain number of points (a circle or a triangle by 3 points, a quadrilateral by 4 points, an ellipse by 5 points, etc.).

Axiom I2 states that a line determined by two points in a plane does not leave the plane; namely, the line and the plane share the same type of curvature. There are other shapes with this property, for instance, an arc or a circle on a spherical surface.

Axiom I3 can be interpreted as follows. A chair with 3 legs never wobbles, even if the legs do not have equal lengths, while a table with 4 legs usually wobbles. This is because the ends of the table legs are not usually 4 coplanar points.

Axiom I4 can be visualized by considering any two pages of the same book, which intersect over the book spine, which is a line.

An equivalent formulation of Axiom I5 is to state the existence of tetrahedrons in the space.

Even if these 5 axioms look pretty straightforward, they can be used to prove some results that do not look that obvious, as shall be seen in the next set of exercises.

Exercise 1.4.1 *Prove that if two distinct lines intersect, then their intersection is at most a point.*

Exercise 1.4.2 *Show that if a line meets a plane and it is not contained in the plane, then their intersection is a point.*

Exercise 1.4.3 *Prove that given two distinct lines which intersect, there is an unique plane that contains the lines.*

Exercise 1.4.4 *Prove that there are two lines which do not intersect.*

Exercise 1.4.5 (Desargues' theorem) *Let ℓ_1, ℓ_2, ℓ_3 be three lines in space passing through a point O. Let A_1, $B_1 \in \ell_1$, A_2, $B_2 \in \ell_2$, A_3, $B_3 \in \ell_3$. Suppose the following lines intersect at*

$$A_1A_2 \cap B_1B_2 = W,$$

$$A_2A_3 \cap B_2B_3 = V,$$

$$A_1A_3 \cap B_1B_3 = U.$$

Prove that the points U, V, W are collinear. See Figure 1.1.

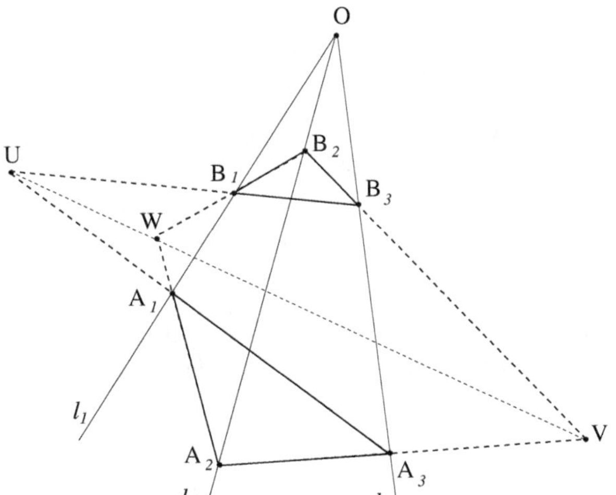

Figure 1.1: The points U, V, W are collinear.

Remark 1.4.6 A metric proof of Desargues' theorem in the case of a planar configuration will be given later using Menelaus theorem.

Exercise 1.4.7 *Let $A_1, \ldots, A_n$ be n points in the plane such that any three of them are collinear. Prove that all the points belong to a line.*

Exercise 1.4.8 *Let $d_1, \ldots, d_n$ be n lines in the space such that any three of them are coplanar. Prove that all the lines belong to the same plane.*

Exercise 1.4.9 *Let $d_1, \ldots, d_n$ be n lines in the space such that any three of them intersect. Prove that all the lines intersect at the same point.*

Exercise 1.4.10 (Gallai-Sylvester) *Given finitely many (at least three) points in the plane, which are not all on one line, show that there is a line which passes through exactly two of these points.*

Exercise 1.4.11 *Given n arbitrary points $A_1, \ldots, A_n$ in the plane, find the maximum number of lines determined by these points.*

Exercise 1.4.12 *Let α, β and γ be three planes which do not have a common point, i.e., $\alpha \cap \beta \cap \gamma = \emptyset$. Assuming that the planes are not parallel, consider the lines $d_1 = \alpha \cap \beta$, $d_2 = \beta \cap \gamma$, and $d_3 = \gamma \cap \alpha$. Show that the lines d_1, d_2 and d_3 are parallel (i.e., $d_i \cap d_j = \emptyset$ for any $i \neq j$).*

Exercise 1.4.13 *Find the number of diagonals of a convex polygon in the plane having N vertices.*

1.5 Metric axioms

This set of axioms assumes, besides the undefined concepts (point, line and plane), also the concept of real number, which is used to measure distances between points and measures of angles. These axioms follow Birkhoff's system, which has the advantage of making the analytic geometry easier to use.

Definition 1.5.1 *Let $\mathcal{P}$ denote the plane. A distance function is a function d defined on pair of points in the plane with the following properties*

D1. $d(A, B) \geq 0$, *for any* $A, B \in \mathcal{P}$,

D2. $d(A, B) = 0$ *if and only if* $A = B$,

D3. $d(A, B) = d(B, A)$,

D4. $d(A, B) + d(B, C) \geq d(A, C)$, *for any* $A, B, C \in \mathcal{P}$.

D1 says that d is nonnegative, while **D2** states that d is nondegenerate. **D3** gives the symmetry of d and **D4** is the triangle inequality. The following definition says that the triangle inequality becomes the identity for collinear points, namely for points situated on the same line.

Definition 1.5.2 *Given three points A, B, C we say that B is* **between** *A and C if A, B, C are collinear, and $d(A, B) + d(B, C) = d(A, C)$.*

The next axiom states that we can measure the distance between any two given points on a line by introducing a system of coordinates on the line.

The ruler axiom. *Any line ℓ has a system of coordinates, i.e. there is a bijective (one-to-one, onto) function $f : \ell \to \mathbb{R}$ such that*

$$|f(A) - f(B)| = d(A, B),$$

for any $A, B \in \ell$. Moreover, given points $O, A \in \ell$ there is a system of coordinates f such that $f(O) = 0$ and $f(A) > 0$.

For instance, in the above figure $f(A) = 7$ and $f(B) = 11$. Then $d(A, B) = |f(A) - f(B)| = |7 - 11| = 4$, see Figure 1.2. The number $f(A)$ is called the coordinate of the point A under the system of coordinates induced by the function f.

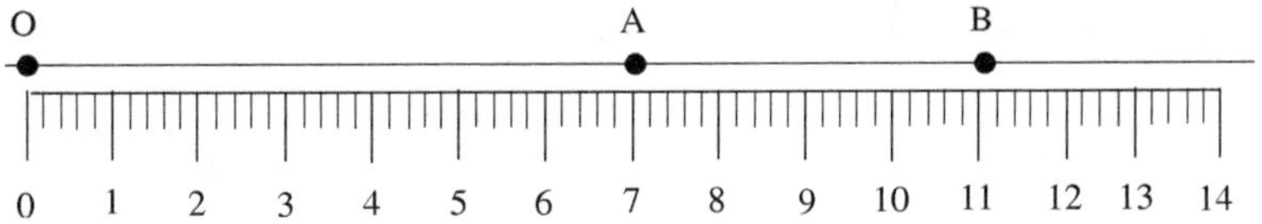

Figure 1.2: The distance between A and B is obtained subtracting the coordinates $f(B)$ and $f(A)$.

Given a distance function d, the system of coordinates is unique. If A is on the right side of the origin O, we choose $f(A) = d(O, A)$. Otherwise $f(A) = -d(O, A)$. The number $f(A)$ is called the *coordinate* of A. On a given line with a fixed origin O, there are two *orientations*. The positive orientation corresponds to $f > 0$, and the negative orientation corresponds to $f < 0$. A *ray* starting at the point M, and pointing into the direction of point N, is defined by

$$(MN = \{A \in MN;\ \mathrm{sign}\big(f(A) - f(M)\big) = \mathrm{sign}\big(f(N) - f(M)\big)\}.$$

The sign function used in the above definition denotes the sign of the difference. Roughly speaking, a ray is the set of points situated at one side of a given point. Sometimes, a ray is also called a *half line*.

A segment AB is defined as the set of points for which we have identity satisfied in the triangle inequality

$$\begin{aligned}[AB] &= \{M \in AB;\ d(A,M) + d(M,B) = d(A,B)\} \\ &= \{M \in AB;\ M \text{ is between } A \text{ and } B\}.\end{aligned}$$

We shall also use the notation (AB) to denote the line segment AB which does not contain the end points A and B.

Definition 1.5.3 *Two segments* $[AB]$ *and* $[A'B']$ *are called* **congruent** *if* $d(A,B) = d(A',B')$. *A point* M *is called a* **midpoint** *of a segment* $[AB]$ *if* M *divides* $[AB]$ *into two congruent segments.*

Exercise 1.5.4 *Prove that each segment has a unique midpoint.*

Exercise 1.5.5 *Let any two points* A, B *in the plane define the function* ρ *by*

$$\rho(A,B) = \frac{d(A,B)}{1 + d(A,B)},$$

where d *is a distance function in the plane. Prove that* ρ *is a distance function in the plane.*

The Euclidean distance

Exercise 1.5.5 shows that there are infinitely many distance functions in the plane. Among the multitude of metrics in the plane, there is one distinguished, in the sense that it is compatible with the human way of perceiving distances in the real world. In order to introduce it, we assume a system of orthogonal coordinates, in which each point is characterized by two real numbers, called the x- and y-coordinates.

Let A and B be two points in the plane with coordinates (x_A, y_A) and (x_B, y_B) and define the function

$$d_E(A,B) = \sqrt{(x_A - x_B)^2 + (y_A - y_B)^2},$$

which is called the *Euclidean distance* between the points A and B.

Exercise 1.5.6 *Show that* d_E *is a distance function, i.e., satisfies properties* **D1**–**D4**.

When we refer to the Euclidean distance, we shall write $|AB|$ instead of $d_E(A,B)$. This is the distance which agrees with human senses and we shall deal with it in Euclidean geometry.

Exercise 1.5.7 *Consider the distance function*

$$\rho(A,B) = \frac{|AB|}{1+|AB|}.$$

(i) Draw the circle centered at the origin with unit radius in the distance ρ, where

$$\mathcal{C}_\rho(O,1) = \{M \in \mathcal{P};\ \rho(O,M) = 1\}.$$

(ii) In which conditions the circle does not exist?

The next sections will introduce other examples of distance functions, which are distinct from the Euclidean distance.

The Taxi-Cab distance

This is the distance traveled by a taxi-cab in a city with a system or orthogonal streets, such as New York or Chicago. If A and B are two points in the plane with coordinates (x_A, y_A) and (x_B, y_B), the function

$$d_T(A,B) = |x_A - x_B| + |y_A - y_B|,$$

is called the *taxi-cab distance* between the points A and B.

Exercise 1.5.8 *Show that d_T is a distance function, i.e., satisfies properties* **D1–D4**.

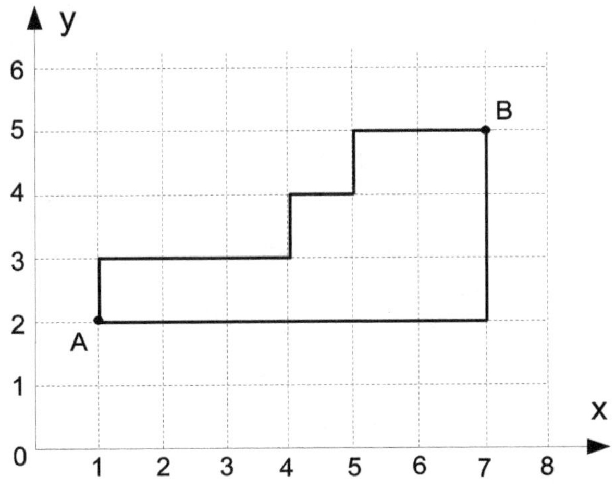

Figure 1.3: The taxi-cab distance between A and B.

Since the Euclidean distance measures the length of the straight segment between the points A and B, while the taxi-cab distance measures the length along a system of orthogonal segments, it makes sense to agree that $d_E(A,B) \le d_T(A,B)$. In the following, we consider an example.

Consider the points $A(1,2)$ and $B(7,5)$, see Figure 1.3. The taxi-cab distance between them is

$$d_T(A,B) = |1-7| + |2-5| = 6+3 = 9,$$

while the Euclidean distance is

$$d_E(A,B) = |AB| = \sqrt{|1-7|^2 + |2-5|^2} = \sqrt{36+9} = \sqrt{45} < 9 = d_T(A,B).$$

Exercise 1.5.9 (*i*) *Using the taxi-cab distance definition, show that* $d_E(A,B) \le d_T(A,B)$, *for any two points* A, B *in the plane.*

(*ii*) *We have* $d_E(A,B) = d_T(A,B)$ *only in one of the following cases:*

$$(a)\ A = B, \qquad (b)\ x_A = x_B, \qquad (c)\ y_A = y_B.$$

Exercise 1.5.10 (*i*) *The points* $A(0,2)$, $B(2,0)$, $C(0,-2)$, $D(-2,0)$, $E(1,1)$, $F(1,-1)$, $G(-1,-1)$, $H(-1,1)$ *are situated on a square of side* $2\sqrt{2}$ *centered at the origin. Show that the taxi-cab distance between the above points and the origin is* 2.

(*ii*) *Draw a unit circle in the taxi-cab distance. How do you explain that it has four corners?*

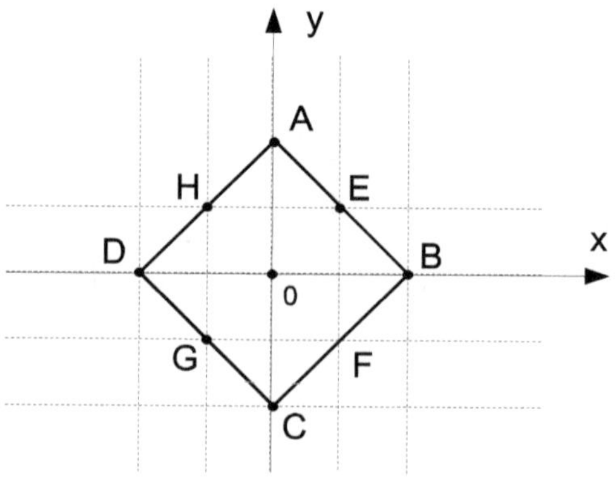

Figure 1.4: The sphere in the taxi-cab distance.

The post office distance

Let P be a fixed point in the plane. For any two points A and B in the plane, let

$$d_P(A,B) = \begin{cases} 0, & \text{if } A = B, \\ |AP| + |PB|, & \text{if } A \neq B \end{cases}$$

be the post office distance between A and B. A letter sent from house A to house B should go through the post office P. See Figure 1.5.

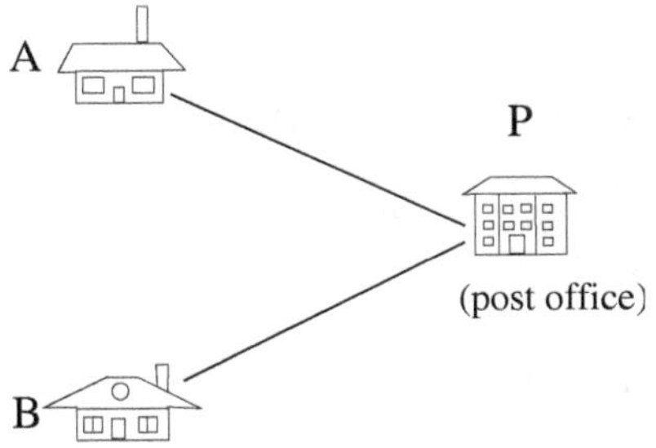

Figure 1.5: A letter goes from A to B passing through the post office P.

Exercise 1.5.11 *Consider the points $A(4,8)$, $B(5,4)$, $P(2,3)$. Find the post office distance between A and B.*

Exercise 1.5.12 *Show that d_P satisfies the properties* **D1–D4** *of a distance function.*

Exercise 1.5.13 (*i*) *Show that $d_E(A,B) \leq d_P(A,B)$.*

(*ii*) *Show that $d_E(A,B) = d_P(A,B)$ if and only if A,B,P are collinear with P between A and B.*

The geometry induced by d_T is called post office geometry. A circle centered at A with radius r in post office geometry is the set of the points M such that $d_P(A,M) = r$.

Exercise 1.5.14 *Let A be a fixed point in the plane and let $P \neq A$ be the post office point. Draw a circle centered at A with radius $r = 2|AP|$ in the post office geometry.*

In the post office geometry a point B is between A and C if

$$d_P(A,B) + d_P(B,C) = d_P(A,C).$$

Exercise 1.5.15 *Given two points A and C, find the set of points B between A and C in post office geometry.*

Exercise 1.5.16 *Let $P \notin \{A, B, C\}$. Show that*

$$d_P(A,C)^2 + d_P(A,B)^2 = d_P(B,C)^2$$

if and only if

$$|AP| + |BP| + |CP| = \frac{|CP| \cdot |BP|}{|AP|}.$$

1.6 Angle measure axioms

An angle is defined as a pair of rays in the plane that start from the same point. If the halves of line are $\ell = (AB$ and $m = (AC$, we shall denote the angle by $\widehat{\ell, m}$ or by $\widehat{BAC}$. Sometimes when there is no possibility of confusion, we may denote the angle just by $\widehat{A}$, see Figure 1.6.

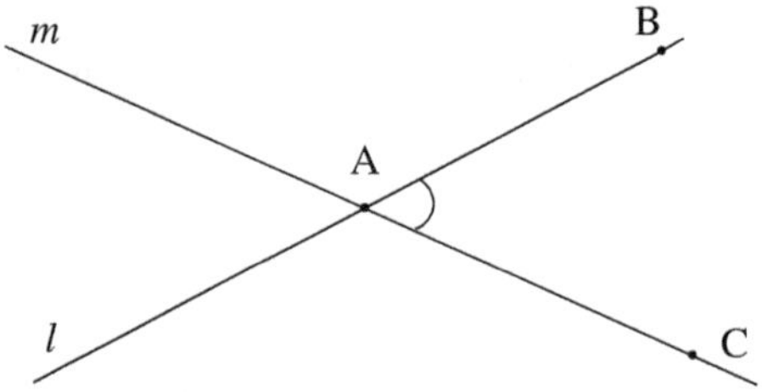

Figure 1.6: The rays $(AB$ and $(AC$ form the angle $\widehat{BAC}$ or $\widehat{A}$.

Let $\mathcal{D}$ be an application which assigns to each point of the plane a unique line and to each line a unique point, such that the incidence is preserved. This is, if a point M belongs to line a, then the line $\mathcal{D}(M)$ passes through the point $\mathcal{D}(a)$. The application $\mathcal{D}$ is called a **duality** of the plane. Each configuration that consists in points and lines will be transformed by duality into a new configuration of lines and points.

In particular, if two points M, L belong to a line a, then the dual configuration will consist in two lines $\ell = \mathcal{D}(L)$ and $m = \mathcal{D}(M)$ passing through the point $A = \mathcal{D}(a)$. This new configuration is called an *angle*, see Figure 1.7.

By the ruler axiom, there is a unique nonegative number $d(L, M)$ associated with the points L, M on the line a, called distance. The dual concept of distance between two points is the **measure** of the angle. The

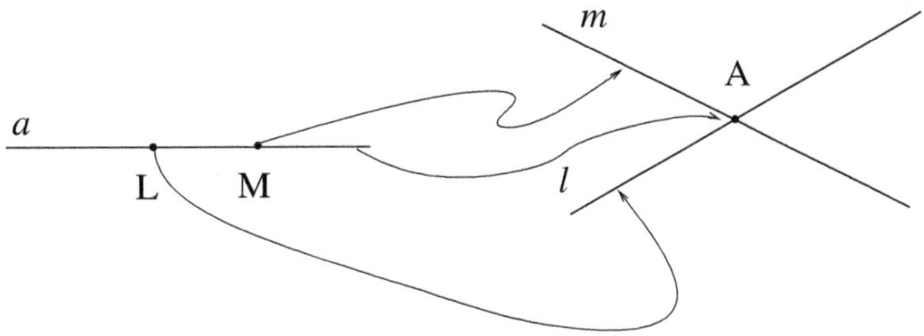

Figure 1.7: The angle born from duality.

dual of the ruler axiom is called the protractor axiom. The name indicates the way the angle can be measured (i.e., using a protractor). This is stated in the following.

The existence of angle measure axiom *Each angle is associated with a unique real number between* 0° *and* 180°, *called its* **measure**.

We shall denote both the angle and its measure by $\widehat{A}$. The reader will understand from the context whether this means one or the other.

Remark 1.6.1 We avoid angles with measures more than 180°.

Definition 1.6.2 *An angle is called* **acute** *if its measure is smaller than* 90°. *If its measure is equal to* 90°, *the angle is called* **right**. *An angle with measure between* 90° *and* 180° *is called* **obtuse**. *If the angle measure is equal to* 180°, *the angle is called* **straight**. *See Figure 1.8.*

If the angle between two lines is a right angle, we say that the lines are **perpendicular**. The axiom dealing with the construction of an angle is given below:

The Protractor Axiom *Given a ray* $(OA$ *and a number* $\theta \in [0^\circ, 180^\circ]$, *there is a ray* $(OB$ *such that the measure of the angle* $\widehat{AOB}$ *is equal to* θ.

The Addition Angle Axiom *Given the angle* $\widehat{AOB}$ *and* $D \in Int\widehat{AOB}$ *the following relation holds among the measures of the angles formed*

$$\widehat{AOB} = \widehat{AOD} + \widehat{DOB}.$$

Definition 1.6.3 *Two angles are called* **complementary** *(respectively* **supplementary)** *if the sum of their measures is* 90° *(respectively* 180°*).*

Exercise 1.6.4 *Show that two angles that are supplementary, or complementary to the same angle have equal measures.*

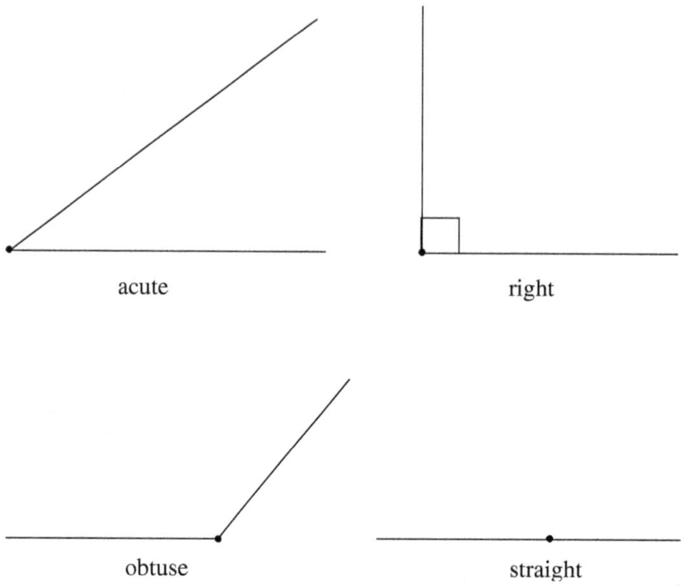

Figure 1.8: Different types of angles.

Exercise 1.6.5 *The angles $\widehat{BAC}$ and $\widehat{B'AC'}$ in Figure 1.9 are called* **vertical** *angles. Show that two vertical angles have the same measure.*

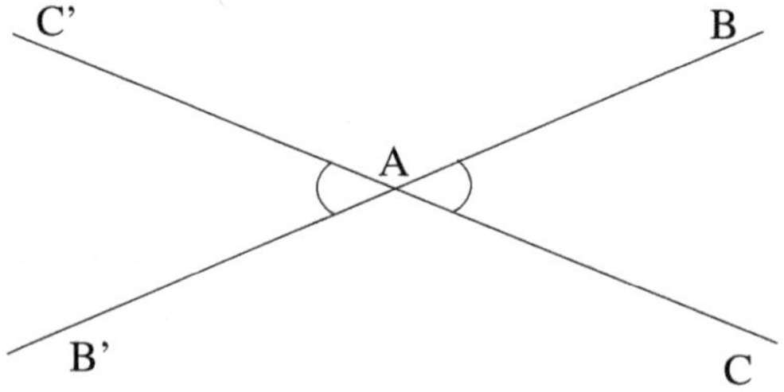

Figure 1.9: $\widehat{BAC}$ and $\widehat{B'AC'}$ are equal vertical angles.

1.7 Plane separation axioms

Before we present the separation of a plane into half planes, we need a few preliminary definitions.

Definition 1.7.1 *A set K is called convex if for any two points $A, B \in K$, the segment $[AB] \subset K$.*

Example 1.7.2 *The interior of a triangle or rectangle, a disk, a half plane, or the entire plane, the interior of an angle are examples of convex sets. A ring, a circular crown, or a boomerang-type shape are not convex sets.*

Exercise 1.7.3 *(i) Let A and B be two convex sets. Prove that $A \cap B$ is convex. Is $A \cup B$ convex?*

(ii) Let $A_1, \ldots, A_n$ be n convex sets. Show that $A_1 \cap \cdots \cap A_n$ is a convex set.

(iii) Show that the statement in (ii) does not hold if the number of the sets is infinite.

Recall from the set theory the definition for the difference of two sets

$$A \backslash B = \{x;\ x \in A \wedge x \notin B\}.$$

Plane separation axiom *Let ℓ be a line lying in the plane $\mathcal{P}$. Then*

$$P \backslash \ell = H_1 \cup H_2,$$

where

S1. *H_1, H_2 are convex sets.*

S2. *$H_1 \cap H_2 = \emptyset$ (disjoint sets).*

S3. *If $A \in H_1$ and $B \in H_2$, then the line ℓ intersects the segment $[AB]$.*

The convex sets H_1, H_2 are called the **half-planes** determined by ℓ. See Figure 1.10.

In some axiomatic systems the separation axiom is replaced by **the postulate of Pasch**. Here we shall prove this postulate as a theorem based on the plane separation axiom.

Theorem 1.7.4 (The postulate of Pasch) *If ℓ is a line passing through an interior point of a side of a triangle, which does not pass through any of the vertices, then ℓ intersects exactly one more side of the triangle at an interior point.*

Proof: Consider a triangle ABC. Assume the line ℓ passes through the interior point $D \in AB$. Then ℓ separates the plane into two half-planes H_1 and H_2, with $A \in H_1$, $B \in H_2$. Now we apply

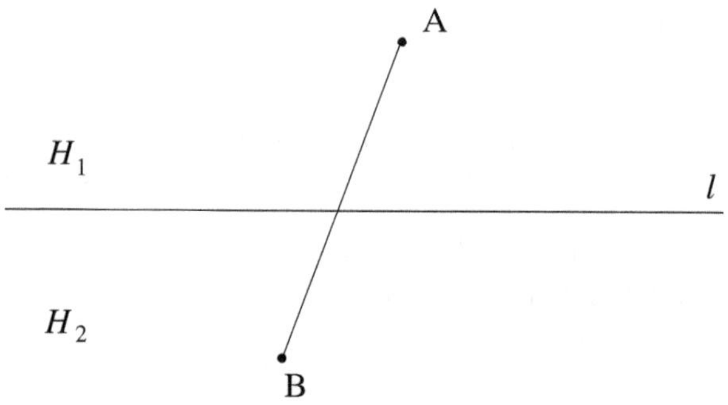

Figure 1.10: The points A and B are separated by the line l.

The Dirichlet principle *Given 2 boats and 3 fishermen, there is a boat that contains at least two fishermen.*[1]
In our case, the boats are H_1 and H_2 and the fishermen are the points A, B, C. There is a half-plane which contains two of the above points. We distinguish two cases, see Figure 1.11.

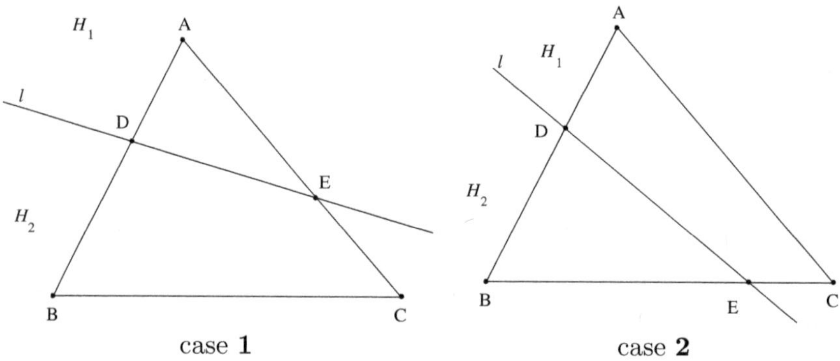

Figure 1.11: The transversal line in the axiom of Pasch.

1. $A \in H_1$, and $B, C \in H_2$. In this case, A and C are separated by the line ℓ and by **S3** it follows that ℓ intersects segment $[AC]$. By convexity of H_2, the segment $[BC]$ is included in H_2 and hence does not intersect ℓ.

[1] Dirichlet's principle represents an axiom in itself. It can be formulated in general like this: "If $n+1$ objects are placed into n boxes, there is at least one box which contains at least two objects." This statement is equivalent to both Zermelo's axiom and Axiom of choice from the set theory.

2. $A, C \in H_1$, and $B \in H_2$. In this case, B and C are separated by the line ℓ and by **S3** the line ℓ intersects segment $[BC]$. As $A, C \in H_1$, by convexity the segment $[AC]$ is included in H_1 and does not intersect ℓ.

Hence the line ℓ intersects exactly one more side of the triangle ΔABC at an interior point E, because E is between A and C. ■

Definition 1.7.5 *The interior of an angle $\widehat{BAC}$ is the intersection of the half-planes H_B and H_C, where H_B is the half-plane determined by AC and B, and H_C is the half-plane determined by AB and C.*

As the intersection of two convex sets is a convex set, it follows that the interior of an angle is a convex set.

Exercise 1.7.6 *Define the interior of a triangle. Prove that it is a convex set.*

Another result equivalent to the Pasch postulate and the plane separation axiom is the **crossbar theorem**.

Exercise 1.7.7 (Crossbar theorem) *If D lies in the interior of angle $\widehat{BAC}$, then AD meets segment $[BC]$ at some interior point E, see Figure* 1.12.

Remark 1.7.8 One may show that the postulate of Pasch, the crossbar theorem and the plane separation axiom are all equivalent. In some axiomatic systems, the plane separation axiom is replaced by the postulate of Pasch.

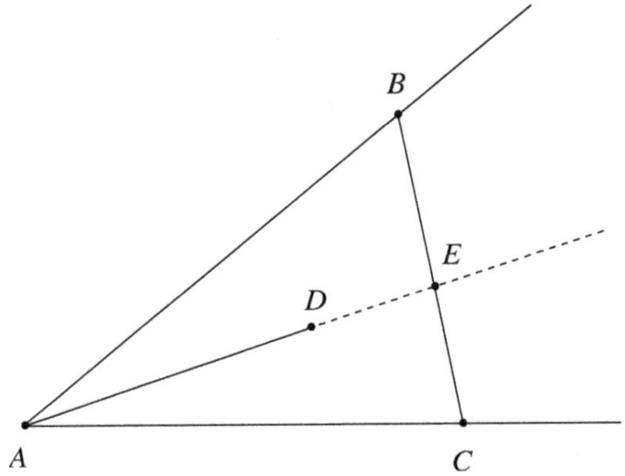

Figure 1.12: The Crossbar theorem.

Exercise 1.7.9 *A line separates the plane into two convex regions. Two lines separate the plane into at most four convex regions. Three lines separate the plane into at most seven convex regions. Let $f(n)$ be the maximum number of regions n arbitrary lines can divide the plane into. Find the function $f(n)$ as a function of n. (Hint: Use induction over n).*

Similar separation considerations work for the 3-dimensional space:

Space separation axiom *Let Π be a plane in the space $\mathcal{S}$. Then*

$$S\backslash\Pi = U_1 \cup U_2,$$

where

S1. *U_1, U_2 are convex sets.*

S2. *$U_1 \cap U_2 = \emptyset$ (disjoint sets).*

S3. *If $A \in U_1$ and $B \in U_2$, then the segment $[AB]$ intersects the plane Π.*

Exercise 1.7.10 *Find the number of space regions the sides of a tetrahedron divides the space into.*

Exercise 1.7.11 *(i) Find the number of plane regions the sides of a square divides the plane into.*
(ii) Find the number of space regions the sides of a cube divides the space into.

Exercise 1.7.12 (Kobon triangles) *Let $T(n)$ be the maximum number of triangles that can be constructed with n lines in the plane. Find $T(n)$ for $n \in \{1, 2, 3, 4\}$. Note that there is no general formula for $T(n)$.*

Exercise 1.7.13 *Let $f(n)$ be the maximum number of space regions n planes can divide the space into. Find $f(n)$ for $n \in \{1, 2, 3\}$.*

Use Dirichlet's principle to solve the following problems.

Exercise 1.7.14 *(i) 5 points belong to the interior of a square with the side equal to 1. Prove that there are two points with the distance smaller than or equal to $\sqrt{2}/2$.*
(ii) 9 points belong to the interior of a cube with the side equal to 1. Prove that there are two points with the distance smaller than or equal to $\sqrt{3}/2$.

Exercise 1.7.15 *Let $n \geq 1$ and consider $n^2 + 1$ points in the interior of a square with the side equal to 1. Prove that there are two points with the distance smaller than or equal to $\sqrt{2}/n$.*

Exercise 1.7.16 *(i) 5 points belong to the interior of an equilateral triangle with the side equal to 1. Prove that there are two points with the distance smaller than or equal to* $1/2$.
(ii) 17 points belong to the interior of an equilateral triangle with the side equal to 1. Prove that there are two points with the distance smaller than or equal to $1/4$.
(iii) Formulate and prove a generalization.

1.8 Congruence cases for triangles

Most proofs of classical geometry are based on congruent triangles. They represent a first robust tool used to prove congruences for angles or segments.

Definition 1.8.1 *Two triangles* ΔABC *and* $\Delta A'B'C'$ *are called* **congruent** *if the lengths of their sides and angles are respectively equal:*

$$|AB| = |A'B'|, \quad |BC| = |B'C'|, \quad |AC| = |A'C'|$$

$$\widehat{A} = \widehat{A'}, \qquad \widehat{B} = \widehat{B'}, \qquad \widehat{C} = \widehat{C'}.$$

We shall write $\Delta ABC \equiv \Delta A'B'C'$. In order to establish the congruence of two triangles, we need to verify 6 identities. If we employ the following axiom, then in order to show that two triangles are congruent, we need to match only three elements: two sides and the angle between them, two angles and the side between them, or three sides.

Axiom (Angle congruence axiom) *Let* ΔABC *and* $\Delta A'B'C'$ *be two triangles such that* $|AB| = |A'B'|$, $|AC| = |A'C'|$ *and* $\widehat{BAC} = \widehat{B'A'C'}$. *Then* $\widehat{ABC} = \widehat{A'B'C'}$.
The above axiom has the following consequence.

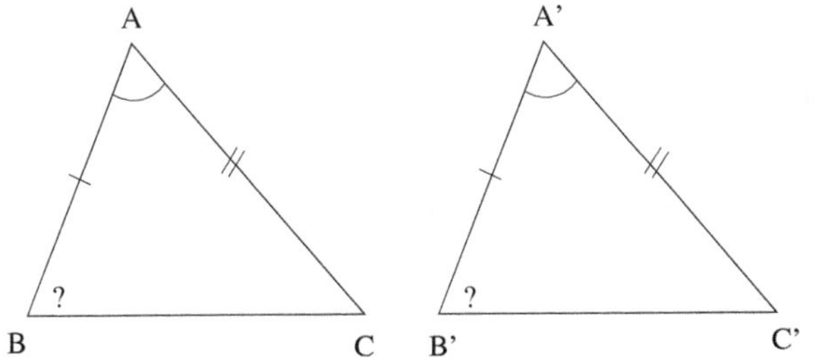

Figure 1.13: Angle congruence axiom.

Proposition 1.8.2 *Let ΔABC and $\Delta A'B'C'$ be two triangles such that $|AB| = |A'B'|$, $|AC| = |A'C'|$, and $\widehat{BAC} = \widehat{B'A'C'}$. Then $\widehat{ACB} = \widehat{A'C'B'}$.*

Theorem 1.8.3 (SAS congruence case) *Let ΔABC and $\Delta A'B'C'$ be two triangles such that $|AB| = |A'B'|$, $|AC| = |A'C'|$ and $\widehat{A} = \widehat{A'}$. Then $\Delta ABC \equiv \Delta A'B'C'$.*

Proof: Applying the above axiom, we get $\widehat{B} = \widehat{B'}$ and $\widehat{C} = \widehat{C'}$. In order to show that the triangles are congruent, we still need to prove that $|BC| = |B'C'|$.
Either $|BC| \leq |B'C'|$ or $|BC| \geq |B'C'|$. Consider the case $|BC| \leq |B'C'|$. Then we can choose a point $D \in B'C'$ such that $|BC| = |B'D|$. See Figure 1.14. Applying the above axiom for triangles ΔABC and $\Delta A'B'D$ yields

$$\widehat{BAC} = \widehat{B'A'D}.$$

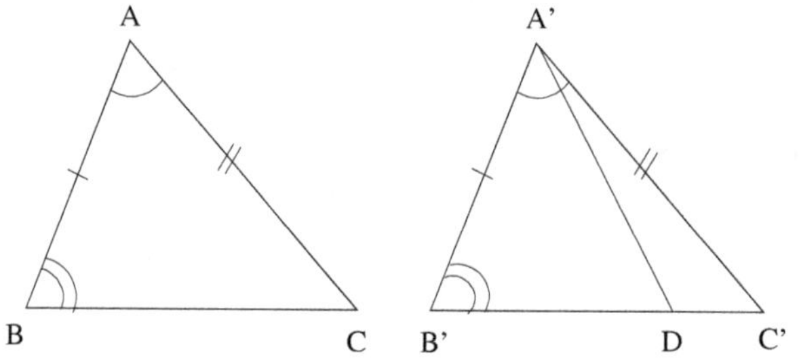

Figure 1.14: SAS congruence case.

Using the hypothesis,

$$\widehat{BAC} = \widehat{B'A'C'} = \widehat{B'A'D} + \widehat{DA'C'}$$

and subtracting the last two identities yields $\widehat{DA'C'} = 0$, and hence $D = C'$ and then $|BC| = |B'C'|$. ∎

Theorem 1.8.4 (ASA congruence case) *Let ΔABC and $\Delta A'B'C'$ be two triangles such that $|AB| = |A'B'|$, $\widehat{A} = \widehat{A'}$, and $\widehat{B} = \widehat{B'}$. Then $\Delta ABC \equiv \Delta A'B'C'$.*

Proof: Either $|BC| \leq |B'C'|$ or $|BC| \geq |B'C'|$. In the following, we choose $|BC| \leq |B'C'|$. Then there is a point D in the segment $[B'C']$ such that

$|BC| = |B'D|$. By the (SAS) congruence case, we have

$$\Delta ABC \equiv \Delta A'B'D \text{ because } \begin{cases} |AB| = |A'B'| & \text{(hypothesis)}, \\ \widehat{B} = \widehat{B'} & \text{(hypothesis)}, \\ |BC| = |B'D| & \text{(construction)}. \end{cases}$$

Hence $\widehat{BAC} = \widehat{B'A'D}$ which implies $\widehat{DA'C'} = 0$, so $D = C'$. Thus $|BC| = |B'C'|$, and $\Delta ABC \equiv \Delta A'B'C'$ by (SAS). ■

The way the triangle congruences axioms are used in practice is as follows. If we need to show that two segments or angles are congruent, we consider them as part of two triangles and prove the congruence of these triangles (using 3 elements, which do not contain the elements we have to prove the congruence for).

The following result is known as *pons asinorum.* Its name comes from the notion that anyone unable to "cross the bridge" in Euclid's school must be an ass.

Theorem 1.8.5 *Let ΔABC be a triangle such that $|AB| = |AC|$. Then $\widehat{B} = \widehat{C}$.*

Proof: By the (SAS) congruence case, we have

$$\Delta ABC \equiv \Delta ACB \text{ because } \begin{cases} |AB| = |AC|, \\ |AC| = |AB|, \\ \widehat{BAC} = \widehat{CAB}, \end{cases}$$

so $\widehat{B} = \widehat{C}$. ■

A triangle with two equal sides is called an *isosceles triangle.*

Exercise 1.8.6 *Prove that a triangle with two equal angles is isosceles.*

Exercise 1.8.7 (SSS congruence case) *Show that if two triangles have equal corresponding sides, then they are congruent.*

Exercise 1.8.8 *Let ΔABC be an isosceles triangle with $|AB| = |AC|$ and denote by M the midpoint of side BC.*

(i) Show that AM is perpendicular on BC.

(ii) Prove that the angles $\widehat{BAM}$ and $\widehat{MAC}$ have the same measure.

1.9 The parallel postulate and its consequences

That, if a straight line falling on two straight lines makes the interior angles on the same side less than two right angles, the two straight lines, if produced indefinitely, meet on that side on which the angles are less than two right angles.

Euclid, c. 300 B.C.

The parallel postulate is the last axiom of Euclidean geometry. For a long time, many mathematicians have tried to show that this postulate is a consequence of the other axioms. Sometimes wrong proofs were given. These proofs were using other statements equivalent to the parallel postulate. It was only in the 1800s that mathematicians realized that this axiom is independent of the others. The geometry that considers the parallel postulate is called Euclidean. The geometries where the negation of the postulate is considered are called non-Euclidean.

The postulate is based on the concept of parallel lines. Two lines ℓ and m, which are situated in the same plane, are called parallel if

$$\ell \cap m = \emptyset, \qquad \text{or} \qquad \ell = m.$$

If ℓ and m are parallel lines, we write $\ell \| m$.

We note that the relation "$\|$" is an *equivalence relation* on the set of lines in the plane $\mathcal{P}$. This means:

1) $\|$ is *reflexive*: $\ell \| \ell$, for any line $\ell \subset \mathcal{P}$,
2) $\|$ is *symmetric*: if $\ell \| m$ then $m \| \ell$, for any $\ell, m \subset \mathcal{P}$, and
3) $\|$ is *transitive*: if $\ell \| m$ and $m \| n$, then $\ell \| n$.

Example 1.9.1 (*i*) *(Friendship relations) We say that A is in friendship relation with B and write $A\mathcal{F}B$ if A is a friend of B. Then $\mathcal{F}$ is not transitive, so $\mathcal{F}$ is not an equivalence relation.*

(ii) (Love relations) We say that a love relation is not transitive or symmetric, so it is not an equivalence relation.

(iii) (Citizenship) Having the same citizenship is an equivalence relation. (Show it!)

Next, we state the last of the Euclidean geometry axioms.

The Parallel postulate: *Given a line ℓ and a point $P \notin \ell$, there is a unique line m through P parallel to ℓ.*

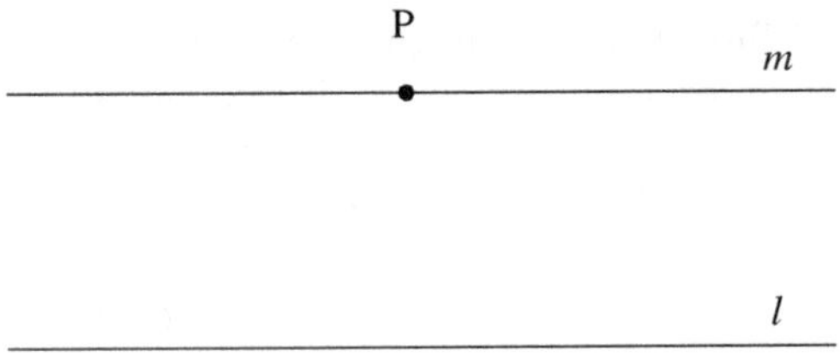

Figure 1.15: The parallel postulate.

The above postulate states two things: the existence and the uniqueness of the parallel line. From this point of view, it looks more like a theorem. This was the reason why mathematicians had a hard time believing that this is another axiom of geometry.

Definition 1.9.2 *A line s that intersects two parallel lines ℓ and m determines the following pairs of angles:*

- *alternate interior pairs:* $(1, 1'), (2, 2')$,
- *alternate exterior pairs:* $(4, 4'), (3, 3')$,
- *correspondent pairs:* $(3, 1'), (2, 4'), (4, 2'), (1, 3')$.

See Figure 1.16(a)*. The line s will be called a secant.*

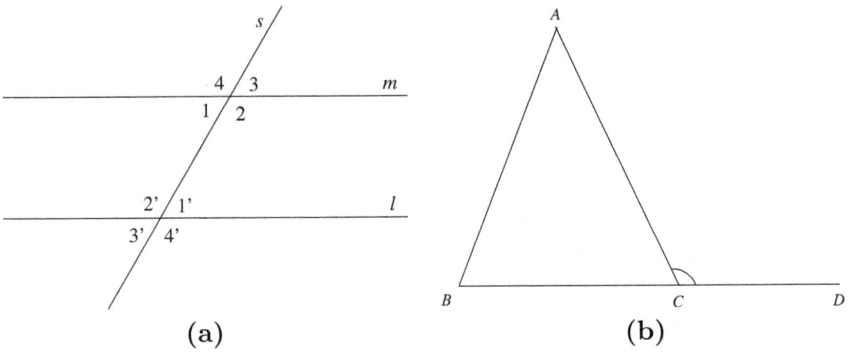

Figure 1.16: (a) Pairs of equal angles; (b) $\widehat{ACD}$ as an exterior angle.

Definition 1.9.3 *Let ΔABC be a triangle and suppose D is a point on the extension of the segment BC such that C is between B and D. Then $\widehat{ACD}$ is called an exterior angle of the given triangle. The angles $\widehat{BAC}$, $\widehat{ABC}$ are called opposite interior angles of $\widehat{ACD}$.*

Theorem 1.9.4 (The exterior angle inequality) *An exterior angle of a triangle is greater than either opposite interior angle:*

$$\widehat{ACD} > \widehat{BAC}, \quad \widehat{ACD} > \widehat{ABC},$$

see Figure 1.16(b).

Proof: The proof uses the following construction, see Figure 1.17(a). Let M be the midpoint of segment AC. Let B' be the point on the line through B and M so that M is the midpoint of BB'. By the (SAS) congruence case

$$\Delta AMB \equiv \Delta CMB' \quad \text{because} \quad \begin{cases} |BM| = |MB'| & \text{(construction)}, \\ \widehat{AMB} = \widehat{B'MC} & \text{(vertical angles)}, \\ |AM| = |MC| & (M \text{ midpoint}). \end{cases}$$

Hence $\widehat{BAM} = \widehat{MCB'}$. As $\widehat{MCB'} < \widehat{ACD}$, it follows that $\widehat{BAM} < \widehat{ACD}$, and so $\widehat{BAC} < \widehat{ACD}$. This proves the first inequality.

In order to prove the second inequality, see Figure 1.17(b). We note that $\widehat{ACD} = \widehat{BCD'}$ as vertical angles and hence it suffices to show that $\widehat{ABC} < \widehat{BCD'}$. We shall perform a construction similar to the one before. Let N

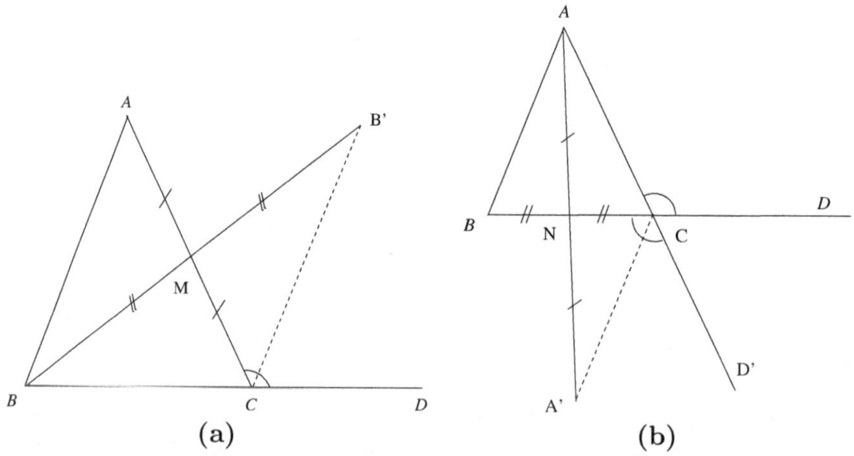

Figure 1.17: The construction needed for the proof of the exterior angle inequality.

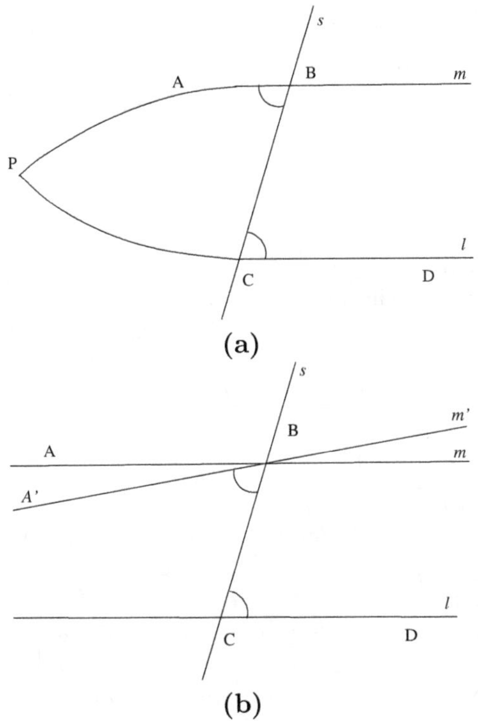

Figure 1.18: For the proof of Theorem 1.9.5: (a) the first part; (b) the second part.

be the midpoint of BC and AA'. We have by the (SAS) congruence case

$$\Delta BAN \equiv \Delta CA'N \quad \text{because} \quad \begin{cases} |AN| = |NA'| & \text{(construction)}, \\ \widehat{ANB} = \widehat{A'NC} & \text{(vertical angles)}, \\ |BN| = |NC| & (N \text{ midpoint}). \end{cases}$$

Hence $\widehat{ABN} = \widehat{NCA'} < \widehat{NCD'} = \widehat{ACD}$. ■

The following theorem is an important tool used in proving that two lines are parallel. The lines are assumed to belong to the same plane.

Theorem 1.9.5 *Two lines are parallel if and only if there is a secant line which intersects them such that the alternate interior pairs of angles are equal.*

Proof: " $\Longleftarrow$ " We shall use proof by contradiction. Let ℓ and m be the lines and s be the secant. Let a pair of equal alternate interior angles be $\widehat{ABC} = \widehat{BCD}$. Assume $\ell \nparallel m$. Then there is a point of intersection $\{P\} = \ell \cap m$. See Figure 1.18(a). The angle $\widehat{BCD}$ is an exterior angle of triangle ΔBPC and Theorem 1.9.4 yields $\widehat{BCD} > \widehat{PBC}$, which is in contradiction with the hypothesis. It follows that $\ell \| m$.

" $\Longrightarrow$ " We need to show that if $\ell \| m$ then $\widehat{ABC} = \widehat{BCD}$. We use proof by contradiction again. Assume $\widehat{ABC} \neq \widehat{BCD}$.
Then we can construct a point A' such that $\widehat{A'BC} = \widehat{BCD}$. See Figure 1.18(b). Let m' be the line determined by the points A' and B. The implication " $\Longleftarrow$ " yields that $m' \| \ell$. Hence through B there are two distinct lines m and m' parallel to ℓ. This is in contradiction with the parallel postulate. It follows that $\widehat{ABC} = \widehat{BCD}$. ∎

Corollary 1.9.6 *The following statements are equivalent:*
(i) The lines ℓ and m are parallel.
(ii) There is a secant which determines a pair of equal alternate interior angles.
(iii) Any secant determines a pair of equal alternate interior angles.
(iv) There is a secant which determines a pair of equal alternate exterior angles.
(v) Any secant determines a pair of equal alternate exterior angles.
(vi) There is a secant which determines a pair of equal correspondent angles.
(vii) Any secant determines a pair of equal correspondent angles.

Exercise 1.9.7 *Prove Corollary 1.9.6.*

In the following, we find the sum of the interior angles of a given triangle. This sum does not depend on the triangle and it is characteristic to Euclidean geometry.

Theorem 1.9.8 *The sum of the angles of any triangle is* $180°$.

Proof: The proof requires the following construction. Consider a line ℓ through A parallel to the base BC. By Theorem 1.9.5, the secants AB and AC determine pairs of equal alternate interior angles.
Then we have

$$\begin{aligned}\widehat{A} + \widehat{B} + \widehat{C} &= \widehat{2} + \widehat{1} + \widehat{3} \\ &= \widehat{BAC} + \widehat{BAX} + \widehat{CAY} = 180°.\end{aligned}$$

∎

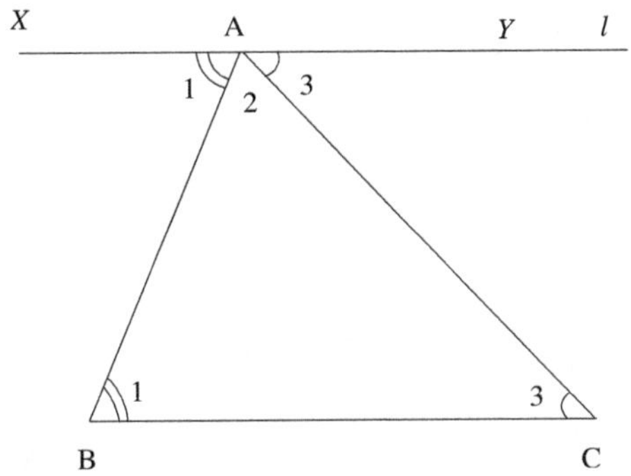

Figure 1.19: The sum of the angles of a triangle is 180°.

Remark 1.9.9 The fact that the sum of the angles of any triangle is 180° is a statement equivalent to the parallel postulate. Theorem 1.9.8 states one of these implications. The converse is proved by considering a line ℓ and a point $A \notin \ell$. We shall construct a line m through A parallel to ℓ as follows. Choose two distinct points $A, B \in \ell$. By the angle measure axiom, choose a point D such that $\widehat{ACB} = \widehat{CAD}$, so that D and B are in opposite half-planes determined by the line AC. Then $AD \| BC$ and choose $m = AD$. Otherwise, let $P = AD \cap BC$ and note that $\widehat{ACB}$ as an exterior angle for ΔAPC is strictly greater than $\widehat{CAD}$, contradiction.

The following is a strong version of Theorem 1.9.4. However, it takes place only in Euclidean geometry.

Theorem 1.9.10 (The exterior angle theorem) *An exterior angle of a triangle is equal to the sum of the opposite interior angles:*

$$\widehat{ACD} = \widehat{BAC} + \widehat{ABC}.$$

Proof: Using Theorem 1.9.8 yields

$$\widehat{ACD} = 180 - \widehat{ACB} = \widehat{BAC} + \widehat{ABC},$$

see Figure 1.20. ■

Remark 1.9.11 The exterior angle identity holds only in the case of Euclidean geometry, since the proof is based on the parallel postulate. However, the exterior angle inequality, see Theorem 1.9.4, holds in general for the case of non-Euclidean geometries.

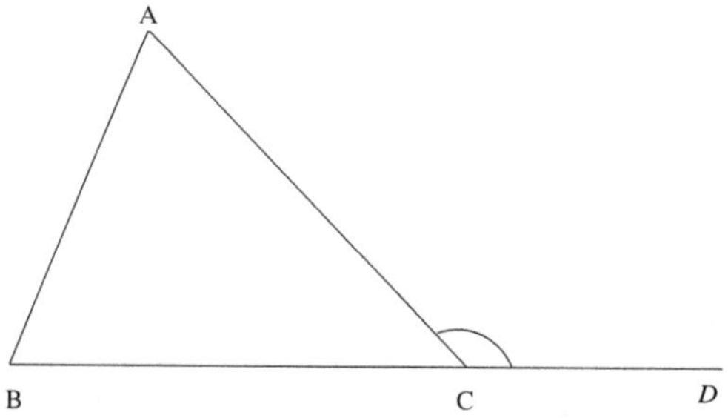

Figure 1.20: The exterior angle identity.

The fact that the sum of the angles of a triangle is 180 is a theorem of Euclidean geometry. One can show that this is equivalent with the parallel postulate. In the case of non-Euclidean triangles, the sum of the measures of angles can be either smaller (hyperbolic geometry) or larger (spherical geometry) and it is not the same for all triangles.

1.10 A paper folding construction

In this book we shall encounter several types of proofs. A proof technique suited for many problems of geometry is the **proof by construction.** This idea consists of considering extra points or lines which complete some triangles of the figure, which are then used to prove some properties. At some point in the Greek school, this was allowed as the only proof technique. The Greek mathematicians were allowed to do their constructions using the unmarked ruler and the compass only. This way, they managed to construct the irrational number $\sqrt{2}$, and the golden number ϕ. However, they were not able to construct the number π. This problem is related to the famous problem of circle quadrature: *Construct a square with the same area with a given circle.* The construction was proven impossible after the discovery that the number π is a transcendental number.[2]

Other two historical construction problems were proven impossible. One of them is the cube duplication. The problem asked to construct a cube with double volume than a given cube, using only the unmarked ruler and the compass. This problem aroused when the Greeks tried to double the size of Apollo's altar. This problem is equivalent with the construction of the number $\sqrt[3]{2}$ by ruler and compass. The construction was proved impossible 2000 years later.

[2]A transcendental number is a number which cannot be a root of a polynomial with integer coefficients.

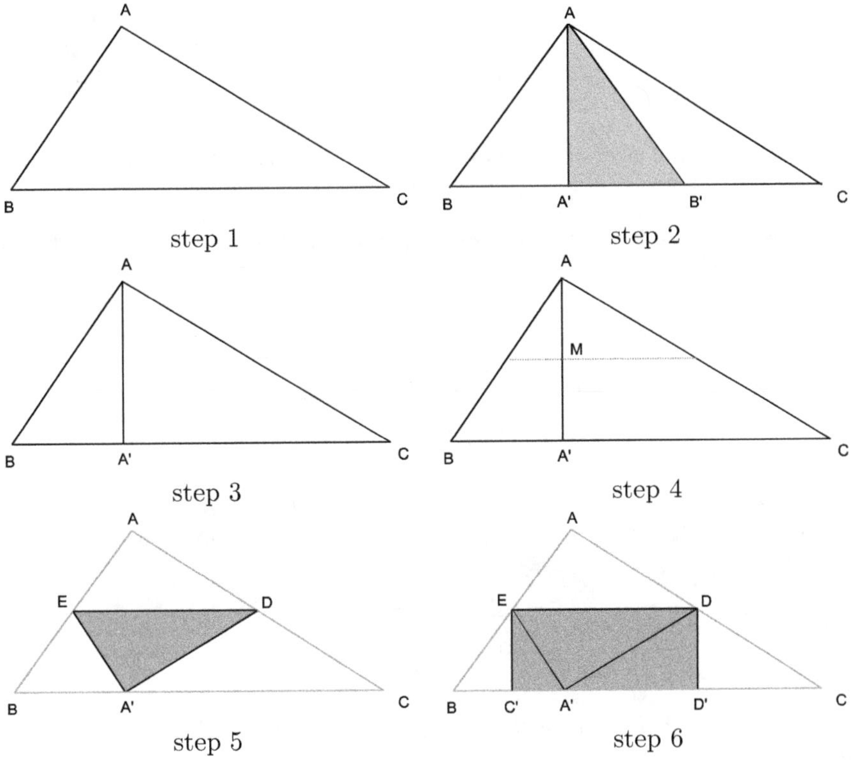

Figure 1.21: Paper folding construction.

Another geometric problem of antiquity was the trisection of an arbitrary given angle using the ruler and the compass. The trisection was proven algebraically impossible in 1800s, after Galois showed that this kind of constructions are possible only when a certain group is solvable.

The conclusion is that not any problem can be solved using geometric constructions. However, there are many classical geometric problems which can be successfully approached by construction.

We shall present next another proof by construction of the fact that the sum of the angles in a triangle is 180° by using a paper folding technique.

Paper folding problem. We shall prove that the sum of the angles of a triangle is 180° using a paper folding construction, see Figure 1.21.

- Step 1. Consider the paper triangle ΔABC.
- Step 2. Fold the triangle such that B overlaps B'.

- Step 3. Unfolding we obtain the altitude AA'.
- Step 4. Find the midpoint M of the segment AA' by folding A over A' and consider the parallel line through M to BC.
- Step 5. Fold over the line CD so A overlaps A'. We obtained triangle $\Delta EDA'$.
- Step 6. Fold B and C over A'. This way we obtain the rectangle $EDD'C'$.

Since $\widehat{EA'D} = \widehat{A}$, $\widehat{EA'B} = \widehat{B}$ and $\widehat{DA'C} = \widehat{C}$, it follows that

$$\widehat{A} + \widehat{B} + \widehat{C} = \widehat{EA'D} + \widehat{EA'B} + \widehat{DA'C} = 180.$$

The above construction is based on the next exercise.

Exercise 1.10.1 *Let AA' be the altitude of triangle ΔABC. Denote by E and D the midpoints of the sides AB and AC. Show the following:*
(i) $\widehat{ADE} = \widehat{EDA'}$ *and* $\widehat{AED} = \widehat{DEA'}$.
(ii) $\widehat{A'DD'} = \widehat{D'DC}$ *and* $\widehat{BCC'} = \widehat{C'CA'}$.
(iii) $\widehat{EBA'} = \widehat{EA'B}$ *and* $\widehat{DCD'} = \widehat{DA'D}$.

Exercise 1.10.2 *Let M be the midpoint of the side BC of ΔABC. Show that*

$$|AM| < (|AB| + |AC|)/2.$$

1.11 Angle measure on the circle

It is easier to square a circle than
to get round a mathematician.
A. De Morgan, 1840

In this section, we shall present a few consequences of the exterior angle theorem regarding angle measures. A circle is the set of points situated at equal distance from a fixed point. We shall denote by $\mathcal{C}(O, R)$ the circle centered at O with radius R. A circle is uniquely determined by its radius and center. We shall investigate several types of angles relative to a given circle, such as central, inscribed, interior and exterior, and compute their measures in terms of the arcs they determine on the circle.

Definition 1.11.1 *Let $A, B, C \in \mathcal{C}(O, R)$. The angle $\widehat{ABC}$ is called an inscribed angle and the angle $\widehat{AOC}$ is called a central angle.*

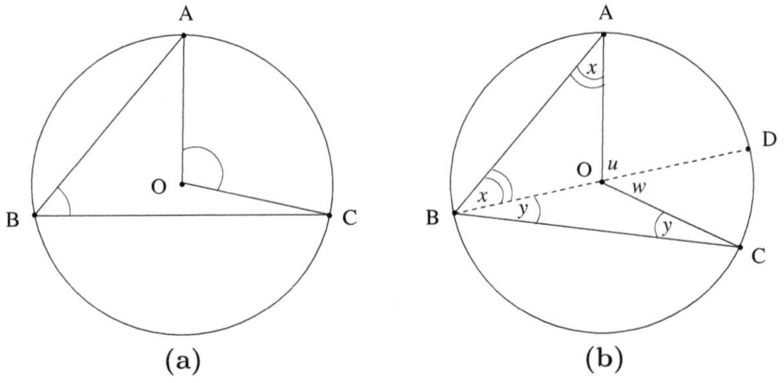

Figure 1.22: Proof of Theorem 1.11.2.

Theorem 1.11.2 *The measure of a central angle is twice as large as the measure of an inscribed angle that cuts the same arc of circle.*

Proof: We shall consider the angles as in the Figure 1.22(a). We shall show that

$$\widehat{AOC} = 2\widehat{ABC}.$$

As $|AO| = |BO| = |CO|$, then the triangles ΔAOB and ΔBOC are isosceles. Then let

$$x = \widehat{BAO} = \widehat{ABO}.$$

Let $u = \widehat{AOD}$ and $w = \widehat{DOC}$. By the exterior angle theorem we have $u = 2x$ and $w = 2y$. Adding yields $u + w = 2(x + y)$, which is equivalent to $\widehat{AOC} = 2\widehat{ABC}$. This proof is consistent with the Figure 1.22(b). ∎

Exercise 1.11.3 *Prove Theorem 1.11.2 in the case when O is not in the interior of the angle* $\widehat{ABC}$.

The arc measure. If the arc measure of a central angle is θ radians then the arc cut by its sides has the length $s = \theta R$, where R is the radius. Then $\theta = s/R$. Rescaling the circle, one may assume that $R = 1$. Then we can say that the arc measure of a central angle is equal to the arc length cut on the circle. Consequently, the arc measure of an inscribed angle is equal to half of the arc cut on the circle. In the case $\theta = 2\pi$ (i.e., 360 degrees) we obtain $s = 2\pi R$, which is the circumference of the circle.

Let AB be a chord on a circle. We shall denote by $arc(AB)$ the measure of the central angle $\widehat{AOB}$, see Figure 1.23(a).

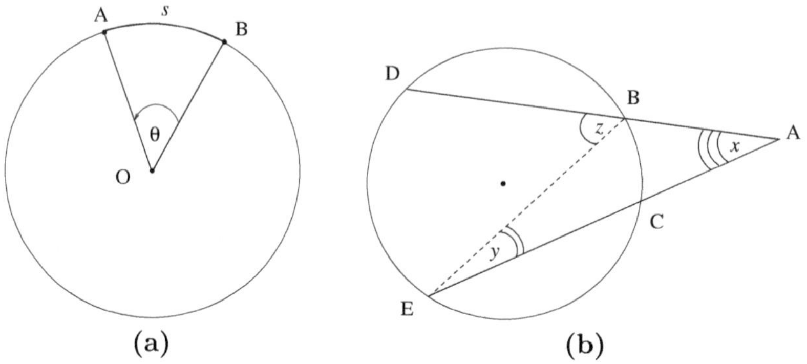

Figure 1.23: (a) The central angle $\widehat{AOB}$; (b) The exterior angle $\widehat{DAE}$.

Proposition 1.11.4 *Let A be a point in the exterior of a circle. Suppose that two lines through A intersect the circle at the points B, D and C, E, respectively. Then*

$$\widehat{DAE} = \frac{arc(DE) - arc(BC)}{2}.$$

Proof: Consider the Figure 1.23(b). The exterior angle theorem for ΔABE yields

$$z = x + y.$$

From the above discussion, $z = arc(DE)/2$ and $y = arc(BC)/2$. Therefore

$$x = z - y = arc(DE)/2 - arc(BC)/2.$$

■

Exercise 1.11.5 *Let A be a point in the interior of a circle. Suppose that two lines passing through A intersect the circle at B, C and D, E, respectively. Prove that*

$$\widehat{BAD} = \frac{arc(BD) + arc(EC)}{2}.$$

Why 360 degrees? Based on the belief that the earth revolves around sun in 360 days, the Sumerians introduced the concept of 360 degrees around a circle, about 3000 B.C. This was based on the belief that the trajectory of the earth around the sun is a perfect circle. Therefore, a 1 degree angle had the meaning of the angle swept by the vectorial radius of the earth in one day. Nowadays, this concept is adopted by convention, and used without any reference to its original meaning. The degree notation uses $\circ$ as a superscript.

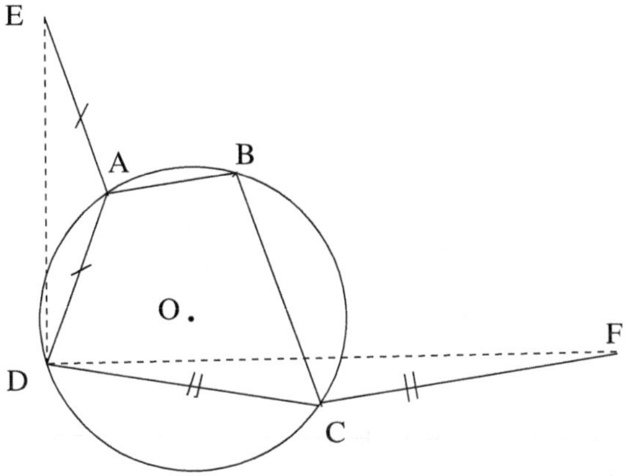

Figure 1.24: For Exercise 1.11.6.

Exercise 1.11.6 *In Figure 1.24, the quadrilateral $ABCD$ is inscribed in a circle. Construct E and F, such that $AE \| BC$, $|AE| = |AD|$, $CF \| AB$, $|CF| = |CD|$. Show the perpendicularity relation $DF \perp DE$.*

Definition 1.11.7 *Let $\mathcal{C}$ be a circle. A line ℓ is called tangent line to the circle if the intersection $\ell \cap \mathcal{C}$ is only one point. If the intersection $\ell \cap \mathcal{C}$ is two points, then the line ℓ is called a secant line.*

Exercise 1.11.8 *Prove that a line cannot intersect a circle at more than two points.*

Exercise 1.11.9 *Let $A \in \mathcal{C}$ be a point on the circle of center O. If ℓ is a line passing through A such that the radius OA is perpendicular to the line ℓ, show that ℓ is a tangent line to the circle $\mathcal{C}$.*

Exercise 1.11.10 *Let $A \in \mathcal{C}$ be a point on the circle of center O. If ℓ is the tangent line to the circle passing through A, show that OA is perpendicular on ℓ.*

Exercise 1.11.11 *Let $A \in \mathcal{C}$ be a point on a circle. Prove the existence and uniqueness of the tangent line to the circle $\mathcal{C}$ passing through A.*

Exercise 1.11.12 *Let A be a point in the exterior of a circle $\mathcal{C}$ and let $T_1, T_2 \in \mathcal{C}$ such that AT_1 and AT_2 are tangent to the circle. Show that $|AT_1| = |AT_2|$.*

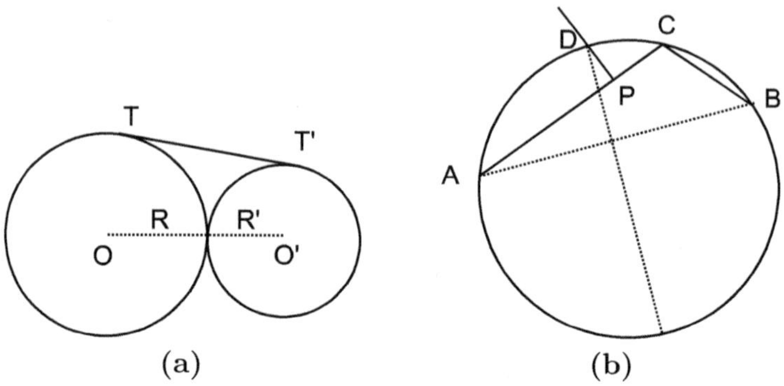

Figure 1.25: (a) For Exercise 1.11.15; (b) For Exercise 1.11.18.

Exercise 1.11.13 *Let AB be a chord on the circle $\mathcal{C}(O,R)$. Let M be the midpoint of AB. Prove that $OM \perp AB$.*

Exercise 1.11.14 *Let AB be a chord on the circle $\mathcal{C}(O,R)$. Prove that the perpendicular line through O to the chord AB passes though the midpoint of AB.*

Exercise 1.11.15 *Let $\mathcal{C}(O,R)$ and $\mathcal{C}(O',R')$ be two tangent exterior circles. Let TT' be their common tangent, see Figure 1.25(a). Find the length of $|TT'|$ in terms of the radii R and R'.*

Exercise 1.11.16 *Let $A, B, A', B' \in \mathcal{C}$ be four points on a circle. Show that $arc(AB) = arc(A'B')$ if and only if $|AB| = |A'B'|$ (equal arcs correspond to equal chords).*

Exercise 1.11.17 *Let ΔABC be an isosceles triangle and $M \in BC$ such that $AM \perp BC$. Show that $|BM| = |MC|$.*

Solved Exercise 1.11.18 (the broken chord property) *Let $\mathcal{C}$ be a circle centered at O and consider the broken chord ACB, see Figure 1.25(b). Let D be the midpoint of the arc AB and P be the foot of the perpendicular from D to the broken chord ACB. Show that*

$$|AP| = |PC| + |CB|.$$

Proof: We construct the point $P' \in AP$ such that $|C'P| = |PC|$, see Figure 1.26. To prove the relation $|AP| = |PC|+|CB|$, we need to show that $|AC'| = |CB|$. This follows from the triangle congruence $\Delta ADC' \equiv \Delta BDC$ by (ASA) case as follows. First, we have the identity of inscribed angles

$$\widehat{DAC'} = \widehat{DAC} = \frac{1}{2}arc(DC) = \widehat{CBD}.$$

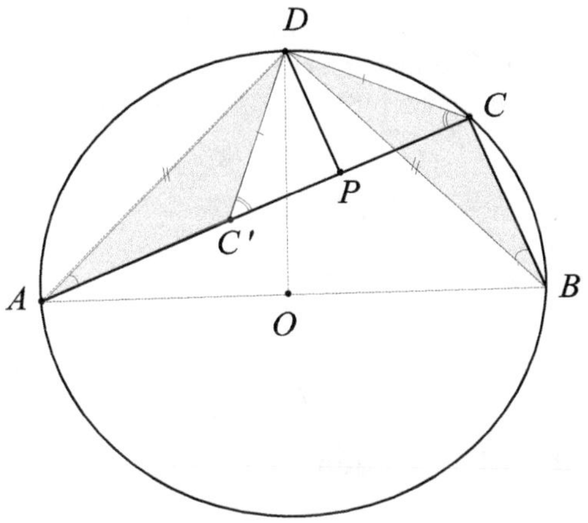

Figure 1.26: The proof of the broken chord theorem.

Then we have $|AD| = |DB|$ as chords corresponding to equal arcs. By construction, $\Delta DCC'$ is isosceles, so its base angles are equal, $\widehat{DCC'} = \widehat{DC'C}$. Then by the exterior angle identity,

$$\widehat{ADC'} = \widehat{DC'P} - \widehat{DAC'} = \widehat{DCC'} - \widehat{DAC'} = \frac{1}{2}(AD) - \frac{1}{2}(DC)$$
$$= \frac{1}{2}(DB) - \frac{1}{2}(DC) = \frac{1}{2}arc(CB) = \widehat{BDC}.$$

■

Exercise 1.11.19 *Consider an equilateral triangle ΔABC. Choose $E \in AB$ and $D \in AC$ such that $|AE| = 2|BE|$ and $|DC| = 2|AD|$. Let $F = CE \cap BD$. Show the following.*

(i) $\widehat{DFC} = 60°$.

(ii) $\widehat{AFC} = 90°$.

Exercise 1.11.20 *Consider a triangle ΔABC inscribed in a circle. Using the property of inscribed angles, show that $\widehat{A} + \widehat{B} + \widehat{C} = 180°$.*

1.12 Useful algebraic inequalities

Sometimes geometric inequalities can be reduced to certain algebraic inequalities. This section introduces a few of these algebraic inequalities.

We shall start with the most remarkable. Let $a, b > 0$ be two real numbers. We consider the arithmetic, geometric and harmonic means of two positive numbers a and b by $A(a,b)$, $G(a,b)$ and $H(a,b)$, respectively, where

$$\begin{aligned} A(a,b) &= \frac{a+b}{2}, \\ G(a,b) &= \sqrt{ab}, \\ H(a,b) &= \frac{1}{A\left(\frac{1}{a}, \frac{1}{b}\right)} = \frac{2}{\frac{1}{a}+\frac{1}{b}} = \frac{2ab}{a+b}. \end{aligned}$$

Proposition 1.12.1 (The inequality of means) *For any $a, b > 0$ the following inequalities hold:*

$$\min\{a,b\} \leq H(a,b) \leq G(a,b) \leq A(a,b) \leq \max\{a,b\}.$$

Proof: Each inequality can be shown by a straightforward computation. For instance, the inequality between the harmonic and the geometric means can be proved as follows

$$H(a,b) \leq G(a,b) \Longleftrightarrow \frac{2ab}{a+b} \leq \sqrt{ab} \Longleftrightarrow \frac{(2ab)^2}{(a+b)^2} \leq ab \Longleftrightarrow$$
$$4ab \leq a^2 + 2ab + b^2 \Longleftrightarrow a^2 + b^2 - 2ab \geq 0 \Longleftrightarrow (a-b)^2 \geq 0,$$

which holds true for all values of a and b, with identity reached for $a = b$. ■

Corollary 1.12.2 *If $x, y > 0$ then*

$$\frac{x}{y} + \frac{y}{x} \geq 2.$$

The equality holds for $x = y$.

Proof: Let $a = \frac{x}{y}$ and $b = \frac{y}{x}$. Then the above relation is equivalent to $a + b \geq 2$. Since $ab = 1$, applying the inequality between the geometric and arithmetic means yields the desired inequality. The equality holds exactly for $a = b$, i.e., for $x = y$. ■

Lemma 1.12.3 *If $x, y, z > 0$ then*

$$(x+y)(y+z)(z+x) \geq 8xyz.$$

Equality holds exactly when $x = y = z$.

Proof: An algebraic computation yields

$$\begin{aligned}\frac{(x+y)(y+z)(z+x)}{xyz} &= \frac{(xy+xz+y^2+yz)(z+x)}{xyz}\\ &= \frac{xyz+xz^2+y^2z+yz^2+x^2y+x^2z+y^2x+yzx}{xyz}\\ &= 1+\frac{z}{y}+\frac{y}{x}+\frac{z}{x}+\frac{x}{z}+\frac{x}{y}+\frac{y}{z}+1\\ &= 1+1+\underbrace{\left(\frac{z}{y}+\frac{y}{z}\right)}_{\geq 2}+\underbrace{\left(\frac{y}{x}+\frac{x}{y}\right)}_{\geq 2}+\underbrace{\left(\frac{z}{x}+\frac{x}{z}\right)}_{\geq 2}\geq 8,\end{aligned}$$

by an application of Corollary 1.12.2. The equality holds when $x = y = z$. ■

We recall that a function is concave if its graph is bent downwards. Equivalently, a function is concave if the chord joining any two points on the graph lies below the graph between these points. For instance, $\ln x$ on $(0, \infty)$, $\sin x$ on $(0, \pi)$, $\cos x$ on $(0, \pi/2)$, $\sin^2 x$ on $(0, \pi)$, and $\ln \sin x$ on $(0, \pi)$ are concave functions.
We also recall the sum and the product notations of n numbers

$$\sum_{k=1}^{n} x_k = x_1 + x_2 + \cdots + x_n, \qquad \prod_{k=1}^{n} x_k = x_1 x_2 \cdots x_n.$$

Proposition 1.12.4 (Jensen's inequality) *Let $f : (a, b) \to \mathbb{R}$ be a concave function. Then for any $x_1, \ldots, x_n \in (a, b)$, we have*

$$f\left(\frac{1}{n}\sum_{k=1}^{n} x_k\right) \geq \frac{1}{n}\sum_{k=1}^{n} f(x_k). \tag{1.12.1}$$

This result is proved by Calculus methods. Heuristically, it states the effect of the curvature of the graph of f with respect to centroids as follows: the coordinate of the centroid of the equally weighted system of points $\{f(x_k)\}$ is smaller than the value of the function f applied to the centroid of the equally weighted system of points $\{x_k\}$.

This behavior is graphically clear in the case of two points, see Figure 1.27. Let AB be a chord with endpoints $A(a, f(a))$ and $B(b, f(b))$ on the graph of $y = f(x)$, with f concave function. The midpoint of AB, denoted by M, has coordinates $\left(\frac{a+b}{2}, \frac{f(a)+f(b)}{2}\right)$. Consider the point C with coordinates $\left(\frac{a+b}{2}, f\left(\frac{a+b}{2}\right)\right)$. The fact that the graph of $y = f(x)$ is bent downward makes the point C to be above the point M, which means that

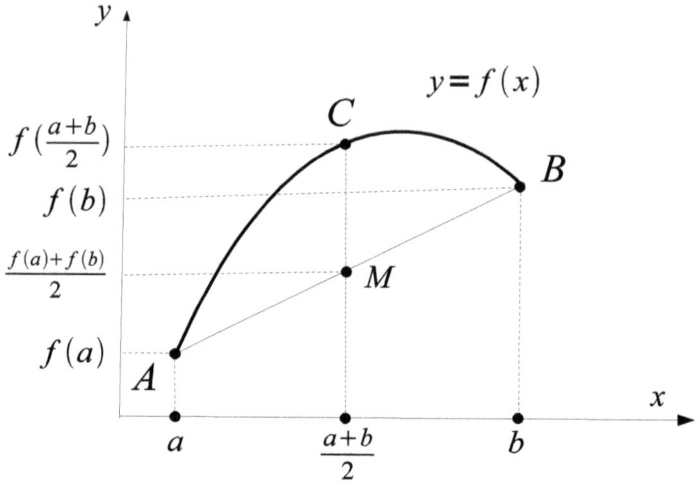

Figure 1.27: The graphical interpretation of Jensen's inequality.

their y-coordinates satisfy the inequality $f\left(\dfrac{a+b}{2}\right) \geq \dfrac{f(a)+f(b)}{2}$. The identity is reached when $C = M$, namely, when the graph of $y = f(x)$ is a straight line.

Remark 1.12.5 There is a similar Jensen's inequality for the case of convex functions. This is obtained by replacing " $\geq$ " in (1.12.1) by " $\leq$ ". However, since in geometric applications we shall use only concave functions, we shall use only inequality (1.12.1).

A few more applications of Jensen's inequality to triangles are as follows. 1. The angles of any triangle ΔABC have values in $(0°, 180°)$. We also know that the sum of their measure is $\widehat{A}+\widehat{B}+\widehat{C} = 180°$. Applying (1.12.1) for the concave function $f(x) = \sin x$ on $(0°, 180°)$ yields

$$\sin\frac{\widehat{A}+\widehat{B}+\widehat{C}}{3} \geq \frac{1}{3}\left(\sin\widehat{A}+\sin\widehat{B}+\sin\widehat{C}\right)$$

which simplifies to

$$\sin\widehat{A}+\sin\widehat{B}+\sin\widehat{C} \leq \frac{3\sqrt{3}}{2}. \tag{1.12.2}$$

We note that the identity is reached for the case of an equilateral triangle, $\widehat{A} = \widehat{B} = \widehat{C} = 60°$.

2. To obtain a similar inequality for products of sines, we apply Jensen's inequality for the concave function $f(x) = \ln \sin x$. Thus

$$\ln \sin \frac{\widehat{A} + \widehat{B} + \widehat{C}}{3} \geq \frac{1}{3}\left(\ln \sin \widehat{A} + \ln \sin \widehat{B} + \ln \sin \widehat{C}\right) \Longleftrightarrow$$

$$\ln \frac{\sqrt{3}}{2} \geq \ln \left(\sin \widehat{A} \sin \widehat{B} \sin \widehat{C}\right)^{1/3} \Longleftrightarrow$$

$$\sin \widehat{A} \sin \widehat{B} \sin \widehat{C} \leq \frac{3\sqrt{3}}{8}, \tag{1.12.3}$$

with identity also reached for the equilateral triangle. We make the remark that the concavity of $f(x) = \ln \sin x$ follows from the second derivative test, $f''(x) = -\dfrac{1}{\sin^2 x} < 0$.

Exercise 1.12.6 *Show that in any triangle ΔABC, we have*

$$\sin^2 \widehat{A} + \sin^2 \widehat{B} + \sin^2 \widehat{C} \leq \frac{9}{4}.$$

Exercise 1.12.7 *Show that in any acute-angled triangle ΔABC, we have*

$$\cos \widehat{A} + \cos \widehat{B} + \cos \widehat{C} \leq \frac{3}{2}.$$

Exercise 1.12.8 *Show that in any triangle ΔABC, we have*

$$\cos \widehat{A} \cos \widehat{B} \cos \widehat{C} \leq \frac{1}{8}.$$

The next result infers the inequality of means from Jensen's inequality.

Proposition 1.12.9 (Generalized inequality of means) *For any positive numbers x_k, $1 \leq k \leq n$, we have*

$$\frac{n}{\displaystyle\sum_{k=1}^{n} \frac{1}{x_k}} \leq \left(\prod_{k=1}^{n} x_k\right)^{1/n} \leq \frac{1}{n} \sum_{k=1}^{n} x_k. \tag{1.12.4}$$

All identities are reached simultaneously for the case $x_1 = \cdots = x_n$.

Proof: Jensen's inequality applied for the concave function $f(x) = \ln x$ yields

$$\ln \left(\frac{1}{n} \sum_{k=1}^{n} x_k\right) \geq \frac{1}{n} \sum_{k=1}^{n} \ln(x_k).$$

Using the properties of the logarithmic function, the right side becomes

$$\frac{1}{n}\sum_{k=1}^{n}\ln(x_k) = \frac{1}{n}\ln\left(\prod_{k=1}^{n}x_k\right) = \ln\left(\prod_{k=1}^{n}x_k\right)^{1/n}.$$

Substituting in the previous inequality, yields

$$\ln\left(\frac{1}{n}\sum_{k=1}^{n}x_k\right) \geq \ln\left(\prod_{k=1}^{n}x_k\right)^{1/n}.$$

Using the injectivity of the logarithmic function, we obtain the following inequality between the geometric and arithmetic means

$$\left(\prod_{k=1}^{n}x_k\right)^{1/n} \leq \frac{1}{n}\sum_{k=1}^{n}x_k. \tag{1.12.5}$$

The first inequality in (1.12.4), namely, the inequality between the harmonic and the geometric means, follows from (1.12.5). To this end, we apply (1.12.5) for $y_k = 1/x_k$

$$\left(\prod_{k=1}^{n}y_k\right)^{1/n} \leq \frac{1}{n}\sum_{k=1}^{n}y_k \Longleftrightarrow \left(\prod_{k=1}^{n}\frac{1}{x_k}\right)^{1/n} \leq \frac{1}{n}\sum_{k=1}^{n}\frac{1}{x_k} \Longleftrightarrow$$

$$\frac{1}{\left(\prod_{k=1}^{n}x_k\right)^{1/n}} \leq \frac{\sum_{k=1}^{n}\frac{1}{x_k}}{n} \Longleftrightarrow \left(\prod_{k=1}^{n}x_k\right)^{1/n} \geq \frac{n}{\displaystyle\sum_{k=1}^{n}\frac{1}{x_k}}.$$

■

Exercise 1.12.10 *Show that for any three positive numbers, x, y, z, we have*

$$\frac{x}{y}+\frac{y}{z}+\frac{z}{x} \geq 3.$$

Another type of algebraic inequality is obtained from the sign of the discriminant of a quadratic polynomial as follows.

Proposition 1.12.11 *Let $f(x) = ax^2 + bx + c$ be a quadratic polynomial with real coefficients, $a, b, c \in \mathbb{R}$. If $f(x) \geq 0$ for all $x \in \mathbb{R}$, then $b^2 \leq 4ac$.*

Proof: Since $ax^2 + bx + c \geq 0$ for all real values of x, the parabola $y = f(x)$ either does not intersect the x-axis, or it is tangent to it. Hence, the quadratic equation $f(x) = 0$ has at most one real root, which means that its discriminant is non-positive, $\Delta = b^2 - 4ac \leq 0$, fact that implies the desired inequality. ■

We make the remark that we also have $a > 0$, namely, the parabola is oriented upwards, and $c > 0$, namely the y-intercept of the parabola is above the x-axis. Therefore $ac > 0$. From here, using the inequality of means, we have

$$|b| \leq 2\sqrt{ac} \leq 2\frac{a+c}{2} = a + c.$$

Proposition 1.12.12 *For any $a, b, c \in \mathbb{R}$, we have*

$$a^2 + b^2 + c^2 \geq ab + bc + ca.$$

Proof: The inequality is equivalent to

$$\begin{aligned} a^2 + b^2 + c^2 \geq ab + bc + ca &\iff \\ a^2 + b^2 - 2ab + c^2 \geq -ab + bc + ca &\iff \\ (a-b)^2 \geq bc + ca - ab - c^2 &\iff \\ (a-b)^2 \geq (a-c)(c-b). \end{aligned}$$

Denoting $x = a - c$, $y = c - b$, we have $x + y = a - b$. Then it suffices to show $(x+y)^2 \geq xy$. This is equivalent to the inequality $x^2 + y^2 + xy \geq 0$, which is obvious for $x = 0$. If $x \neq 0$, then this becomes $x^2(1 + \frac{y}{x} + (\frac{y}{x})^2) \geq 0$. Let $u = \frac{y}{x}$. Since the quadratic function $f(u) = u^2 + u + 1$ does not have any real roots (it has a negative discriminant), then its sign is constant. Since $f(0) = 1 > 0$, then $u^2 + u + 1 > 0$. Hence, $x^2 + y^2 + xy \geq 0$, with equality when $x = y = 0$.

Proof variant: A more elegant proof uses Rolle's theorem. To this end, we consider the following polynomial

$$P(x) = (x-a)(x-b)(x-c) = x^3 - (a+b+c)x^2 + (ab+bc+ca)x + abc,$$

which has three real roots, a, b, and c. By Rolle's theorem, its derivative

$$P'(x) = 3x^2 - 2(a+b+c)x + (ab+bc+ca)$$

has two real roots, so its discriminant is non-negative

$$\begin{aligned} \Delta = 4(a+b+c)^2 - 4 \cdot 3 \cdot (ab+bc+ca) \geq 0 &\iff \\ (a+b+c)^2 \geq 3 \cdot (ab+bc+ca) &\iff \\ a^2 + b^2 + c^2 + 2(ab+bc+ca) \geq 3 \cdot (ab+bc+ca) &\iff \\ a^2 + b^2 + c^2 \geq ab + bc + ca, \end{aligned}$$

which shows the inequality. ■

Other algebraic inequalities are provided by the boundedness of trigonometric functions, such as $\sin x \leq 1$ and $\cos x \leq 1$, for all $x \in \mathbb{R}$. For instance, the law of cosines states that $\cos \widehat{A} = \frac{b^2+c^2-a^2}{2bc}$. Applying that $\cos \widehat{A} \leq 1$ implies the inequality $\frac{b^2+c^2-a^2}{2bc} \leq 1$, where $\{a, b, c\}$ denote the side lengths of a triangle ΔABC. After cross multiplication, this inequality can be written equivalently as $b^2 + c^2 - 2bc \leq a^2$, or, after using the binomial formula and taking the square root, $|b - c| \leq a$. It is interesting that we obtained a geometric inequality, which is a particular form of the triangle inequality.

It is worth noting that many geometric inequalities get reduced to the triangle inequality. This is stated by an axiom which defines the distance in the plane and can be restated as follows:

If A, B and C are three points in the plane, then $|AB| + |BC| \geq |AC|$, with identity reached in the case when the points belong to the same line, with B between points A and C.

Sometimes, a slightly more elaborate version is more suitable:

Proposition 1.12.13 *Let $P_1, P_2, \ldots, P_n$ be points in the plane. Then*

$$|P_1P_n| \leq |P_1P_2| + |P_2P_3| + \cdots + |P_{n-1}P_n|.$$

Proof: The proof is done by induction over n. For $n = 3$, we obtain the usual triangle inequality. We assume the statement valid for some $n \geq 3$, namely

$$\mathcal{S}_n : \quad |P_1P_n| \leq |P_1P_2| + |P_2P_3| + \cdots + |P_{n-1}P_n|$$

holds. Adding in both sides $|P_nP_{n+1}|$ and applying the triangle inequality, $\mathcal{S}_n$ implies

$$|P_1P_{n+1}| \leq |P_1P_n| + |P_nP_{n+1}| \leq |P_1P_2| + |P_2P_3| + \cdots + |P_{n-1}P_n| + |P_nP_{n+1}|.$$

Hence, we proved the statement

$$\mathcal{S}_{n+1} : \quad |P_1P_{n+1}| \leq |P_1P_2| + |P_2P_3| + \cdots + |P_{n-1}P_n| + |P_nP_{n+1}|$$

which resolves the induction step. By the method of mathematical induction, the statement $\mathcal{S}_n$ holds true for any $n \geq 3$. ∎

Chapter 2

Triangle Geometry

Let no one ignorant of geometry enter here.
Plato's academy entrance

The metric approach used by this chapter is mainly based on trigonometric computation. Thus, we shall start with a brief review of the basic trigonometric functions and their properties.

2.1 A review of trigonometric functions

Let ΔABC be a right triangle with $\widehat{A} = 90°$, see Figure 2.1(a). As the sum of the angles in a triangle is $180°$, then $\widehat{B} + \widehat{C} = 90°$. The definitions of the basic trigonometric functions using triangle ΔABC are as follows:

$$\begin{aligned}\sin\widehat{C} &= \frac{\text{opposite}}{\text{hypotenuse}} = \frac{c}{a}, \quad \cos\widehat{C} = \frac{\text{adjacent}}{\text{hypotenuse}} = \frac{b}{a},\\ \tan\widehat{C} &= \frac{\text{opposite}}{\text{adjacent}} = \frac{c}{b}, \quad \cot\widehat{C} = \frac{1}{\tan\widehat{C}} = \frac{b}{c}.\end{aligned}$$

We note that

$$\begin{aligned}\sin\widehat{C} = \frac{c}{a} = \cos\widehat{B} = \cos(90° - \widehat{C}),\\ \cos\widehat{C} = \frac{b}{a} = \sin\widehat{B} = \sin(90° - \widehat{C}).\end{aligned}$$

The above trigonometric functions can be extended to $(0°, 180°)$, see Figures 2.1(b), (c) and (d). The function sine is increasing on $(0°, 90°)$, and is decreasing on $(90°, 180°)$. The function cosine is decreasing on the whole

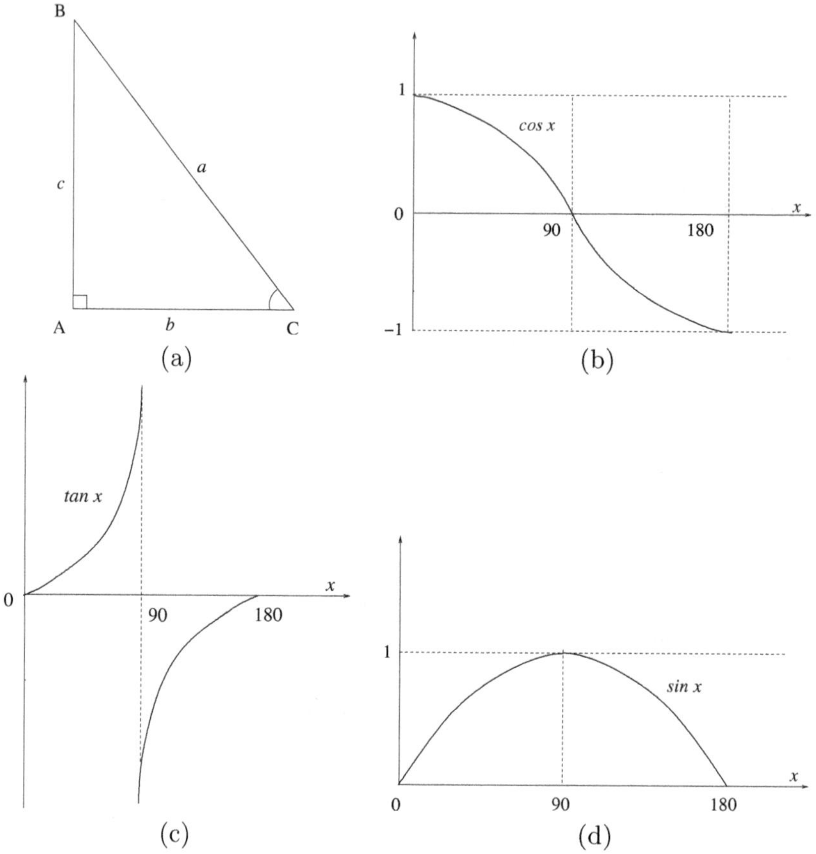

Figure 2.1: Graphs of trigonometric functions.

interval and has negative values on $(90°, 180°)$. The function tangent is increasing, having a vertical asymptote at $x = 90°$.

The facts concerning monotonicity are given by the following table.

x	$0°$		$90°$		$180°$
$\sin x$	0	↗	1	↘	0
$\cos x$	1	↘	0	↘	-1
$\tan x$	0	↗	$+\infty \| -\infty$	↗	0
$\cot x$	$+\infty$	↘	0	↘	$-\infty$

Proposition 2.1.1 *We have*
(i) $\sin^2 x + \cos^2 x = 1$ *(the fundamental identity of trigonometry)*

(ii) $\sin(90° - x) = \cos x$, $\cos(90° - x) = \sin x$
(iii) $0 < \sin x \leq 1$, $-1 < \cos x < 1$, *for any* $x \in (0, 180°)$.

Proof: *(i)* Consider a right triangle with $\widehat{A} = 90$ and $\widehat{C} = x$. Using the definition of the sin and cos functions and the Pythagorean theorem we have

$$\sin^2 x + \cos^2 x = \left(\frac{c}{a}\right)^2 + \left(\frac{b}{a}\right)^2 = \frac{c^2 + b^2}{a^2} = \frac{a^2}{a^2} = 1.$$

The fundamental identity of trigonometry and the Pythagorean theorem are equivalent.
(ii) This comes from the definition of the trigonometric functions as quotients of lengths of sides.
(iii) See the above monotonicity table. ∎

Exercise 2.1.2 *Show that*

$$(i) \qquad 1 + \tan^2 x = \frac{1}{\cos^2 x}$$

$$(ii) \qquad 1 + \cot^2 x = \frac{1}{\sin^2 x}.$$

The following identities transform the sums into products:

Proposition 2.1.3 *We have*

$$\begin{aligned}
\cos(x + y) &= \cos x \cos y - \sin x \sin y, \\
\cos(x - y) &= \cos x \cos y + \sin x \sin y, \\
\sin(x + y) &= \sin x \cos y + \cos x \sin y, \\
\sin(x - y) &= \sin x \cos y - \cos x \sin y, \text{ and} \\
\tan(x + y) &= \frac{\tan x + \tan y}{1 - \tan x \tan y}.
\end{aligned}$$

Remark 2.1.4 The above proposition shows that the trigonometric functions are not additive. This means that $\sin(x+y)$ is not equal to $\sin x + \sin y$! The proof of the first four relations follow from the trigonometric circle and the definitions of sine and cosine functions as projections on the axes.

Corollary 2.1.5 *We have*

$$\begin{aligned}
\cos x \cos y &= \frac{1}{2}\big[\cos(x+y) + \cos(x-y)\big], \\
\sin x \cos y &= \frac{1}{2}\big[\sin(x+y) + \sin(x-y)\big], \\
\sin x \sin y &= \frac{1}{2}\big[\cos(x-y) - \cos(x+y)\big], \text{and} \\
\cos x \sin y &= \frac{1}{2}\big[\sin(x+y) - \sin(x-y)\big].
\end{aligned}$$

Proof: Add or subtract two by two the equations in Proposition 2.1.3, either the first two or the last two equations. ■

Corollary 2.1.6 *The following formulas of transforming the products into sums hold:*

$$\begin{aligned}
2\cos\frac{\alpha+\beta}{2}\cos\frac{\alpha-\beta}{2} &= \cos\alpha + \cos\beta, \\
2\sin\frac{\alpha+\beta}{2}\cos\frac{\alpha-\beta}{2} &= \sin\alpha + \sin\beta, \\
2\sin\frac{\alpha+\beta}{2}\sin\frac{\alpha-\beta}{2} &= \cos\beta - \cos\alpha, \text{ and} \\
2\cos\frac{\alpha+\beta}{2}\sin\frac{\alpha-\beta}{2} &= \sin\alpha - \sin\beta.
\end{aligned}$$

Proof: Substitute $\alpha = x+y$ and $\beta = x-y$ in the Corollary 2.1.5. ■

Making $x = y$ in Proposition 2.1.3 yields the formula for the double angle.

Corollary 2.1.7 *We have*

$$\begin{aligned}
(i) \qquad \cos 2x &= \cos^2 x - \sin^2 x = 1 - 2\sin^2 x = 2\cos^2 x - 1, \text{and} \\
(ii) \qquad \sin 2x &= 2\sin x \cos x.
\end{aligned}$$

Proof: The first identity in (i) and (ii) comes from the substitution $x = y$ in Proposition 2.1.3. By the fundamental identity of trigonometry

$$\begin{aligned}
\cos^2 x - \sin^2 x &= 1 - \sin^2 x - \sin^2 x = 1 - 2\sin^2 x, \text{and} \\
\cos^2 x - \sin^2 x &= \cos^2 x - (1 - \cos^2 x) = 2\cos^2 x - 1,
\end{aligned}$$

which proves the second and the third identity of (i). ■

Exercise 2.1.8 *For any x, where $0° < x < 90°$, prove that*

$$\begin{aligned}
(i) \qquad \sin x &= \sqrt{\frac{1-\cos 2x}{2}}, \\
(ii) \qquad \cos x &= \sqrt{\frac{1+\cos 2x}{2}}, \\
(iii) \qquad \tan x &= \sqrt{\frac{1-\cos 2x}{1+\cos 2x}}.
\end{aligned}$$

How do the above formulas change for $90° < x < 180°$?

Exercise 2.1.9 *Let AB be a chord on the circle $\mathcal{C}(O, R)$. Show that $|AB| = 2R\sin\frac{\widehat{AOB}}{2}$.*

Exercise 2.1.10 *Find formulas for $\sin(x+y+z)$ and $\cos(x+y+z)$ in terms of trigonometric functions of x, y and z.*

Exercise 2.1.11 *Prove the formula*

$$\tan(x+y+z) = \frac{\tan x + \tan y + \tan z - \tan x \tan y \tan z}{1 - \tan x \tan y - \tan y \tan z - \tan z \tan x}.$$

Exercise 2.1.12 *Using Exercise 2.1.11 show that in any triangle ΔABC, we have*

$$\tan\widehat{A} + \tan\widehat{B} + \tan\widehat{C} = \tan\widehat{A}\tan\widehat{B}\tan\widehat{C}.$$

Exercise 2.1.13 *Show that*

$$\tan^{-1}(1) + \tan^{-1}(2) + \tan^{-1}(3) = 180°.$$

An application of the previous formulas is to transform the product of cosines into sums of squares of sines and cosines.

Proposition 2.1.14 *In any triangle ΔABC, we have*

$$1 + \cos\widehat{A}\cos\widehat{B}\cos\widehat{C} = \frac{1}{2}\left(\sin^2\widehat{A} + \sin^2\widehat{B} + \sin^2\widehat{C}\right) \qquad (2.1.1)$$

$$1 - 2\cos\widehat{A}\cos\widehat{B}\cos\widehat{C} = \cos^2\widehat{A} + \cos^2\widehat{B} + \cos^2\widehat{C}. \qquad (2.1.2)$$

Proof: We shall use a trigonometric computation that uses repeatedly the "product to sums" formula

$$\cos x \cos y = \frac{1}{2}\left(\cos(x+y) + \cos(x-y)\right).$$

Thus, we have

$$
\begin{aligned}
\left(\cos\widehat{A}\cos\widehat{B}\right)\cos\widehat{C} &= \frac{1}{2}\left(\cos\left(\widehat{A}+\widehat{B}\right)+\cos\left(\widehat{A}-\widehat{B}\right)\right)\cos\widehat{C} \\
&= \frac{1}{2}\left(\cos\left(\widehat{A}+\widehat{B}\right)\cos\widehat{C}+\cos\left(\widehat{A}-\widehat{B}\right)\cos\widehat{C}\right) \\
&= \frac{1}{4}\left[\cos\left(\widehat{A}+\widehat{B}+\widehat{C}\right)+\cos\left(\widehat{A}+\widehat{B}-\widehat{C}\right)\right. \\
&\quad \left.+\cos\left(\widehat{A}-\widehat{B}+\widehat{C}\right)+\cos\left(\widehat{A}-\widehat{B}-\widehat{C}\right)\right].
\end{aligned}
$$

Using $\widehat{A}+\widehat{B}+\widehat{C}=180°$ and $\cos(180°-x)=-\cos x$, we obtain

$$
\begin{aligned}
\cos\widehat{A}\cos\widehat{B}\cos\widehat{C} &= \frac{1}{4}\left(-1-\cos\left(2\widehat{C}\right)-\cos\left(2\widehat{B}\right)-\cos\left(2\widehat{A}\right)\right) \\
&= -\frac{1}{4}\left(1+1-2\sin^2\widehat{C}+1-2\sin\widehat{B}+1-2\sin\widehat{A}\right) \\
&= -1+\frac{1}{2}\left(\sin^2\widehat{A}+\sin^2\widehat{B}+\sin^2\widehat{C}\right),
\end{aligned}
$$

where we used the double angle formula $\cos(2x)=1-2\sin^2 x$.

To show formula (2.1.2), we shall proceed similarly as in the computation that led to (2.1.1), using at the end the formula $\cos(2x)=2\cos^2 x-1$ as follows

$$
\begin{aligned}
\cos\widehat{A}\cos\widehat{B}\cos\widehat{C} &= \frac{1}{4}\left(-1-\cos\left(2\widehat{C}\right)-\cos\left(2\widehat{B}\right)-\cos\left(2\widehat{A}\right)\right) \\
&= -\frac{1}{4}\left(1+2\cos^2\widehat{C}-1+2\cos^2\widehat{B}-1+2\cos^2\widehat{A}-1\right) \\
&= \frac{1}{2}-\frac{1}{2}\left(\cos^2\widehat{A}+\cos^2\widehat{B}+\cos^2\widehat{C}\right),
\end{aligned}
$$

which is equivalent to (2.1.2). ■

Solved Exercise 2.1.15 *Show that in any triangle ΔABC, the following identity holds*

$$\cos\widehat{A}+\cos\widehat{B}+\cos\widehat{C}=1+4\sin\frac{\widehat{A}}{2}\sin\frac{\widehat{B}}{2}\sin\frac{\widehat{C}}{2}.$$

Proof: Using the formula of transforming products of sines into difference of cosines, we have

$$\begin{aligned}
\sin\frac{\widehat{A}}{2}\sin\frac{\widehat{B}}{2}\sin\frac{\widehat{C}}{2} &= \frac{1}{2}\left(\cos\frac{\widehat{A}-\widehat{B}}{2}-\cos\frac{\widehat{A}+\widehat{B}}{2}\right)\sin\frac{\widehat{C}}{2}\\
&= \frac{1}{2}\cos\frac{\widehat{A}-\widehat{B}}{2}\sin\frac{\widehat{C}}{2}-\frac{1}{2}\cos\frac{\widehat{A}+\widehat{B}}{2}\sin\frac{\widehat{C}}{2}\\
&= \frac{1}{4}\left(\sin\frac{\widehat{A}-\widehat{B}+\widehat{C}}{2}+\sin\frac{\widehat{C}-\widehat{A}+\widehat{B}}{2}\right)\\
&\quad -\frac{1}{4}\left(\sin\frac{\widehat{A}+\widehat{B}+\widehat{C}}{2}+\sin\frac{\widehat{C}-\widehat{A}-\widehat{B}}{2}\right)\\
&= \frac{1}{4}\left(\sin\left(90-\widehat{B}\right)+\sin\left(90-\widehat{A}\right)-1+\sin\left(90-\widehat{C}\right)\right)\\
&= \frac{1}{4}\left(\cos\widehat{B}+\cos\widehat{A}-1+\cos\widehat{C}\right),
\end{aligned}$$

which after the separation of the sum of cosines leads to the desired relation. ■

Exercise 2.1.16 *Show that in any triangle ΔABC, we have*

$$\sin\widehat{A}+\sin\widehat{B}+\sin\widehat{C}=4\cos\frac{\widehat{A}}{2}\cos\frac{\widehat{B}}{2}\cos\frac{\widehat{C}}{2}.$$

Solved Exercise 2.1.17 *Show that*

$$\sin(u+y)\sin(u-y)=\sin^2 u-\sin^2 y.$$

Proof: Using difference of squares formula as well as the formulas for transforming the sums and differences into products, we have

$$\begin{aligned}
\sin^2 u-\sin^2 y &= (\sin u-\sin y)(\sin u+\sin y)\\
&= 2\cos\frac{u+y}{2}\sin\frac{u-y}{2}\cdot 2\sin\frac{u+y}{2}\cos\frac{u-y}{2}\\
&= 2\sin\frac{u-y}{2}\cos\frac{u-y}{2}\cdot 2\sin\frac{u+y}{2}\cos\frac{u+y}{2}\\
&= \sin(u+y)\sin(u-y).
\end{aligned}$$

■

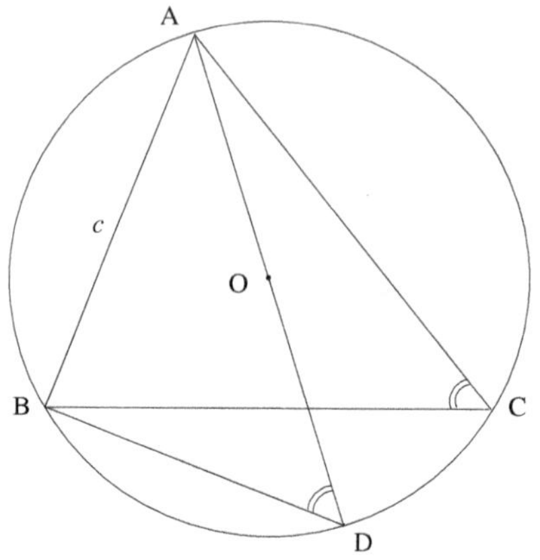

Figure 2.2: The proof of law of sines.

2.2 The law of sines

The relationship between the lengths of the sides and the sines of the opposite angles in a triangle is called *the law of sines.* Theorems in Euclid on lengths of chords contain the same ideas as the law of sines. This law was also known to Ptolemy.

We shall consider in the following a triangle ΔABC with side lengths a, b, c and angle measures $\widehat{A}$, $\widehat{B}$, $\widehat{C}$. Let R denote the radius of the circumcircle of ΔABC.

Theorem 2.2.1 (The law of sines) *We have*

$$\frac{a}{\sin\widehat{A}} = \frac{b}{\sin\widehat{B}} = \frac{c}{\sin\widehat{C}} = 2R.$$

Proof: Consider the triangle ΔABC inscribed in a circle with radius R. A diameter is drawn with one endpoint at A terminating at D. The right triangle ΔABD is created. Using the right triangle definitions of sine yields $\sin\widehat{ADB} = \dfrac{|AB|}{|AD|}$. The angles $\widehat{ACB}$ and $\widehat{ADB}$ are equal, as both are inscribed angles and are cutting the same arc. Therefore they have equal sines. Substituting, we get $\sin C = \dfrac{|AB|}{|AD|} = \dfrac{c}{2R}$ since AD is a diameter with length $2R$. This can be written as

$$\frac{c}{\sin\widehat{C}} = 2R.$$

Since the choice of angle C was arbitrary, we could show that the same holds for each side and opposite angle pair. We shall call this procedure *cyclic permutation*. Hence we obtain two analogous formulas

$$\frac{a}{\sin \widehat{A}} = 2R, \quad \frac{b}{\sin \widehat{B}} = 2R.$$

■

Exercise 2.2.2 *Let ΔABC be a right triangle with $\widehat{A} = 90$, $|AB| = 3$, $|AC| = 4$. Find $|AM|$, where M is the midpoint of BC.*

In triangle geometry, we often use the semiperimeter of a triangle instead of the perimeter. The explanation consists of the fact that the trigonometric formulas developed using the semiperimeter look simpler. The semiperimeter is defined by the relation

$$p = \frac{1}{2}(a + b + c).$$

Exercise 2.2.3 *Show that in any triangle, we have*

$$\max\{a, b, c\} \leq 2R.$$

Exercise 2.2.4 *Show that in any triangle*

$$p < 3R.$$

Why do we have a strict inequality?

Exercise 2.2.5 *Show that*

$$\sin \widehat{A} + \sin \widehat{B} + \sin \widehat{C} = \frac{p}{R}.$$

Exercise 2.2.6 *Let ΔABC be a triangle with $2a = b + c$. Show that*

$$2 \sin \frac{\widehat{A}}{2} = \cos \frac{\widehat{B} - \widehat{C}}{2}.$$

2.3 The area of a triangle

Consider the triangle ABC and let AA' be the altitude from the vertex A with length h_a. Let h_b and h_c be the lengths of the altitudes from vertices B and C.

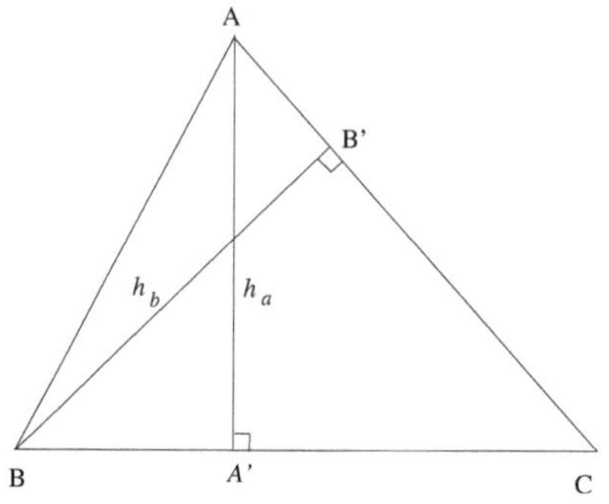

Figure 2.3: The altitudes of a triangle.

Theorem 2.3.1 *The product of the altitudes and the lengths of the corresponding sides is the same for all altitudes:*

$$ah_a = bh_b = ch_c. \tag{2.3.3}$$

Proof: Consider Figure 2.3. Using the definition of sine and using the right triangles $\Delta ACA'$ and $\Delta CBB'$, we have $h_a = b\sin\widehat{C}$ and $h_b = a\sin\widehat{C}$. Multiplying by a and respectively by b yields $ah_a = ab\sin\widehat{C} = bh_b$. Hence $ah_a = bh_b$ as they are equal to the same expression. By cyclic permutations we have $ah_a = ch_c$ and $ch_c = bh_b$. Then (2.3.3) obviously follows. ■

Definition 2.3.2 *The area of the triangle ΔABC is defined by*

$$\mathcal{A}_{ABC} = \frac{1}{2}ah_a = \frac{1}{2}bh_b = \frac{1}{2}ch_c.$$

Proposition 2.3.3 *Let $\mathcal{T}$ be the set of all triangles in the plane. The area function $\mathcal{A} : \mathcal{T} \to \mathbb{R}$ has the following properties.*
(i) $\mathcal{A} > 0$;
(ii) Let A' be a point in the interior of the segment BC of the triangle ABC. Then

$$\mathcal{A}_{ABC} = \mathcal{A}_{ABA'} + \mathcal{A}_{AA'C}.$$

Proof: Relation (i) is obvious. To show (ii), we apply the definition of the area in triangles ΔABC, $\Delta ABA'$ and $\Delta AA'C$:

$$\begin{aligned}\mathcal{A}_{ABC} &= \frac{1}{2}ah_a = \frac{1}{2}(|BA'| + |A'C|)h_a \\ &= \frac{1}{2}|BA'|h_a + \frac{1}{2}|A'C|h_a = \mathcal{A}_{ABA'} + \mathcal{A}_{AA'C},\end{aligned}$$

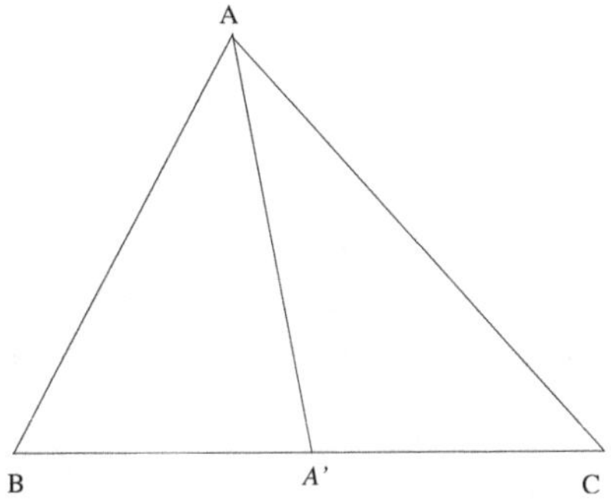

Figure 2.4: The median AA'.

where we used that triangles $\Delta ABA'$ and $\Delta AA'C$ share the same altitude from vertex A. ■

Remark 2.3.4 We note that in the case when A' is the midpoint of BC, the areas of triangles $\Delta ABA'$ and $\Delta ACA'$ become equal.

Exercise 2.3.5 *Show that the altitudes h_a and h_b cannot be parallel, see Figure 2.3.*

Exercise 2.3.6 *Let M be a point inside the triangle ΔABC. Show that*

$$\mathcal{A}_{ABC} = \mathcal{A}_{ABM} + \mathcal{A}_{BCM} + \mathcal{A}_{ACM}.$$

Exercise 2.3.7 *Let M be a point outside the triangle ΔABC. Assuming that M is inside the angle $\widehat{BAC}$, show that*

$$\mathcal{A}_{ABC} = \mathcal{A}_{ABM} - \mathcal{A}_{BCM} + \mathcal{A}_{ACM}.$$

Remark 2.3.8 The previous properties can be stated equivalently by saying that the area is an additive measure. Other examples of measure functions are the length, volume, defect, etc. Another example will be given below.

Another area measure example Consider a grid of equidistant dots and a polygon (convex or concave) with the vertices situated at the dots of the grid. Let the distance between the lines of the grid be 1 unit. Denote by S the area of the polygon (measured in square units), by N the number of dots of the grid situated in the interior of the polygon, and by k the number of the dots of the grid situated on the sides of the polygon (including the vertices).

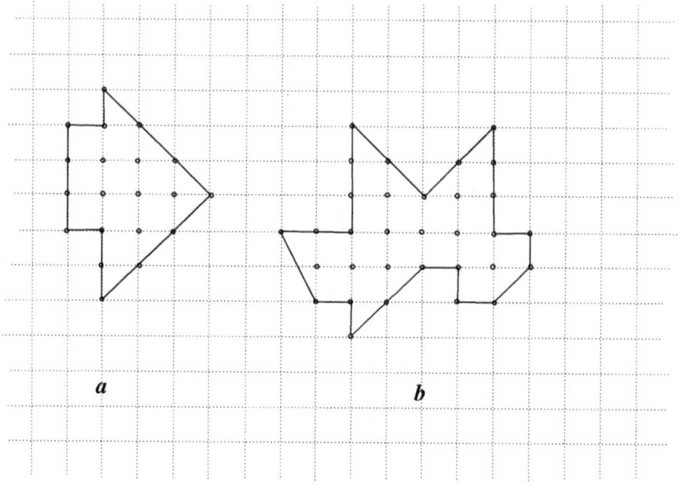

Figure 2.5: Polygons on the grid.

In the Figure 2.5(a) a simple counting yields

$$\begin{aligned} S &= 9+6\cdot\frac{1}{2}=12, \\ N &= 6, \text{and} \\ k &= 14. \end{aligned}$$

We note that

$$N+\frac{k}{2}-1=6+7-1=12=S.$$

In Figure 2.5(b) a careful counting provides

$$\begin{aligned} S &= 19.5 \\ N &= 9 \\ k &= 23. \end{aligned}$$

We can verify that

$$N+\frac{k}{2}-1=9+\frac{23}{2}-1=19.5=S.$$

Exercise 2.3.9 (*i*) *Show that for any polygon the following inequality holds*

$$N+\frac{k}{2}>1.$$

(*ii*) *Show that if a polygon* $\mathcal{P}$ *is partitioned into two polygons* $\mathcal{P}_1$ *and* $\mathcal{P}_2$, *then we have*

$$N + \frac{k}{2} = \left(N_1 + \frac{k_1}{2} - 1\right) + \left(N_2 + \frac{k_2}{2} - 1\right).$$

(*iii*) *(Fick's formula) The area of any polygon with the vertices situated in the dots of a unit grid is given by*

$$S = N + \frac{k}{2} - 1.$$

Remark 2.3.10 Exercise 2.3.9(i) states that the area is positive, i.e., $S > 0$. Exercise 2.3.9(ii) states that the function $S = N + \frac{k}{2} - 1$ is additive, i.e., $S = S_1 + S_2$.

2.4 More formulas for the area of a triangle

There are many formulas for the area of a triangle, involving different triangle elements, such as sides, angles, inradius, circumradius, semiperimeter, etc. This way, the area can be seen as a common base for different triangle elements, and can be regarded as the source of many useful metrical relations.

Proposition 2.4.1 *The area of the triangle ABC is given by*

$$\mathcal{A}_{ABC} = \frac{1}{2}ab\sin\widehat{C} = \frac{1}{2}ac\sin\widehat{B} = \frac{1}{2}bc\sin\widehat{A}.$$

Proof: Using the definition of the area and $h_a = c\sin B$, we have

$$\mathcal{A}_{ABC} = \frac{1}{2}ah_a = \frac{1}{2}ac\sin\widehat{B}.$$

By cyclic permutations we obtain the other two formulas. ■

The following formula for the area involves four elements, the side lengths and the circumradius.

Proposition 2.4.2 *We have*

$$\mathcal{A}_{ABC} = \frac{abc}{4R}.$$

Proof: From the law of sines

$$\frac{c}{\sin\widehat{C}} = 2R \Longrightarrow \sin\widehat{C} = \frac{c}{2R}.$$

Substituting, yields

$$\begin{aligned}\mathcal{A}_{ABC} &= \frac{1}{2}ab\sin\widehat{C} = \frac{1}{2}ab\frac{c}{2R} \\ &= \frac{abc}{4R}.\end{aligned}$$

■

A formula for the area of triangle ABC involving the angles and the circumradius is given in the following.

Proposition 2.4.3 *The area of* ΔABC *is given by*

$$\mathcal{A}_{ABC} = 2R^2 \sin\widehat{A}\sin\widehat{B}\sin\widehat{C}.$$

Proof: From the law of sines

$$\sin\widehat{A} = \frac{a}{2R}, \quad \sin\widehat{B} = \frac{b}{2R}, \quad \sin\widehat{C} = \frac{c}{2R}.$$

Then

$$2R^2 \sin\widehat{A}\sin\widehat{B}\sin\widehat{C} = 2R^2\frac{a\cdot b\cdot c}{(2R)^3} = \frac{abc}{4R},$$

which is the formula for the area given by Proposition 2.4.2. ■

2.5 The incircle of a triangle

Since the incircle of a triangle is related to the concept of angle bisector, we shall define this concept first and state its main properties.

Definition 2.5.1 *Let* $\widehat{AOB}$ *be an angle and* P *be a point in the interior of the angle. Then* OP *is called the angle bisector of angle* $\widehat{AOB}$ *if* $\widehat{AOP} = \widehat{POB}$.

In the following, we shall denote by $d(P, \ell)$ the distance from the point P to the line ℓ.

Theorem 2.5.2 *Let* $\widehat{AOB}$ *be an angle and* P *be a point in the interior of this angle. Then* OP *is the angle bisector if and only if* $d(P, OA) = d(P, OB)$.

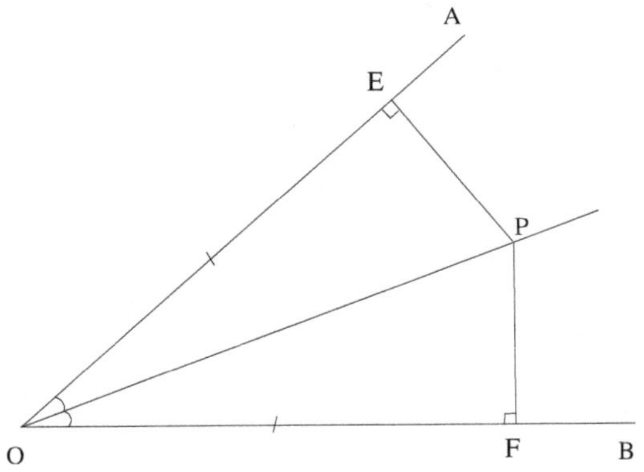

Figure 2.6: The angle bisector OP.

Proof: Let $E \in OA$ and $F \in OB$ be the orthogonal projections of P on OA and OB, respectively. Then $d(P, OA) = |PE|$ and $d(P, OB) = |PF|$.
" $\Longrightarrow$ " If OP is the angle bisector, then the following triangles are congruent by the (SAS) case:

$$\Delta OPE \equiv \Delta OPF \text{ because } \begin{cases} OP & \text{(common side)}, \\ |OE| = |OF| & \text{(both equal to } |OP| \cos \widehat{EOP}\text{)}, \\ \widehat{AOP} = \widehat{POB} & \text{(definition of angle bisector)}. \end{cases}$$

" $\Longleftarrow$ " If $|EP| = |PF|$, the above triangles are congruent by the (SAS) case. This is left as an exercise for the reader. ■

Theorem 2.5.3 *The angle bisectors of a triangle intersect at a point I, which belongs to the interior of the triangle.*

Proof: Let I be the intersection point of the angle bisectors from B and C. The existence of I is a consequence of the crossbar theorem. Then applying the implication " $\Longrightarrow$ " of Theorem 2.5.2 yields

$$\begin{aligned} d(I, BC) &= d(I, AB), \text{and} \\ d(I, BC) &= d(I, AC). \end{aligned}$$

By transitivity, we have $d(I, AB) = d(I, AC)$. Applying the implication " $\Longleftarrow$ " of Theorem 2.5.2, it follows that I belongs to the angle bisector of angle $\widehat{A}$. ■

Corollary 2.5.4 *Let $r = d(I, AB) = d(I, BC) = d(I, AC)$. Then the circle centered at I with radius r is tangent to the sides AB, BC, CA at some interior points of the sides.*

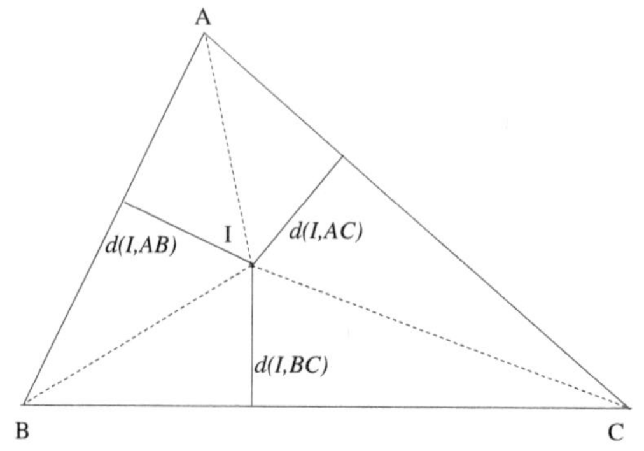

Figure 2.7: The incircle I as the intersection of the angle bisectors.

Proof: It is a consequence of Theorem 2.5.2 and of the fact that the radius of a circle is perpendicular to any tangent. ∎

Definition 2.5.5 *The circle $\mathcal{C}(I, r)$ centered at I and of radius r is called the incircle of triangle ABC. The point I is called the incenter and r is called the inradius.*

Exercise 2.5.6 *A triangle has the side lengths equal to* 3, 4 *and* 5.

(i) Find the inradius r of the triangle.

(ii) Find the circumradius R.

Exercise 2.5.7 *An isosceles triangle ΔABC has $|AB| = |AC| = 8$ and $|BC| = 6$. Find the inradius r of the triangle.*

Finding the contact points

Let A', B' and C' be the points where the incircle is tangent to the sides BC, CA and AB, respectively. We have shown before that $\Delta BIC' \equiv \Delta BIA'$. Then $|BA'| = |BC'|$. In a similar way, $|CA'| = |CB'|$ and $|AC'| = |AB'|$. Let $x = |BA'| = |BC'|$, $y = |A'C| = |CB'|$ and $z = |AB'| = |AC'|$. Then x, y, z are the solutions of the following linear system

$$\begin{aligned} x + y &= a, \\ z + y &= b, \\ z + x &= c. \end{aligned}$$

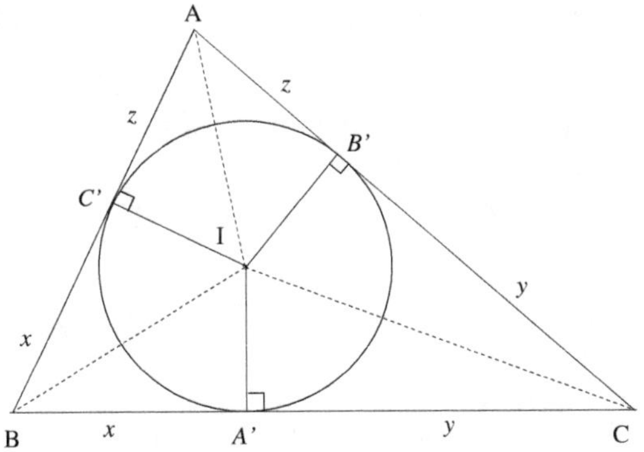

Figure 2.8: The contact points of the incircle.

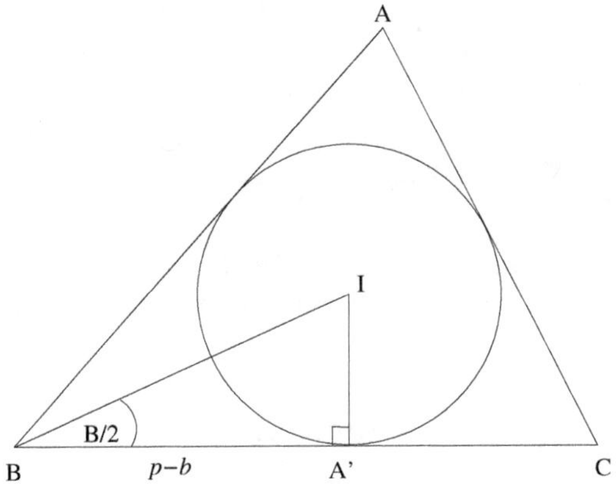

Figure 2.9: The distance $|IA'|$.

Standard methods of linear algebra show that the above system has a unique solution given by

$$\begin{aligned} x &= \frac{-b+a+c}{2} = \frac{a+b+c}{2} - b = p - b, \\ y &= \frac{-c+a+b}{2} = \frac{a+b+c}{2} - c = p - c, \\ z &= \frac{-a+b+c}{2} = \frac{a+b+c}{2} - a = p - a. \end{aligned}$$

The distances from the incenter to the vertices

As the radius IA' is perpendicular to the tangent BC, the triangle BIA' is right with $\widehat{BA'I} = 90°$. Using the definition of the sine function, we have

$$\sin\frac{\widehat{B}}{2} = \frac{r}{|BI|} \Longrightarrow |BI| = \frac{r}{\sin\frac{\widehat{B}}{2}}.$$

In a similar way, we obtain

$$|AI| = \frac{r}{\sin\frac{\widehat{A}}{2}}, \qquad |CI| = \frac{r}{\sin\frac{\widehat{C}}{2}}.$$

Exercise 2.5.8 *Show that*

$$\frac{1}{|AI|} + \frac{1}{|BI|} + \frac{1}{|CI|} = \frac{1}{r}\left(\sin\frac{\widehat{A}}{2} + \sin\frac{\widehat{B}}{2} + \sin\frac{\widehat{C}}{2}\right).$$

Exercise 2.5.9 *Show that*

$$\sin^2\frac{\widehat{B}}{2} = \frac{r^2}{r^2 + (p-b)^2}.$$

Exercise 2.5.10 *A triangle has the sides lengths equal to* 5, 7 *and* 9. *Find the distances between the contact points of incircle and the vertices of the triangle.*

A formula for the area involving inradius

The following formula for the area involves two elements, the inradius and the semiperimeter.

Theorem 2.5.11 *Let $\mathcal{A}$ be the area of the triangle ΔABC and let $p = \frac{1}{2}(a+b+c)$ be the semiperimeter. Then*

$$\mathcal{A} = pr.$$

Proof: Using the additivity property of the area (see statement 2 of Proposition 2.3.3), and the definition of the area of a triangle yields

$$\begin{aligned}\mathcal{A} &= \mathcal{A}_{BIC} + \mathcal{A}_{CIA} + \mathcal{A}_{AIB} = \frac{1}{2}ra + \frac{1}{2}rb + \frac{1}{2}rc \\ &= \frac{1}{2}(a+b+c)r = pr.\end{aligned}$$

■

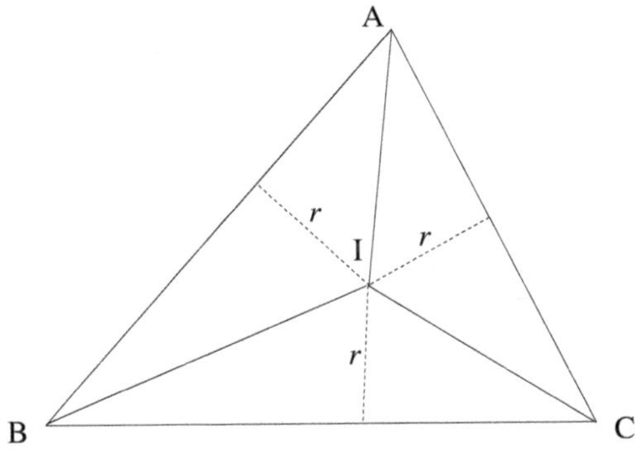

Figure 2.10: The segments IA, IB, IC divide the triangle ΔABC into three triangles.

Exercise 2.5.12 *Let h_a, h_b, h_c and r denote the altitudes and the inradius of the triangle ΔABC. Prove that*

$$\frac{1}{h_a} + \frac{1}{h_b} + \frac{1}{h_c} = \frac{1}{r}.$$

Exercise 2.5.13 *Prove the inequality*

$$h_a h_b h_c \geq 27r^3.$$

Exercise 2.5.14 *Let ΔABC be an equilateral triangle and $M \in Int\, \Delta ABC$ be an arbitrary point. Prove that*

$$dist(M, BC) + dist(M, AB) + dist(M, CA)$$

does not depend on the position of the point M.

Exercise 2.5.15 *Let $ABCD$ be a trapezoid with $AB \| DC$. Let $O = AC \cap BD$ be the intersection of the diagonals. Show that triangles ΔAOD and ΔBOC have equal areas.*

2.6 Excircles and excenters

The incircle is tangent to the sides of the triangle ABC at some interior points. There are three more circles tangent to the triangle sides, with two contact points on the extensions of the sides and one contact point in the interior of the side. See Figure 2.11.

The centers I_a, I_b, I_c of these circles are called excenters and the radii r_a, r_b and r_c are called exradii.

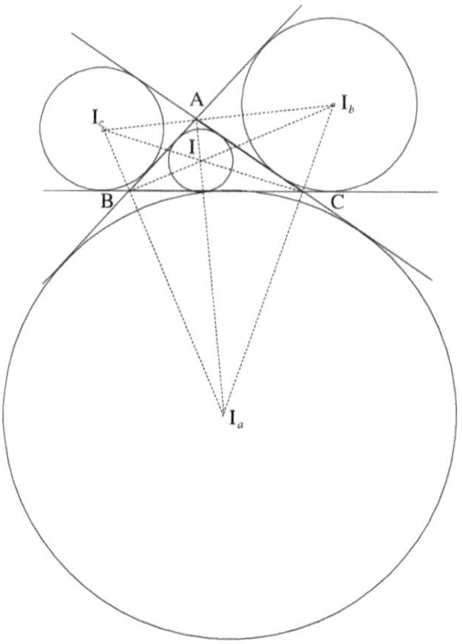

Figure 2.11: Excircles and incircle of a triangle.

Exercise 2.6.1 *Prove that the area of a triangle can be written as*

$$\mathcal{A} = (p-a)r_a = (p-b)r_b = (p-c)r_c.$$

Exercise 2.6.2 *Prove that*

$$\frac{1}{r_a} + \frac{1}{r_b} + \frac{1}{r_c} = \frac{1}{r}.$$

Exercise 2.6.3 *Prove that*

$$\mathcal{A} = \sqrt{r r_a r_b r_c}.$$

Exercise 2.6.4 *Show the inequalities*
(i) $r_a r_b r_c \geq 27r^3$
(ii) $\mathcal{A} \geq 3\sqrt{3}r^2$.

Exercise 2.6.5 *Show the following inequalities*

(i) $\dfrac{1}{h_a h_b} + \dfrac{1}{h_b h_c} + \dfrac{1}{h_c h_a} \leq \dfrac{1}{3r^2}$

(ii) $\dfrac{1}{r_a r_b} + \dfrac{1}{r_b r_c} + \dfrac{1}{r_c r_a} \leq \dfrac{1}{3r^2}$.

Exercise 2.6.6 *Show that in any triangle, we have*

$$\frac{r_a}{r_b} = \frac{a-b+c}{b+c-a}.$$

Exercise 2.6.7 *Let* 3, 4, 5 *be the side lengths of triangle* ΔABC. *Find the exradii* r_a, r_b, *and* r_c.

Exercise 2.6.8 *Show that in a triangle, we have* $2a = b + c$ *if and only if* $r_a = 3r$.

2.7 The circumcircle of a triangle

The circumcircle of a triangle is related to the concept of perpendicular bisector. We shall define this concept next.

Definition 2.7.1 *The perpendicular bisector of a segment* AB *is the line perpendicular to* AB *through the midpoint of* AB. *See Figure 2.12(a).*

Lemma 2.7.2 *Let* AB *be a segment. The point* P *belongs to the perpendicular bisector of* AB *if and only if* $|PA| = |PB|$.

Proof: " $\Longrightarrow$ " Let M be the midpoint of AB. Then $\Delta AMP \equiv \Delta BMP$ by the (SAS) congruence case (PM is a common side, $|AM| = |MB|$ and $\widehat{PMA} = \widehat{PMB} = 90°$). Hence the hypotenuses of these triangles will have equal lengths $|AP| = |PB|$.
" $\Longleftarrow$ " As $|AP| = |PB|$ (hypothesis) and $|AM| = |MB|$ (M is a midpoint), then $\Delta AMP \equiv \Delta BMP$ by the (SSS) case. Then $\widehat{AMP} = \widehat{BMP}$. As

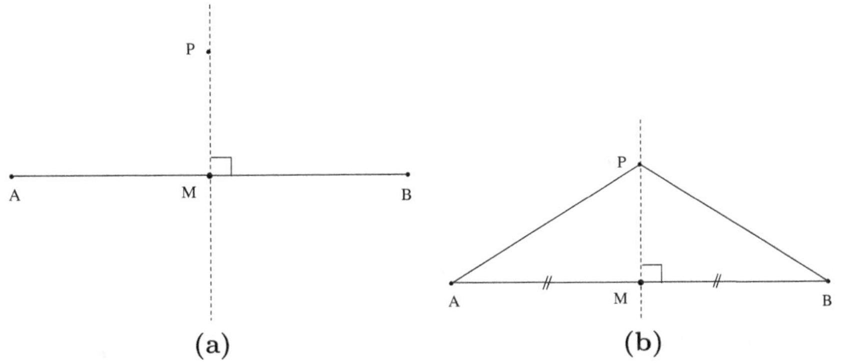

Figure 2.12: (a) The perpendicular bisector; (b) The distances $|PA|$ and $|PB|$.

$\widehat{AMP}+\widehat{BMP}=180°$, it follows that $\widehat{AMP}=90°$, $\widehat{BMP}=90°$ and hence $PM \perp AB$. ■

Theorem 2.7.3 *The perpendicular bisectors of the sides of a triangle intersect at a point O, called the circumcenter.*

Proof: Let A', B', C' be the midpoints of the sides BC, AC and AB, respectively. Let p_a, p_b, p_c be the perpendicular bisectors of segments BC, AC and AB, respectively. Then p_a and p_b will intersect at a point. Otherwise, $p_a \| p_b$ and, as $AC \perp p_b$, $AC \perp p_a$, and hence $AC \| BC$, because $BC \perp p_a$, which is a contradiction. Hence p_a and p_b are not parallel. Let $p_a \cap p_b = \{O\}$. Lemma 2.7.2 yields

$$\begin{aligned} O \in p_a &\implies |OB| = |OC|, \text{and} \\ O \in p_b &\implies |OA| = |OC|. \end{aligned}$$

By transitivity $|OB| = |OA|$ and hence, using the implication " $\Longleftarrow$ " of Lemma 2.7.2 yields $O \in p_c$. It follows that $p_a \cap p_b \cap p_c = \{O\}$. ■

Definition 2.7.4 *The circumcenter O is the point situated at equal distance from the triangle vertices A, B and C. The quantity*

$$R = |OA| = |OB| = |OC|$$

is called the circumradius. The circle centered at O with radius R is called the circumcircle of ΔABC. See the Figure 2.13(b).

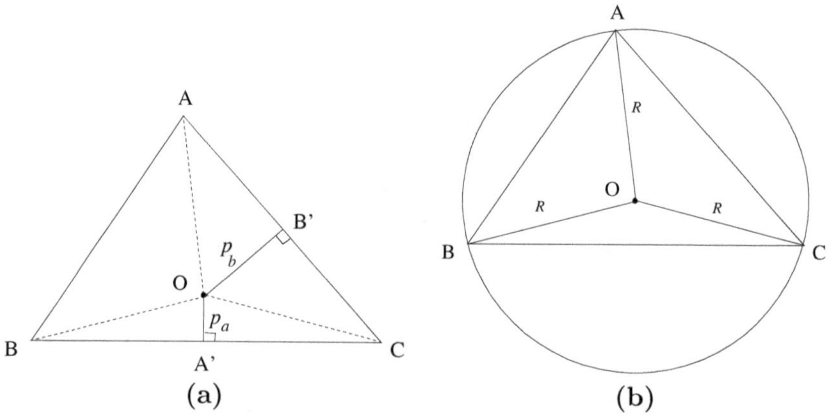

Figure 2.13: (a) The circumcenter O as the intersection of the perpendicular bisectors; (b) The distances $|OA| = |OB| = |OC|$.

Exercise 2.7.5 *Let I and O denote the incenter and circumcenter of ΔABC, respectively. Show that*

$$\widehat{IAO} = \frac{|\widehat{B} - \widehat{C}|}{2}.$$

Exercise 2.7.6 *An isosceles triangle ΔABC has the sides $|AB| = |AC| = 9$ and $|BC| = 3$. Find its circumradius R.*

Exercise 2.7.7 *The ancient temple of Osiris is situated at equal distances from three cities: Anapur, Bengapur and Cyprus. A monk from the temple is planning a journey through the above three cities. Find the plan of the journey such that the following conditions are satisfied:*
(i) the monk passes through each city exactly once, and
(ii) the distance traveled is the minimum possible.

Exercise 2.7.8 *(i) With the usual notations, prove that*

$$4R = r_a + r_b + r_c - r.$$

(ii) Show the inequality

$$r_a + r_b + r_c \geq 9r,$$

with the identity reached for the equilateral triangle.

The distances from the circumcenter to the sides

The next result provides formulas for the distances from the circumcenter of a triangle and its sides. They will be useful later for proving the existence of the Euler line as well as for finding an area formula of the triangle involving the perimeter of the orthic triangle.

Proposition 2.7.9 *The distances from the circumcenter O to the sides BC, CA, AB are equal to $d_a = R\cos\widehat{A}$, $d_b = R\cos\widehat{B}$ and $d_c = R\cos\widehat{C}$, respectively.*

Proof: Let D be the orthogonal projection of the circumcenter O on the side AB, and let $d_c = d(O, AB) = |OD|$. A diameter is drawn with one endpoint at A terminating at A'. The right triangle $\Delta AA'C$ is created with $\widehat{ACA'} = 90°$. Because $\widehat{ACA'}$ is inscribed into a semicircle, we have $x = \widehat{BAO} = \widehat{BCA'}$ as inscribed angles, cutting the same arc. Then $x = 90° - \widehat{C}$. Applying the definition of sine in ΔOAD yields

$$\begin{aligned} \sin x &= \frac{d_c}{|AO|} = \frac{d_c}{R} \\ \Longrightarrow d_c &= R\sin x = R\sin(90° - \widehat{C}) = R\cos\widehat{C}. \end{aligned}$$

■

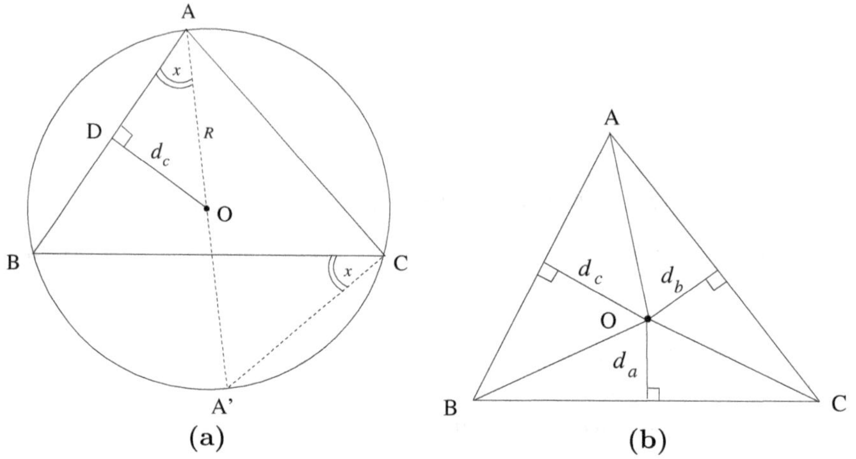

Figure 2.14: (a) The distance d_c from O to AB; (b) The distances d_a, d_b, d_c.

It is worth noting that a variant of proof uses the Pythagorean theorem in ΔAOD as follows

$$\begin{aligned} d_c^2 &= R^2 - \frac{c^2}{4} = R^2 - R^2 \sin^2 \widehat{C} \\ &= R^2 \left(1 - \sin^2 \widehat{C}\right) = R^2 \cos^2 \widehat{C}. \end{aligned}$$

We used the Law of Sines formula $c = 2R \sin \widehat{C}$ and the fundamental identity of trigonometry. Taking the square root yields the desired result.

Proposition 2.7.9 has several consequences as follows.

Corollary 2.7.10 *The area of an acute triangle is given by*

$$\mathcal{A} = \frac{1}{2} R \left(a \cos \widehat{A} + b \cos \widehat{B} + c \cos \widehat{C}\right).$$

Proof: The circumcenter $O \in IntABC$ because the triangle is acute. (Justify why!) The additivity property for the area function yields

$$\begin{aligned} \mathcal{A} &= \mathcal{A}_{BOC} + \mathcal{A}_{COA} + \mathcal{A}_{AOB} \\ &= \frac{1}{2} d_a \cdot a + \frac{1}{2} d_b \cdot b + \frac{1}{2} d_c \cdot c \\ &= \frac{1}{2} \left(Ra \cos \widehat{A} + Rb \cos \widehat{B} + Rc \cos \widehat{C}\right) \\ &= \frac{1}{2} R \left(a \cos \widehat{A} + b \cos \widehat{B} + c \cos \widehat{C}\right). \end{aligned}$$

■

Remark 2.7.11 We shall show in a later section that the quantities $a_O = a\cos\widehat{A}$, $b_O = b\cos\widehat{B}$, $c_O = c\cos\widehat{C}$ are the lengths of the sides of the orthic triangle. Its semiperimeter is $p_O = (a_O + b_O + c_O)/2$. Then Corollary 2.7.10 yields the formula

$$\mathcal{A} = Rp_O. \tag{2.7.4}$$

Using $\mathcal{A} = pr$, it follows that $Rp_O = pr$ or

$$\frac{R}{r} = \frac{p}{p_O}. \tag{2.7.5}$$

In the next section, we shall prove the Euler inequality $R/r \geq 2$. From this will follow the inequality $p \geq 2p_O$.

Corollary 2.7.12 *The area of a triangle can be expressed as*

$$\mathcal{A} = \frac{1}{2}R^2\left(\sin 2\widehat{A} + \sin 2\widehat{B} + \sin 2\widehat{C}\right).$$

Proof: The law of sines and Corollary 2.7.10 yield

$$\begin{aligned}
\mathcal{A} &= \frac{1}{2}R\left(a\cos\widehat{A} + b\cos\widehat{B} + c\cos\widehat{C}\right)\\
&= \frac{1}{2}R\left(2R\sin\widehat{A}\cos\widehat{A} + 2R\sin\widehat{B}\cos\widehat{B} + 2R\sin\widehat{C}\cos\widehat{C}\right)\\
&= \frac{1}{2}R\left(R\sin 2\widehat{A} + R\sin 2\widehat{B} + R\sin 2\widehat{C}\right)\\
&= \frac{1}{2}R^2\left(\sin 2\widehat{A} + \sin 2\widehat{B} + \sin 2\widehat{C}\right).
\end{aligned}$$

■

Corollary 2.7.13 *If A, B, C are the angle measures of a triangle, then the following identity holds*

$$\sin 2\widehat{A} + \sin 2\widehat{B} + \sin 2\widehat{C} = 4\sin\widehat{A}\sin\widehat{B}\sin\widehat{C}.$$

Proof: Using the law of sines and Proposition 2.4.2 yields

$$\begin{aligned}
\mathcal{A} &= \frac{abc}{4R} = \frac{\left(2R\sin\widehat{A}\right)\left(2R\sin\widehat{B}\right)\left(2R\sin\widehat{C}\right)}{4R}\\
&= 2R^2\sin\widehat{A}\sin\widehat{B}\sin\widehat{C}.
\end{aligned}$$

Using Corollary 2.7.12 yields

$$\begin{aligned}
\frac{1}{2}R^2\left(\sin 2\widehat{A} + \sin 2\widehat{B} + \sin 2\widehat{C}\right) &= \mathcal{A}\\
&= 2R^2\sin\widehat{A}\sin\widehat{B}\sin\widehat{C}.
\end{aligned}$$

Multiplying by $2/R^2$ yields the desired result. ■

Exercise 2.7.14 *Show that among all triangles inscribed in a given circle $\mathcal{C}(\mathcal{O},\mathcal{R})$ having a fixed base BC and a variable vertex A, the isosceles triangle has the maximum area.*

Exercise 2.7.15 *(i) Show that among all triangles inscribed in a given circle $\mathcal{C}(\mathcal{O},\mathcal{R})$, the equilateral triangle has the maximum area.*

(ii) The product $\sin\widehat{A}\sin\widehat{B}\sin\widehat{C}$ *reaches its maximum in the case of the equilateral triangle.*

2.8 Euler's inequality

After this section, we shall be able to answer the following question: *Can you construct a triangle inscribed in a circle of radius $R = 10$ and circumscribed to a circle of radius $r = 6$?*
This question was asked first by Leonhard Euler around 1720. Even though there is a geometric way to solve the problem, in this section we shall give an algebraic solution.

The proof of the following result uses some algebraic inequalities. A proof based on purely geometric concepts will be given later on page 152.

Theorem 2.8.1 (Euler's inequality) *In any triangle $R \geq 2r$; the identity holds exactly for equilateral triangles.*

Proof: Let $\mathcal{A}$ denote the area of the triangle ABC. Using the formulas

$$\mathcal{A} = \frac{abc}{4R}, \qquad \mathcal{A} = pr$$

of Proposition 2.4.2 and Theorem 2.5.11 yields

$$R = \frac{abc}{4\mathcal{A}}, \qquad 2r = 2\frac{\mathcal{A}}{p}.$$

Dividing, we have

$$\frac{R}{2r} = \frac{abc}{4\mathcal{A}} \cdot \frac{p}{2\mathcal{A}} = \frac{abc\,p}{8\mathcal{A}^2}.$$

Using Heron's formula (see Section 2.17),

$$\mathcal{A} = \sqrt{p(p-a)(p-b)(p-c)}$$

and Lemma 1.12.3 yields

$$\begin{aligned}\frac{R}{2r} &= \frac{abc\,p}{8p(p-a)(p-b)(p-c)} = \frac{abc}{8(p-a)(p-b)(p-c)} \\ &= \frac{1}{8}\frac{(x+y)(y+z)(x+z)}{xyz} \geq \frac{1}{8}\cdot 8 = 1,\end{aligned}$$

where $x = p-a$, $y = p-b$, $z = p-c$, and $x+y = p-a+p-b = 2p-a-b = c$, $x+z = b$ and $y+z = a$. Lemma 1.12.3 also shows that equality holds exactly when $x = y = z$, which means that $a = b = c$, i.e., ΔABC is an equilateral triangle. ∎

Returning to the question at the beginning of this section, we see that since $R = 10 < 2 \cdot 6 = 2r$, the answer is "no".

2.9 Van Schouten's theorem

The following result gives an unexpected property of the equilateral triangle. It is one of the most beautiful and elegant theorems of classical geometry. The property was published in 1646 by Francis Van Schouten (1615–1666). An extension of this theorem, due to Dimitrie Pompeiu (1925), will be presented in a future chapter.

Theorem 2.9.1 *Let M be a point on the arc AC of the circumcircle of the equilateral triangle ΔABC. Then*

$$|MA| + |MC| = |MB|.$$

Proof: We shall give a trigonometric proof. Let $x = \widehat{ABM} = \widehat{ACM}$ and let R be the circumradius of ΔABC, see Figure 2.15(a). The law of sines

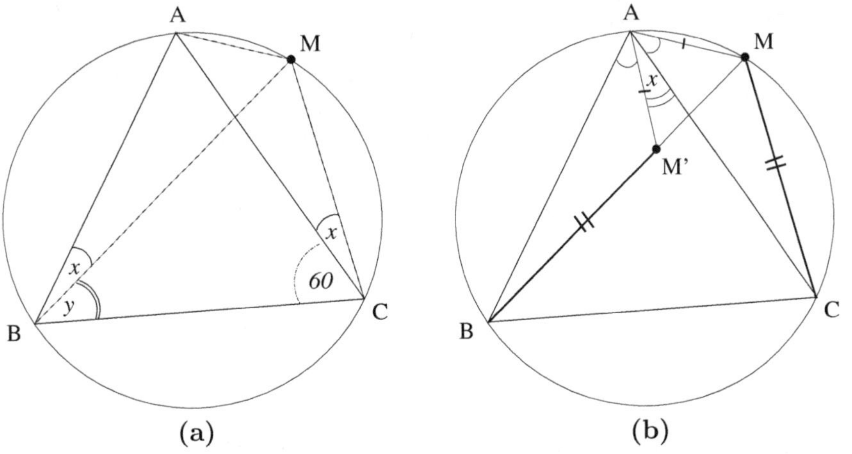

Figure 2.15: Van Schouten theorem: (a) the trigonometric proof; (b) the construction proof.

applied to ΔAMC and ΔMCB yields

$$\frac{|MA|}{\sin x} = 2R, \quad \frac{|MC|}{\sin y} = 2R, \quad \frac{|MB|}{\sin(x+60^\circ)} = 2R.$$

Therefore the relation $|MB| = |MA| + |MC|$ is equivalent to

$$\begin{aligned} 2R\sin(x+60^\circ) &= 2R\sin x + 2R\sin y \iff \\ \sin(x+60^\circ) &= \sin x + \sin(60^\circ - x) \iff \\ \sin x\cos 60^\circ + \cos x \sin 60^\circ &= \sin x + \sin 60^\circ \cos x - \sin x \cos 60^\circ \iff \\ \sin x \cos 60^\circ &= \sin x - \sin x \cos 60^\circ \iff \\ 2\sin x \cos 60^\circ &= \sin x, \end{aligned}$$

which holds true since $\cos 60^\circ = 1/2$. ■

The trigonometric proof is not always the shortest or the most elegant. We next present a proof based on an ingenious construction.

Proof involving construction: Choose $M' \in BM$ such that $|MA| = |MM'|$, see Figure 2.15(b). As $\widehat{AMM'} = \widehat{ACB} = 60^\circ$, then $\Delta AMM'$ is isosceles and has all angles equal to 60°. It follows that it is equilateral. (Justify why!) Hence $|MA| = |M'A| = |MM'|$.

Let $x = \widehat{M'AC}$. Then $\widehat{BAM'} + x = 60^\circ = x + \widehat{CAM}$. Hence $\widehat{BAM'} = \widehat{CAM}$. As $AB = AC$ (because ΔABC is equilateral), we have $\Delta AM'B \equiv \Delta AMC$ by the (SAS) congruence case. Hence $|BM'| = |MC|$, and $|MA| + |MC| = |MM'| + |BM'| = |BM|$.

We make the remark that if M does not belong to the circumcircle of ΔABC, we have $|MA| + |MC| > |BM|$, which is called Pompeiu's inequality. This follows from the following metrical relation.

Proposition 2.9.2 *Let M be a point inside the angle $\widehat{ABC}$ of the equilateral triangle ΔABC. Then*

$$|MB|^2 = (|MA| + |MC|)^2 - 4|AM|\,|MC|\cos^2\left(30^\circ + \frac{\widehat{AMC}}{2}\right).$$

Proof: The proof uses a construction together with the law of cosines. Figure 2.16 shows a 60° rotation about the point C. Thus, the points B, A, M became A, A' and M', respectively. Therefore, $\Delta AA'C$ and $\Delta MM'C$ are equilateral triangles (Why?) and $|MB| = |AM'|$. Then it suffices to

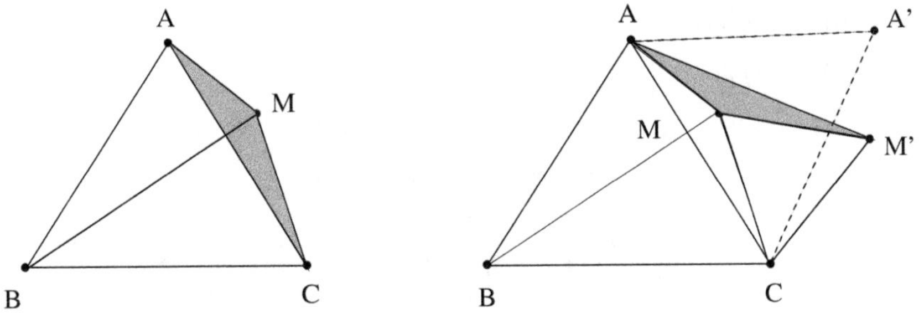

Figure 2.16: Pompeiu's theorem.

compute $|AM'|$. Applying the law of cosines in $\Delta AMM'$ yields

$$\begin{aligned}
|AM'|^2 &= |AM|^2 + |MM'|^2 - 2|AM|\,|MM'|\cos\widehat{AMM'} \\
&= |AM|^2 + |MC|^2 - 2|AM|\,|MC|\cos\left(360^\circ - 60^\circ - \widehat{AMC}\right) \\
&= |AM|^2 + |MC|^2 + 2|AM|\,|MC| \\
&\quad - 2|AM|\,|MC|\left(1 + \cos\left(60^\circ + \widehat{AMC}\right)\right) \\
&= (|AM| + |MC|)^2 - 2|AM|\,|MC|\left(1 + \cos\left(60^\circ + \widehat{AMC}\right)\right) \\
&= (|AM| + |MC|)^2 - 4|AM|\,|MC|\cos^2\left(30^\circ + \frac{\widehat{AMC}}{2}\right).
\end{aligned}$$

■

It is worth noting that for $\widehat{AMC} = 120^\circ$ the cosine term vanishes. This corresponds to the case when M belongs to the circumcircle of ΔABC and the previous result recovers van Schouten's theorem.

Solved Exercise 2.9.3 *An equilateral triangle ΔABC is inscribed in a circle of radius R. Show that for any point M on the circumcircle of ΔABC, we have*

$$|MA|^2 + |MB|^2 + |MC|^2 = 6R^2.$$

Proof: We assume, without the loss of generality, that the point C belongs to the arc AC and consider the notations from Figure 2.15(a). Using the law of sines, we have

$$\begin{aligned}
&|MA|^2+|MB|^2+|MC|^2\\
&=(2R\sin x)^2+(2R\sin(60^\circ+x))^2+(2R\sin(60^\circ-x))^2\\
&=4R^2\left(\sin^2 x+\sin^2(60^\circ+x)+\sin^2(60^\circ-x)\right)\\
&=4R^2\left[\sin^2 x+\left(\frac{\sqrt{3}}{2}\cos x+\frac{1}{2}\sin x\right)^2+\left(\frac{\sqrt{3}}{2}\cos x-\frac{1}{2}\sin x\right)^2\right]\\
&=4R^2\frac{6}{4}(\sin^2 x+\cos^2 x)=6R^2.
\end{aligned}$$

■

Solved Exercise 2.9.4 *Let M be the midpoint of the arc AB on the circle $\mathcal{C}(O,R)$ and P an arbitrary point on the arc AB. Show that*

$$|PA|\cdot|PB|+|PM|^2=|MA|^2.$$

Proof: We shall provide a trigonometric solution. Let $\theta=\widehat{AOB}$, $\frac{\theta}{2}=\widehat{AOM}=\widehat{MOB}$ and $x=\widehat{MOP}$, see Figure 2.17. With the notations $y=x/2$ and $u=\theta/4$, we have

$$|PA|=\frac{R}{2}\sin\frac{x+\frac{\theta}{2}}{2}=\frac{R}{2}\sin(y+u),\ |PB|=\frac{R}{2}\sin\frac{\frac{\theta}{2}-x}{2}=\frac{R}{2}\sin(u-y)$$

$$|PM|=\frac{R}{2}\sin\frac{x}{2}=\frac{R}{2}\sin y,\qquad |MA|=\frac{R}{2}\sin\frac{\theta}{4}=\frac{R}{2}\sin u.$$

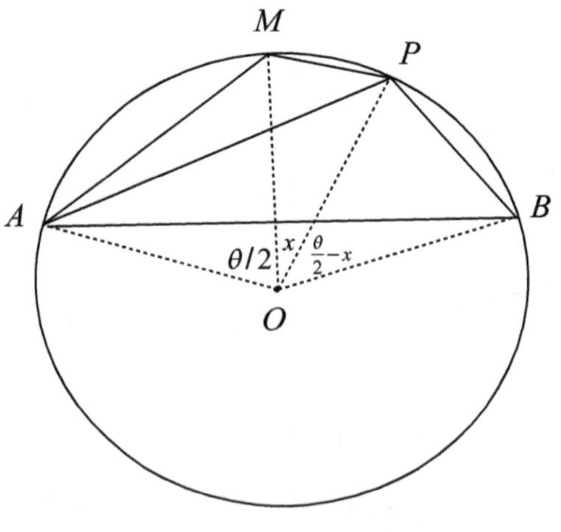

Figure 2.17: For Exercise 2.9.4.

Substituting these relations in $|PA| \cdot |PB| + |PM|^2 = |MA|^2$ we obtain the trigonometric relation $\sin(y+u)\sin(u-y) + \sin^2 y = \sin^2 u$, which holds true by Exercise 2.1.17. ■

2.10 A proof by contradiction

We shall exemplify the proof by contradiction using the following problem:

Prove that in any non-isosceles triangle ΔABC the angle bisector from A and the perpendicular bisector of side BC intersect outside of ΔABC.

Proof: First, we need to ensure that the angle bisector from A and the perpendicular bisector of side BC intersect.

By contradiction, assuming that they do not intersect, then they must be parallel. Thus we obtain that the angle bisector from A is perpendicular on BC. Let A' be the midpoint of BC. Then we can show that $\Delta AA'B \equiv \Delta AA'C$ by (ASA) congruence case. Therefore $|AB| = |AC|$, which contradicts that ΔABC is not isosceles. Hence, there is an intersection point between the angle bisector from A and the perpendicular bisector of side BC, which will be denoted by P.

Next, we shall show that P is outside of ΔABC. We proceed again by contradiction, by assuming that P is inside of ΔABC. Then its orthogonal projections, A', E and F onto the sides BC, AB and AC are located inside the sides, see Figure 2.18.

By (SAS) congruence case, we have $\Delta APE \equiv \Delta APF$ since

- AP is a common side
- $\widehat{EAP} = \widehat{PAE}$ since AP is an angle bisector
- $\widehat{APE} = \widehat{APF}$ since they have equal complements.

Then $|AE| = |AF|$ and $|PE| = |PF|$.

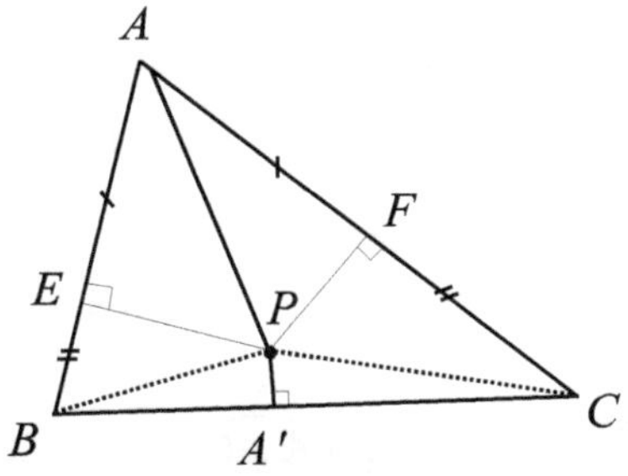

Figure 2.18: The case when P is inside of ΔABC.

Since we also have

- $|PE| = |PF|$ from the previous triangle congruence
- $|PB| = |PC|$ since P is a point on the perpendicular bisector of BC.
- $\widehat{BEP} = \widehat{CFP} = 90°$,

the right triangles $\Delta PEB \equiv \Delta PFC$. Therefore, $|BE| = |FC|$.

Now, writing $|AB| = |BE|+|AE|$ and $|AC| = |AF|+|FC|$ and using the aforementioned equal segments, we obtain $|AB| = |AC|$, which contradicts that ΔABC is not isosceles.

Hence, the intersection point P must be outside of ΔABC.

2.11 Similar triangles

To Thales the primary question was not "What do we know?" but "How do we know it?"

Aristotle

Besides the use of congruence of triangles, the similarity of triangles provides another very powerful tool in the Euclidean geometry. This section deals with the concept of similar triangles.

The following theorem is a consequence of the law of sines. It is also known in the literature under the name of Thales' theorem and it is equivalent to the (SAS) similarity case.

Theorem 2.11.1 *Let ΔABC be a triangle. Suppose that B' and C' are distinct points so that either $B' \in AB$ and $C' \in AC$ or $A \in BB'$ and $A \in CC'$. Then*

$$1) \qquad B'C' \| BC \Longrightarrow \frac{|AB'|}{|AB|} = \frac{|AC'|}{|AC|} = \frac{|B'C'|}{|BC|},$$

$$2) \qquad \frac{|AB'|}{|AB|} = \frac{|AC'|}{|AC|} \Longrightarrow B'C' \| BC.$$

Proof: We shall prove this in the case when the points B', C' belong to the interior of the sides AB, AC.

1) Applying the law of sines in $\Delta AB'C'$ and ΔABC yields

$$\frac{|B'C'|}{\sin \widehat{B'AC'}} = \frac{|AC'|}{\sin \widehat{AB'C'}} = \frac{|AB'|}{\sin \widehat{AC'B'}}$$
$$\frac{|BC|}{\sin \widehat{BAC}} = \frac{|AC|}{\sin \widehat{ABC}} = \frac{|AB|}{\sin \widehat{ACB}}.$$

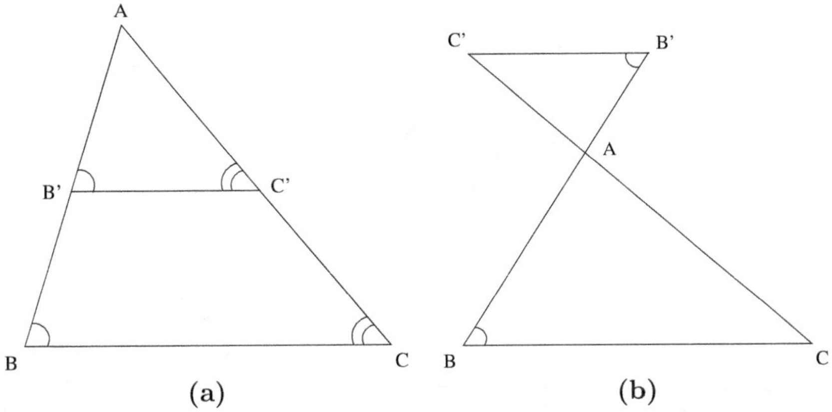

Figure 2.19: Thales theorem: (a) B' and C' interior points; (b) B' and C' exterior points.

Dividing the above two identities yields

$$\frac{|B'C'|}{|BC|} \cdot \frac{\sin \widehat{BAC}}{\sin \widehat{B'AC'}} = \frac{|AC'|}{|AC|} \cdot \frac{\sin \widehat{ABC}}{\sin \widehat{AB'C'}} = \frac{|AB'|}{|AB|} \cdot \frac{\sin \widehat{ACB}}{\sin \widehat{AC'B}'}. \tag{2.11.6}$$

As $B'C' \| BC$, we have the congruent corresponding angles

$$\widehat{ABC} = \widehat{AB'C'} \qquad \text{and} \qquad \widehat{ACB} = \widehat{AC'B'}.$$

Hence (2.11.6) reduces to

$$\frac{|B'C'|}{|BC|} = \frac{|AC'|}{|AC|} = \frac{|AB'|}{|AB|},$$

which implies statement 1.

2) Consider the left side of statement 2 satisfied. Construct a line ℓ through B' parallel to BC and let $C'' = \ell \cap AC$. See Figure 2.20.
Using statement 1 proved before yields

$$\frac{|AB'|}{|AB|} = \frac{|AC''|}{|AC|}.$$

Comparing with the left side of statement 2 yields

$$\frac{|AC'|}{|AC|} = \frac{|AC''|}{|AC|},$$

which implies $|AC'| = |AC''|$ and hence $C' = C''$. ■

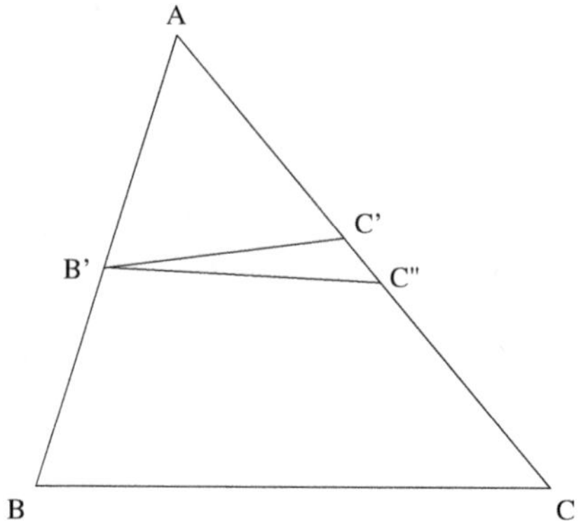

Figure 2.20: See the proof of Theorem 2.11.1.

As a consequence, we have the following result.

Corollary 2.11.2 (The midpoint connector theorem) *Let B', C' be the midpoints of the sides AC and AB of the triangle ABC. Then*

(*i*) $B'C' \| BC$

(*ii*) $|B'C'| = \frac{1}{2}|BC|$.

Proof: The left side of statement 2 of Theorem 2.11.1 is equal to 1/2. Statement 2 of Theorem 2.11.1 gives (*i*) and statement 1 gives (*ii*). ∎

Exercise 2.11.3 *Let B' be the midpoint of AB and $C' \in AC$ such that $B'C' \| BC$. Show that*

(*i*) *C' is the midpoint of AC*

(*ii*) $|B'C'| = \frac{1}{2}|BC|$.

Exercise 2.11.4 *Let $ABCD$ be a convex quadrilateral and we denote by M, N, P, Q, the midpoints of the sides DA, AB, BC, CD, respectively. Prove that $MNPQ$ is a parallelogram.*

Definition 2.11.5 *Two triangles ΔABC and $\Delta A'B'C'$ are called* **similar** *if*

$$\frac{|A'B'|}{|AB|} = \frac{|A'C'|}{|AC|} = \frac{|B'C'|}{|BC|}.$$

We write $\Delta ABC \sim \Delta A'B'C'$.

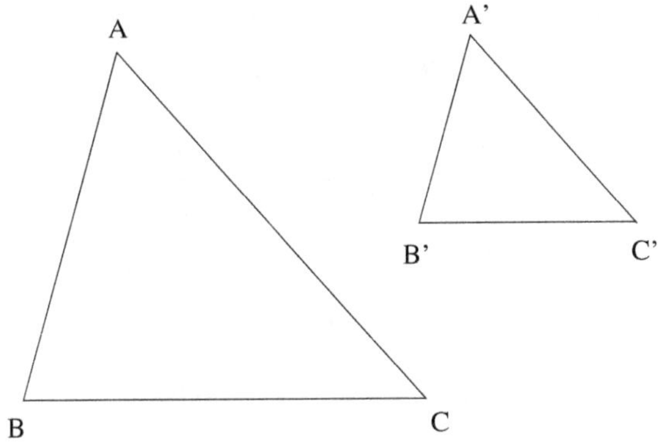

Figure 2.21: Similar triangles.

The following result is a consequence of the law of cosines. Even if this law will be stated on page 110, we shall use it here, so that we have all the similarity cases presented together in the same section. We note that the law of cosines is based on the Pythagorean relation, which in turn, can be proved using the concept of area, see page 88.

Proposition 2.11.6 *Two triangles ΔABC and $\Delta A'B'C'$ are similar if and only if they have the same angle measures.*

Proof: "$\Longrightarrow$" Let $\Delta ABC \sim \Delta A'B'C'$. Choose k such that

$$\frac{1}{k} = \frac{|A'B'|}{|AB|} = \frac{|A'C'|}{|AC|} = \frac{|B'C'|}{|BC|}.$$

Then

$$|AB| = k|A'B'|, \quad |BC| = k|B'C'|, \quad |AC| = k|A'C'|.$$

The law of cosines, see Theorem 2.16.1, yields

$$\begin{aligned}\cos\widehat{A} &= \frac{|AB|^2 + |AC|^2 - |BC|^2}{2|AB|\cdot|AC|} = \frac{k^2|A'B'|^2 + k^2|A'C'|^2 - k^2|B'C'|^2}{2k|A'B'|\cdot k|A'C'|} \\ &= \frac{k^2\left[|A'B'|^2 + |A'C'|^2 - |B'C'|^2\right]}{k^2\left[2|A'B'|\cdot|A'C'|\right]} = \frac{|A'B'|^2 + |A'C'|^2 - |B'C'|^2}{2|A'B'|\cdot|A'C'|} \\ &= \cos\widehat{A'}.\end{aligned}$$

As $0 < \widehat{A}, \widehat{A'} < 180^\circ$ and the cosine function is one-to-one on $[0^\circ, 180^\circ]$, from $\cos\widehat{A} = \cos\widehat{A'}$ it follows that $\widehat{A} = \widehat{A'}$. In a similar way, $\widehat{B} = \widehat{B'}$ and $\widehat{C} = \widehat{C'}$.

"$\Longleftarrow$" Suppose $\widehat{A} = \widehat{A'}$, $\widehat{B} = \widehat{B'}$, $\widehat{C} = \widehat{C'}$. Then $\sin\widehat{A} = \sin\widehat{A'}$, $\sin\widehat{B} = \sin\widehat{B'}$, $\sin\widehat{C} = \sin\widehat{C'}$. The law of sines applied for triangles ΔABC and $\Delta A'B'C'$ yields

$$\frac{|BC|}{\sin\widehat{A}} = \frac{|AC|}{\sin\widehat{B}} = \frac{|AB|}{\sin\widehat{C}}$$

$$\frac{|B'C'|}{\sin\widehat{A'}} = \frac{|A'C'|}{\sin\widehat{B'}} = \frac{|A'B'|}{\sin\widehat{C'}}.$$

Dividing, we get

$$\frac{|BC|}{|B'C'|} = \frac{|AC|}{|A'C'|} = \frac{|AB|}{|A'B'|},$$

i.e., $\Delta ABC \sim \Delta A'B'C'$. ■

Similarity cases

Consider two triangles ΔABC and $\Delta A'B'C'$. There are three similarity cases for triangles: angle-angle (AA), side-angle-side (SAS) and side-side-side (SSS). The (SSS) similarity case is in Definition 2.11.5. The other two similarity cases are given in the following propositions.

Proposition 2.11.7 (AA) *If* $\widehat{B} = \widehat{B'}$ *and* $\widehat{C} = \widehat{C'}$ *then* $\Delta ABC \sim \Delta A'B'C'$.

Proof: $\widehat{A} = 180 - \widehat{B} - \widehat{C} = 180 - \widehat{B'} - \widehat{C'} = \widehat{A'}$. Proposition 2.11.6 yields similar triangles. ■

Proposition 2.11.8 (SAS) *If* $\widehat{A} = \widehat{A'}$ *and*

$$\frac{|A'B'|}{|AB|} = \frac{|A'C'|}{|AC|}, \tag{2.11.7}$$

then $\Delta ABC \sim \Delta A'B'C'$.

Proof: Without loss of generality, we may assume that $|A'B'| \leq |AB|$. Hence $|A'C'| \leq |AC|$. (Why?) Choose $B'' \in AB$ and $C'' \in AC$ such that $|AB''| = |A'B'|$ and $|AC''| = |A'C'|$. Then $\Delta AB''C'' \equiv \Delta A'B'C'$, by the (SAS) congruence case. Then (2.11.7) becomes

$$\frac{|AB''|}{|AB|} = \frac{|AC''|}{|AC|}.$$

From Theorem 2.11.1, it follows that $B''C''\|BC$. As the corresponding angles are equal, it follows that triangles ΔABC and $\Delta AB''C''$ have the same angle measures and hence they are similar by the (AA) case. Hence

$$\Delta ABC \sim \Delta AB''C'' \equiv \Delta A'B'C'.$$

It follows that $\Delta ABC \sim \Delta A'B'C'$. ■

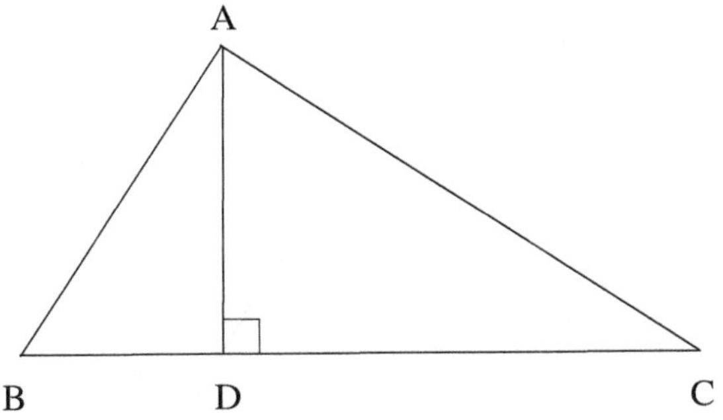

Figure 2.22: $|AB|$ is a geometric mean of $|BD|$ and $|BC|$.

2.12 The Pythagorean theorem in a right triangle

The following result deals with a side length as a geometric mean.

Proposition 2.12.1 *Let AD be the altitude of triangle ΔABC. Then $\widehat{A} = 90°$ if and only if $|AB| = \sqrt{|BD| \cdot |BC|}$.*

Proof: " $\Longrightarrow$ " Let $\widehat{A} = 90°$. Then the similarity case (AA) yields $\Delta ABD \sim \Delta CBA$. Hence

$$\frac{|AB|}{|CB|} = \frac{|BD|}{|BA|} \Longrightarrow |AB|^2 = |CB| \cdot |BD|.$$

" $\Longleftarrow$ " If $|AB| = \sqrt{|BD| \cdot |BC|}$ then $\dfrac{|AB|}{|CB|} = \dfrac{|BD|}{|BA|}$ and as $\widehat{ABD} = \widehat{ABC}$, the similarity case (SAS) yields $\Delta ABD \sim \Delta CBA$. Hence $\widehat{BAC} = \widehat{BDA} = 90°$ by Theorem 2.11.6. ■

Theorem 2.12.2 (Pythagorean theorem) *If ΔABC is a right triangle with $\widehat{A} = 90°$, then $|BC|^2 = |AB|^2 + |AC|^2$.*

Proof: Using Proposition 2.12.1, we have

$$\begin{aligned} |AB|^2 &= |BD| \cdot |BC| \\ |AC|^2 &= |DC| \cdot |BC|. \end{aligned}$$

Summing yields $|AB|^2 + |AC|^2 = (|BD| + |DC|) \cdot |BC| = |BC|^2$. ■

We make the remark on the following geometric interpretation of the previous proof. Using Figure 2.23, we notice that the area of the square

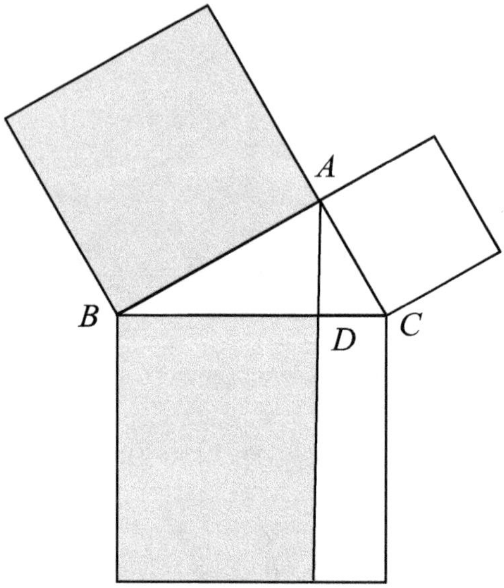

Figure 2.23: The square and rectangle that are shaded have the same area.

constructed along the hypothenuse can be decomposed into two rectangles, each having the area equal to the area of one the squares constructed along the legs.

Exercise 2.12.3 *Let AD be an altitude of the right triangle ΔABC with $\widehat{A} = 90°$. Show that*
(i) $|AD| = \sqrt{|BD| \cdot |DC|}$,
(ii) $|AD| \leq \dfrac{|BC|}{2}$,
(iii) $|AD| < \dfrac{1}{2}(|AB| + |AC|)$, *and*
(iv) $\dfrac{1}{|AD|^2} = \dfrac{1}{|AC|^2} + \dfrac{1}{|AB|^2}$.

Proofs of the Pythagorean relation using areas

The following two proofs of the Pythagorean relation use the concept of area, its additivity properties and the binomial formula.

1. We consider four copies of a right triangle with legs a and b and hypothenuse c. We set them as in Figure 2.24(a). They form a large square

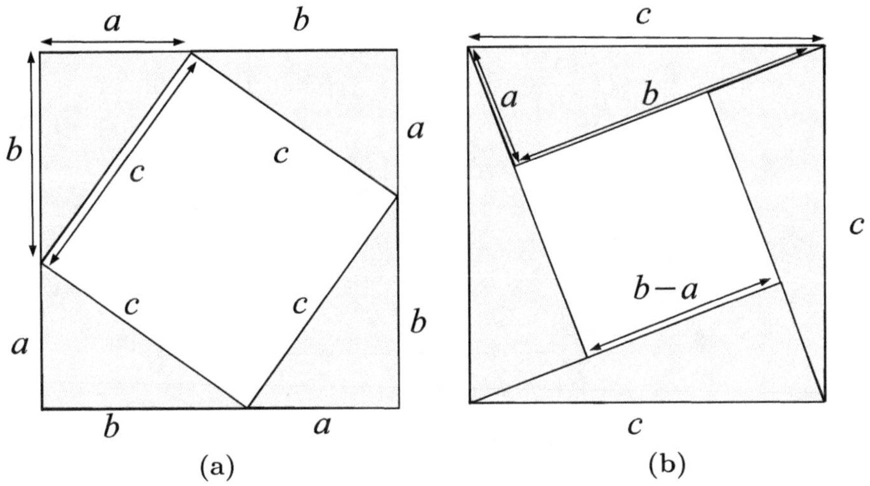

Figure 2.24: Two proofs of the Pythagorean relation using areas.

with side size $a+b$ and an inner square with side size c. Writing that the area of the large square is the sum of the areas of the four right triangles and the inner square yields

$$(a+b)^2 = 4\frac{ab}{2} + c^2.$$

Using the binomial formula, after cancelation, we obtain $c^2 = a^2+b^2$, which is the Pythagorean relation.

2. We consider again four copies of a right triangle with legs a and b and hypothenuse c. We set them now as in Figure 2.24(b). They form a large square with side size c and an inner square with side size $b-a$. Writing again that the area of the large square is the sum of the areas of the four right triangles and the inner square, we obtain

$$c^2 = 4\frac{ab}{2} + (b-a)^2,$$

which after a straightforward computation leads again to the Pythagorean relation.

2.13 The orthocenter

Let ABC be a triangle and let $A_1 \in BC$, $B_1 \in AC$, $C_1 \in AB$, such that $AA_1 \perp BC$, $BB_1 \perp AC$, $CC_1 \perp AB$. We recall that AA_1, BB_1, CC_1 are

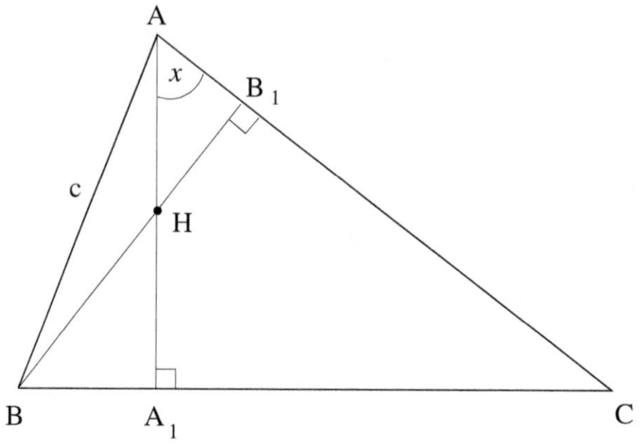

Figure 2.25: The orthocenter H.

called the altitudes of ΔABC. A figure in the case of an acute triangle is given in Figure 2.25.

Theorem 2.13.1 *The altitudes AA_1, BB_1, CC_1 intersect at a point H, called the orthocenter of the ΔABC.*

Proof: The length of the altitude $|AA_1|$ is given by

$$|AA_1| = c \sin \widehat{B}. \tag{2.13.8}$$

The lines AA_1 and BB_1 intersect at a point, because otherwise $AA_1 \| BB_1$ and as $AC \perp BB_1$, $BC \perp AA_1$ we have $AC \| BC$, which is a contradiction. Let $AA_1 \cap BB_1 = \{H\}$. In a similar way, let $CC_1 \cap AA_1 = \{H'\}$. If we show that

$$\frac{|AH|}{|AA_1|} = \frac{|AH'|}{|AA_1|} \tag{2.13.9}$$

then it follows $H = H'$.

Let $x = \widehat{HAB_1} = 90° - \widehat{C}$. See Figure 2.25. In ΔAHB_1, we have

$$\begin{aligned} |AH| &= \frac{|AB_1|}{\cos x} = \frac{|AB_1|}{\cos(90° - C)} \\ &= \frac{|AB_1|}{\sin \widehat{C}}. \end{aligned}$$

Using ΔABB_1,

$$|AB_1| = c \cos \widehat{A}.$$

Substituting yields

$$|AH| = \frac{c\cos\widehat{A}}{\sin\widehat{C}}. \tag{2.13.10}$$

Dividing (2.13.10) and (2.13.8) yields

$$\frac{|AH|}{|AA_1|} = \frac{c\cos\widehat{A}}{c\sin\widehat{C}\sin\widehat{B}} = \frac{\cos\widehat{A}}{\sin C\sin\widehat{B}}. \tag{2.13.11}$$

Swapping the roles of BB_1 and CC_1 and replacing H by H' yields a similar relation

$$\frac{|AH'|}{|AA_1|} = \frac{\cos\widehat{A}}{\sin\widehat{C}\sin\widehat{B}}. \tag{2.13.12}$$

Comparing relations (2.13.11) and (2.13.12) yields (2.13.9) and hence $H = H'$. ■

We shall provide below another proof involving a construction.

Proof: Consider parallel lines through the vertices A, B and C, which form the triangle $\Delta A'B'C'$, see Figure 2.26. Since $ACBC'$ and $BACA'$ are parallelograms, we have $|AC| = |BC'|$ and $|AC| = |BA'|$, see Exercise 2.13.10. By transitivity, we get $|BC'| = |BA'|$, so B is the midpoint of $A'C'$. Since $AC \| A'C'$, the altitude from B to AC is perpendicular on $A'C'$. This way, the altitude from B of triangle ΔABC become the perpendicular bisector of $A'C'$. The same argument applies to the other sides of the triangle. The altitudes of ΔABC becomes perpendicular bisectors for $\Delta A'B'C'$. Since the perpendicular bisectors of any triangle intersect at a point, see Theorem 2.7.3, it follows that the altitudes of triangle ΔABC intersect at a point. ■

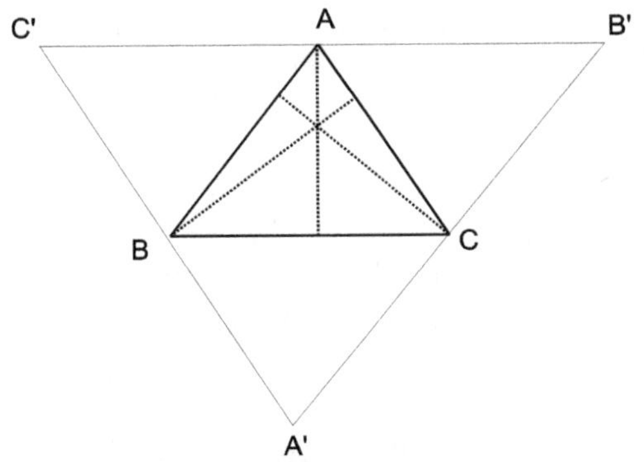

Figure 2.26: The altitudes of ΔABC are perpendicular bisectors of $\Delta A'B'C'$.

Remark 2.13.2 The orthocenter of triangle ΔABC is the circumcenter of triangle $\Delta A'B'C'$.

Solved Exercise 2.13.3 (*i*) *Let ΔABC be a triangle with acute base angles, $\widehat{B} < 90°$, $\widehat{C} < 90°$. Show that the altitude from the vertex A is situated inside the triangle ΔABC.*

(*ii*) *Show that ΔABC is an acute-angled triangle if and only if its orthocenter, H, belongs to the interior of the triangle.*

Proof: (*i*) Let $AA_1 \perp BC$, with $A_1 \in BC$. By contradiction, we assume that A_1 is not between B and C. Let's say that A_1 is on the left side of B. Then $\widehat{ABC}$ is an exterior angle in ΔAA_1B, so $\widehat{ABC} > \widehat{AA_1B} = 90°$, which contradicts that $\widehat{B} < 90°$. A similar argument for the case when A_1 is on the right side of B leads to the contradictory inequality $\widehat{C} > 90°$. Hence, $A_1 \in (BC)$. Using that the interior of a triangle is a convex set, we obtain that (AA_1) is inside triangle ΔABC.
(*ii*) " $\Longrightarrow$ " Let ΔABC be an acute-angled triangle. From part (*i*) the altitude $(AA_1) \subset Int(\Delta ABC)$. Since $H \in (AA_1)$, it follows that H is an interior point of the triangle.
" $\Longleftarrow$ " Assume that $H \in Int(\Delta ABC)$. Let $A_1 = AH \cap BC$. Then $AA_1 \perp BC$, and then ΔABA_1 and AA_1C are right triangles. Since a right triangle has two acute angles, it follows that $\widehat{B} < 90°$ and $\widehat{C} < 90°$. Similarly, starting from another altitude we obtain $\widehat{A} < 90°$, and hence ΔABC is an acute-angled triangle.

As a remark, in an acute-angled triangle the orthic triangle has the vertices inside the sides of ΔABC. ■

Solved Exercise 2.13.4 *Let O denote the circumcenter of ΔABC. Prove that the point $O \in Int(\Delta ABC)$ if and only if ΔABC is an acute-angled triangle.*

Proof: " $\Longrightarrow$ " Let $O \in Int(\Delta ABC)$, so the points A and O are on the same side of the line BC. Extend AO until it intersects the circle the second time at A''. The point A'' is on the other side of BC than A and O. Join the points A'' with B and C. By construction, $\widehat{ACB} < \widehat{ACA''} = \frac{1}{2}arc(ABA'') = 90°$, so $\widehat{C} < 90°$. A similar reason applies to the other angles.
" $\Longleftarrow$ " Let ΔABC be an acute-angled triangle. By contradiction, we assume $O \notin Int(\Delta ABC)$. Let BB'' be the diameter from B. Then $\widehat{BAC} > \widehat{BAB''} = 90°$, which contradicts the fact that the angle $\widehat{A}$ is acute. ■

The next property was found by Lazare Carnot (1753–1823), one of the most important personalities of the French Revolution and a mathematician as well.

Theorem 2.13.5 (Carnot theorem) *Let H be the orthocenter of ΔABC. Then, the circumcircles of ΔBHC, ΔAHC and ΔAHB have the same radii as the circumcircle of ΔABC*

Proof: We shall assume the case of an acute triangle. See Figure 2.27. Let R_A, R_B, R_C be the circumradii of ΔBHC, ΔAHC and ΔAHB, respectively. The idea of the proof is to apply the law of sines to the above triangles. From ΔBB_1C and ΔCC_1B, we have

$$\begin{aligned} x &= \widehat{B_1BC} = 90^\circ - \widehat{C}, \\ y &= \widehat{C_1CB} = 90^\circ - \widehat{B}. \end{aligned}$$

In ΔBHC, we have

$$\begin{aligned} \widehat{BHC} &= 180^\circ - x - y \\ &= 180^\circ - (90^\circ - \widehat{C}) - (90^\circ - \widehat{B}) \\ &= 180^\circ - 90^\circ + \widehat{C} - 90^\circ + \widehat{B} \\ &= \widehat{B} + \widehat{C} = 180^\circ - \widehat{A}. \end{aligned}$$

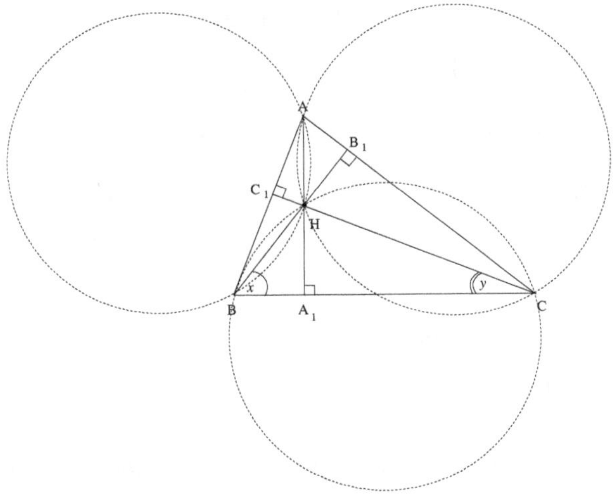

Figure 2.27: Carnot's theorem.

The law of sines applied in ΔBHC and ΔABC yields

$$\begin{aligned} 2R_A &= \frac{|BC|}{\sin \widehat{BHC}} = \frac{a}{\sin(180^\circ - \widehat{A})} \\ &= \frac{a}{\sin \widehat{A}} = 2R. \end{aligned}$$

Hence $R_A = R$. Similarly, $R_B = R$ and $R_C = R$. ■

Proposition 2.13.6 *We have*

$$|AH| = 2R\cos \widehat{A}, \qquad |BH| = 2R\cos \widehat{B}, \qquad |CH| = 2R\cos \widehat{C}.$$

Proof: The law of sines in ΔBHC yields

$$\frac{|BH|}{\sin y} = \frac{|CH|}{\sin x} = 2R.$$

It follows that

$$\begin{aligned} |BH| &= 2R\sin y = 2R\sin(90^\circ - \widehat{B}) \\ &= 2R\cos \widehat{B}. \end{aligned}$$

In a similar way we find the other two relations. ■

Exercise 2.13.7 *Prove that a triangle ΔABC is obtuse if and only if the orthocenter $H \notin Int\, ABC$.*

Solved Exercise 2.13.8 (Carnot's relation) *Prove that in any triangle we have*

$$|AH| + |BH| + |CH| = 2(R + r).$$

Proof: Using Proposition 2.13.6 and Exercise 2.1.15, we have

$$\begin{aligned} |AH| + |BH| + |CH| &= 2R\left(\cos \widehat{A} + \cos \widehat{B} + \cos \widehat{C}\right) \\ &= 2R\left(1 + 4\sin\frac{\widehat{A}}{2}\sin\frac{\widehat{B}}{2}\sin\frac{\widehat{C}}{2}\right) = 2R\left(1 + \frac{r}{R}\right) \\ &= 2(R + r). \end{aligned}$$

■

Exercise 2.13.9 *If O and H denote the circumcenter and the orthocenter of ΔABC, show the inequality*

$$|AO| + |BO| + |CO| \geq |AH| + |BH| + |CH|.$$

Exercise 2.13.10 *A quadrilateral $ABCD$ is called a parallelogram if $AB \| CD$ and $BC \| AD$. Show that in a parallelogram we have the following.*

(i) $|AB| = |CD|$ *and* $|BC| = |AD|$.

(ii) The diagonals AC and BD bisect each other.

Exercise 2.13.11 *Let $ABCD$ be a quadrilateral. Show that if $|AB| = |CD|$ and $|BC| = |AD|$, then $ABCD$ is a parallelogram.*

Exercise 2.13.12 *A quadrilateral $ABCD$ is called a rhombus if $|AB| = |BC| = |CD| = |DA|$. Show that a rhombus is a parallelogram.*

Exercise 2.13.13 *Show that in a rhombus the diagonals are perpendicular. Is the reciprocal true?*

Exercise 2.13.14 *Show that the angle bisectors of the interior angles of a parallelogram intersect at the vertices of a rectangle.*

Exercise 2.13.15 *Consider the acute-angled triangle ΔABC. Let O_A, O_B, O_C be the circumcenters of ΔBHC, ΔAHC and ΔAHB, respectively. See Figure* 2.28. *Show the following:*

(i) The circumradius of $\Delta O_A O_B O_C$ is equal to the circumradii of ΔBHC, ΔAHC and ΔAHB,

(ii) The circumcenter of $\Delta O_A O_B O_C$ is the orthocenter of ΔABC,

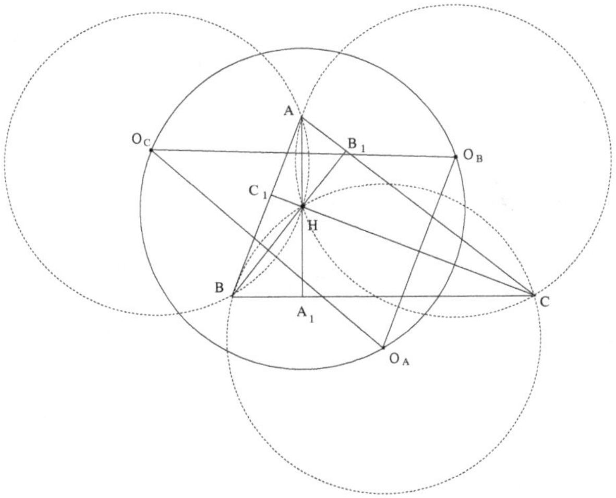

Figure 2.28: Carnot circles.

(iii) $\Delta O_A O_B O_C \equiv \Delta ABC$, *and*

(iv) $\widehat{O_B H O_C} = 2\widehat{A}$, $\widehat{O_C H O_A} = 2\widehat{B}$, $\widehat{O_A H O_B} = 2\widehat{C}$.

Exercise 2.13.16 (Tzitzeica, 1904) *Consider three equal circles passing through the same point. Let A, B, C be the points where the circles intersect two by two. Show that there is a circle of the same size as the initial circles passing though the points A, B, C. Make the relation with the Carnot circles given by Figure* 2.28.

2.14 Geometry and Physics

In this section, we shall present a few geometry problems related to several branches of Physics.

Electricity

We shall focus our concern on the relation given in Exercise 2.6.2

$$\frac{1}{r_a} + \frac{1}{r_b} + \frac{1}{r_c} = \frac{1}{r}. \tag{2.14.13}$$

The very same functional relation holds in an electrical circuit among three parallel resistors, r_a, r_b and r_c, under a voltage V, see Figure 2.29(a). The equivalent circuit is given in Figure 2.29(b), which replaces the three resistors by an equivalent resistor, r. The algebraic relation relating r with r_a, r_b and r_c is given by (2.14.13). This is a consequence of Ohm's formula applied in both circuits. First, Ohm's formula applied to the circuit given in Figure 2.29(b) relates the voltage, V, the current intensity, I and the

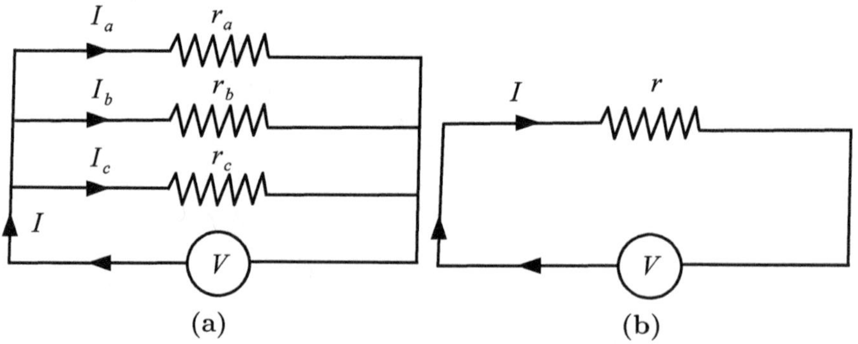

Figure 2.29: (a) Parallel resistors; (b) Equivalent resistor.

resistance, r, by formula $V = rI$. By Kirchhoff's formula the intensities add algebraically as $I = I_a + I_b + I_c$. On each branch we can apply Ohm's formula obtaining relations $V = r_a I_a$, $V = r_b I_b$, and $V = r_c I_c$, where we used that all resistors share the same voltage. Expressing resistances in terms of intensities and voltage, we have

$$\frac{1}{r} = \frac{I}{V} = \frac{I_a}{V} + \frac{I_b}{V} + \frac{I_c}{V} = \frac{1}{r_a} + \frac{1}{r_b} + \frac{1}{r_c},$$

which retrieves (2.14.13).

To obtain the geometric counterpart, we shall substitute the area for voltage, $\mathcal{A} = V$, the semiperimeter for total intensity, $p = I$, and the inradius, r, for the total resistance. Then the relation $\mathcal{A} = pr$ becomes the analog of the Ohm's law. To continue, we substitute $p-a$, $p-b$ and $p-c$ for the intensities I_a, I_b and I_c, which verify Kirchoff's formula, $p = (p-q)+(p-b)+(p-c)$. Considering the resistances r_a, r_b and r_c as exradii, by Exercise 2.6.1 the triangle area is given by $\mathcal{A} = (p-a)r_a = (p-b)r_b = (p-c)r_c$, which represents the analog relations of $V = I_a r_a = I_b r_b = I_c r_c$. Then the aforementioned relation becomes

$$\frac{1}{r} = \frac{p}{\mathcal{A}} = \frac{p-a}{\mathcal{A}} + \frac{p-b}{\mathcal{A}} + \frac{p-c}{\mathcal{A}} = \frac{1}{r_a} + \frac{1}{r_b} + \frac{1}{r_c},$$

fact that proves Exercise 2.14.13.

Mechanics

We consider the following problem: *Let ΔABC be a triangle with centroid G. Show that there is at least one side of ΔABC such that the orthogonal projection of G falls inside that side.*

We shall supply first a geometric proof. We denote by AA_1 and AA' the altitude and the median from A, respectively, see Figure 2.30(a). We assume the largest angle of the triangle ΔABC is located at A. Consequently, the altitude AA_1 is interior to the triangle and hence A_1 falls inside the side BC. (Why?) Since the centroid G belongs to AA', then its projection G_1 on BC is located between A_1 and A'. Using that both A_1 and A' are inside of BC, it follows that the point G_1 is also located inside the side BC.

An interesting non-computational proof variant involves the following mechanical approach. By contradiction, we assume that none of the orthogonal projections of G fall inside any of the sides. Considering the triangle as a slab of material of constant density, the previous assumption states that the projection of the centroid falls outside the stability basis of the triangle. Then by the law of dynamics the triangle will fall, performing a

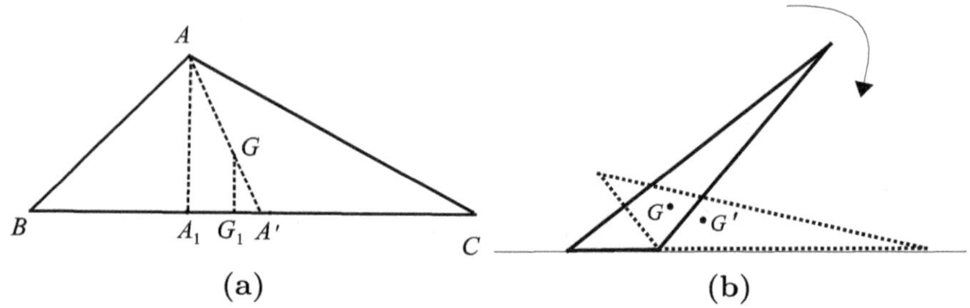

Figure 2.30: (a) The projection of the centroid on the side BC; (b) The triangle rotates to set the centroid from position G into a lower potential position G'.

rotation, see Figure 2.30(b). As the centroid projection is always outside the basis, the triangle will fall forever, in the virtue of getting set into a state of lower gravitational potential. Since a perpetuum mobile does not exist, the initial assumption must be false, and hence the triangle rotation must stop in the configuration where the projection of G falls inside a basis.

It is worth noting that this type of mechanical reasoning can be easily adapted to any convex polygon, or more general, to any convex polyhedron in 3 dimensions.

Optics

The reflected ray If a light beam reaches a mirror, the angle between the incident beam and the mirror is called the **incidence angle**. The angle between the reflected beam and the plane of the mirror is called the **reflection angle**. One of the law of optics states

$$\boxed{\text{incidence angle} = \text{reflection angle}}.$$

Exercise 2.14.1 *Let $AO)$ and $(OB$ be a pair of incidence-reflected rays, and let $(ON$ denote the normal at the plane of the mirror. Show that $(ON$ is the angle bisector of angle $\widehat{AOB}$. See Figure 2.31*(a).

The light triangle If the sides of the acute-angled triangle ΔABC are mirrors, any light beam will reflect inside the triangle. If the ray's trajectory is closed, then we have a light triangle, see Figure 2.31(b). The triangle $\Delta A'B'C'$ is a light triangle if and only if the reflection identity holds for each side

$$\widehat{C'A'B} = \widehat{B'A'C}, \quad \widehat{A'B'C} = \widehat{C'B'A}, \quad \widehat{AC'B'} = \widehat{A'C'B}. \qquad (2.14.14)$$

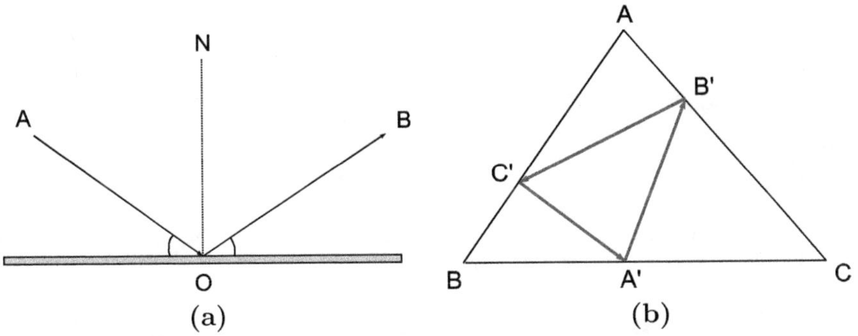

Figure 2.31: (a) Incidence and reflection angles; (b) Light triangle.

Since by Fermat's principle the light travels following the shortest path, it follows that the triangle $\Delta A'B'C'$ satisfying property (2.14.14) is the triangle with the smallest perimeter among all the triangles inscribed in ΔABC. Later we shall prove this principle. We shall also show that there is a unique triangle with this property and it can be constructed by joining the feet of the altitudes of ΔABC. This means that AA', BB' and CC' are altitudes in ΔABC.

The billiard problem The law of reflection applies also in the case of a billiard with a perfect elastic wall. A ball which strikes the wall bounces back such that the incidence and the reflection angles are equal, see Figure 2.32(a). The billiard problem can be stated as:

Given two points A and B in the same half plane with boundary line d, find a point $C \in d$ such that AC and CB make equal angles with d.

The solution of this problem is related to a variational problem, which we shall pursue in the following. To this end, we shall emphasize first some properties of an isosceles trapezoid based on Figure 2.32(b).

Lemma 2.14.2 *Let $ABB'A'$ be an isosceles trapezoid with $|AB| = |A'B'|$ and denote by d the vertical symmetry axis. Let C be the point where the diagonals intersect. Then*

(i) $C \in d$,

(ii) $\widehat{ACU} = \widehat{BCV}$, and

(iii) Among all the points $M \in d$, the sum $|AM| + |BM|$ is minimum for the point $M = C$.

Proof: (*i*) The symmetry axis d is the perpendicular bisector of both AA' and BB'. Because of the symmetry, we have $|CA'| = |CA|$ and

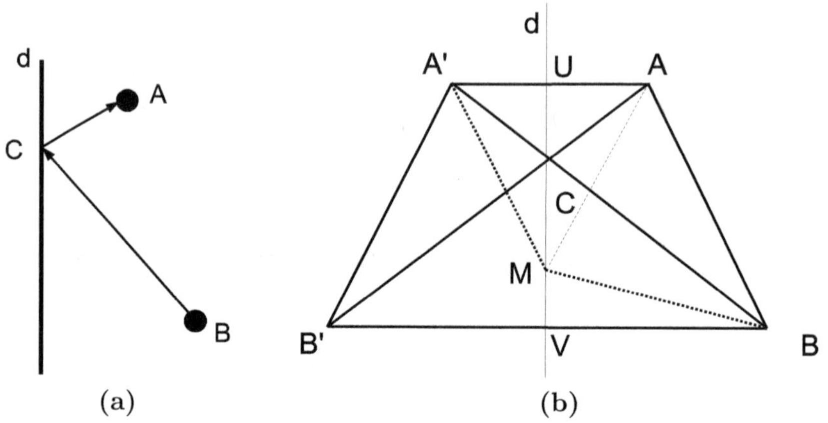

Figure 2.32: The billiard problem.

$|CB'| = |CB|$, and hence C belongs to the perpendicular bisector d, see Lemma 2.7.2.

(*ii*) By (SAS) congruence case we have $\Delta ACU \equiv \Delta A'CU$, because

$$\begin{cases} |AU| = |A'U|, & \text{by symmetry;} \\ |BV| = |B'V|, & \text{by symmetry;} \\ \widehat{AUV} = \widehat{A'UV}, & AA' \perp d. \end{cases}$$

Hence $\widehat{A'CU} = \widehat{ACU}$. Since $\widehat{A'CU} = \widehat{BCV}$ (vertical angles), by transitivity we get $\widehat{A'CU} = \widehat{BCV}$.

(*iii*) Consider an arbitrary point $M \in d \backslash \{C\}$. We need to show that

$$|AC| + |CB| < |AM| + |BM|.$$

Using that $|AM| = |A'M|$ and $|AC| = |A'C|$, applying the triangle inequality in $\Delta A'MB$ yields

$$\begin{aligned} |AC| + |CB| &= |A'C| + |CB| = |A'B| < |A'M| + |MB| \\ &= |AM| + |MB|. \end{aligned}$$

■

The previous result shows that the law of reflection is equivalent to a variational problem as follows.

Corollary 2.14.3 (Fermat's principle) *Let A and B be two fixed points on the same side of a line d and M be a variable point on d. The point*

M satisfies the minimum value of the sum $|MA| + |MB|$ if and only if the incidence angle between MA and d equals the reflection angle between MB and d.

The minimality of the sum $|MA|+|MB|$ is equivalent to the fact that the ball travels between A and B along the path of shortest time. Consequently, the minimum distance traveled by the ball between A and B is equal to the length of the diagonal of the isosceles trapezoid $ABB'A'$. This can also be regarded as the trajectory of a light beam starting at A, reflecting on the mirror d and arriving at B; this is done by following the trajectory for which the incidence and the reflection angles are equal.

Going back to the initial billiard problem, using Lemma 2.14.2, point C can be found in the following way:

- Construct the symmetric point A' of A with respect to the line d;
- Let C be the intersection of the line d with the segment $A'B$. The intersection is assured by the plane separation axiom applied to points A' and B;
- Point C can also be obtained using the triangle similarity $\Delta BVC \sim \Delta AUC$:
$$\frac{|CU|}{|CV|} = \frac{|AU|}{|BV|}.$$

Exercise 2.14.4 *Consider the points $A(1, 4)$ and $B(5, 1)$. Find a point M on the y-axis such that the sum $|AM| + |BM|$ is minimum.*

Exercise 2.14.5 *Consider the points $A(1, 4)$ and $B(5, 1)$. Find a point M on the y-axis such that the sum $|AM|^2 + |BM|^2$ is minimum. Do you get the same point as in Exercise 2.14.4? Why?*

Exercise 2.14.6 *Consider two points A, B in the plane of a circle $\mathcal{C}$. Find the point M on $\mathcal{C}$ such that the sum $|AM| + |BM|$ is minimum. Consider the following cases: both points are outside the circle; both points are inside; and one point is outside and one is inside.*

The law of reflection applies also in the case of billiards with curved walls. In this case, the ball reflects such that the incidence and the reflection angles with respect to the tangent line to the wall are equal. A distinguished case is played by the elliptical billiards.

Definition 2.14.7 *Let F_1 and F_2 be two given points. A ellipse with the focal points F_1 and F_2 is the set of points in the plane such that*

$$\mathcal{E}_c = \{M; |MF_1| + |MF_2| = c\}, \text{ with } c > 0.$$

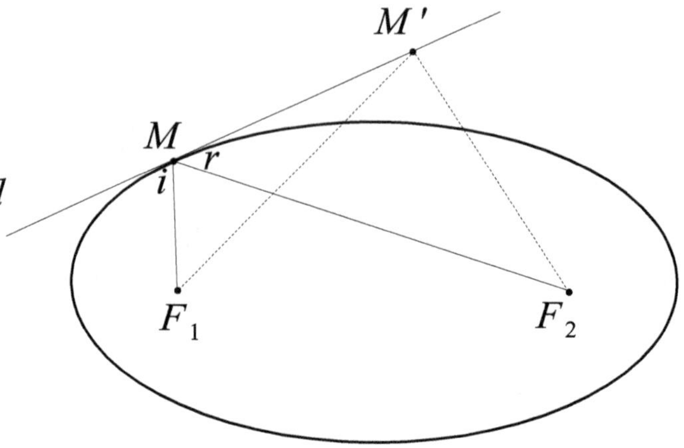

Figure 2.33: The reflection property of an ellipse: $i = r$.

All ellipses in the family $(\mathcal{E}_c)_{c>0}$ have the same focal points. If $c' > c$, then $\mathcal{E}_c$ lies inside $\mathcal{E}_{c'}$. If $F_1 = F_2$, then $(\mathcal{E}_c)_{c>0}$ becomes a family of concentric circles of diameter c.

The next result deals with the reflection property of an ellipse, see Figure 2.33.

Proposition 2.14.8 *Let M be a point on the ellipse $\mathcal{E}$ and d be the tangent line to $\mathcal{E}$ at M. Then the lines F_1M and F_2M make equal angles with the tangent d.*

Proof: It suffices to show that the point M satisfies the minimality condition provided by Corollary 2.14.3. To this end, we consider another point $M' \in d$. Since d is a tangent line and $\mathcal{E}$ is convex, then M' is an exterior point to the ellipse. Let $|MF_1| + |MF_2| = c$. Then $|M'F_1| + |M'F_2| = c'$ with $c' > c$. Then we have the inequality

$$|MF_1| + |MF_2| < |M'F_1| + |M'F_2|, \quad \forall M' \in d \backslash \{M\}.$$

Therefore, the tangency point M satisfies the minimum property and by Corollary 2.14.3, it follows the equality between the specified angles. ■

Consequently, if a billiard ball, initially located at F_1, is struck in any direction on a billiard table of elliptical shape, then it will bounce back hitting the ball situated at F_2.

Exercise 2.14.9 *Consider a billiard table of elliptical shape with focal points F_1 and F_2. A ball, initially located at F_1, is struck into the billiard wall.*

Assuming that the ball moves forever, what will happen to its trajectory in the long run?

Exercise 2.14.10 *Under the hypotheses of Exercise 2.14.9, we consider a second ball, initially located at F_2, which is struck into the billiard wall simultaneously with the first ball. Do the balls struck each other? Assume the balls have non-negligible radii.*

The existence of closed trajectories of balls in convex billiards is a complex problem of ergodic theory. However, the case of a triangular billiard is of distinguished nature as it relates to the classical notion of orthic triangle. We shall study this problem in the next section.

2.15 The orthic triangle

Let AA_1, BB_1, CC_1 be the altitudes of ΔABC. The triangle $\Delta A_1B_1C_1$ is called the **orthic** triangle of ΔABC. The trigonometric computations in this section will be made under the assumption that the triangle ΔABC is an acute triangle, i.e., the measure of all angles is less than $90°$.
In the following, we shall compute the elements of the orthic triangle $\Delta A_1B_1C_1$ in terms of the elements of triangle ABC.

The sides of the orthic triangle

Denote the side lengths of the orthic triangle by $a_0 = B_1C_1$, $b_0 = A_1C_1$, and $c_0 = A_1B_1$. The points A, B_1, H and C_1 belong to a circle because

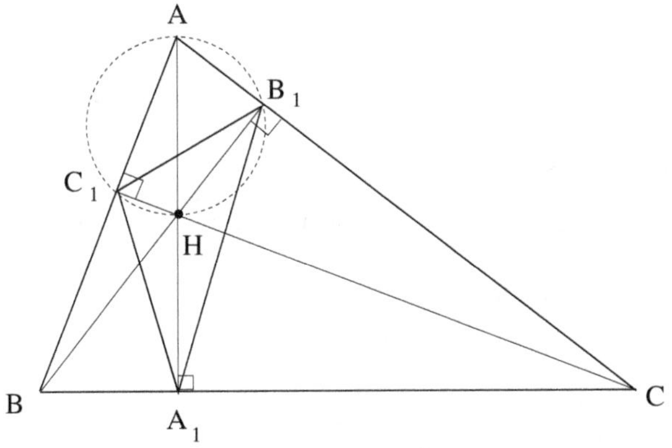

Figure 2.34: The orthic triangle $\Delta A_1B_1C_1$.

ΔAB_1H and ΔAC_1H are right triangles and with the common hypotenuse AH whose length is the diameter of their circumcircle.

The angle made by the diagonal B_1C_1 of the quadrilateral AB_1HC_1 with the side AC_1 is equal to the angle made by the other diagonal AH with the opposite side HB_1 (why?):

$$\begin{aligned}\widehat{B_1C_1A} &= \widehat{AHB_1} = \widehat{BHA_1} \text{ (vertical angles)}\\ &= 90^\circ - \widehat{HBC} \text{ (}\Delta HBA_1 \text{ is a right triangle)}\\ &= 90^\circ - \widehat{B_1BC}\\ &= \widehat{C} \text{ (from } \Delta BB_1C\text{)}.\end{aligned}$$

Hence $\widehat{B_1C_1A} = \widehat{ACB}$. Similarly we have $\widehat{C_1B_1A} = \widehat{ABC}$. It follows that the triangles ΔAB_1C_1 and ΔABC are similar because they have the same angles. Hence

$$\frac{|AC|}{|AC_1|} = \frac{|BC|}{|B_1C_1|} = \frac{|AB|}{|AB_1|} \iff \frac{b}{|AC_1|} = \frac{a}{a_0} = \frac{c}{|AB_1|}.$$

Then one side of the orthic triangle is given by

$$a_0 = \frac{a}{c}|AB_1| = \frac{a}{c} \cdot c\cos\widehat{A} = a\cos\widehat{A}, \tag{2.15.15}$$

where we have used the fact that ΔABB_1 is a right triangle. Similarly, we obtain $b_0 = b\cos\widehat{B}$ and $c_0 = c\cos\widehat{C}$.

The semiperimeter of the orthic triangle

Using the law of sines in ΔABC, we have

$$a = 2R\sin\widehat{A}, \qquad b = 2R\sin\widehat{B}, \qquad c = 2R\sin\widehat{C},$$

where R is the circumradius of ΔABC. Then

$$a_0 = a\cos\widehat{A} = 2R\sin\widehat{A}\cos\widehat{A} = R\sin 2\widehat{A}.$$

Using similar relations and Corollary 2.7.13 yields

$$\begin{aligned}p_0 &= \frac{1}{2}(a_0 + b_0 + c_0) = \frac{1}{2}\left(R\sin 2\widehat{A} + R\sin 2\widehat{B} + R\sin 2\widehat{C}\right)\\ &= \frac{R}{2}\left(\sin 2\widehat{A} + \sin 2\widehat{B} + \sin 2\widehat{C}\right)\\ &= 2R\sin\widehat{A}\sin\widehat{B}\sin\widehat{C}.\end{aligned}$$

Hence the semiperimeter of the orthic triangle is given by

$$p_0 = 2R \sin \widehat{A} \sin \widehat{B} \sin \widehat{C}. \tag{2.15.16}$$

The law of sines in ΔABC provides

$$\sin \widehat{A} = \frac{a}{2R}, \qquad \sin \widehat{B} = \frac{b}{2R}, \qquad \sin \widehat{C} = \frac{c}{2R}.$$

Substituting in (2.15.16) yields

$$\begin{aligned} p_0 &= 2R \cdot \frac{a}{2R} \cdot \frac{b}{2R} \cdot \frac{c}{2R} \\ &= \frac{abc}{4R^2} = \frac{1}{R}\frac{abc}{4R} \\ &= \frac{1}{R}\mathcal{A}_{ABC}, \end{aligned}$$

where we used Proposition 2.4.2. We obtain another formula for the area of a triangle.

Proposition 2.15.1 (Tzitzeica's formula) *The area of ΔABC is given by*

$$\mathcal{A}_{ABC} = p_0 R,$$

where R is the circumradius of ΔABC and p_0 is the semiperimeter of the orthic triangle.

Corollary 2.15.2 *Let r and p denote the inradius and the semiperimeter of ΔABC. Then*

$$\frac{R}{r} = \frac{p}{p_0}.$$

Proof: Using $\mathcal{A}_{ABC} = pr$ from Theorem 2.5.11, Tzitzeica's formula yields

$$p_0 R = pr \Longrightarrow \frac{R}{r} = \frac{p}{p_0}.$$

■

Corollary 2.15.3 *We have $p \geq 2p_0$.*

Proof: Using Euler's inequality $\frac{R}{r} \geq 2$, Corollary 2.15.2 yields $\frac{p}{p_0} \geq 2$. ■

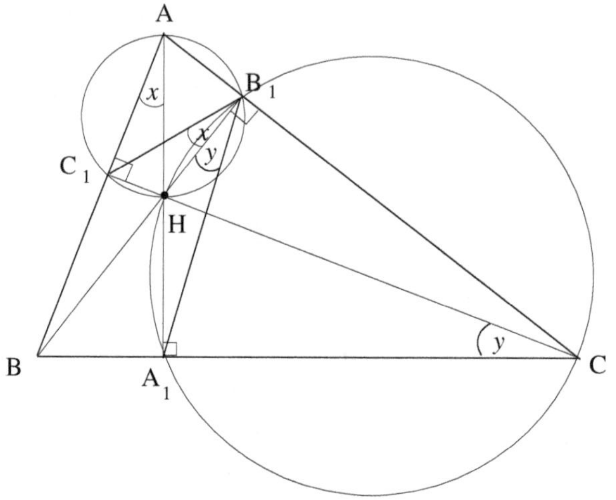

Figure 2.35: The angle of the orthic triangle $\Delta A_1B_1C_1$.

The angles of the orthic triangle

In this paragraph, we suppose that ΔABC is an acute triangle. The quadrilaterals AB_1HC_1 and B_1CA_1H can be inscribed into circles of diameters AH and HC, respectively. The justification is left for the reader.
Hence, we have

$$\begin{aligned} x &= \widehat{C_1AH} = \widehat{C_1B_1H} \text{ and} \\ y &= \widehat{HB_1A_1} = \widehat{HCA_1}. \end{aligned}$$

Then

$$\begin{aligned} \widehat{C_1B_1A_1} &= x + y = \widehat{C_1B_1H} + \widehat{HB_1A_1} \\ &= \widehat{C_1AH} + \widehat{HCA_1} \\ &= \widehat{BAA_1} + \widehat{C_1CB} \\ &= \left(90 - \widehat{B}\right) + \left(90 - \widehat{B}\right) \\ &= 180 - 2\widehat{B}. \end{aligned}$$

In a similar way, we find the other two angles $\widehat{B_1C_1A_1}$ and $\widehat{C_1A_1B_1}$. Thus, we arrived to the following result:

Proposition 2.15.4 *The measure of the angles for the orthic triangle of an acute triangle* ΔABC *are* $180° - 2\widehat{A}$, $180° - 2\widehat{B}$, *and* $180° - 2\widehat{C}$.

Corollary 2.15.5 *The altitudes of ΔABC are angle bisectors for the orthic triangle $\Delta A_1B_1C_1$. The orthocenter H of ΔABC is the incenter of $\Delta A_1B_1C_1$.*

Proof: We have shown that $x = 90° - \widehat{B} = y$, and hence BB_1 bisects the angle $\widehat{C_1B_1A_1}$. ■

Corollary 2.15.6 *The orthic triangle $\Delta A_1B_1C_1$ is a light triangle, i.e., we have*

$$\widehat{A_1B_1C} = \widehat{C_1B_1A}, \qquad \widehat{C_1A_1B} = \widehat{B_1A_1C}, \qquad \widehat{B_1C_1A} = \widehat{A_1C_1B}.$$

Equivalently stated, the orthic triangle represents a closed trajectory of a ball in a triangular billiard. In fact, this is the only triangle trajectory with this property. The uniqueness is not clear at the moment, but it follows from the argument used in Fagnano's theorem proof in Section 2.19.

Exercise 2.15.7 *Show that if the triangle ΔABC has a right angle at A, then $H = A$.*

Exercise 2.15.8 *Let ΔABC be a triangle that has an obtuse angle at A. Find the angles of the orthic triangle in terms of the angles of ΔABC.*

Exercise 2.15.9 *Show that $\Delta ABC \sim \Delta A_1B_1C_1$ if and only if ΔABC is an equilateral triangle.*

The circumradius of the orthic triangle

Let a_0, b_0, c_0 and $\widehat{A_0}$, $\widehat{B_0}$, $\widehat{C_0}$ be the lengths of the sides and the angles of the orthic triangle. The law of sines yields

$$\frac{a_0}{\sin \widehat{A_0}} = \frac{b_0}{\sin \widehat{B_0}} = \frac{c_0}{\sin \widehat{C_0}} = 2R_0.$$

Using Proposition 2.15.4 and the law of sines in ΔABC,

$$\begin{aligned} 2R_0 &= \frac{a_0}{\sin \widehat{A_0}} = \frac{a \cos \widehat{A}}{\sin(180 - 2\widehat{A})} \\ &= \frac{a \cos \widehat{A}}{\sin(2\widehat{A})} = \frac{a \cos \widehat{A}}{2 \sin \widehat{A} \cos \widehat{A}} \\ &= \frac{1}{2} \frac{a}{\sin \widehat{A}} = \frac{1}{2} \cdot 2R = R. \end{aligned}$$

We have arrived at the following result.

Proposition 2.15.10 *The circumradius of the orthic triangle is half of the circumradius of the initial triangle*

$$R_0 = \frac{R}{2}.$$

Exercise 2.15.11 *Prove that the incenter of the orthic triangle is the orthocenter of the initial triangle.*

Exercise 2.15.12 *Prove that triangles ΔABC, ΔAB_1C_1, ΔA_1BC_1 and ΔA_1B_1C are similar.*

The area of the orthic triangle

The area of the orthic triangle will be denoted by $\mathcal{A}_0$. Proposition 2.4.2 yields

$$\begin{aligned}
\mathcal{A}_0 &= \frac{a_0b_0c_0}{4R_0} = \frac{a\cos\widehat{A}\, b\cos\widehat{B}\, c\cos\widehat{C}}{4\dfrac{R}{2}} \\
&= 2\frac{abc\cos\widehat{A}\cos\widehat{B}\cos\widehat{C}}{4R} = \frac{abc}{4R}\cdot 2\cos\widehat{A}\cos\widehat{B}\cos\widehat{C} \\
&= \mathcal{A}_{ABC}\cdot 2\cos\widehat{A}\cos\widehat{B}\cos\widehat{C}.
\end{aligned}$$

Hence

$$\frac{\mathcal{A}_0}{\mathcal{A}_{ABC}} = 2\cos\widehat{A}\cos\widehat{B}\cos\widehat{C}. \qquad (2.15.17)$$

The inradius of the orthic triangle

Theorem 2.5.11, Proposition 2.15.1 and (2.15.17) yields

$$r_0 = \frac{\mathcal{A}_0}{p_0} = R\frac{\mathcal{A}_0}{p_0R} = R\frac{\mathcal{A}_0}{\mathcal{A}_{ABC}} = 2R\cos\widehat{A}\cos\widehat{B}\cos\widehat{C}.$$

Hence

$$r_0 = 2R\cos\widehat{A}\cos\widehat{B}\cos\widehat{C}. \qquad (2.15.18)$$

Inequalities from the orthic triangle

The expression $\cos A\cos B\cos C$ which appears in formulas (2.15.17) and (2.15.18) is bounded from above. The following result improves on this statement. The proof uses the theory of quadratic equations to show a geometric inequality.

Lemma 2.15.13 *In any triangle the following inequality holds:*

$$\cos\widehat{A}\cos\widehat{B}\cos\widehat{C} \leq \frac{1}{8}.$$

Proof: Let $w = \cos\widehat{A}\cos\widehat{B}\cos\widehat{C}$ and $x = \cos\widehat{C}$. Using Corollary 2.1.5 yields

$$\begin{aligned} w &= \frac{1}{2}\left(\cos\left(\widehat{A}+\widehat{B}\right)+\cos\left(\widehat{A}-\widehat{B}\right)\right)\cos\widehat{C} \\ &= \frac{1}{2}\left(\cos\left(180-\widehat{C}\right)+\cos\left(\widehat{A}-\widehat{B}\right)\right)\cos\widehat{C} \\ &= \frac{1}{2}\left(-\cos\widehat{C}+\cos\left(\widehat{A}-\widehat{B}\right)\right)\cos\widehat{C} \\ &= \frac{1}{2}\left(-x+\cos\left(\widehat{A}-\widehat{B}\right)\right)x \\ &= \frac{1}{2}\left(-x^2+x\cos\left(\widehat{A}-\widehat{B}\right)\right). \end{aligned}$$

Then x is a solution of the quadratic equation

$$2w = -x^2 + x\cos\left(\widehat{A}-\widehat{B}\right),$$

which can also be written as

$$x^2 - x\cos\left(\widehat{A}-\widehat{B}\right) + 2w = 0.$$

As the solution is real ($x = \cos C$ is a real number), the discriminant of the equation has to be non-negative:

$$\begin{aligned} 0 &\leq \Delta = \cos^2\left(\widehat{A}-\widehat{B}\right) - 4\cdot 2w \Longleftrightarrow \\ 8w &\leq \cos^2\left(\widehat{A}-\widehat{B}\right) \leq 1 \Longrightarrow \\ w &\leq \frac{1}{8}. \end{aligned}$$

■

Remark 2.15.14 The identity in the above inequality is reached exactly when $\cos A = \cos\widehat{B} = \cos\widehat{C} = \frac{1}{2}$, i.e., when $\widehat{A} = \widehat{B} = \widehat{C} = 60$, which is the case of an equilateral triangle. There is also a proof using Jensen's inequality, see Exercise 1.12.8.

Proposition 2.15.15 *We have*

$$\mathcal{A}_0 \leq \frac{1}{4}\mathcal{A}_{ABC}.$$

Proof: Using Lemma 2.15.13 and formula (2.15.17), we have

$$\begin{aligned} \mathcal{A}_0 &= 2\mathcal{A}_{ABC} \cos\widehat{A}\cos\widehat{B}\cos\widehat{C} \\ &\le 2\mathcal{A}_{ABC}\frac{1}{8} = \frac{1}{4}\mathcal{A}_{ABC}. \end{aligned}$$

■

Definition 2.15.16 *Let ABC be a triangle and M, N, P be the midpoints of its sides. The triangle ΔMNP is called the* **median triangle**.

Exercise 2.15.17 *Let $\mathcal{A}_{med}$ denote the area of the median triangle. Show that*

$$\mathcal{A}_0 \le \mathcal{A}_{med}.$$

Proposition 2.15.18 *We have $r_0 \le \dfrac{R}{4}$.*

Proof: Using Lemma 2.15.13 and (2.15.18), we have

$$r_0 = 2R\cos\widehat{A}\cos\widehat{B}\cos\widehat{C} \le 2R\frac{1}{8} = \frac{R}{4}.$$

■

Remark 2.15.19 Using $R_0 = R/2$ yields $2r_0 \le R_0$, which is Euler's inequality for the orthic triangle.

2.16 Law of cosines

Besides the law of sines, the law of cosines plays an important role in metric Euclidean geometry. It represents a functional relation between the side lengths and the cosines of the angles. This relation has many applications in computing lengths and angles in triangle and quadrilateral geometry. A few applications will be presented in this section. The law of cosines might also be considered as a generalization of the Pythagorean theorem since it is an extension of this theorem to any angle, not necessarily right.

Theorem 2.16.1 (The law of cosines) *Let a, b, c denote the lengths of the sides of an arbitrary triangle ΔABC. Then*

$$a^2 = b^2 + c^2 - 2bc\cos\widehat{A}.$$

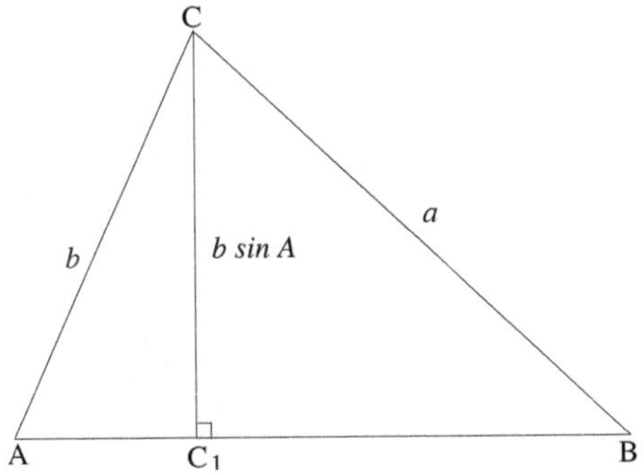

Figure 2.36: The case $\widehat{A} < 90°$.

Proof: *The case* $0° < \widehat{A} < 90°$: Let CC_1 be the altitude from the vertex C. Then C_1 is between A and B. See Figure 2.36. From triangle ΔACC_1 we have $|CC_1| = b \sin \widehat{A}$ and $|C_1B| = |AB| - |AC_1| = c - b\cos\widehat{A}$. The Pythagorean theorem for ΔCC_1B with $\widehat{C_1} = 90°$ yields $|BC|^2 = |CC_1|^2 + |C_1B|^2$, which is equivalent to

$$\begin{aligned}
a^2 &= \left(b\sin\widehat{A}\right)^2 + \left(c - b\cos\widehat{A}\right)^2 \\
&= b^2\sin^2\widehat{A} + c^2 - 2bc\cos\widehat{A} + b^2\cos^2\widehat{A} \\
&= b^2\left(\sin^2\widehat{A} + \cos^2\widehat{A}\right) + c^2 - 2bc\cos\widehat{A} \\
&= b^2 + c^2 - 2bc\cos\widehat{A}.
\end{aligned}$$

The case $\widehat{A} = 90°$: Then $\cos\widehat{A} = 0°$ and we have $a^2 = b^2 + c^2$, which is exactly the Pythagorean theorem.

The case $90° < \widehat{A} < 180°$: Let $u = 180° - \widehat{A}$. In this case, A is between C_1 and B. The Pythagorean theorem applied to ΔBCC_1 yields $|BC|^2 = |CC_1|^2 + |C_1B|^2$, which can be written as

$$\begin{aligned}
a^2 &= (b\sin u)^2 + (b\cos u + c)^2 \\
&= b^2\sin^2 u + b^2\cos^2 u + 2bc\cos u + c^2 \\
&= b^2(\sin^2 u + \cos^2 u) + 2bc\cos u + c^2 \\
&= b^2 + c^2 + 2bc\cos u \\
&= b^2 + c^2 + 2bc\cos\left(180° - \widehat{A}\right) \\
&= b^2 + c^2 - 2bc\cos\widehat{A}.
\end{aligned}$$

■

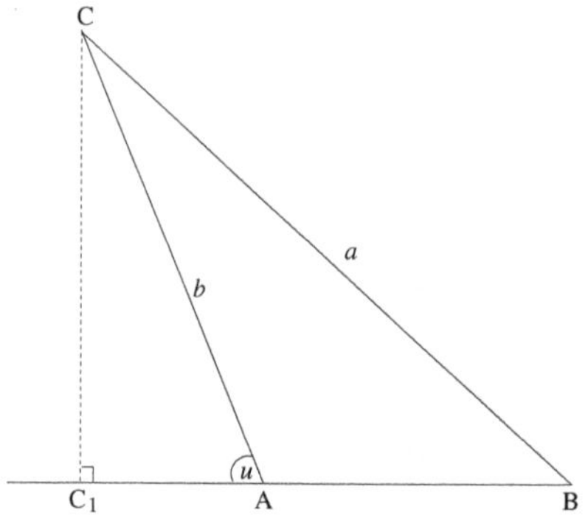

Figure 2.37: The case $\widehat{A} > 90°$.

Corollary 2.16.2 *The cosines of the angles of a triangle ΔABC can be expressed in terms of the side lengths a, b, c:*

$$\begin{aligned}\cos\widehat{A} &= \frac{b^2+c^2-a^2}{2bc},\\ \cos\widehat{B} &= \frac{a^2+c^2-b^2}{2ac}, \text{ and}\\ \cos\widehat{C} &= \frac{a^2+b^2-c^2}{2ab}.\end{aligned}$$

Exercise 2.16.3 *Show that in any triangle we have*

$$bc\cos\widehat{A} + ca\cos\widehat{B} + ab\cos\widehat{C} = \frac{1}{2}(a^2+b^2+c^2).$$

The following result expresses the cosines of the half angles in terms of the side lengths.

Theorem 2.16.4 *In any triangle ABC, we have*

$$\begin{aligned}1)\quad \cos\frac{\widehat{A}}{2} &= \sqrt{\frac{p(p-a)}{bc}},\\ 2)\quad \sin\frac{\widehat{A}}{2} &= \sqrt{\frac{(p-b)(p-c)}{bc}}, \text{ and}\\ 3)\quad \tan\frac{\widehat{A}}{2} &= \sqrt{\frac{(p-b)(p-c)}{p(p-a)}}.\end{aligned}$$

Proof: 1) Exercise 2.1.8 and Corollary 2.16.2 yield

$$\begin{aligned}\cos\frac{\widehat{A}}{2} &= \sqrt{\frac{1+\cos\widehat{A}}{2}} = \sqrt{\frac{1}{2}+\frac{1}{2}\cos\widehat{A}}\\ &= \sqrt{\frac{1}{2}+\frac{1}{2}\frac{b^2+c^2-a^2}{2bc}} = \sqrt{\frac{2bc+b^2+c^2-a^2}{4bc}}\\ &= \sqrt{\frac{(b+c)^2-a^2}{4bc}} = \sqrt{\frac{(b+c-a)(b+c+a)}{2\cdot 2\cdot bc}}\\ &= \sqrt{\frac{(p-a)p}{bc}}.\end{aligned}$$

The proofs of 2) and 3) are left as an exercise for the reader. ■

Exercise 2.16.5 *Prove parts* 2) *and* 3) *of Theorem 2.16.4.*

2.17 Heron's formula

A consequence of the above result is the well-known formula of Heron, which expresses the area of a triangle in terms of its side lengths.

Proposition 2.17.1 (Heron's formula) *The area of a triangle with sides' lengths a, b, c is given by*

$$\mathcal{A} = \sqrt{p(p-a)(p-b)(p-c)}.$$

Proof: We follow the notations in Figure 2.38, where I is the incenter and $IB_1 \perp AC$. Let r denote the inradius.
We know that $|AB_1| = p-a$ and $r = |IB_1| = (p-a)\tan\frac{\widehat{A}}{2}$. Then the area of triangle ABC is

$$\begin{aligned}\mathcal{A} &= pr = p(p-a)\tan\frac{\widehat{A}}{2} = p(p-a)\sqrt{\frac{(p-b)(p-c)}{p(p-a)}}\\ &= \sqrt{p(p-a)(p-b)(p-c)}.\end{aligned}$$

■

Proof variant: One can also prove Heron's formula starting from the area formula involving two sides and an angle and using Theorem 2.16.4

$$\begin{aligned}\mathcal{A} &= \frac{1}{2}bc\sin\widehat{A} = \frac{1}{2}bc\cdot 2\sin\frac{\widehat{A}}{2}\cos\frac{\widehat{A}}{2}\\ &= bc\sqrt{\frac{(p-b)(p-c)}{bc}}\sqrt{\frac{p(p-a)}{bc}} = \sqrt{p(p-a)(p-b)(p-c)}.\end{aligned}$$

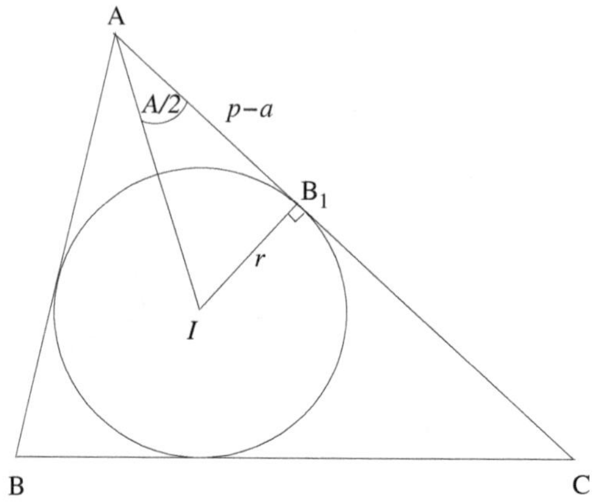

Figure 2.38: See the proof of Heron's formula.

Exercise 2.17.2 *Show that the inradius is given by*

$$r = (p-a)\tan\frac{\widehat{A}}{2} = (p-b)\tan\frac{\widehat{B}}{2} = (p-c)\tan\frac{\widehat{C}}{2}.$$

Exercise 2.17.3 *Show that the exradii are given by*

$$r_a = p\tan\frac{\widehat{A}}{2}, \quad r_b = p\tan\frac{\widehat{B}}{2}, \quad r_c = p\tan\frac{\widehat{C}}{2}.$$

Exercise 2.17.4 *Prove the following relations:*

(*i*) $\cos\frac{\widehat{A}}{2}\cos\frac{\widehat{B}}{2}\cos\frac{\widehat{C}}{2} = \frac{p}{4R}$, *and*

(*ii*) $\sin\widehat{A} + \sin\widehat{B} + \sin\widehat{C} = 4\cos\frac{\widehat{A}}{2}\cos\frac{\widehat{B}}{2}\cos\frac{\widehat{C}}{2}$.

Heron's formula has many consequences in the metric geometry of a triangle. For instance, one of them is to express the area of a triangle in terms of inradius and angles.

Proposition 2.17.5 *We have*

$$\mathcal{A} = r^2\cot\frac{\widehat{A}}{2}\cot\frac{\widehat{B}}{2}\cot\frac{\widehat{C}}{2}. \tag{2.17.19}$$

Proof: Using Exercise 2.17.2, we express the cube of the inradius as

$$r^3 = (p-a)\tan\frac{\widehat{A}}{2}(p-b)\tan\frac{\widehat{B}}{2}(p-c)\tan\frac{\widehat{C}}{2}.$$

Then multiplying by the semiperimeter, p, and using the area formula $\mathcal{A} = pr$ and Heron's formula, yields

$$pr \cdot r^2 = p(p-a)(p-b)(p-c) \tan\frac{\widehat{A}}{2} \tan\frac{\widehat{B}}{2} \tan\frac{\widehat{C}}{2} \Longleftrightarrow$$
$$\mathcal{A}r^2 = \mathcal{A}^2 \tan\frac{\widehat{A}}{2} \tan\frac{\widehat{B}}{2} \tan\frac{\widehat{C}}{2}.$$

Solving for $\mathcal{A}$ leads to the desired formula. ■

As a consequence, we obtain a relation between the inradius and circumradius.

Proposition 2.17.6 *We have*

$$r = 4R \sin\frac{\widehat{A}}{2} \sin\frac{\widehat{B}}{2} \sin\frac{\widehat{C}}{2}. \tag{2.17.20}$$

Proof: By Proposition 2.4.3, we have $\mathcal{A} = 2R^2 \sin\widehat{A} \sin\widehat{B} \sin\widehat{C}$. Equating against the area relation given by (2.17.19) and using the double sine angle formula yields

$$r^2 \cot\frac{\widehat{A}}{2} \cot\frac{\widehat{B}}{2} \cot\frac{\widehat{C}}{2} = 2R^2 \sin\widehat{A} \sin\widehat{B} \sin\widehat{C} \Longleftrightarrow$$
$$r^2 \frac{\cos\frac{\widehat{A}}{2} \cos\frac{\widehat{B}}{2} \cos\frac{\widehat{C}}{2}}{\sin\frac{\widehat{A}}{2} \sin\frac{\widehat{B}}{2} \sin\frac{\widehat{C}}{2}} = 2^4 R^2 \sin\frac{\widehat{A}}{2} \cos\frac{\widehat{A}}{2} \sin\frac{\widehat{B}}{2} \cos\frac{\widehat{B}}{2} \sin\frac{\widehat{C}}{2} \cos\frac{\widehat{C}}{2} \Longleftrightarrow$$
$$r^2 = 4^2 R^2 \sin^2\frac{\widehat{A}}{2} \sin^2\frac{\widehat{B}}{2} \sin^2\frac{\widehat{C}}{2}.$$

■

Euler's inequality $R \geq 2r$ implies the following inequality:

Corollary 2.17.7 *We have*

$$\sin\frac{\widehat{A}}{2} \sin\frac{\widehat{B}}{2} \sin\frac{\widehat{C}}{2} \leq \frac{1}{8}.$$

Remark 2.17.8 We can arrive to the identity (2.17.20) directly, using the formula for the sine of the half angle and then using Heron's formula.

Another consequence of the law of cosines is given next.

Proposition 2.17.9 *Let ABC be a triangle. The following inequalities hold*

$$|b-c| < a, \qquad |a-c| < b, \qquad |a-b| < c.$$

Proof: As $\widehat{A} \neq 0$, then $\cos\widehat{A} \neq 1$. Using $\cos\widehat{A} < 1$, the law of cosines provides

$$b^2 + c^2 - a^2 < 2bc \iff b^2 - 2bc + c^2 < a^2 \iff (b-c)^2 < a^2 \iff |b-c| < a.$$

In particular, $b - c < a$, and so $b < a + c$. The other inequalities can be proved by cyclic permutations of the sides of the triangle ΔABC. ■

2.18 The area of the contact triangle

Let A', B' and C' be the contact points of the incircle of ΔABC with the sides BC, AC and AB, respectively. Denote the sides of the *contact triangle* $\Delta A'B'C'$ by $a' = |B'C'|$, $b' = |A'C'|$ and $c' = |A'B'|$. We shall compute these sides by applying the law of cosines in triangles $\Delta AB'C'$, $\Delta BA'C'$ and $\Delta CA'B'$ and the fact that $|AC'| = |AB'| = p-a$, $|BC'| = |BA'| = p-b$ and $|CA'| = |CB'| = p-c$. Thus, we have

$$\begin{aligned} a'^2 &= |AC'|^2 + |AB'|^2 - 2|AC'|\cdot|AB'|\cos\widehat{A} \\ &= 2(p-a)^2(1-\cos\widehat{A}) = 4(p-a)^2\sin^2\frac{\widehat{A}}{2}. \end{aligned}$$

By cyclic permutations, we obtain

$$a' = 2(p-a)\sin\frac{\widehat{A}}{2}, \qquad b' = 2(p-b)\sin\frac{\widehat{B}}{2}, \qquad c' = 2(p-c)\sin\frac{\widehat{C}}{2}.$$

Since the circumradius of $\Delta A'B'C'$ is the inradius of ΔABC, then

$$\begin{aligned} \mathcal{A}_{A'B'C'} &= \frac{a'b'c'}{4r} = 2\frac{p(p-a)(p-b)(p-c)}{pr}\sin\frac{\widehat{A}}{2}\sin\frac{\widehat{B}}{2}\sin\frac{\widehat{C}}{2} \\ &= 2\mathcal{A}_{ABC}\cdot\sin\frac{\widehat{A}}{2}\sin\frac{\widehat{B}}{2}\sin\frac{\widehat{C}}{2} \\ &= \frac{r}{2R}\mathcal{A}_{ABC}, \end{aligned}$$

where we used Heron's formula for the area of a triangle and identity (2.17.20).

Using Euler's inequality, $R \geq 2r$, we obtain $\mathcal{A}_{A'B'C'} \leq \frac{1}{4}\mathcal{A}_{ABC}$. Since the area of the median triangle of ΔABC (the triangle obtained joining the midpoints of its sides) is equal to $\frac{1}{4}\mathcal{A}_{ABC}$, it follows that the area of the contact triangle is less than or equal to the area of the median triangle.

In certain applications, it is useful to express the symmetric quantities $a^2+b^2+c^2$ and $ab+bc+ca$ in terms of r, R and p. This can be accomplished as an application of the Heron's formula.

Proposition 2.18.1 *Let a, b, c denote the side lengths of a triangle. If p, r, R denote respectively the semiperimeter, the inradius and the circumradius of the triangle, then we have*

(i) $ab + bc + ca = p^2 + r^2 + 4Rr$

(ii) $a^2 + b^2 + c^2 = 2p^2 - 2r^2 - 8Rr.$

Proof: (*i*) Expressing the area of the triangle in two ways, $\mathcal{A} = pr$, and by Heron's formula $\mathcal{A} = \sqrt{p(p-a)(p-b)(p-c)}$, then equating the expressions, taking the square and expanding the parenthesis and dividing by p yields

$$pr^2 = p^3 - (a+b+c)p^2 + (ab+bc+ca)p - abc.$$

Substituting $abc = 4R\mathcal{A} = 4Rpr$ in the previous relation and then dividing by p, we get

$$r^2 = p^2 - (a+b+c)p + (ab+bc+ca) - 4Rr.$$

Rearranging, we obtain $ab + bc + ca = r^2 + p^2 + 4Rr$.

(*ii*) Using the binomial formula and part (*i*), we can express the sum of squares as

$$\begin{aligned} a^2+b^2+c^2 &= (a+b+c)^2 - 2(ab+bc+ca) = 4p^2 - 2(r^2+p^2+4Rr) \\ &= 2p^2 - 2r^2 - 8Rr, \end{aligned}$$

which proves the desired relation. ■

Exercise 2.18.2 *Show that in any triangle, we have the inequality* $p \geq \sqrt{3r(r+4R)}$.

Exercise 2.18.3 *Show the inequality* $p \geq \sqrt{3r(r_a+r_b+r_c)}$.

2.19 Fagnano's problem

The next result was proposed and solved by Toschi di Fagnano in 1775. The proof method described in the following belongs to L. Fejér.

Theorem 2.19.1 *The triangle of smallest perimeter inscribed into an acute-angled triangle ABC is the orthic triangle of triangle ABC.*

Proof: We shall perform the proof in two steps. In the first step, we consider one vertex fixed and the other two variable. In the second step, we determine the position of the fixed vertex such that the perimeter reaches its minimum value.

Step 1: Let P be a fixed point inside the segment BC and Q and S variable points on the sides AC and AB, respectively. We shall find the positions of S and Q such that the perimeter of triangle ΔPSQ is minimum. To this end, we construct the points P' and P'' as the symmetric points of point P with respect to sides AC and AB, respectively, see Figure 2.39(a). This construction implies that $\Delta SPP''$ and $\Delta QPP'$ are isosceles triangles. Consequently,

$$|PQ| + |QS| + |SP| = |P'Q| + |QS| + |SP''|,$$

i.e., the perimeter of ΔPQS is equal to the length broken segment $P'QSP''$. Since the endpoints P' and P'' are fixed, the positions of S and Q such that the length of the broken segment $P'QSP''$ is minimum are the intersections S' and Q' of the line $P'P''$ with the sides AB and AC. This fact is a consequence of the triangle inequality

$$|P'Q| + |QS| + |SP''| \geq |P'Q'| + |Q'S'| + |S'P''| = |P'P''|.$$

Step 2: Each position of the point P determines unique optimal positions for S and Q as determined by Step 1. Now we shall find the position of P such that the length $|P'P''|$ is minimum. We use Figure 2.39(b), where $S = S'$ and $Q = Q'$. We first note that $|AP| = |AP'| = |AP''|$, fact which is a consequence of the construction of P' and P'' as symmetric points of P with respect to the sides. Another consequence of the symmetry is that $\widehat{P'AC} = \widehat{CAP}$ and $\widehat{PAB} = \widehat{BAP''}$, which implies

$$\widehat{P'AP''} = \widehat{P'AC} + \widehat{CAP} + \widehat{PAB} + \widehat{BAP''} = 2\widehat{BAC} = 2\widehat{A},$$

which is independent of the position of the point P. The law of cosines applied in the isosceles triangle $\Delta AP'P''$ yields

$$\begin{aligned} |P'P''|^2 &= |AP'|^2 + |AP''|^2 - 2|AP'|\,|AP''| \cos \widehat{P'AP''} \\ &= 2|AP|^2 - 2|AP|^2 \cos\left(2\widehat{A}\right) = 2|AP|^2 \left(1 - \cos\left(2\widehat{A}\right)\right) \\ &= 4|AP|^2 \sin^2 \widehat{A}. \end{aligned} \tag{2.19.21}$$

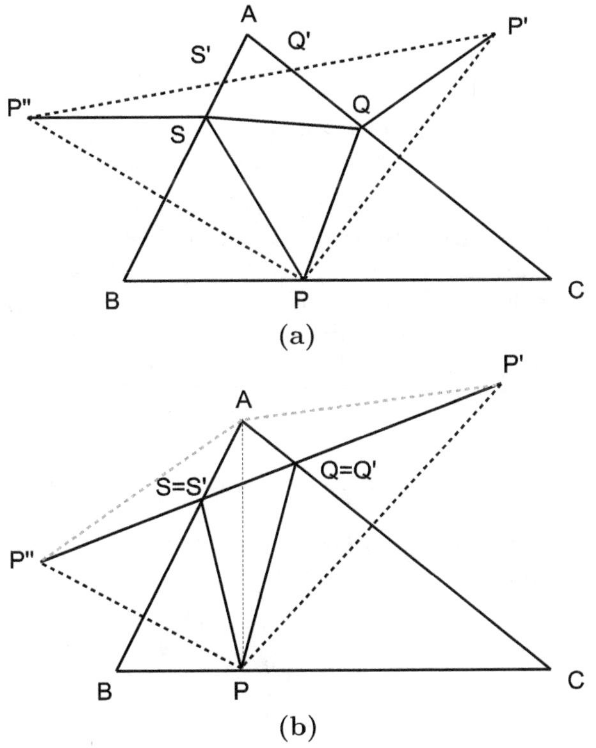

Figure 2.39: Fagnano's theorem proof: (a) P fixed on BC; (b) P variable on BC.

Therefore $|P'P''| = 2|AP|\sin\widehat{A}$ is minimum whenever $|AP|$ is. This is achieved in the case when AP is the altitude from A. It follows from the previous proof that the inscribed triangle ΔPQS of minimum perimeter thus obtained is unique. Due to this uniqueness, we shall get to the same triangle if we start from Q or S, instead of P, we obtain also that BQ and CS are altitudes from B and C. Hence, ΔPQS is the orthic triangle of ΔABC. ■

The previous proof produces the semiperimeter of the orthic triangle in terms of an altitude and its correspondent angle as

$$p_0 = \frac{1}{2}|P'P''| = |AP|\sin\widehat{A} = h_a \sin\widehat{A}. \tag{2.19.22}$$

Using $\sin\widehat{B} = \dfrac{h_a}{c}$ and the law of sines $c = 2R\sin\widehat{C}$, the aforementioned formula becomes

$$p_0 = h_a \sin \widehat{A} = 2R \sin \widehat{A} \sin \widehat{B} \sin \widehat{C}. \tag{2.19.23}$$

We note that we could have got to the same formula directly, equating two area formulas, $\mathcal{A} = p_0 R$ and $\mathcal{A} = 2R^2 \sin \widehat{A} \sin \widehat{B} \sin \widehat{C}$.

Using Fagnano's theorem, Theorem 2.19.1, and Fermat's principle, Corollary 2.14.3, we can show that the orthic triangle is a light triangle (or equivalently, a closed trajectory of a ball in a triangular billiard). To this end, we consider the points S and Q fixed, while P is variable on BC, such that $|PQ| + |QS| + |SP|$ is minimum. This means $|PQ| + |SP|$ is minimum, so by Corollary 2.14.3, we get $\widehat{SPB} = \widehat{QPC}$. The triangle PQS of minimum perimeter should satisfy these type of angle identities for all vertices

$$\widehat{SPB} = \widehat{QPC}, \qquad \widehat{SQA} = \widehat{PQC}, \qquad \widehat{ASQ} = \widehat{PSB}.$$

Exercise 2.19.2 *Starting from formula (2.19.22), show that in any acute-angled triangle we have*

$$\frac{1}{h_a} + \frac{1}{h_b} + \frac{1}{h_c} = \frac{1}{r}.$$

Exercise 2.19.3 *Use the minimality property of the orthic triangle to show that in any acute-angled triangle we have:*

(i) $4 \sin \widehat{A} \sin \widehat{B} \sin \widehat{C} \le \sin \widehat{A} + \sin \widehat{B} + \sin \widehat{C}$

(ii) $a \cos \widehat{A} + b \cos \widehat{B} + c \cos \widehat{C} \le p.$

Exercise 2.19.4 *Is Fagnano's theorem valid if* $\widehat{A} = 90°$*? What about the case when* $\widehat{A} > 90°$*?*

2.20 Keyl's problem

This section deals with a geometric variational problem, which is somehow related to Fagnano's problem, but is more sophisticated, in the sense that its solution requires besides geometric methods also calculus techniques of minimization. The problem was posed by Keyl[1] in 1985.

Problem: *Consider an acute-angled triangle* ΔABC *and* D *a variable point on the side* $[BC]$. *Let* D_1 *and* D_1 *be the symmetric points of* D *with*

[1]Eugeniu Keyl was the author's high school math teacher. He posed this problem when the author was a high-school freshman. The solution included here is the author's.

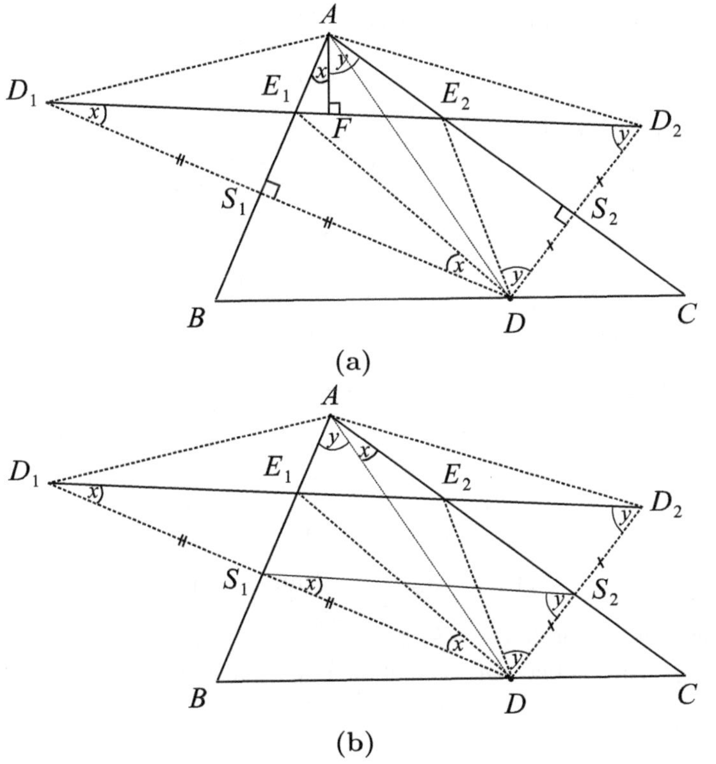

Figure 2.40: The construction for Keyl's problem.

respect to the sides AB and AC, respectively. Let $E_1 = D_1D_2 \cap AB$ and $E_2 = D_1D_2 \cap AC$. Find the position of the point D such that $|E_1E_2|$ is minimum.

We shall show that this problem has a unique solution. To this end, we shall start with a few notations. Let S_1 and S_2 be the orthogonal projections of D on the sides AB and AC, respectively, and denote $x = \widehat{E_1DS_1}$ and $y = \widehat{E_2DS_2}$, see Figure 2.40(a). Due to the construction symmetry, triangles ΔE_1DD_1 and ΔE_2DD_2 are isosceles, so $\widehat{E_1D_1S_1} = x$ and $\widehat{E_2D_2S_2} = y$. Let F be the orthogonal projection of A on E_1E_2. We note that F is between E_1 and E_2, as AF is an altitude in the acute-angled triangle ΔAE_1E_2 ($\widehat{A} < 90°$ by hypothesis, $\widehat{AE_1E_2} = \widehat{D_1E_1S_1} = 90° - x < 90°$ and $\widehat{AE_2E_1} = \widehat{D_2E_2S_2} = 90° - y < 90°$). Therefore, we shall compute the length $|E_1E_2|$ as $|E_1E_2| = |E_1F| + |FE_2|$.

By an angle inspection, we note that $\widehat{E_1AF} = x$ and $\widehat{E_2AF} = y$. Then

$$|E_1F| = |AF| \tan x, \quad |E_2F| = |AF| \tan y,$$

and then adding yields

$$|E_1E_2| = |E_1F| + |FE_2| = |AF|(\tan x + \tan y). \tag{2.20.24}$$

To find $|AF|$, we shall express the area of triangle ΔAD_1D_2 in two ways and use that $|D_1D_2| = 2|AD| \sin \widehat{A}$, see relation (2.19.21), as follows

$$\begin{aligned}
\frac{1}{2}|AF|\,|D_1D_2| &= \frac{1}{2}|AD_1|\,|AD_2| \sin \widehat{D_1AD_2} \iff \\
2|AF|\,|AD| \sin \widehat{A} &= |AD|^2 \sin(2\widehat{A}) \iff \\
|AF| &= \frac{2|AD|^2 \sin \widehat{A} \cos \widehat{A}}{2|AD| \sin \widehat{A}} \iff |AF| = |AD| \cos \widehat{A}. \quad (2.20.25)
\end{aligned}$$

Substituting (2.20.25) into (2.20.24) yields

$$|E_1E_2| = |AD| \cos \widehat{A}\,(\tan x + \tan y). \tag{2.20.26}$$

Since the angle $\widehat{A}$ is fixed, and $x = \widehat{A} - y$, it suffices to express $|AD|$ in terms of y. To this end, we shall show first that $\widehat{BAD} = x$ and $\widehat{DAC} = y$, see Figure 2.40(b).

Since S_1S_2 is a midpoint connector line in triangle ΔDD_1D_2, we have $S_1S_2 \| D_1D_2$. Then via equal corresponding angles, we have

$$\widehat{S_1S_2D} = \widehat{E_2D_2D} = y.$$

Since $\widehat{AS_1D} = \widehat{AS_2D} = 90°$, the quadrilateral AS_1DS_2 is cyclic (as inscribed in a circle of diameter AD) and hence the angle made by one diagonal with a side is equal to the angle made by the other diagonal with the opposite side, namely $\widehat{BAD} = \widehat{S_1S_2D}$. Therefore, $\widehat{BAD} = y$. Similarly, we can show that[2] $\widehat{DAC} = x$.

In order to express $|AD|$ in terms of angle y, we apply the law of sines in triangle ΔABD

$$\frac{|AD|}{\sin \widehat{B}} = \frac{|AB|}{\sin(180° - \widehat{B} - y)},$$

which implies

$$|AD| = \frac{c \sin \widehat{B}}{\sin(\widehat{B} + y)}. \tag{2.20.27}$$

[2]Equivalently, we can say that AF and AD are isogonals with respect to the angle $\widehat{BAC}$.

Substituting (2.20.27) into (2.20.26) yields

$$\begin{aligned}|E_1E_2| &= \frac{c\sin\widehat{B}}{\sin(\widehat{B}+y)}\cos\widehat{A}\,(\tan x+\tan y)\\ &= \underbrace{c\sin\widehat{B}\cos\widehat{A}}_{\text{constant}}\,\frac{\tan(\widehat{A}-y)+\tan y}{\sin(\widehat{B}+y)}.\end{aligned}$$

Therefore, $|E_1E_2|$ is minimum whenever the expression

$$f(y)=\frac{\tan(\widehat{A}-y)+\tan y}{\sin(\widehat{B}+y)},\qquad y\in(0,\widehat{A})$$

is minimum, with $0<\widehat{A}<90°$. To find the value of y for which $f(y)$ is minimum, we shall solve the equation $f'(y)=0$. This equation writes as

$$\left[\frac{1}{\cos^2 y}-\frac{1}{\cos^2\left(\widehat{A}-y\right)}\right]\sin\left(\widehat{B}+y\right)+\left(\tan\left(\widehat{A}-y\right)+\tan y\right)\cos\left(\widehat{B}+y\right)=0.$$

Multiplying by $\cos^2 y\cos^2(\widehat{A}-y)$ and using the difference of squares formula, we obtain

$$\begin{aligned}&\left(\cos\left(\widehat{A}-y\right)+\cos y\right)\left(\cos\left(\widehat{A}-y\right)-\cos y\right)\sin\left(\widehat{B}+y\right)+\cos y\cos\\ &\quad\left(\widehat{A}-y\right)\left[\cos y\sin\left(\widehat{A}-y\right)+\cos\left(\widehat{A}-y\right)\sin y\right]\cos\left(\widehat{B}+y\right)=0,\end{aligned}$$

which is equivalent to

$$\begin{aligned}&2\cos\frac{\widehat{A}}{2}\cos\left(\frac{\widehat{A}}{2}-y\right)2\sin\frac{\widehat{A}}{2}\sin\left(\frac{\widehat{A}}{2}-y\right)\sin\left(\widehat{B}+y\right)\\ &\qquad+\cos y\cos\left(\widehat{A}-y\right)\sin\widehat{A}\cos\left(\widehat{B}+y\right)=0.\end{aligned}$$

After using the formula of the sine of double angle, we get

$$\sin\widehat{A}\sin\left(\widehat{A}-2y\right)\sin\left(\widehat{B}+y\right)+\cos y\cos\left(\widehat{A}-y\right)\sin\widehat{A}\cos\left(\widehat{B}+y\right)=0.$$

Dividing by $\sin\widehat{A}$ and transforming $\cos y\cos(\widehat{A}-y)$ into sums of cosines yields

$$2\sin\left(\widehat{A}-2y\right)\sin\left(\widehat{B}+y\right)+\left[\cos\widehat{A}+\cos\left(\widehat{A}-2y\right)\right]\cos\left(\widehat{B}+y\right)=0,$$

which is equivalent to the following equation in y

$$\frac{2\sin\left(\widehat{A}-2y\right)}{\cos\widehat{A}+\cos\left(\widehat{A}-2y\right)}+\cot\left(\widehat{B}+y\right)=0. \tag{2.20.28}$$

We shall show that equation (2.20.28) has a unique solution $y\in(0,90^\circ)$. To show this, we consider the function $g:(0,90^\circ)\to\mathbb{R}$

$$g(y)=\frac{2\sin\left(\widehat{A}-2y\right)}{\cos\widehat{A}+\cos\left(\widehat{A}-2y\right)}+\cot\left(\widehat{B}+y\right).$$

We first note that

$$g(0)=\tan\widehat{A}+\cot\widehat{B}>0$$

since $\widehat{A},\widehat{B}\in(0,90^\circ)$. Next, we compute the value of $g(y)$ for values of y close to but smaller than 90°. For $\epsilon>0$ small enough, we have

$$\begin{aligned}g(90^\circ-\epsilon)&=\frac{2\sin\left(\widehat{A}-2(90^\circ-\epsilon)\right)}{\cos\widehat{A}+\cos\left(\widehat{A}-2(90^\circ-\epsilon)\right)}+\cot\left(\widehat{B}+(90^\circ-\epsilon)\right)\\&=\frac{-2\sin\left(\widehat{A}+2\epsilon\right)}{\cos\widehat{A}-\cos\left(\widehat{A}+2\epsilon\right)}+\cot\left(\widehat{B}+90^\circ-\epsilon\right).\end{aligned}$$

For ϵ close to zero, by continuity, the difference $\cos\widehat{A}-\cos(\widehat{A}+2\epsilon)$ tends to zero with positive values. Consequently, $g(90^\circ-\epsilon)<0$ for ϵ positive small. In fact, the value of $g(90^\circ-\epsilon)$ can be made as negatively large as we wish for ϵ small enough.

Since $g(y)$ is a continuous function, which starts with positive values and ends with negative values, it should take the intermediary value of zero for some values of y between 0 and 90°. This fact shows the existence of solutions for equation (2.20.28). The solution uniqueness follows from the fact that the function $g(y)$ is strictly decreasing on $(0,90^\circ)$, since it is a sum of two decreasing functions, $g(y)=g_1(y)+g_2(y)$, where

$$\begin{aligned}g_1(y)&=\frac{2\sin\left(\widehat{A}-2y\right)}{\cos\widehat{A}+\cos\left(\widehat{A}-2y\right)}\\g_2(y)&=\cot\left(\widehat{B}+y\right).\end{aligned}$$

The fact that g_1 and g_2 are decreasing follows from their derivative signs

$$g_1'(y) = \frac{-8\left(1+\cos\widehat{A}\cos\left(\widehat{A}-2y\right)\right)}{\left(\cos\widehat{A}+\cos\left(\widehat{A}-2y\right)\right)^2} < 0$$

$$g_2'(y) = \frac{-1}{\sin^2\left(\widehat{B}+y\right)} < 0.$$

Hence, equation (2.20.28) has a unique solution $y \in (0, 90°)$. This corresponds to a unique value $x = \widehat{A} - y$, and hence to a unique position of the point D on BC.

Remark 2.20.1 In the case of a general triangle, the optimal position of the point D does not coincide with any of the points studied before, namely, AD is not the angle bisector, neither the median, or the altitude, or the symmedian from the vertex A.

Exercise 2.20.2 *We consider the case of a triangle ΔABC with $\widehat{A} = 90°$. Show that $|E_1E_2| = 0$ for any position of the point D on the side BC.*

Exercise 2.20.3 *Let ΔABC be an equilateral triangle. Show that $|E_1E_2|$ is minimum if and only if D is the midpoint of BC. Find this distance in terms of the triangle side.*

Exercise 2.20.4 (*i*) *Find an expression of the area of triangle ΔAFD in terms of y.*
(*ii*) *Show that $\mathcal{A}_{AFD}$ is minimum if and only if AD is an angle bisector.*

Exercise 2.20.5 *Let AD be the altitude from the vertex A in triangle ΔABC. Find the distance $|E_1E_2|$ in terms of the triangle elements.*

Exercise 2.20.6 *Let AD be the angle bisector from the vertex A in triangle ΔABC. Find the distance $|E_1E_2|$ in terms of the triangle sides.*

2.21 Stewart's relation and applications

Definition 2.21.1 *A segment that joins a vertex of a triangle with a point on the opposite side is called a cevian.*[3]

The medians, angle bisectors, and altitudes are examples of cevians. Theorem 2.21.2 provides a useful formula to compute lengths of cevians.

[3] After the name of Giovanni Ceva, Italian mathematician 1647–1734.

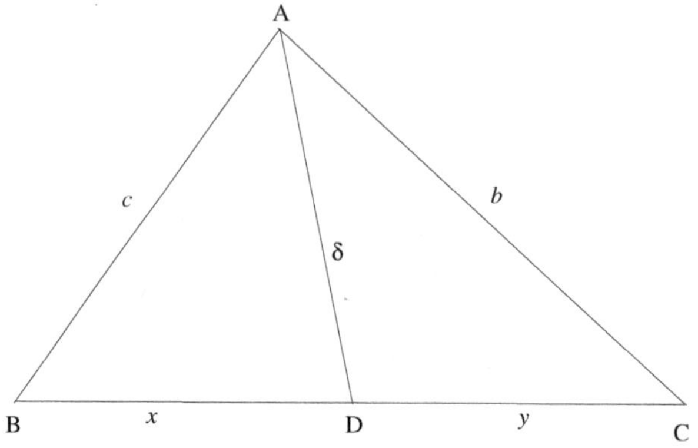

Figure 2.41: See Stewart's relation.

Theorem 2.21.2 (Stewart's relation) *Let ΔABC be an arbitrary triangle. Consider the point $D \in BC$ between B and C. Then*

$$|AD|^2 \cdot |BC| = |AB|^2 \cdot |DC| + |AC|^2 \cdot |BD| - |BD| \cdot |DC| \cdot |BC|. \quad (2.21.29)$$

Proof: Let $c = |AB|$, $b = |AC|$, $x = |BD|$, $y = |DC|$, $a = |BC|$ and $\delta = |AD|$. The relation (2.21.29) becomes

$$\delta^2 \cdot a = c^2 y + b^2 x - axy. \quad (2.21.30)$$

The law of cosines applied to triangles ΔABC and ΔABD yields

$$\begin{aligned} b^2 &= c^2 + a^2 - 2ac\cos\widehat{B}, \text{and} \\ \delta^2 &= c^2 + x^2 - 2cx\cos\widehat{B}. \end{aligned}$$

Multiplying the first of the above relations by $-x$ and the second by a yields

$$\begin{aligned} -b^2 x &= -c^2 x - a^2 x + 2acx\cos\widehat{B}, \text{and} \\ a\delta^2 &= c^2 a + x^2 a - 2acx\cos\widehat{B}. \end{aligned}$$

Adding,

$$\begin{aligned} a\delta^2 - b^2 x &= c^2(a - x) + xa(x - a) \\ &= c^2 y + xa(-y) \\ \Longrightarrow a\delta^2 &= b^2 x + c^2 y - xay, \end{aligned}$$

which is (2.21.30). ■

Recall that a median is the cevian that joins a vertex of a triangle with the midpoint of the opposite side. Stewart's relation can be used in the computation of the length of a median.

Theorem 2.21.3 (The median theorem) *Let m_a denote the length of the median corresponding to the vertex A. Then*

$$m_a^2 = \frac{b^2 + c^2}{2} - \frac{a^2}{4}. \tag{2.21.31}$$

Proof: Applying Stewart's relation with $x = y = \frac{a}{2}$ yields

$$a\, m_a^2 = b^2\, \frac{a}{2} + c^2\, \frac{a}{2} - \frac{a}{2}\, a\, \frac{a}{2}.$$

Divide by a to get (2.21.31). ■

Exercise 2.21.4 *Show that*

$$m_a^2 = \frac{a^2}{4} + bc \cos \widehat{A},$$

where m_a denotes the length of the median from vertex A.

Exercise 2.21.5 *Show that*

$$m_a^2 + m_b^2 + m_c^2 = \frac{3}{4}(a^2 + b^2 + c^2),$$

where m_a, m_b, m_c denote the lengths of the medians of ΔABC.

Exercise 2.21.6 *Prove that a triangle ABC has $\widehat{A} = 90°$ if and only if $m_a = \frac{a}{2}$.*

Exercise 2.21.7 *Let A' be the point where the incircle is tangent to the side BC of ΔABC. Find the length of AA' in terms of the side lengths a, b, c.*

Exercise 2.21.8 *(i) Using the median theorem, show that $m_a < \frac{1}{2}(b + c)$.*
(ii) Show that $m_a + m_b + m_c < 2p$.
(iii) Show that $m_a + m_b + m_c \leq \frac{3}{2}(a^2 + b^2 + c^2)^{1/2}$.
(iv) Show the inequality $\frac{3}{2}p \leq m_a + m_b + m_c$.

2.22 The angle bisector theorem

In order to prove Steiner's theorem, we need the following result.

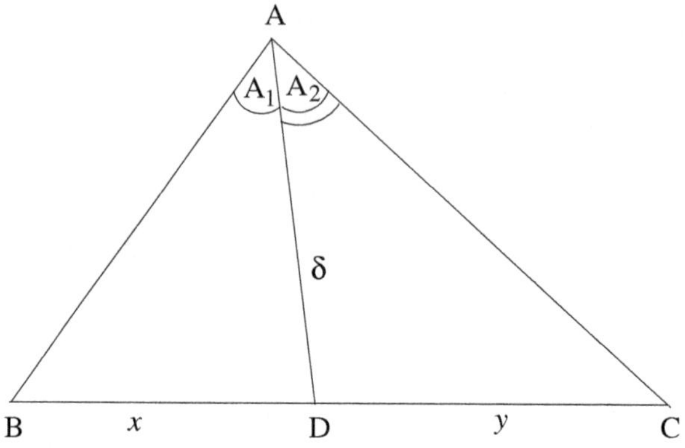

Figure 2.42: See Lemma 2.22.1.

Lemma 2.22.1 *Let AD be a cevian of* ΔABC *with D between B and C. Then*

$$\frac{|BD|}{|DC|} = \frac{\sin \widehat{A_1}}{\sin \widehat{A_2}} \cdot \frac{c}{b},$$

where $\widehat{A_1} = \widehat{BAD}$ *and* $\widehat{A_2} = \widehat{DAC}$.

Proof: Let $\delta = |AD|$. The law of sines applied to ΔABD and ΔADC yields

$$\frac{|BD|}{\sin \widehat{A_1}} = \frac{\delta}{\sin \widehat{B}} \Longrightarrow |BD| = \frac{\delta}{\sin \widehat{B}} \sin \widehat{A_1} \tag{2.22.32}$$

$$\frac{|DC|}{\sin \widehat{A_2}} = \frac{\delta}{\sin \widehat{C}} \Longrightarrow |DC| = \frac{\delta}{\sin \widehat{C}} \sin \widehat{A_2}. \tag{2.22.33}$$

Dividing relations (2.22.32) and (2.22.33) yields

$$\begin{aligned}
\frac{|BD|}{|DC|} &= \frac{\delta}{\sin \widehat{B}} \sin \widehat{A_1} \cdot \frac{\sin \widehat{C}}{\delta} \frac{1}{\sin \widehat{A_2}} \\
&= \frac{\sin \widehat{A_1} \sin \widehat{C}}{\sin \widehat{A_2} \sin \widehat{B}} = \frac{\sin \widehat{A_1}}{\sin \widehat{A_2}} \frac{c/(2R)}{b/(2R)} \\
&= \frac{\sin \widehat{A_1}}{\sin \widehat{A_2}} \cdot \frac{c}{b}.
\end{aligned}$$

■

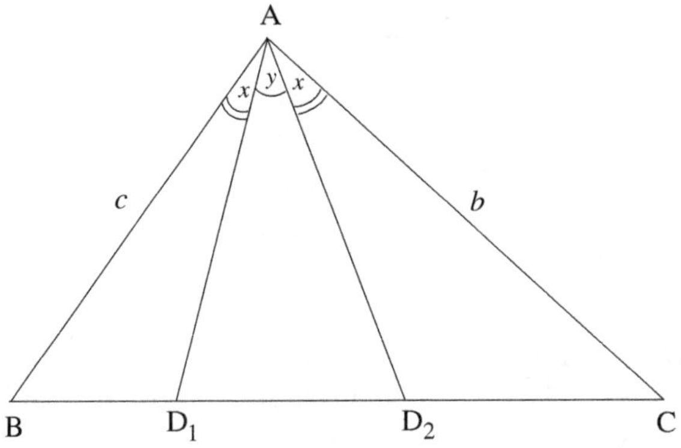

Figure 2.43: Isogonal cevians.

Remark 2.22.2 A shorter proof variant bypasses the law of sines by using areas:

$$\frac{|BD|}{|DC|} = \frac{\mathcal{A}_{ABD}}{\mathcal{A}_{ADC}} = \frac{\frac{1}{2}c\delta \sin \widehat{A_1}}{\frac{1}{2}b\delta \sin \widehat{A_2}} = \frac{c}{b} \cdot \frac{\sin \widehat{A_1}}{\sin \widehat{A_2}}.$$

One of the most important applications of Lemma 2.22.1 is the following result.

Theorem 2.22.3 (The angle bisector theorem) *The cevian AD is an angle bisector of ΔABC if and only if*

$$\frac{|BD|}{|DC|} = \frac{|AB|}{|AC|}. \tag{2.22.34}$$

Proof: AD is an angle bisector if and only if $\widehat{A_1} = \widehat{A_2}$. Applying Lemma 2.22.1 yields (2.22.34). ■

Definition 2.22.4 *Two cevians AD_1 and AD_2 are called isogonal if*

$$\widehat{BAD_1} = \widehat{D_2AC}.$$

Theorem 2.22.5 (Steiner's relation) *Let AD_1 and AD_2 be two isogonal cevians in ΔABC. Then*

$$\frac{|BD_1|}{|D_1C|} \cdot \frac{|BD_2|}{|D_2C|} = \left(\frac{|AB|}{|AC|}\right)^2.$$

Proof: Applying Lemma 2.22.1 for the cevians AD_1 and AD_2 yields

$$\begin{aligned}\frac{|BD_1|}{|D_1C|} &= \frac{\sin x}{\sin(x+y)}\cdot\frac{c}{b}\\ \frac{|BD_2|}{|D_2C|} &= \frac{\sin(x+y)}{\sin x}\cdot\frac{c}{b}.\end{aligned}$$

Multiplying yields

$$\begin{aligned}\frac{|BD_1|}{|D_1C|}\cdot\frac{|BD_2|}{|D_2C|} &= \frac{\sin x}{\sin(x+y)}\cdot\frac{\sin(x+y)}{\sin x}\cdot\left(\frac{c}{b}\right)^2\\ &= \left(\frac{c}{b}\right)^2.\end{aligned}$$

Another proof: The following proof involves areas. We have from Definition 2.3.2 and Proposition 2.4.1 that

$$\begin{aligned}\frac{c\,|AD_1|}{b\,|AD_2|} &= \frac{\mathcal{A}_{BAD_1}}{\mathcal{A}_{CAD_2}} = \frac{|BD_1|\,h_a}{|D_2C|\,h_a} = \frac{|BD_1|}{|D_2C|}\\ \frac{c\,|AD_2|}{b\,|AD_1|} &= \frac{\mathcal{A}_{BAD_2}}{\mathcal{A}_{CAD_1}} = \frac{|BD_2|\,h_a}{|D_1C|\,h_a} = \frac{|BD_2|}{|D_1C|}.\end{aligned}$$

Multiplying yields

$$\left(\frac{c}{b}\right)^2 = \frac{|BD_1|}{|D_2C|}\cdot\frac{|BD_2|}{|D_1C|}.$$

■

Exercise 2.22.6 *Let ΔABC be a triangle and let $D, E \in BC$ such that $|BD| = |DE| = |EC|$. Show that if $|AD| = |AE|$, then ΔABC is isosceles, with $|AB| = |AC|$.*

Definition 2.22.7 *The isogonal cevian of the median is called symmedian.*

We shall find the length of the symmedian using Steiner's relation.

Proposition 2.22.8 *Let m_a and s_a denote the lengths of the median and symmedian from vertex A of ΔABC. Then*

$$s_a = \frac{2bc}{b^2+c^2}\,m_a.$$

Proof: Let AD_2 and AD_1 be the median and the symmedian, respectively. As D_2 is the midpoint of BC, Steiner's relation $\frac{|BD_1|}{|D_1C|}\cdot\frac{|BD_2|}{|D_2C|} = \left(\frac{c}{b}\right)^2$

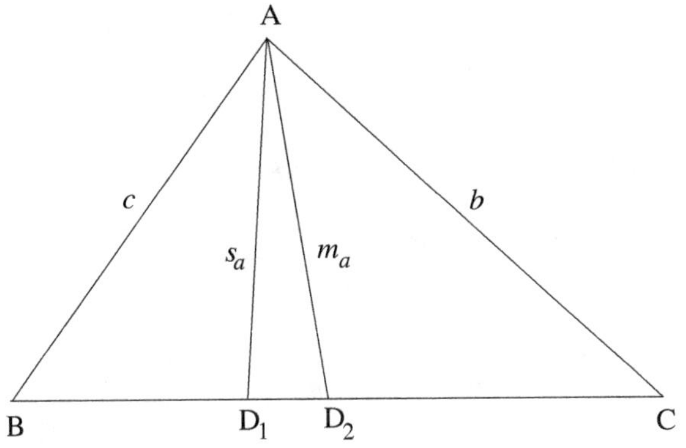

Figure 2.44: The median and the symmedian.

becomes

$$\frac{|BD_1|}{|D_1C|} = \left(\frac{c}{b}\right)^2 \implies \frac{|BD_1| + |D_1C|}{|D_1C|} = \frac{c^2+b^2}{b^2}$$
$$\implies \frac{a}{|D_1C|} = \frac{c^2+b^2}{b^2} \implies |D_1C| = \frac{ab^2}{c^2+b^2}. \tag{2.22.35}$$

As $\widehat{BAD_2} = \widehat{CAD_1}$, we have

$$\begin{aligned}\frac{|D_1C|}{|BD_2|} &= \frac{|D_1C|\,h_a}{|BD_2|\,h_a} = \frac{\mathcal{A}_{AD_1C}}{\mathcal{A}_{ABD_2}} \\ &= \frac{s_a\, b\, \sin \widehat{CAD_1}}{m_a\, c\, \sin \widehat{BAD_2}} = \frac{s_a}{m_a} \cdot \frac{b}{c}.\end{aligned}$$

Hence

$$\frac{|D_1C|}{|BD_2|} = \frac{s_a}{m_a}\frac{b}{c}.$$

Using (2.22.35) and $|BD_2| = a/2$ yields

$$\frac{ab^2}{c^2+b^2} \cdot \frac{2}{a} = \frac{s_a}{m_a}\frac{b}{c} \implies \frac{s_a}{m_a} = \frac{2bc}{c^2+b^2}.$$

■

Exercise 2.22.9 *Prove that we have $s_a \leq m_a$ with identity exactly when $b = c$.*

Exercise 2.22.10 *Prove that*

$$\frac{s_a}{m_a} + \frac{s_b}{m_b} + \frac{s_c}{m_c} \le 3.$$

Exercise 2.22.11 *Show that the symmedian length is given by*

$$s_a^2 = \frac{b^2c^2}{b^2+c^2}\left(2 - \frac{a^2}{b^2+c^2}\right).$$

Exercise 2.22.12 *Show that in any triangle* $b^2 + c^2 > \frac{1}{2}a^2$.

Definition 2.22.13 *Let AA' be a cevian in triangle ΔABC. Then AA' is called a cevian of rank k if*

$$\frac{|BA'|}{|A'C|} = \left(\frac{c}{b}\right)^k.$$

Remark 2.22.14 The cevian of rank $k = 0$ is the median and Theorem 2.22.3 implies that the cevian of rank $k = 1$ is the angle bisector.

Proposition 2.22.15 *The cevian of rank $k = 2$ is the symmedian.*

Proof: Let AD_1 be the symmedian. We have shown in (2.22.35) that

$$|D_1C| = \frac{ab^2}{b^2+c^2}.$$

In a similar way

$$|BD_1| = \frac{ac^2}{b^2+c^2}.$$

Then

$$\frac{|BD_1|}{|D_1C|} = \frac{ac^2}{b^2+c^2} \cdot \frac{b^2+c^2}{ab^2} = \left(\frac{c}{b}\right)^2 \Longrightarrow k = 2.$$

∎

Exercise 2.22.16 *Let ℓ_k be the length of the cevian of rank k from the vertex A. Express ℓ_k in terms of the triangle side lengths a, b, c. Show that in the cases $k = 0, 1, 2$, we obtain the formulas for the lengths of the median, angle bisector and symmedian.*

Exercise 2.22.17 *What is the rank of the isogonal of a cevian of rank k?*

Exercise 2.22.18 *What is the length of the isogonal of a cevian of rank k in terms of the triangle side lengths?*

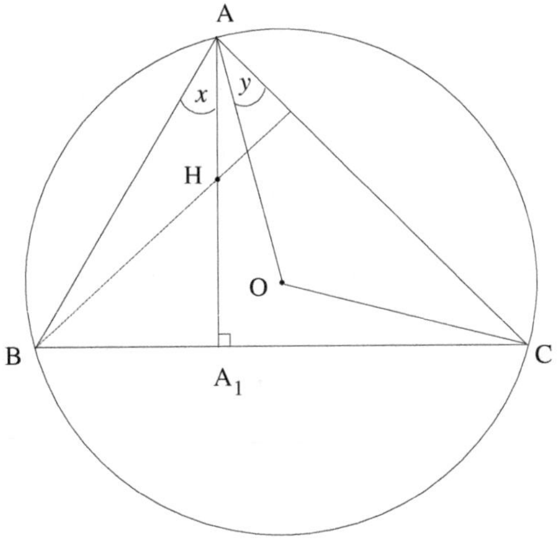

Figure 2.45: AH and AO are isogonal cevians.

2.23 An example of isogonal cevians

Let H, O denote the orthocenter and the circumcenter of the acute triangle ΔABC, respectively. We next show that AH and AO are isogonal cevians. Let $x = \widehat{BAH}$ and $y = \widehat{OAC}$. We have $x = 90° - \widehat{B}$, since ΔABA_1 is a right triangle. As a central angle, $\widehat{AOC} = 2\widehat{ABC} = 2\widehat{B}$. As ΔAOC is isosceles ($|AO| = |OC|$), we have

$$\begin{aligned} y &= \frac{180 - \widehat{AOC}}{2} = 90° - \frac{1}{2}\widehat{AOC} \\ &= 90° - \widehat{B} = x. \end{aligned}$$

Hence $x = y$, and the cevians AH and AO are isogonal.

Let AA_1 be the altitude from vertex A. See Figure 2.46. Then $|BA_1| = c\cos B$, $|A_1C| = b\cos C$. Let U be the intersection of BC with the extension of AO. Applying Steiner's relation yields

$$\begin{aligned} \frac{|BA_1|}{|A_1C|}\frac{|BU|}{|UC|} &= \left(\frac{c}{b}\right)^2 \Longleftrightarrow \\ \frac{c\cos\widehat{B}}{b\cos\widehat{C}}\frac{|BU|}{|UC|} &= \frac{c^2}{b^2} \Longleftrightarrow \\ \frac{|BU|}{|UC|} &= \frac{c}{b}\frac{\cos\widehat{C}}{\cos\widehat{B}} = \frac{c_0}{b_0}, \end{aligned}$$

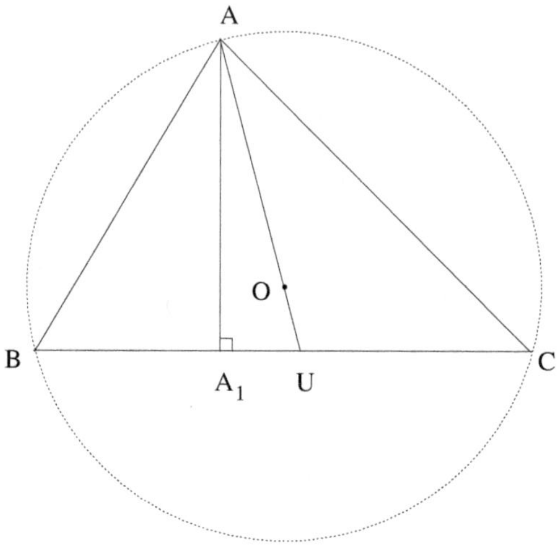

Figure 2.46: AU is the isogonal of the altitude AA_1.

where a_0, b_0, c_0 are the lengths of the sides of the orthic triangle from Section 2.15.
Elementary operations with fractions yield

$$\begin{aligned}
\frac{|BU|}{|UC|} = \frac{c_0}{b_0} &\iff \frac{|BU| + |UC|}{|UC|} = \frac{c_0 + b_0}{b_0} \\
&\iff \frac{a}{|UC|} = \frac{c_0 + b_0}{b_0} \\
&\iff |UC| = \frac{ab_0}{c_0 + b_0}.
\end{aligned}$$

Then we can compute

$$\begin{aligned}
|BU| &= a - |UC| = a - \frac{ab_0}{c_0 + b_0} \\
&= a\left(1 - \frac{b_0}{c_0 + b_0}\right) \\
&= a\frac{c_0 + b_0 - b_0}{c_0 + b_0} = \frac{ac_0}{c_0 + b_0}.
\end{aligned}$$

We have

$$|UC| = \frac{ab_0}{c_0 + b_0}, \qquad |BU| = \frac{ac_0}{c_0 + b_0}.$$

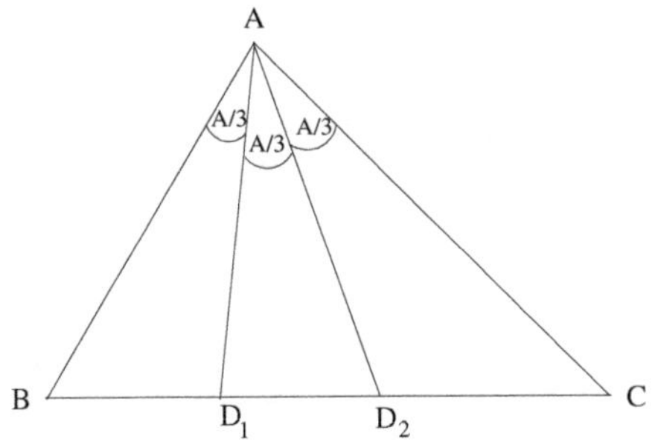

Figure 2.47: The trisectors of angle $\widehat{A}$.

Stewart's relation yields

$$\begin{aligned} a\,|AU|^2 &= c^2\,|UC| + b^2\,|BU| - |BU|\cdot|UC|\cdot a \\ &= c^2\cdot\frac{ab_0}{c_0+b_0} + b^2\frac{ac_0}{c_0+b_0} - \frac{ac_0}{c_0+b_0}\cdot\frac{ab_0}{c_0+b_0}\cdot a. \end{aligned}$$

Dividing by a, we get the length of AU:

$$|AU|^2 = \frac{(b_0c^2+c_0b^2)(b_0+c_0) - a^2b_0c_0}{(c_0+b_0)^2}.$$

Exercise 2.23.1 *Compute the lengths of the trisectors of the angle $\widehat{A}$ of triangle ΔABC in terms of its angles and side lengths, see Figure 2.47 (there are two trisectors of angle A which trisect the angle in three equal angles).*

2.24 The length of the angle bisector

Let $i_a = |AD|$ be the length of the angle bisector from the vertex A. Let $x = |BD|$ and $y = |DC|$. The angle bisector theorem yields

$$\frac{x}{y} = \frac{c}{b} \Longrightarrow bx = cy.$$

Using $x = a - y$ yields

$$\begin{aligned} b(a-y) = cy \quad &\Longleftrightarrow \quad ba - by = cy \\ &\Longleftrightarrow \quad ba = y(c+b). \end{aligned}$$

Then $y = \dfrac{ba}{b+c}$ and $x = \dfrac{ca}{b+c}$.

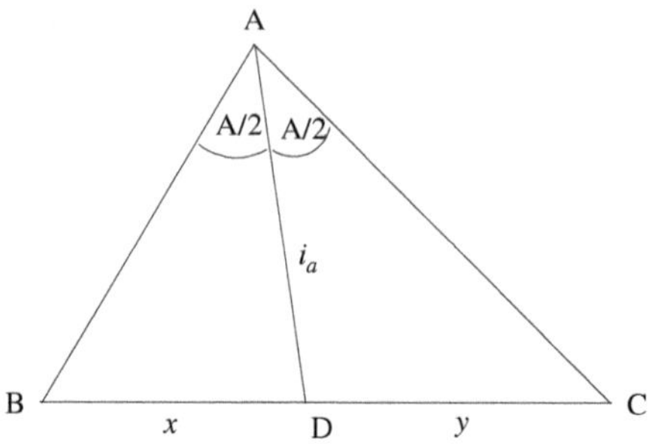

Figure 2.48: The angle bisector AD.

Stewart's relation yields

$$\begin{aligned} a \cdot i_a^2 &= c^2\, y + b^2\, x - a\, x\, y \\ &= c^2 \frac{ba}{b+c} + b^2 \frac{ca}{b+c} - a \frac{ca}{b+c} \cdot \frac{ba}{b+c}. \end{aligned}$$

Dividing by a and factoring out $\dfrac{1}{b+c}$ yields

$$\begin{aligned} i_a^2 &= \frac{1}{b+c}\left[c^2 b + b^2 c - \frac{ca \cdot ba}{b+c}\right] \\ &= \frac{1}{b+c}\left[bc(b+c) - \frac{bc\, a^2}{b+c}\right] \\ &= \frac{bc}{(b+c)^2}\left[(b+c)^2 - a^2\right] \\ &= \frac{bc}{(b+c)^2}(b+c-a)(b+c+a) \\ &= \frac{4bc}{(b+c)^2}(p-a)p. \end{aligned}$$

Hence the length of the angle bisector from the vertex A is

$$i_a = \frac{2}{b+c}\sqrt{bcp(p-a)}. \tag{2.24.36}$$

Proposition 2.24.1 *The length of the angle bisector from the vertex A is*

$$i_a = \frac{2bc}{b+c} \cos \frac{\widehat{A}}{2}.$$

Proof: From (2.24.36) and formula 1) of Theorem 2.16.4, we get

$$\begin{aligned} i_a &= \frac{2}{b+c}\sqrt{bcp(p-a)} = \frac{2}{b+c}\sqrt{(bc)^2\frac{p(p-a)}{bc}} \\ &= \frac{2bc}{b+c}\cos\frac{\widehat{A}}{2}. \end{aligned}$$

■

Corollary 2.24.2 *We have*

$$i_a < \frac{2bc}{b+c} = \frac{2}{\dfrac{1}{c}+\dfrac{1}{b}}.$$

Exercise 2.24.3 *Prove that in any triangle we have $i_a < \sqrt{bc}$. Does the identity $i_a = \sqrt{bc}$ ever hold?*

Exercise 2.24.4 *Using the formulas for the angle bisector and median, show that $i_a \le m_a$. When is the identity reached?*

Exercise 2.24.5 *Prove that in a right triangle ΔABC with $\widehat{A} = 90$, we have*

$$i_a = \frac{bc\sqrt{2}}{b+c}.$$

Exercise 2.24.6 *Let ΔABC be a right triangle ($\widehat{A} = 90°$) with side lengths 3, 4, 5. Find the length of the angle bisector i_a.*

Exercise 2.24.7 *Let ΔABC be a right triangle with sides length 3, 4, 5. Find the length of the symmedian s_a.*

Exercise 2.24.8 *Let H and I be the orthocenter and the incenter of triangle ABC. Show that:*

(i) $\widehat{HAI} = \dfrac{|\widehat{B} - \widehat{C}|}{2}$

(ii) $h_a = i_a \cos\dfrac{\widehat{B} - \widehat{C}}{2}$

(iii) If the altitude and the bisector starting from the same vertex have the same length then the triangle is isosceles.

The following proposition is needed later.

Proposition 2.24.9 *Let $x, y, z \geq 0$. Then*

$$\sqrt{x} + \sqrt{y} + \sqrt{z} \leq \sqrt{3(x+y+z)}. \tag{2.24.37}$$

Proof: Let $u = \sqrt{x}, v = \sqrt{y}, w = \sqrt{z}$. Then (2.24.37) becomes

$$u + v + w \leq \sqrt{3(u^2+v^2+w^2)} \Longleftrightarrow$$
$$(u+v+w)^2 \leq 3(u^2+v^2+w^2) \Longleftrightarrow$$
$$u^2+v^2+w^2+2uv+2vw+2uw \leq u^2+v^2+w^2+2(u^2+v^2+w^2) \Longleftrightarrow$$
$$2uv+2vw+2uw \leq 2(u^2+v^2+w^2),$$

which holds because

$$\begin{aligned} 2uv &\leq u^2+v^2 \Longleftrightarrow (u-v)^2 \geq 0 \\ 2vw &\leq v^2+w^2 \Longleftrightarrow (v-w)^2 \geq 0 \\ 2uw &\leq u^2+w^2 \Longleftrightarrow (u-w)^2 \geq 0. \end{aligned}$$

■

Proposition 2.24.10 *The sum of the lengths of the sides of a triangle satisfies the inequality*

$$i_a + i_b + i_c \leq \frac{\sqrt{3}}{2}(a+b+c). \tag{2.24.38}$$

Proof: The inequality between the arithmetic and geometric means can be written as $2\sqrt{bc} \leq b+c$. This and (2.24.36) give

$$i_a = \frac{2}{b+c}\sqrt{bcp(p-a)} = \frac{2\sqrt{bc}}{b+c}\sqrt{p(p-a)} \leq \sqrt{p(p-a)}.$$

By cyclic permutations, we have

$$\begin{aligned} i_a &\leq \sqrt{p(p-a)} \\ i_b &\leq \sqrt{p(p-b)} \\ i_c &\leq \sqrt{p(p-c)}. \end{aligned}$$

Let $x = p-a$, $y = p-b$ and $z = p-c$. Adding and using Proposition 2.24.9, yields

$$\begin{aligned} i_a + i_b + i_c &\leq \sqrt{p}\left(\sqrt{p-a}+\sqrt{p-b}+\sqrt{p-c}\right) \\ &= \sqrt{p}(\sqrt{x}+\sqrt{y}+\sqrt{z}) \leq \sqrt{p}\sqrt{3(x+y+z)} \\ &= \sqrt{p}\sqrt{3(p-a+p-b+p-c)} \\ &= \sqrt{p}\sqrt{3(3p-2p)} = \sqrt{p}\sqrt{3p} \\ &= \sqrt{3}p = \frac{\sqrt{3}}{2}(a+b+c). \end{aligned}$$

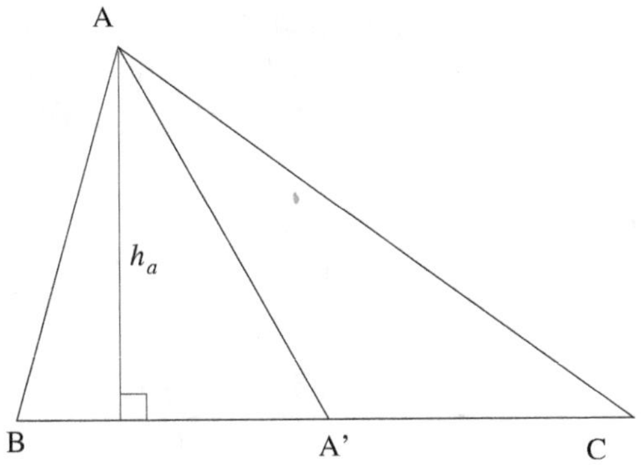

Figure 2.49: The median AA' bisects the triangle in two equal areas.

∎

Remark 2.24.11 In fact, a stronger version of inequality (2.24.38) holds, given by

$$i_a + i_b + m_c \leq \frac{\sqrt{3}}{2}(a + b + c),$$

as proved by L. Panaitopol [65] in 1980. This is the tightest it can get, since considering two medians and a bisector, or even three medians violates the inequality, see Exercises 2.24.12 and 2.24.13.

Exercise 2.24.12 *Let ΔABC be a right triangle with sides $a = \sqrt{10}$, $b = 1$ and $c = 1$. Show that*

$$m_a + m_b + m_c > \frac{\sqrt{3}}{2}(a + b + c).$$

Exercise 2.24.13 *Let ΔABC be a right triangle with sides $a = 1$, $b = \sqrt{10}$ and $c = 3$. Show that*

$$i_a + m_b + m_c > \frac{\sqrt{3}}{2}(a + b + c).$$

In a triangle, the equality of two cevians of the same type implies the fact that the triangle is isosceles. This is not hard to show for the case of altitudes, and medians, see Exercises 2.24.15 and 2.24.16. In the case of angle bisectors, the statement validation requires some more involved computations.

Theorem 2.24.14 (Steiner-Lehmus) *Two angle bisectors in a triangle have the same length if and only if the triangle is isosceles.*

Proof: " $\Longrightarrow$ " Let ΔABC be a triangle with side lengths a, b and c and consider two equal angle bisectors, $i_a = i_b$. We shall show that $a = b$ by an algebraic computation using the formula for the length of an angle bisector. We have

$$\begin{aligned}
i_a = i_b \Longleftrightarrow i_a^2 = i_b^2 \Longleftrightarrow \\
bc\left[1-\left(\frac{a}{b+c}\right)^2\right] = ac\left[1-\left(\frac{b}{a+c}\right)^2\right] \Longleftrightarrow \\
b-a = \frac{ba^2}{(b+c)^2} - \frac{ab^2}{(a+c)^2} \Longleftrightarrow \\
b-a = ab\left[\frac{a}{(b+c)^2} - \frac{b}{(a+c)^2}\right] \Longleftrightarrow
\end{aligned}$$

$$\begin{aligned}
(b-a)(b+c)^2(a+c)^2 = ab\big[a(a+c)^2 - b(b+c)^2\big] \Longleftrightarrow \\
(b-a)(b+c)^2(a+c)^2 = ab(a-b)\big(a^2+b^2+c^2+ab+2c(a+b)\big) \Longleftrightarrow \\
(b-a)\big[(b+c)^2(a+c)^2 + ab\big(a^2+b^2+c^2+ab+2c(a+b)\big)\big] = 0 \Longleftrightarrow \\
a = b,
\end{aligned}$$

since the term in the brackets is strictly positive.
" $\Longleftarrow$ " We assume the triangle is isosceles, with $a = b$. Then following the previous computation backwards, we obtain $i_a = i_b$. ■

Exercise 2.24.15 *Prove that* $h_a = h_b$ *if and only if* $a = b$.

Exercise 2.24.16 *Use the median theorem to prove that* $m_a = m_b$ *if and only if* $a = b$.

Exercise 2.24.17 *Show the inequality*

$$\sin\widehat{A}\sin\widehat{B}\sin\widehat{C} \le \frac{1}{4}\Big(\sin\widehat{A} + \sin\widehat{B} + \sin\widehat{C}\Big).$$

2.25 The length of an altitude

Since the altitude of a triangle is related to the concept of area, one simple way of computing its length is using this relationship as follows:

$$h_a = \frac{2\mathcal{A}}{a} = \frac{2}{a}\cdot\frac{abc}{4R} = \frac{bc}{2R}. \tag{2.25.39}$$

Furthermore, using the law of sines we can express the altitude length as

$$h_a = 2R\sin\widehat{B}\sin\widehat{C}. \tag{2.25.40}$$

Exercise 2.25.1 *Let h_a, h_b and h_c be the altitudes of ΔABC. Show that*

$$h_a + h_b + h_c = 2r + \frac{p^2 + r^2}{2R}.$$

Exercise 2.25.2 (*i*) *Show that* $h_a = \dfrac{2pr}{a}$.
(*ii*) *Prove the inequality* $h_a + h_b + h_c > 6r$.

Exercise 2.25.3 *Show that in any triangle ΔABC, we have the inequality*

$$\frac{1}{\sin\widehat{B}\sin\widehat{C}} + \frac{1}{\sin\widehat{C}\sin\widehat{A}} + \frac{1}{\sin\widehat{A}\sin\widehat{B}} \geq 4.$$

The altitude length can also be computed using Stewart's relation in ΔABC. With the following notation setup

$$h_a = |AD|, \quad x = |BD| = c\cos\widehat{B}, \quad y = |DC| = b\cos\widehat{C}$$

we have

$$h_a^2 a = c^2 y + b^2 x - axy.$$

Solving for the altitude yields

$$\begin{aligned} h_a^2 &= \frac{c^2 y}{a} + \frac{b^2 x}{a} - xy \\ &= \frac{bc}{a}\left[c\cos\widehat{C} + b\cos\widehat{B} - a\cos\widehat{B}\cos\widehat{C}\right]. \end{aligned} \tag{2.25.41}$$

Formula (2.25.41) is not very eye catching. We shall simplify it by showing that the expression in the parenthesis is just the semiperimeter of the orthic triangle. For this, it suffices to prove the identity

$$c\cos\widehat{C} + b\cos\widehat{B} - a\cos\widehat{B}\cos\widehat{C} = \frac{1}{2}\left(a\cos\widehat{A} + b\cos\widehat{B} + c\cos\widehat{C}\right). \tag{2.25.42}$$

This is equivalent to

$$c\cos\widehat{C} + b\cos\widehat{B} = a\left(\cos\widehat{A} + 2\cos\widehat{B}\cos\widehat{C}\right).$$

Using the law of cosines, we obtain the following relation in a, b and c

$$\begin{aligned} &\frac{c(b^2 + a^2 - c^2)}{2ba} + \frac{b(a^2 + c^2 - b^2)}{2ac} \\ &= a\left(\frac{b^2 + c^2 - a^2}{2bc} + 2\frac{a^2 + c^2 - b^2}{2ac} \cdot \frac{b^2 + a^2 - c^2}{2ba}\right). \end{aligned}$$

A straightforward algebraic computation, which is left for the reader, shows that the previous identity holds true. Denoting the semiperimeter of the orthic triangle by

$$p_0 = \frac{1}{2}\left(a\cos\widehat{A} + b\cos\widehat{B} + c\cos\widehat{C}\right),$$

formula (2.25.41) becomes

$$h_a = \sqrt{\frac{bc}{a}p_0}. \tag{2.25.43}$$

Remark 2.25.4 We have obtained two formulas for h_a. Now, equating formulas (2.25.43) and (2.25.39) we obtain

$$\frac{\sqrt{abc}}{2R} = \sqrt{p_0} \iff \frac{abc}{4R^2} = p_0 \iff \mathcal{A} = Rp_0,$$

which recovers a formula for the triangle area in terms of its circumradius and semiperimeter of the orthic triangle, see Proposition 2.15.1.

Solved Exercise 2.25.5 *Show that in any triangle ΔABC, we have*
(i) $a^2 + b^2 + c^2 = 8R^2 + 4Rr_0$, *and*
(ii) $8R^2 < a^2 + b^2 + c^2 \leq 9R^2$,
where r_0 is the inradius of the orthic triangle.

Proof: (a) From the formulas $|OH|^2 = 9R^2 - (a^2 + b^2 + c^2)$ and $|OH|^2 = R^2 - 4Rr_0$, see Corollary 2.28.3, equating we obtain relation (i).
(b) Since $|OH|^2 \geq 0$, we obtain $a^2 + b^2 + c^2 \leq R^2$. From part (i) we also have $a^2 + b^2 + c^2 = 8R^2 + 4Rr_0 > 8R^2$. We note that this inequality is strict, as both R and r_0 are always positive. ■

2.26 The centroid

The centroid plays the role of center of mass for a triangle. A triangle ΔABC has three medians AA', BB' and CC', where A', B', C' are the midpoints of the sides BC, AC, AB, respectively.

Lemma 2.26.1 *Any median divides the triangle into two triangles of the same area. That is*

$$\mathcal{A}_{ABA'} = \mathcal{A}_{AA'C} = \frac{1}{2}\mathcal{A}_{ABC}.$$

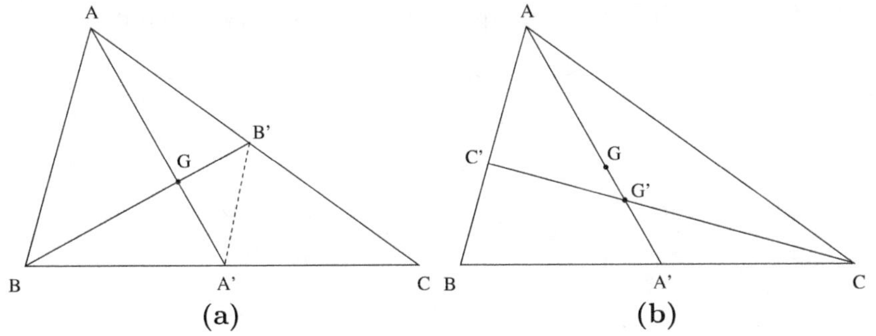

Figure 2.50: (a) Existence of centroid; (b) Uniqueness of centroid.

Proof: Let h_a be the length of the altitude corresponding to the side BC of triangle ΔABC. Then $\Delta ABA'$ and $\Delta AA'C$ have the same altitude corresponding to the side BC.

$$\begin{aligned} \mathcal{A}_{ABA'} &= \frac{1}{2}|BA'| \cdot h_a = \frac{1}{2}|A'C| \cdot h_a \quad (A' \text{ midpoint}) \\ &= \mathcal{A}_{AA'C}. \end{aligned}$$

■

The following property deals with the existence and the position of the centroid.

Theorem 2.26.2 *The medians AA', BB' and CC' intersect at a point G, called the* **centroid**. *The point G has the property*

$$\frac{|AG|}{|GA'|} = \frac{|BG|}{|GB'|} = \frac{|CG|}{|GC'|} = 2.$$

Proof: Using the crossbar theorem, Exercise 1.7.7, it follows that AA' intersects BB'. Let $AA' \cap BB' = \{G\}$. As B' and A' are midpoints, by the midpoint connector theorem, Corollary 2.11.2, we have

$$A'B' \| AB \text{ and } |A'B'| = \frac{1}{2}|AB|.$$

By the (AA) similarity case

$$\Delta AGB \sim \Delta A'GB'.$$

Then

$$\frac{|AG|}{|GA'|} = \frac{|BG|}{|GB'|} = \frac{|AB|}{|A'B'|} = 2.$$

Let $CC' \cap AA' = \{G'\}$. See Figure 2.50(b). A similar computation yields

$$\frac{|AG'|}{|G'A'|} = 2.$$

Then

$$\frac{|AG|}{|GA'|} = \frac{|AG'|}{|G'A'|} \iff \frac{|AG|}{|AG| + |GA'|} = \frac{|AG'|}{|AG'| + |G'A'|} \iff$$
$$\frac{|AG|}{|AA'|} = \frac{|AG'|}{|AA'|} \iff |AG| = |AG'|$$

and hence $G = G'$, which completes the proof. ∎

Exercise 2.26.3 *Let G be the centroid of triangle ΔABC. Show that*

$$\mathcal{A}_{AGB} = \mathcal{A}_{BGC} = \mathcal{A}_{AGC}.$$

Exercise 2.26.4 *Let M be a point in the interior of triangle ΔABC such that*

$$\mathcal{A}_{AMB} = \mathcal{A}_{BMC} = \mathcal{A}_{AMC}.$$

Show that M is the centroid of triangle ΔABC.

Remark 2.26.5 The point G is situated at the intersections of the cevians which divides the triangle into triangles of equal areas. For this reason the point G is also called the center of mass of the triangle.

2.27 The power of a point and applications

This section introduces an invariant, called the power of a point with respect to a circle. It can be used as a computation tool for distances between two points, especially if one of the points is the center of the circle.

Definition and properties

Let $\mathcal{C}(O,\rho)$ be a circle of center O and radius ρ. Let

$$Int\, \mathcal{C}(O,\rho) = \{M; |OM| < \rho\}$$

be the interior of the circle and

$$Ext\, \mathcal{C}(O,\rho) = \{M; |OM| > \rho\}$$

be the exterior of the circle.

Definition 2.27.1 *Let A be a point in the plane and let $\mathcal{C}(O,\rho)$ be a circle centered at O with radius ρ. The* **power of the point** *A with respect to $\mathcal{C}(O,\rho)$ is defined as*

$$Power(A) = \begin{cases} |AO|^2 - \rho^2 & \textit{if } A \in Ext\,\mathcal{C}(O,\rho), \\ 0 & \textit{if } A \in \mathcal{C}(O,\rho), \\ \rho^2 - |AO|^2 & A \in Int\,\mathcal{C}(O,\rho). \end{cases}$$

We note the power of a point is always a non-negative number which vanishes on the circle.

Theorem 2.27.2 *Let ℓ be an arbitrary line through A which intersects the circle $\mathcal{C}(O,\rho)$ at M and N. Then $|AM| \cdot |AN| = Power(A)$.*

Proof: We shall investigate three cases:
(*i*) Case $A \in Ext\,\mathcal{C}(O,\rho)$:
Suppose that the line through A and O meets $\mathcal{C}(O,\rho)$ at O_1 and O_2. See Figure 2.51(a). Then $\widehat{NO_2A} = 180° - \widehat{NMO_1}$ (angles inscribed in circles) and $\widehat{AMO_1} = 180° - \widehat{NMO_1}$ (supplementary angles).
Hence $\widehat{NO_2A} = \widehat{AMO_1}$. As $\widehat{MAO_1} = \widehat{O_2AN}$, the (AA) similarity case yields

$$\Delta ANO_2 \sim \Delta AO_1M.$$

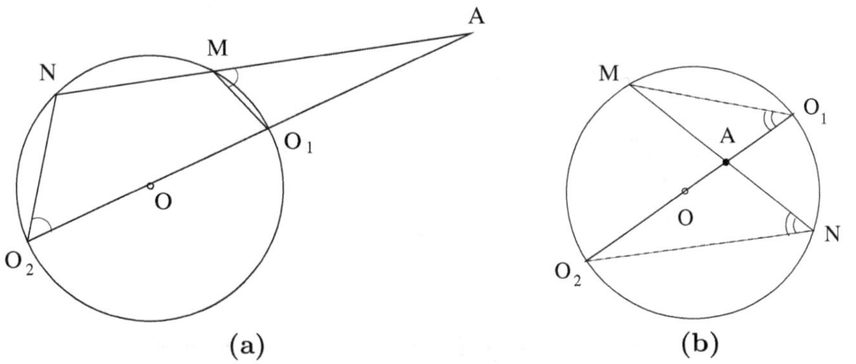

Figure 2.51: (a) $A \in Ext\,\mathcal{C}(O,\rho)$; (b) $A \in Int\,\mathcal{C}(O,\rho)$.

Hence

$$\frac{|AN|}{|AO_1|} = \frac{|AO_2|}{|AM|} \Longrightarrow |AN| \cdot |AM| = |AO_1| \cdot |AO_2|$$
$$= (|AO| - |O_1O|)(|AO| + |OO_2|)$$
$$= (|AO| - \rho)(|AO| + \rho)$$
$$= |AO|^2 - \rho^2 = \text{Power}(A).$$

(*ii*) Case $A \in Int\,\mathcal{C}(O, \rho)$: In this case, $\widehat{MO_1O_2} = \widehat{MNO_2}$ (inscribed angles cutting the same arc) and $\widehat{MAO_1} = \widehat{NAO_2}$ (equal vertical angles). Then $\Delta ANO_2 \sim \Delta AO_1M$ by the similarity case (AA). Therefore

$$\begin{aligned} |AN| \cdot |AM| &= |AO_1| \cdot |AO_2| \\ &= (\rho - |AO|)(\rho + |AO|) \\ &= \rho^2 - |AO|^2 \\ &= \text{Power(A)}. \end{aligned} \tag{2.27.44}$$

(*iii*) Case $A \in \mathcal{C}(O, \rho)$: In this case, both the power and the product vanish. ■

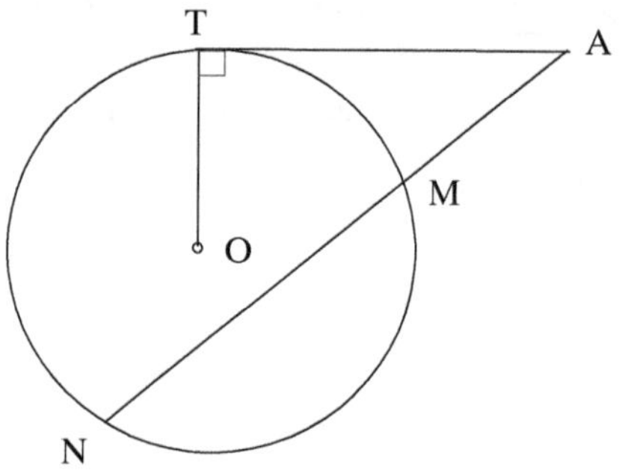

Figure 2.52: The case $A \in \mathcal{C}(O, \rho)$.

Corollary 2.27.3 *Let $T \in \mathcal{C}(O, \rho)$ and $A \notin Int\,\mathcal{C}(O, \rho)$ such that AT is tangent to $\mathcal{C}(O, \rho)$. Then*

$$|AM| \cdot |AN| = |AT|^2.$$

Proof: The Pythagorean theorem in ΔATO and Theorem 2.27.2 yield

$$|AT|^2 = |OA|^2 - |OT|^2 = |OA|^2 - \rho^2 = \text{Power}(A).$$

■

Corollary 2.27.4 (the two-chord-theorem) *When two chords of a circle intersect, the product of the length of the segments formed on one chord equals that on the other chord. That is* $|AM| \cdot |AN| = |AO_1| \cdot |AO_2|$.

Proof: The relation follows from (2.27.44). See also Figure 2.51(b). ∎

Exercise 2.27.5 (the reciprocal of the power of a point) *Let* MN *and* BC *be two distinct lines such that* $MN \cap BC = \{A\}$ *and* $|AM| \cdot |AN| = |AB| \cdot |AC|$. *Show that the points* M, N, B, *and* C *belong to a circle.*

Exercise 2.27.6 *Consider the circle of center* O *and the chord* AB. *Let* M *be a point on* AB *such that* $|AM| = 5$ *and* $|MB| = 3$. *If* $|MO| = 1$, *find the radius of the circle.*

Exercise 2.27.7 *Consider a chord* AB *in a circle and let* M *be its midpoint. Let* XY *be another chord passing through* M *such that* $|AM| = |XM|$. *Show that*

(i) $|AB| = |XY|$,

(ii) $\widehat{AXB} = 90^\circ$, *and*

(iii) $AXBY$ *is a rectangle.*

Exercise 2.27.8 *Let* A, B, C *be three points in the interior of a circle* $\mathcal{C}$. *Let* $\{M, N\}$, $\{P, Q\}$, $\{R, S\}$ *be the points where the extensions of the sides* AB, BC *and* AC *respectively intersect the circle* $\mathcal{C}$. *Denote* $x = |AR|$, $y = |CS|$, $z = |BN|$, $u = |CP|$, $v = |QB|$, $t = |AM|$, $a = |BC|$, $b = |AC|$, $c = |AB|$, *see Figure 2.53(a). Prove the identity*

$$a(u - v) + b(x - y) + c(z - t) = 0.$$

Exercise 2.27.9 *Let* ΔABC *be a triangle and denote by* A' *the point where the side* BC *is tangent to the inner circle. The line* AA' *intersects the second time the inner circle at* A'', *see Figure 2.53(b). Compute* $|AA''|$.

2.28 Computing distances between remarkable centers

The power of a point is useful for computing distances between different remarkable centers of a triangle. We shall compute the distances $|OI|$ and $|OH|$, which were computed for the first time by Leonhard Euler in the 1700s.

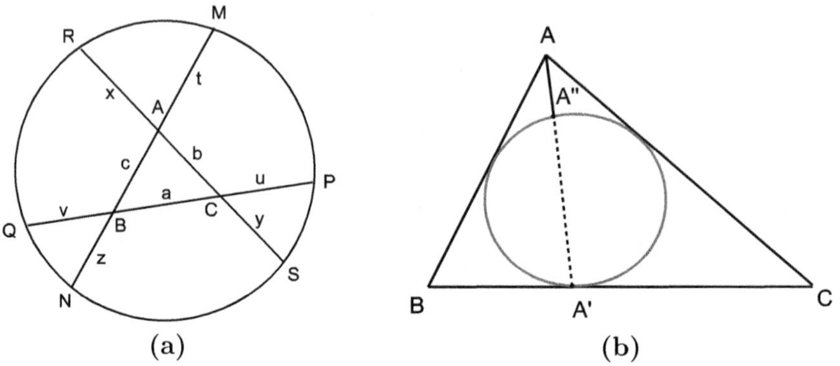

Figure 2.53: (a) See Exercise 2.27.8; (b) For Exercise 2.27.9.

Proposition 2.28.1 *Let O be the circumcenter and H be the orthocenter of ΔABC. Then*

$$|OH| = R\sqrt{1 - 8\cos\widehat{A}\cos\widehat{B}\cos\widehat{C}}.$$

Proof: We shall assume that triangle ΔABC is acute. The obtuse case is left for the reader as an exercise.

Let AA_1 be the altitude from the vertex A. As $H \in AA_1 \subset Int\,\mathcal{C}(O,R)$, it follows that H is an interior point of the circumcircle.
Then $\text{Power}(H) = R^2 - |OH|^2$ and hence

$$|OH|^2 = R^2 - \text{Power}(H). \tag{2.28.45}$$

We shall compute Power(H) using the chord which passes through points A and H. Let A_2 be the point where the altitude AA_1 intersects the circle a second time. By Theorem 2.27.2, the power of H can be expressed as

$$\text{Power}(H) = |AH| \cdot |HA_2|. \tag{2.28.46}$$

From Proposition 2.13.6 we know that $|AH| = 2R\cos A$. We shall compute $|HA_2|$ showing first that $|HA_1| = |A_1A_2|$ and computing $|HA_1|$ from ΔHBA_1. By the (ASA) congruence case, we have

$$\Delta HBA_1 \equiv \Delta A_2BA_1 \text{ because } \begin{cases} BA_1 & \text{(common side),} \\ \widehat{HA_1B} = \widehat{BA_1A_2} & \text{(right angles),} \\ \widehat{HBA_1} = \widehat{A_1BA_2} & \text{because} \end{cases}$$

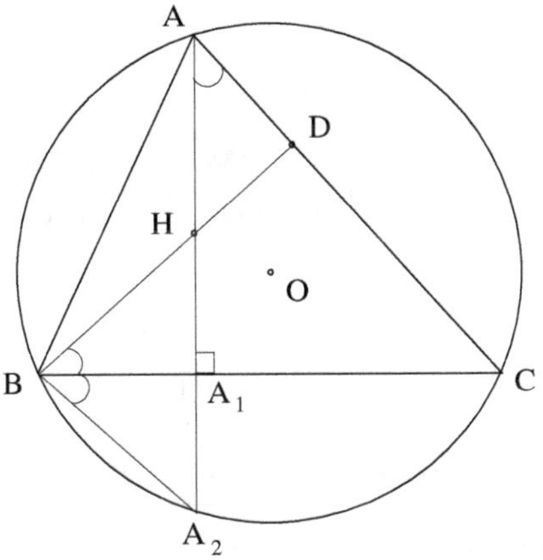

Figure 2.54: The power of H.

$$\begin{aligned}\widehat{HBA_1} &= 90° - \widehat{ACB} \text{ (from } \Delta DCB) \\ &= \widehat{A_1AC} \text{ (from } \Delta AA_1C) \\ &= \widehat{CBA_2} \text{ (same arc measure)} \\ &= \widehat{A_1BA_2}.\end{aligned}$$

Hence $|HA_1| = |A_1A_2|$ and then

$$|HA_2| = 2|HA_1|. \tag{2.28.47}$$

From ΔHBA_1 we have

$$\begin{aligned}\frac{|HA_1|}{|BH|} &= \sin \widehat{HBA_1} = \sin\left(90° - \widehat{C}\right) = \cos \widehat{C} \\ \Longrightarrow &\quad |HA_1| = |BH| \cos \widehat{C} = 2R \cos \widehat{B} \cos \widehat{C}, \quad (2.28.48)\end{aligned}$$

the last step coming from Proposition 2.9.3. Substituting in (2.28.47) yields

$$|HA_2| = 2|HA_1| = 4R \cos \widehat{B} \cos \widehat{C}.$$

Substituting in (2.28.46) yields

$$\begin{aligned}\text{Power}(H) &= |AH| \cdot |HA_2| \\ &= 2R \cos \widehat{A} \cdot 4R \cos \widehat{B} \cos \widehat{C} \\ &= 8R^2 \cos \widehat{A} \cos \widehat{B} \cos \widehat{C},\end{aligned}$$

and hence (2.28.45) becomes

$$\begin{aligned} |OH|^2 &= R^2 - \text{Power}(H) \\ &= R^2 - 8R^2 \cos\widehat{A}\cos\widehat{B}\cos\widehat{C} \\ &= R^2\left(1 - 8\cos\widehat{A}\cos\widehat{B}\cos\widehat{C}\right). \end{aligned}$$

■

Remark 2.28.2 As $|OH| \geq 0$, it follows that

$$\cos\widehat{A}\cos\widehat{B}\cos\widehat{C} \leq 1/8,$$

which is the inequality proved in Section 2.15 by algebraic methods.

Corollary 2.28.3 *Let r_0 denote the inradius of the orthic triangle. Then*

$$|OH|^2 = R^2 - 4Rr_0.$$

Proof: Substitute $r_0 = 2R\cos\widehat{A}\cos\widehat{B}\cos\widehat{C}$, which comes from (2.15.18), in Proposition 2.28.1. ■

Sometimes, it is more useful to express the distance $|OH|$ in terms of side lengths rather than angles.

Corollary 2.28.4 *We have*

$$|OH|^2 = 9R^2 - (a^2 + b^2 + c^2). \tag{2.28.49}$$

Proof: By the law of sines, we can express the sides as $a^2 = 4R^2\sin^2\widehat{A}$, $b^2 = 4R^2\sin^2\widehat{B}$, $c^2 = 4R^2\sin^2\widehat{C}$. Using formula (2.1.1), we obtain

$$\begin{aligned} 9R^2 - (a^2 + b^2 + c^2) &= 9R^2 - 4R^2\left(\sin^2\widehat{A} + \sin^2\widehat{B} + \sin^2\widehat{C}\right) \\ &= 9R^2 - 4R^2\left(2 + 2\cos\widehat{A}\cos\widehat{B}\cos\widehat{C}\right) \\ &= R^2\left(1 - 8\cos\widehat{A}\cos\widehat{B}\cos\widehat{C}\right) = |OH|^2. \end{aligned}$$

■

Theorem 2.28.5 (Euler) *Let O and I be the circumcenter and the incenter of ΔABC. Then*

$$|OI| = \sqrt{R(R - 2r)}.$$

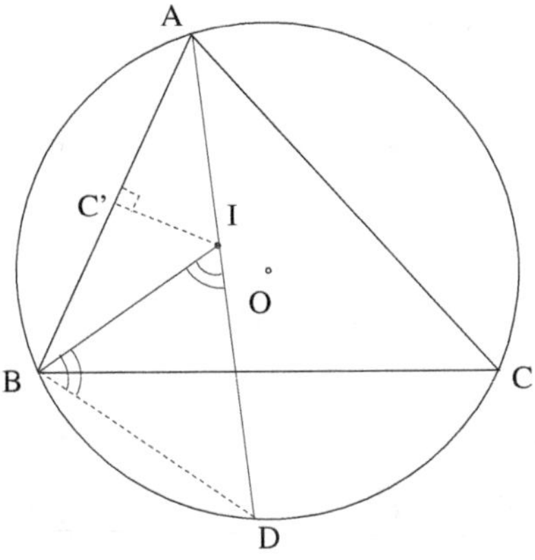

Figure 2.55: Finding distance $|OI|$.

Proof: Let D be the point where the extension of AI intersects the circumcircle for the second time. We have

$$\begin{aligned} \text{Power}(I) = R^2 - |OI|^2 \Longrightarrow |OI|^2 &= R^2 - \text{Power}(I) \\ &= R^2 - |AI| \cdot |ID|. \quad (2.28.50) \end{aligned}$$

We shall compute the lengths of the segments AI and ID.
Let C' be the orthogonal projection of I on the side AB. In $\Delta AIC'$ we have

$$\sin \frac{\widehat{A}}{2} = \sin \widehat{BAI} = \frac{|IC'|}{|AI|} = \frac{r}{|AI|} \Longrightarrow |AI| = \frac{r}{\sin \frac{\widehat{A}}{2}}. \quad (2.28.51)$$

We shall show that ΔDBI is isosceles. Using that $\widehat{DIB}$ is an exterior angle in ΔABI yields

$$\widehat{DIB} = \widehat{IAB} + \widehat{IBA} = \frac{\widehat{A}}{2} + \frac{\widehat{B}}{2}. \quad (2.28.52)$$

As BI is an angle bisector

$$\begin{aligned} \widehat{IBD} &= \widehat{IBC} + \widehat{CBD} \\ &= \widehat{IBC} + \widehat{DAC} \text{ (inscribed angles cutting the same arc)} \\ &= \frac{\widehat{B}}{2} + \frac{\widehat{A}}{2}. \quad (2.28.53) \end{aligned}$$

Comparing (2.28.52) and (2.28.53) yields $\widehat{DIB} = \widehat{IBD}$. Hence triangle ΔDBI is an isosceles triangle, and therefore $|ID| = |BD|$. Thus it suffices to compute $|BD|$. This is done using the law of sines in triangle ΔABD:

$$\frac{|BD|}{\sin\frac{\widehat{A}}{2}} = 2R \Longrightarrow |BD| = 2R\sin\frac{\widehat{A}}{2}.$$

Hence

$$|ID| = 2R\sin\frac{\widehat{A}}{2}.$$

Then (2.28.50), (2.28.51) and this give

$$\begin{aligned}|OI|^2 &= R^2 - |AI|\cdot|ID| = R^2 - \frac{r}{\sin\frac{\widehat{A}}{2}}\cdot 2R\sin\frac{\widehat{A}}{2}\\ &= R^2 - 2Rr = R(R-2r).\end{aligned}$$

■

Corollary 2.28.6 (Euler's inequality) *In any triangle, we have $R \geq 2r$.*

Corollary 2.28.7 *A triangle is equilateral if and only if $O = I$.*

Proof: $O = I \Longleftrightarrow |OI| = 0 \Longleftrightarrow R = 2r \Longleftrightarrow \Delta ABC$ is equilateral by Theorem 2.8.1. ■

It is interesting to note that a converse of the Euler relation holds. This will nicely answer the question at the beginning of Section 2.8. More precisely, we have the following statement:

Proposition 2.28.8 *Consider the interior circles, $\mathcal{C}(O,R)$ and $\mathcal{C}(I,r)$, of centers O, I and radii R and r, which satisfy relation*

$$|OI|^2 = R(R-2r).$$

Then there are infinitely many triangles inscribed in $\mathcal{C}(O,R)$ and circumscribed to $\mathcal{C}(I,r)$.

Proof: Euler's relation implies $|OI| < R$, so $\mathcal{C}(I,r)$ is inside $\mathcal{C}(O,R)$. We choose an arbitrary point $A \in \mathcal{C}(O,R)$ and let B be the point where the tangent line from A to $\mathcal{C}(I,r)$ intersects the circle $\mathcal{C}(O,R)$. Consider the tangent line from B to $\mathcal{C}(I,r)$, which intersects $\mathcal{C}(O,R)$ at point C. Then let A_1 be the point where the tangent line from C to $\mathcal{C}(I,r)$ intersects $\mathcal{C}(O,R)$. For arbitrary interior circles $\mathcal{C}(O,R)$ and $\mathcal{C}(I,r)$, the points A and A_1 are in general distinct, see Figure 2.56. We shall show that if Euler's relation

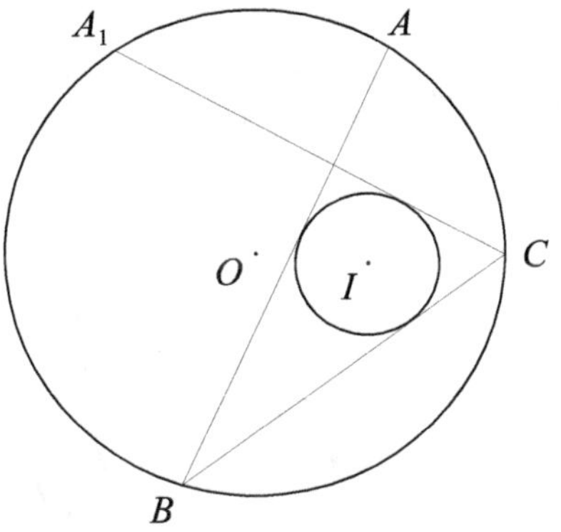

Figure 2.56: In general, the points A and A_1 are distinct.

holds, $|OI|^2 = R(R - 2r)$, then $A = A_1$. Since this will hold for any initial point A, we shall obtain infinitely many triangles ΔABC circumscribed to $\mathcal{C}(I, r)$ and inscribed into $\mathcal{C}(O, R)$. The proof will follow a few steps.

Step 1: Let A' and A'' be the points where AI and A_1I intersect the circle $\mathcal{C}(O, R)$ for the second time, see Figure 2.57(a). We shall evaluate the power of the point I with respect to $\mathcal{C}(O, R)$ using Euler's relation as follows:

$$Power(I) = R^2 - |OI|^2 = R^2 - R(R - 2r) = 2Rr.$$

The power can also be evaluated using the two-chords theorem as

$$Power(I) = |AI| \cdot |IA'| = |A_1I| \cdot |IA''|.$$

Therefore, it follows that

$$|AI| \cdot |IA'| = 2Rr, \quad |A_1I| \cdot |IA''| = 2Rr. \tag{2.28.54}$$

Step 2: At this stage we shall prove that $|A'B| = |IA'|$ and $|A''C| = |IA''|$, see Figure 2.57(a). To this end we shall employ the following construction. Let S be the contact point between AB and $\mathcal{C}(I, r)$, and let B_2 be the opposite diametral point of B in $\mathcal{C}(O, R)$. We note that $\Delta B_2A'B \sim \Delta ASI$ by (AA) similarity case as follows:

(i) $\widehat{B_2A'B} = 90°$ as being an angle inscribed in a semicircle and $\widehat{ASI} = 90°$ as the radius IS is perpendicular on the tangent AB; hence $\widehat{B_2A'B} = \widehat{ASI}$.

(ii) $\widehat{SAI} = \widehat{BAA'} = \frac{1}{2}arc(BA')$, as angles cutting the same arc on $\mathcal{C}(O, R)$.

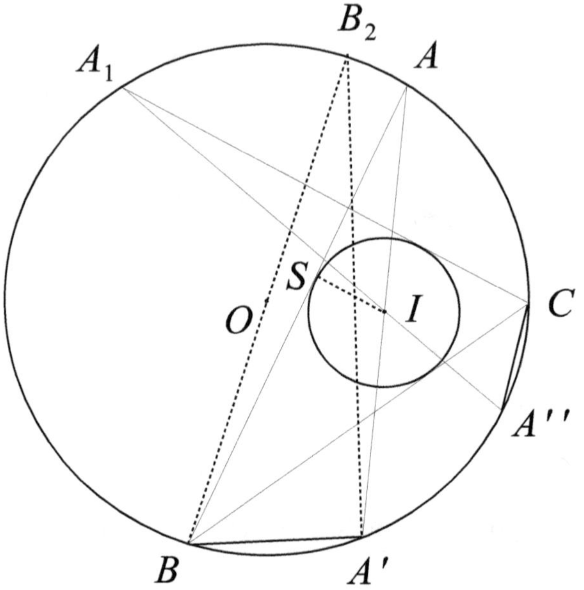

(a)

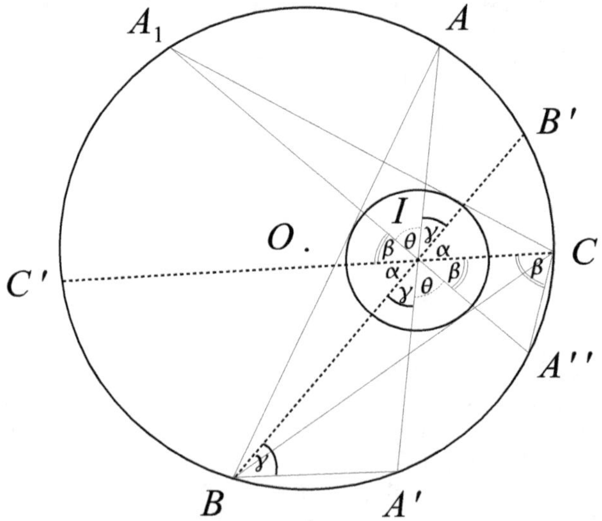

(b)

Figure 2.57: (a) For Step 1 and Step 2; (b) For Step 3.

Writing the similarity rations, $\dfrac{|B_2B|}{|A'B|} = \dfrac{|AI|}{|IS|}$, and using that $|B_2B| = 2R$ and $|IS| = r$, cross multiplying yields $|AI| \cdot |A'B| = 2Rr$. Comparing with the first relation of (2.28.54) yields $|A'B| = |IA'|$.

A similar construction (involving the projection of I on A_1C and the opposite diametral point of C) provides the identity $|A''C| = |IA''|$.

Step 3: We shall show that $arc(A'A'') = 0$, which implies that $A' = A''$ and hence $A = A_1$. This will be proved by an angle addition as in the following. Let B' and C' be the intersections of BI and CI with $\mathcal{C}(O, R)$ for the second time, see Figure 2.57(b). First, for the sake of simplicity, we make the angle notations $\alpha = \widehat{C'IB} = \widehat{B'IC}$, $\beta = \widehat{A_1IC'} = \widehat{A''IC}$, $\gamma = \widehat{BIA'} = \widehat{AIB'}$, $\theta = \widehat{A'IA''} = \widehat{A_1IA}$. Each of these angles can be evaluated in terms of the arcs on the circle $\mathcal{C}(O, R)$. Thus

$$\alpha = \frac{1}{2}(arc(B'C) + arc(BC')), \text{and}$$

$$\theta = \frac{1}{2}(arc(A'A'') + arc(A_1A)),$$

since they are interior angles. Using *Step 2*, we obtain that triangles $\Delta A'IB$ and $\Delta A''IC$ are isosceles, so $\widehat{IBA'} = \gamma$, $\widehat{ICA''} = \beta$. Using the measure of inscribed angles we have

$$\beta = \widehat{C'CA''} = \frac{1}{2}arc(A''C'), \text{and}$$

$$\gamma = \widehat{B'BA'} = \frac{1}{2}arc(A'B').$$

Substituting into $\alpha + \beta + \gamma + \theta = 180°$ we obtain

$$\frac{arc(B'C) + arc(BC')}{2} + \frac{arc(A'A'') + arc(A_1A)}{2} + \frac{arc(A''C')}{2} + \frac{arc(A'B')}{2} = 180°,$$

which by using that the sum of the measures of the entire circle is 360°, we obtain

$$\frac{1}{2}(360° + 2\,arc(A'A'')) = 180°,$$

which implies $arc(A'A'') = 0$. This implies $A = A_1$, which ends the proof. ∎

Remark 2.28.9 If $|OI|^2 \neq R(R - 2Rr)$, then $A \neq A_1$, regardless of the position of the initial point A. In this case the reader may try the more challenging computation of the arc measure $arc(AA_1)$ in terms of the distance $d = |OI|$ and the radii R and r.

Exercise 2.28.10 *Show that* $Power(I) = \dfrac{abc}{a+b+c}$.

Let O_0 and I_0 be the circumcenter and the incenter of the orthic triangle of ΔABC. Let H be the orthocenter of ΔABC. We have shown in Corollary 2.10.5 that $I_0 = H$. Using Theorem 2.28.5 for the orthic triangle yields

$$\begin{aligned} |O_0H|^2 &= |O_0I|_0^2 = R_0^2 - 2R_0r_0 \\ &= \left(\frac{R}{2}\right)^2 - 2\frac{R}{2}r_0 \\ &= \frac{1}{4}\left(R^2 - 4Rr_0\right) \\ &= \left(\frac{|OH|}{2}\right)^2, \end{aligned}$$

where we used $R_0 = R/2$ from Proposition 2.10.8. We have arrived at the following result.

Proposition 2.28.11 *We have*

$$|O_0H| = \frac{|OH|}{2}.$$

Remark 2.28.12 We shall show later that O_0 is the midpoint of OH.

Exercise 2.28.13 *Find the power of the centroid G of ΔABC with respect to the circumcircle in terms of the sides of the triangle.*

Exercise 2.28.14 *Let AA_1, BB_1, CC_1 be the altitudes of triangle ΔABC. Show that we have*

$$|HA| \cdot |HA_1| = |HB| \cdot |HB_1| = |HC| \cdot |HC_1|,$$

where H denotes the orthocenter.

Solved Exercise 2.28.15 *Let $\mathcal{C}(O,R)$ be a circle and let A, B be two points on the same diameter such that $|AO| = |OB|$. Consider an arbitrary point $M \in \mathcal{C}(O,R)$ and let N, P be the points where the lines MA and MB intersects the circle $\mathcal{C}(O,R)$ second time.*

Show that the sum $\dfrac{1}{|AN|^2} + \dfrac{1}{|BP|^2}$ *is independent of the position of M on the circle $\mathcal{C}(O,R)$.*

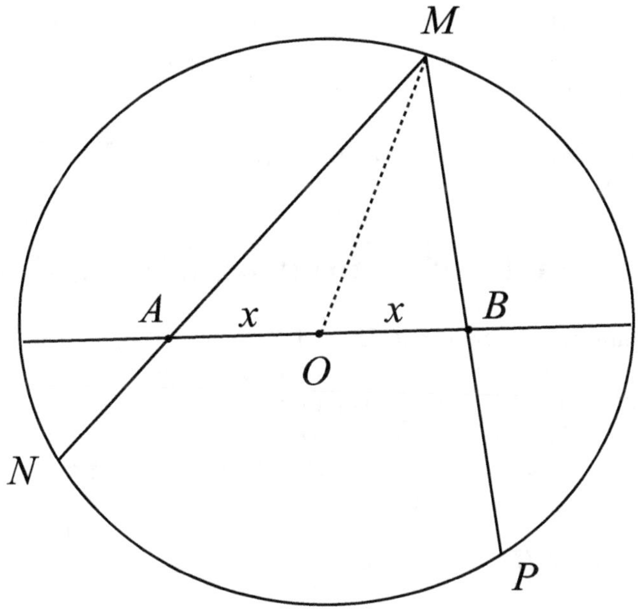

Figure 2.58: For Exercise 2.28.15.

Proof: Let $x = |AO| = |OB|$, see Figure 2.58. Then $Power(A) = R^2 - x^2 = |AM| \cdot |AN|$ and $Power(B) = R^2 - x^2 = |BM| \cdot |BP|$, so $|AN| = \frac{R^2 - x^2}{|AM|}$ and $|BP| = \frac{R^2 - x^2}{|BM|}$. Then we have

$$\frac{1}{|AN|^2} + \frac{1}{|BP|^2} = \frac{|AM|^2 + |BM|^2}{(R^2 - x^2)^2}. \tag{2.28.55}$$

The median theorem in ΔMAB yields

$$R^2 = |OM|^2 = \frac{1}{2}(|AM|^2 + |BM|^2) - \frac{|AB|^2}{4},$$

so $|AM|^2 + |BM|^2 = 2R^2 + 2x^2$. Substituting in (2.28.55) we obtain

$$\frac{1}{|AN|^2} + \frac{1}{|BP|^2} = \frac{2(R^2 + x^2)}{(R^2 - x^2)^2},$$

which is independent of the position of M. ∎

Exercise 2.28.16 *Let P and Q be two points inside of circle $\mathcal{C}(O,R)$ and let us denote $\alpha = \widehat{POQ}$. Show that the distance between P and Q can be expressed in terms of their powers as*

$$|PQ|^2 = 2R^2 - Pow(P) - Pow(Q) - 2\cos\alpha\sqrt{(R^2 - Pow(P))(R^2 - Pow(Q))}.$$

2.29 Variants of computing $|OH|$ and $|OI|$

We have computed distances $|OH|$ and $|OI|$ in a triangle using the power of points H and I with respect to the circumcircle of ΔABC. We shall provide next some ad-hoc methods based on the Pythagorean theorem and trigonometric relations to compute $|OH|$ and $|OI|$. We shall arrive at formulas similar to the ones obtained using the power of a point.

The distance $|OH|$

We consider a triangle ΔABC and we shall assume $\widehat{A} \geq \widehat{B} \geq \widehat{C}$, see Figure 2.59. Let $AA_1 \perp BC$ and denote by A' the midpoint of the side BC. Then we shall find $|OH|$ from the trapezoid $HOA'A_1$, where we know three of its sides

$$\begin{aligned}
|HA_1| &= 2R\cos\widehat{B}\cos\widehat{C},\\
|OA'| &= R\cos\widehat{A},\\
|BA_1| &= c\cos\widehat{B}, \text{and}\\
|A_1A'| &= \frac{a}{2} - c\cos\widehat{B}.
\end{aligned}$$

To this end we consider the point $O' \in HA_1$ such that $OO' \| BC$. Then Pythagorean theorem applied in triangle $\Delta HOO'$ provides

$$\begin{aligned}
|OH|^2 &= |OO'|^2 + |HO'|^2 = |A_1A'|^2 + (|HA_1| - |OA'|)^2\\
&= \left(\frac{a}{2} - c\cos\widehat{B}\right)^2 + \left(2R\cos\widehat{B}\cos\widehat{C} - R\cos\widehat{A}\right)^2\\
&= \frac{a^2}{4} + c^2\cos^2\widehat{B} - ac\cos\widehat{B} + 4R^2\cos^2\widehat{B}\cos^2\widehat{C} + R^2\cos^2\widehat{A}\\
&\quad - 4R^2\cos\widehat{A}\cos\widehat{B}\cos\widehat{C}.
\end{aligned}$$

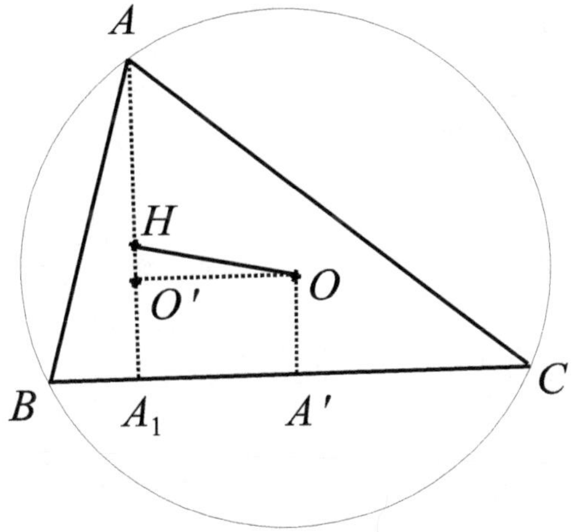

Figure 2.59: Computing the distance $|OH|$.

Using the law of sines, $a = 2R\sin\widehat{A}$, $c = 2R\sin\widehat{C}$, collecting terms yields

$$\begin{aligned}|OH|^2 &= R^2\left(\sin^2\widehat{A} + \cos^2\widehat{A}\right) + 4R^2\cos^2\widehat{B}\left(\sin^2\widehat{C} + \cos^2\widehat{C}\right)\\&\quad - 4R^2\sin\widehat{A}\sin\widehat{B}\sin\widehat{C} - 4R^2\cos\widehat{A}\cos\widehat{B}\cos\widehat{C}\\&= R^2 + 4R^2\cos\widehat{B}\left(\cos\widehat{B} - \sin\widehat{A}\sin\widehat{C}\right)\\&\quad - 4R^2\cos\widehat{A}\cos\widehat{B}\cos\widehat{C}. \qquad (2.29.56)\end{aligned}$$

The expression in the parenthesis can be computed as

$$\begin{aligned}\cos\widehat{B} - \sin\widehat{A}\sin\widehat{C} &= \cos\left(180 - \left(\widehat{A} + \widehat{C}\right)\right) - \sin\widehat{A}\sin\widehat{C}\\&= -\cos\left(\widehat{A} + \widehat{C}\right) - \sin\widehat{A}\sin\widehat{C}\\&= -\cos\widehat{A}\cos\widehat{C}.\end{aligned}$$

Substituting in (2.29.56) yields

$$\begin{aligned}|OH|^2 &= R^2 - 4R^2\cos\widehat{A}\cos\widehat{B}\cos\widehat{C} - 4R^2\cos\widehat{A}\cos\widehat{B}\cos\widehat{C}\\&= R^2 - 8R^2\cos\widehat{A}\cos\widehat{B}\cos\widehat{C},\end{aligned}$$

which agrees with a previous formula obtained using the power of H.

The distance $|OI|$

We consider a triangle ΔABC with $\widehat{B} \geq \widehat{C}$. Since by Exercise 2.7.5 we have $\widehat{IAO} = \frac{\widehat{B}-\widehat{C}}{2}$, an application of the law of cosines in triangle ΔAIO provides

$$\begin{aligned} |OI|^2 &= |AI|^2 + |AO|^2 - 2|AI| \cdot |AO| \cos \frac{\widehat{B}-\widehat{C}}{2} \\ &= \frac{r^2}{\sin^2 \frac{\widehat{A}}{2}} + R^2 - 2\frac{r}{\sin \frac{\widehat{A}}{2}} R \cos \frac{\widehat{B}-\widehat{C}}{2} \\ &= R^2 - \frac{2Rr}{\sin \frac{\widehat{A}}{2}} \left(\cos \frac{\widehat{B}-\widehat{C}}{2} - \frac{1}{2}\frac{r}{R}\frac{1}{\sin \frac{\widehat{A}}{2}} \right). \end{aligned} \tag{2.29.57}$$

Using relation (2.17.20)

$$r = 4R \sin \frac{\widehat{A}}{2} \sin \frac{\widehat{B}}{2} \sin \frac{\widehat{C}}{2}$$

the expression in the parenthesis can be computed as

$$\begin{aligned} \cos \frac{\widehat{B}-\widehat{C}}{2} - \frac{1}{2}\frac{r}{R}\frac{1}{\sin \frac{\widehat{A}}{2}} &= \cos \frac{\widehat{B}}{2} \cos \frac{\widehat{C}}{2} + \sin \frac{\widehat{B}}{2} \sin \frac{\widehat{C}}{2} - 2 \sin \frac{\widehat{B}}{2} \sin \frac{\widehat{C}}{2} \\ &= \cos \frac{\widehat{B}}{2} \cos \frac{\widehat{C}}{2} - \sin \frac{\widehat{B}}{2} \sin \frac{\widehat{C}}{2} \\ &= \cos \frac{\widehat{B}+\widehat{C}}{2} = \cos \frac{180 - \widehat{A}}{2} = \cos \left(90^\circ - \frac{\widehat{A}}{2} \right) \\ &= \sin \frac{\widehat{A}}{2}. \end{aligned}$$

Substituting in (2.29.57), after simplification we obtain the desired formula $|OI|^2 = R^2 - 2Rr = R(R - 2R)$.

Remark 2.29.1 By Exercise 2.15.11 the incenter of the orthic triangle, denoted by I_0, is the orthocenter of ΔABC, namely, $I_0 = H$. Since the circumcenter of the orthic triangle is $O_0 = O_9$, the center of the Euler's circle of ΔABC, then Euler's relation applied to the orthic triangle provides

$$|O_9 I_0|^2 = R_0(R_0 - 2r_0), \tag{2.29.58}$$

where R_0, r_0 are the circumradius and the inradius of the orthic triangle.

Using that $R_0 = R/2$ and Proposition 2.28.11 yields

$$\begin{aligned} |O_9 I_0|^2 &= |O_9 H|^2 = \frac{1}{4}|OH|^2 \\ &= \frac{1}{4}R^2\left(1 - 8\cos\widehat{A}\cos\widehat{B}\cos\widehat{C}\right). \end{aligned} \tag{2.29.59}$$

Equating (2.29.58) and (2.29.59), we obtain

$$\frac{R}{2}\left(\frac{R}{2} - 2r_0\right) = \frac{1}{4}R^2\left(1 - 8\cos\widehat{A}\cos\widehat{B}\cos\widehat{C}\right), \tag{2.29.60}$$

which after simplifications becomes

$$r_0 = 2R\cos\widehat{A}\cos\widehat{B}\cos\widehat{C},$$

which recovers formula (2.15.18).

2.30 Euler's line

In the 1720s, Euler proposed the following problem:

Given three points H, G and O, find a triangle such that these points are the orthocenter, the centroid and the circumcenter of the triangle.

After 42 years, he had shown that the triangle can be constructed only if the points O, G, H are collinear and some proportionality relation holds.

Theorem 2.30.1 *Let O, G, H be the circumcenter, the centroid and the orthocenter of a triangle. Then G is between O and H and*

$$|OH| = 3|OG|.$$

Proof: We shall assume ΔABC acute. The obtuse case is left for the reader. Let A' be the midpoint of BC and AA_1 be the altitude from vertex A. Let $AA' \cap HO = \{Q\}$. The segments AH and OA' are parallel because both are perpendicular to the same segment BC (O is on the perpendicular bisector of BC). Then $\widehat{HAQ} = \widehat{QA'O}$ (alternate interior angles) and $\widehat{AQH} = \widehat{OQA'}$ (vertical angles). Then by the (AA) similarity case we have

$$\Delta AHQ \sim \Delta A'OQ.$$

Then

$$\frac{|AH|}{|A'O|} = \frac{|AQ|}{|QA'|} = \frac{|HQ|}{|QO|}.$$

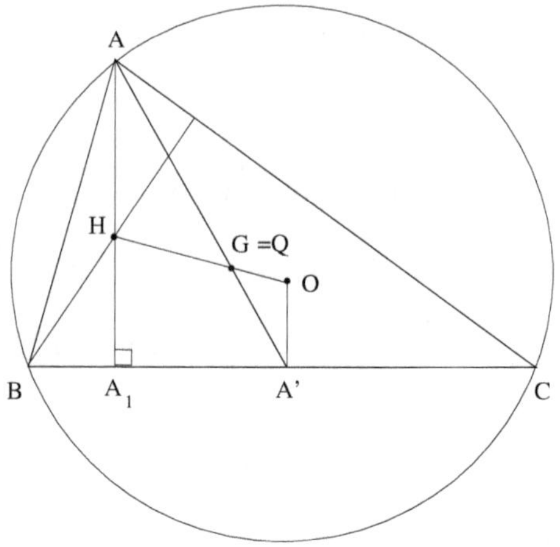

Figure 2.60: The median AA' bisects the triangle in two equal areas.

We have from Propositions 2.13.6 and 2.7.9 that $|AH| = 2R\cos\widehat{A}$ and $|OA'| = R\cos\widehat{A}$. Therefore $\frac{|AH|}{|A'O|} = 2$ and hence $\frac{|AQ|}{|QA'|} = 2$. Hence Q is a point on the median situated at one third from the base and two thirds from the vertex. It follows that $Q = G$, the centroid. We have

$$\frac{|HQ|}{|QO|} = \frac{|HG|}{|OG|} = 2 \Longrightarrow |HG| = 2|OG|.$$

■

2.31 Euler's circle

The following propositions will be used in the proof of Euler's circle theorem.

Proposition 2.31.1 *Let ΔABC be a right triangle with $\widehat{A} = 90$. Then the length of the median from A is half the hypotenuse length.*

Proof: Consider the circumcircle of ΔABC. Then BC is a diameter and the circumcenter is A', the midpoint of BC. The median AA' is a radius in this circle and hence it is half of its diameter. That is $|AA'| = |BC|/2$. ■

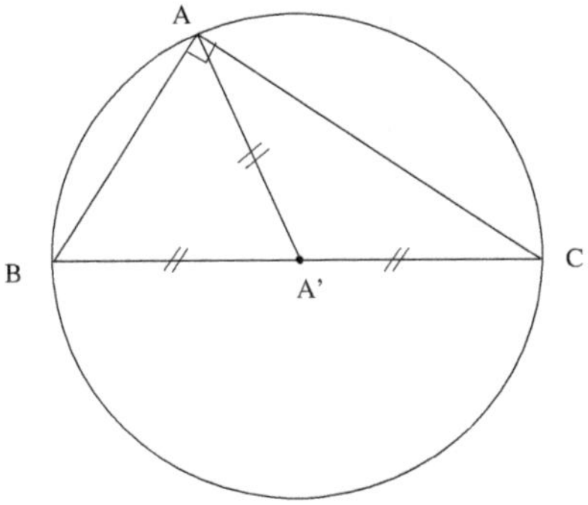

Figure 2.61: Right triangle inscribed in a semicircle.

Another proof: We shall give a proof based on metric relations. Substituting the Pythagoras relation $a^2 = b^2 + c^2$ into the median theorem, Theorem 2.21.3 yields

$$m_a^2 = \frac{b^2 + c^2}{2} - \frac{a^2}{4} = \frac{a^2}{2} - \frac{a^2}{4} = \frac{a^2}{4},$$

and hence $m_a = a/2$.

Definition 2.31.2 *A trapezoid $ABCD$ ($AB \| CD$) is called* **isosceles** *if the nonparallel sides are congruent, i.e., $|AD| = |BC|$.*

Proposition 2.31.3 *The trapezoid $ABCD$ ($AB \| CD$) is isosceles if and only if $\widehat{ADC} = \widehat{BCD}$.*

Proof: Let $AA_1 \perp DC$ and $BB_1 \perp DC$, with $A_1, B_1 \in DC$. Then ABB_1A_1 is a rectangle and hence $|AA_1| = |BB_1|$. See Figure 2.62(a).
" $\Longrightarrow$ " Assume $AD = BC$. Then by the (SSS) congruence case

$$\Delta AA_1D \equiv \Delta BB_1C \text{ because } \begin{cases} |AA_1| = |BB_1| & (ABB_1A_1 \text{ rectangle}), \\ |AD| = |BC| & (\text{hypothesis}), \\ |DA_1| = |B_1C| & \text{because} \end{cases}$$

$$|DA_1| = \sqrt{|AD|^2 - |AA_1|^2} = \sqrt{|BC|^2 - |BB_1|^2} = |B_1C|.$$

Hence $\widehat{ADC} = \widehat{BCD}$.
" $\Longleftarrow$ " If $\widehat{ADC} = \widehat{BCD}$, using $|AA_1| = |BB_1|$ yields

$$|AD| = |AA_1| / \sin \widehat{ADC} = |BB_1| / \sin \widehat{BCD} = |BC|.$$

■

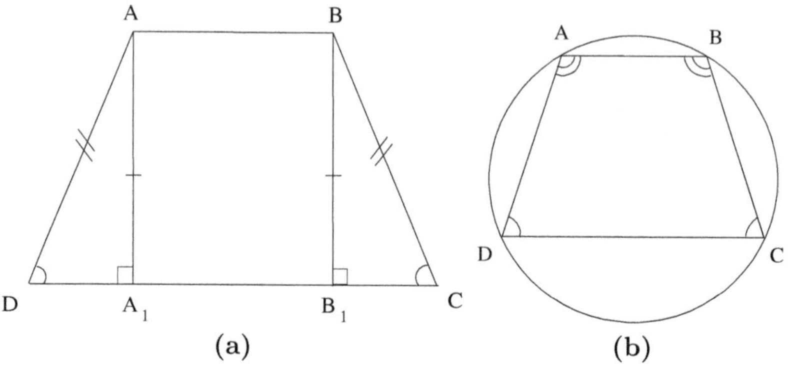

Figure 2.62: (a) Isosceles trapezoid; (b) Isosceles trapezoid inscribed in a circle.

Exercise 2.31.4 *Show that in a trapezoid ($AB\|CD$) we have $\widehat{ADC} = \widehat{BCD}$ if and only if $\widehat{DAB} = \widehat{ABC}$.*

A quadrilateral is called **cyclic** if it can be inscribed in a circle.

Proposition 2.31.5 *A trapezoid is cyclic if and only if it is isosceles.*

Proof: We shall use Figure 2.62(b).
"$\Longrightarrow$" As $AB\|CD$ then $\widehat{A}+\widehat{D} = 180° = \widehat{B}+\widehat{C}$. If $ABCD$ is cyclic then $\widehat{A}+\widehat{C} = \widehat{D}+\widehat{B} = 180°$. Subtracting, yields

$$\widehat{A}+\widehat{D}-\left(\widehat{A}+\widehat{C}\right) = 0 \Longleftrightarrow \widehat{D} = \widehat{C}.$$

"$\Longleftarrow$" If $ABCD$ is isosceles then $\widehat{D} = \widehat{C}$. As $\widehat{A}+\widehat{D} = 180°$ and $\widehat{B}+\widehat{C} = 180°$ it follows that $\widehat{A} = \widehat{B}$. Then

$$360° = \widehat{A}+\widehat{B}+\widehat{C}+\widehat{D} = 2\left(\widehat{A}+\widehat{C}\right).$$

Then $\widehat{A}+\widehat{C} = 180°$, and hence A, B, C, D belong to a circle. ∎

In the next two propositions, AA_1, BB_1, CC_1 denote the altitudes of ΔABC. Let A', B', C' be the midpoints of the sides BC, CA and AB, respectively. Let A'', B'', C'' denote the midpoints of the segments AH, BH, CH, where H denotes the orthocenter of ΔABC. Then $\Delta A_1B_1C_1$ is the orthic triangle and $\Delta A'B'C'$ is the median triangle of ΔABC.

Proposition 2.31.6 *Triangles $\Delta A'B'C'$ and $\Delta A_1B_1C_1$ have the same circumcircle.*

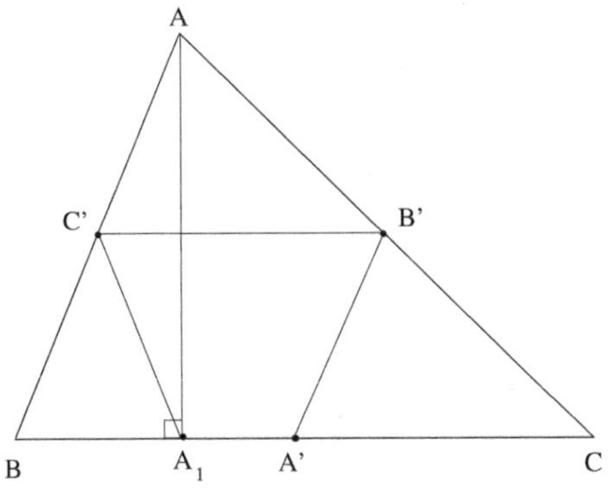

Figure 2.63: The isosceles trapezoid $A'A_1C'B'$.

Proof: Let $\mathcal{C}(A'B'C')$ be the circumcircle of $\Delta A'B'C'$. We shall show that A_1, B_1, $C_1 \in \mathcal{C}(A'B'C')$. This proves the result because a circle is determined by three points. We prove first that $A_1A'B'C'$ is an isosceles trapezoid. As C' and B' are the midpoints of AB and AC, then $B'C' \| BC$ (by the midpoint connector theorem) and hence $A_1A'B'C'$ is a trapezoid. As A' and B' are midpoints, the midpoint connector theorem, Corollary 2.11.2 yields

$$|A'B'| = \frac{1}{2}|AB|.$$

In the right triangle AA_1B, the cevian A_1C' is a median. Then by Proposition 2.31.1

$$|A_1C'| = \frac{1}{2}|AB|.$$

Hence $|A'B'| = \frac{1}{2}|AB| = |A_1C'|$ and then $A_1A'B'C'$ is an isosceles trapezoid. Applying Proposition 2.31.5, it follows that $A_1A'B'C'$ is cyclic and hence $A_1 \in \mathcal{C}(A'B'C')$.
In a similar way, we can show that $B_1 \in \mathcal{C}(A'B'C')$ and $C_1 \in \mathcal{C}(A'B'C')$. ■

Proposition 2.31.7 *Triangles $\Delta A_1B_1C_1$ and $\Delta A''B''C''$ have the same circumcircle.*

Proof: We shall show that A'', B'', $C'' \in \mathcal{C}(A_1B_1C_1)$. As B_1C'' is a median of the right triangle ΔB_1HC, by Proposition 2.31.1

$$|B_1C''| = |C''C| = \frac{1}{2}|HC|.$$

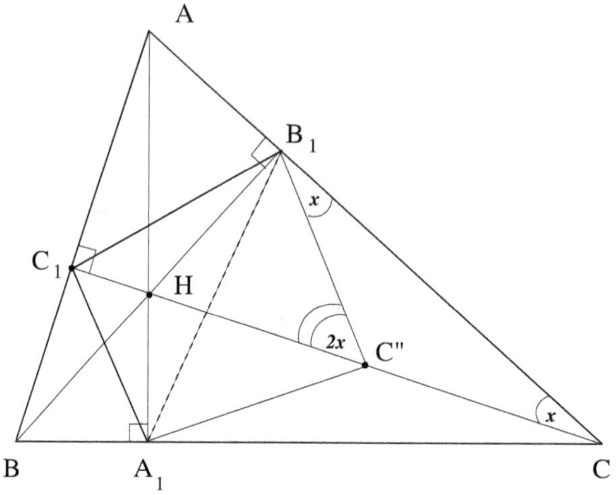

Figure 2.64: See the Euler's circle theorem.

It follows that $\Delta B_1C''C$ is isosceles. Let

$$x = \widehat{C''B_1C} = \widehat{C''CB_1}.$$

The exterior angle theorem for $\Delta B_1C''C$ yields

$$\widehat{B_1C''C_1} = 2x = 2(90^\circ - \widehat{A}) = 180^\circ - 2\widehat{A}. \tag{2.31.61}$$

As $\Delta A_1B_1C_1$ is the orthic triangle, Proposition 2.15.4 implies that

$$\widehat{C_1A_1B_1} = 180^\circ - 2\widehat{A}. \tag{2.31.62}$$

Comparing (2.31.61) and (2.31.62) yields

$$\widehat{B_1C''C_1} = \widehat{C_1A_1B_1}.$$

Hence $C_1B_1C''A_1$ is a cyclic quadrilateral. Then $C'' \in \mathcal{C}(A_1B_1C_1)$. In a similar way A'', $B'' \in \mathcal{C}(A_1B_1C_1)$. ■

Theorem 2.31.8 (Euler's circle theorem) *The points A_1, B_1, C_1, A', B', C', A'', B'', C'' belong to the same circle.*

Proof: From Propositions 2.31.6 and 2.31.7 we have respectively

$$\mathcal{C}(A_1B_1C_1) = \mathcal{C}(A'B'C'),$$
$$\mathcal{C}(A_1B_1C_1) = \mathcal{C}(A''B''C''),$$

and hence these three circumcircles are equal. ■

Definition 2.31.9 *The above circle is called the* **Euler circle** *of* ΔABC.

Exercise 2.31.10 *Let $ABCD$ be a trapezoid with $AB \| DC$ and let M and N be the midpoints of the segments AD and BC, respectively. Prove that*

(*i*) $MN \| AB$, *and*

(*ii*) $|MN| = \frac{1}{2}(|AB| + |DC|)$.

Exercise 2.31.11 (Midpoint connector theorem for trapezoids) *Let $ABCD$ be a trapezoid with $AB \| DC$ and let M be the midpoint of the segment AD. Prove that the parallel through M to AB intersects the segment BC at its midpoint.*

Exercise 2.31.12 *Show that the center of Euler's circle is the midpoint of the segment OH, where O and H denote the circumcenter and orthocenter, respectively.*

2.32 Intersection and collinearity

We shall introduce a few theorems, due to Menelaus, Ceva and Oppel, which can be used to prove intersection and collinearity of lines and points from the metric point of view. We shall discuss several remarkable results as applications of these important theorems.

2.32.1 The theorem of Menelaus

Consider a line ℓ which does not pass through any of the vertices of ΔABC. By Pasch's postulate, Theorem 1.7.4, if ℓ intersects one side of the triangle at an interior point, then ℓ will also intersect one more side at an interior point.

The following theorem is due to the Greek mathematician Menelaus (about 100 A.C.) and it gives a powerful technique to prove collinearity.

Theorem 2.32.1 (Menelaus) *Suppose that a line which does not pass through any of the vertices of ΔABC intersects the lines BC, CA, AB at the points A', B', C', respectively. Then the points A', B', C' are collinear if and only if*

$$\frac{|A'B|}{|A'C|} \cdot \frac{|B'C|}{|B'A|} \cdot \frac{|C'A|}{|C'B|} = 1. \tag{2.32.63}$$

Proof: We shall consider the following two configurations:

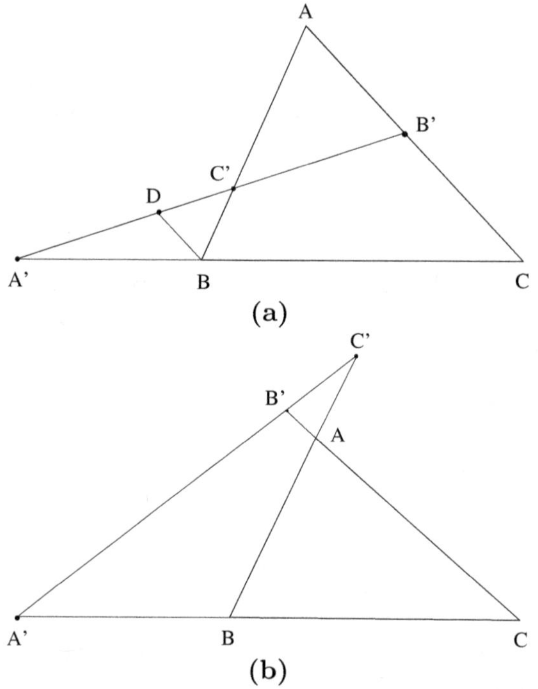

Figure 2.65: Configurations for Menelaus theorem.

(*i*) The configuration in Figure 2.65(a) in which exactly two of the points A', B', C' are situated in the interiors of sides and one point belongs to the exterior of a side.

(*ii*) The configuration in Figure 2.65(b) in which all three are exterior points belonging to the extensions of the sides.

We shall prove the theorem for the first configuration. The second is left to the reader.

" $\Longrightarrow$ " We consider $A'B'C'$ collinear and we shall show that (2.32.63) holds.
The line through B parallel to AC intersects $A'C'$ at a point D. Then the following pairs of triangles are similar:

$$\Delta A'BD \sim \Delta A'CB' \quad \text{and} \quad \Delta BDC' \sim \Delta AB'C'.$$

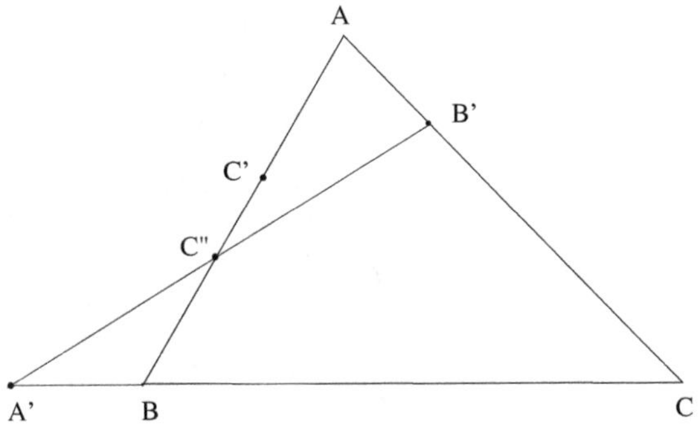

Figure 2.66: The uniqueness of point C.

Therefore we have

$$\frac{|A'B|}{|A'C|} = \frac{|BD|}{|CB'|} \quad \text{and} \quad \frac{|C'A|}{|C'B|} = \frac{|AB'|}{|BD|},$$

and hence

$$\begin{aligned} \frac{|A'B|}{|A'C|} \cdot \frac{|B'C|}{|B'A|} \cdot \frac{|C'A|}{|C'B|} &= \frac{|BD|}{|CB'|} \cdot \frac{|B'C|}{|B'A|} \cdot \frac{|AB'|}{|BD|} \\ &= \frac{|BD|}{|CB'|} \cdot \frac{|CB'|}{|B'A|} \cdot \frac{|B'A|}{|BD|} = 1. \end{aligned}$$

" $\Longleftarrow$ " We shall prove this by contradiction. Assume (2.32.63) is satisfied but the points A', B', C' are not collinear. Then let $C'' = A'B' \cap AB$. By Pasch's postulate it follows that C'' is an interior point of the segment AB. Using the implication " $\Longrightarrow$ " proved above, as A', B', C'' are collinear points, we have

$$\frac{|A'B|}{|A'C|} \cdot \frac{|B'C|}{|B'A|} \cdot \frac{|C''A|}{|C''B|} = 1.$$

On the other hand, by hypothesis

$$\frac{|A'B|}{|A'C|} \cdot \frac{|B'C|}{|B'A|} \cdot \frac{|C'A|}{|C'B|} = 1.$$

Dividing the last two relations yields

$$\begin{aligned} \frac{|C''A|}{|C''B|} = \frac{|C'A|}{|C'B|} \quad &\Longleftrightarrow \quad \frac{|C''A|}{|C''A| + |C''B|} = \frac{|C'A|}{|C'A| + |C'B|} \Longleftrightarrow \\ \frac{|C''A|}{|AB|} = \frac{|C'A|}{|AB|} \quad &\Longleftrightarrow \quad |C''A| = |C'A| \Longleftrightarrow C' = C'', \end{aligned}$$

and hence A', B', C' are collinear. ■

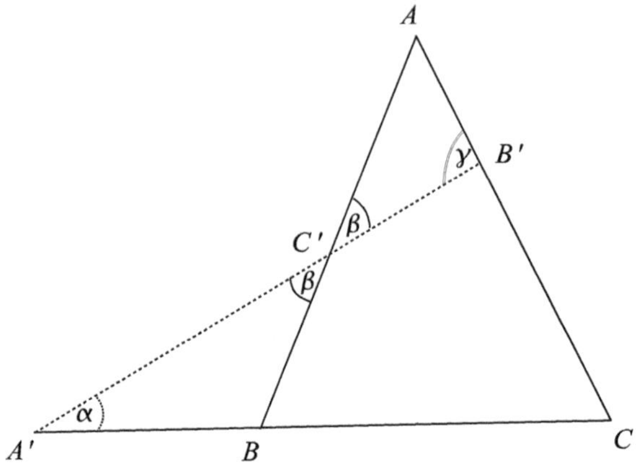

Figure 2.67: For the proof of Menelaus theorem using areas.

We note that the direct implication of Menelaus' theorem can also be proved using the concept of area.[4] We shall refer to Figure 2.67.

Considering the angle notations $\alpha = \widehat{C'A'B}$, $\beta = \widehat{BC'A'} = \widehat{AC'B'}$, and $\gamma = \widehat{AB'C'}$ we have

$$\frac{|A'B|}{|A'C|} \cdot \frac{|CB'|}{|B'A|} \cdot \frac{|AC'|}{|C'B|} = \frac{|A'B|}{|A'C|} \cdot \frac{|CB'|}{|B'A|} \cdot \frac{|AC'|}{|C'B|} \cdot \frac{|A'C'|}{|A'C'|} \cdot \frac{|C'B'|}{|C'B'|} \cdot \frac{|A'B'|}{|A'B'|}$$
$$= \frac{|A'B| \cdot |A'C'|}{|C'B| \cdot |A'C'|} \cdot \frac{|AC'| \cdot |C'B'|}{|B'C'| \cdot |AB'|} \cdot \frac{|CB'| \cdot |A'B'|}{|A'C| \cdot |A'B'|}$$
$$= \frac{|A'B| \cdot |A'C'| \cdot \sin\alpha}{|C'B| \cdot |A'C'| \cdot \sin\beta} \cdot \frac{|AC'| \cdot |C'B'| \cdot \sin\beta}{|B'C'| \cdot |AB'| \cdot \sin\gamma} \cdot \frac{|CB'| \cdot |A'B'| \cdot \sin(180^\circ - \gamma)}{|A'C| \cdot |A'B'| \cdot \sin\alpha}$$
$$= \frac{\mathcal{A}_{A'BC'}}{\mathcal{A}_{A'BC'}} \cdot \frac{\mathcal{A}_{AB'C'}}{\mathcal{A}_{AB'C'}} \cdot \frac{\mathcal{A}_{CA'B'}}{\mathcal{A}_{CA'B'}} = 1.$$

Exercise 2.32.2 *Using Menelaus theorem show that any two medians of a triangle intersect at* $2/3$ *of the vertices and* $1/3$ *from the bases.*

Exercise 2.32.3 *Consider a triangle* ΔABC *and the points* $B' \in [AC]$, $C' \in [AB]$ *and* $A' \in (CB$ *such that* $|B'C| = 2$, $|B'A| = 1$, $|C'A| = 2$, $|C'B| = 2$, $|BC| = 5$, $|A'B| = 5$, $|A'C| = 10$. *Show that the points* A', B' *and* C' *are collinear.*

[4]There are several results where a proof involving similarity can be replaced by a proof involving areas.

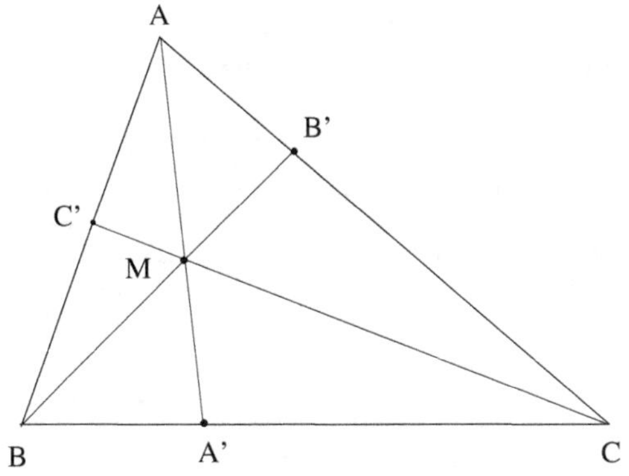

Figure 2.68: Van Aubel theorem.

2.32.2 Applications of the Menelaus theorem

The Van Aubel theorem and applications

Theorem 2.32.4 (Van Aubel) *Let AA', BB', CC' be cevians of ΔABC which intersect at a point $M \in Int\ ABC$. Then*

$$\frac{|AC'|}{|C'B|} + \frac{|AB'|}{|B'C|} = \frac{|AM|}{|MA'|}. \tag{2.32.64}$$

Proof: Applying the Menelaus theorem to triangles $\Delta ABA'$ and $\Delta AA'C$ cut by the lines $C'MC$ and BMB' respectively, yields

$$\frac{|CB|}{|CA'|} \cdot \frac{|MA'|}{|MA|} \cdot \frac{|C'A|}{|C'B|} = 1 \implies \frac{|C'A|}{|C'B|} = \frac{|MA|}{|MA'|} \cdot \frac{|CA'|}{|CB|}$$

$$\frac{|BC|}{|BA'|} \cdot \frac{|MA'|}{|MA|} \cdot \frac{|B'A|}{|B'C|} = 1 \implies \frac{|B'A|}{|B'C|} = \frac{|MA|}{|MA'|} \cdot \frac{|BA'|}{|BC|}.$$

Adding, we have

$$\begin{aligned} \frac{|C'A|}{|C'B|} + \frac{|B'A|}{|B'C|} &= \frac{|MA|}{|MA'|}\left(\frac{|CA'|}{|CB|} + \frac{|BA'|}{|BC|}\right) \\ &= \frac{|MA|}{|MA'|} \cdot \frac{|BC|}{|BC|} = \frac{|MA|}{|MA'|}, \end{aligned}$$

which is relation (2.32.64). ■

Corollary 2.32.5 *Let $M \in Int\, ABC$ be the intersection point of the cevians AA', BB' and CC'. Then the following inequality holds*

$$\frac{AM}{MA'} + \frac{BM}{MB'} + \frac{CM}{MC'} \geq 6.$$

Proof: Using Van Aubel's theorem for cevians AA', BB', CC' and the point M yields

$$\begin{aligned}
\frac{|AM|}{|MA'|} &= \frac{|AC'|}{|C'B|} + \frac{|AB'|}{|B'C|} \\
\frac{|BM|}{|MB'|} &= \frac{|BC'|}{|C'A|} + \frac{|BA'|}{|A'C|} \\
\frac{|CM|}{|MC'|} &= \frac{|CA'|}{|A'B|} + \frac{|CB'|}{|B'A|}.
\end{aligned}$$

Adding and applying Lemma 1.12.2 yields

$$\begin{aligned}
\frac{|AM|}{|MA'|} + \frac{|BM|}{|MB'|} + \frac{|CM|}{|MC'|} = {} & \underbrace{\left(\frac{|AC'|}{|C'B|} + \frac{|BC'|}{|C'A|}\right)}_{\geq 2} \\
& + \underbrace{\left(\frac{|AB'|}{|B'C|} + \frac{|CB'|}{|B'A|}\right)}_{\geq 2} + \underbrace{\left(\frac{|BA'|}{|A'C|} + \frac{|CA'|}{|A'B|}\right)}_{\geq 2} \geq 6.
\end{aligned}$$

■

Exercise 2.32.6 *Let $M \in Int\, ABC$. Prove that*

$$\frac{|AM|}{|MA'|} + \frac{|BM|}{|MB'|} + \frac{|CM|}{|MC'|} = 6$$

if and only if $M = G$, where G is the centroid of the triangle.

Solved Exercise 2.32.7 *Let M be a point in the interior of triangle ΔABC and denote by A', B', C' the intersections between AM, BM, CM with the sides BC, AC and AB, respectively. Prove that*

$$\frac{|MA'|}{|AA'|} + \frac{|MB'|}{|BB'|} + \frac{|MC'|}{|CC'|} = 1.$$

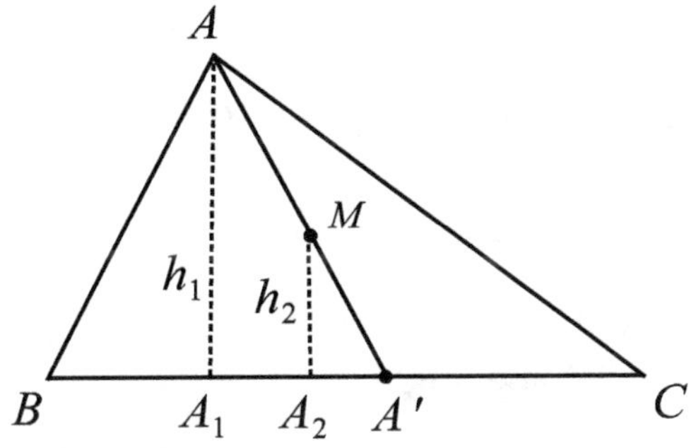

Figure 2.69: For Exercise 2.32.7.

Proof: Let A_1 and A_2 be the orthogonal projections of A and M on the side BC, see Figure 2.69. Using the triangles similarity $\Delta A'MA_2 \sim \Delta A'AA_1$ and the area definition, we obtain

$$\frac{|MA'|}{|AA'|} = \frac{|MA_2|}{|AA_1|} = \frac{h_2}{h_1} = \frac{ah_2/2}{ah_1/2} = \frac{\mathcal{A}_{BMC}}{\mathcal{A}_{ABC}}.$$

Similar reasons provide $\frac{|MB'|}{|BB'|} = \frac{\mathcal{A}_{AMC}}{\mathcal{A}_{ABC}}$ and $\frac{|MC'|}{|CC'|} = \frac{\mathcal{A}_{AMB}}{\mathcal{A}_{ABC}}$. Summing and using the area additivity yields

$$\frac{|MA'|}{|AA'|} + \frac{|MB'|}{|BB'|} + \frac{|MC'|}{|CC'|} = \frac{\mathcal{A}_{BMC} + \mathcal{A}_{AMC} + \mathcal{A}_{AMB}}{\mathcal{A}_{ABC}} = \frac{\mathcal{A}_{ABC}}{\mathcal{A}_{ABC}} = 1.$$

■

Solved Exercise 2.32.8 *Show that in any acute-angled triangle the following relation holds*

$$\cot\widehat{A}\cot\widehat{B} + \cot\widehat{B}\cot\widehat{C} + \cot\widehat{C}\cot\widehat{A} = 1.$$

Proof: Let $M = H$, the orthocenter. Since ΔABC is an acute-angled triangle, $H \in Int(ABC)$. Then we have equation (2.28.48), and

$$|AA'| = h_a = c\sin\widehat{B} = 2R\sin\widehat{C}\sin\widehat{B},$$

by the laws of sines. Then

$$\frac{|HA'|}{|AA'|} = \frac{\cos\widehat{B}\cos\widehat{C}}{\sin\widehat{C}\sin\widehat{B}} = \cot\widehat{B}\cot\widehat{C}$$

and two more analog relations. Applying Exercise 2.32.7 yields

$$\cot\widehat{B}\cot\widehat{C}+\cot\widehat{C}\cot\widehat{A}+\cot\widehat{A}\cot\widehat{B}=\frac{|HA'|}{|AA'|}+\frac{|HB'|}{|BB'|}+\frac{|HC'|}{|CC'|}=1.$$

■

Finding the length of the angle bisector

Another application of the Van Aubel theorem is to find the length of the angle bisector. Let $M = I$ be the incenter. Let $a = |BC|, b = |AC|, c = |AB|$. The Van Aubel theorem implies that $\frac{|AI|}{|IA'|} = \frac{|AC'|}{|C'B|} + \frac{|AB'|}{|B'C|}$. We rewrite $\frac{|AC'|}{|C'B|}$ and $\frac{|AB'|}{|B'C|}$ using the angle bisector theorem, Theorem 2.22.3, to get

$$\begin{aligned}\frac{|AI|}{|IA'|} &= \frac{b}{a}+\frac{c}{a}=\frac{b+c}{a} \Longleftrightarrow \frac{|AI|}{|AI|+|IA'|}=\frac{b+c}{a+b+c} \Longleftrightarrow \\ \frac{|AI|}{i_a} &= \frac{b+c}{2p},\end{aligned}$$

where $i_a = |AA'|$ is the length of the angle bisector. Hence

$$i_a = \frac{2p\cdot|AI|}{b+c}. \tag{2.32.65}$$

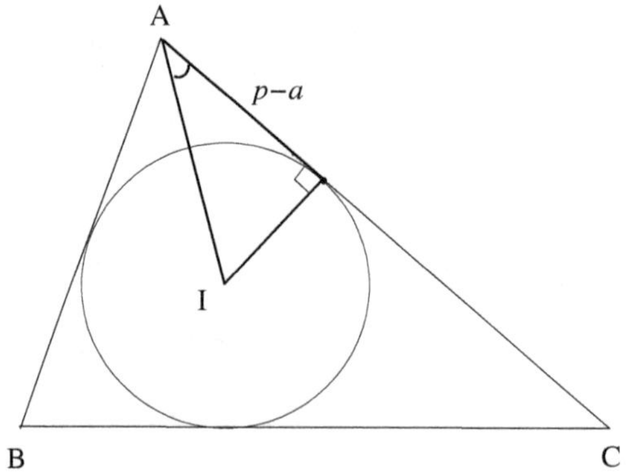

Figure 2.70: The angle bisector AI.

From Figure 2.70, we have

$$\cos\frac{\widehat{A}}{2} = \frac{p-a}{|AI|} \Longrightarrow |AI| = \frac{p-a}{\cos\frac{\widehat{A}}{2}}.$$

Substituting in (2.32.65) yields

$$i_a = \frac{2p}{b+c}\cdot\frac{p-a}{\cos\frac{\widehat{A}}{2}}. \tag{2.32.66}$$

Using $\cos\frac{\widehat{A}}{2} = \sqrt{\frac{p(p-a)}{bc}}$ from Theorem 2.16.4, the relation (2.32.66) becomes

$$i_a = \frac{2p(p-a)}{b+c}\sqrt{\frac{bc}{p(p-a)}} = \frac{2\sqrt{bc}}{b+c}\sqrt{p(p-a)},$$

which is the length of the angle bisector.

Exercise 2.32.9 *Show that for any acute triangle ΔABC the following inequality holds:*

$$\frac{\cos\widehat{A}}{\cos\widehat{B}\cos\widehat{C}} + \frac{\cos\widehat{B}}{\cos\widehat{A}\cos\widehat{C}} + \frac{\cos\widehat{C}}{\cos\widehat{A}\cos\widehat{B}} \geq 6.$$

Exercise 2.32.10 *Use Van Aubel's theorem to prove that the centroid G lies at 2/3 from the vertices and 1/3 from the base of each side of a triangle.*

Exercise 2.32.11 *Let K be the point of intersection of symmedians of a right triangle ΔABC, with $\widehat{A} = 90°$. Use Van Aubel's theorem to show that K is the midpoint of the symmedian from A.*

The Newton-Gauss line

The following lemma is necessary in order to prove the Newton-Gauss theorem.

Lemma 2.32.12 *Let AD be a cevian of triangle ΔABC with D between B and C. Let E, F and M be the midpoints of the segments AB, AC and AD, respectively. Then*

(i) The points E, M, F are collinear, and

(ii) $\frac{|EM|}{|EF|} = \frac{|BD|}{|BC|}$.

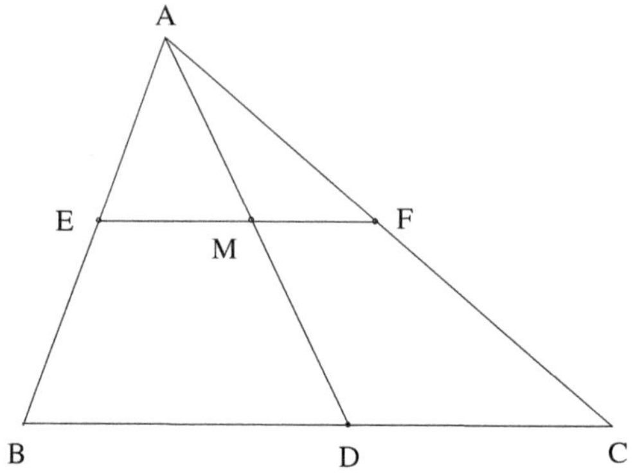

Figure 2.71: See Lemma 2.32.12.

Proof: (*i*) Using the midpoint connector theorem in triangles ΔABD and ΔADC implies that $EM \| BC$ and $MF \| BC$. The parallel postulate (uniqueness part) assures that the lines EM and MF are the same, i.e., E, M, F are collinear.
(*ii*) By the midpoint connector theorem we have

$$|EM| = \frac{1}{2}|BD| \text{ and } |EF| = \frac{1}{2}|BC|.$$

Dividing yields

$$\frac{|EM|}{|EF|} = \frac{|BD|}{|BC|}.$$

■

Theorem 2.32.13 (Newton-Gauss) *Let $ABCD$ be a quadrilateral such that the opposite sides are not parallel. Let $E = AB \cap CD$ and $F = AD \cap BC$. Then the midpoints of the segments AC, BD and EF are collinear.*

Proof: We shall assume $ABCD$ convex. Let M, N, P be the midpoints of the segments AC, BD and EF respectively. Let X, Y, Z be the midpoints of BC, CE and BE respectively. Applying Lemma 2.32.12 to ΔACE, ΔBEF and ΔBED yields the following triplets of collinear points

$$(M, X, Y), \qquad (P, Y, Z), \qquad (Z, X, N).$$

Hence M, N and P belong to the extensions of the sides of ΔXYZ. By the Menelaus theorem, the points M, N, P are collinear if and only if

$$\frac{|PY|}{|PZ|} \cdot \frac{|MX|}{|MY|} \cdot \frac{|NZ|}{|NX|} = 1. \tag{2.32.67}$$

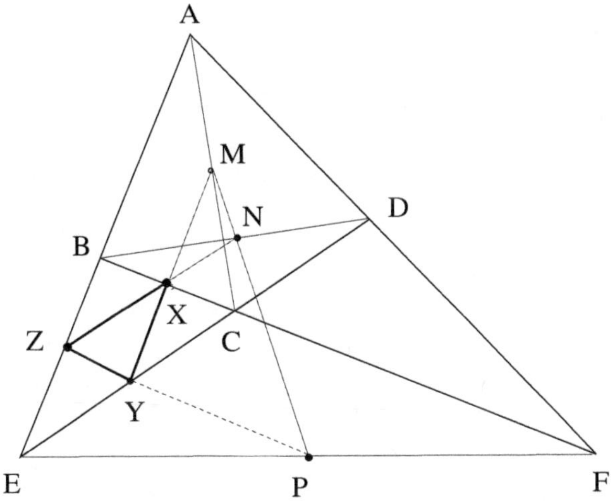

Figure 2.72: The Newton-Gauss line MNP.

Applying Lemma 2.32.12, part (*ii*) yields

$$\frac{|PY|}{|PZ|} = \frac{|FC|}{|FB|}, \qquad \frac{|MX|}{|MY|} = \frac{|AB|}{|AE|}, \qquad \frac{|NZ|}{|NX|} = \frac{|DE|}{|DC|}.$$

Substituting on the left side of (2.32.67) yields

$$\frac{|PY|}{|PZ|} \cdot \frac{|MX|}{|MY|} \cdot \frac{|NZ|}{|NX|} = \frac{|FC|}{|FB|} \cdot \frac{|AB|}{|AE|} \cdot \frac{|DE|}{|DC|} = 1,$$

by the Menelaus theorem applied to ΔBEC and line AF. Hence the identity (2.32.67) is proved and the points M, N, P are collinear. ∎

The line MNP is called the **Newton-Gauss line** of the quadrilateral $ABCD$.

Desargues' theorem

Theorem 2.32.14 (Desargues) *Let ABC and DEF be two triangles. Let $X = AC \cap DF$, $Y = AB \cap DE$ and $Z = BC \cap EF$ and $O = AD \cap CF$. The following two sentences are equivalent:*

i) The lines AD, BE and CF intersect at the point O.

ii) The points X, Y, Z are collinear.

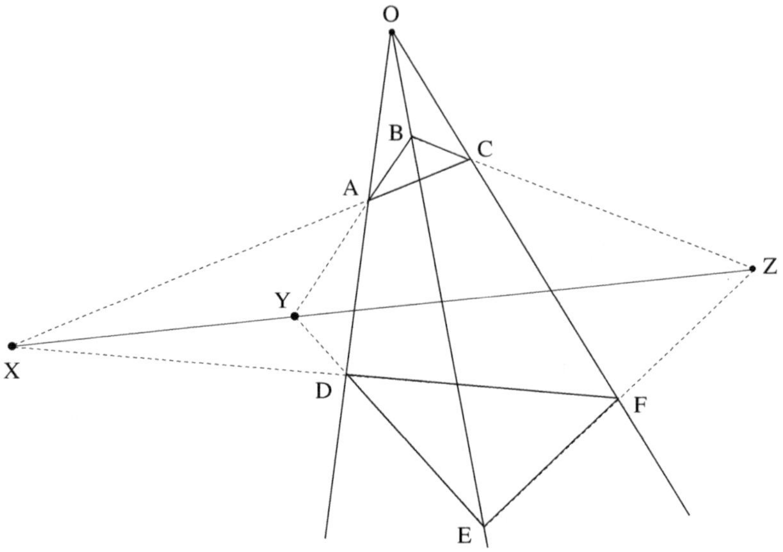

Figure 2.73: Desargues' theorem.

Proof: *i)* $\Longrightarrow$ *ii)* Menelaus theorem applied to triangles ΔOBC, ΔOAC, ΔOAB cut by the secants EFZ, XDF and respectively YDE yields

$$\frac{|ZB|}{|ZC|}\cdot\frac{|FC|}{|FO|}\cdot\frac{|EO|}{|EB|} = 1 \Longrightarrow \frac{|ZB|}{|ZC|} = \frac{|FO|}{|FC|}\cdot\frac{|EB|}{|EO|} \quad (2.32.68)$$

$$\frac{|XC|}{|XA|}\cdot\frac{|FO|}{|FC|}\cdot\frac{|DA|}{|DO|} = 1 \Longrightarrow \frac{|XC|}{|XA|} = \frac{|FC|}{|FO|}\cdot\frac{|DO|}{|DA|} \quad (2.32.69)$$

$$\frac{|YA|}{|YB|}\cdot\frac{|EB|}{|EO|}\cdot\frac{|DO|}{|DA|} = 1 \Longrightarrow \frac{|YA|}{|YB|} = \frac{|EO|}{|EB|}\cdot\frac{|DA|}{|DO|}. \quad (2.32.70)$$

Multiplying relations (2.32.68), (2.32.69), (2.32.70) term by term yields

$$\frac{|ZB|}{|ZC|}\cdot\frac{|XC|}{|XA|}\cdot\frac{|YA|}{|YB|} = \frac{|FO|}{|FC|}\frac{|EB|}{|EO|}\cdot\frac{FC}{|FO|}\frac{|DO|}{|DA|}\cdot\frac{|EO|}{|EB|}\frac{|DA|}{|DO|} = 1.$$

Applying the Menelaus theorem to ΔABC yields that X, Y, Z are collinear.

Before proving the second implication we shall adopt the following definitions.

The point O is called the **omology center** of ΔABC and ΔDEF and the line XYZ is called the **omology axis** of ΔABC and ΔDEF.

$i) \Longleftarrow ii)$ The point X is the omology center of ΔAYD and ΔCZF. The implication $i) \Longrightarrow ii)$ implies that the pairs of sides (AY, CZ), (AD, CF), (YD, FZ) intersect at three collinear points, which are B, O, E. Hence BE, AD, CF intersect at a point, which is O. ∎

Exercise 2.32.15 *What happens with Desargues' theorem in the case when the lines AD, BE, CF are parallel?*

Exercise 2.32.16 *Let $\Delta A_1B_1C_1$ be the orthic triangle of ΔABC. Suppose the following lines intersect: $BC \cap B_1C_1 = \{A''\}$, $AC \cap A_1C_1 = \{B''\}$, $AB \cap A_1B_1 = \{C''\}$. Prove that the points A'', B'', C'' are collinear.*

Exercise 2.32.17 *Let A', B', C' be the contact points where the incircle is tangent to the sides BC, AC and AB of ΔABC. Show that the pairs of lines $(AB, A'B')$, $(AC, A'C')$ and $(BC, B'C')$ intersect at three collinear points, provided none of the pairs are parallel lines.*

Pascal's theorem

Blaise Pascal started to work on geometry at the age of 12. He discovered that the sum of the angles of a triangle is 180 even before he read Euclid's work. At the age of 14, Blaise Pascal started to accompany his father to Mersenne's meetings where he met Desargues. At the age of 16, Pascal presented a paper at one of the meetings in June 1639. It contained a number of projective geometry theorems, including the property of cyclic hexagrams presented in the following.

Pascal is the inventor of the first digital calculator. This device was used in helping his father with his work collecting taxes. Blaise worked on it for 3 years between 1642 and 1645. The device resembled a mechanical calculator of the 1940s.

Theorem 2.32.18 (Pascal) *Let $\mathcal{C}$ be a circle and A, B, C, D, E, $F \in \mathcal{C}$. Let $M = AB \cap DE$, $N = EF \cap BC$ and $P = AF \cap CD$. Then the points M, N, P are collinear.*

Proof: We shall consider the Figure 2.74. Applying the Menelaus theorem

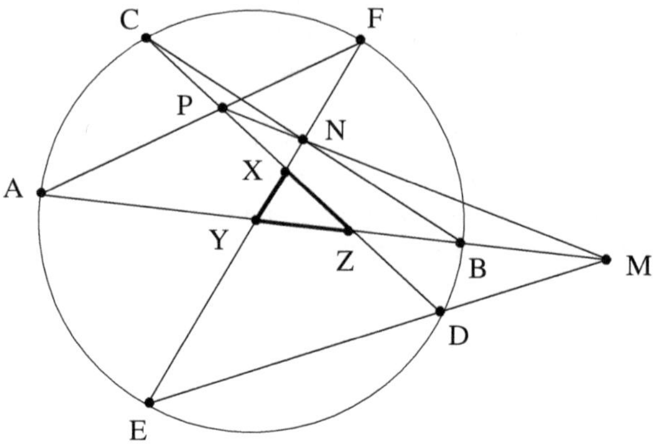

Figure 2.74: Pascal's hexagram.

to ΔXYZ cut by the secants EDM, APF and BNC yields

$$\frac{|MY|}{|MZ|}\cdot\frac{|DZ|}{|DX|}\cdot\frac{|EX|}{|EY|}=1 \implies \frac{|MY|}{|MZ|}=\frac{|DX|}{|DZ|}\cdot\frac{|EY|}{|EX|}$$

$$\frac{|AY|}{|AZ|}\cdot\frac{|PZ|}{|PX|}\cdot\frac{|FX|}{|FY|}=1 \implies \frac{|PZ|}{|PX|}=\frac{|AZ|}{|AY|}\cdot\frac{|FY|}{|FX|}$$

$$\frac{|BY|}{|BZ|}\cdot\frac{|NX|}{|NY|}\cdot\frac{|CZ|}{|CX|}=1 \implies \frac{|NX|}{|NY|}=\frac{|BZ|}{|BY|}\cdot\frac{|CX|}{|CZ|}.$$

Multiplying term by term yields

$$\begin{aligned}\frac{|MY|}{|MZ|}\cdot\frac{|PZ|}{|PX|}\cdot\frac{|NX|}{|NY|} &= \frac{|DX|}{|DZ|}\frac{|EY|}{|EX|}\cdot\frac{|AZ|}{|AY|}\frac{|FY|}{|FX|}\cdot\frac{|BZ|}{|BY|}\frac{|CX|}{|CZ|}\\ &= \frac{|DX|\cdot|CX|}{|FX|\cdot|EX|}\cdot\frac{|EY|\cdot|FY|}{|BY|\cdot|AY|}\cdot\frac{|AZ|\cdot|BZ|}{|DZ|\cdot|CZ|}\\ &= \frac{\text{Power}(X)\cdot\text{Power}(Y)\cdot\text{Power}(Z)}{\text{Power}(X)\cdot\text{Power}(Y)\cdot\text{Power}(Z)}=1.\end{aligned}$$

Applying the Menelaus theorem to ΔXYZ shows that M, N, P are collinear.

■

Corollary 2.32.19 *If the opposite sides of a cyclic hexagon are not parallel, then they intersect at three collinear points.*

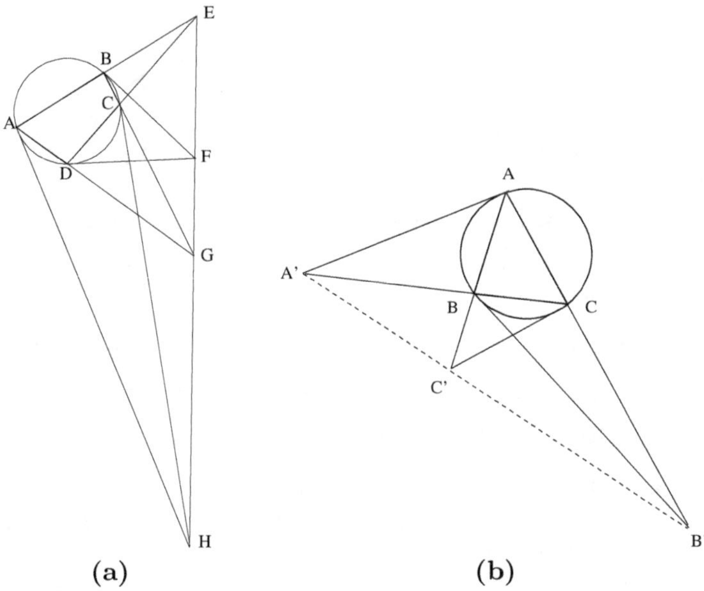

Figure 2.75: (a) See Corollary 2.32.20; (b) See Corollary 2.32.21.

Corollary 2.32.20 *Let Q be a cyclic quadrilateral. Then the opposite sides and the tangents to the circumcircle at the opposite vertices of Q intersect at four collinear points.*

Corollary 2.32.21 *The tangent lines to the circumcircle of a triangle at the vertices intersect the opposite sides at three collinear points.*

Exercise 2.32.22 *Prove Corollary 2.32.20. See Figure 2.75(a).*

Exercise 2.32.23 *Prove Corollary 2.32.21. See Figure 2.75(b).*

Remark 2.32.24 The French mathematician Jean Victor Poncelet (1788–1867) found a generalization of the Menelaus theorem for any polygon. For instance, in the case of a quadrilateral the theorem becomes:

Theorem 2.32.25 *Suppose that a line intersects the sides A_1A_2, A_2A_3, A_3A_4 and A_4A_1 of the quadrilateral $A_1A_2A_3A_4$ at points M_1, M_2, M_3, and M_4. Then*

$$\frac{|M_1A_1|}{|M_1A_2|} \cdot \frac{|M_2A_2|}{|M_2A_3|} \cdot \frac{|M_3A_3|}{|M_3A_4|} \cdot \frac{|M_4A_4|}{|M_4A_1|} = 1.$$

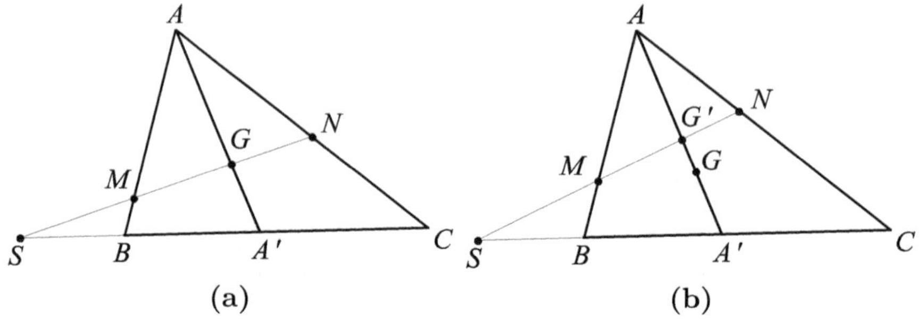

Figure 2.76: (a) The direct implication. (b) The converse implication.

Line passing through the centroid

Proposition 2.32.26 *Let ΔABC be a triangle with centroid G. A line ℓ intersects the sides AB and AC at M and N, respectively. Then the line ℓ passes through G if and only if*

$$\frac{|MB|}{|MA|} + \frac{|NC|}{|NA|} = 1. \tag{2.32.71}$$

Proof: " $\Longrightarrow$ " Let $G \in \ell$. We distinguish two cases: (i) $\ell \| BC$ and (ii) $\ell \nparallel BC$.
(i) If $\ell \| BC$ then $MN \| BC$ and by Thales' theorem we get

$$\frac{|MB|}{|MA|} = \frac{|NC|}{|NA|} = \frac{|GA'|}{|GA|} = \frac{1}{2},$$

so (2.32.71) obviously holds.
(ii) If $\ell \nparallel BC$ let $\ell \cap BC = \{S\}$, see Figure 2.76(a). Applying Menelaus theorem in $\Delta ABA'$ with transversal SMG yields

$$\frac{|SB|}{|SA'|} \cdot \frac{|GA'|}{|GA|} \cdot \frac{|MA|}{|MB|} = 1. \tag{2.32.72}$$

Also, Menelaus theorem applied in $\Delta AA'C$ with transversal SGN provides

$$\frac{|SA'|}{|SC|} \cdot \frac{|NC|}{|NA|} \cdot \frac{|AG|}{|GA'|} = 1. \tag{2.32.73}$$

Using that the centroid G verifies the relation $\dfrac{|AG|}{|GA'|} = 2$, relations (2.32.72) and (2.32.73) imply

$$\frac{|MB|}{|MA|} = \frac{|SB|}{2|SA'|}$$
$$\frac{|NC|}{|NA|} = \frac{|SC|}{2|SA'|}.$$

Adding we obtain

$$\frac{|MB|}{|MA|} + \frac{|NC|}{|NA|} = \frac{|SB| + |SC|}{2|SA'|} = 1,$$

where we used that A' is the midpoint of BC.
" $\Longleftarrow$ " We assume that relation (2.32.71) holds and we need to show $G \in \ell$. To this end, let $G' = MN \cap AA'$ and denote $\eta = \frac{|AG'|}{|G'A'|}$, see Figure 2.76(b). It suffices to prove that $\eta = 2$, fact that will imply $G' = G$.
A computation similar with the one done in the implication " $\Longrightarrow$ " provides

$$\frac{|MB|}{|MA|} = \frac{|SB|}{\eta|SA'|}, \qquad \frac{|NC|}{|NA|} = \frac{|SC|}{\eta|SA'|}.$$

Adding these relations and using (2.32.71) we obtain

$$1 = \frac{|MB|}{|MA|} + \frac{|NC|}{|NA|} = \frac{|SB| + |SC|}{\eta|SA'|} = \frac{2|SA'|}{\eta|SA'|} = \frac{2}{\eta},$$

which implies $\eta = 2$, and hence $G = G'$. ■

Corollary 2.32.27 *In the hypotheses of Proposition 2.32.26, the following inequality holds*

$$\frac{|AM|}{|MB|} + \frac{|AN|}{|NC|} > 3.$$

Proof: By Lemma 1.12.2 we have

$$\frac{|AM|}{|MB|} + \frac{|MB|}{|AM|} \geq 2, \qquad \frac{|AN|}{|NC|} + \frac{|NC|}{|AN|} \geq 2,$$

with both equalities reached for both M and N midpoints. Adding and regrouping terms, we have

$$\Big(\frac{|AM|}{|MB|} + \frac{|AN|}{|NC|}\Big) + \underbrace{\Big(\frac{|MB|}{|AM|} + \frac{|NC|}{|AN|}\Big)}_{=1} \geq 4,$$

which implies the desired inequality. We note that the inequality is always strict. Otherwise, both points M and N would be midpoints and then in this case $\frac{|MB|}{|AM|} + \frac{|NC|}{|AN|} = 2 \neq 1$, which is a contradiction. ■

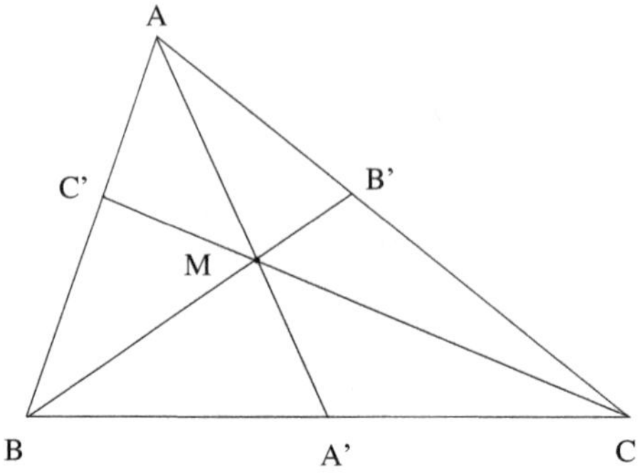

Figure 2.77: Ceva's theorem.

2.32.3 Ceva's theorem

Giovanni Ceva (1647–1734) was an Italian mathematician who worked on synthetic geometry and mechanics. In his honor, the segments which join a vertex with a point on the opposite side of a triangle are called **cevians**. The following result is known as Ceva's theorem and it gives a powerful method to prove intersection of cevians.

Theorem 2.32.28 (Ceva) *Let $A' \in BC$, $B' \in AC$ and $C' = AB$ be interior points of the sides of the triangle ΔABC. Then the cevians AA', BB', CC' intersect if and only if*

$$\frac{|A'B|}{|A'C|} \cdot \frac{|B'C|}{|B'A|} \cdot \frac{|C'A|}{|C'B|} = 1. \qquad (2.32.74)$$

Proof: "$\Longrightarrow$" Assume $AA' \cap BB' \cap CC' = \{M\}$.

The Menelaus theorem applied to $\Delta ABA'$ and $\Delta ACA'$ cut by the secants CMC' and BMB', respectively, yields

$$\frac{|CA'|}{|CB|} \cdot \frac{|C'B|}{|C'A|} \cdot \frac{|MA|}{|MA'|} = 1, \text{and}$$
$$\frac{|BC|}{|BA'|} \cdot \frac{|MA'|}{|MA|} \cdot \frac{|B'A|}{|B'C|} = 1.$$

Multiplying term by term yields

$$\begin{aligned} 1 &= \frac{|CA'|}{|CB|}\cdot\frac{|C'B|}{|C'A|}\cdot\frac{|MA|}{|MA'|}\cdot\frac{|BC|}{|BA'|}\cdot\frac{|MA'|}{|MA|}\cdot\frac{|B'A|}{|B'C|} \\ &= \frac{|CA'|}{|BA'|}\cdot\frac{|C'B|}{|C'A|}\cdot\frac{|B'A|}{|B'C|} \\ &= \left[\frac{|A'B|}{|A'C|}\cdot\frac{|B'C|}{|B'A|}\cdot\frac{|C'A|}{|C'B|}\right]^{-1}, \end{aligned}$$

which leads to (2.32.74).
" $\Longleftarrow$ " We shall use a proof by contradiction.
Applying crossbar theorem, the cevians BB' and CC' intersect at a point M'. Let $\{A''\} = AM' \cap BC$. The point A'' is between B and C by the crossbar theorem. Applying the implication " $\Longrightarrow$ " of the proof, we have

$$\frac{|A''B|}{|A''C|}\cdot\frac{|B'C|}{|B'A|}\cdot\frac{|C'A|}{|C'B|} = 1. \tag{2.32.75}$$

On the other hand, we have by hypothesis

$$\frac{|A'B|}{|A'C|}\cdot\frac{|B'C|}{|B'A|}\cdot\frac{|C'A|}{|C'B|} = 1. \tag{2.32.76}$$

Dividing (2.32.75) and (2.32.76) yields

$$\frac{|A''B|}{|A''C|} = \frac{|A'B|}{|A'C|},$$

which can be written as

$$\frac{|A''B|}{|A''B|+|A''C|} = \frac{|A'B|}{|A'B|+|A'C|} \Longleftrightarrow \frac{|A''B|}{|BC|} = \frac{|A'B|}{|BC|} \Longleftrightarrow A' = A''.$$

Hence AA' passes through M', and hence AA', BB', CC' intersect at a point. ■

Remark 2.32.29 If A', B', C' are not interior points of the sides BC, CA, AB then the implication " $\Longleftarrow$ " might fail. Consider A' between B and C and let B', C' be outside of AC and AB, respectively, such that $BB' \| AA' \| CC'$. Then the relation (2.32.74) holds but AA', BB', CC' do not intersect.

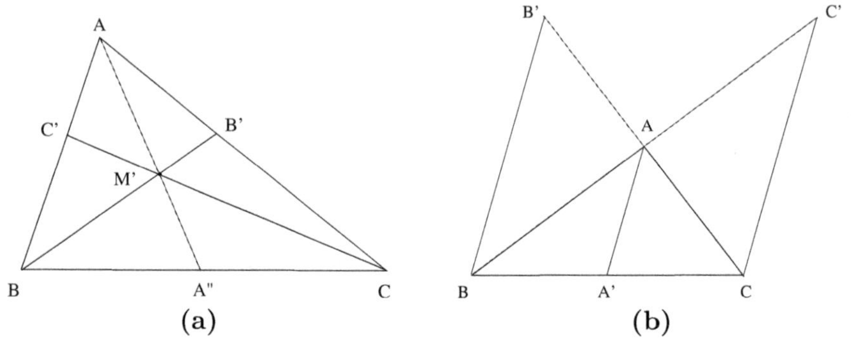

Figure 2.78: (a) Proof of Ceva's theorem; (b) Counterexample for Ceva's theorem.

In order to show this, let $\lambda = \frac{A'B}{A'C}$. By Thales' theorem, Theorem 2.11.1 applied to $\Delta BB'C$ we have

$$\frac{|B'C|}{|B'A|} = \frac{|BC|}{|BA'|} = \frac{|BA'| + |A'C|}{|BA'|} = 1 + \frac{|A'C|}{|BA'|} = 1 + \frac{1}{\lambda} = \frac{1+\lambda}{\lambda}.$$

On the other side we have

$$\frac{|C'A|}{|C'B|} = \frac{|CA'|}{|CB|} = \frac{|CA'|}{|CA'| + |A'B|} = \frac{1}{1 + \frac{|A'B|}{|CA'|}} = \frac{1}{1+\lambda}.$$

Multiplying yields

$$\frac{|A'B|}{|A'C|} \cdot \frac{|B'C|}{|B'A|} \cdot \frac{|C'A|}{|C'B|} = \lambda \cdot \frac{1+\lambda}{\lambda} \cdot \frac{1}{1+\lambda} = 1,$$

and hence (2.32.74) holds.

We shall next give a proof of the forward implication of Ceva's theorem using areas instead of Menelaus theorem. The proof is based on the following result.

Lemma 2.32.30 *Let AD be a cevian of the triangle ABC with D between B and C. Then*

$$\frac{|BD|}{|DC|} = \frac{\mathcal{A}_{ABD}}{\mathcal{A}_{ADC}}.$$

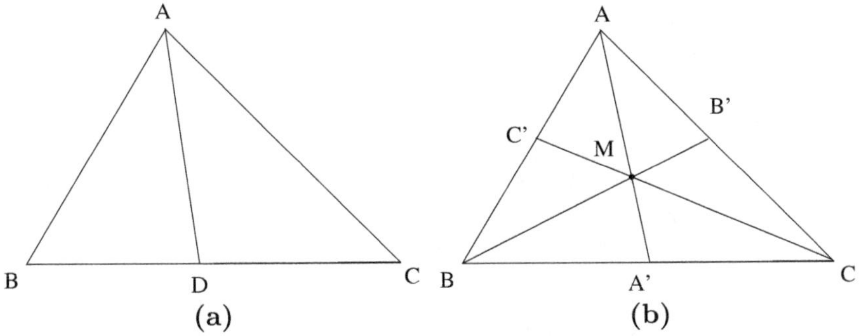

Figure 2.79: (a) Cevian AD; (b) Cevians intersecting at M.

Proof: Let h_a denote the length of the altitude from the vertex A. Using the definition of the area of a triangle, we have

$$\frac{|BD|}{|DC|} = \frac{\frac{1}{2}|BD| \cdot h_a}{\frac{1}{2}DC \cdot h_a} = \frac{\mathcal{A}_{ABD}}{\mathcal{A}_{ADC}}.$$

■

Proof of Ceva's theorem using areas: " $\Longrightarrow$ " We assume AA', BB' and CC' intersect at a point. Applying Lemma 2.32.30 to ΔABC and ΔMBC yields

$$\begin{aligned} \frac{|A'B|}{|A'C|} &= \frac{\mathcal{A}_{AA'B}}{\mathcal{A}_{AA'C}} = \frac{\mathcal{A}_{MA'B}}{\mathcal{A}_{MA'C}} \\ &= \frac{\mathcal{A}_{AA'B} - \mathcal{A}_{MA'B}}{\mathcal{A}_{AA'C} - \mathcal{A}_{MA'C}} = \frac{\mathcal{A}_{MAB}}{\mathcal{A}_{MAC}}. \end{aligned}$$

We have also used the fact that if $\frac{a}{b} = \frac{c}{d}$, then $\frac{a}{b} = \frac{c}{d} = \frac{a-c}{b-d}$, which is left for the reader as an exercise.
By cyclic permutations we arrive at the following identities

$$\frac{|A'B|}{|A'C|} = \frac{\mathcal{A}_{MAB}}{\mathcal{A}_{MAC}}, \qquad \frac{|B'C|}{|B'A|} = \frac{\mathcal{A}_{MBC}}{\mathcal{A}_{MAB}}, \qquad \frac{|C'A|}{|C'B|} = \frac{\mathcal{A}_{MAC}}{\mathcal{A}_{MBC}}.$$

Multiplying the above identities yields

$$\frac{|A'B|}{|A'C|} \cdot \frac{|B'C|}{|B'A|} \cdot \frac{|C'A|}{|C'B|} = \frac{\mathcal{A}_{MAB}}{\mathcal{A}_{MAC}} \cdot \frac{\mathcal{A}_{MBC}}{\mathcal{A}_{MAB}} \cdot \frac{\mathcal{A}_{MAC}}{\mathcal{A}_{MBC}} = 1.$$

" $\Longleftarrow$ " By contradiction, we assume that AA', BB' and CC' do not intersect at a point. Let $BB' \cap CC' = \{M\}$. Denote by A'' the point of

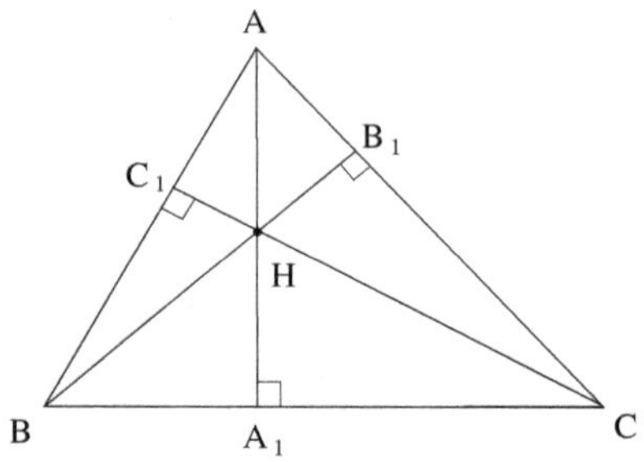

Figure 2.80: The altitudes intersecting at H.

intersection of cevian AM with the side BC. Since AA'', BB' and CC' intersect, applying part " $\Longrightarrow$ " we have

$$\frac{|A''B|}{|A''C|} \cdot \frac{|B'C|}{|B'A|} \cdot \frac{|C'A|}{|C'B|} = 1$$

Comparing to relation (2.32.74) yields

$$\frac{|A''B|}{|A''C|} = \frac{|A'B|}{|A'C|}.$$

This is equivalent to

$$\frac{|A''B|}{|A''C| + |A''B|} = \frac{|A'B|}{|A'C| + |A'B|} \Longleftrightarrow \frac{|A''B|}{|BC|} = \frac{|A'B|}{|BC|},$$

which implies $|A''B| = |A'B|$, so that $A' = A''$. Hence, AA' passes through the point M.

2.32.4 Applications of Ceva's theorem

Proposition 2.32.31 *The altitudes of a triangle intersect at a point.*

Proof: Let AA_1, BB_1, CC_1 be the altitudes of ΔABC. Then we have

$$\frac{|A_1B|}{|A_1C|} = \frac{c\cos\widehat{B}}{b\cos\widehat{C}}, \qquad \frac{|B_1C|}{|B_1A|} = \frac{a\cos\widehat{C}}{c\cos\widehat{A}}, \qquad \frac{|C_1A|}{|C_1B|} = \frac{b\cos\widehat{A}}{a\cos\widehat{B}}.$$

Multiplying term by term yields

$$\frac{|A_1B|}{|A_1C|} \cdot \frac{|B_1C|}{|B_1A|} \cdot \frac{|C_1A|}{|C_1B|} = \frac{c\cos\widehat{B}}{b\cos\widehat{C}} \cdot \frac{a\cos\widehat{C}}{c\cos\widehat{A}} \cdot \frac{b\cos\widehat{A}}{a\cos\widehat{B}} = 1.$$

The proposition now follows from Ceva's theorem. ∎

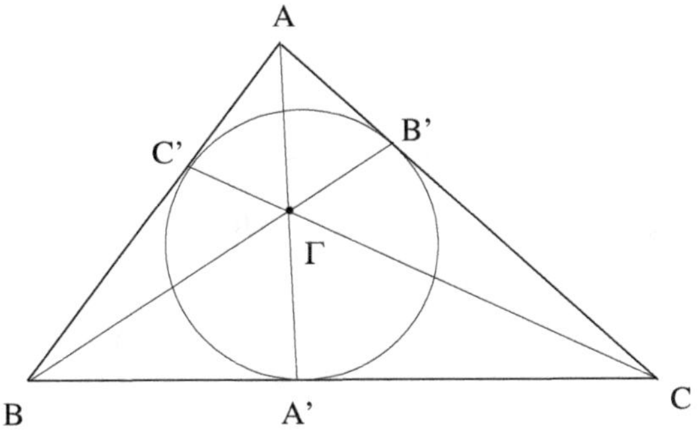

Figure 2.81: The Gergonne point Γ.

Exercise 2.32.32 *Using Ceva's theorem show that the medians of a triangle intersect.*

Exercise 2.32.33 *Using Ceva's theorem show that the angle bisectors of a triangle intersect.*

The next result is due to the French mathematician J. D. Gergonne (1771–1859).

Theorem 2.32.34 *Let A', B', C' be the contact points between the incircle and the sides BC, CA, AB of ΔABC. Then the cevians AA', BB', CC' intersect at a point Γ, called the Gergonne point.*

Proof: Using that the length of the tangents from an exterior point to the incircle are equal, we have $|BA'| = |BC'|$, $|A'C| = |CB'|$, $|AB'| = |AC'|$. Then

$$\frac{|A'B|}{|A'C|} \cdot \frac{|B'C|}{|B'A|} \cdot \frac{|C'A|}{|C'B|} = \frac{|A'B|}{|C'B|} \cdot \frac{|B'C|}{|CA'|} \cdot \frac{|C'A|}{|AB'|} = 1.$$

Applying Ceva's theorem yields the desired result. ■

Exercise 2.32.35 (Nagel) *Let A'', B'', C'' be the contact points between the excircles of ΔABC and the sides BC, CA, AB. Show that the cevians AA'', BB'', CC'' intersect.*

We next present a trigonometric variant of Ceva's theorem.

Proposition 2.32.36 *Let $A' \in BC$, $B' \in AC$ and $C' = AB$ be interior points of the sides of the triangle ΔABC. Then the cevians AA', BB', CC' intersect if and only if*

$$\frac{\sin \widehat{A_1}}{\sin \widehat{A_2}} \cdot \frac{\sin \widehat{B_1}}{\sin \widehat{B_2}} \cdot \frac{\sin \widehat{C_1}}{\sin \widehat{C_2}} = 1.$$

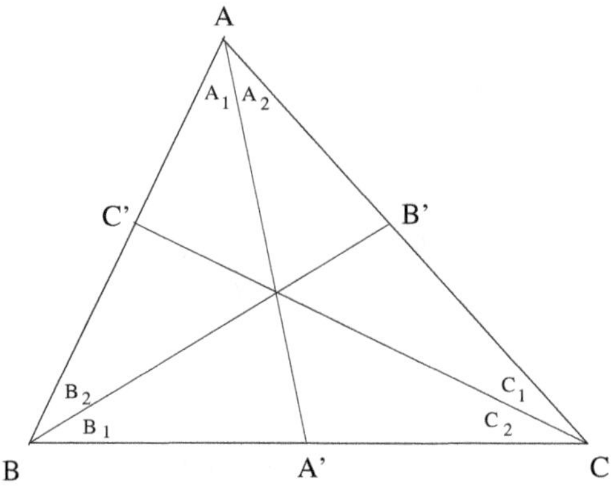

Figure 2.82: Ceva with angles.

Proof: Applying Lemma 2.32.30 yields

$$\frac{|A'B|}{|A'C|} = \frac{\mathcal{A}_{ABA'}}{\mathcal{A}_{ACA'}} = \frac{|AB|\cdot\sin\widehat{A_1}\cdot|AA'|}{|AC|\cdot\sin\widehat{A_2}\cdot|AA'|} = \frac{|AB|}{|AC|}\cdot\frac{\sin\widehat{A_1}}{\sin\widehat{A_2}}$$

$$\frac{|B'C|}{|B'A|} = \frac{\mathcal{A}_{BCB'}}{\mathcal{A}_{BAB'}} = \frac{|BC|\cdot\sin\widehat{B_1}\cdot|BB'|}{|BA|\cdot\sin\widehat{B_2}\cdot|BB'|} = \frac{|BC|}{|BA|}\cdot\frac{\sin\widehat{B_1}}{\sin\widehat{B_2}}$$

$$\frac{|C'A|}{|C'B|} = \frac{\mathcal{A}_{ACC'}}{\mathcal{A}_{BCC'}} = \frac{|AC|\cdot\sin\widehat{C_1}\cdot|CC'|}{|BC|\cdot\sin\widehat{C_2}\cdot|CC'|} = \frac{|AC|}{|BC|}\cdot\frac{\sin\widehat{C_1}}{\sin\widehat{C_2}}.$$

Multiplying term by term yields

$$\begin{aligned}\frac{|A'B|}{|A'C|}\cdot\frac{|B'C|}{|B'A|}\cdot\frac{|C'A|}{|C'B|} &= \frac{|AB|}{|AC|}\cdot\frac{|BC|}{|BA|}\cdot\frac{|AC|}{|BC|}\cdot\frac{\sin\widehat{A_1}}{\sin\widehat{A_2}}\cdot\frac{\sin\widehat{B_1}}{\sin\widehat{B_2}}\cdot\frac{\sin\widehat{C_1}}{\sin\widehat{C_2}}\\ &= \frac{\sin\widehat{A_1}}{\sin\widehat{A_2}}\cdot\frac{\sin\widehat{B_1}}{\sin\widehat{B_2}}\cdot\frac{\sin\widehat{C_1}}{\sin\widehat{C_2}}.\end{aligned}$$

This proposition now follows from Ceva's theorem. ∎

Isogonal points

Definition 2.32.37 *The lines AA', AA'' are called* **isogonal** *if $\widehat{BAA'} = \widehat{A''AC}$ and A', $A'' \in BC$ are both interior or both exterior points.*

Example 2.32.1 *The angle bisector is a self-isogonal cevian.*

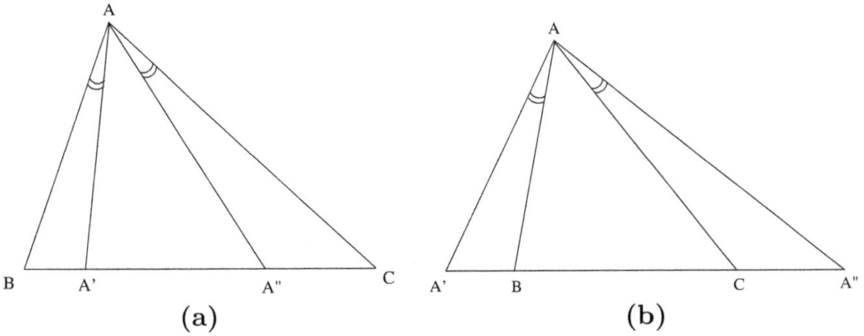

Figure 2.83: (a) Interior isogonals; (b) Exterior isogonals.

Example 2.32.2 *Let O and H be the circumcenter and the orthocenter, respectively, of ΔABC. Then AO, AH are isogonal (see Section 2.23).*

Proposition 2.32.38 *Let A', $A'' \in BC$, B', $B'' \in AC$, C', $C'' \in AB$ such that (AA', AA''), (BB', BB''), (CC', CC'') are isogonal pairs of cevians. Then AA', BB', CC' intersect if and only if AA'', BB'', CC'' intersect.*

Proof: By isogonality we have

$$\begin{aligned} \frac{\sin \widehat{BAA'}}{\sin \widehat{A'AC}} &= \frac{\sin \widehat{A''AC}}{\sin \widehat{BAA''}}, \\ \frac{\sin \widehat{CBB'}}{\sin \widehat{B'BA}} &= \frac{\sin \widehat{B''BA}}{\sin \widehat{CBB''}}, \text{ and} \\ \frac{\sin \widehat{ACC'}}{\sin \widehat{C'CB}} &= \frac{\sin \widehat{C''CB}}{\sin \widehat{ACC''}}. \end{aligned}$$

Multiplying yields

$$\frac{\sin \widehat{BAA'}}{\sin \widehat{A'AC}} \cdot \frac{\sin \widehat{CBB'}}{\sin \widehat{B'BA}} \cdot \frac{\sin \widehat{ACC'}}{\sin \widehat{C'CB}} = \frac{\sin \widehat{A''AC}}{\sin \widehat{BAA''}} \cdot \frac{\sin \widehat{B''BA}}{\sin \widehat{CBB''}} \cdot \frac{\sin \widehat{C''CB}}{\sin \widehat{ACC''}}.$$

Applying Proposition 2.32.36 yields

$$AA', BB', CC' \text{ intersect} \Longleftrightarrow AA'', BB'', CC'' \text{ intersect}.$$

■

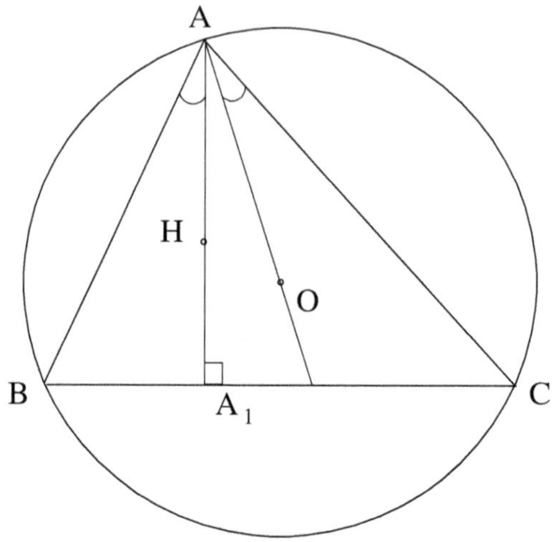

Figure 2.84: Isogonal cevians.

Definition 2.32.39 *Let $\{M'\} = AA' \cap BB' \cap CC'$ and $\{M''\} = AA'' \cap BB'' \cap CC''$, assuming that their intersection is not empty. The points M' and M'' are called* **isogonal conjugate points**.

Exercise 2.32.40 *Consider a triangle ΔABC in the plane $\mathcal{P}$. Let $\mathcal{J} : \mathcal{P} \to \mathcal{P}$ be the transform of the plane which assigns each point M' its isogonal conjugate point: $\mathcal{J}(M') = M''$. Let $I : \mathcal{P} \to \mathcal{P}$ be the identity function given by $I(M) = M$. Show the following:*

(i) $\mathcal{J}^2 = I$ *(involution),*

(ii) $\mathcal{J}(Int\ ABC) \subset Int\ ABC$,

(iii) $\mathcal{J}(O) = H, \mathcal{J}(H) = O$, *and*

(iv) $\mathcal{J}(I) = I$,

where O, H, I denote the circumcenter, orthocenter and incenter, respectively.

Exercise 2.32.41 *Show that the restriction of the isogonal transform $\mathcal{J}$: $Int\ ABC \to \mathcal{P}$ defined above is one-to-one. Is it onto?*

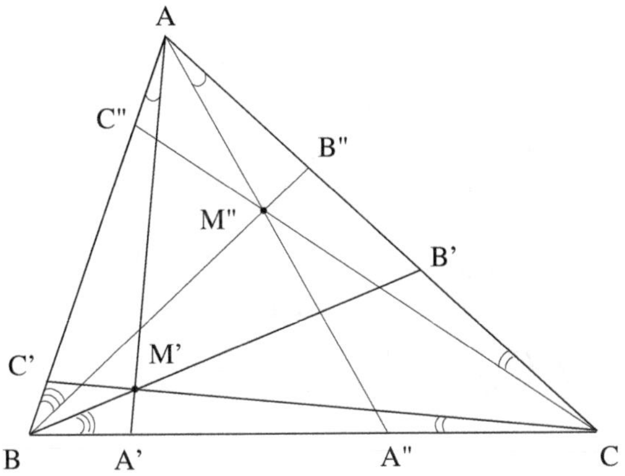

Figure 2.85: Isogonal points.

Lemoine's point

We recall that the symmedian is the isogonal cevian of the median (see Definition 2.22.7). As the medians of a triangle intersect at a point, the centroid G, Proposition 2.32.38 implies that the symmedians intersect at the isogonal point of G, called the **Lemoine's point** and denoted by K.

Remark 2.32.42 The isogonal points of O and H are H and O, respectively. One might think that the isogonal transform of the Euler's line OH is OH itself. However, the isogonal transform of the Euler line is an equilateral hyperbola circumscribed to ΔABC, passing through O and H.

An equivalent way of constructing the isogonal conjugate is given by the next result.

Theorem 2.32.43 *Let P and M be two isogonal conjugate points in triangle ΔABC. Let P_1, P_2 and P_3 be the symmetric points of P with respect to the sides BC, CA and AB of ΔABC. Then M is the circumcenter of triangle $\Delta P_1P_2P_3$.*

Proof: Since BP_3 and BP_1 are the symmetrics of BP with respect to BA and BC, respectively, we have

$$|BP| = |BP_3| = |BP_1|, \tag{2.32.77}$$

$$\widehat{P_3BA} = \widehat{ABP} = \alpha, \qquad \widehat{PBC} = \widehat{CBP_1} = x + \alpha \tag{2.32.78}$$

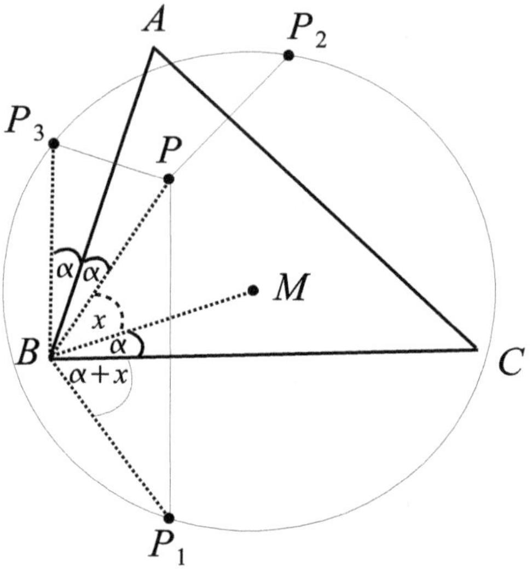

Figure 2.86: For Theorem 2.32.43.

where $x = \widehat{PBM}$, see Figure 2.86. Also, by isogonality, we have $\widehat{MBC} = \alpha = \widehat{PBA}$. Then we have the congruent triangles $\Delta BMP_3 \equiv \Delta BMP_1$ by the (SAS) congruence case as follows:

- BM is a common side
- $|BP_3| = |BP_1|$ by (2.32.77)
- $\widehat{P_3BM} = 2\alpha + x = \widehat{MBP_1}$ by (2.32.78).

Then $|MP_1| = |MP_3|$. By a similar procedure we obtain $|MP_1| = |MP_2|$. Therefore the point M is located at an equal distance from the points P_1, P_2 and P_3, and hence it is the circumcenter of $\Delta P_1P_2P_3$. ∎

Exercise 2.32.44 *Let H be the orthocenter of ΔABC. Use Theorem 2.32.43 to show that the circumcenter O is the isogonal conjugate of H.*

Exercise 2.32.45 (*i*) *Let O_A, O_B and O_C be the symmetric points of the circumcenter O of ΔABC with respect to the sides BC, AC and AB, respectively. Find the circumcenter of $\Delta O_AO_BO_C$.*

(*ii*) *Let G_A, G_B and G_C be the symmetric points of the centroid G of ΔABC with respect to the sides BC, AC and AB, respectively. Find the circumcenter of $\Delta G_AG_BG_C$.*

2.32.5 Isogonal conjugate points in normal coordinates

Let M be a point in the plane of triangle ΔABC and let's denote the distances from M to the sides by $x_a = d(M, BC)$, $x_b = d(M, AC)$, and $x_c = d(M, AB)$. We make the following sign convention: we consider the distance x_a positive if the point M belongs to the same half-plane determined by BC as the point A, and negative otherwise. We make the same sign convention for x_b and x_c. The numbers (x_a, x_b, x_c) are called the *normal coordinates* of M with respect to ΔABC.

Example 2.32.3 (*i*) *The vertices A, B and C have the normal coordinates $(h_a, 0, 0)$, $(0, h_b, 0)$ and $(0, 0, h_c)$, respectively, where h_a, h_b and h_c denote the lengths of the altitudes.*
(*ii*) *The incenter, I, of ΔABC has the normal coordinates (r, r, r), where r denotes the triangle inradius. The excenters I_a, I_b and I_c have the normal coordinates given respectively by $(-r_a, r_a, r_a)$, $(r_b, -r_b, r_b)$, and $(r_c, r_c, -r_c)$.*
(*iii*) *The circumcenter, O, of ΔABC has the normal coordinates*

$$(R\cos\widehat{A},\ R\cos\widehat{B},\ R\cos\widehat{C}).$$

(*iv*) *The orthocenter H of ΔABC has the normal coordinates*

$$(2R\cos\widehat{B}\cos\widehat{C},\ 2R\cos\widehat{A}\cos\widehat{C},\ 2R\cos\widehat{A}\cos\widehat{B}).$$

Exercise 2.32.46 (*i*) *Find the normal coordinates of the centroid of ΔABC.*
(*ii*) *Let O_9 denote the center of the Euler circle of ΔABC. Find the normal coordinates of O_9.*

Exercise 2.32.47 *Show that the sum of the normal coordinates of any point in the plane with respect to a given equilateral triangle is constant.*

The normal coordinates are not independent. Any one of the coordinates can be expressed in terms of the other two. This follows from the following linear constraint

$$ax_a + bx_b + cx_c = 2\mathcal{A}, \tag{2.32.79}$$

which is a consequence of the additivity property of areas. The proof of relation (2.32.79) has to be addressed in two cases: (*i*) the case $M \in Int(\Delta ABC)$; (*ii*) the case $M \notin Int(\Delta ABC)$.

For instance, in the case (*i*), splitting the area of ΔABC in three areas yields

$$\mathcal{A} = \mathcal{A}_{BMC} + \mathcal{A}_{AMC} + \mathcal{A}_{ABM} = \frac{1}{2}ax_a + \frac{1}{2}bx_b + \frac{1}{2}cx_c,$$

which is equivalent to the desired constraint. The proof in the case (*ii*) is using similar area additivity properties, taken into account that a negative normal coordinate corresponds to an area subtraction.

Consequently, using the area formula $\mathcal{A} = \frac{1}{2}ch_c$, relation (2.32.79) provides the dependency of one normal coordinate in terms of the others

$$x_c = h_c - \frac{a}{c}x_a - \frac{b}{c}x_b.$$

We are interested to study how normal coordinates transform under an isogonal inversion. For this we need the following result:

Lemma 2.32.48 *Let P and Q be two points on two isogonal lines with respect to the angle $\widehat{AOB}$. Let P_1, Q_1 and P_2, Q_2 be the orthogonal projections of P and Q on the lines OA and OB, respectively. Then*

$$\frac{|PP_1|}{|PP_2|} = \frac{|QQ_2|}{|QQ_1|}. \tag{2.32.80}$$

Proof: We consider the case when P and Q belong to the interior of angle $\widehat{AOB}$, see Figure 2.87. Let $\alpha = \widehat{AOP} = \widehat{QOB}$ and $x = \widehat{POQ}$. Then expressing $|PP_1|$, $|PP_2|$, $|QQ_1|$ and $|QQ_2|$ from the right triangles ΔPOP_1, ΔPOP_2, ΔQQ_1O and ΔQQ_2O yields

$$\begin{aligned}
|PP_1| &= |PO|\sin\alpha,\\
|PP_2| &= |PO|\sin(\alpha + x),\\
|QQ_1| &= |QO|\sin(\alpha + x), \text{and}\\
|QQ_2| &= |QO|\sin\alpha.
\end{aligned}$$

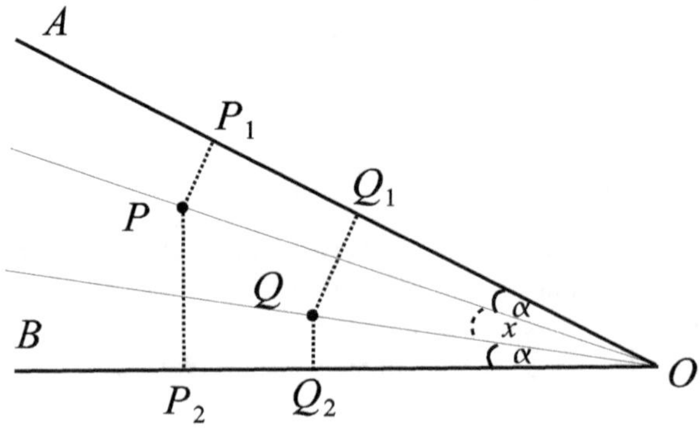

Figure 2.87: For Lemma 2.32.48.

Then

$$|PP_1| \cdot |QQ_1| = |PO| \cdot |QO| \sin\alpha \sin(\alpha + x) = |PP_2| \cdot |QQ_2|,$$

which implies the identity (2.32.80). ■

Consider two points, M and M', which are isogonal conjugate in triangle ΔABC, and let (x_a, x_b, x_c) and (x'_a, x'_b, x'_c) be their normal coordinates. Applying Lemma 2.32.48 three times, once for each angle of ΔABC, we obtain the following equal products of normal coordinates

$$k = x_a x'_a = x_b x'_b = x_c x'_c, \tag{2.32.81}$$

where $k = k(M, M')$ depends on the position of the pair (M, M') and is subject to be found. Therefore the normal coordinates of M' and M are related by

$$x'_a = \frac{k}{x_a}, \qquad x'_b = \frac{k}{x_b}, \qquad x'_c = \frac{k}{x_c}. \tag{2.32.82}$$

Substituting into the constraint (2.32.79)

$$ax'_a + bx'_b + cx'_c = 2\mathcal{A}$$

yields

$$k = \frac{2\mathcal{A}}{\dfrac{a}{x_a} + \dfrac{b}{x_b} + \dfrac{c}{x_c}} = \frac{ax_a + bx_b + cx_c}{\dfrac{a}{x_a} + \dfrac{b}{x_b} + \dfrac{c}{x_c}}. \tag{2.32.83}$$

Substituting into (2.32.82) we obtain the following equations of an isogonal inversion with respect to ΔABC

$$\begin{aligned}
x'_a &= \frac{1}{x_a} \frac{ax_a + bx_b + cx_c}{\dfrac{a}{x_a} + \dfrac{b}{x_b} + \dfrac{c}{x_c}} \\
x'_b &= \frac{1}{x_b} \frac{ax_a + bx_b + cx_c}{\dfrac{a}{x_a} + \dfrac{b}{x_b} + \dfrac{c}{x_c}} \\
x'_c &= \frac{1}{x_c} \frac{ax_a + bx_b + cx_c}{\dfrac{a}{x_a} + \dfrac{b}{x_b} + \dfrac{c}{x_c}}.
\end{aligned}$$

Remark 2.32.49 Due to the symmetry of relation (2.32.81), we can also express

$$k = \frac{2\mathcal{A}}{\dfrac{a}{x'_a} + \dfrac{b}{x'_b} + \dfrac{c}{x'_c}},$$

which combined with the previous expression of k yields the identity

$$\frac{a}{x_a} + \frac{b}{x_b} + \frac{c}{x_c} = \frac{a}{x'_a} + \frac{b}{x'_b} + \frac{c}{x'_c}. \qquad (2.32.84)$$

Example 2.32.4 *(i) If $M = I$, the incenter, then $M' = I$ is also the incenter, and using Example 2.32.3 we obtain $k = x_a x'_a = r^2$.*
(ii) If $M = H$, the orthocenter, then $M' = O$, the circumcenter, and Example 2.32.3 provides $x_a = 2R\cos\widehat{B}\cos\widehat{C}$ and $x'_a = R\cos\widehat{A}$. Therefore we have $k = 2R^2\cos\widehat{A}\cos\widehat{B}\cos\widehat{C}$.

Exercise 2.32.50 *Using that the Lemoine point K is the isogonal conjugate of the centroid, find the normal coordinates of K.*

Solved Exercise 2.32.51 *Show that the normal coordinates (x_a, x_b, x_c) of the circumcenter, O, satisfy*

$$x_a + x_b + x_c = R + r.$$

Proof:

$$x_a + x_b + x_c = R\cos\widehat{A} + R\cos\widehat{B} + R\cos\widehat{C} = R\left(\cos\widehat{A} + \cos\widehat{B} + \cos\widehat{C}\right)$$
$$= R\left(1 + 4\sin\frac{\widehat{A}}{2}\sin\frac{\widehat{B}}{2}\sin\frac{\widehat{C}}{2}\right) = R\left(1 + \frac{r}{R}\right) = R + r,$$

where we used Exercise 2.1.15. ■

Definition 2.32.52 *Two cevians AA', AA'' are* **isotomic** *if $|BA'| = |A''C|$ and A', A'' are both interior or both on the extension of segment BC.*

Exercise 2.32.53 *Let (AA', AA''), (BB', BB''), (CC', CC'') be three isotomic pairs of cevians with A', $A'' \in BC$, B', $B'' \in AC$, C', $C'' \in AB$. Prove that the cevians AA', BB', CC' intersect if and only if AA'', BB'', CC'' intersect.*

The harmonic conjugate

Now we shall present a fixed point property, which is a consequence of Menelaus' and Ceva's theorems, see Figure 2.88(b).

Proposition 2.32.54 *Let ABC be a triangle, and let $D \in BC$ be an interior point. For any point $M \in AD$ let $N = AC \cap BM$ and $P = CM \cap AB$. Then the line PN passes through a fixed point.*

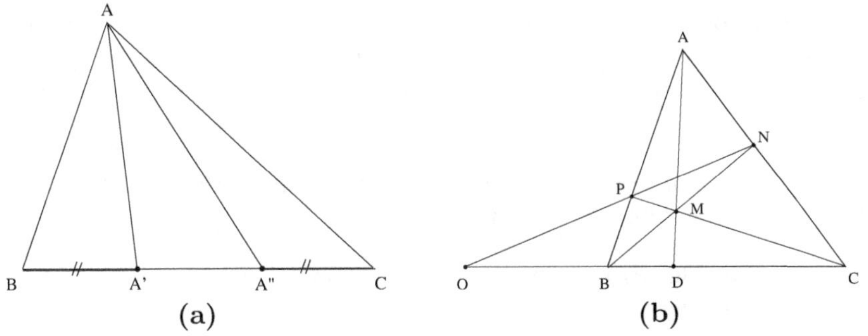

Figure 2.88: (a) Isotomic points; (b) Harmonic conjugate points.

Proof: Let $\{O\} = PN \cap BC$ for some fixed position of the point M. We shall show that O is the fixed point. It suffices to prove that the quotient $\frac{|OB|}{|OC|}$ does not depend on the position of the point M.
Applying the Menelaus theorem to ΔABC cut by the secant OPN yields

$$\frac{|OB|}{|OC|} \cdot \frac{|NC|}{|NA|} \cdot \frac{|PA|}{|PB|} = 1 \Longrightarrow \frac{|OB|}{|OC|} = \frac{|NA|}{|NC|} \cdot \frac{|PB|}{|PA|}. \tag{2.32.85}$$

Ceva's theorem applied to ΔABC yields

$$\frac{|DB|}{|DC|} \cdot \frac{|NC|}{|NA|} \cdot \frac{|PA|}{|PB|} = 1 \Longrightarrow \frac{|BD|}{|DC|} = \frac{|NA|}{|NC|} \cdot \frac{|PB|}{|PA|}. \tag{2.32.86}$$

Comparing (2.32.85) and (2.32.86) we get

$$\frac{|OB|}{|OC|} = \frac{|BD|}{|DC|}, \tag{2.32.87}$$

and hence $\frac{|OB|}{|OC|}$ does not depend of the position of M. ■

Remark 2.32.55 The point O is independent of A because of (2.32.87).

Definition 2.32.56 *The point O is called the* **harmonic conjugate** *of D with respect to the segment BC.*

Exercise 2.32.57 *Let A_1, A_2, A_3, A_4 be four collinear points in this order. Prove that A_1 is the harmonic conjugate of A_3 if and only if*

$$\frac{|A_1A_2|}{|A_1A_4|} : \frac{|A_2A_3|}{|A_3A_4|} = \frac{|A_1A_2|}{|A_1A_4|} \cdot \frac{|A_3A_4|}{|A_2A_3|} = 1.$$

Exercise 2.32.58 *Let H, O_9, G, O be the orthocenter, the center of the Euler circle, the centroid and the circumcenter of triangle ΔABC, respectively. Show that H is the harmonic conjugate of G with respect to O_9O.*

Exercise 2.32.59 *Let AA' and AA'' be the interior and the exterior angle bisector of the angle A of triangle ABC, respectively. Prove that A'' is the harmonic conjugate of A' with respect to BC.*

Exercise 2.32.60 *Prove that there is no point harmonic conjugate to the midpoint of the segment BC.*

2.32.6 Oppel's theorem and applications

This section provides a useful tool to prove the concurrence of three lines normal to the sides of a triangle.

Theorem 2.32.61 (Oppel) *Let A', B' and C' be points on the sides BC, CA and AB of ΔABC. Then the perpendiculars at A', B' and C' on the triangle sides are concurrent at a point if and only if*

$$|BA'|^2 + |CB'|^2 + |AC'|^2 = |A'C|^2 + |B'A|^2 + |C'B|^2. \qquad (2.32.88)$$

Proof: " $\Longrightarrow$ " Let M be the intersection point of the perpendiculars at A', B' and C' on the triangle sides, see Figure 2.89(a). The Pythagorean relation provides

$$\begin{aligned} |BA'|^2 &= |BM|^2 - |MA'|^2, \text{and} \\ |A'C|^2 &= |MC|^2 - |MA'|^2, \end{aligned}$$

which after subtraction yields

$$|BA'|^2 - |A'C|^2 = |BM|^2 - |MC|^2.$$

Similarly, we obtain two more relations

$$\begin{aligned} |CB'|^2 - |B'A|^2 &= |MC|^2 - |MA|^2, \text{and} \\ |AC'|^2 - |C'B|^2 &= |MA|^2 - |BM|^2. \end{aligned}$$

Adding the last three relations and canceling in pairs we obtain

$$\begin{aligned} &|BA'|^2 - |A'C|^2 + |CB'|^2 - |B'A|^2 + |AC'|^2 - |C'B|^2 \\ &\qquad = |BM|^2 - |MC|^2 + |MC|^2 - |MA|^2 + |MA|^2 - |BM|^2 = 0, \end{aligned}$$

which implies relation (2.32.88).

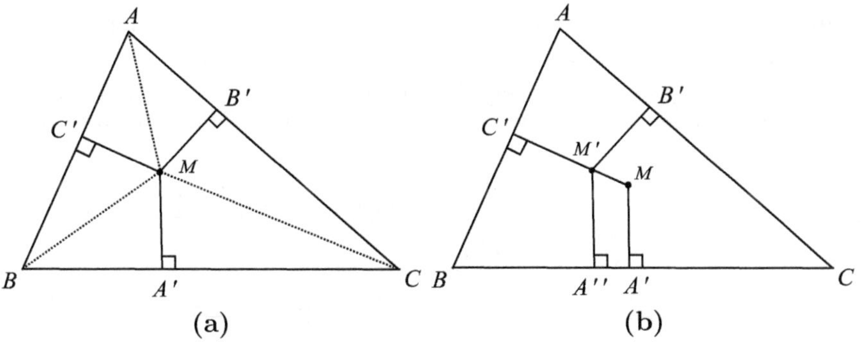

Figure 2.89: (a) For the direct implication of Oppel's theorem. (b) For the converse of Oppel's theorem.

" $\Longleftarrow$ " We consider relation (2.32.88) satisfied. By contradiction, we assume that the perpendiculars at A', B' and C' on the triangle sides are not concurrent at a point, see Figure 2.89(b). Let M be the intersection point of the perpendiculars at A' and C' and M' the intersection point of the perpendiculars at C' and B'. Let A'' be the orthogonal projection of M' on BC. Applying part " $\Longrightarrow$ " for both points M and M' yields

$$|BA'|^2 + |CB'|^2 + |AC'|^2 = |A'C|^2 + |B'A|^2 + |C'B|^2$$
$$|BA''|^2 + |CB'|^2 + |AC'|^2 = |A''C|^2 + |B'A|^2 + |C'B|^2.$$

Subtracting and canceling terms yields

$$\begin{aligned} |BA'|^2 - |BA''|^2 = |A'C|^2 - |A''C|^2 \Longleftrightarrow \\ |BA'|^2 - |A'C|^2 = |BA''|^2 - |A''C|^2 \end{aligned} \tag{2.32.89}$$

The rest of the proof is based on the configuration given by Figure 2.89(b). Factorizing the previous relation we have

$$(\underbrace{|BA'| + |A'C|}_{=|BC|})(|BA'| - |A'C|) = (\underbrace{|BA''| + |A''C|}_{=|BC|})(|BA''| - |A''C|)$$

and dividing by BC yields

$$\begin{aligned} |BA'| - |A'C| = |BA''| - |A''C| \Longleftrightarrow \\ |BA'| - |BA''| = |A'C| - |A''C| \Longleftrightarrow \\ |A''A'| = -|A''A'| \Longleftrightarrow 2|A''A'| = 0 \end{aligned}$$

so $A' = A''$, and hence $M = M'$. ■

Remark 2.32.62 There are several cases (when A' or A'' do not belong to the interior of the side BC), which are not covered by the previous proof. Actually, all cases can be considered simultaneously if we employ the use of oriented segments as follows. Relation (2.32.89) can be written in terms of inner product[5] as

$$\begin{aligned}
|BA'|^2 - |A'C|^2 = |BA''|^2 - |A''C|^2 &\iff \\
(\overrightarrow{BA'} + \overrightarrow{A'C}) \cdot (\overrightarrow{BA'} - \overrightarrow{A'C}) = (\overrightarrow{BA''} + \overrightarrow{A''C}) \cdot (\overrightarrow{BA''} - \overrightarrow{A''C}) &\iff \\
\overrightarrow{BC} \cdot (\overrightarrow{BA'} - \overrightarrow{A'C}) = \overrightarrow{BC} \cdot (\overrightarrow{BA''} - \overrightarrow{A''C}) &\iff \\
\overrightarrow{BC} \cdot (\overrightarrow{BA'} - \overrightarrow{A'C} - \overrightarrow{BA''} + \overrightarrow{A''C}) = 0 &\iff \\
\overrightarrow{BC} \cdot (\overrightarrow{A'B} + \overrightarrow{BA''} + \overrightarrow{A'C} + \overrightarrow{CA''}) = 0 &\iff \\
\overrightarrow{BC} \cdot (\overrightarrow{A'A''} + \overrightarrow{A'A''}) = 0 &\iff \\
\overrightarrow{BC} \cdot \overrightarrow{A'A''} = 0 \iff \epsilon |BC|\,|A'A''| = 0,&
\end{aligned}$$

with $\epsilon \in \{\pm 1\}$, from where $|A'A''| = 0$, which implies $A' = A''$. Hence, $M = M'$ and the proof is finished.

Remark 2.32.63 Sometimes it is useful to use a variant of Oppel's theorem that states relation (2.32.88) in terms of the midpoints A_0, B_0 and C_0 of the sides BC, AC and AB. We start with the evaluation of the difference of squares using inner product as follows

$$\begin{aligned}
|BA'|^2 - |A'C|^2 &= (\overrightarrow{BA'} + \overrightarrow{A'C})(\overrightarrow{BA'} - \overrightarrow{A'C}) \\
&= \overrightarrow{BC} \cdot (\overrightarrow{BA_0} + \overrightarrow{A_0A'} - \overrightarrow{A'A_0} - \overrightarrow{A_0C}) \\
&= \overrightarrow{BC} \cdot (\underbrace{\overrightarrow{BA_0} + \overrightarrow{CA_0}}_{=0} + 2\overrightarrow{A_0A'}) = 2\overrightarrow{BC} \cdot \overrightarrow{A_0A'}.
\end{aligned}$$

Then

$$|BA'|^2 - |A'C|^2 = 2\overrightarrow{BC} \cdot \overrightarrow{A_0A'}.$$

A similar computation leads to two more relations

$$|CB'|^2 - |B'A|^2 = 2\overrightarrow{CA} \cdot \overrightarrow{B_0B'}, \text{and}$$

$$|AC'|^2 - |C'B|^2 = 2\overrightarrow{AB} \cdot \overrightarrow{C_0C'}.$$

[5]The inner product of two collinear oriented segments is the product of their lengths times ϵ, where $\epsilon = +1$ if they have the same orientations and $\epsilon = -1$ if they have opposite orientations.

Adding the last three identities yields

$$|BA'|^2 - |A'C|^2 + |CB'|^2 - |B'A|^2 + |AC'|^2 - |C'B|^2$$
$$= 2(\overrightarrow{BC} \cdot \overrightarrow{A_0A'} + \overrightarrow{CA} \cdot \overrightarrow{B_0B'} + \overrightarrow{AB} \cdot \overrightarrow{C_0C'}).$$

Hence, relation (2.32.88) can be stated equivalently as

$$\overrightarrow{BC} \cdot \overrightarrow{A_0A'} + \overrightarrow{CA} \cdot \overrightarrow{B_0B'} + \overrightarrow{AB} \cdot \overrightarrow{C_0C'} = 0. \qquad (2.32.90)$$

Remark 2.32.64 The triangle $\Delta A'B'C'$ obtained by projecting the point M on the triangle sides is called the *pedal triangle* of M with respect to ΔABC. Thus, Oppel's theorem provides a necessary and sufficient condition for a triangle $\Delta A'B'C'$ to be a pedal triangle.

In the following we shall encounter a few remarkable cases.

1. The case $M = O$, the circumcenter. In this case $A' = A_0$, $B' = B_0$ and $C' = C_0$ are midpoints and relation (2.32.90) is obviously satisfied. Hence, the perpendicular bisectors of the sides intersect at the points O. Equivalently, the pedal triangle of the point O is the median triangle $\Delta A'B'C'$.

2. The case $M = I$, the incenter. Relation (2.32.90) holds since $|BA'| = |BC'|$, $|A'C| = |CB'|$ and $|AB'| = |AC'|$. We obtain that the perpendiculars at A', B' and C' intersect at the point I. Thus, the pedal triangle of the point I is the contact triangle $\Delta A'B'C'$.

3. The case $M = H$, the orthocenter. Since $|BA'| = c\cos\widehat{B}$ and $|A'C| = b\cos\widehat{C}$, so using the law of cosines we obtain

$$\begin{aligned}
|BA'|^2 - |A'C|^2 &= c^2\cos^2\widehat{B} - b^2\cos^2\widehat{C} \\
&= c^2\left(\frac{a^2+c^2-b^2}{2ac}\right)^2 - b^2\left(\frac{b^2+a^2-c^2}{2ba}\right)^2 \\
&= \frac{1}{4}\left\{\frac{(a^2+c^2-b^2)^2}{a^2} - \frac{(b^2+a^2-c^2)^2}{a^2}\right\} \\
&= \frac{4a^2(c^2-b^2)}{4a^2} = c^2 - b^2.
\end{aligned}$$

Then relation (2.32.90) is equivalent to

$$(c^2 - b^2) + (b^2 - a^2) + (a^2 - c^2) = 0,$$

which is obviously satisfied. Therefore, Oppel's theorem implies now that the altitudes are concurrent. Also, the pedal triangle of the orthocenter, H, is the orthic triangle.

The next result deals with a concurrency property related to two triangles symmetric with respect to an axis.

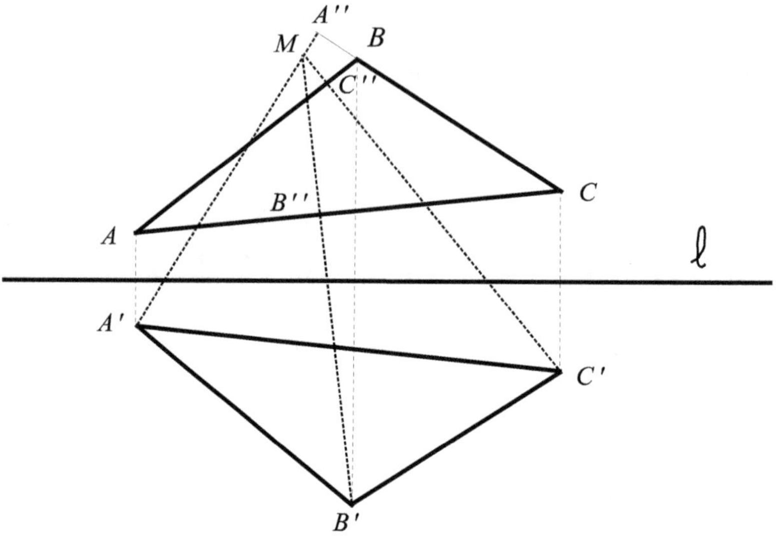

Figure 2.90: The lines $A'A''$, $B'B''$ and $C'C''$ intersect at the point M.

Proposition 2.32.65 *Let ΔABC and $\Delta A'B'C'$ be two triangles that are symmetric with respect to the line ℓ. Let A'', B'' and C'' be the orthogonal projections of A', B' and C' on the sides BC, CA and AB of ΔABC. Then the lines $A'A''$, $B'B''$ and $C'C''$ are intersecting at a point.*

Proof: First, using the symmetry with respect to ℓ we have that $ACC'A'$ is an isosceles trapezoid, so its diagonals are congruent, $|AC'| = |A'C|$, see Figure 2.90. Similarly, we have $|AB'| = |A'B|$ and $|BC'| = |B'C|$. An application of the Pythagorean theorem provides

$$\begin{aligned}
|BA''|^2 - |A''C|^2 &= (|BA'|^2 - |A'A''|^2) - (|A'C|^2 - |A'A''|^2) \\
&= |BA'|^2 - |A'C|^2, \\
|CB''|^2 - |B''A|^2 &= (|B'C|^2 - |B'B''|^2) - (|B'A|^2 - |B'B''|^2) \\
&= |B'C|^2 - |B'A|^2, \text{and} \\
|AC''|^2 - |C''B|^2 &= (|C'A|^2 - |C'C''|^2) - (|BC'|^2 - |CC'|^2) \\
&= |C'A|^2 - |BC'|^2.
\end{aligned}$$

Adding and using the aforementioned symmetry congruences, we cancel terms in pairs and obtain

$$\begin{aligned}|BA''|^2 - |A''C|^2 + |CB''|^2 - |B''A|^2 + |AC''|^2 - |C''B|^2 \\ = |BA'|^2 - |A'C|^2 + |B'C|^2 - |B'A|^2 + |C'A|^2 - |BC'|^2 = 0.\end{aligned}$$

By Theorem 2.32.61 the lines $A'A''$, $B'B''$ and $C'C''$ are concurrent. ∎

Remark 2.32.66 The aforementioned concurrency property is symmetric in the sense that if A_1, B_1 and C_1 are the orthogonal projections of A, B and C on the sides $B'C'$, $C'A'$ and $A'B'$, then the lines AA_1, BB_1 and CC_1 are also intersecting at a point. In general, this point is not the same as the point provided by Proposition 2.32.65. These two intersection points are symmetric with respect to the line ℓ.

The next application of the Oppel theorem deals with the orthopol of a triangle with respect to a line.

Proposition 2.32.67 *Let A', B', C' be the orthogonal projections of the vertices A, B, and C of ΔABC on a line ℓ. Let A'', B'', C'' be the orthogonal projections of the points A', B', and C' on the sides BC, CA and AB, respectively. Then $A'A''$, $B'B''$ and $C'C''$ are intersecting at a point (called the orthopol of ΔABC with respect to the line ℓ).*

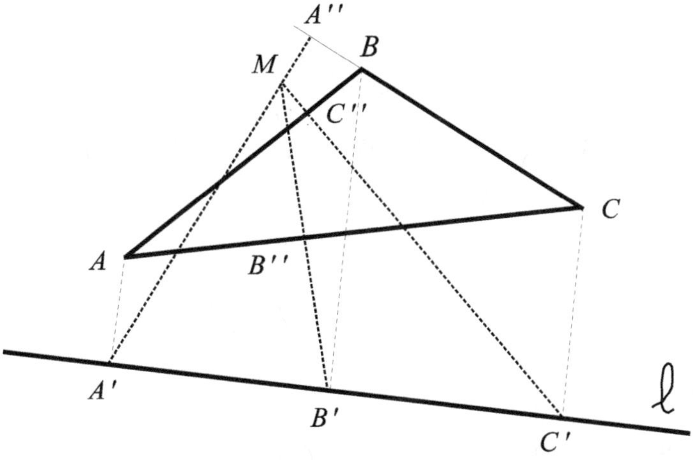

Figure 2.91: The lines $A'A''$, $B'B''$ and $C'C''$ intersect at the orthopol of ΔABC with respect to the line ℓ.

Proof: We shall refer to Figure 2.91. A repetitive application of the Pythagorean relation yields

$$|A''C|^2 = |A'C|^2 - |A'A''|^2 = |A'C'|^2 + |CC'|^2 - |A'A''|^2, \text{ and}$$
$$|A''B|^2 = |A'B|^2 - |A'A''|^2 = |A'B'|^2 + |BB'|^2 - |A'A''|^2.$$

Subtracting we obtain

$$|A''C|^2 - |A''B|^2 = |A'C'|^2 - |A'B'|^2 + |CC'|^2 - |BB'|^2.$$

We also have two more similar formulas

$$|C''B|^2 - |C''A|^2 = |C'B'|^2 - |C'A'|^2 + |BB'|^2 - |AA'|^2, \text{ and}$$

$$|B''A|^2 - |B''C|^2 = |B'A'|^2 - |B'C'|^2 + |AA'|^2 - |CC'|^2.$$

Adding the last three relations and canceling in pairs yields

$$|A''C|^2 - |A''B|^2 + |C''B|^2 - |C''A|^2 + |B''A|^2 - |B''C|^2 = 0.$$

By Oppel's theorem, Theorem 2.32.61, we obtain that $A'A''$, $B'B''$ and $C'C''$ are concurrent. ■

Exercise 2.32.68 *Where is the orthopol located in the case when the line ℓ is one of the triangle sides?*

The next application provides an equivalent way of constructing the isogonal inversion map.

Proposition 2.32.69 *Let A', B', C' be the orthogonal projections of a point M on the sides BC, CA and AB of ΔABC. The circle $\mathcal{C}(A', B', C')$ passing through A', B' and C' intersects the sides BC, CA and AB the second time at A'', B'' and C''. Then the perpendiculars on the sides at A'', B'' and C'' intersect at a point M'.*

Proof: In order to apply Oppel's theorem, it suffices to show that relation (2.32.90) holds, which in this case becomes

$$\overrightarrow{BC} \cdot \overrightarrow{A_0A''} + \overrightarrow{CA} \cdot \overrightarrow{B_0B''} + \overrightarrow{AB} \cdot \overrightarrow{C_0C''} = 0, \tag{2.32.91}$$

where A_0, B_0 and C_0 denote the midpoints of the sides, see Figure 2.92. Let A_1, B_1, C_1 be the midpoints of the chords $A'A''$, $B'B''$ and $C'C''$, respectively. Since the perpendiculars at the midpoints of the chords intersect at the center of the circle, applying (2.32.90) we have

$$\overrightarrow{BC} \cdot \overrightarrow{A_0A_1} + \overrightarrow{CA} \cdot \overrightarrow{B_0B_1} + \overrightarrow{AB} \cdot \overrightarrow{C_0C_1} = 0.$$

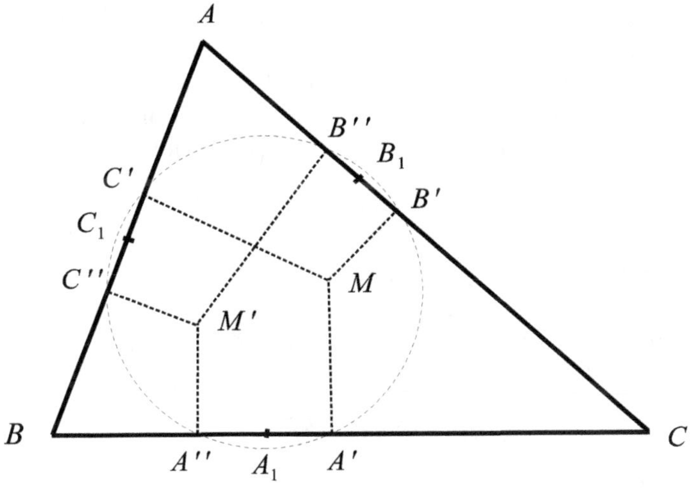

Figure 2.92: The points M and M' are isogonal conjugate.

Using the midpoint conditions $2\overrightarrow{A_0A_1} = \overrightarrow{A_0A'} + \overrightarrow{A_0A''}$, $2\overrightarrow{B_0B_1} = \overrightarrow{B_0B'} + \overrightarrow{B_0B''}$, $2\overrightarrow{C_0C_1} = \overrightarrow{C_0C'} + \overrightarrow{C_0C''}$, the previous identity becomes

$$\overrightarrow{BC} \cdot (\overrightarrow{A_0A'} + \overrightarrow{A_0A''}) + \overrightarrow{CA} \cdot (\overrightarrow{B_0B'} + \overrightarrow{B_0B''}) + \overrightarrow{AB} \cdot (\overrightarrow{C_0C'} + \overrightarrow{C_0C''}) = 0,$$

which after regrouping yields

$$\overrightarrow{BC} \cdot \overrightarrow{A_0A'} + \overrightarrow{CA} \cdot \overrightarrow{B_0B'} + \overrightarrow{AB} \cdot \overrightarrow{C_0C'} = \overrightarrow{BC} \cdot \overrightarrow{A_0A''} + \overrightarrow{CA} \cdot \overrightarrow{B_0B''} + \overrightarrow{AB} \cdot \overrightarrow{C_0C''}.$$

Since $\Delta A'B'C'$ is the pedal triangle of the point M, the left side of the previous expression vanishes, which implies that the right side also vanishes, fact that shows that (2.32.91) holds. ■

Remark 2.32.70 Proposition 2.32.69 is equivalently to: A circle intersects the sides BC, CA and AB of ΔABC at the pair of points (A', A''), $(B'B'')$ and (C', C''). Then $\Delta A'B'C'$ is a pedal triangle if and only if $\Delta A''B''C''$ is also a pedal triangle.

Corollary 2.32.71 *Assume the hypotheses of Proposition 2.32.69 are satisfied. Then the points M and M' are isogonal conjugate.*

Proof: Let M'' be the isogonal conjugate of M. By contradiction, we assume that $M'' \neq M'$. Then by Theorem 2.36.1 the orthogonal projections of M and M'' on the sides of ΔABC are 6 concyclic points. The circle that contains these points has three points in common with the circle passing through the points $\{A', B', C', A'', B'', C''\}$. Since a circle is unique determined by three points, these two circles coincide, so the projections of M'' on the sides are $\{A'', B'', C''\}$. Since by Theorem 2.32.91 the perpendiculars at A'', B'' and C'' intersect at M', it follows that $M' = M''$. ■

2.33 The isogon center

Definition 2.33.1 *Let T be an interior point of the acute-angled triangle ΔABC. The point T is called the isogon center of ΔABC if*

$$\widehat{ATB} = \widehat{BTC} = \widehat{CTA} = 120^\circ, \tag{2.33.92}$$

namely, each side is seen from T under the same angle.

Since the size of an object is influenced by the angle it subtends in the observer viewpoint, then an observer that is situated at the isogon center, T, perceives all sides of the triangle as equal in size. See Figure 2.93(a).

Proposition 2.33.2 *The isogon center is unique.*

Proof: We need to show that there is only one point T inside ΔABC such that relation (2.33.92) holds. We assume by contradiction, that there is another point T' inside ΔABC such that

$$\widehat{AT'B} = \widehat{BT'C} = \widehat{CT'A} = 120^\circ. \tag{2.33.93}$$

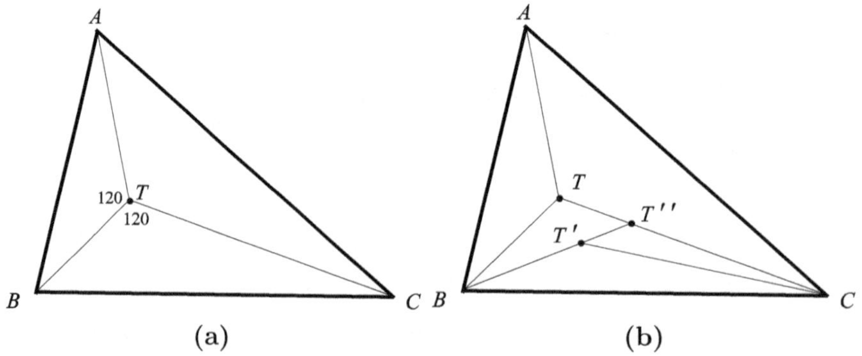

Figure 2.93: (a) The isogon center T; (b) Uniqueness of the isogon center.

Then the point T' belongs to the interior or to the sides of one of the triangles ΔABT, ΔBTC, or ΔATC.

For instance, if $T' \in \operatorname{Int} \Delta BTC$, then using the exterior angle inequality, Theorem 1.9.4, we have $\widehat{BT'C} > \widehat{BT''C} > \widehat{BTC} = 120°$, see Figure 2.93(b), which contradicts relation (2.33.93). We conclude that $T = T'$. ■

Proposition 2.33.3 *Let T be the isogon center of an acute-angled triangle ΔABC. Then*

$$|TA| + |TB| + |TC| = \left(\frac{1}{2}(a^2 + b^2 + c^2) + 2\sqrt{3}\mathcal{A}\right)^{1/2}, \qquad (2.33.94)$$

where $\mathcal{A}$ denotes the area of triangle ΔABC.

Proof: Instead of computing separately each of the distances $|TA|$, $|TB|$, and $|TC|$ and then summing them up, we shall compute the entire sum using van Schouten theorem as follows.

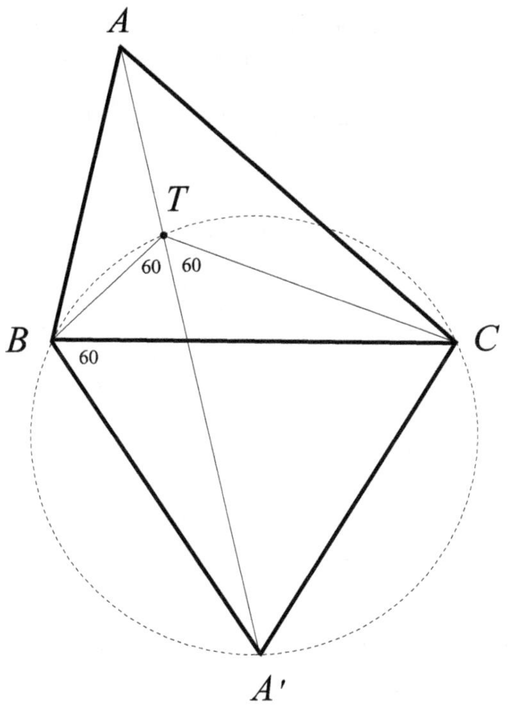

Figure 2.94: The intersection of AA' with the circle $\mathcal{C}(BCA')$ is the isogon center.

We start by constructing the equilateral triangle $\Delta BCA'$ on the exterior of the side BC, see Figure 2.94. Let T be the point where the segment AA' intersects the circumcircle of $\Delta BCA'$. We show that T is the isogon center by computing the angles with vertex at T

$$\begin{aligned}
\widehat{ATB} &= 180^\circ - \widehat{BTA'} = 180^\circ - \frac{1}{2}arc(BA') = 180^\circ - \widehat{BCA'} \\
&= 180^\circ - 60^\circ = 120^\circ, \\
\widehat{ATC} &= 180^\circ - \widehat{CTA'} = 180^\circ - \frac{1}{2}arc(CA') = 180^\circ - \widehat{CBA'} \\
&= 180^\circ - 60^\circ = 120^\circ, \text{and} \\
\widehat{BTC} &= \widehat{BTA'} + \widehat{CTA'} = 60^\circ + 60^\circ = 120^\circ.
\end{aligned}$$

By van Schouten theorem, Theorem 2.9.1, we have $|TB| + |TC| = |TA'|$. Therefore

$$|TA| + |TB| + |TC| = |TA| + |TA'| = |AA'|.$$

Then, it suffices to compute the length $|AA'|$. To this end, we apply the law of cosines in triangle $\Delta ABA'$ together with some trigonometric formulas

$$\begin{aligned}
|AA'|^2 &= |AB|^2 + |BA'|^2 - 2\,|AB|\,|BA'| \cos \widehat{ABA'} \\
&= c^2 + a^2 - 2ac\cos(\widehat{B} + 60^\circ) \\
&= c^2 + a^2 - 2ac\left(\frac{1}{2}\cos \widehat{B} - \frac{\sqrt{3}}{2}\sin \widehat{B}\right) \\
&= a^2 + c^2 - ac\frac{a^2 + c^2 - b^2}{2ac} + ac\sqrt{3}\sin \widehat{B} \\
&= \frac{1}{2}(a^2 + b^2 + c^2) + 2\sqrt{3}\,\frac{1}{2}ac\sin \widehat{B} \\
&= \frac{1}{2}(a^2 + b^2 + c^2) + 2\sqrt{3}\mathcal{A},
\end{aligned}$$

which proves the result. ∎

Exercise 2.33.4 *Let ΔABC be a right triangle, with $\widehat{A} = 90^\circ$.*

(i) Show that the triangle has an isogon center and denote it by T.

(ii) Find the sum $|TA| + |TB| + |TC|$ in terms of $|AB|$ and $|AC|$.

In the following we shall compute the distances from the isogon center, T, to vertices in terms of the elements of triangle ΔABC. Denoting $\alpha =$

$|TA|$, $\beta = |TB|$, and $\gamma = |TC|$, then the law of cosines applied to triangles ΔATB, ΔBTC and ΔATC yields

$$\frac{\alpha^2 + \beta^2 - c^2}{2\alpha\beta} = -\frac{1}{2}, \quad \frac{\alpha^2 + \gamma^2 - b^2}{2\alpha\gamma} = -\frac{1}{2}, \quad \frac{\gamma^2 + \beta^2 - a^2}{2\gamma\beta} = -\frac{1}{2}.$$

These relations can be written equivalently as

$$\begin{aligned} \alpha^2 + \beta^2 + \alpha\beta &= c^2, &(2.33.95)\\ \beta^2 + \gamma^2 + \beta\gamma &= a^2, \text{and} &(2.33.96)\\ \gamma^2 + \alpha^2 + \alpha\gamma &= b^2. &(2.33.97) \end{aligned}$$

Together with $\delta = \alpha + \beta + \gamma$, they form four equations that can be solved for α, β and γ. To this end, we subtract the equations (2.33.95)–(2.33.97) two by two to obtain formulas for the differences as

$$\alpha - \gamma = \frac{c^2 - a^2}{\delta}, \quad \gamma - \beta = \frac{b^2 - c^2}{\delta}, \quad \beta - \alpha = \frac{a^2 - b^2}{\delta}.$$

Now we write the relation $\alpha + \gamma + \beta = \delta$ under the equivalent form

$$(\alpha - \gamma) + 2(\gamma - \beta) + 3\beta = \delta$$

and use the above differences formulas to get

$$3\beta = \delta + \frac{a^2 + c^2 - 2b^2}{\delta} = \delta + \frac{a^2 + b^2 + c^2 - 3b^2}{\delta},$$

so

$$\beta = \frac{1}{3}\delta + \frac{a^2 + b^2 + c^2}{3\delta} - \frac{b^2}{\delta}.$$

A similar reasoning leads to

$$\alpha = \frac{1}{3}\delta + \frac{a^2 + b^2 + c^2}{3\delta} - \frac{a^2}{\delta}, \quad \gamma = \frac{1}{3}\delta + \frac{a^2 + b^2 + c^2}{3\delta} - \frac{c^2}{\delta},$$

where $\delta = \left(\frac{1}{2}(a^2 + b^2 + c^2) + 2\sqrt{3}\mathcal{A}\right)^{1/2}$ by equation (2.33.94). The triplet (α, β, γ) determine uniquely the point T inside the triangle ΔABC.

The isogon center is also called the *Torricelli point* of triangle ΔABC. The reason is the following result.

Theorem 2.33.5 (Torricelli circles) *Let ΔABC be an acute-angled triangle and construct in exterior the equilateral triangles $\Delta BCA'$, $\Delta AB'C$ and $\Delta ABC'$.*

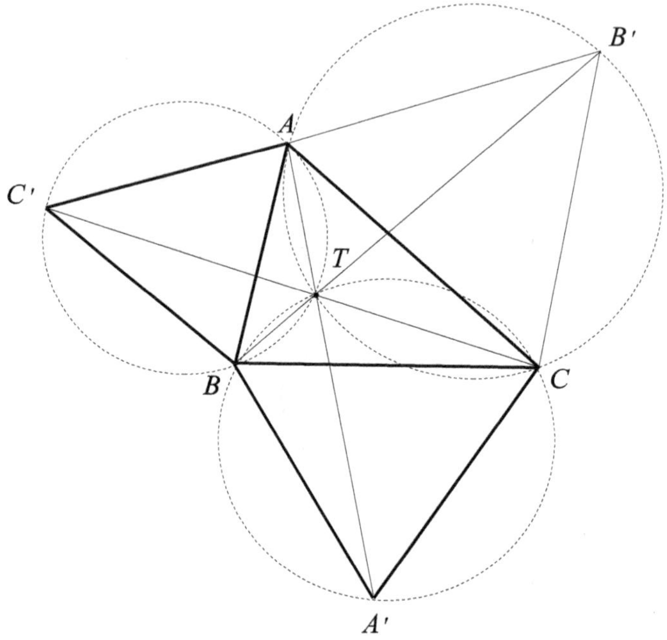

Figure 2.95: Torricelli circles and the isogon point.

1) *The circumcircles of triangles $\Delta BCA'$, $\Delta AB'C$ are passing through the isogon center, T.*

2) *The lines AA', BB' and CC' intersect at the isogon center, T.*

3) *The segments $|AA'|$, $|BB'|$ and $|CC'|$ are equal.*

Proof: We shall show the parts 1) and 2) simultaneously. In the proof of Proposition 2.33.3, we showed that the segment AA' intersects the circumcircle of $\Delta BCA'$ at the isogon center, T. Since the same argument can be applied for the segments BB', CC' and the circumcircles of $\Delta AB'C$ and $\Delta ABC'$, respectively, then using the uniqueness of the isogon center, it follows that T is the intersection point of AA', BB' and CC' as well as the intersection of the aforementioned circles, see Figure 2.95.
3) From the previous proof we have $|AA'| = |TA| + |TB| + |TC|$. Since similar arguments apply for $|BB'|$ and $|CC'|$, we obtain $|AA'| = |BB'| = |CC'|$. ∎

The following result produces the solution for the Fermat's problem that consists of finding the point which minimizes the sum of distances between the point and the triangle vertices.

Theorem 2.33.6 *Let ΔABC be an acute-angled triangle and M be an interior point. Then the point M for which the sum*

$$|MA| + |MB| + |MC| \tag{2.33.98}$$

is minimum is the isogon center of the triangle ΔABC.

Proof: The proof is a consequence of the triangle inequality as follows. Since $\Delta BCA'$ is equilateral, by Pompeiu's inequality, Proposition 2.9.2, for any point M in the plane we have

$$|MB| + |MC| \geq |MA'|,$$

Then applying the triangle inequality in $\Delta AMA'$ we have

$$|MA| + |MB| + |MC| \geq |MA| + |MA'| \geq |AA'|. \tag{2.33.99}$$

Since $|AA'|$ does not depend on M, it follows that the minimum of the sum (2.33.98) is obtained when both inequalities in (2.33.99) become identities:

(i) The first one occurs when M belongs to the circumcircle of $\Delta BCA'$, via van Schouten theorem, Theorem 2.9.1.

(ii) The second inequality holds when M belongs to AA'.

Hence, both inequalities occur when M is located the intersection between AA' and the circumcircle, ie., M is the isogon center. ∎

The physical interpretation of the minimum property of the isogon center consists of Plateau's experiment: If some vertical pegs are placed at the vertices A, B and C and the triangle is immersed in soap water, then a minimal soap surface will get formed joining the vertices with the point T. From this point of intersection of the soap films each side can be seen under an angle of 120 degrees.

The analytic solution: For comparison reasons, we shall attempt to show that the minimality of the sum (2.33.98) is reached for $M = T$ using the method of Lagrange multipliers. To this end, we assume that the sum (2.33.98) is minimum and show that $\widehat{AMB} = \widehat{BMC} = \widehat{AMC}$.

We shall use the following notations: $\alpha = |AM|$, $\beta = |BM|$, $\gamma = |CM|$ and $\omega_a = \widehat{BMC}$, $\omega_b = \widehat{AMC}$, $\omega_c = \widehat{AMB}$.

The sum (2.33.98) is computed using the binomial formula

$$(\alpha + \beta + \gamma)^2 = \alpha^2 + \beta^2 + \gamma^2 + 2\alpha\beta + 2\beta\gamma + 2\gamma\alpha. \tag{2.33.100}$$

The sum of squares follows from a cyclic application of the law of cosines in triangles ΔABM, ΔBMC, ΔCMA as follows

$$\alpha^2 + \beta^2 = c^2 + 2\alpha\beta \cos \omega_c,$$

$$\beta^2 + \gamma^2 = a^2 + 2\beta\gamma \cos \omega_a, \text{ and}$$

$$\gamma^2 + \alpha^2 = b^2 + 2\alpha\gamma \cos \omega_b.$$

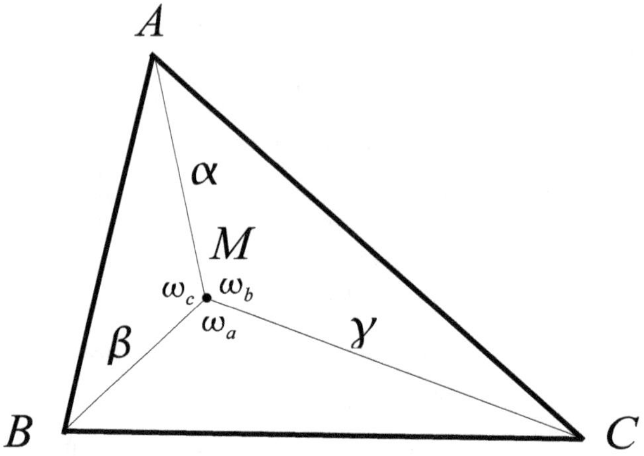

Figure 2.96: Angles and distances induced by the point M.

Taking the sum yields

$$\alpha^2 + \beta^2 + \gamma^2 = \frac{1}{2}(a^2 + b^2 + c^2) + \alpha\beta\cos\omega_c + \beta\gamma\cos\omega_a + \alpha\gamma\cos\omega_b.$$

Substituting in (2.33.100) we obtain

$$(\alpha+\beta+\gamma)^2 = \frac{a^2 + b^2 + c^2}{2} + \alpha\beta(2+\cos\omega_c) + \beta\gamma(2+\cos\omega_a) + \alpha\gamma(2+\cos\omega_b).$$

Consequently, the minimum of $\alpha + \beta + \gamma$ is reached simultaneously with the minimum of the expression

$$F(\alpha, \beta, \gamma) = \alpha\beta(2 + \cos\omega_c) + \beta\gamma(2 + \cos\omega_a) + \alpha\gamma(2 + \cos\omega_b).$$

Since the area of ΔABC can be split into the areas of ΔATB, ΔBTC, and ΔATC, then α, β and γ satisfy the constraint

$$\beta\gamma\sin\omega_a + \alpha\gamma\sin\omega_b + \alpha\beta\sin\omega_c = 2\mathcal{A}.$$

Adding the constraint to the function F, we obtain the Lagrangian function with multiplier λ

$$L = F(\alpha, \beta, \gamma) + \lambda(\beta\gamma\sin\omega_a + \alpha\gamma\sin\omega_b + \alpha\beta\sin\omega_c).$$

The minimum condition implies

$$\frac{\partial L}{\partial \alpha} = \frac{\partial L}{\partial \beta} = \frac{\partial L}{\partial \gamma} = 0,$$

which becomes

$$\beta(2+\cos\omega_c+\lambda\sin\omega_c)+\gamma(2+\cos\omega_b+\lambda\sin\omega_b)=0,$$
$$\gamma(2+\cos\omega_a+\lambda\sin\omega_a)+\alpha(2+\cos\omega_c+\lambda\sin\omega_c)=0, \text{ and}$$
$$\beta(2+\cos\omega_a+\lambda\sin\omega_a)+\alpha(2+\cos\omega_b+\lambda\sin\omega_b)=0.$$

Since this is a linear system with nontrivial solution (α, β, γ), its coefficients matrix is singular, namely the following determinant vanishes

$$\begin{vmatrix} 0 & 2+\cos\omega_c+\lambda\sin\omega_c & 2+\cos\omega_b+\lambda\sin\omega_b \\ 2+\cos\omega_c+\lambda\sin\omega_c & 0 & 2+\cos\omega_a+\lambda\sin\omega_a \\ 2+\cos\omega_b+\lambda\sin\omega_b & 2+\cos\omega_a+\lambda\sin\omega_a & 0 \end{vmatrix}=0,$$

or, equivalently

$$(2+\cos\omega_a+\lambda\sin\omega_a)(2+\cos\omega_b+\lambda\sin\omega_b)(2+\cos\omega_c+\lambda\sin\omega_c)=0.$$

Solving for λ we obtain three possible values, which in fact are equal since the multiplier is unique

$$\frac{2+\cos\omega_a}{\sin\omega_a}=\frac{2+\cos\omega_b}{\sin\omega_b}=\frac{2+\cos\omega_c}{\sin\omega_c}=-\lambda. \tag{2.33.101}$$

Then the angles ω_a, ω_b and ω_c can take those values for which relation (2.33.101) is satisfied. To study this, we take the function $f : (0, 180°) \to \mathbb{R}$ defined by $f(x) = \dfrac{2+\cos x}{\sin x}$ and rewrite (2.33.101) as

$$f(\omega_a)=f(\omega_b)=f(\omega_c). \tag{2.33.102}$$

The graph of $f(x)$ is given by Figure 2.97. It has vertical asymptotes at $x \in \{0°, 180°, 360°\}$, f is convex on $(0°, 180°)$, with a minimum at $x = 120°$ and f is concave on $(180°, 360°)$, with a maximum at $x = 240°$.

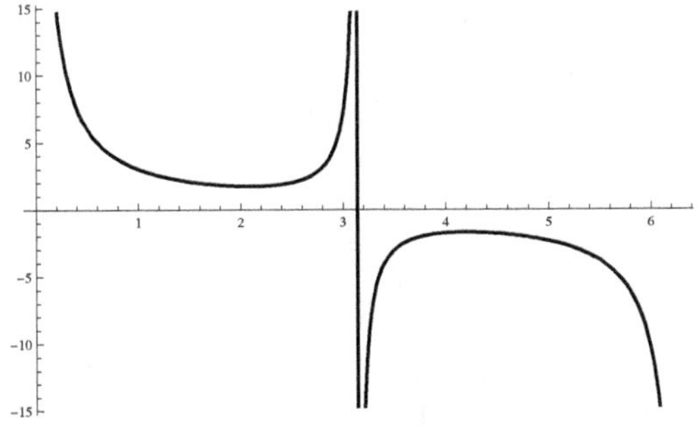

Figure 2.97: The graph of $f(x) = \dfrac{2+\cos x}{\sin x}$, $x \in (0, 360°)$.

Since $\omega_a, \omega_b, \omega_c \in (0°, 180°)$, and the maximum number of intersections between any horizontal line $y = -\lambda$ and the graph $y = f(x)$ is at most two distinct points, it follows that at least two of the angles have equal values, say $\omega_a = \omega_b$.
Therefore:

(*i*) either all angles are equal, and hence $\omega_a = \omega_b = \omega_c = 120$ and the proof ends here, or

(*ii*) $\omega_a = \omega_b \neq \omega_c$. We shall show that this case does not hold. From $\omega_a + \omega_b + \omega_c = 360$ we get $\omega_c = 360 - 2\omega_a$ and since $\omega_c < 180$ it follows that $\omega_a > 90$. From (2.33.102) we have

$$f(\omega_a) = f(\omega_c) = f(360 - 2\omega_a) = f(-2\omega_a) = -f(2\omega_a),$$

so ω_a is a solution of the trigonometric equation

$$\frac{2 + \cos x}{\sin x} = -\frac{2 + \cos(2x)}{\sin(2x)}.$$

Using $x > 90°$, this can be solved as

$$\frac{2 + \cos x}{\sin x} = -\frac{1 + 2\cos^2 x}{2\sin x \cos x} \iff 4\cos^2 x + 4\cos x + 1 = 0 \iff$$
$$2\cos x + 1 = 0 \iff x = 120°.$$

Therefore, $\omega_a = \omega_b = 120$, which also implies that $\omega_c = 120$. Hence, we are always in case (*i*), which ends the proof.

2.34 A mechanical interpretation

We shall consider three pulleys situated at the vertices A, B and C from which we hang vertically unit weights. The strings are all binding at a point T, inside the triangle ABC. We ask for the position of the point T such that the system of weights is in equilibrium, see Figure 2.98.

To this end, we consider the force tensions at the point T pointing in the direction of triangle vertices and denote them by u_A, u_B and u_C, respectively. The equilibrium corresponds to a zero sum of vectors, namely $u_A + u_B + u_C = 0$. Besides this, we have also the condition of equal magnitude of tensions given by $|u_A| = |u_B| = |u_C| = 1$. We shall compute the magnitude of the sum $|u_B + u_C|$ in two ways. By Newton's rule of composition of forces, we have $|u_B + u_C|^2 = |u_B|^2 + |u_C|^2 + 2\cos\theta = 2 + 2\cos\theta$, where θ is the measure of the angle made by u_B and u_C. Then using $|u_B + u_C|^2 = |-u_A|^2 = 1$, we obtain the equation $2 + 2\cos\theta = 1$ with solution $\theta = 120°$. We arrive to the same angle measure if we start from any two other vectors. Hence, T is the isogon center of ΔABC.

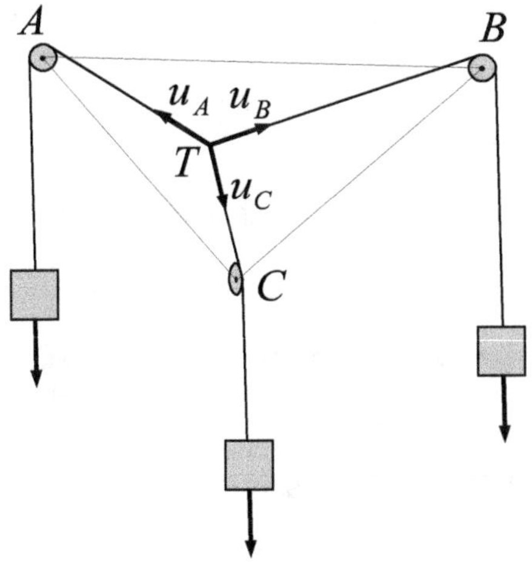

Figure 2.98: The physical interpretation of the isogon point.

2.35 Emerging inequalities

The minimality of the sum of the distances between the isogon center, T, and the vertices of an acute-angled triangle ΔABC implies the inequality

$$|TA| + |TB| + |TC| \leq |MA| + |MB| + |MC| \qquad (2.35.103)$$

for any point M in the plane. Taking certain particular cases of points M and using identity (2.33.94) will lead to a series of inequalities as follows.

1. *The case $M = O$ (the circumcenter)* In this case $|OA| = |OB| = |OC| = R$, the circumradius. Then (2.35.103) and (2.33.94) imply

$$\frac{1}{2}(a^2 + b^2 + c^2) + 2\sqrt{3}\mathcal{A} \leq 9R^2.$$

2. *The case $M = H$ (the orthocenter)* By Carnot's relation we have $|HA| + |HB| + |HC| = 2(R + r)$. The associated inequality in this case is

$$\frac{1}{2}(a^2 + b^2 + c^2) + 2\sqrt{3}\mathcal{A} \leq 4(R + r)^2.$$

3. *The case $M = G$ (the centroid)* Using that G is located at 2/3 of the vertices along each median, we obtain

$$\frac{1}{2}(a^2 + b^2 + c^2) + 2\sqrt{3}\mathcal{A} \leq \frac{2}{3}(m_a + m_b + m_c).$$

4. *The case $M = I$ (the incenter)* Using $|AI| = r/\sin\frac{\widehat{A}}{2}$ we obtain

$$\frac{1}{2}(a^2+b^2+c^2) + 2\sqrt{3}\mathcal{A} \le r\left(\csc\frac{\widehat{A}}{2} + \csc\frac{\widehat{B}}{2} + \csc\frac{\widehat{C}}{2}\right).$$

Sometimes, if we express the sum of squares of the sides by the formula

$$a^2+b^2+c^2 = 2p^2 - 2r^2 - 8Rr,$$

the left side can simplify. For instance, in the last case after using the aforementioned formula and simplifying by r, we obtain

$$\frac{p^2}{r} + 2\sqrt{3}p - r - 4R \le \csc\frac{\widehat{A}}{2} + \csc\frac{\widehat{B}}{2} + \csc\frac{\widehat{C}}{2}.$$

2.36 A generalization of the Euler's circle

The distinguished role played by the circumcenter, O, and the orthocenter, H, in the case of the Euler's circle can be played by any pair of isogonal conjugate points as follows.

Theorem 2.36.1 *Let M and P be two isogonal conjugate points in ΔABC. Then the orthogonal projections of M and P onto the sides of ΔABC are 6 cyclic points.*

Let M_1, M_2, M_3 and P_1, P_2, P_3 be the orthogonal projections of M and P on the lines BC, AC and AB, respectively. We shall prove that the points P_1, P_3, M_1, M_3 and P_1, P_2, M_1, M_2 belong to the same circle $\mathcal{C}$. To this end we shall employ the reciprocal of the power of a point property, see Exercise 2.27.5.

Thus, showing that P_1, M_1, M_2, M_3 belong to the same circle $\mathcal{C}$ is equivalent to the property

$$|BM_1|\cdot|BP_1| = |BP_3|\cdot|BM_3|. \tag{2.36.104}$$

Each of the previous segment lengths can be computed from some right triangles, see Figure 2.99, as

$$\begin{aligned}
|BM_1| &= |BM|\cos\widehat{MBM_1} = |BM|\cos(\alpha+\theta),\\
|BP_1| &= |BP|\cos\widehat{PBP_1} = |BP|\cos\alpha,\\
|BP_3| &= |BP|\cos\widehat{PBP_3} = |BP|\cos(\alpha+\theta), \text{and}\\
|BM_3| &= |BM|\cos\widehat{MBM_3} = |BM|\cos\alpha.
\end{aligned}$$

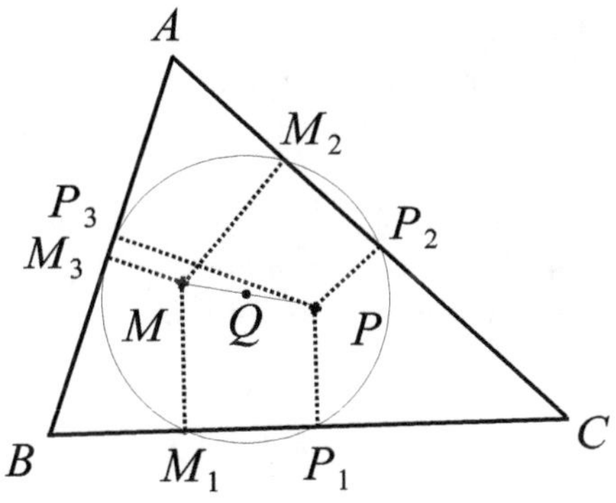

Figure 2.99: The projections of M and P onto the sides of ΔABC are 6 cyclic points.

Then relation (2.36.104) can be verified by direct substitution, each side being equal to the expression $|BM|\cdot|BP|\cos\alpha\cos(\alpha+\theta)$. The center of the circle $\mathcal{C}$ is obtained by intersecting the perpendicular bisectors of the chords M_1P_1 and P_3M_3. Both of these chords will pass through the midpoint of the segment MP, which we denote by Q. This follows from applying the midpoint connector theorem for trapezoids MM_1P_1P and MM_3P_3P, see Exercise 2.31.11. Therefore, the points P_1, M_1, P_3, M_3 belong to the circle $\mathcal{C}$ centered at Q and having radius $\rho = |QP_1|$.

A similar proof shows that the points P_1, P_2, M_1, M_2 belong to a circle. Since this circle is also centered at Q and has radius $\rho = |QP_1|$, then it is the same as the circle $\mathcal{C}$.

Therefore, the points P_1, P_3, M_1, M_3, P_1, P_2, M_1, M_2 belong to the same circle. The center of this circle is located at the midpoint of the segment MP.

Remark 2.36.2 (i) In the case $M = H$, the orthocenter, and $P = O$, the circumcircle, we obtain the classical Euler's circle.
(ii) Considering $M = G$, the centroid, and $P = K$, the Lemoine's point (the symedian center), we obtain that the projections of the centroid and the Lemoine's point onto the sides of ΔABC are 6 cyclic points.

2.37 Backwards proofs

We shall present a method of proof which works the argument backwards. Even if this is not very often seen in geometry, we shall include here two examples, one in which we need to find an angle, and the other where we show that a triangle is equilateral.

In the first problem, where we need to find the angle, we show first that the solution of the problem is unique. Then in order to show the existence, we just make an educated guess for the angle measure and verify its veridicity.

The second problem, which is the classical Morley's theorem, the argument is worked backwards, starting from an equilateral triangle and ending to the initial triangle (or a similar copy of it).

(i) The square problem

Let M be a point in the interior of the square $ABCD$ such that $\widehat{MAB} = \widehat{MBA} = 15°$, see Figure 2.100. Find the measure of the angle $\widehat{DMC}$.

Solution: We first observe that the measure of the angle $\widehat{DMC}$ is determined in an unique way by the measure of angles $\widehat{MAB}$ and $\widehat{ABM}$. This comes from the construction of an angle axiom: The rays $(AM$ and $(BM$ are unique such that $\widehat{BAM} = \widehat{ABM} = 15°$. The point M is unique as the intersection of the rays $(AM$ and $(BM$.

Now we shall make an educated guess for the angle $\widehat{DMC}$. To this end, we let $\widehat{DMC} = 60°$. Due to symmetry reasons, the triangle ΔDMC is isosceles, and hence it is in fact equilateral (Why?). Hence $|DM| = |MC| = |DC|$. It follows that $|MC| = |BC|$ and $|MD| = |AD|$. Hence the triangles

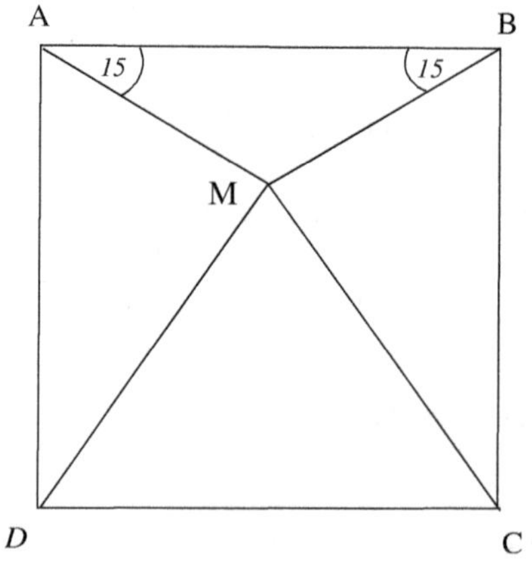

Figure 2.100: Is ΔMDC equilateral?

ΔDMA and ΔCMB are isosceles. Since $\widehat{MCD} = \widehat{MDC} = 60°$, then

$$\widehat{MCB} = 90° - \widehat{MCD} = 90° - 60° = 30°,$$
$$\widehat{MDA} = 90° - \widehat{MDC} = 90° - 60° = 30°.$$

Using that ΔDMA and ΔCMB are isosceles triangles, we have

$$\widehat{MBC} = \frac{1}{2}(180° - \widehat{MCB}) = \frac{1}{2}(180° - 30°) = 75°,$$

$$\widehat{MAD} = \frac{1}{2}(180° - \widehat{MDA}) = \frac{1}{2}(180° - 30°) = 75°.$$

Hence

$$\widehat{ABM} = 90° - \widehat{MBC} = 90° - 75° = 15°,$$

$$\widehat{MAB} = 90° - \widehat{DAM} = 90° - 75° = 15°.$$

Since we arrived at $\widehat{BAM} = \widehat{ABM} = 15°$, it follows that our guess $\widehat{DMC} = 60°$ is the correct one.

Remark 2.37.1 The square problem from page 220 can also be approached using a proof by construction as follows. Let ΔNAB be an equilateral triangle constructed in the exterior of the square on the side AB, see Figure 2.101. By symmetry reasons $NM \perp AB$. Since also $AD \perp AB$, then it follows that $AD \| NM$. In order to show that $ANMD$ is a parallelogram, it

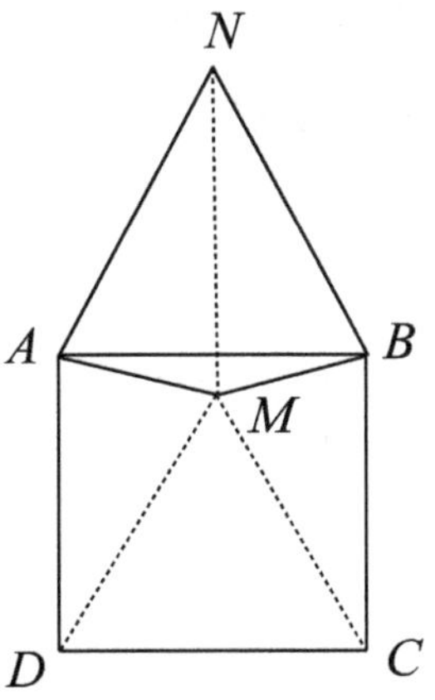

Figure 2.101: The construction of an equilateral triangle on the side AB.

suffices to show that $|AD| = |NM|$. We start by evaluating the base angles in ΔNAM

$$\widehat{NAM} = \widehat{NAB} + \widehat{BAM} = 60^\circ + 15^\circ = 75^\circ, \text{and}$$
$$\widehat{NMA} = \frac{1}{2}\widehat{AMB} = \frac{1}{2}(180^\circ - 2 \cdot 15^\circ) = 75^\circ,$$

so ΔNAM is isosceles, with $|AN| = |NM|$. Using $|AN| = |AB| = |AD|$, it follows that $|NM| = |AD|$, and hence $ANMD$ is a parallelogram. Consequently, the other pair of opposite sides are equal, $|AN| = |DM|$. Since $|AN| = |AB| = |DC|$, it follows that $|DM| = |DC|$. Using a symmetric reasoning we obtain $|CM| = |DC|$. This implies that ΔMDC is an equilateral triangle, and hence $\widehat{DMC} = 60^\circ$.

Exercise 2.37.2 *Solve the same problem as above with the condition* $\widehat{MAB} = \widehat{MBA} = 45^\circ$.

Solved Exercise 2.37.3 *Let I be the incenter of ΔABC and M a point on the angle bisector from A. Show that*

$$M = I \Longleftrightarrow \widehat{BMC} = 90^\circ + \frac{\widehat{A}}{2}.$$

Proof: " $\Longrightarrow$ " We evaluate in ΔBMC

$$\widehat{BMC} = 180^\circ - \frac{\widehat{B}}{2} - \frac{\widehat{C}}{2} = 180^\circ - \frac{180^\circ - \widehat{A}}{2} = 90^\circ + \frac{\widehat{A}}{2}.$$

" $\Longleftarrow$ " Let $\widehat{BMC} = 90^\circ + \frac{\widehat{A}}{2}$ and we assume, by contradiction, that $M \neq I$. We denote by AA' the angle bisector from A. We assume $M \in (AI)$. By the exterior angle inequality we get

$$\widehat{BMC} = \widehat{BMI} + \widehat{IMC} < \widehat{BIA'} + \widehat{A'IC} = \widehat{BIC} = 90^\circ + \frac{\widehat{A}}{2},$$

so $\widehat{BMC} < 90^\circ + \frac{\widehat{A}}{2}$, which contradicts the hypothesis. Assuming $M \in (IA')$ we obtain $\widehat{BMC} > 90^\circ + \frac{\widehat{A}}{2}$, which also contradicts hypothesis. It follows that $M = I$. ■

(ii) Morley's Theorem

Theorem 2.37.4 (Morley) *The three points of intersection of the adjacent trisectors of the angles of any triangle form an equilateral triangle.*

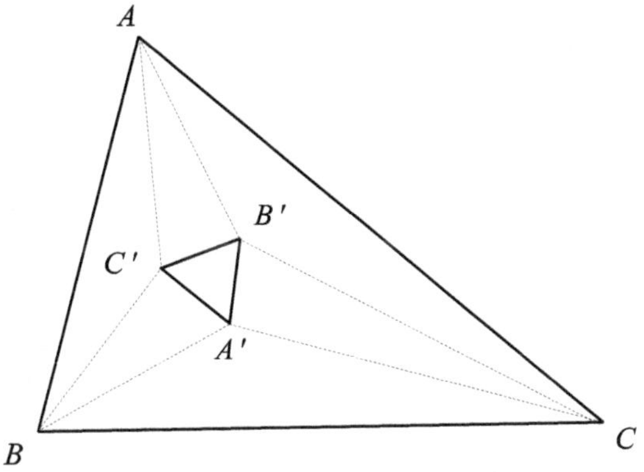

Figure 2.102: Morley's theorem.

Proof: Let ΔABC be the given triangle and A', B', C' be the intersections of the adjacent trisectors from vertices $\{B, C\}$, $\{A, C\}$, and $\{A, B\}$, see Figure 2.102. We need to show that $|A'B'| = |B'C'| = |A'C'|$.

This proof is performed backwards, namely, we start from a given equilateral triangle $\Delta A'B'C'$ and construct the initial triangle ΔABC. We choose the following angle measures

$$\alpha = 60^\circ - \frac{1}{3}\widehat{A}, \quad \beta = 60^\circ - \frac{1}{3}\widehat{B}, \quad \gamma = 60^\circ - \frac{1}{3}\widehat{C},$$

where $\{\widehat{A}, \widehat{B}, \widehat{C}\}$ are the angles of ΔABC, see Figure 2.103. We note that $\alpha + \beta + \gamma = 120^\circ$. This implies that $\alpha + \beta + \gamma + 60^\circ = 180^\circ$, fact that will be useful shortly. We construct in the exterior of triangle $\Delta A'B'C'$ isosceles triangles $\Delta A''B'C'$, $\Delta B''A'C'$ and $\Delta C''A'B'$ such that the base angles have measure given by

$$\widehat{A''C'B'} = \widehat{A''B'C'} = \alpha, \widehat{B''A'C'} = \widehat{B''C'A'} = \beta, \widehat{C''A'B'} = \widehat{C''B'A'} = \gamma.$$

We consider next the intersection points $\{A\} = B''C' \cap C''B'$, $\{B\} = A''C' \cap C''A'$, and $\{C\} = A''B' \cap B''A'$. We shall show that $\{AC', AB'\}$, $\{BC', BA'\}$ and $\{CB', CA'\}$ are the angle trisectors from A, B and C of triangle ΔABC.

It suffices to show that BA' and CA' are two adjacent trisectors, the rest having a similar proof. To this end we shall show that BA' and CA' are angle bisectors in $\Delta A''BC$. To show this it suffices to prove that the point A' is the incenter of triangle $\Delta A''BC$. This fact will follow from Exercise 2.37.3 as follows.

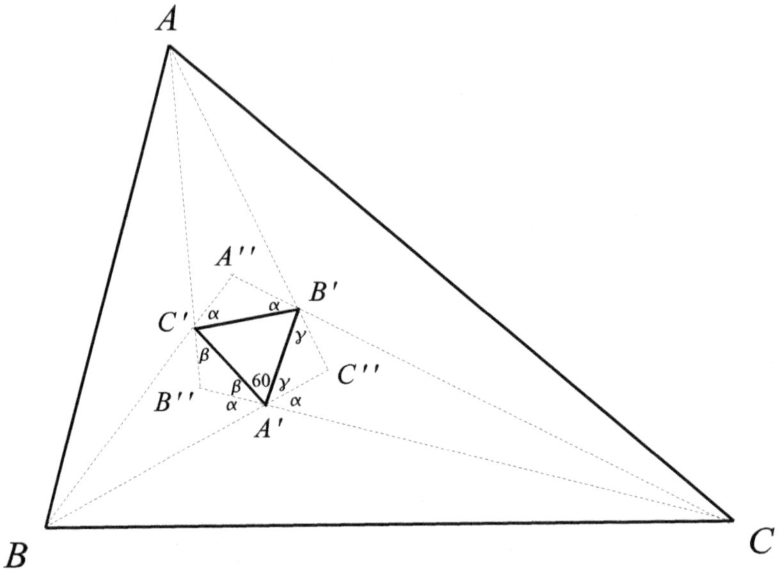

Figure 2.103: The proof of Morley's theorem.

First, due to the construction symmetry, we note that $A''A'$ is a symmetry axis of the hexagon $A''B'C''A'B''C'$. This fact implies that $A''A'$ is an angle bisector of $\widehat{BA''C}$. Next we shall evaluate $\widehat{BA'C}$ in terms of $\widehat{BA''C}$. We have

$$\frac{\widehat{BA''C}}{2} = \frac{\widehat{C'A''B'}}{2} = \frac{180^\circ - 2\alpha}{2} = 90^\circ - \alpha.$$

The aforementioned condition, $\alpha+\beta+\gamma+60^\circ = 180^\circ$, implies that $\widehat{C''A'C} = \widehat{B''A'B} = \alpha$. Then

$$\widehat{BA'C} = 360^\circ - (\alpha + \beta + \gamma + 60^\circ) - \alpha = 360^\circ - 180^\circ - \alpha$$
$$= 180^\circ - \alpha = 90^\circ + (90^\circ - \alpha) = 90^\circ + \frac{\widehat{BA''C}}{2}.$$

Then Exercise 2.37.3 implies that A' is the incenter of $\Delta BA''C$ and hence

$$\widehat{C'BA'} = \widehat{A'BC}, \qquad \widehat{A'CB} = \widehat{A'CB'}. \tag{2.37.105}$$

A similar reasoning shows that C' and B' are incenters for triangles $\Delta AC''B$ and $\Delta AB''C$, respectively, so

$$\widehat{ABC'} = \widehat{C'BB''}, \qquad \widehat{ACB'} = \widehat{B'CA'}. \tag{2.37.106}$$

Identities (2.37.105) and (2.37.106) imply that BA' and CA' are trisectors. A similar argument applies for the other trisectors.

We note that the triangle ΔABC recovered by the aforementioned construction, starting from the equilateral triangle $\Delta A'B'C'$, is similar with the initial triangle ΔABC (as having the same angle measures). Applying a scaling transformation we obtain the initial triangle. This scaling transformation scales also the initial equilateral triangle. ■

Chapter 3

Quadrilateral Geometry

The goal of this chapter is to understand the quadrilateral geometry and to extend some results from the triangle geometry to the new framework. Even if many formulas and concepts are similar for both triangles and quadrilaterals, there are also important differences between them. A triangle is determined by its side lengths, while a quadrilateral is not. Consequently, not all quadrilaterals will have a circumcircle or an incircle. However, all do have an area and a centroid. A quadrilateral also comes with new elements, such as Miquel's point or Aubert's line.

3.1 Basic properties of quadrilaterals

Four points in the plane, any three of them noncollinear, determine a **quadrilateral**. If A, B, C, D are the points, $ABCD$ is a quadrilateral if the pairs of segments (AB, DC) and (AD, BC) do not intersect, see Figure 3.1(a).
The points A, B, C, D are called *vertices* and the segments AB, BC, CD, DA are called *sides*. The segments AC, DB are called *diagonals*.

A quadrilateral $ABCD$ is called **convex** if $Int\, ABCD$ is a convex set. See Figure 3.1(c). A quadrilateral which is not convex is called concave, see Figure 3.1(d).

Definition 3.1.1 *A quadrilateral $ABCD$ is called* **cyclic** *if there is a circle $\mathcal{C}$ passing through the points A, B, C, D.*

Theorem 3.1.2 *Any cyclic quadrilateral is convex.*

Proof: We start with a triangle ΔABC and let α_A, α_B, α_C be the arcs which do not contain A, B, C, respectively. As $ABCD$ is cyclic, the point

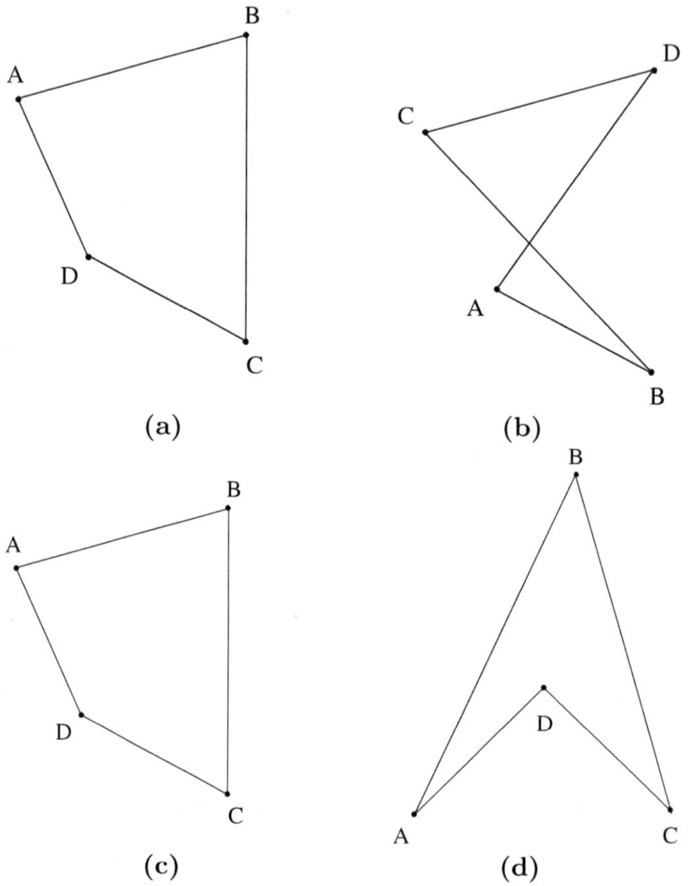

Figure 3.1: (a) $ABCD$ quadrilateral; (b) $ABCD$ not quadrilateral; (c) Convex quadrilateral; (d) Concave quadrilateral.

D may belong to α_A, α_B or α_C, see Figure 3.2. If $D \in \alpha_C$, then the segments $DC \cap AB \neq \emptyset$ and hence $ABCD$ is not a quadrilateral.
The same for $D \in \alpha_A$. By exclusion, it follows that $D \in \alpha_B$. Then C and D belong to the same half plane defined by the line AB. We write the interior of the quadrilateral as the intersection of the interiors of two angles as

$$Int\,ABCD = Int\,\widehat{ABC} \cap Int\,\widehat{ADC}.$$

Using that the interior of an angle is a convex set (why?), and the intersection of two convex sets is also convex, it follows that $ABCD$ is a convex quadrilateral. ■

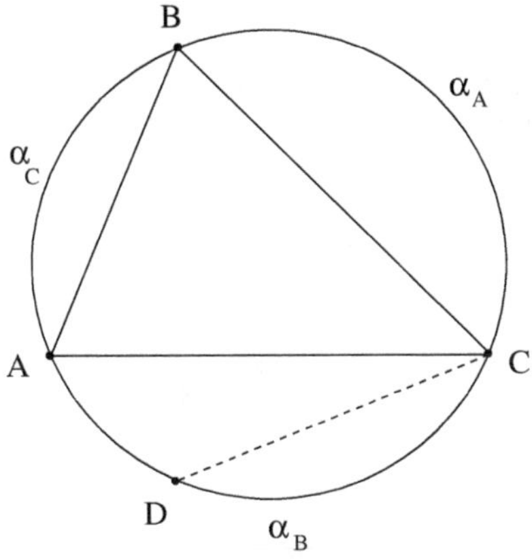

Figure 3.2: The case $D \in \alpha_B$.

The following two theorems are important tools in showing that a quadrilateral is cyclic.

Theorem 3.1.3 *Let $ABCD$ be a convex quadrilateral. Then $ABCD$ is cyclic if and only if*

$$\widehat{ABC} + \widehat{ADC} = 180°.$$

Proof: " $\Longrightarrow$ " Using angles inscribed in a circle, Proposition 1.11.4 yields

$$\begin{aligned}\widehat{ABC} + \widehat{ADC} &= \frac{1}{2}\Big(arc(AB) + arc(BC)\Big) + \frac{1}{2}\Big(arc(CD) + arc(DA)\Big) \\ &= \frac{1}{2} \cdot 360° = 180°.\end{aligned}$$

" $\Longleftarrow$ " From the convexity condition the point D belongs to a half plane defined by AC that does not contain B.
Let $\mathcal{C}$ be the circle passing through A, B and C. Assume $D \notin \mathcal{C}$. Then $D \in Int\,\mathcal{C}$ or $D \in Ext\,\mathcal{C}$.

If $D \in Int\,\mathcal{C}$, let $E = CD \cap \mathcal{C}$. As $ABCE$ is cyclic, applying the implication " $\Longrightarrow$ " of the proof yields

$$\widehat{ABC} + \widehat{AEC} = 180°.$$

As $\widehat{ABC} + \widehat{ADC} = 180°$, subtracting the last two relations yields $\widehat{ADC} = \widehat{AEC}$, which is a contradiction because exterior angle inequality, Theorem 1.9.4, implies that $\widehat{ADC} > \widehat{AEC}$.
The case $D \in Ext\,\mathcal{C}$ leads to a similar contradiction. ∎

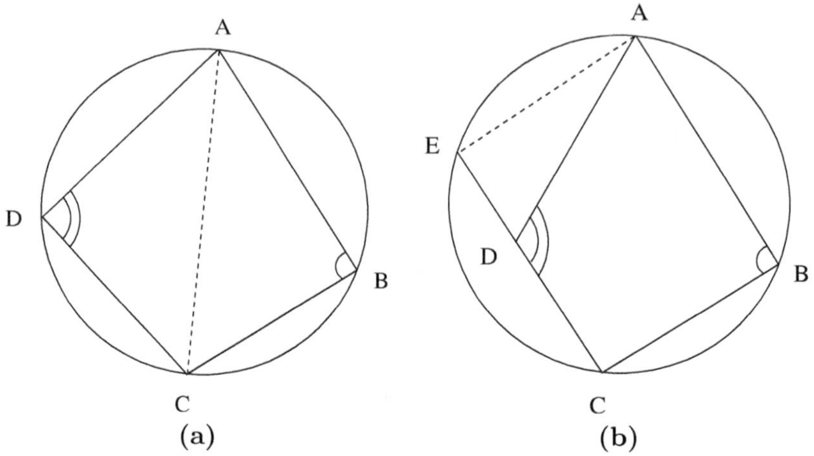

Figure 3.3: (a) Cyclic quadrilateral; (b) The reciprocal of Theorem 3.1.3.

Remark 3.1.4 Without the convexity condition the above theorem is not necessarily true. Consider Figure 3.4, the nonconvex quadrilateral $ABCD$. We have $\widehat{ABC} + \widehat{ADC} = 180°$ but the quadrilateral is not cyclic.

We recall that a rhombus is a parallelogram having all sides equal.

Exercise 3.1.5 (*i*) *Prove that a parallelogram is a cyclic quadrilateral if and only if it is a rectangle.*

(*ii*) *Prove that a rhombus is a cyclic quadrilateral if and only if it is a square.*

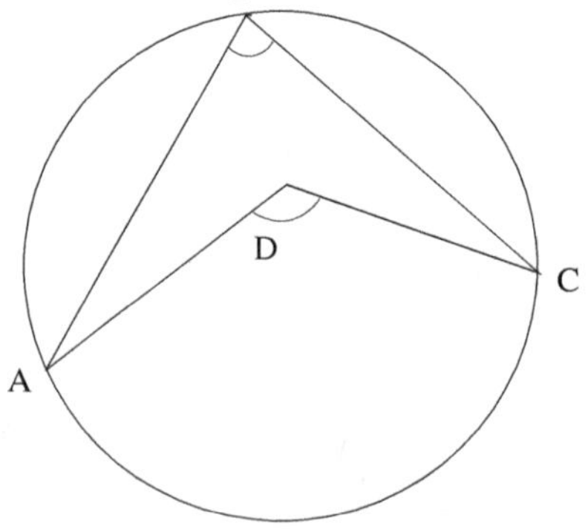

Figure 3.4: A counterexample.

The next theorem has already been used in the proof of Proposition 2.31.6. We included it here for the sake of completeness.

Theorem 3.1.6 *Let $ABCD$ be a convex quadrilateral. Then $ABCD$ is cyclic if and only if $\widehat{ABD} = \widehat{ACD}$, i.e., the angle between a diagonal and one of the sides is equal to the angle between the other diagonal and the opposite side.*

Proof: " $\Longrightarrow$ " We consider Figure 3.5(a). As angles inscribed in the same circle, we have according to Theorem 1.11.4 that

$$\widehat{ABD} = \frac{1}{2} arc(AD) = \widehat{ACD}.$$

" $\Longleftarrow$ " We prove this by contradiction. We shall refer to the Figure 3.5(b). Assume $\widehat{ABD} = \widehat{ACD}$ and $ABCD$ not cyclic. Let $\mathcal{C}$ be the circumcircle of ΔABD. Then $C \notin \mathcal{C}$. Then $C \in Int\,\mathcal{C}$ or $C \in Ext\,\mathcal{C}$. Suppose $C \in Ext\,\mathcal{C}$. Then let $C' = AC \cap \mathcal{C}$. Hence $ABC'D$ is cyclic. Using the implication " $\Longrightarrow$ " proved before, it follows that

$$\widehat{ABD} = \widehat{AC'D}.$$

From the exterior angle theorem we have

$$\widehat{AC'D} > \widehat{ACD}.$$

Comparing the last two displays yields

$$\widehat{ABD} > \widehat{ACD},$$

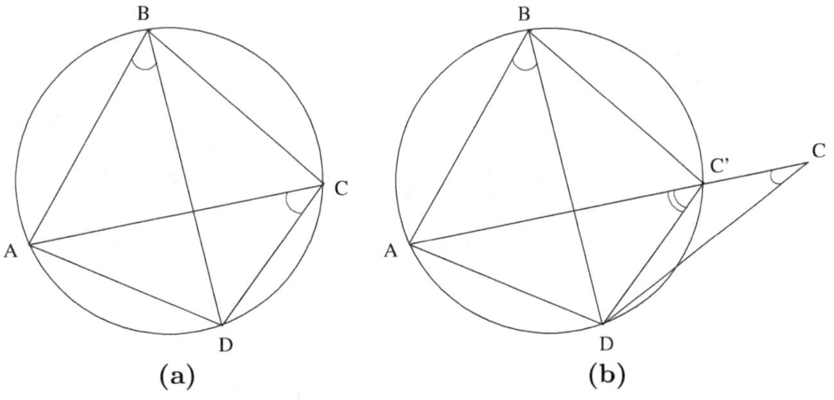

Figure 3.5: See Theorem 3.1.6.

which is a contradiction. Hence $ABCD$ is cyclic. The case $C \in Int\, \mathcal{C}$ is left for the reader. ∎

Remark 3.1.7 The convexity condition is essential. See a counterexample in Figure 3.6, where we have $\widehat{ABD} = \widehat{ACD}$ but the quadrilateral is not cyclic.

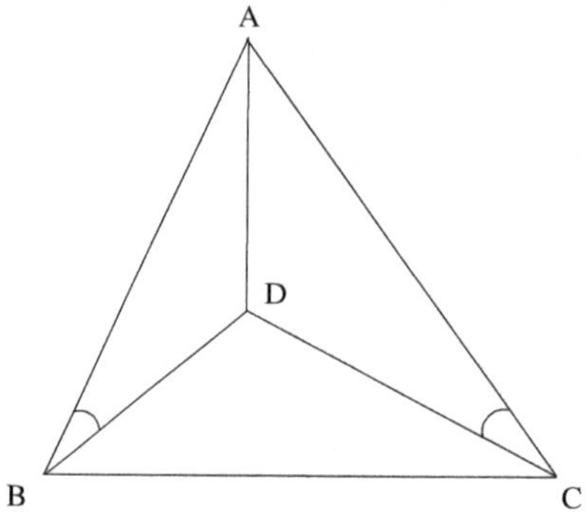

Figure 3.6: A counterexample.

Exercise 3.1.8 *A trapezoid is cyclic if and only if it is isosceles.*

Solved Exercise 3.1.9 *Show that in any convex quadrilateral the sum of diagonals is larger than the quadrilateral semiperimeter and smaller than its perimeter, i.e.,*

$$\frac{1}{2}(a+b+c+d) < \delta_1 + \delta_2 < a+b+c+d.$$

Proof: Let $ABCD$ be a convex quadrilateral with side lengths a, b, c, d and diagonals lengths δ_1 and δ_2. Let O be the point of intersection of diagonals, see Figure 3.7(a). The triangle inequality applied in triangles formed by two sides and one diagonal yields $\delta_1 < a+b$, $\delta_1 < c+d$, $\delta_2 < b+c$ and $\delta_2 < a+d$. Summing these four inequalities and taking half, we obtain $\delta_1 + \delta_2 < a+b+c+d$. For the second inequality, we apply the triangle inequality in the four triangles formed by O and two vertices: $a < |BO| + |AO|$, $b < |BO| + |CO|$, $c < |CO| + |DO|$, $d < |AO| + |DO|$. Summing and using $\delta_1 = |AC| = |AO| + |CO|$ and $\delta_2 = |BD| = |BO| + |DO|$ yields $a+b+c+d < 2(\delta_1 + \delta_2)$, which means $(a+b+c+d)/2 < \delta_1 + \delta_2$. ∎

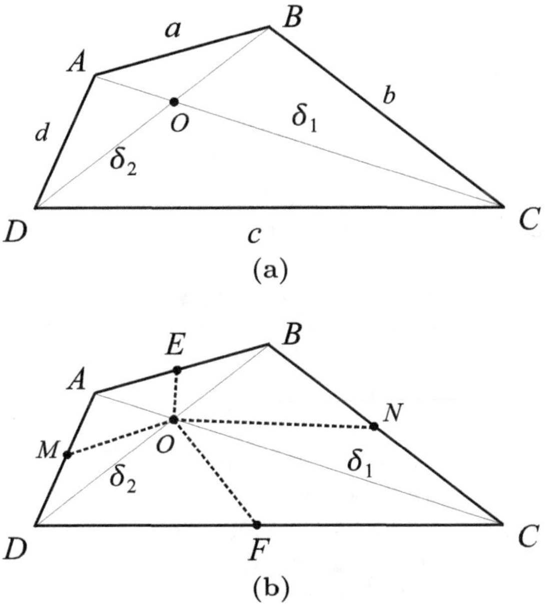

Figure 3.7: (a) For Exercise 3.1.9; (b) For Exercise 3.1.10.

Solved Exercise 3.1.10 *Let $ABCD$ be a convex quadrilateral and denote by E, F, M, N the midpoints of the sides AB, CD, AD and BC, respectively. If $O = AC \cap BD$, show that*

$$|OE| + |OF| + |OM| + |ON| < |AC| + |BD|.$$

Proof: Since OE is a median in triangle ΔOAB, by Exercise 2.21.8, part (i), we have $|OE| < \frac{1}{2}(|OA| + |OB|)$. Similarly, we have $|ON| < \frac{1}{2}(|OB| + |OC|)$, $|OF| < \frac{1}{2}(|OC| + |OD|)$, and $|OM| < \frac{1}{2}(|OA| + |OD|)$. Summing we obtain

$$|OE|+|OF|+|OM|+|ON| < (|OA|+|OC|)+(|OB|+|OD|) = |AC|+|BD|.$$

■

Solved Exercise 3.1.11 *Consider the same notations as in Exercise 3.1.10. Prove that*

$$|MN| + |EF| < |AC| + |BD|.$$

Proof: By triangle inequalities in ΔOMN and ΔOEF we have $|MN| < |OM|+|ON|$ and $|EF| < |OE|+|OF|$. Summing and using Exercise 3.1.10

yields

$$|MN| + |EF| < |OM| + |ON| + |OE| + |OF| < |AC| + |BD|.$$

■

Solved Exercise 3.1.12 *Show that a parallelogram ABCD is a rhombus if and only if the diagonals AC and BD are perpendicular.*

Proof: Let $O = AC \cap BD$. Since the diagonals in a parallelogram bisect, then O is the midpoint of AC, so BO is a median in ΔABC. If $ABCD$ is a rhombus, then $|AB| = |BC|$, so ΔABC becomes isosceles. Then $\Delta ABO \equiv \Delta CBO$ by (SSS) congruence case. Then $\widehat{BOA} = \widehat{BOC}$ and since $\widehat{BOA} + \widehat{BOC} = 90°$, then $BO \perp AC$, or equivalently, $BD \perp AC$.

Conversely, we assume now that the diagonals AC and BD are perpendicular. Then $\widehat{BOA} = \widehat{BOC} = 90°$. Since O is the midpoint of AC, then $\Delta ABO \equiv \Delta CBO$ by (SAS). Therefore, $|AB| = BC|$. Since the opposite sides of a parallelogram are congruent, it follows that all the sides of $ABCD$ are equal, and hence it is a rhombus. ■

3.2 Applications of cyclic quadrilaterals

3.2.1 Simson's line

The next theorem was found by W. Wallace (1768–1843) in 1799. Servois (1767–1847) attributes the theorem to R. Simson (1687–1768) who passed away before the theorem was discovered!

Theorem 3.2.1 *A point M belongs to the circumcircle of ΔABC if and only if the orthogonal projections of M on the sides AB, BC, CA or on their extensions are three collinear points.*

Proof: " $\Longrightarrow$ " We shall consider the case when M belongs to the arc AC. The other cases have a similar proof and are left for the reader. Let P, Q, S be the orthogonal projections of the point M on the lines AB, AC, BC respectively, see Figure 3.8(a). The proof idea is to notice that we have three cyclic quadrilaterals, $ABCM$, $MQSC$ and $MPAQ$; then, using relations between angles we show that $\widehat{PQS} = 180°$.

The point M belongs to the circumcircle of ΔABC if and only if the quadrilateral $ABCM$ is cyclic, which is equivalent to

$$\widehat{BAM} + \widehat{BCM} = 180°. \tag{3.2.1}$$

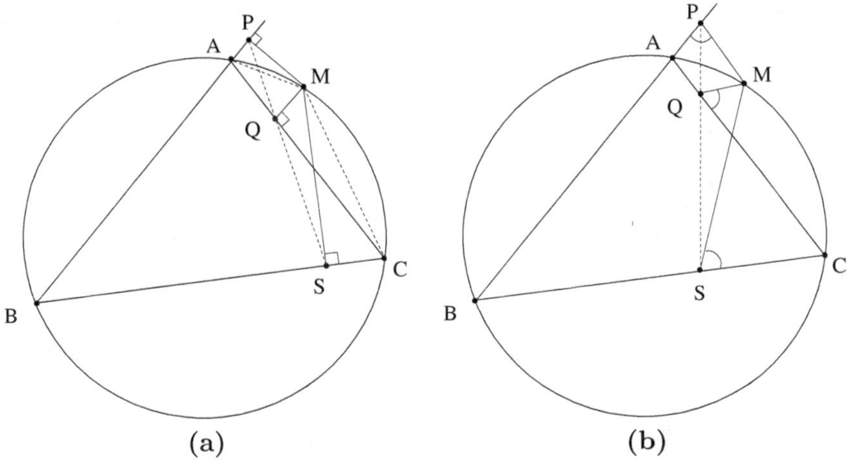

Figure 3.8: (a) Simson's line; (b) generalized Simson's line.

As $\widehat{CQM} = \widehat{MSC} = 90°$, then $MQSC$ is cyclic and hence $\widehat{BCM} = 180° - \widehat{SQM}$. Using $\widehat{MPA} + \widehat{MQA} = 90° + 90° = 180°$ the quadrilateral $MPAQ$ is cyclic and hence $\widehat{MAP} = \widehat{PQM}$.
Then relation (3.2.1) is equivalent to

$$\begin{aligned}(180° - \widehat{MAP}) + (180° - \widehat{SQM}) &= 180° \Longleftrightarrow \\ (180° - \widehat{PQM}) + (180° - \widehat{SQM}) &= 180° \Longleftrightarrow \\ 180° - \widehat{PQM} - \widehat{SQM} &= 0 \Longleftrightarrow \\ \widehat{PQM} + \widehat{SQM} &= 180°, \qquad (3.2.2)\end{aligned}$$

which means that P, Q, S are collinear.
" $\Longleftarrow$ " We can just work our way back through the algebraic relations starting from (3.2.2) until we reach relation (3.2.1), which means that $ABCM$ is cyclic, and hence M belongs to the circle.

Or, we can proceed by contradiction as follows. We assume that P, Q and S are collinear (with Q between P and S), but M does not belong to the circumcircle of ΔABC. Let's assume that M is inside the circle. Then, the extension of $(QM$ intersects the circle at M'. Let P', $Q' = Q$, and S' be the orthogonal projections of the point M' on the lines AB, AC, BC respectively. By part " $\Longrightarrow$ " of the proof we have that P', Q, and S' are collinear. Since the points P' and S' are on the same side of the line PQS, then $\widehat{P'QS'} < \widehat{PQS} = 180°$, which contradicts the collinearity. A similar contradiction is obtained if we assume that M is outside the circle.
■

Definition 3.2.2 *The line PQS is called the Simson line of the triangle ΔABC and the point M.*

Remark 3.2.3 The French mathematician L. Carnot (1753–1823) formulated the theorem in the following more general framework:

Let M be a point on the circumcircle of ΔABC. Then the projections of same angle and in the same sense of M on the sides AB, BC, CA or on their extensions are three collinear points. See Figure 3.8(b).

Exercise 3.2.4 *Prove the previous statement.*

3.2.2 The Miquel point

The following result was proposed by the well-known mathematician J. Steiner in the French journal Annales des Gergonne in 1827–1828. Someone published in the same year a solution to the problem, signing with the name *Miquel.* The property states that four circles have a point in common. The main tool used in the proof are angle relations in cyclic quadrilaterals.

Theorem 3.2.5 *Let $ABCD$ be a convex quadrilateral such that the opposite sides are not parallel. Let $E = AB \cap CD$ and $F = BC \cap AD$. Then the circumcircles of triangles AED, ABF, BCE, and CDF intersect at a point M, called the* **Miquel** *point.*

Proof: Let BM be the common chord of the circumcircles of ΔABF and ΔBCE. We shall show that the point M belongs to the circumcircles of ΔCDF and ΔADE. We need to show that the quadrilaterals $CDFM$ and $AEMD$ are cyclic.
Using that $ABMF$ and $BCME$ are cyclic quadrilaterals yields

$$\begin{aligned}
\widehat{CMF} &= \widehat{AMF} - \widehat{AMC} \\
&= \widehat{ABF} - \widehat{AMC} = \widehat{ABF} - \left(\widehat{BMC} - \widehat{BMA}\right) \\
&= \widehat{ABF} + \widehat{BMA} - \widehat{BMC} \\
&= \widehat{ABF} + \widehat{BMA} - \widehat{BEC} \\
&= \widehat{ABF} + \widehat{BFA} - \widehat{BEC} \\
&= \left(180^\circ - \widehat{BAF}\right) - \widehat{BEC} \\
&= 180^\circ - \left(\widehat{BAF} + \widehat{BEC}\right) \\
&= \widehat{ADE} = 180^\circ - \widehat{CDF}.
\end{aligned}$$

Hence $\widehat{CMF} + \widehat{CDF} = 180^\circ$, i.e., the quadrilateral $CDFM$ is cyclic.
In a similar way $AEMD$ is cyclic, which completes the proof. ∎

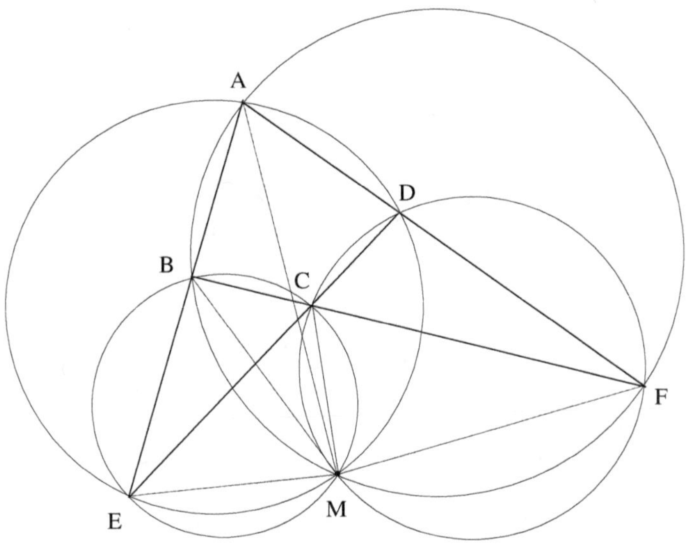

Figure 3.9: The Miquel point M.

Corollary 3.2.6 *The Miquel point has the same Simson line with respect to the triangles ΔAED, ΔABF, ΔBCE, ΔCDF.*

Proof: Any of the above two triangles have two sides in common. The Simson lines for any two triangles will be the same because they have two points in common. Hence all four Simson's lines coincide. ■

3.2.3 Properties of the Simson line

The following result deals with the direction of the Simson line.

Proposition 3.2.7 *Let M be a point on the circumcircle of ΔABC. The perpendicular line from M to BC intersects the circumcircle again at the point M'. Then the Simson line of M with respect to ΔABC is parallel to AM'.*

Proof: Let D, E, F be the projections of M on BC, AC, AB, respectively. Using that two inscribed angles cutting the same arc are equal yields

$$\widehat{BAM'} = \widehat{BMM'}. \tag{3.2.3}$$

As $MD \perp BC$ and $MF \perp AB$ then $FBDM$ is cyclic and hence

$$\widehat{BMD} = \widehat{DFB}. \tag{3.2.4}$$

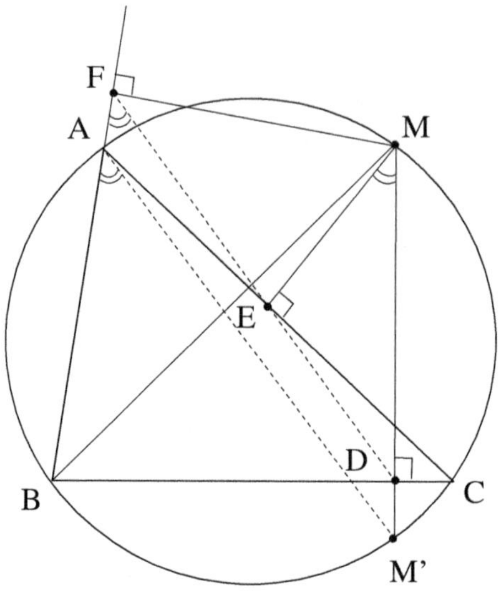

Figure 3.10: See Proposition 3.2.7.

Comparing (3.2.3) and (3.2.4) yields

$$\widehat{BAM'} = \widehat{BMM'} = \widehat{BMD} = \widehat{DFB}.$$

We obtain two equal alternate interior angles and hence $AM' \| FD$. ■

Corollary 3.2.8 *Given a triangle ABC and a direction in the plane there is a point M on the circumcircle of ΔABC such that the Simson line of M is parallel with the given direction.*

We shall next calculate the angle between two Simson lines.

Proposition 3.2.9 *Let M, N be two points on the circumcircle of ΔABC. Denote by s_M, s_N the Simson lines of M, N with respect to ΔABC. Then*

$$\widehat{s_M, s_N} = \frac{1}{2} arc(MN).$$

In particular, MN is a diameter of the circumcircle if and only if their Simson lines are perpendicular.

Proof: Using Proposition 3.2.7 we have $s_M \| AM'$ and $s_N \| AN'$. Therefore

$$\widehat{s_M, s_N} = \widehat{M'AN'} = \frac{1}{2} arc(M'N') = \frac{1}{2} arc(MN),$$

because $|MN| = |M'N'|$ (symmetry reasons). ■

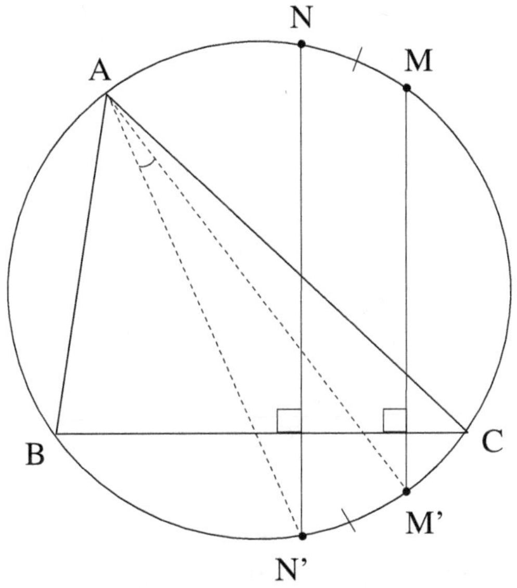

Figure 3.11: See Proposition 3.2.7.

The next proposition deals with the position of the Simson line.

Proposition 3.2.10 *The Simson line of the point M with respect to ΔABC passes through the midpoint of the segment HM, where H is the orthocenter of ΔABC.*

Proof: Let $A_1 \in BC$, $AA_1 \perp BC$ and let A_2 be the point where the altitude AA_1 intersects the circumcircle again. Let $D \in BC$, $MD \perp BC$ and let M' be the point where MD intersects the circumcircle again. Let $N = BC \cap MA_2$, $P = AM' \cap MA_2$ and $R = AM' \cap BC$.

The proof will be done in a few steps.

Step 1: We show $HA_1 = A_1A_2$.
This follows from the following triangle congruence by case (ASA):

$$\Delta BHA_1 \equiv \Delta BA_2A_1 \text{ because } \begin{cases} BA_1 & \text{(common side)} \\ \widehat{BA_1H} = \widehat{BA_1A_2} = 90^\circ & \\ \widehat{HBA_1} = \widehat{A_1BA_2} & \text{because} \end{cases}$$

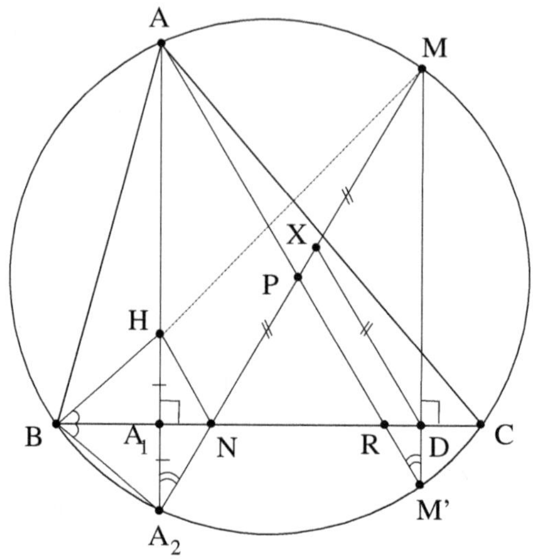

Figure 3.12: See Proposition 3.2.10.

$$\begin{aligned}\widehat{HBA_1} &= 90^\circ - \widehat{BCA}\\ &= 90^\circ - \widehat{BA_2A} \text{ (inscribed angles)}\\ &= \widehat{A_1BA_2}.\end{aligned}$$

Step 2: We prove $\widehat{HNA_1} = \widehat{A_1NA_2}$.
This follows from the following triangle congruence by case (SAS):

$$\Delta HNA_1 \equiv \Delta A_2NA_1 \text{ because } \begin{cases} A_1N & \text{(common side)}\\ \widehat{HA_1N} = \widehat{NA_1A_2} = 90^\circ & \\ |HA_1| = |A_1A_2| & \text{(by step 1)}\end{cases}$$

Step 3: We show $HN \| AM'$.
The lines HN, AM' intersected by the secant BC form the corresponding angles $\widehat{BNH}$ and $\widehat{BRA}$. We shall show that $\widehat{BNH} = \widehat{BRA}$.

$$\begin{aligned}\widehat{BNH} = \widehat{A_1NH} &= \widehat{A_1NA_2}\\ &= 90^\circ - \widehat{A_1A_2N} = 90^\circ - \widehat{AA_2M}\\ &= 90^\circ - \widehat{AM'M} \text{ (inscribed angles)}\\ &= 90^\circ - \widehat{RM'D} = \widehat{DRM'}\\ &= \widehat{NRP} \text{ (vertical angles)}\\ &= \widehat{BRA}.\end{aligned}$$

Step 4: We show that the Simson line of M is parallel to HN.
From Proposition 3.2.7, the Simson line of M is parallel to AM'. Using step 3 yields the desired result.

Step 5: Let X be the point where the Simson line of M intersects MN. Then we prove that $NX = XM$.
The triangle ΔXND is isosceles with $XN = XD$ because

$$\begin{aligned}\widehat{XND} &= \widehat{A_1NA_2} \text{ (vertical angles)}\\ &= \widehat{HNA_1} \text{ (step 2)}\\ &= \widehat{PRN} \text{ (step 3)}\\ &= \widehat{XDN} \text{ (step 4).}\end{aligned}$$

As ΔMND has a right angle at D, then the triangle ΔXDM is isosceles because

$$\widehat{XMD} = 90° - \widehat{XND} = 90° - \widehat{XDN} = \widehat{XDM}.$$

Hence $|XD| = |XM|$. Using $|XD| = |NX|$ yields $|NX| = |XM|$.

Step 6: We show that the Simson line of M intersects HM at its midpoint. The Simson line of M is the line XD. Using that X is the midpoint of MN and $XD \| HN$, the midpoint connector theorem applied to ΔMHN yields that XD intersects the side HM at its midpoint. ■

Corollary 3.2.11 *The Simson line of the point M is the line parallel to AM' through the midpoint of HM.*

3.2.4 Aubert's line of a quadrilateral

Theorem 3.2.12 *Let $ABCD$ be a convex quadrilateral such that the opposite sides are not parallel. Let $E = AB \cap CD$ and $F = BC \cap AD$. Then the orthocenters of triangles ΔAED, ΔABF, ΔBCE, and ΔCDF are four collinear points. See Figure 3.13*(a).

Proof: By Corollary 3.2.6 the Simson lines of the Miquel point M with respect to ΔAED, ΔABF, ΔBCE, and ΔCDF coincide. Denote the common Simson line by **s**. By Proposition 3.2.10 the line **s** contains the midpoints of the segments H_1M, H_2M, H_3M, H_4M, where H_i, $i = 1, \ldots, 4$ are the orthocenters of ΔBCE, ΔABF, ΔAED, and ΔCDF. By the midpoint connector theorem the line **s** is parallel to H_1H_2, H_2H_3, H_3H_4, H_1H_4, and hence H_1, H_2, H_3, H_4 are collinear. See Figure 3.13(b). They belong to a line parallel to the Simson line of the Miquel point. ■

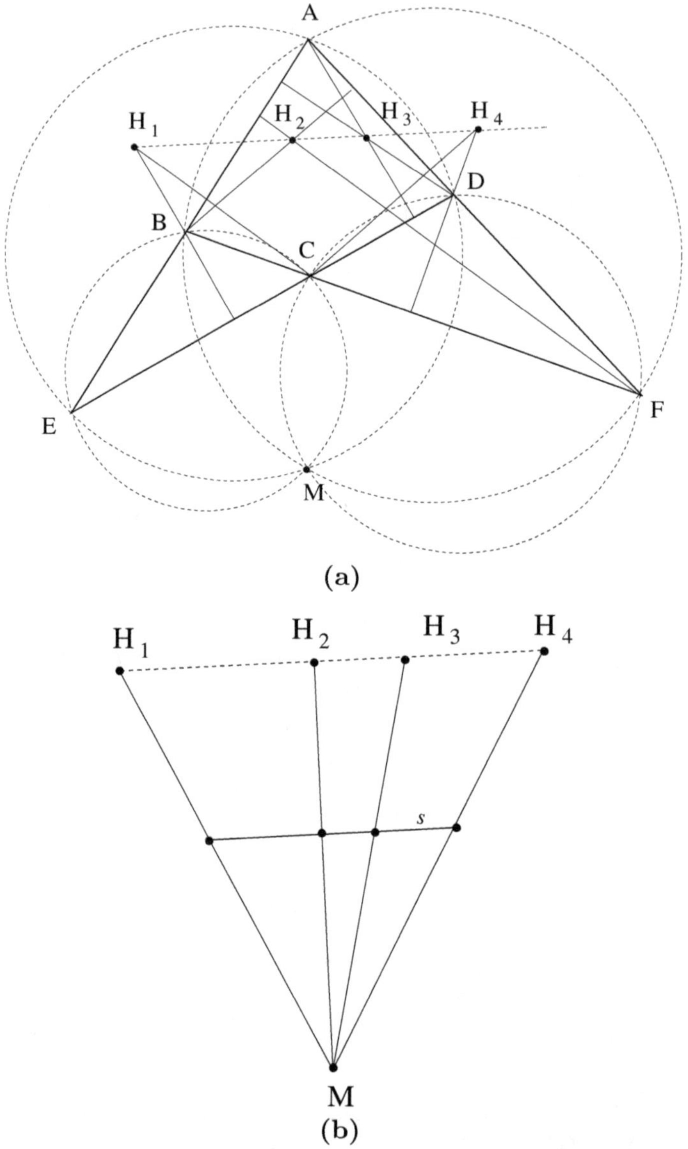

Figure 3.13: (a) Aubert's line; (b) See the proof of Theorem 3.2.12.

The line containing the orthocenters H_1, H_2, H_3 and H_4 is called the *Aubert's line* of the quadrilateral $ABCD$.

3.2.5 Ptolemy's theorems

The Greek astronomer Claudius Ptolemy (120–190 C.E.) discovered two interesting formulas relating the sides and the diagonals of a cyclic quadrilateral. These relations are presented in the following. They provide relations for the product and quotient of diagonals in terms of the quadrilateral sides. Consequently, they provide close form formulas for the diagonals of a quadrilateral in terms of its sides.

Theorem 3.2.13 (Ptolemy's first theorem) *Let $ABCD$ be a cyclic quadrilateral. Then*

$$|AB| \cdot |DC| + |AD| \cdot |BC| = |AC| \cdot |BD|.$$

Proof: Choose a point M between A and C such that $\widehat{ABM} = \widehat{DBC}$. By the (AA) similarity case we have

$$\Delta ABM \sim \Delta DBC \text{ because } \begin{cases} \widehat{DBC} = \widehat{ABM} & \text{(construction)}, \\ \widehat{BDC} = \widehat{BAM} & \text{(cutting the same arc } BC\text{)}. \end{cases}$$

Hence

$$\frac{|AB|}{|DB|} = \frac{|BM|}{|BC|} = \frac{|AM|}{|DC|} \Longrightarrow |AB| \cdot |DC| = |AM| \cdot |DB|. \tag{3.2.5}$$

By the (AA) similarity case we also have

$$\Delta ABD \sim \Delta MBC \text{ because } \begin{cases} \widehat{ADB} = \widehat{ACB} & \text{(cutting the same arc } AB\text{)}, \\ \widehat{ABD} = \widehat{MBC} & \text{because} \end{cases}$$

$$\widehat{ABD} = \widehat{ABM} + \widehat{MBD} = \widehat{MBD} + \widehat{DBC} = \widehat{MBC}.$$

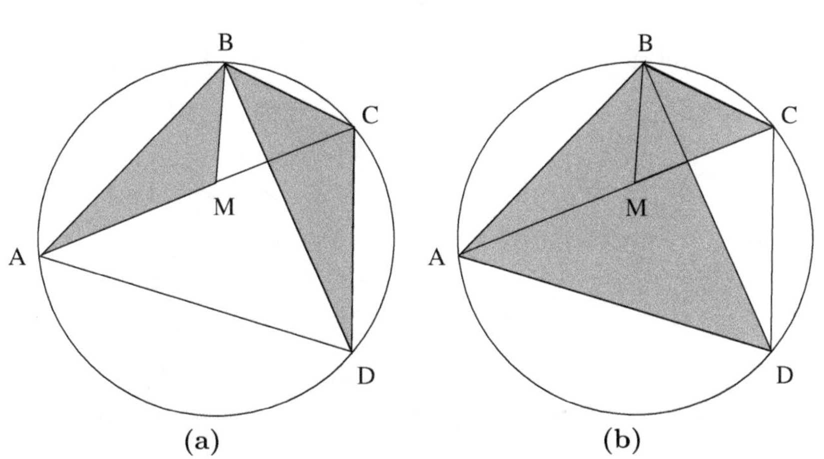

(a) (b)

Figure 3.14: See the proof of Ptolemy's first theorem.

Hence

$$\frac{|AB|}{|MB|} = \frac{|BD|}{|BC|} = \frac{|AD|}{|MC|} \Longrightarrow |BC| \cdot |AD| = |BD| \cdot |MC|. \qquad (3.2.6)$$

Summing (3.2.5) and (3.2.16) yields

$$\begin{aligned} |AB| \cdot |DC| + |BC| \cdot |AD| &= |AM| \cdot |DB| + |BD| \cdot |MC| \\ &= |DB| \cdot (|MC| + |MA|) = |DB| \cdot |AC|. \end{aligned}$$

■

The next exercise states a converse of the Ptolemy's first theorem.

Exercise 3.2.14 (*i*) *Let $ABCD$ be a convex quadrilateral. Show that*

$$|AB| \cdot |DC| + |AD| \cdot |BC| \geq |AC| \cdot |BD|.$$

(*ii*) *Show that the previous inequality becomes identity if and only if the quadrilateral $ABCD$ is cyclic.*

Exercise 3.2.15 *Show that in the case when ΔABC is an equilateral triangle we obtain Van Schouten's theorem, i.e., $|DA|+|DC| = |DB|$, for any point D on the arc AC.*

Theorem 3.2.16 (Ptolemy's second theorem) *Let $ABCD$ be a cyclic quadrilateral. Then*

$$\frac{|AC|}{|BD|} = \frac{|AD| \cdot |AB| + |BC| \cdot |CD|}{|AB| \cdot |BC| + |AD| \cdot |DC|}. \qquad (3.2.7)$$

Proof: We shall refer to Figure 3.15. Let R be the radius of the circumcircle of $ABCD$. Writing the area of the quadrilateral in two ways yields

$$\begin{aligned} \mathcal{A}_{ABCD} &= \mathcal{A}_{ABC} + \mathcal{A}_{ADC} = \frac{|AB| \cdot |BC| \cdot |AC|}{4R} + \frac{|AD| \cdot |DC| \cdot |AC|}{4R} \\ \mathcal{A}_{ABCD} &= \mathcal{A}_{ABD} + \mathcal{A}_{BCD} = \frac{|AD| \cdot |AB| \cdot |BD|}{4R} + \frac{|BC| \cdot |CD| \cdot |BD|}{4R}. \end{aligned}$$

Hence

$$\begin{aligned} |AB| \cdot |BC| \cdot |AC| + |AD| \cdot |DC| \cdot |AC| &= |AD| \cdot |AB| \cdot |BD| \\ &\quad + |BC| \cdot |CD| \cdot |BD| \\ \Longleftrightarrow |AC|(|AB| \cdot |BC| + |AD| \cdot |DC|) &= |BD|(|AD| \cdot |AB| + |BC| \cdot |CD|) \\ \Longleftrightarrow \frac{|AC|}{|BD|} &= \frac{|AD| \cdot |AB| + |BC| \cdot |CD|}{|AB| \cdot |BC| + |AD| \cdot |DC|}. \end{aligned}$$

■

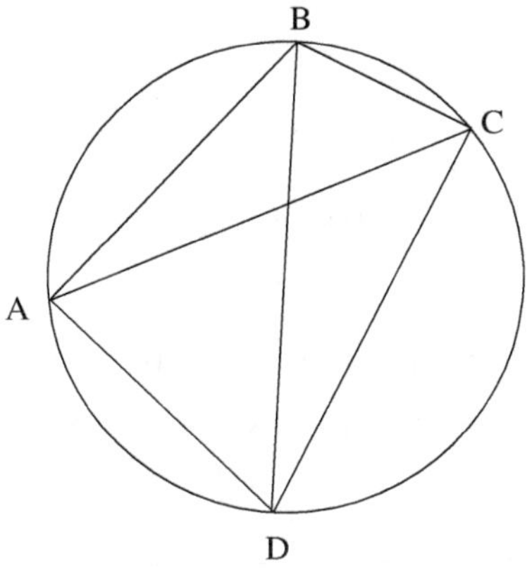

Figure 3.15: For Ptolemy's second theorem.

Corollary 3.2.17 *Let D be a point on the arc AC of the circumcircle of an equilateral triangle ABC. Then*

$$|AD| \cdot |DC| = |BD|^2 - \ell^2,$$

where ℓ is the length of the sides of ΔABC.

Proof: Ptolemy's second theorem yields

$$\frac{\ell}{|BD|} = \frac{|AD| \cdot \ell + \ell \cdot |CD|}{\ell^2 + |AD| \cdot |DC|} \Longleftrightarrow$$

$$\frac{1}{|BD|} = \frac{|AD| + |CD|}{\ell^2 + |AD| \cdot |DC|} = \frac{|BD|}{\ell^2 + |AD| \cdot |DC|},$$

where we used Van Schouten's theorem, Theorem 2.7.4. Hence

$$|BD|^2 = \ell^2 + |AD| \cdot |DC|,$$

which is equivalent to the desired relation. ■

Exercise 3.2.18 *Find the position of a point D on the arc AC of the circumcircle of equilateral triangle ABC such that the product $AD \cdot DC$ is maximum.*

Solved Exercise 3.2.19 *Let $ABCD$ be a convex quadrilateral and consider the distances $d_A = d(A, BD)$, $d_C = d(C, BD)$, $d_B = d(B, AC)$, and $d_D = d(D, AC)$ from the vertices to nonadjacent diagonals. Prove that*

$$\frac{|AC|}{|BD|} = \frac{d_A + d_C}{d_B + d_D}.$$

Proof: We write the quadrilateral area in two equivalent way as follows

$$\mathcal{A}_{ABCD} = \mathcal{A}_{ABC} + \mathcal{A}_{ACD} = \frac{1}{2}|AC|\, d_B + \frac{1}{2}|AC|\, d_D = \frac{1}{2}|AC|(d_B + d_D), \text{ and}$$

$$\mathcal{A}_{ABCD} = \mathcal{A}_{ABD} + \mathcal{A}_{BCD} = \frac{1}{2}|BD|\, d_A + \frac{1}{2}|BD|\, d_C = \frac{1}{2}|BD|(d_A + d_C).$$

Equating yields

$$|AC|(d_B + d_D) = |BD|(d_A + d_C),$$

which is equivalent to the desired conclusion. ■

Solved Exercise 3.2.20 *Let $ABCD$ be a cyclic quadrilateral and consider the notations $a = |AB|$, $b = |BC|$, $c = |CD|$, $d = |AD|$, $\delta_1 = |AC|$, and $\delta_2 = |BD|$. Prove that*

$$ab + cd = 2R\delta_2 \sin\phi,$$

where ϕ is the angle made by the diagonals.

Proof: Using the additivity of areas and the formula $\mathcal{A}_{ABCD} = \frac{1}{2}\delta_1\delta_2 \sin\phi$ we have

$$\begin{aligned}
\mathcal{A}_{ABD} + \mathcal{A}_{ACD} + \mathcal{A}_{ABC} + \mathcal{A}_{BCD} &= 2\mathcal{A}_{ABCD} \iff \\
\frac{ad\delta_2}{4R} + \frac{cd\delta_1}{4R} + \frac{ab\delta_1}{4R} + \frac{bc\delta_2}{4R} &= 2\mathcal{A}_{ABCD} \iff \\
\delta_1(ab + cd) + \delta_2(bc + ad) &= 4R\delta_1\delta_2 \sin\phi.
\end{aligned}$$

Dividing by δ_1 and using the second Ptolemy's theorem yields

$$\begin{aligned}
(ab + cd) + \frac{\delta_2}{\delta_1}(bc + ad) &= 4R\delta_2 \sin\phi \iff \\
(ab + cd) + \frac{ab + cd}{bc + ad}(bc + ad) &= 4R\delta_2 \sin\phi \iff \\
ab + cd &= 2R\delta_2 \sin\phi.
\end{aligned}$$

■

Solved Exercise 3.2.21 *Let M be the midpoint of the arc AB on the circle $\mathcal{C}(O, R)$ and P an arbitrary point on the arc AB. Using Ptolemy's theorem show that*

$$|PA| \cdot |PB| + |PM|^2 = |MA|^2.$$

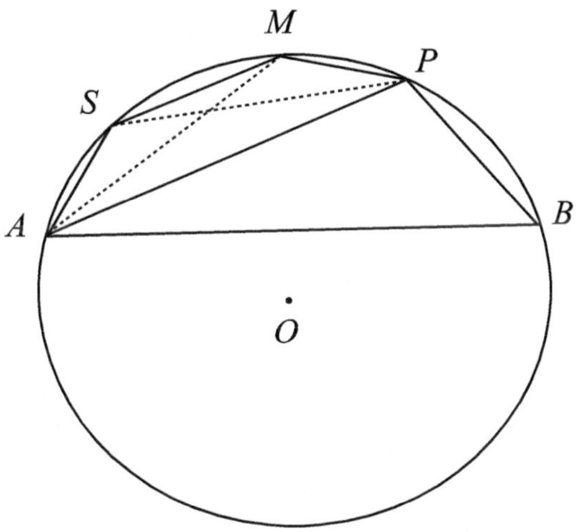

Figure 3.16: For the Exercise 3.2.21.

Proof: Take S on the circle $\mathcal{C}(O, R)$ such that $MS \| PA$, see Figure 3.16. Then $ASMP$ is a trapezoid, which is cyclic, so it is isosceles (see Proposition 2.31.5), so $|MP| = |SA|$ and has equal diagonals, $|SP| = |AM|$. Also,

$$arc(PB) = arc(MB) - arc(MP) = arc(MA) - arc(SA) = arc(SM),$$

so $|PB| = |SM|$, as to equal arcs correspond equal chords. Ptolemy's theorem, Theorem 3.2.13, applied to the cyclic quadrilateral $ASMP$ yields

$$|AP| \cdot |SM| + |AS| \cdot |MP| = |SP| \cdot |AM| \iff |AP| \cdot |PB| + |MP|^2 = |AM|^2.$$

■

3.2.6 Pappus' theorem and applications

Theorem 3.2.22 (Pappus) *If a quadrilateral is cyclic, then the product of the distances of a point on the circumcircle to two opposite sides is equal to the product of distances to the other two opposite sides.*

Proof: Let P be a point on the circumcircle of a cyclic quadrilateral $ABCD$ and denote by E, G, H and F the orthogonal projections of P on the sides AB, CD, AD and BC, respectively, see Figure 3.17. The distances from P to the sides are given by $d(P, AB) = |PE|$, $d(P, CD) = |PG|$,

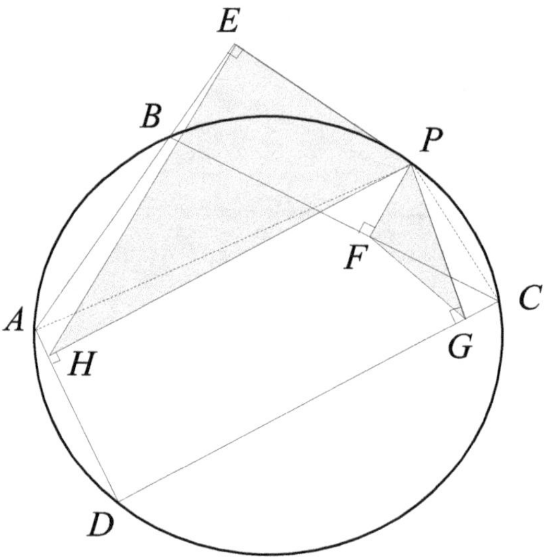

Figure 3.17: For the Pappus theorem.

$d(P, AD) = |PH|$, and $d(P, BC) = |PF|$. The relation we need to prove is equivalent to

$$\frac{|PE|}{|PF|} = \frac{|PH|}{|PG|}. \tag{3.2.8}$$

To show this it suffices to verify the triangle similarity $\Delta PEH \sim \Delta PFG$. This follows from the (AA) similarity case. To this end we have $\widehat{EPH} = \widehat{FPG}$ because

$$\begin{aligned}\widehat{EPH} &= 180° - \widehat{EAD} \ \text{(since } PEAH \text{ is cyclic)}\\ &= 180° - \widehat{BAD} = \widehat{BCD} \ \text{(since } ABCD \text{ is cyclic)}\\ &= \widehat{FCG} = \widehat{FPG} \ \text{(since } PFGC \text{ is cyclic).}\end{aligned}$$

The second pair of congruent angles are $\widehat{EHP} = \widehat{PGF}$. This follows from

$$\begin{aligned}\widehat{EHP} &= \widehat{EAP} \ \text{(since } EAHP \text{ is cyclic)}\\ &= \widehat{BAP} = \widehat{BCP} \ (= \tfrac{1}{2}arc(BP))\\ &= \widehat{FCP} = \widehat{PGF} \ \text{(since } PFGC \text{ is cyclic).}\end{aligned}$$

Hence $\Delta PEH \sim \Delta PFG$, which implies relation (3.2.8). ■

We shall provide some applications of Pappus' theorem to a triangle. To this end we shall consider the case when $A = D$, namely, when the quadrilateral $ABCD$ degenerates to a triangle ΔABC. We note that the quadrilateral side AD becomes in this case the tangent line at A to the circumcircle of ΔABC. If P is a point on the circumcircle of ΔABC, then we denote by E, G, F the orthogonal projections of P onto the sides AB, AC and BC, respectively. Let H be the orthogonal projection of P on the tangent line at A, see Figure 3.18(a). Then Pappus' theorem provides

$$|PF| \cdot |PH| = |PG| \cdot |PE|. \tag{3.2.9}$$

We also note that the projection points E, G and F are collinear, as they belong to the Simpson's line of the point P with respect to ΔABC. This setup is useful in the proof of the following result.

Proposition 3.2.23 *Let A', B', C' be the contact points of the inner circle of ΔABC with the sides BC, AC and BC, respectively. Let P be an arbitrary point on the inner circle of ΔABC. Then the product of the normal distances from M to the sides of ΔABC is equal to the product of the normal distances from P to the sides of $\Delta A'B'C'$.*

Proof: For the sake of simplicity we denote $\alpha = |PF|$, $\alpha' = |PH|$, $\beta = |PG|$ and $\gamma = |PE|$. Then (3.2.9) becomes $\alpha\alpha' = \beta\gamma$. Two other similar relations are $\beta\beta' = \alpha\gamma$ and $\gamma\gamma' = \alpha\beta$, where $\beta' = d(P, \text{tangent at } B)$ and $\gamma' = d(P, \text{tangent at } C)$. Then multiplying the previous three relations and taking a square root, we obtain $\alpha\beta\gamma = \alpha'\beta'\gamma'$, which is the desired relation. See Figure 3.18(b). ■

Remark 3.2.24 The previous relation is valid in the particular case when P is the Feuerbach point of ΔABC, see page 412 for details.

3.3 The centroid of a quadrilateral

Theorem 3.3.1 (Varignon) *Let $ABCD$ be a convex quadrilateral, and let E, F, M, N, X, Y be the midpoints of the sides AB, BC, CD, DA and diagonals BD, AC respectively. Then the segments NF, EM and XY intersect at a point K, which is the midpoint of the segments NF, EM and XY, see Figure 3.19.*

Proof: The midpoint connector theorem applied to ΔABC, ΔADC yields

$$\begin{aligned} EF \| AC, \ |EF| &= \frac{1}{2}|AC|, \text{and} \\ NM \| AC, \ |NM| &= \frac{1}{2}|AC|. \end{aligned}$$

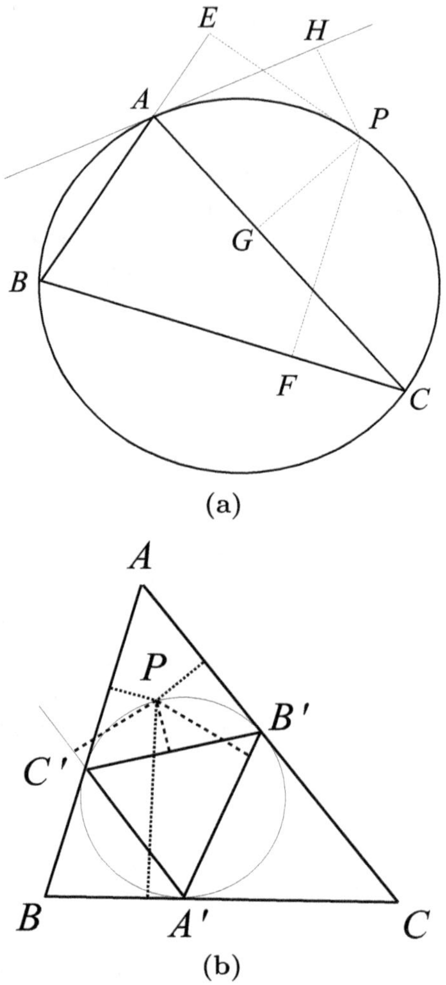

Figure 3.18: (a) For relation (3.2.9); (b) For Proposition 3.2.23.

Then $EF \| NM$ and $|EF| = |NM|$. Hence $EFMN$ is a parallelogram. The diagonals NF and EM bisect (Why?). Let $\{K\} = EM \cap NF$.

Applying the midpoint connector theorem to ΔABD and ΔABC yields

$$\begin{aligned} NX \| AB, \ |NX| &= \frac{1}{2}|AB|, \text{and} \\ YF \| AB, \ |YF| &= \frac{1}{2}|AB|. \end{aligned}$$

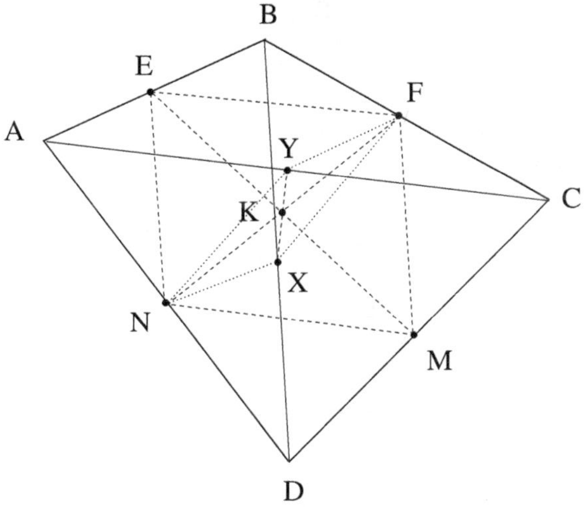

Figure 3.19: The centroid of a quadrilateral.

It follows that $YFXN$ is a parallelogram and hence the diagonals YX and NF bisect. Their intersection point is K. ∎

Definition 3.3.2 *The segments EM, NF, XY are called* **bimedians** *of the quadrilateral. Their intersection point K is called the* **centroid** *of the quadrilateral.*

Remark 3.3.3 Unlike for the triangle case, in the case of a quadrilateral the centroid does not represent the center of mass or center of gravity of the entire solid quadrilateral. It does represent just the centroid of the system of points $\{A, B, C, D\}$, equally weighted.

Exercise 3.3.4 *Let $ABCD$ be a trapezoid with $AB \| CD$ and $|AB| < |CD|$. Show that one of the bimedians does not bisect the trapezoid in two trapezoids of equal areas.*

We shall next compute the length of the third bimedian XY. The result is due to Euler, and its proof is an application of the median theorem.

Theorem 3.3.5 (Euler) *Consider the assumptions and notations from Varignon's theorem. We have*

$$4|XY|^2 = |AB|^2 + |BC|^2 + |CD|^2 + |DA|^2 - |AC|^2 - |BD|^2. \quad (3.3.10)$$

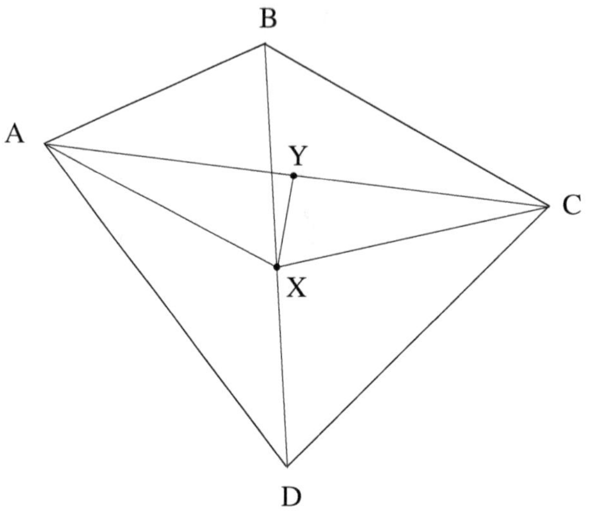

Figure 3.20: For the proof of Theorem 3.3.5.

Proof: The median theorem, Theorem 2.21.3, applied to ΔACX, ΔBAD, ΔBCD yields

$$\begin{aligned}|XY|^2 &= \frac{1}{2}(|AX|^2 + |CX|^2) - \frac{1}{4}|AC|^2 \\ &= \frac{1}{2}\Big[\frac{1}{2}(|AB|^2 + |AD|^2) - \frac{1}{4}|BD|^2 + \frac{1}{2}(|CB|^2 + |CD|^2) - \frac{1}{4}|BD|^2\Big] \\ &\quad - \frac{1}{4}|AC|^2 \\ &= \frac{1}{4}(|AB|^2 + |AD|^2 + |CB|^2 + |CD|^2) - \frac{1}{4}|BD|^2 - \frac{1}{4}|AC|^2,\end{aligned}$$

which yields (3.3.10). See Figure 3.20. ∎

Exercise 3.3.6 *Let $ABCD$ be a convex quadrilateral and F, N be the midpoints of the sides BC, AD. Show that*

$$4|NF|^2 = |AB|^2 + |CD|^2 + |AD|^2 + |BD|^2 + |AC|^2 - |BC|^2 - |AD|^2.$$

Exercise 3.3.7 *Let E, F, M, N, X, Y be the midpoints of the sides AB, BC, CD, DA and diagonals BD and AC of the convex quadrilateral*

ABCD. Prove that

$$\begin{aligned} |EM|^2 + |NF|^2 &= \frac{1}{2}(|BD|^2 + |AC|^2), \textit{ and} \\ |EM|^2 + |NF|^2 + |XY|^2 &= \frac{1}{4}(|AB|^2 + |BC|^2 + |CD|^2 \\ &\quad + |DA|^2 + |BD|^2 + |AC|^2). \end{aligned}$$

Exercise 3.3.8 *Let $ABCD$ be a convex quadrilateral. Show that*

$$|AB|^2 + |BC|^2 + |CD|^2 + |DA|^2 \geq |AC|^2 + |BD|^2,$$

with identity if and only if $ABCD$ is a parallelogram.

The next exercise is an extension of Exercise 1.10.2 from the triangle to quadrilaterals.

Exercise 3.3.9 *Let M and N be the midpoints of sides AD and BC of the convex quadrilateral $ABCD$.*

(i) Show the inequality $|MN| \leq (|AB| + |DC|)/2$.

(ii) Prove that we have $|MN| = (|AB| + |DC|)/2$ if and only if $AB \| DC$.

Exercise 3.3.10 *Let $ABCD$ be a convex quadrilateral.*

(i) Show the inequality $2|AB| \cdot |CD| + |AD|^2 + |BC|^2 \geq |AC|^2 + |BD|^2$.

(ii) Prove that the inequality in part (i) becomes identity if and only if $AB \| DC$.

Solved Exercise 3.3.11 *Let $ABCD$ be a convex quadrilateral and E, F, M, N be the midpoints of the sides AB, DC, AD and BC, respectively. Show that the bimedians EF and MN are perpendicular if and only if the quadrilateral has equal diagonals.*

Proof: An application of the midpoint connector theorem shows that $ENFM$ is a parallelogram. We assume $EF \perp MN$. Then the parallelogram $ENFM$ has perpendicular diagonals, so it is a rhombus, i.e., all its sides are equal, see Exercise 3.1.12. In particular, $|EN| = |NF|$. Since from the midpoint connector theorem $|EN| = \frac{1}{2}|AC|$ and $|NF| = \frac{1}{2}|BD|$, we obtain $|AC| = |BD|$. We assume now that $|AC| = |BD|$. Then $|EN| = |NF|$ and the parallelogram $ENFM$ becomes a rhombus by Exercise 3.1.12 and then its diagonals are perpendicular, $EF \perp MN$. ■

3.4 The anticenter of a cyclic quadrilateral

In this section we shall introduce a new center for a quadrilateral, which is an analogue of the orthocenter of a triangle. This property holds for cyclic quadrilaterals only.

Theorem 3.4.1 *For every side of $ABCD$ we construct a line through its midpoint and perpendicular to the opposite side. Then these lines intersect at a point, called the* **anticenter**.

Proof: Let O be the center of the circumcircle. It is known that the centroid K is the midpoint of the bimedian FN. Let H be a point on OK such that $|HK| = |KO|$. Then $HFON$ is a quadrilateral with the property that the diagonals bisect, and hence it is a parallelogram (Why?). Therefore

$$HF \| ON, \quad NH \| OF. \tag{3.4.11}$$

Using that the line passing through the center of a circle and the midpoint of a chord is perpendicular to the chord, yields

$$ON \perp AD, \quad OF \perp BC. \tag{3.4.12}$$

Now (3.4.11) and (3.4.12) yield

$$HF \perp AD, \quad NH \perp BC.$$

In a similar way, using the parallelogram $EHMO$ yields

$$HM \perp AB, \quad HE \perp DC.$$

■

We make the remark that the anticenter is also called the *Mathot point* of a cyclic quadrilateral.

Exercise 3.4.2 *Consider a cyclic quadrilateral $ABCD$. Show the following:*
(i) The lines through the midpoint of each diagonal and perpendicular to the other diagonal pass through the anticenter.
(ii) Let X, Y be the midpoints of the diagonals AC, BD and let $\{Z\} = AC \cap BD$. If X, Y, Z are distinct points, the anticenter of $ABCD$ is the orthocenter of ΔXYZ.

Remark 3.4.3 The analogue of the Euler's line for a cyclic quadrilateral is the line determined by O, K, and H. Hence, it makes sense to compute the distances between these important points. This will be done in the next section.

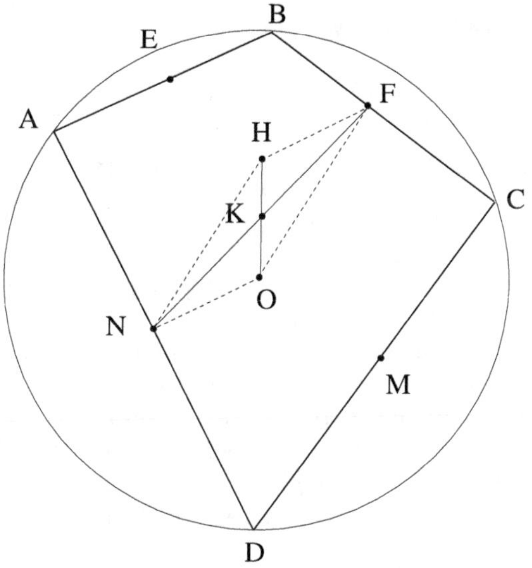

Figure 3.21: The anticenter H. The line OKH is an analog of the Euler line for quadrilaterals.

3.5 The distance $|OH|$

We start by computing the distance between the circumcenter, O, and the anticenter, H, using the median theorem in triangle ΔONF, see Figure 3.21. To this end we need first to express the sides $|OF|$ and $|ON|$ in terms of the quadrilateral sides $a = |AB|$, $b = |BC|$, $c = |CD|$, $d = |AD|$, diagonals $\delta_1 = |AC|$, $\delta_2 = |BD|$, and circumradius, R. Applying the Pythagorean theorem in the right triangles ΔOFC and ΔOND we have

$$|OF|^2 = R^2 - \frac{b^2}{4}, \qquad |ON|^2 = R^2 - \frac{d^2}{4}.$$

The median theorem applied in triangle ΔONF yields

$$\begin{aligned}|OK|^2 &= \frac{1}{2}(|ON|^2 + |OF|^2) - \frac{1}{4}|NF|^2 = \frac{1}{2}\left(R^2 + R^2 - \frac{b^2}{4} - \frac{d^2}{4}\right) - \frac{1}{4}|NF|^2 \\ &= R^2 - \frac{1}{8}(b^2 + d^2) - \frac{1}{16}(a^2 + c^2 - b^2 - d^2 + \delta_1^2 + \delta_2^2) \\ &= R^2 - \frac{1}{16}(a^2 + b^2 + c^2 + d^2 + \delta_1^2 + \delta_2^2).\end{aligned}$$

Now, since $|OH| = 2|OK|$ we obtain

$$|OH|^2 = 4R^2 - \frac{1}{4}(a^2 + b^2 + c^2 + d^2 + \delta_1^2 + \delta_2^2). \tag{3.5.13}$$

Since $|OH|^2 \geq 0$, we obtain the following inequality

$$16R^2 \geq a^2 + b^2 + c^2 + d^2 + \delta_1^2 + \delta_2^2, \tag{3.5.14}$$

which holds for any cyclic quadrilateral. The inequality becomes an equality in the case of a rectangle.

To show this we note that the equality holds in (3.5.14) if and only if $O = H$, or equivalently, $O = K$. This corresponds to the case when the line NF becomes a diameter that is perpendicular on both AD and BC, which implies $AD \| BC$. Similarly, we obtain $AB \| DC$, so that $ABCD$ becomes a parallelogram. Since the rectangle is the only cyclic parallelogram, see Exercise 3.5.1, then we obtain the desired conclusion.

Exercise 3.5.1 *Show that if a parallelogram is cyclic, then it is a rectangle.*

3.6 Cocyclic quadrilaterals

Definition 3.6.1 *A quadrilateral $ABCD$ is called* **cocyclic** *if it is circumscribed to a circle, i.e., there is a circle $\mathcal{C}$ tangent to the quadrilateral sides at interior points of the sides.*

We note that sometimes cocyclic quadrilaterals are also called *tangential* quadrilaterals.

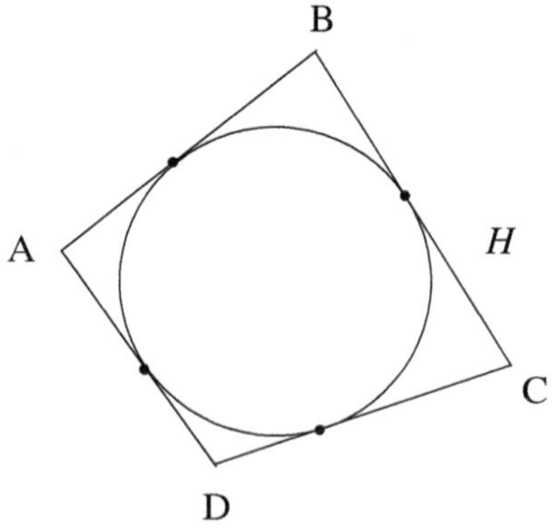

Figure 3.22: A cocyclic quadrilateral.

Theorem 3.6.2 *If a quadrilateral is cocyclic, then it is convex.*

Proof: Let $ABCD$ be a quadrilateral circumscribed to a circle $\mathcal{C}$. The side AB is tangent to $\mathcal{C}$. Let $\mathcal{H}$ denote the half-plane determined by AB which contains $\mathcal{C}$. Then $\mathcal{H}$ contains the contact points between the circle $\mathcal{C}$ and the sides BC and AD. So the segments BC and AD are included in $\mathcal{H}$ and hence $C, D \in \mathcal{H}$. ■

Not all quadrilaterals are circumscribed to a circle. We shall present an analogue of Theorem 2.5.3.

Theorem 3.6.3 *The quadrilateral $ABCD$ is circumscribed to a circle if and only if the angle bisectors of the angles A, B, C and D intersect at a point.*

Proof: " $\Longrightarrow$ " See Figure 3.23. Let I be the center of the incircle of $ABCD$. Let E, F, M, N be the orthogonal projections of I on the sides AB, BC, CD and DA respectively. As $|IE| = |IF| = |IM| = |IN|$, Theorem 2.5.2 yields that I belongs to the angle bisectors of angles A, B, C, D.

" $\Longleftarrow$ " Let I be the point where the angle bisectors intersect. Let E, F, M, N be the orthogonal projections of I on the sides. By Theorem 2.5.2 we have

$$|IE| = |IF| = |IM| = |IN|.$$

Consider the circle $\mathcal{C}(I, r)$ with $r = IE$. Because $IE \perp AB$, $IF \perp BC$, $IM \perp DC$, $IN \perp AD$, it follows that $\mathcal{C}(I, r)$ is tangent to the sides of the quadrilateral $ABCD$. ■

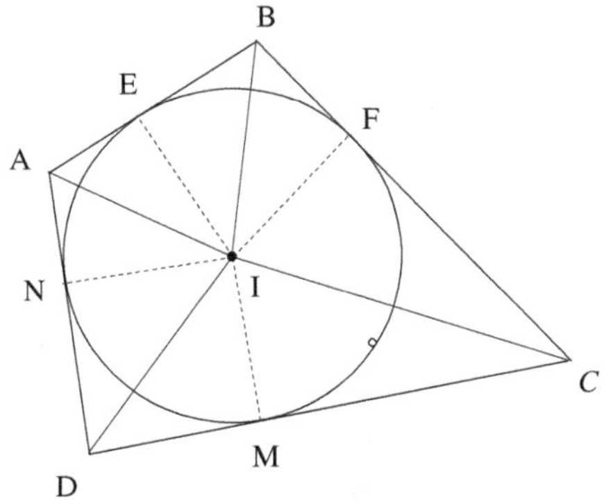

Figure 3.23: The angle bisectors of a cocycle quadrilateral.

Proposition 3.6.4 *Let a, b, c, d be the lengths of the sides of the cocyclic quadrilateral $ABCD$. If r denotes the inradius, then*

$$\mathcal{A}_{ABCD} = \frac{1}{2}(a+b+c+d)\cdot r.$$

Proof: Let I be the incenter. Using the additivity of area yields

$$\begin{aligned}\mathcal{A}_{ABCD} &= \mathcal{A}_{ABI} + \mathcal{A}_{BIC} + \mathcal{A}_{CID} + \mathcal{A}_{DIA}\\ &= \frac{1}{2}|AB|\cdot|EI| + \frac{1}{2}|BC|\cdot|IF| + \frac{1}{2}|DC|\cdot|IM| + \frac{1}{2}|AD|\cdot|IN|\\ &= \frac{1}{2}(a+b+c+d)\cdot r.\end{aligned}$$

■

Theorem 3.1.3 shows that a quadrilateral is cyclic if and only if the sum of the opposite angles is the same. The following theorem is an analogue for the cocyclic quadrilaterals, where the role of the opposite angles is replaced by the role of the opposite sides.

Theorem 3.6.5 (Pithot) *Let $ABCD$ be a convex quadrilateral. Then $ABCD$ can be circumscribed to a circle if and only if*

$$|AB| + |CD| = |AD| + |BC|.$$

Proof: " $\Longrightarrow$ " See Figure 3.24. Let M, N, P, Q be the contact points between the circle and the sides AB, BC, CD and DA respectively. As the lengths of the tangent segments from a point to a circle are equal, we have

$$|AM| = |AQ|,\ \ |BM| = |BN|,\ \ |CP| = |CN|,\ \ |DQ| = |DP|.$$

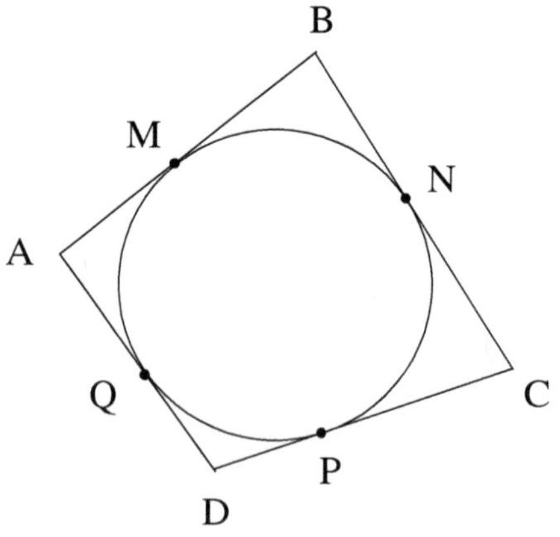

Figure 3.24: For Theorem 3.6.5.

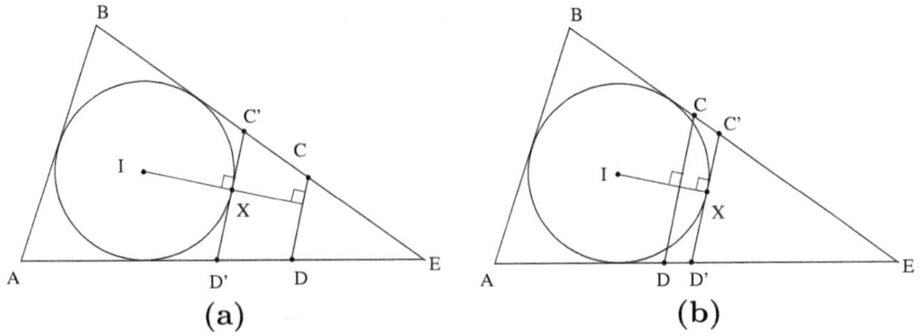

Figure 3.25: For the proof of Pithot's theorem.

Hence

$$\begin{aligned}|AB| + |CD| &= |AM| + |MB| + |CP| + |PD| \\ &= |AQ| + |BN| + |CN| + |DQ| \\ &= |AQ| + |DQ| + |BN| + |CN| \\ &= |AD| + |BC|.\end{aligned}$$

" $\Longleftarrow$ " Assume that $|AB| + |CD| = |AD| + |BC|$. We shall show that the quadrilateral is circumscribed to a circle.
Let $\{E\} = AD \cap BC$, and let $\mathcal{C}$ be the incircle of ΔABE. If CD is tangent to $\mathcal{C}$, then we obtain the desired result. Assume that CD is not tangent to $\mathcal{C}$. Let I be the incenter of ΔABE. The line from I perpendicular to CD intersects the circle $\mathcal{C}$ at two points. Let X be the point such that the tangent line at X to $\mathcal{C}$ separates points I and E. Let C', D' be the intersections with BE and AE of the tangent line to $\mathcal{C}$ at X. As $ABC'D'$ has an incircle, using the implication " $\Longrightarrow$ " proved before, yields

$$|AB| + |C'D'| = |AD'| + |BC'|.$$

Subtracting from the relation

$$|AB| + |CD| = |AD| + |BC|$$

in the situation of Figure 3.25(a) yields

$$\begin{aligned}|CD| - |C'D'| &= |AD| + |BC| - |AD'| - |BC'| \\ &= (|AD| - |AD'|) + (|BC| - |BC'|) \\ &= |D'D| + |C'C|.\end{aligned}$$

It follows that $|CD| = |C'D'| + |D'D| + |C'C|$, which contradicts the following inequalities obtained from the triangle inequality:

$$|C'D'| + |D'D| + |C'C| > |D'D| + |CD'| > |CD|.$$

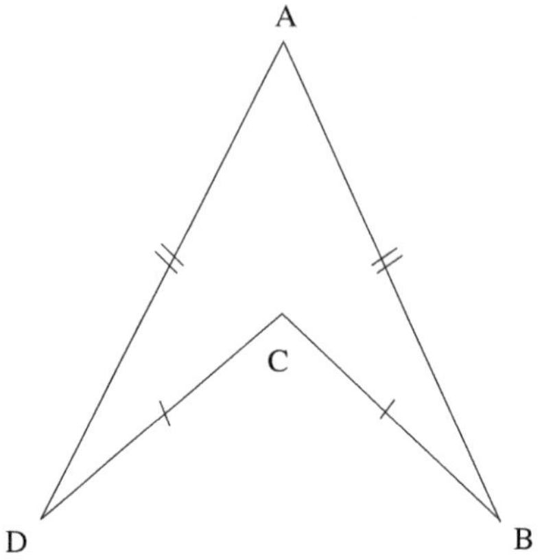

Figure 3.26: A counterexample.

In the case of Figure 3.25(b) we have

$$\begin{aligned} |CD| - |C'D'| &= -|DD'| - |CC'| \\ \Longrightarrow |CD| &= |DD'| + |CC'| + |C'D'|, \end{aligned}$$

which leads to a similar contradiction.

Hence CD is tangent to the circle $\mathcal{C}$. ■

Remark 3.6.6 If the quadrilateral is not convex the conclusion might fail. In Figure 3.26 we have $|AB| = |AD|$, $|BC| = |CD|$. Then $|AB| + |CD| = |AD| + |BC|$, but $ABCD$ cannot be circumscribed to a circle.

3.6.1 Newton's theorems

The description of right lines and circles upon which geometry is founded, belongs to mechanics. Geometry does not teach us to draw these lines, but requires them to be drawn.

Isaac Newton, 1687

Isaac Newton (1642–1727) is well-known for his formulation of the principles of mechanics and for inventing calculus. In classical geometry he is present with the following theorem.

Theorem 3.6.7 (Newton) *Let $ABCD$ be a quadrilateral circumscribed to a circle. Let M, N, P, Q be the contact points as usual. Then the segments AC, BD, MP, QN intersect at a point.*

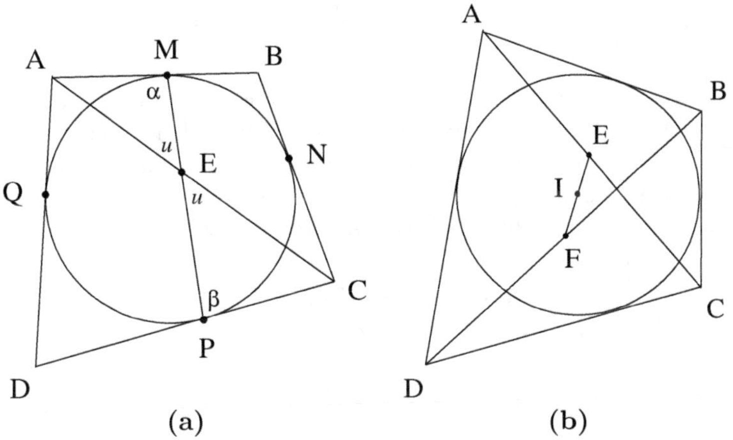

Figure 3.27: (a) Newton's theorem; (b) Newton's line.

Proof: We consider Figure 3.27(a). Let $\alpha = \widehat{AMP}$, $\beta = \widehat{MPC}$. Then $\alpha + \beta = 180$, because

$$\alpha = \widehat{AMP} = \frac{1}{2}arc(MQ) + \frac{1}{2}arc(QP) = \widehat{MPD} = 180 - \beta.$$

Let $\{E\} = MP \cap AC$. The law of sines applied to ΔAME and ΔCEP yields

$$\frac{|AE|}{\sin\alpha} = \frac{|AM|}{\sin u}, \qquad \frac{|EC|}{\sin\beta} = \frac{|PC|}{\sin u},$$

where $u = \widehat{AEM} = \widehat{PEC}$. Because $\alpha = 180 - \beta$, $\sin\alpha = \sin\beta$, and so dividing the above identities yields

$$\frac{|AE|}{|EC|} = \frac{|AM|}{|PC|}. \tag{3.6.15}$$

Let $E' = NQ \cap AC$. In a similar way, we obtain

$$\frac{|AE'|}{|E'C|} = \frac{|AQ|}{|NC|}. \tag{3.6.16}$$

Using $|AM| = |AQ|$, $|CN| = |CP|$ (the lengths of the tangents from a point to a circle are equal), from (3.6.15) and (3.6.16) yields

$$\frac{|AE|}{|EC|} = \frac{|AE'|}{|E'C|} \iff \frac{|AE|}{|AE| + |EC|} = \frac{|AE'|}{|AE'| + |E'C|}$$

$$\iff \frac{|AE|}{|AC|} = \frac{|AE'|}{|AC|} \Longrightarrow E = E'.$$

Hence $E \in AC \cap NQ \cap MP$. In a similar way, one may show that $E \in BD \cap NQ \cap MP$. ■

Let E, F be the midpoints of the diagonals. Then the line passing through E and F is called the **Newton's line.** It is worth to notice that this concept makes sense in the case when $E \neq F$, namely when the quadrilateral is not a parallelogram. In the next few problems we assume that the quadrilateral is general enough such that its Newton's line exists.

Exercise 3.6.8 *Let E be the midpoint of the diagonal AC of the convex quadrilateral $ABCD$. Show that*

$$\mathcal{A}_{AED} + \mathcal{A}_{BEC} = \frac{1}{2}\mathcal{A}_{ABDC}.$$

Exercise 3.6.9 *Consider a convex quadrilateral $ABCD$. Show that a point M belongs to Newton's line if and only if*

$$\mathcal{A}_{AMB} + \mathcal{A}_{CMD} = \mathcal{A}_{BMC} + \mathcal{A}_{AMD}.$$

Exercise 3.6.10 (Newton's theorem) *Prove that the center of the circle inscribed in a quadrilateral lies on the line joining the midpoints of the quadrilateral's diagonals, see Figure 3.27(b).*

Solved Exercise 3.6.11 *Show that there are infinitely many noncongruent quadrilaterals circumscribed to a given circle.*

Proof: Let a, b, c, d denote the quadrilateral sides lengths and x, y, z, u be the lengths of the segments made by the contact points on the sides. Since the length of the tangents from a point to a circle are equal, we have

$$\begin{aligned} x + y &= a, \\ y + z &= b, \\ z + u &= c, \text{ and} \\ u + x &= d. \end{aligned}$$

The matrix of this linear system is degenerate, and the solution depends on a real parameter t as

$$x = a - t, \quad y = t, \quad z = b - t, \quad u = d - a + t,$$

with $a - d < t < \min\{a, b\}$. ■

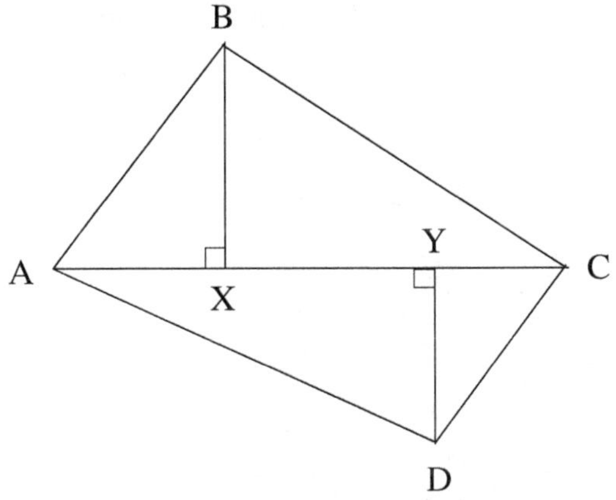

Figure 3.28: See the proof of Theorem 3.7.2.

3.7 Orthodiagonal quadrilaterals

Definition 3.7.1 *A convex quadrilateral $ABCD$ is called* **orthodiagonal** *if the diagonals AC and BD are perpendicular.*

Theorem 3.7.2 *The convex quadrilateral $ABCD$ is orthodiagonal if and only if*

$$|AB|^2 + |CD|^2 = |BC|^2 + |AD|^2.$$

Proof: Let X and Y be the orthogonal projections of the points B and D on the diagonal AC. The quadrilateral $ABCD$ is orthodiagonal if and only if $BD \perp AC$. This is equivalent with $XY = 0$ or $X = Y$. Hence it suffices to estimate the distance XY as a function of the sides of the quadrilateral.

Using triangles ΔABX and ΔDCY, we have

$$\cos \widehat{BAX} = \frac{|AX|}{|AB|} \quad \text{and} \quad \cos \widehat{YCD} = \frac{|YC|}{|CD|}.$$

The law of cosines applied to ΔABC and ΔACD yields

$$\begin{aligned}
|BC|^2+|AD|^2 &= |AB|^2+|AC|^2-2|AB|\cdot|AC|\cdot\cos\widehat{BAC}\\
&\quad +|AC|^2+|CD|^2-2|AC|\cdot|CD|\cdot\cos\widehat{ACD}\\
&= |AB|^2+|AC|^2-2|AB|\cdot|AC|\cdot\frac{|AX|}{|AB|}\\
&\quad +|AC|^2+|CD|^2-2|AC|\cdot|CD|\cdot\frac{|YC|}{|CD|}\\
&= |AB|^2+|AC|^2-2|AC|\cdot|AX|+|AC|^2+|CD|^2\\
&\quad -2|AC|\cdot|YC|\\
&= |AB|^2+|CD|^2+2|AC|^2-2|AC|(|AX|+|YC|)\\
&= |AB|^2+|CD|^2+2|AC|(|AC|-|AX|-|YC|)\\
&= |AB|^2+|CD|^2+2|AC|\cdot|XY|.
\end{aligned}$$

Therefore

$$|BC|^2+|AD|^2=|AB|^2+|CD|^2+2|AC|\cdot|XY|, \tag{3.7.17}$$

and hence $X=Y$ if and only if $|BC|^2+|AD|^2=|AB|^2+|CD|^2$. ∎

Proposition 3.7.3 *The area of an orthodiagonal quadrilateral is*

$$\mathcal{A}=\frac{1}{2}\delta_1\delta_2,$$

where δ_1*,* δ_2 *are the lengths of the diagonals.*

Proof: Let $ABCD$ be an orthodiagonal quadrilateral and $AC\cap BD=\{M\}$. Then

$$\begin{aligned}
\mathcal{A}=\mathcal{A}_{ABC}+\mathcal{A}_{ACD} &= \frac{1}{2}|BM|\cdot|AC|+\frac{1}{2}|DM|\cdot|AC|\\
&= \frac{1}{2}|AC|(|BM|+|DM|)\\
&= \frac{1}{2}|AC|\cdot|BD|=\frac{1}{2}\delta_1\delta_2.
\end{aligned}$$

∎

3.8 The area of a quadrilateral

The following formula expresses the area of a quadrilateral in terms of the diagonal lengths and the angle between the diagonals. It can be considered as the natural generalization of the result provided by Proposition 3.7.3.

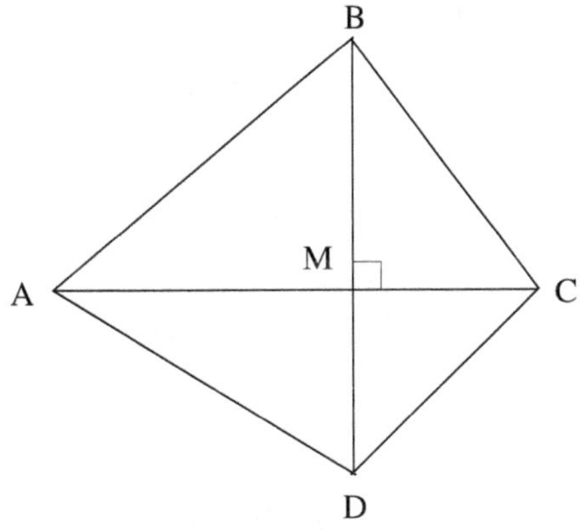

Figure 3.29: Orthodiagonal quadrilateral.

Proposition 3.8.1 *Denote by δ_1, δ_2 the lengths of the diagonals of a convex quadrilateral $ABCD$. Let φ be the acute angle between the diagonals. Then the area of $ABCD$ is given by*

$$\mathcal{A} = \frac{1}{2}\delta_1\delta_2 \sin\varphi.$$

Proof: Let $\{M\} = AC \cap BD$. Then

$$\begin{aligned}
\mathcal{A} &= \mathcal{A}_{ABM} + \mathcal{A}_{MBC} + \mathcal{A}_{CMD} + \mathcal{A}_{AMD} \\
&= \frac{1}{2}|AM| \cdot |BM| \sin\varphi + \frac{1}{2}|BM| \cdot |MC| \sin(180 - \varphi) \\
&\quad + \frac{1}{2}|MC| \cdot |MD| \sin\varphi + \frac{1}{2}|AM| \cdot |MD| \sin(180 - \varphi) \\
&= \frac{1}{2}\Big[|BM|(|AM| + |MC|)\sin\varphi + |MD|(|MC| + |AM|)\sin\varphi\Big] \\
&= \frac{1}{2}(|BM| \cdot |AC| \sin\varphi + |MD| \cdot |AC| \sin\varphi) \\
&= \frac{1}{2}|AC|(|BM| + |MD|)\sin\varphi \\
&= \frac{1}{2}|AC| \cdot |BD| \sin\varphi \\
&= \frac{1}{2}\delta_1\delta_2 \sin\varphi.
\end{aligned}$$

■

The angle between the diagonals depends on the lengths of the sides. This relation is provided by the following lemma.

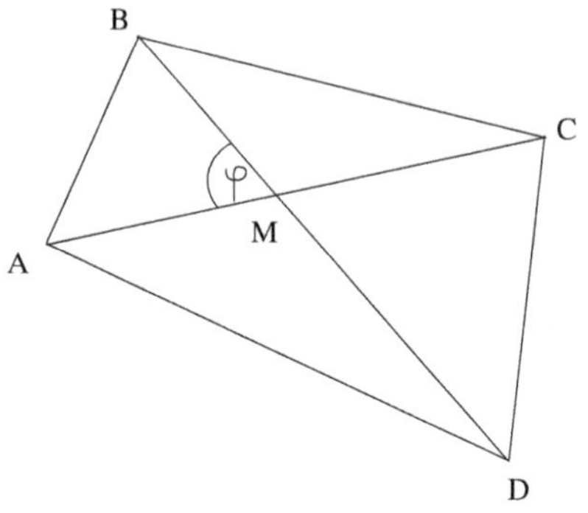

Figure 3.30: The angle φ.

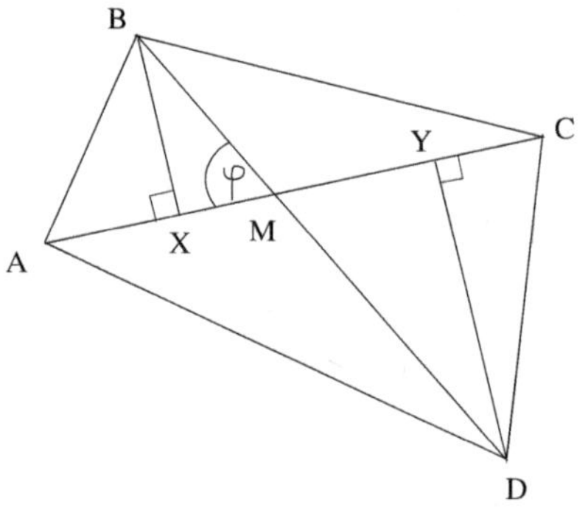

Figure 3.31: See the proof of Lemma 3.8.2.

Lemma 3.8.2 *Let φ be the acute angle between the diagonals AC and BD of the convex quadrilateral $ABCD$. Then*

$$\cos\varphi = \frac{\left||AB|^2 - |BC|^2 + |CD|^2 - |DA|^2\right|}{2|AC|\cdot|BD|}. \tag{3.8.18}$$

Proof: Let X, Y be the orthogonal projections of B, D on the diagonal AC. Then expressing $\cos\varphi$ from ΔBMX and ΔDMY yields

$$\cos\varphi = \frac{|XM|}{|BM|} = \frac{|MY|}{|MD|} = \frac{|XM| + |MY|}{|BM| + |MD|} = \frac{|XY|}{|BD|}.$$

Formula (3.7.17) yields

$$|XY| = \frac{|BC|^2 + |AD|^2 - |AB|^2 - |CD|^2}{2|AC|}.$$

Hence

$$\cos\varphi = \frac{|XY|}{|BD|} = \frac{|BC|^2 + |AD|^2 - |AB|^2 - |CD|^2}{2|AC| \cdot |BD|}.$$

■

The following theorem expresses the area of a quadrilateral in terms of the sides and diagonals.

Theorem 3.8.3 *Let $ABCD$ be a convex quadrilateral. Let $a = AB$, $b = BC$, $c = CD$, $d = AD$ and $\delta_1 = AC$, $\delta_2 = BD$. Then the area of the quadrilateral is given by*

$$\mathcal{A} = \frac{1}{4}\left[4\delta_1^2\delta_2^2 - (a^2 - b^2 + c^2 - d^2)^2\right]^{1/2}. \tag{3.8.19}$$

Proof: Using Lemma 3.8.2 and Proposition 3.8.1 yields

$$\begin{aligned}\mathcal{A} &= \frac{1}{2}\delta_1\delta_2 \sin\varphi = \frac{1}{2}\delta_1\delta_2\sqrt{1 - \cos^2\varphi} \\ &= \frac{1}{2}\delta_1\delta_2\sqrt{1 - \frac{(a^2 - b^2 + c^2 - d^2)^2}{4\delta_1^2\delta_2^2}} \\ &= \frac{1}{4}\sqrt{4\delta_1^2\delta_2^2 - (a^2 - b^2 + c^2 - d^2)^2}.\end{aligned}$$

■

3.8.1 Heron's formula for a cyclic quadrilateral

In the case of cyclic quadrilaterals, the area can be expressed using a formula analogous to Heron's formula for the area of a triangle.

Theorem 3.8.4 (Heron's formula) *Let a, b, c, d be the side lengths of a cyclic quadrilateral. Let $p = (a + b + c + d)/2$ denote the semiperimeter. Then the quadrilateral area is given by*

$$\mathcal{A} = \sqrt{(p - a)(p - b)(p - c)(p - d)}.$$

Proof: Relation (3.8.19) can be written as

$$16\mathcal{A}^2 = 4\delta_1^2\delta_2^2 - (a^2 - b^2 + c^2 - d^2)^2. \tag{3.8.20}$$

Substituting Ptolemy's relation $\delta_1\delta_2 = ac + bd$, see Theorem 3.2.13, in (3.8.20) yields

$$\begin{aligned}16\mathcal{A}^2 &= 4(ac+bd)^2 - (a^2 - b^2 + c^2 - d^2)^2\\ &= (2ac + 2bd + a^2 - b^2 + c^2 - d^2)(2ac + 2bd - a^2 + b^2 - c^2 + d^2)\\ &= [(a^2 + 2ac + c^2) - (b^2 - 2bd + d^2)][(b^2 + 2bd + d^2) - (a^2 - 2ac + c^2)]\\ &= [(a+c)^2 - (b-d)^2][(b+d)^2 - (a-c)^2]\\ &= (a+c-b+d)(a+c+b-d)(b+d+a-c)(b+d-a+c)\\ &= (2p-2b)(2p-2d)(2p-2c)(2p-2a).\end{aligned}$$

Hence

$$\mathcal{A}^2 = (p-a)(p-b)(p-c)(p-d). \qquad \blacksquare$$

Exercise 3.8.5 *Prove that if a cyclic quadrilateral is circumscribed to a circle of radius r, then*

$$r = \frac{\sqrt{(p-a)(p-b)(p-c)(p-d)}}{p}.$$

3.8.2 Archimedes' formula

There was more imagination in the head of Archimedes than in that of Homer.

Voltaire

Archimedes' formula is the generalization of the Heron's formula for any convex quadrilateral.

Theorem 3.8.6 (Archimedes' formula) *The area of a convex quadrilateral $ABCD$ with sides lengths a, b, c, d is given by*

$$\mathcal{A}^2 = (p-a)(p-b)(p-c)(p-d) - abcd\cos^2\frac{\widehat{B}+\widehat{D}}{2},$$

where $p = (a+b+c+d)/2$.

Proof: Using the additivity of area

$$\mathcal{A} = \mathcal{A}_{ABC} + \mathcal{A}_{ADC} = \frac{1}{2}ab\sin\widehat{B} + \frac{1}{2}cd\sin\widehat{D}$$

and hence $2\mathcal{A} = ab\sin\widehat{B} + cd\sin\widehat{D}$. Therefore

$$\begin{aligned}4\mathcal{A}^2 &= (ab\sin\widehat{B} + cd\sin\widehat{D})^2\\ &= a^2b^2\sin^2\widehat{B} + c^2d^2\sin^2\widehat{D} + 2abcd\sin\widehat{B}\sin\widehat{D}. \qquad (3.8.21)\end{aligned}$$

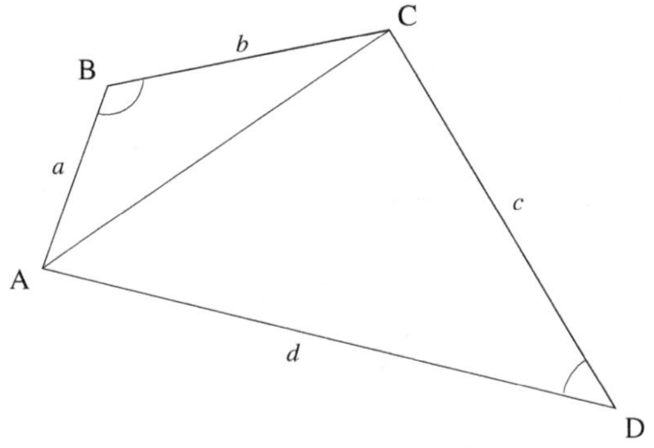

Figure 3.32: See the proof of Theorem 3.8.6.

The law of cosines applied to ΔABC and ΔACD yields

$$\begin{aligned} |AC|^2 &= a^2 + b^2 - 2ab\cos\widehat{B}, \text{ and} \\ |AC|^2 &= c^2 + d^2 - 2cd\cos\widehat{D}. \end{aligned}$$

Thus

$$\begin{aligned} a^2 + b^2 - 2ab\cos\widehat{B} &= c^2 + d^2 - 2cd\cos\widehat{D} \iff \\ a^2 + b^2 - c^2 - d^2 &= 2ab\cos\widehat{B} - 2cd\cos\widehat{D} \implies \\ (a^2 + b^2 - c^2 - d^2)^2 &= 4a^2b^2\cos^2\widehat{B} + 4c^2d^2\cos^2\widehat{D} - 8abcd\cos\widehat{B}\cos\widehat{D}. \end{aligned}$$

Multiplying (3.8.21) by 4 yields

$$16\mathcal{A}^2 = 4a^2b^2\sin^2\widehat{B} + 4c^2d^2\sin^2\widehat{D} + 8abcd\sin\widehat{B}\sin\widehat{D}.$$

Adding the last two identities and using formula $\cos(\widehat{B}+\widehat{D}) = \cos\widehat{B}\cos\widehat{D} - \sin\widehat{B}\sin\widehat{D}$ yields

$$16\mathcal{A}^2 + (a^2 + b^2 - c^2 - d^2)^2 = 4a^2b^2 + 4c^2b^2 - 8abcd\cos\left(\widehat{B} + \widehat{D}\right) \iff$$

$$\begin{aligned} 16\mathcal{A}^2 &= -(a^2 + b^2 - c^2 - d^2)^2 + (4a^2b^2 + 4c^2d^2 + 8abcd) - 8abcd \\ &\quad -8abcd\cos\left(\widehat{B} + \widehat{D}\right) \\ &= -(a^2 + b^2 - c^2 - d^2)^2 + (2ab + 2cd)^2 - 8abcd\left(1 + \cos\left(\widehat{B} + \widehat{D}\right)\right) \\ &= (2ab + 2cd + a^2 + b^2 - c^2 - d^2)(2ab + 2cd - a^2 - b^2 + c^2 + d^2) \\ &\quad -16abcd\cos^2\frac{\widehat{B} + \widehat{D}}{2} \\ &= \left[(a+b)^2 - (c-d)^2\right]\left[(c+d)^2 - (a-b)^2\right] - 16abcd\cos^2\frac{\widehat{B} + \widehat{D}}{2} \end{aligned}$$

$$\begin{aligned}
&= (a+b+c-d)(a+b-c+d)(c+d+a-b)(c+d-a+b) \\
&\quad -16abcd\cos^2\frac{\widehat{B}+\widehat{D}}{2} \\
&= (2p-2d)(2p-2c)(2p-2b)(2p-2a) - 16abcd\cos^2\frac{\widehat{B}+\widehat{D}}{2} \\
&= 16(p-a)(p-b)(p-c)(p-d) - 16abcd\cos^2\frac{\widehat{B}+\widehat{D}}{2}.
\end{aligned}$$

■

3.8.3 The largest area of a quadrilateral

We shall present in this section an application of Archimedes' formula. Given the side lengths a, b, c, d, there are infinitely many quadrilaterals with these side lengths. We are interested in finding the one with the largest area.

Theorem 3.8.7 *A cyclic quadrilateral Q has the largest area of all quadrilaterals with sides of the same length as those of Q.*

Proof: The largest area should be reached for a convex quadrilateral (Justify why!). Using Archimedes' formula

$$\begin{aligned}
\mathcal{A}^2 &= (p-a)(p-b)(p-c)(p-d) - abcd\cos^2\frac{\widehat{B}+\widehat{D}}{2} \\
&\le (p-a)(p-b)(p-c)(p-d).
\end{aligned}$$

The maximum is reached when $\cos^2\dfrac{\widehat{B}+\widehat{D}}{2} = 0$ which means $\widehat{B}+\widehat{D} = 180°$, i.e., for the cyclic quadrilateral. ■

The next theorem shows that there is only one cyclic quadrilateral of given side lengths.

Theorem 3.8.8 *Two cyclic quadrilaterals with equal side lengths in order are congruent.*

Proof: The (SSS) congruence condition for triangles implies that it suffices to prove that the lengths of the sides of a cyclic quadrilateral determine uniquely the lengths of the diagonals. Let a, b, c, d be the side lengths. Ptolemy's theorems, Theorems 3.2.5 and 3.2.16 imply that the lengths of diagonals satisfy the following system of equations

$$\begin{aligned}
\frac{x}{y} &= \frac{ab+cd}{ad+bc}, \\
xy &= ac+bd.
\end{aligned}$$

The above system has an unique solution given by

$$x = \sqrt{\frac{(ab+cd)(ac+bd)}{ad+bc}}, \text{ and}$$

$$y = \sqrt{\frac{(ac+bd)(ad+bc)}{ab+cd}}.$$

■

Corollary 3.8.9 *Given the sides a, b, c, d in order, there is at most only one quadrilateral of the largest area with these sides.*

3.9 Metric relations in quadrilaterals

3.9.1 Bretschneider's relation

The following is the generalization of Ptolemy's formula for any convex quadrilateral.

Theorem 3.9.1 (Bretschneider) *In any convex quadrilateral we have, using the usual notation,*

$$\delta_1^2\delta_2^2 = a^2c^2 + b^2d^2 - 2abcd\cos(\widehat{B}+\widehat{D}).$$

Proof: We shall write the area of the quadrilateral in two ways. Archimedes' formula and relation (3.8.20) yield

$$\begin{aligned} 16\mathcal{A}^2 &= 16(p-a)(p-b)(p-c)(p-d) - 16abcd\cos^2\frac{\widehat{B}+\widehat{D}}{2} \\ &= 4\delta_1^2\delta_2^2 - (a^2-b^2+c^2-d^2)^2. \end{aligned}$$

Subtracting we obtain

$$\begin{aligned} 4\delta_1^2\delta_2^2 &= (a^2-b^2+c^2-d^2)^2 + 16(p-a)(p-b)(p-c)(p-d) \\ &\quad -16abcd\cos^2\frac{\widehat{B}+\widehat{D}}{2}. \end{aligned} \tag{3.9.22}$$

The second term in the right side can be computed as follows:

$$\begin{aligned} &16(p-a)(p-b)(p-c)(p-d) \\ = \ &(a+c-b+d)(a+c+b-d)(b+d+a-c)(b+d-a+c) \\ = \ &[(a+c)^2-(b-d)^2][(b+d)^2-(a-c)^2] \\ = \ &[(a^2+2ac+c^2)-(b^2-2bd+d^2)][(b^2+2bd+d^2)-(a^2-2ac+c^2)] \\ = \ &4(ac+bd)^2-(a^2-b^2+c^2-d^2)^2. \end{aligned}$$

Substituting in (3.9.22) yields

$$\begin{aligned} 4\delta_1^2\delta_2^2 &= (a^2-b^2+c^2-d^2)^2+4(ac+bd)^2-(a^2-b^2+c^2-d^2)^2 \\ &\quad -16abcd\cos^2\frac{\widehat{B}+\widehat{D}}{2} \\ &= 4(ac+bd)^2-16abcd\cos^2\frac{\widehat{B}+\widehat{D}}{2}. \end{aligned}$$

Then

$$\begin{aligned} \delta_1^2\delta_2^2 &= (ac+bd)^2-4abcd\cos^2\frac{\widehat{B}+\widehat{D}}{2} \\ &= a^2c^2+b^2d^2+2abcd-4abcd\cos^2\frac{\widehat{B}+\widehat{D}}{2} \\ &= a^2c^2+b^2d^2+2abcd\left(1-2\cos^2\frac{\widehat{B}+\widehat{D}}{2}\right) \\ &= a^2c^2+b^2d^2-2abcd\cos(\widehat{B}+\widehat{D}), \end{aligned}$$

where we used the trigonometric formula $\cos 2x = 2\cos^2 x - 1$. ■

Corollary 3.9.2 (Ptolemy's theorem) *A convex quadrilateral is cyclic if and only if*

$$\delta_1\delta_2 = ac+bd.$$

Proof: Theorem 3.1.3 shows that a convex quadrilateral $ABCD$ is cyclic if and only if $\widehat{B}+\widehat{D}=180$, which means $\cos(\widehat{B}+\widehat{D})=-1$. In this case, Bretschneider's relation becomes

$$\delta_1^2\delta_2^2 = a^2c^2+b^2d^2+2abcd = (ac+bd)^2, \tag{3.9.23}$$

which is equivalent with the desired relation. ■

Corollary 3.9.3 *Two opposite angles in a convex quadrilateral are complementary if and only if*

$$\delta_1^2\delta_2^2 = a^2c^2+b^2d^2.$$

Corollary 3.9.4 (Pythagorean theorem) *Let a, b, c be the side lengths of a right triangle with hypothenuse length c. Then*

$$a^2+b^2=c^2.$$

Proof: Consider a rectangle $ABCD$ with $AB=DC=b$ and $AD=BC=a$. The diagonal length is $|AC|=|BD|=c$. Ptolemy's relation becomes $c^2=a^2+b^2$. See Figure 3.33. ■

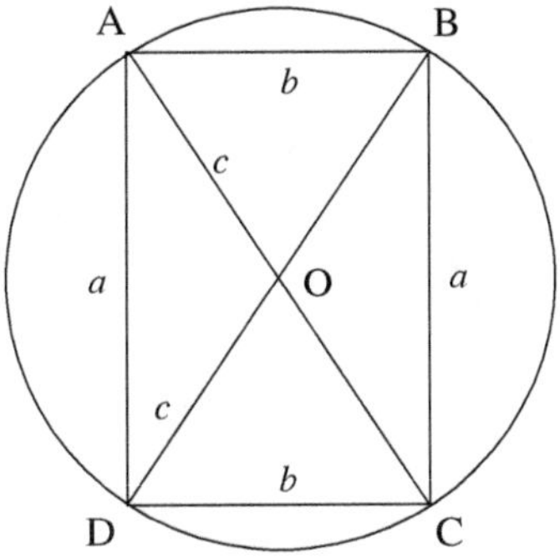

Figure 3.33: Rectangle inscribed in circle.

3.9.2 Carlitz' identity for bicentric quadrilaterals

The shortest path between two truths in the real domain passes through the complex domain.
Jacques Hadamard

A quadrilateral which has both a circumcircle that passes through each vertex, and an incircle that is tangent to each side is called **bicentric**. Let O and I be the centers of the circumcircle and incircle, respectively. We have shown in Theorem 2.14.8 that in a triangle Euler's relation holds: $|OI|^2 = R^2 - 2Rr$. The goal of this section is to extend this relation to bicentric quadrilaterals. The following result is due to the American mathematician Leonard Carlitz (1907–1999).

Theorem 3.9.5 *Let a, b, c, d be the side lengths of a bicentric quadrilateral. Then the distance between the circumcenter and the incenter is given by*

$$|OI|^2 = R^2 - 2Rr \cdot \mu,$$

where

$$\mu = \sqrt{\frac{(ab+cd)(ad+bc)}{(a+c)^2(ac+bd)}} \tag{3.9.24}$$

and R, r denote the circumradius and, respectively, the inradius.

Proof: Let E be the point where the line AI intersects the circumcircle. Let F be the orthogonal projection of I on the side AB, which is the contact point between the incircle and the side AB. Let $a = AB$, $b = BC$, $c = CD$, $d = DA$ and $\alpha = arc(AB)$, $\beta = arc(BC)$, $\gamma = arc(CD)$, $\delta = arc(DA)$. Let ES be a diameter of the circumcircle.

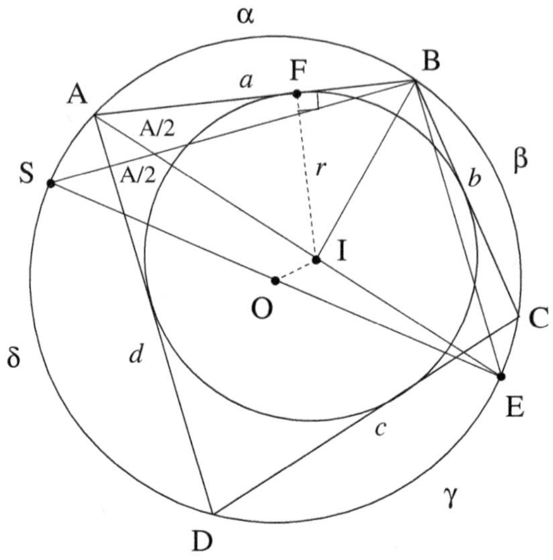

Figure 3.34: See the proof of Theorem 3.9.5.

The proof uses the power of the point I with respect to the circumcircle. See Section 2.29. We have

$$|AI| \cdot |IE| = \text{Power}(I) = R^2 - |OI|^2 \Longrightarrow |OI|^2 = R^2 - |AI| \cdot |IE|. \quad (3.9.25)$$

We shall show that

$$|AI| \cdot |IE| = 2Rr \cdot \mu. \quad (3.9.26)$$

The (AA) similarity case yields

$$\Delta SBE \sim \Delta AFI \text{ because } \begin{cases} \widehat{IAF} = \frac{1}{2} arc(BE) = \widehat{BSE}, \\ \widehat{AFI} = 90 = \widehat{SBE}. \end{cases}$$

Hence

$$\frac{|AI|}{|SE|} = \frac{|FI|}{|BE|} \Longleftrightarrow \frac{|AI|}{2R} = \frac{r}{|BE|} \Longrightarrow |AI| = \frac{2Rr}{|BE|}.$$

Then

$$|AI| \cdot |IE| == 2Rr \cdot \frac{|IE|}{|BE|}, \quad (3.9.27)$$

From relations (3.9.27) and (3.9.26) it follows that we still need to show that

$$\frac{|IE|}{|BE|} = \mu, \quad (3.9.28)$$

with μ given by (3.9.24).

We shall next find the measures of the angles $\widehat{IBE}$ and $\widehat{EIB}$ in terms of the arc measures α and γ:

$$\begin{aligned}
\widehat{IBE} &= 180^\circ - \widehat{BEI} - \widehat{BIE} \\
&= 180^\circ - \widehat{BEI} - \left(\widehat{IAB} + \widehat{ABI}\right) \\
&= 180^\circ - \frac{\alpha}{2} - \frac{A}{2} - \frac{B}{2} \\
&= 180^\circ - \frac{\alpha}{2} - \frac{\beta+\gamma}{4} - \frac{\gamma+\delta}{4} \\
&= \frac{180^\circ \cdot 4 - 2\alpha - \beta - \gamma - \gamma - \delta}{4} \\
&= \frac{2 \cdot 360^\circ - 2\overbrace{(\alpha+\beta+\gamma+\delta)}^{=360^\circ} + \beta + \delta}{4} \\
&= \frac{\beta+\delta}{4} = \frac{360^\circ - (\alpha+\gamma)}{4} \\
&= 90^\circ - \frac{\alpha+\gamma}{4}. \qquad (3.9.29)
\end{aligned}$$

$$\begin{aligned}
\widehat{EIB} &= 180^\circ - \widehat{BEI} - \widehat{IBE} \\
&= 180^\circ - \frac{\alpha}{2} - \left(90^\circ - \frac{\alpha+\gamma}{4}\right) \quad \text{(see (3.9.29))} \\
&= 90^\circ - \frac{\alpha}{2} + \frac{\alpha+\gamma}{4} \\
&= 90^\circ + \frac{-2\alpha+\alpha+\gamma}{4} = 90^\circ + \frac{\gamma-\alpha}{4} \\
&= 90^\circ - \frac{\alpha-\gamma}{4}.
\end{aligned}$$

The law of sines in ΔBIE yields

$$\frac{|IE|}{|BE|} = \frac{\sin \widehat{IBE}}{\sin \widehat{EIB}} = \frac{\sin(90^\circ - \frac{\alpha+\gamma}{4})}{\sin(90^\circ - \frac{\alpha-\gamma}{4})} = \frac{\cos\frac{\alpha+\gamma}{4}}{\cos\frac{\alpha-\gamma}{4}}.$$

Hence (3.9.28) becomes

$$\frac{\cos\frac{\alpha+\gamma}{4}}{\cos\frac{\alpha-\gamma}{4}} = \mu.$$

We shall next compute μ in terms of the angles α and γ. Using

$$a = 2R\sin\frac{\alpha}{2}, \quad b = 2R\sin\frac{\beta}{2}, \quad c = 2R\sin\frac{\gamma}{2}, \quad d = 2R\sin\frac{\delta}{2},$$

we have

$$ab + cd = 4R^2 \left(\sin\frac{\alpha}{2} \sin\frac{\beta}{2} + \sin\frac{\gamma}{2} \sin\frac{\delta}{2} \right). \tag{3.9.30}$$

As $\alpha + \beta + \gamma + \delta = 360^\circ$ we get $\dfrac{\alpha+\beta}{2} = 180^\circ - \dfrac{\gamma+\delta}{2}$ and hence

$$\cos\frac{\alpha+\beta}{2} = -\cos\frac{\gamma+\delta}{2}. \tag{3.9.31}$$

Using (3.9.31) and Corollary 2.1.5, the parenthesized expression in right side of relation (3.9.30) becomes

$$\begin{aligned}
& \sin\frac{\alpha}{2} \sin\frac{\beta}{2} + \sin\frac{\gamma}{2} \sin\frac{\delta}{2} \\
= {} & \frac{1}{2}\left[\cos\frac{\alpha-\beta}{2} - \cos\frac{\alpha+\beta}{2} + \cos\frac{\gamma-\delta}{2} - \cos\frac{\gamma+\delta}{2}\right] \\
= {} & \frac{1}{2}\left[\cos\frac{\alpha-\beta}{2} + \cos\frac{\gamma-\delta}{2}\right] \\
= {} & \frac{1}{2}\left[-\cos\left(180^\circ - \frac{\beta-\alpha}{2}\right) + \cos\frac{\gamma-\delta}{2}\right] \\
= {} & \frac{1}{2}\left[-\cos\frac{360^\circ+\alpha-\beta}{2} + \cos\frac{\gamma-\delta}{2}\right] \\
= {} & \frac{1}{2}\left[-\cos\frac{(\alpha+\beta+\gamma+\delta)+\alpha-\beta}{2} + \cos\frac{\gamma-\delta}{2}\right] \\
= {} & \frac{1}{2}\left[-\cos\frac{(\alpha+\gamma)+(\alpha+\delta)}{2} + \cos\frac{(\alpha+\gamma)-(\alpha+\delta)}{2}\right] \\
= {} & \sin\frac{\alpha+\gamma}{2} \sin\frac{\alpha+\delta}{2}.
\end{aligned}$$

Substituting back in (3.9.30) yields

$$ab + cd = 4R^2 \sin\frac{\alpha+\gamma}{2} \sin\frac{\alpha+\delta}{2}.$$

In a similar way we have

$$\begin{aligned}
ad + bc &= 4R^2 \sin\frac{\alpha+\gamma}{2} \sin\frac{\alpha+\beta}{2} \\
ac + bd &= 4R^2 \sin\frac{\alpha+\beta}{2} \sin\frac{\alpha+\delta}{2} \\
a + c &= 4R \sin\frac{\alpha+\gamma}{4} \cos\frac{\alpha-\gamma}{4}.
\end{aligned}$$

Then

$$\begin{aligned}\mu^2 &= \frac{(ab+cd)(ad+bc)}{(a+c)^2(ac+bd)}\\ &= \frac{4R^2\sin\frac{\alpha+\gamma}{2}\sin\frac{\alpha+\delta}{2}\cdot 4R^2\sin\frac{\alpha+\gamma}{2}\sin\frac{\alpha+\beta}{2}}{4\cdot 4R^2\sin^2\frac{\alpha+\gamma}{4}\cos^2\frac{\alpha-\gamma}{4}\cdot 4R^2\sin\frac{\alpha+\beta}{2}\sin\frac{\alpha+\delta}{2}}\\ &= \frac{\sin^2\frac{\alpha+\gamma}{2}}{4\sin^2\frac{\alpha+\gamma}{4}\cos^2\frac{\alpha-\gamma}{4}}\\ &= \frac{4\sin^2\frac{\alpha+\gamma}{4}\cos^2\frac{\alpha+\gamma}{4}}{4\sin^2\frac{\alpha+\gamma}{4}\cos^2\frac{\alpha-\gamma}{4}} = \left[\frac{\cos\frac{\alpha+\gamma}{4}}{\cos\frac{\alpha-\gamma}{4}}\right]^2.\end{aligned}$$

■

The following result is an analog of Euler's inequality from triangle geometry.

Corollary 3.9.6 *Let a, b, c, d be the lengths of the sides of a bicentric quadrilateral in order. Then*

$$R \geq 2r\sqrt{\frac{(ab+cd)(ad+bc)}{(a+c)^2(ac+bd)}}.$$

Remark 3.9.7 The Hungarian mathematician L. Fejes Tóth proved in 1948 that for a bicentric quadrilateral

$$R \geq r\sqrt{2}.$$

Exercise 3.9.8 *Prove the second Bretschneider relation*

$$\frac{\delta_1^2}{\delta_2^2} = \frac{a^2b^2+c^2d^2-2abcd\cos(\widehat{A}+\widehat{C})}{a^2d^2+b^2c^2-2abcd\cos(\widehat{B}+\widehat{D})},$$

and show that the quadrilateral is cyclic if and only if

$$\frac{\delta_1}{\delta_2} = \frac{ab+cd}{ad+bc}.$$

Exercise 3.9.9 *Prove that if the quadrilateral $ABCD$ is circumscribed to a circle then*

(i) $ab\sin^2\frac{\widehat{B}}{2} = cd\sin^2\frac{\widehat{D}}{2}$,

(ii) $\mathcal{A} = \sqrt{abcd} \cdot \sin \frac{\widehat{A} + \widehat{C}}{2}$, *and*

(iii) If ABCD is cyclic, then $\mathcal{A} = \sqrt{abcd}$.

Exercise 3.9.10 *Let* φ *be the angle between the diagonals of a convex quadrilateral. Then the area is given by*

$$\mathcal{A} = \frac{1}{4}|a^2 - b^2 + c^2 - d^2| \tan \varphi.$$

Exercise 3.9.11 (Fermat's problem for polygons) *(i) Let ABCD be a convex quadrilateral. Show that there is a unique point P in the plane such that the sum*

$$|PA| + |PB| + |PC| + |PD|$$

is minimum. Find the point.

(ii) Let $A_1A_2A_3A_4A_5A_6$ *be a hexagon circumscribed to a circle. Show that there is a unique point P in the plane such that the sum* $\sum_{i=1}^{6} |A_iP|$ *is minimum. Find the point.*

Solved Exercise 3.9.12 *Let O be the circumcenter of the cyclic quadrilateral ABCD and assume* $O \in Int(ABCD)$. *Let M be the point of the intersection of its diagonals and denote by P, Q, R, S the orthogonal projections of M onto the sides AB, BC, CD and DA, respectively.*

(i) Show that the points P, Q, R, S belong to the interior of the quadrilateral sides.

(ii) Show that the quadrilateral PQRS satisfies the "billiard property" (the incidence and the reflection angles are equal).

Proof: (*i*) It suffices to show that $P \in (AB)$, the others being similar. We have $\widehat{ABM} = \widehat{ABD} = \frac{1}{2}arc(AD) = \frac{\widehat{AOD}}{2} < \frac{180}{2} = 90^\circ$. Similarly, $\widehat{BAM} < 90^\circ$. Applying Exercise 2.13.3, part (*i*), to triangle ΔABM yields that P is between A and B.

(*ii*) It suffices to show $\widehat{SPM} = \widehat{MPQ}$, the other identities following by a similar argument, see Figure 3.35. We note that $APMS$ and $PBQM$ are cyclic quadrilaterals since a pair of two opposite angles have the sum equal to $90^\circ + 90^\circ = 180$. Using properties of cyclic quadrilaterals, we have

$$\widehat{SPM} = \widehat{MAS} = \widehat{CAD} = \frac{1}{2}arc(DC) = \widehat{DBC} = \widehat{MBQ} = \widehat{MPQ}.$$

We make the note that the quadrilateral $PQRS$ has the smallest perimeter among all quadrilaterals which are inscribed in $ABCD$. This is an analog of Fagnano's theorem. ∎

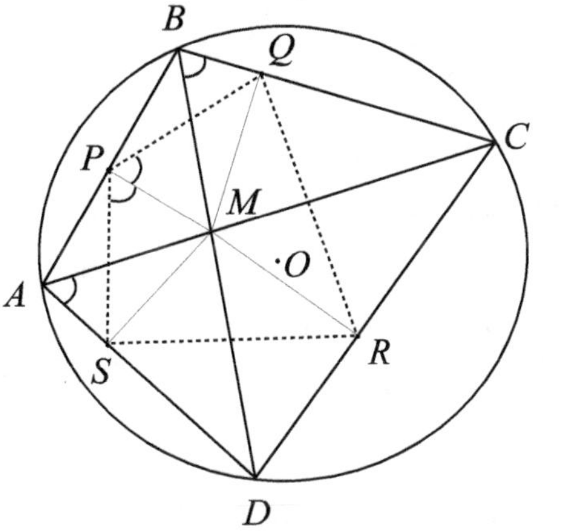

Figure 3.35: For Exercise 3.9.12.

Solved Exercise 3.9.13 *Let $ABCD$ be a trapezoid with $AB \| DC$. We denote by M and N the midpoints of the sides AD and BC, respectively, and by O the point of intersection of diagonals. Let E and F be the points of intersection of the line parallel to AB passing through O with the sides AD and BC, respectively.*

(i) Prove that O is the midpoint of EF;

(ii) Show that $|EF| = \dfrac{2|AB| \cdot |DC|}{|AB| + |DC|}$;

(iii) Show the inequality $|EF| \leq MN$.

Proof: (*i*) We use Figure 3.36. From $\Delta AOB \sim \Delta COD$ we have $\frac{|DO|}{|OB|} = \frac{|CO|}{|OA|} = \frac{|DC|}{|AB|}$, or equivalently, $\frac{|DO|}{|DO|+|OB|} = \frac{|CO|}{|CO|+|OA|} = \frac{|DC|}{|DC|+|AB|}$, or $\frac{|DO|}{|DB|} = \frac{|CO|}{|CA|} = \frac{|DC|}{|DC|+|AB|}$. Starting from $\Delta DEO \sim \Delta DAB$ we obtain $\frac{|EO|}{|AB|} = \frac{|DO|}{|DB|} = \frac{|DC|}{|DC|+|AB|}$, so $|EO| = \frac{|DC| \cdot |AB|}{|DC|+|AB|}$. Using the similarity $\Delta OCF \sim \Delta ACB$ we obtain $\frac{|OF|}{|AB|} = \frac{|OC|}{|AC|} = \frac{|DC|}{|DC|+|AB|}$, so $|OF| = \frac{|DC| \cdot |AB|}{|DC|+|AB|}$. It follows that $|EO| = |OF|$, i.e., O is the midpoint of EF.

(*ii*) $|EF| = |EO| + |OF| = 2|EO| = \frac{2|DC| \cdot |AB|}{|DC|+|AB|}$.

(*iii*) Since $MN|$ is a midpoint connector in a trapezoid, by Exercise 2.31.10, we have $|MN| = \frac{1}{2}(|AB| + |DC|)$. Using that the harmonic mean is smaller than the arithmetic mean, see page 44, we obtain $|EF| \leq |MN|$. The equality is reached when $|AB| = |CD|$. ∎

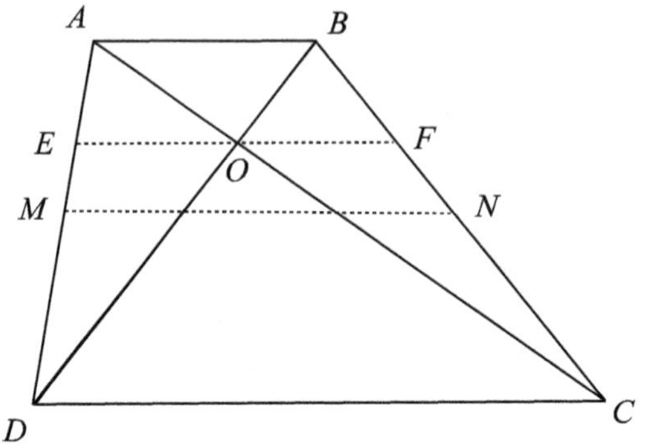

Figure 3.36: For Exercise 3.9.13.

3.10 The physical centroid

The concept of physical centroid of a system of weights placed at different points in the plane consists in considering the entire weight system concentrated at only one point — the centroid. We shall start with the definition of the physical centroid of a system of two weights.

Definition 3.10.1 *Two weights of sizes w_1 and w_2 are applied at two points A_1 and A_2. The point G situated between A_1 and A_2 is called the physical centroid of the system of weights $\{w_1, w_2\}$ if $x_1w_1 = x_2w_2$, where $x_1 = |A_1G|$ and $x_2 = |A_2G|$.*

The mechanical interpretation of Definition 3.10.1 is that the momentum of two parallel forces of sizes w_1 and w_2 applied at A_1 and A_2 stay in equilibrium. This is equivalent to the fact that the forces have equal angular momenta about the point G, so the system stays in equilibrium, see Figure 3.37.

Inductively, we can define the physical centroid, G, of a system of n weights, $\{w_1, w_2, \ldots, w_{n-1}, w_n\}$ applied to the coplanar points A_1, $A_2, \ldots, A_n$, as follows. Let B be the physical centroid of the system of $n-1$ weights $\{w_1, w_2, \ldots, w_{n-1}\}$ given by the induction hypothesis. Then G is defined to be the physical centroid of the system $\{w, w_n\}$ applied at B and A_n, where $w = \sum_{k=1}^{n-1} w_k$.

The only issue with this practical construction is the uniqueness of the point G. To this end we need to show that regardless of how the subsets are nested during the construction, for instance

$$\{w_1, w_2\} \subset \cdots \subset \{w_1, w_2, \ldots, w_{n-1}\} \subset \{w_1, w_2, \ldots, w_{n-1}, w_n\},$$

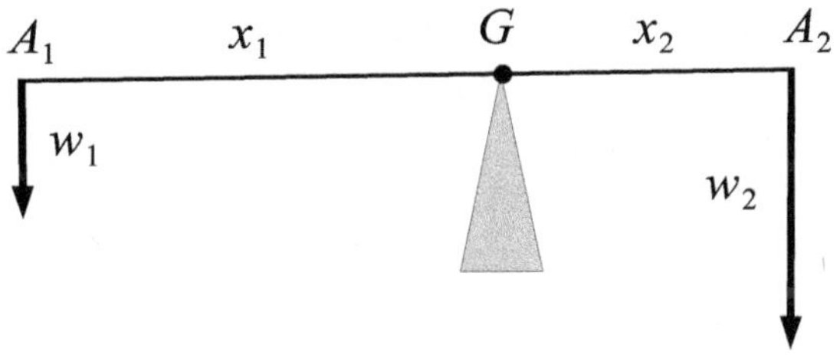

Figure 3.37: The center of mass G of the system of weights $\{w_1, w_2\}$.

we end up with the same centroid. This follows either from the plausible physical hypothesis that the centroid of a system of n weights is unique, or from an inductive vectorial computation showing that we end up always with the same reference vector

$$\overrightarrow{OG} = \frac{w_1\overrightarrow{OA_1} + \cdots + w_n\overrightarrow{OA_n}}{w_1 + \cdots + w_n},$$

where O is a given reference point.

The concept of physical centroid can be extended to a continuum domain of points, D, with a weight $w(x)$ associated to each point $x \in D$. Even if finding the centroid in this case requires more advanced integral calculus techniques, we shall consider in the following the case of an uniform weight distribution on D and take into account a few particular shapes of D with axial symmetry.

We start by accepting that the physical centroid of a line segment AB considered as a continuum set of points with an uniform weight distribution is the midpoint of the segment. This is physically plausible, since the weight of the first half segment is equal to the weight of the second half. Thus, the weight of the segment AB can be concentrated in its midpoint, M, with a weight proportional to the segment length, $|AB|$.

Proposition 3.10.2 *Let D be a planar shape with the physical centroid G.*
(i) If D has an axis of symmetry, ℓ, then $G \in \ell$.
(ii) If D has two distinct axes of symmetry, ℓ_1 and ℓ_2, then $\{G\} = \ell_1 \cap \ell_2$.

Proof: (*i*) The symmetry line ℓ divides the domain D into two domains D_1 and D_2, symmetric with respect to ℓ. Their centroids are located at G_1 and G_2, respectively, which are also symmetric with respect to ℓ. The centroid of D is the centroid of the system of two equal weights, $w_1 = area(D_1)$, $w_2 = area(D_2)$, located at G_1 and G_2. Hence, the centroid G is located at the midpoint of G_1G_2, which belongs to ℓ.

(*ii*) By part (*i*), the centroid G belongs to both ℓ_1 and ℓ_2, so it is located at their intersection point. ■

As a consequence, the centroids of planar shapes in the form of a disk, rectangle, or equilateral triangle is located at their centers.

Exercise 3.10.3 *Let D be a planar shape with the physical centroid G. Show that if D has three axes of symmetry, ℓ_1, ℓ_2, and ℓ_3, then*

$$\{G\} = \ell_1 \cap \ell_2 = \ell_2 \cap \ell_3 = \ell_3 \cap \ell_1.$$

Exercise 3.10.4 *Let D be a planar shape with two axes of symmetry, ℓ_1 and ℓ_2, such that $\ell_1 \| \ell_2$. Show that $\ell_1 = \ell_2$.*

3.10.1 Centroids of a triangle

When talking about the centroid of a triangle from the physical point of view we have a few variant choices. Let G_0 be the center of mass of the point ensemble $\{A, B, C\}$ formed by unit masses centered at the vertices A, B and C. Let G_1 be the center of mass of the bars ensemble $\{AB, BC, CA\}$ formed by homogeneous bars joint at the vertices A, B and C. Finally, let's denote by G_2 the center of mass of a homogeneous triangular slab with vertices at A, B and C. For an arbitrary triangle these three variants do not necessarily coincide. Their relation will be discussed in this section.

Let G denote the geometric centroid, which is the intersection of the triangle medians, as introduced in Section 2.26.

Proposition 3.10.5 *For any triangle we have $G_0 = G_2 = G$.*

Proof: We start by proving $G_0 = G$. Consider unit masses concentrated at each of the points $\{A, B, C\}$. The pair of points $\{B, C\}$ can be assimilated with one point of mass 2 situated at the midpoint between B and C. Let's denote this point by A'. Then the centroid of the ensemble of points $\{A, B, C\}$ is the centroid of the pair of points $\{A', A\}$, so $G_0 \in AA'$. Since A' weights as 2 and A weights as 1, then G_0 is located twice as close to A' than to A, namely at 1/3 from A' and 2/3 from A. This point is the point of intersection of the medians. Hence $G_0 = G$.

We shall show next that $G_2 = G$. To this end, we consider a variable segment EF parallel to BC, with $E \in (AB)$ and $F \in (AC)$, see Figure 3.38(a). We consider the mass of the segment EF concentrated at its midpoint, M. We shall show that M describes the median AA', where A' is the midpoint of BC. Let $EF \cap AA' = \{M'\}$. Then Thales' theorem applied to $\Delta ABA'$ and $\Delta AA'C$ yields the equal ratios

$$\frac{|EM'|}{|BA'|} = \frac{|AM'|}{|AA'|} = \frac{|M'F|}{|A'C|}.$$

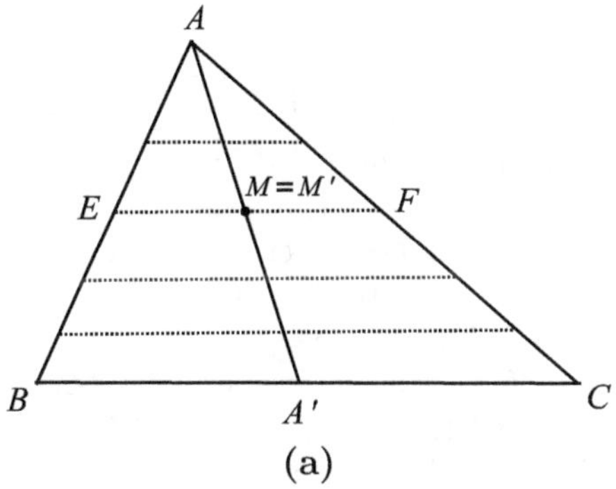

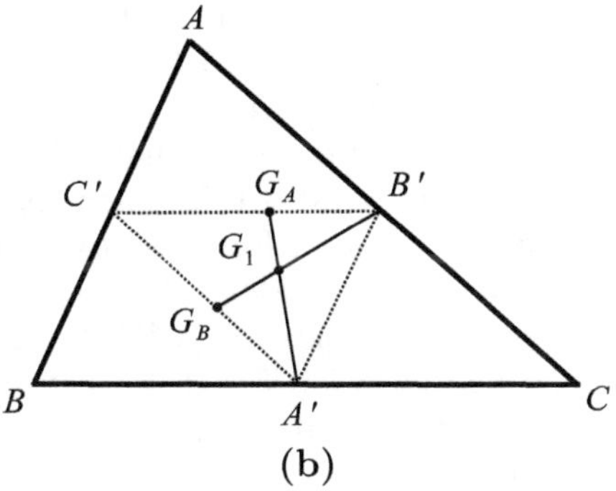

Figure 3.38: (a) G_2 belongs to AA'; (b) G_1 is the incenter of the median triangle $\Delta A'B'C'$.

Since $|BA'| = |A'C|$ we obtain $|EM'| = |M'F|$, so $M' = M$, the midpoint of EF. Hence, when EF varies, the point M describes the median AA'. The centroid G_2, which is the centroid of all points M together with some weight, must belong to this line. A similar approach shows that G_2 belongs to the other two medians, BB' and CC', fact that implies that $G_2 = G$ is the geometric centroid of ΔABC. ■

Proposition 3.10.6 *For any triangle ΔABC, the centroid G_1 is the incenter of the median triangle of ΔABC.*

Proof: Let A', B' and C' be the midpoints of BC, CA and AB, respectively. The homogeneous bar BC can be concentrated to the middle of the segment, A', with the weight proportional to the length $|BC| = a$. Similarly, the segments AB and AC can be concentrated to their midpoints, C' and B', with weights proportional to $c = |AB|$ and $b = |AC|$, respectively, see Figure 3.38(b).

Let G_A be the centroid of the pair of points $\{B', C'\}$ associated with the system of weights $\{b, c\}$. Then the centroid G_1 belongs to the segment $A'G_A$. By the properties of the centroid of two weights, we have

$$\frac{|G_AB'|}{|G_AC'|} = \frac{c}{b} = \frac{|A'B'|}{|A'C'|},$$

where the last equality used the midpoint connector property, $|A'C'| = b/2$ and $|A'B'| = c/2$. Then the angle bisector theorem, Theorem 2.22.3, implies that $A'G_A$ is the angle bisector of $\widehat{B'A'C'}$. Similarly, G_1 belongs to the other two angle bisectors, $B'G_B$ and $C'G_C$ of the median triangle $\Delta A'B'C'$. Hence, G is the incenter of the triangle $\Delta A'B'C'$. ■

Since in general, the centroid G of the interior of ΔABC does not coincide with the centroid G_1 of the boundary of ΔABC, it makes sense to compute the distance $|GG_1|$. Instead of performing a long trigonometric computation, we notice that triangles $\Delta A'B'C'$ and ΔABC are similar with similarity ratio 1/2. Then the distance between the inner center and geometric centroid for these two triangles must be under the same ratio, namely $|I_m G_m| = \frac{1}{2}|IG|$, where I_m, I and G_m, G are the inner centers and geometric centroids for $\Delta A'B'C'$ and ΔABC, respectively.

By Proposition 3.10.6 we have $G_1 = I_m$. We also notice that $G_m = G$, since the support lines for the medians of $\Delta A'B'C'$ and ΔABC are respectively the same. Therefore, $|G_1G| = \frac{1}{2}|IG|$. Using one of the formulas for $|IG|$, Equation (7.11.15), we obtain

$$|G_1G|^2 = \frac{1}{4}\left(\frac{1}{3}(bc + ca + ab) - 4rR - \frac{1}{9}(a^2 + b^2 + c^2)\right).$$

3.10.2 Centroids of a quadrilateral

Similarly with the triangle case, we have several variants of physical centroids associated with a quadrilateral $ABCD$. Let G_0 be the center of mass of the point ensemble $\{A, B, C, D\}$ formed by unit masses centered at the vertices A, B, C and D. Let G_1 be the center of mass of the bars ensemble $\{AB, BC, CD, DA\}$ formed by homogeneous bars joint at the vertices A, B, C and D. We also denote by G_2 the center of mass of a homogeneous quadrilateral slab with vertices at A, B, C and D.

As usual, let K denote the quadrilateral centroid, which is the intersection of its bimedians, see Theorem 3.3.1.

Proposition 3.10.7 *In any quadrilateral $G_0 = K$. In general, $G_1 \neq K$ and $G_2 \neq K$.*

Proof: We first show $G_0 = K$. Let E and F be the midpoints of AB and DC, respectively. We split the system of points $\{A, B, C, D\}$ into the union of two subsystems, $\{A, B\}$ and $\{C, D\}$. The centroids of $\{A, B\}$ and $\{C, D\}$ are the points E and F, respectively. Each of them count with a weight 2 (the sum of two unit weights). Then the centroid of the system $\{A, B, C, D\}$ is the midpoint of EF, since each E and F have the same weight. Since the midpoint of the bimedian EF is the point K (see Varignon's theorem, Theorem 3.3.1), we obtain the desired identity, $G_0 = K$, see Figure 3.39(a).

Next we show that $G_1 \neq K$ for a quadrilateral with unequal sides. We consider the sides AB, BC, CD and DA concentrated at its midpoints, E, M, F and N, with weights a, b, c, and d, respectively. We split the ensemble $\{E, M, F, N\}$ into $\{E, F\}$ and $\{M, N\}$. The centroid of the pair

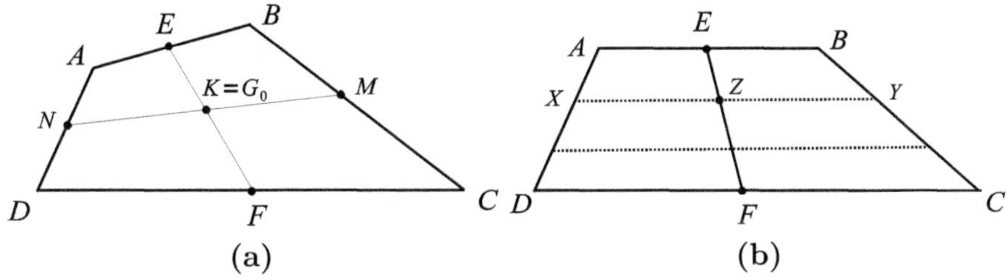

Figure 3.39: (a) The centroid G_0 lies at the intersection of bimedians; (b) The midpoint of XY spans the bimedian EF.

$\{E, F\}$, denoted by U, belongs to the line segment EF, but it is not its midpoint since $a \neq c$. Similarly, centroid of the pair $\{M, N\}$, denoted by V, belongs to the line segment EF, but it is not its midpoint either. Therefore, $K \notin \{U, V\}$, and hence $K \notin (EF)$. Since the centroid G_1 is located between E and F, it follows that $G_1 \neq K$.

Now we show that, in general, $G_2 \neq K$. It suffices to prove this for the particular case when the quadrilateral is a trapezoid. We assume $AB \| BC$ with $|AB| < |DC|$, see Figure 3.39(b). Consider the variable segment XY, with $X \in (AD)$, $Y \in (BC)$ and $XY \| AB$. The homogeneous weighted segment XY has the centroid concentrated at its midpoint, Z. Now, when X and Y are variable, the segment XY sweeps the entire trapezoid. During this process, the point Z describes the bimedian EF. Since the centroid G_2 is obtained aggregating the points Z, then $G_2 \in EF$. Since the weight of Z increases from E to F, as the length $|XY|$ does the same, it follows that G_2 cannot be located at the midpoint of EF, so $G_2 \neq K$. ∎

Part II

Complex Numbers Approach

Chapter 4

Complex Numbers in Geometry

Complex numbers have been introduced in mathematics as an extension of the set of real numbers as early as the 16th century in order to solve quadratic equations. For instance, let's consider for instance the equation $x^2 - 4 = 0$, which can be factorized as $(x - 2)(x + 2) = 0$. From here the solutions are $x = \pm 2$. However, there are many other equations, such as $x^2 + 4 = 0$, which do not factorize over real numbers. In order to obtain solutions for this equation, one has to introduce "imaginary numbers".[1] If i is defined as a number with the property that $i^2 = -1$, then $x = \pm 2i$ are the solutions of the aforementioned equation. A verification of this fact is immediate: $(2i)^2 + 4 = 4i^2 + 4 = 4(-1) + 4 = 0$.

Complex numbers start to be useful when mathematicians realized that formulas for solving quadratic, cubic and quadratic equations work also in the case when square roots of negative numbers are involved. It was interesting to realize that these expressions provided roots which could not be found by the use of real numbers.

This chapter deals with an approach of plane geometry using complex numbers. Each point in the plane is associated with a complex number and each operation with complex numbers corresponds to a certain geometric transform. Many geometric results shall be recovered by the use of complex numbers.

[1] This term was introduced by René Descartes.

4.1 Algebraic properties of complex numbers

Let $\mathbb{R}$ denote the set of real numbers and define the set of complex numbers by

$$\mathbb{C} = \{z;\ z = a + ib,\ a, b \in \mathbb{R},\ i = \sqrt{-1}\}.$$

The numbers $a = Re(z)$ and $b = Im(z)$ are called the *real* and the *imaginary* parts of the complex number z. If $z_1 = a_1 + ib_1$ and $z_2 = a_2 + ib_2$ are two complex numbers then $z_1 = z_2$ if and only if $a_1 = a_2$ and $b_1 = b_2$. This property can be also stated in a different way by saying that $\{1, i\}$ is a basis for the vectorial space $\mathbb{C}$.

In the following we shall discuss operations with complex numbers. The properties of addition and multiplication will endow $\mathbb{C}$ with a field structure. This means that the addition, subtraction, product and division of two non-zero complex numbers is still a complex number.

Addition and subtraction Let $z_1 = a_1 + ib_1$ and $z_2 = a_2 + ib_2$ be two complex numbers. The addition and subtraction are defined on components, *i.e.*,

$$\begin{aligned} z_1 + z_2 &= (a_1 + a_2) + i(b_1 + b_2), \text{and} \\ z_1 - z_2 &= (a_1 - a_2) + i(b_1 - b_2). \end{aligned}$$

One can easily verify the associativity and commutativity of addition, *i.e.*,

$$\begin{aligned} (z_1 + z_2) + z_3 &= z_1 + (z_2 + z_3) \\ z_1 + z_2 &= z_2 + z_1. \end{aligned}$$

The zero element is defined by $0 = 0 + i \cdot 0$ and the opposite of $z = a + ib$ is $-z = -a - ib$. This can be stated saying that $(\mathbb{C}, +)$ is an *abelian group*.

Multiplication The product of the complex numbers $z_1 = a_1 + ib_1$ and $z_2 = a_2 + ib_2$ is defined by

$$\begin{aligned} z_1 z_2 &= (a_1 + ib_1)(a_2 + ib_2) \\ &= a_1 a_2 + i a_1 b_2 + i b_1 a_2 + i^2 b_1 b_2 \\ &= (a_1 a_2 - b_1 b_2) + i(a_1 b_2 + b_1 a_2), \end{aligned}$$

where we used that $i^2 = -1$. One can verify by direct computation the associativity and commutativity of multiplication

$$\begin{aligned} (z_1 z_2) z_3 &= z_1 (z_2 z_3) \\ z_1 z_2 &= z_2 z_1. \end{aligned}$$

The unit element is $1 = 1 + i \cdot 0$. In the following we shall show that each complex number $z \neq 0$ has an inverse z^{-1}. This means there is a complex number z^{-1} such that

$$zz^{-1} = z^{-1}z = 1.$$

We need to define first a few new concepts. If $z = a + ib$ is a complex number, its *conjugate* is

$$\bar{z} = a - ib.$$

The *modulus* of z is defined by $|z| = \sqrt{a^2 + b^2}$. We note that z and $\bar{z}$ have the same modulus $|z| = |\bar{z}|$.

Proposition 4.1.1 (Properties of the modulus) *The following properties hold*

(*i*) *The modulus is multiplicative:* $|z_1 z_2| = |z_1| \cdot |z_2|, \forall z_1, z_2 \in \mathbb{C}$.

(*ii*) *The modulus commutes with powers:* $|z^n| = |z|^n, \forall z \in \mathbb{C}, n \in \mathbb{N}$.

(*iii*) *The product between a number and its conjugate is a real number:* $z\bar{z} = |z|^2$.

(*iv*) *If* $z \in \mathbb{R}$, *then* $|z|$ *is the usual absolute value of a real number.*

(*v*) *Triangle inequality:* $|z_1 + z_2| \leq |z_1| + |z_2|, \forall z_1, z_2 \in \mathbb{C}$. *The identity is reached if and only if* $z_2 = tz_1$, *with* $t \in \mathbb{R}$.

Proof: (*i*) It suffices to show the squares are equal

$$\begin{aligned} |z_1 z_2|^2 &= |(a_1 a_2 - b_1 b_2) + i(a_1 b_2 + b_1 a_2)|^2 \\ &= (a_1 a_2 - b_1 b_2)^2 + (a_1 b_2 + b_1 a_2)^2 \\ &= a_1^2 a_2^2 + b_1^2 b_2^2 + a_1^2 b_2^2 + b_1^2 a_2^2 = (a_1^2 + b_1^2)(a_2^2 + b_2^2) \\ &= |z_1|^2 |z_2|^2. \end{aligned}$$

(*ii*) It follows from (*i*) inductively over n.

(*iii*) Let $z = a + ib$. The difference of squares formula yields

$$z\bar{z} = (a + ib)(a - ib) = a^2 - i^2 b^2 = a^2 + b^2 = |z|^2 = |\bar{z}|^2.$$

(*iv*) Let $z \in \mathbb{R}$, $z = a + i \cdot 0$. The definition of the modulus yields $|z| = \sqrt{a^2 + 0^2} = |a|$.

(*v*) Let $z_1 = a_1 + ib_1$, $z_2 = a_2 + ib_2$ be two complex numbers. Raising to the square the inequality $|z_1 + z_2| \leq |z_1| + |z_2|$ yields

$$\begin{aligned} |z_1 + z_2|^2 &\leq |z_1|^2 + |z_2|^2 + 2|z_1 z_2| \Longleftrightarrow \\ (a_1 + a_2)^2 + (b_1 + b_2)^2 &\leq (a_1^2 + b_1^2) + (a_2^2 + b_2^2) + 2\sqrt{(a_1^2 + b_1^2)(a_2^2 + b_2^2)} \\ \Longleftrightarrow a_1 a_2 + b_1 b_2 &\leq \sqrt{(a_1^2 + b_1^2)(a_2^2 + b_2^2)}. \end{aligned}$$

If the left side is negative, the inequality is obviously satisfied. Otherwise, we take the square and obtain

$$(a_1a_2 + b_1b_2)^2 \leq (a_1^2 + b_1^2)(a_2^2 + b_2^2),$$

which after an algebraic manipulation yields

$$0 \leq (a_1b_2 - a_2b_1)^2,$$

which holds true. The identity holds if $a_1b_2 = a_2b_1$, *i.e.*, the components (a_1, b_1) and (a_2, b_2) are proportional, or $z_2 = tz_1$, with $t \in \mathbb{R}$. ■

Exercise 4.1.2 *Show that for any z_1, $z_2 \in \mathbb{C}\backslash\{0\}$ we have*

$$|z_1 - z_2| \geq |z_1| - |z_2|,$$

with identity if and only of z_1 and z_2 are proportional, i.e., $z_1 = tz_2$, $t \in \mathbb{R}$.

The inverse of a complex number Let $z = a + ib$ be a non-zero complex number. Part (*iii*) of Proposition 4.1.1 provides $z\bar{z} = |z|^2$. Dividing by $|z|^2$ yields $z\frac{\bar{z}}{|z|^2} = 1$. Hence the inverse of the complex number z is

$$z^{-1} = \frac{\bar{z}}{|z|^2} = \frac{1}{a^2 + b^2}(a - ib). \tag{4.1.1}$$

Sometimes we denote the inverse by $\frac{1}{z} = z^{-1}$.

Division of complex numbers Let $z_1, z_2 \in \mathbb{C}$, $z_2 \neq 0$. The quotient z_1/z_2 is defined as

$$\frac{z_1}{z_2} = z_1 \cdot \frac{1}{z_2}.$$

The previous properties state that $\mathbb{C}\backslash\{0\}$ is an abelian group with the multiplication of complex numbers. Since $\mathbb{C}$ is a group with addition, we can state that $(\mathbb{C}, +, \cdot)$ has a field structure.[2] All the operations (addition, subtraction, multiplication and division) work on $\mathbb{C}$ in a similar way they work on $\mathbb{R}$.

Other examples of fields A field is a set which is closed with respect to addition, subtraction, multiplication and division by non-zero numbers. The set of real numbers $\mathbb{R}$ and the set of rational numbers $\mathbb{Q}$ are well-known fields. If a number does not belong to the field, then all the linear

[2]A field is a ring with all non-zero elements invertible.

combinations with coefficients in the field form another field called the extension of the initial one. For instance $\sqrt{3} \notin \mathbb{Q}$, and the set

$$\mathbb{Q}[\sqrt{3}] = \{a + b\sqrt{3};\ a, b \in \mathbb{Q}\}$$

is closed with respect to addition, subtraction, multiplication and division by non-zero expressions. $\mathbb{Q}[\sqrt{3}]$ becomes a field, which is the extension of the field $\mathbb{Q}$ of rational numbers.

In a similar way, we can construct the field of complex numbers as an extension of the field of real numbers $\mathbb{R}$. Since $\sqrt{-1}$ does not belong to $\mathbb{R}$, we can construct the extension

$$\mathbb{C} = \mathbb{R}[-1] = \{a + ib;\ \ a, b \in \mathbb{R},\ i = \sqrt{-1}\},$$

which is the field of complex numbers.

Exercise 4.1.3 *Compute the following expressions*

$$(i)\ \frac{4}{3+i}, \qquad (ii)\ 2(1+i)^2 + (1-i)^3, \qquad (iii)\ \frac{(1+i)^3}{(1-i)^2}, \qquad (iv)\ (1+i)^5.$$

Exercise 4.1.4 *For any complex number z show that*

(i) $z + \bar{z} = 2Re(z)$,

(ii) $Re(z) = Re(\bar{z})$, $Im(z) = -Im(\bar{z})$,

(iii) $Re(iz) = -Im(z)$, $Im(iz) = Re(z)$, *and*

(iv) $\left|\frac{1}{z}\right| = \frac{1}{|z|}$, $z \neq 0$.

Exercise 4.1.5 *(i) Show that in general the modulus is not additive*

$$|z_1 + z_2| \neq |z_1| + |z_2|.$$

(ii) For which complex numbers z_1, z_2 do we have

$$|z_1 + z_2| = |z_1| + |z_2|?$$

Exercise 4.1.6 *Solve the following equations*

$$(i)\ \ z = \bar{z}, \qquad (ii)\ \ z = -z, \qquad (iii)\ \ (\bar{z})^2 = z, \qquad (iv)\ \ z^3 = \bar{z}.$$

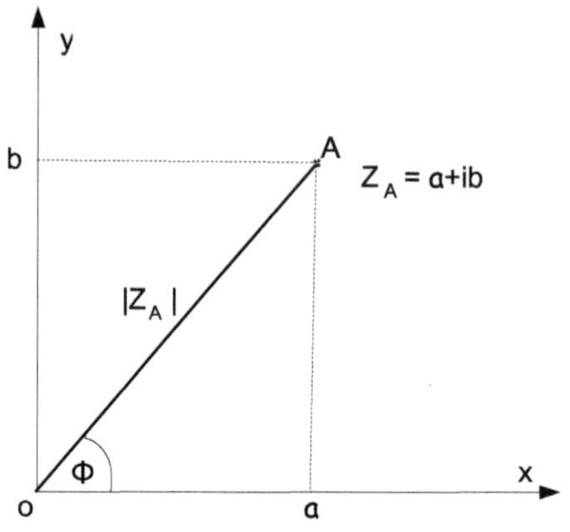

Figure 4.1: The image point A and the affix z_A in the complex plane.

4.2 Geometric interpretation of complex numbers

The complex plane We shall denote by $\mathcal{P}$ and $\mathbb{C}$ the plane and the set of complex numbers, respectively. Consider the function $f : \mathcal{P} \to \mathbb{C}$, $f(A) = z_A$, which associates with each point A with coordinates (a, b) the complex number $z_A = a + ib$. The complex number z_A is called the *affix* of the point A. The point A is called the *image* of the complex number z_A. Since the function f is one-to-one and onto, the plane $\mathcal{P}$ can be identified through the bijection f with the set $\mathbb{C}$ of complex numbers. Hence, instead of working with points, we shall work with complex numbers. The goal of this chapter is to translate the geometric properties into algebraic properties for complex numbers. The aforementioned planar representation of complex numbers is also known under the name of *Argand diagram.*[3]

Since the Pythagorean theorem yields $|OA|^2 = a^2 + b^2 = |z_A|^2$, see Figure 4.1, it follows that the distance between the origin O and the point A is the modulus of the affix z_A. The angle formed by OA with the horizontal axis is denoted with ϕ and is called the *argument* of the complex number z_A. The angle ϕ is measured counter-clock-wise. The argument function assigns with each non-zero complex number the angle ϕ, *i.e.*, $\arg : \mathbb{C}\backslash\{0\} \to [0, 2\pi)$, with $\arg(z) = \phi$, where

$$Re(z) = |z| \cos\phi, \qquad Im(z) = |z| \sin\phi. \tag{4.2.2}$$

[3]Named after the French mathematician Jean-Robert Argand (1768–1822).

The trigonometric form of complex numbers Since the plane can be parametrized by polar coordinates, the complex numbers can also be represented in trigonometric form involving the polar coordinates (ρ, ϕ), where $\rho = |z|$ and $\phi = \arg(z)$. Using (4.2.2) yields

$$z \quad = \quad a + ib = \rho(\cos\phi + i\sin\phi). \tag{4.2.3}$$

We shall refer to formula (4.2.3) as the *trigonometric form* of complex numbers.

Exercise 4.2.1 *Consider the complex numbers in trigonometric form* $z_1 = \rho_1(\cos\phi_1 + i\sin\phi_1)$, $z_2 = \rho_2(\cos\phi_2 + i\sin\phi_2)$. *Show that* $z_1 = z_2$ *if and only if* $\rho_1 = \rho_2$ *and* $\phi_1 = \phi_2$.

In the following we shall discuss the main operations with complex numbers (addition, subtraction, multiplication and division) from the geometric point of view.

Addition Let A_1, A_2 be two points in the plane and let $z_1 = a_1 + ib_1$ and $z_2 = a_2 + ib_2$ be the corresponding affixes. Let P be a point in the plane such that the quadrilateral OA_1PA_2 is a parallelogram. It can be seen from Figure 4.2(a) that the affix of P is

$$z_P = (a_1 + a_2) + i(b_1 + b_2) = z_1 + z_2.$$

Hence, the complex numbers are added using the *parallelogram rule*.

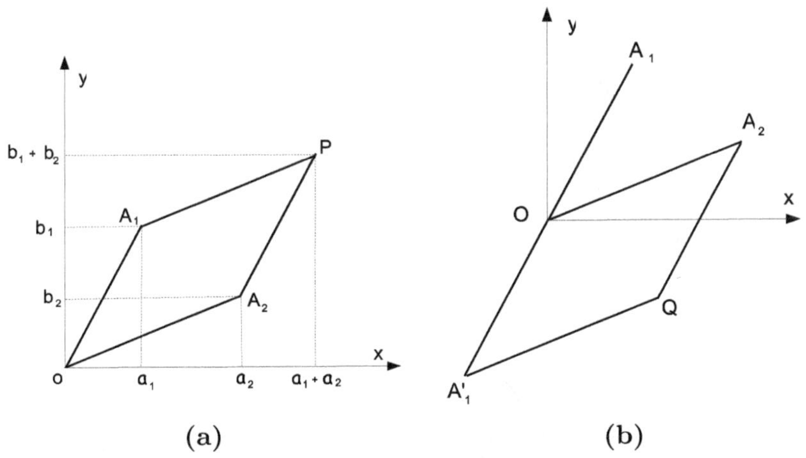

Figure 4.2: The addition and subtraction of complex numbers.

Subtraction Let A_1' be the symmetric of point A_1 with respect to the origin. Its affix is $-z_1$. Since $z_2 - z_1 = z_2 + (-z_1)$, we shall add z_2 and $-z_1$ by the parallelogram rule. Let Q be a point in the plane such that $OA_1'QA_2$ is a parallelogram. Then the affix of the point Q is the difference of the complex numbers z_1 and z_2, see Figure 4.2(b).

Multiplication Consider the points A_1 and A_2 with affixes

$$z_1 = \rho_1(\cos\phi_1 + i\sin\phi_1), \qquad z_2 = \rho_2(\cos\phi_2 + i\sin\phi_2).$$

A computation provides

$$\begin{aligned} & (\cos\phi_1 + i\sin\phi_1)(\cos\phi_2 + i\sin\phi_2) \\ = \; & (\cos\phi_1\cos\phi_2 - \sin\phi_1\sin\phi_2) + i(\cos\phi_1\sin\phi_2 + \sin\phi_1\cos\phi_2) \\ = \; & \cos(\phi_1+\phi_2) + i\sin(\phi_1+\phi_2), \end{aligned}$$

we get

$$\begin{aligned} z_1 z_2 &= \rho_1\rho_2(\cos\phi_1 + i\sin\phi_1)(\cos\phi_2 + i\sin\phi_2) \\ &= \rho_1\rho_2\big(\cos(\phi_1+\phi_2) + i\sin(\phi_1+\phi_2)\big). \end{aligned}$$

We obtained the following multiplication rule:

The product of two complex numbers is obtained by multiplying the modulus and adding the arguments.

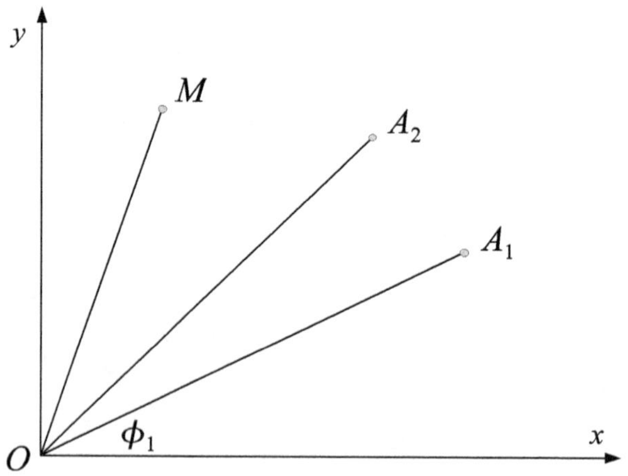

Figure 4.3: The multiplication of two complex numbers: $\phi_1 = \widehat{A_1Ox}$, $\phi_2 = \widehat{A_2Ox} = \widehat{MOA_1}$.

In particular, if $|z_1| = |z_2| = 1$, then the product $z_1 z_2$ corresponds to a point M obtained by a rotation of the point A_1 with angle ϕ_2, see Figure 4.3.

The inverse Let $z = \rho(\cos\phi + i\sin\phi)$. According to (4.1.1), the inverse will be

$$z^{-1} = \frac{\bar{z}}{|z|^2} = \frac{1}{\rho}(\cos\phi - i\sin\phi) = \frac{1}{\rho}\big(\cos(-\phi) + i(\sin(-\phi)\big).$$

Exercise 4.2.2 (*i*) *Let* $z \in \mathbb{C}$ *with* $|z| = 1$. *Find* z^{-1}.

(*ii*) *Find the inverse of the complex number* $\frac{\sqrt{2}}{2}(1-i)$.

The powers of a complex number It is more easily to rise a complex number to a power if it is written in trigonometric form. If $z = \rho(\cos\phi + i\sin\phi)$, then inductively we have

$$\begin{aligned} z^2 &= \rho^2(\cos 2\phi + i\sin 2\phi) \\ z^3 &= \rho^3(\cos 3\phi + i\sin 3\phi) \\ \dots &\quad \dots\dots\dots\dots\dots\dots \\ z^n &= \rho^n(\cos n\phi + i\sin n\phi), \quad \forall n \in \mathbb{N}. \end{aligned}$$

The complex unit circle The *unit circle* in the plane

$$C(O,1) = \{M \in \mathcal{P};\ |OM| = 1\}$$

corresponds to the set of complex numbers with unit modulus:

$$\mathbb{U} = \{z \in \mathbb{C};\ |z| = 1\} = \{\cos\phi + i\sin\phi;\ \phi \in [0, 2\pi)\}.$$

Since $|z_1 z_2| = |z_1| \cdot |z_2|$, the set $\mathbb{U}$ is closed with respect to the multiplication of complex numbers. This means that the product of two complex numbers on the unit circle is a complex number on the same circle. Using $z^{-1} = \bar{z}$, it follows that $\mathbb{U}$ is also closed with respect to the inverse. Equivalently, we can say that $(\mathbb{U}, \cdot)$ forms a subgroup of the multiplicative complex group $(\mathbb{C}, \cdot)$.

The roots of unity Let $n \in \mathbb{N}$ be a natural number. A solution of the equation $z^n = 1$ is called an *nth root of unity.* An obvious root is $\omega_0 = 1$. There are also other roots besides 1. The previous equation can be solved more easily if the complex number is represented in the trigonometric form. To this end, we consider the following complex number

$$\omega_1 = \cos\frac{2\pi}{n} + i\sin\frac{2\pi}{n}.$$

Then the nth power of ω_1 is

$$\begin{aligned}\omega_1^n &= \left(\cos\frac{2\pi}{n} + i\sin\frac{2\pi}{n}\right)^n \\ &= \cos\frac{2n\pi}{n} + i\sin\frac{2n\pi}{n} = \cos 2\pi + i\sin 2\pi \\ &= 1.\end{aligned}$$

Hence ω_1 is another root of unity. Now consider $\omega_2 = \omega_1^2 = \cos\frac{4\pi}{n} + i\sin\frac{4\pi}{n}$. Since

$$\omega_2^n = (\omega_1^2)^n = (\omega_1^n)^2 = 1^2 = 1,$$

it follows that ω_2 is also a root of unity. In general, let $\omega_k = \omega_1^k$, $k \in \{0, 1, 2, \ldots, n-1\}$. Then

$$\omega_k^n = (\omega_1^k)^n = (\omega_1^n)^k = 1^k = 1,$$

so we have obtained the following n roots of unity

$$\begin{aligned}\omega_0 &= 1 \\ \omega_1 &= \cos\frac{2\pi}{n} + i\sin\frac{2\pi}{n} \\ \omega_2 &= \cos\frac{4\pi}{n} + i\sin\frac{4\pi}{n} \\ \cdots &\cdots \cdots\cdots\cdots \\ \omega_k &= \cos\frac{2k\pi}{n} + i\sin\frac{2k\pi}{n} \\ \cdots &\cdots \cdots\cdots\cdots \\ \omega_{n-1} &= \cos\frac{2(n-1)\pi}{n} + i\sin\frac{2(n-1)\pi}{n}.\end{aligned}$$

From the Fundamental Theorem of Algebra, the polynomial equation $z^n - 1 = 0$ has exactly n roots (either complex or real). Since we already found n distinct roots, it follows that all the roots are given by $\omega_0 = 1, \omega_1, \ldots, \omega_{n-1}$. These are called the nth roots of unity.

Proposition 4.2.3 *(i) The product of two nth roots of unity is an nth root of unity.*

(ii) Let $\Omega_n = \{\omega_0, \omega_1, \ldots, \omega_{n-1}\}$. Then Ω_n is a group with respect to multiplication. (Ω_n is called the cyclic group with n elements).

Proof: (*i*) Let $\omega_k, \omega_{k'} \in \Omega_n$. Then

$$\omega_k\omega_{k'} = \omega_1^k\omega_1^{k'} = \omega_1^{k+k'}.$$

If $k + k' < n$, then $\omega_1^{k+k'} \in \Omega_n$. Otherwise, $n \leq k + k' < 2n$, or $0 \leq k + k' - n < n$, and then

$$\omega_1^{k+k'} = \omega_1^{k+k'} \cdot 1 = \omega_1^{k+k'} \omega^{-n} = \omega_1^{k+k'-n} \in \Omega_n.$$

Hence Ω_n is closed with respect to multiplication.

(ii) Since $\omega_k \omega_{n-k} = \omega_1^k \omega_1^{n-k} = \omega_1^n = 1$, it follows that $\omega_k^{-1} = \omega_{n-k}$, and hence Ω_n is closed with respect to the inverse. It follows that Ω_n is a subgroup of $\mathbb{U}$, and hence it is a group. ■

Geometrical interpretation of the roots of unity The nth roots of unity are affixes of points on the unit circle. Let A_k be the image in the plane of the complex number ω_k. Using the geometric interpretation of multiplication, it follows that the angle between the ray OA_k and the x-axis is $\frac{2\pi}{n}$. The angle between two consecutive rays OA_k and OA_{k+1} is also $\frac{2\pi}{n}$. It follows that the points $A_0, A_1, \ldots, A_{n-1}$ are the vertices of a regular polygon, see Figure 4.4.

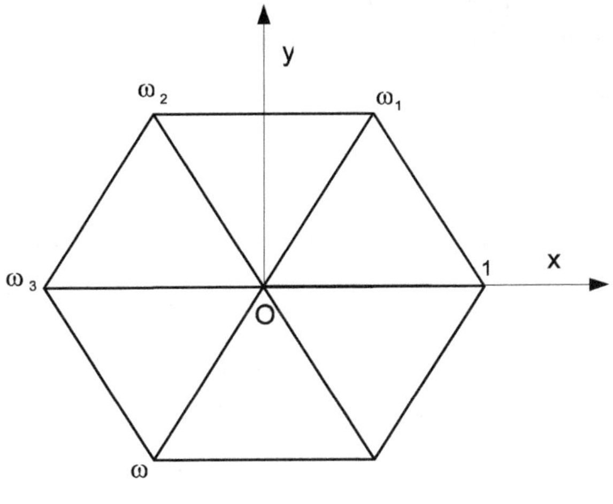

Figure 4.4: The images of the roots of unity form a regular polygon.

Exercise 4.2.4 *Find the roots of unity for* $n = 3$ *and draw the corresponding equilateral triangle.*

Exercise 4.2.5 *Find the roots of unity for* $n = 4$ *and draw the corresponding square.*

Exercise 4.2.6 *Show that the sum of the nth roots of unity is zero:*

$$\omega_0 + \omega_1 + \omega_2 + \cdots + \omega_{n-1} = 0.$$

Exercise 4.2.7 *Show that any 3rd root of unity different than* 1 *satisfies the equation*

$$z^2 + z + 1 = 0.$$

Exercise 4.2.8 *Let z_1, z_2, z_3 be three numbers on the unit circle, i.e., $|z_1| = |z_2| = |z_3| = 1$, such that*

$$z_1 + z_2 + z_3 = 0.$$

Let A_1, A_2, A_3 be the images in the plane of the complex numbers z_1, z_2, z_3. Show that $\Delta A_1A_2A_3$ is an equilateral triangle.

Exercise 4.2.9 *Let n be a natural number. Using complex numbers and the sum of a geometric series, find formulas for the telescopic sums*

$$1 + \cos\phi + \cos 2\phi + \cdots + \cos n\phi$$
$$\sin\phi + \sin 2\phi + \cdots + \sin n\phi.$$

Euler's formula This paragraph will deal with a very useful representation of complex numbers. Let $z = \rho(\cos\phi + i\sin\phi)$ be a complex number written in trigonometric form. Using the Mac-Lauren series

$$\begin{aligned}
\cos x &= \sum_{n\geq 0}(-1)^n \frac{x^{2n}}{(2n)!}, \\
\sin x &= \sum_{n\geq 0}(-1)^n \frac{x^{2n+1}}{(2n+1)!}, \text{ and} \\
e^x &= \sum_{n\geq 0}\frac{x^n}{n!},
\end{aligned}$$

and the powers of i

$$i^{4k} = 1, \qquad i^{4k+1} = i, \qquad i^{4k+2} = -1, \qquad i^{4k+3} = -i,$$

we can write

$$\begin{aligned}
&\cos\phi + i\sin\phi \\
= &\sum_{n\geq 0}(-1)^n\frac{\phi^{2n}}{(2n)!} + i\sum_{n\geq 0}(-1)^n\frac{\phi^{2n+1}}{(2n+1)!} \\
= &\sum_{k\geq 0}\left(\frac{\phi^{4k}}{(4k)!} - \frac{\phi^{4k+2}}{(4k+2)!}\right) + i\sum_{k\geq 0}\left(\frac{\phi^{4k+1}}{(4k+1)!} - \frac{\phi^{4k+3}}{(4k+3)!}\right) \\
= &\sum_{k\geq 0}\frac{\phi^{4k}}{(4k)!} + i\sum_{k\geq 0}\frac{\phi^{4k+1}}{(4k+1)!} + (-1)\sum_{k\geq 0}\frac{\phi^{4k+2}}{(4k+2)!} - i\sum_{n\geq 0}\frac{\phi^{4k+3}}{(4k+3)!} \\
= &\sum_{k\geq 0}\frac{(i\phi)^{4k}}{(4k)!} + \sum_{k\geq 0}\frac{(i\phi)^{4k+1}}{(4k+1)!} + \sum_{k\geq 0}\frac{(i\phi)^{4k+2}}{(4k+2)!} + \sum_{n\geq 0}\frac{(i\phi)^{4k+3}}{(4k+3)!} \\
= &\sum_{n\geq 0}\frac{(i\phi)^n}{n!} = e^{i\phi}.
\end{aligned}$$

We have arrived at the following result:

Theorem 4.2.10 (Euler) *If z is a complex number with $|z| = \rho$ and $arg(z) = \phi$, then*

$$z = \rho e^{i\phi}. \tag{4.2.4}$$

We shall refer to formula (4.2.4) as Euler's representation of complex numbers. Multiplication and division formulas make more sense when dealing with this representation. Also, this representation is preferred when proving geometric properties using complex numbers.

Let $z_1 = \rho_1 e^{i\phi_1}$ and $z_2 = \rho_2 e^{i\phi_2}$ be two complex numbers in Euler's representation. Then

$$\begin{aligned}
z_1 z_2 &= \rho_1 e^{i\phi_1}\cdot\rho_2 e^{i\phi_2} = \rho_1\rho_2 e^{i(\phi_1+\phi_2)} \\
z_1/z_2 &= \rho_1 e^{i\phi_1}/(\rho_2 e^{i\phi_2}) = (\rho_1/\rho_2)\, e^{i(\phi_1-\phi_2)}.
\end{aligned}$$

For instance, we have $e^{i\pi} = -1$, $e^{i\pi/2} = i$, $e^{2i\pi} = 1$. Also, the nth roots of unity can be written in Euler's representation as

$$\omega_0 = e^{\frac{0\cdot i}{n}} = 1,\quad \omega_1 = e^{\frac{2\pi i}{n}},\quad \ldots,\quad \omega_k = e^{\frac{2k\pi i}{n}},\quad \ldots,\quad \omega_{n-1} = e^{\frac{2(n-1)\pi i}{n}}.$$

4.3 Basic geometric concepts

In the following a few useful basic geometric concepts are translated in the language of complex numbers.

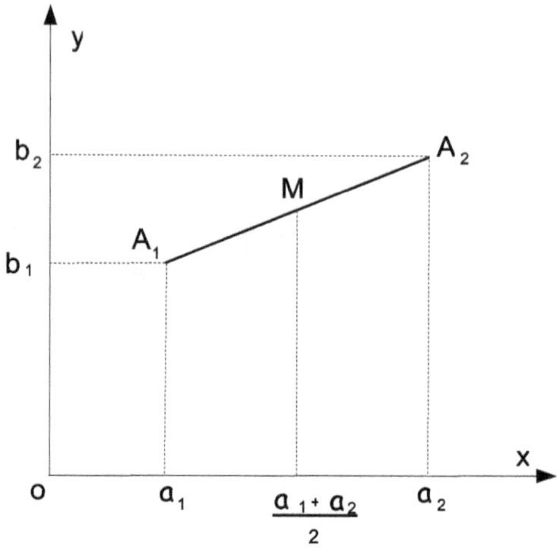

Figure 4.5: The affix of the midpoint of the segment A_1A_2 is $z_M = (z_1 + z_2)/2$.

The distance between two points Let A_1 and A_2 be two points in the plane, with affixes $z_1 = a_1 + ib_1$ and $z_2 = a_2 + ib_2$, see Figure 4.5. The Pythagorean theorem yields

$$|A_1A_2|^2 = (a_2 - a_1)^2 + (b_2 - b_1)^2 = |(a_2 - a_1) + i(b_2 - b_1)|^2 = |z_2 - z_1|^2.$$

Hence, the distance between two points is equal with the modulus of the difference of their affixes:

$$|A_1A_2| = |z_2 - z_1|.$$

The midpoint of a line segment Let M be the midpoint of the line segment A_1A_2. Let $z_M = a_M + ib_M$ be its affix. Using the midpoint connector theorem in a trapezoid, see Figure 4.5, yields

$$a_M = \frac{1}{2}(a_1 + a_2), \qquad b_M = \frac{1}{2}(b_1 + b_2).$$

Hence the affix of the midpoint of a line segment is the average of the end points affixes:

$$z_M = \frac{1}{2}(z_1 + z_2).$$

Exercise 4.3.1 *The quadrilateral $ABCD$ is a parallelogram if and only if*

$$z_A + z_C = z_B + z_D.$$

Exercise 4.3.2 *Let $ABCD$ be a quadrilateral. Show that the midpoints of the sides AB, BC, CD, DA are the vertices of a parallelogram.*

Exercise 4.3.3 *Let $ABCD$ be a quadrilateral. Let Q, M, N, P be the midpoints of the sides AB, BC, CD, DA. Denote by L and S the midpoints of the diagonals AC and BD.*

(i) Show that the line segments QN, PM and LS intersect at the same point G.

(ii) Show that G is the common midpoint of the above line segments.

The angle between two rays Let A_1, A_2 be two points with affixes z_1, z_2. Consider the Euler's representation

$$z_1 = \rho_1 e^{i\phi_1}, \qquad z_2 = \rho_2 e^{i\phi_2}.$$

Dividing, yields $\dfrac{z_2}{z_1} = \dfrac{\rho_2}{\rho_1} e^{i(\phi_2 - \phi_1)}$ and then $\arg\dfrac{z_2}{z_1} = \phi_2 - \phi_1$. Hence, we obtain the following expression for the angle

$$\widehat{A_2OA_1} = \arg\frac{z_2}{z_1}, \tag{4.3.5}$$

see Figure 4.6(a). Now let M be another point in the plane with affix z_M. We are interested in describing the angle $\widehat{A_2MA_1}$, see Figure 4.6(b). Let

$$|MA_1| = |z_1 - z_M| = r_1, \qquad |MA_2| = |z_2 - z_M| = r_2.$$

Then in the Euler representation, we have

$$z_2 - z_M = r_2 e^{i\theta_2}, \qquad z_1 - z_M = r_1 e^{i\theta_1},$$

where $\theta_1 = \widehat{A_1Mx}$ and $\theta_2 = \widehat{A_2Mx}$. Dividing, yields

$$\frac{z_2 - z_M}{z_1 - z_M} = \frac{r_2}{r_1} e^{i(\theta_2 - \theta_1)}.$$

Hence

$$\widehat{A_2MA_1} = \theta_2 - \theta_1 = \arg\frac{z_2 - z_M}{z_1 - z_M}. \tag{4.3.6}$$

The above formula works in the case $\theta_2 > \theta_1$. Otherwise, we need to use absolute values to obtain

$$\widehat{A_2MA_1} = |\theta_2 - \theta_1| = \left|\arg\frac{z_2 - z_M}{z_1 - z_M}\right|. \tag{4.3.7}$$

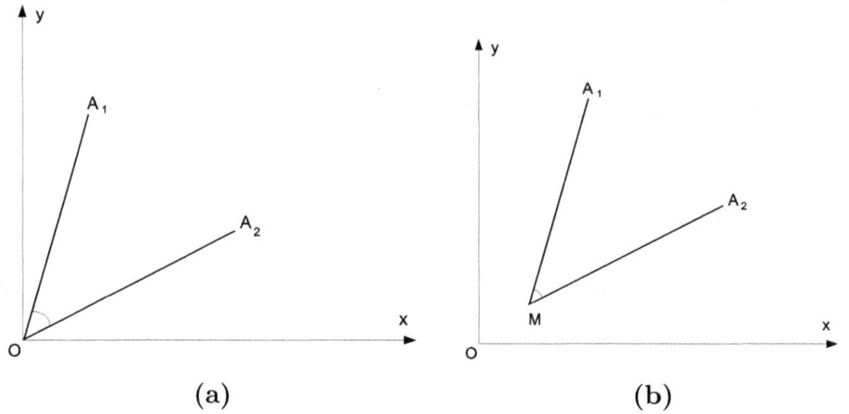

Figure 4.6: The angle between two rays.

The equation of a line If the points A_1, A_2, M are collinear, then making $\widehat{A_2MA_1} = \theta_2 - \theta_1 = \pi$ in the above formula, yields

$$\frac{z_2 - z_M}{z_1 - z_M} = \frac{r_2}{r_1}e^{i\pi} = -\frac{r_2}{r_1} = t \in \mathbb{R}.$$

Solving for z_2 yields the equation

$$z_2 = tz_1 + (1-t)z_M,$$

i.e., z_2 is a convex combination[4] of z_1 and z_M. In conclusion, the equation of a line passing through the points A and B with affixes z_A and z_B is

$$z = tz_A + (1-t)z_B, \qquad t \in \mathbb{R}. \tag{4.3.8}$$

The midpoint connector theorem We shall present a straightforward application of the above formulas. Let ΔABC be a triangle and denote by M and N the midpoints of the sides AB and AC, respectively. The relations between the affixes are

$$z_N = \frac{1}{2}(z_A + z_C), \qquad z_M = \frac{1}{2}(z_A + z_B).$$

Subtracting and taking the modulus yields $|z_N - z_M| = \frac{1}{2}|z_C - z_B|$, and then we obtain $|MN| = \frac{1}{2}|BC|$, *i.e.*, the length of the midpoint connector segment is half the length of the base, see Figure 4.7.

[4]In a convex combination the sum of the coefficients is 1.

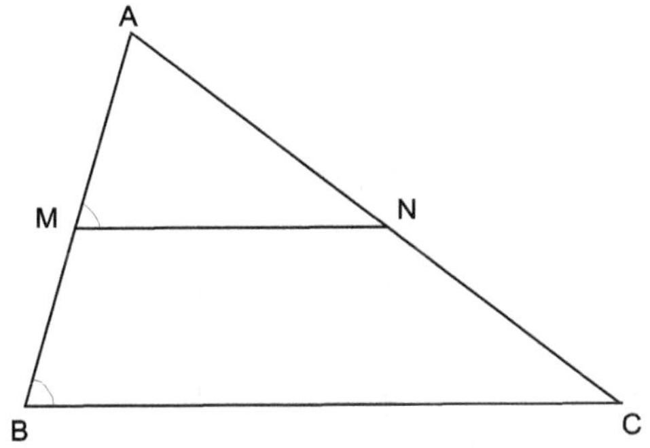

Figure 4.7: The midpoint connector MN.

On the other side, we have

$$\begin{aligned}
\frac{z_A - z_M}{z_N - z_M} &= \frac{z_A - \frac{1}{2}z_A - \frac{1}{2}z_B}{\frac{1}{2}z_A + \frac{1}{2}z_C - \frac{1}{2}z_A - \frac{1}{2}z_B} = \frac{z_A - z_B}{z_C - z_B} \Longrightarrow \\
\arg\frac{z_A - z_M}{z_N - z_M} &= \arg\frac{z_A - z_B}{z_C - z_B} \Longleftrightarrow \\
\widehat{AMN} &= \widehat{ABC}.
\end{aligned}$$

Since the alternate interior angles are equal, it follows that the lines MN and BC are parallel. Hence, the midpoint connector line is parallel with the base.

The rotation centered at the origin Let M and M' be two points such that

$$z_{M'} = z_M e^{i\theta}.$$

Since $|OM'| = |z_{M'}| = |z_M|\cdot|e^{i\theta}| = |z_M| = |OM|$, and $\widehat{M'OM} = \arg\frac{z_{M'}}{z_M} = \arg e^{i\theta} = \theta$, it follows that M' is obtained from M by a rotation of angle θ centered at the origin.

The rotation centered at a point A Consider a rotation centered at A of angle θ, which sends M into M'. Then the relation among the affixes of M, M' and A is

$$z_{M'} = z_A + (z_M - z_A)e^{i\theta}. \tag{4.3.9}$$

This can be shown by rewriting the above relation as

$$z_{M'} - z_A = (z_M - z_A)e^{i\theta}$$

and taking the modulus

$$|z_{M'} - z_A| = |z_M - z_A|.$$

This shows that $|M'A| = |MA|$. Since

$$\widehat{M'AM} = \arg \frac{z_{M'} - z_M}{z_M - z_A} = \arg e^{i\theta} = \theta,$$

it follows that (4.3.9) describes a rotation centered at A of angle θ.

Equilateral triangle Consider an equilateral triangle ΔABC. Since A is obtained from B by a rotation of an angle of 60 degrees around B, we can write

$$z_A = z_B + (z_C - z_B)e^{i\frac{\pi}{3}}. \tag{4.3.10}$$

The above relation is also equivalent with

$$\frac{z_A - z_B}{z_C - z_B} = e^{i\frac{\pi}{3}}. \tag{4.3.11}$$

Translation The equation

$$z_{M'} = z_A + z_M \tag{4.3.12}$$

describes a translation of M into M' with a vector $\overrightarrow{OA}$. All geometric properties are preserved by translations. Many times we shall take advantage of this property to assume that the origin is in one of the given points.

Symmetry with respect to a point A Let M' be the symmetric of the point M with respect to A. Then A is the midpoint of MM' and then

$$z_A = \frac{1}{2}(z_M + z_{M'}).$$

Solving for $z_{M'}$ yield the equation of the symmetry

$$z_{M'} = 2z_A - z_M. \tag{4.3.13}$$

In the following we shall present a few solved exercises.

Solved Exercise 4.3.4 *Let* $z_1, z_2 \in \mathbb{C}$. *Show that*

$$|2z_1 - z_2| = |2z_2 - z_1| \Longleftrightarrow |z_1| = |z_2|.$$

Proof: Let $z_1 = a_1 + ib_1$ and $z_2 = a_2 + ib_2$. We have the following computation

$$\begin{aligned} |2z_1 - z_2|^2 &= |2(a_1 + ib_1) - (a_2 + ib_2)|^2 = |(2a_1 - a_2) + i(2b_1 - b_2)|^2 \\ &= (2a_1 - a_2)^2 + (2b_1 - b_2)^2. \end{aligned} \quad (4.3.14)$$

Similarly we have

$$|2z_2 - z_1|^2 = (2a_2 - a_1)^2 + (2b_2 - b_1)^2. \quad (4.3.15)$$

Using (4.3.14) and (4.3.15) yields

$$\begin{aligned} |2z_1 - z_2|^2 &= |2z_2 - z_1|^2 \Longleftrightarrow \\ (2a_1 - a_2)^2 + (2b_1 - b_2)^2 &= (2a_2 - a_1)^2 + (2b_2 - b_1)^2 \\ a_1^2 + b_1^2 &= a_2^2 + b_2^2 \Longleftrightarrow \\ |z_1|^2 &= |z_2|^2 \Longleftrightarrow \\ |z_1| &= |z_2|. \end{aligned}$$

■

Remark 4.3.5 The geometric significance of Exercise 4.3.4 is as follows. Let A_1 and A_2 be the images of the complex numbers z_1 and z_2. Then $2z_1 - z_2$ is the affix of the symmetric of A_2 with respect to A_1 and $2z_2 - z_1$ is the affix of the symmetric of A_1 with respect to A_2. Let's denote these points by M and N. Then the exercise states that $|OA_1| = |OA_2|$ if and only if $|OM| = |ON|$, where O denotes the origin. More generally, if ΔOMN is a triangle and $A_1, A_2 \in (MN)$ such that $|MA_1| = |A_1A_2| = |A_2N|$, then ΔOMN is isosceles if and only if ΔOA_1A_2 is also isosceles.

Solved Exercise 4.3.6 *Let ΔABC be a triangle and B' and C' the midpoints of the sides AC and AB, respectively. Show that $|AB| = |AC|$ if and only if $|BB'| = |CC'|$. (This can be stated equivalently by saying that a triangle is isosceles if and only if it has two equal medians).*

Proof: Let z_A, z_B, z_C, $z_{B'}$, $z_{C'}$ be the affixes of A, B, C, B', C', respectively. We have

$$\begin{aligned} |BB'| &= |z_{B'} - z_B| = \left|\frac{z_A + z_C}{2} - z_B\right| \\ &= \frac{1}{2}|z_A + z_C - 2z_B|. \end{aligned}$$

Similarly, we can get

$$|CC'| = \frac{1}{2}|z_A + z_B - 2z_C|.$$

Denote $z_1 = z_A - z_B$, $z_2 = z_A - z_C$, and $z_3 = z_B - z_C$. Then $|BB'| = |CC'|$ if and only if

$$\begin{aligned} |z_A + z_C - 2z_B| &= |z_A + z_B - 2z_C| \iff \\ |(z_A - z_B) + (z_C - z_B)| &= |(z_A - z_C) + (z_B - z_C)| \iff \\ |z_1 - z_3| &= |z_2 + z_3|. \end{aligned}$$

Since $z_1 - z_2 = z_A - z_B - z_A + z_C = -z_3$, the above relation becomes

$$|2z_1 - z_2| = |2z_2 - z_1|.$$

Using Exercise 4.3.4 yields $|z_1| = |z_2|$, which becomes $|z_A - z_B| = |z_A - z_C|$. This is equivalent to $|AB| = |AC|$. ■

Solved Exercise 4.3.7 *Let $z_1, z_2 \in \mathbb{C}$. Show that*

$$Re(z_1 z_2) = Re(z_1)\, Re(z_2) - Im(z_1)\, Im(z_2) \tag{4.3.16}$$
$$Im(z_1 z_2) = Re(z_1)\, Im(z_2) + Re(z_2)\, Im(z_1). \tag{4.3.17}$$

Proof: Consider the trigonometric representations

$$z_1 = \cos\theta_1 + i\sin\theta_1, \qquad z_2 = \cos\theta_2 + i\sin\theta_2.$$

Then $z_1 z_2 = \cos(\theta_1+\theta_2) + i\sin(\theta_1+\theta_2)$. Then relations (4.3.16) and (4.3.17) become the following well-known trigonometric formulas

$$\begin{aligned} \cos(\theta_1 + \theta_2) &= \cos\theta_1\cos\theta_2 - \sin\theta_1\sin\theta_2 \\ \sin(\theta_1 + \theta_2) &= \cos\theta_1\sin\theta_2 + \cos\theta_2\sin\theta_1. \end{aligned}$$

■

The next exercise will be used in the sequel several times.

Solved Exercise 4.3.8 *Show that*

(i) $1 - e^{i\theta} = 2\sin\frac{\theta}{2}\, e^{i\frac{\theta-\pi}{2}}$, *and*

(ii) $|1 - e^{i\theta}| = 2|\sin\frac{\theta}{2}|$.

Proof: (i) Applying usual trigonometric formulas we have

$$\begin{aligned}1-e^{i\theta} &= 1-(\cos\theta+i\sin\theta)=(1-\cos\theta)-i\sin\theta\\ &= 2\sin^2\frac{\theta}{2}-2i\sin\frac{\theta}{2}\cos\frac{\theta}{2}=2\sin\frac{\theta}{2}\left(\sin\frac{\theta}{2}-i\cos\frac{\theta}{2}\right)\\ &= 2\sin\frac{\theta}{2}\left(\cos\frac{\pi-\theta}{2}-i\sin\frac{\pi-\theta}{2}\right)\\ &= 2\sin\frac{\theta}{2}\left(\cos\frac{\theta-\pi}{2}+i\sin\frac{\theta-\pi}{2}\right)\\ &= 2\sin\frac{\theta}{2}e^{i\frac{\theta-\pi}{2}}.\end{aligned}$$

(ii) It comes obviously from (i) using that $|e^{i\frac{\theta-\pi}{2}}|=1$. ■

Solved Exercise 4.3.9 *Let* $\epsilon=1+i\sqrt{3}/2$. *Show that*

(i) $|1-\epsilon|=|\epsilon|=1$,

(ii) $|1-\bar{\epsilon}|=|\bar{\epsilon}|=1$, *and*

(iii) $\left|\dfrac{\epsilon}{1-\epsilon}\right|=1$.

Proof: (i) Since $\epsilon=1+i\sqrt{3}/2=e^{i\frac{\pi}{3}}$, then $|\epsilon|=1$, and using Exercise 4.3.8 part (ii) we have

$$|1-\epsilon|=|1-e^{i\frac{\pi}{3}}|=2\sin\frac{\pi}{6}=2\cdot\frac{1}{2}=1.$$

(ii) Using the properties of conjugation and (i) yields

$$|1-\bar{\epsilon}|=|\overline{1-\epsilon}|=|1-\epsilon|=1.$$

(iii) Using the properties of modulus we have

$$\left|\frac{\epsilon}{1-\epsilon}\right|=\frac{|\epsilon|}{|1-\epsilon|}=\frac{1}{1}=1.$$

■

Remark 4.3.10 We note that ϵ, $\frac{1}{\epsilon}$, $1-\epsilon$, $\frac{1}{1-\epsilon}$, $\frac{\epsilon-1}{\epsilon}$, and $\frac{\epsilon}{\epsilon-1}$ belong to the unit circle.

Exercise 4.3.11 *Find the complex numbers* z *such that* $|z|=|1-z|=1$.

Proof: Let $z = a + ib$. Since $|z| = 1$, we have $a^2 + b^2 = 1$. Using the definition of the modulus, we have

$$|1 - z| = |1 - a + ib| = \sqrt{(1-a)^2 + b^2} = \sqrt{1 - 2a + a^2 + b^2} = \sqrt{2 - 2a}.$$

Since $|1 - z| = 1$, yields $a = 1/2$. Then $b^2 = 1 - a^2 = 3/4$, and hence $b = \pm\sqrt{3}/2$. The solutions of the equations are $1/2 \pm i\sqrt{3}/2$, which are ϵ and $\overline{\epsilon}$.

Proof variant: Since $|z| = 1$, we can write $z = e^{i\theta}$. Then by Exercise 4.3.8 part (ii) we have

$$|1 - z| = |1 - e^{i\theta}| = 2|\sin\frac{\theta}{2}| = 1,$$

so $\sin\frac{\theta}{2} = \pm 1/2$ and hence $\theta = \pm\pi/3$. The corresponding solutions are

$$\begin{aligned} \epsilon &= e^{i\pi/3} = \cos\frac{\pi}{3} + i\sin\frac{\pi}{3} = \frac{1}{2} + i\frac{\sqrt{3}}{2}, \\ \overline{\epsilon} &= e^{-i\pi/3} = \cos\frac{\pi}{3} - i\sin\frac{\pi}{3} = \frac{1}{2} - i\frac{\sqrt{3}}{2}. \end{aligned}$$

■

The length of the side of a regular polygon with n vertices Let $A_0A_1\ldots A_{n-1}$ be a regular polygon inscribed in the unit circle, and denote by ℓ_n the length of its side. If $e^{i\phi_j}$ is the affix of A_j, then multiplying by $e^{-i\phi_1}$ we perform a rotation of the polygon into a polygon with the first vertex starting at $(1, 0)$. The other vertices will have the affixes $\omega_k = e^{i\frac{2\pi k}{n}}$. The length of the side is equal to the distance between the first two vertices

$$\ell_n = |\omega_0 - \omega_1| = |1 - e^{i\frac{2\pi}{n}}| = 2|\sin\frac{\pi}{n}|,$$

see Exercise 4.3.8. For instance, the side of an equilateral triangle inscribed in the unit circle is

$$\ell_3 = 2\sin\frac{\pi}{3} = 2\frac{\sqrt{3}}{2} = \sqrt{3},$$

while the side of the regular hexagon is

$$\ell_3 = 2\sin\frac{\pi}{6} = 2\frac{1}{2} = 1.$$

Exercise 4.3.12 *Let A', B', C' be the midpoints of the sides BC, AC and AB respectively. Let B'' and C'' be the symmetric points of A' with respect to B' and C'. Prove that A belongs to the line $B''C''$.*

Exercise 4.3.13 *Let M and N be the midpoints of the sides AD and BC of a quadrilateral $ABCD$. Show that*

$$|MN| \leq \frac{1}{2}(|AB| + |DC|).$$

Solved Exercise 4.3.14 *Let $P_0P_1 \dots P_{n-1}$ be a regular polygon inscribed in a circle of radius $r = 1$.*
(i) Find the product $|P_0P_1| \cdot |P_0P_2| \cdot \dots \cdot |P_0P_{n-1}|$.
(ii) Find the sum $|P_0P_1| + |P_0P_2| + \dots + |P_0P_{n-1}|$.

Proof: (*i*) Consider the polynomial function $f(x) = x^{n-1}+x^{n-2}+\dots+x+1$. Let $x_k = e^{i\frac{2\pi}{n}k}$, $1 \leq k \leq n-1$. Since x_k satisfy the equation $x^n - 1 = 0$ and we have $f(x) = \frac{x^n-1}{x-1}$ for $n \neq 1$, then $f(x_k) = 0$, so x_k are the zeros of $f(x)$. Therefore, we can factorize $f(x) = (x-x_1)(x-x_2)\cdots(x-x_{n-1})$. Using the formula of the polynomial, we have $f(1) = n$. Then using the factorization formula, we have

$$f(1) = (1-x_1)(1-x_2)\cdots(1-x_{n-1}) = \prod_{k=1}^{n-1}(1-e^{i\frac{2\pi}{n}k}).$$

Since $f(1)$ is real, we have

$$n = f(1) = |\prod_{k=1}^{n-1}(1-e^{i\frac{2\pi}{n}k})| = \prod_{k=1}^{n-1}|1-e^{i\frac{2\pi}{n}k}| = \prod_{k=1}^{n-1}|P_0P_k|,$$

where we considered the complex numbers $\{1, e^{i\frac{2\pi}{n}k}\}$ associated with the vertices $\{P_0, P_k\}$.
(*ii*) We have

$$\sum_{k=1}^{n-1}|P_0P_k| = \sum_{k=1}^{n-1}|1-e^{i\frac{2\pi}{n}k}| = \sum_{k=1}^{n-1}2|\sin\frac{k\pi}{n}| = 2\sum_{k=1}^{n-1}\sin\frac{k\pi}{n} = 2\cot\frac{\pi}{2n},$$

where we used the formula $\sum_{k=1}^{n-1}\sin\frac{k\pi}{n} = \cot\frac{\pi}{2n}$. This can be shown as in the following. Consider the sums

$$S_1 = 1 + \cos\frac{\pi}{n} + \cos\frac{2\pi}{n} + \dots + \cos\frac{(n-1)\pi}{n}$$

$$S_2 = 0 + \sin\frac{\pi}{n} + \sin\frac{2\pi}{n} + \dots + \sin\frac{(n-1)\pi}{n},$$

and compute $S_1 + iS_2$. Denoting $z = \cos\frac{\pi}{n} + i\sin\frac{\pi}{n} = e^{i\frac{\pi}{n}}$, using Moivre's formula and the formula of a geometric sum, we have

$$\begin{aligned}
S_1 + iS_2 &= 1 + e^{i\frac{\pi}{n}} + e^{i\frac{2\pi}{n}} + \cdots + e^{i\frac{(n-1)\pi}{n}} \\
&= 1 + z + z^2 + \cdots z^{n-1} = \frac{1-z^n}{1-z} = \frac{2}{1-z} \\
&= \frac{2(1-\bar{z})}{|1-z|^2} = \frac{2(1-e^{-\frac{\pi}{n}i})}{|1-e^{i\frac{\pi}{n}}|^2} = \frac{2-2e^{-\frac{\pi}{n}i}}{4\sin^2\frac{\pi}{2n}} \\
&= \frac{2(1-\cos\frac{\pi}{n})}{4\sin^2\frac{\pi}{2n}} + i\frac{2\sin\frac{\pi}{n}}{4\sin^2\frac{\pi}{2n}} = 1 + i\frac{4\sin\frac{\pi}{2n}\cos\frac{\pi}{2n}}{4\sin^2\frac{\pi}{2n}} = 1 + i\cot\frac{\pi}{2n}.
\end{aligned}$$

Identifying the real and imaginary parts yields $S_1 = 1$ and $S_2 = \cot\frac{\pi}{2n}$. ■

Remark 4.3.15 One can use some standard Calculus techniques to show that

$$\lim_{x\to\infty}\frac{1}{x}\cot\frac{\pi}{2x} = \frac{2}{\pi}.$$

Using this limit, we obtain

$$\lim_{n\to\infty}\frac{1}{n}\sum_{k=1}^{n-1}|P_0P_k| = \lim_{n\to\infty}\frac{2}{n}\cot\frac{\pi}{2n} = \frac{4}{\pi}.$$

Solved Exercise 4.3.16 *Let A, B and C be three points in the plane with affixes z_A, z_B and z_C, respectively, and denote $\omega = e^{\frac{2\pi}{3}i}$.*

(i) Show that if ΔABC is an equilateral triangle then either $\omega^2 z_A + \omega z_B + z_C = 0$, or $\omega z_A + \omega^2 z_B + z_C = 0$.

(ii) If either $\omega^2 z_A + \omega z_B + z_C = 0$, or $\omega z_A + \omega^2 z_B + z_C = 0$, show that ΔABC is equilateral.

Proof: (*i*) Let M be the center of the equilateral triangle ΔABC and O denote the origin. Translating the triangle ΔABC by the vector $\overrightarrow{MO}$ we obtain the equilateral triangle $\Delta A'B'C'$ with vertices given by affixes $z_{A'} = z_A - z_M$, $z_{B'} = z_B - z_M$, $z_{C'} = z_C - z_M$. Let $z_{C'} = \rho e^{i\theta}$. Then $z_{A'} = z_{C'}\omega$ and $z_{B'} = z_{C'}\omega^2$. Using $\omega^3 = 1$, a computation shows that

$$\begin{aligned}
\omega z_{A'} + \omega^2 z_{B'} + z_{C'} &= \omega^2 z_{C'} + \omega^2\omega^2 z_{C'} + z_{C'} = z_{C'}(\omega^2 + \omega^4 + 1) \\
&= z_{C'}\underbrace{(\omega^2 + \omega + 1)}_{=0} = 0.
\end{aligned}$$

Using this identity, we have

$$\begin{aligned} z_A\omega + z_B\omega^2 + z_C &= (z_{A'} + z_M)\omega + (z_{B'} + z_M)\omega^2 + (z_{C'} + z_M) \\ &= \underbrace{(\omega z_{A'} + \omega^2 z_{B'} + z_{C'})}_{=0} + z_M\underbrace{(\omega + \omega^2 + 1)}_{=0} = 0. \end{aligned}$$

Considering the triangle with a different orientation (obtained by swapping the positions of A and B), we obtain $\omega^2 z_A + \omega z_B + z_C = 0$.
(ii) We assume $z_A\omega + z_B\omega^2 + z_C = 0$. Using $1 + \omega + \omega^2 = 0$, we also have $z_C\omega + z_C\omega^2 + z_C = 0$. Subtracting term by term, we obtain $z_A - z_C = -\omega(z_B - z_C)$. Then take the modulus

$$|z_A - z_C| = |-\omega| \cdot |z_B - z_C| = |z_B - z_C|,$$

so $|AC| = |BC$. Similarly, subtracting $z_A\omega + z_A\omega^2 + z_A = 0$, we obtain $|AC| = |AB|$. Hence ΔABC is equilateral. ■

Solved Exercise 4.3.17 *Let ΔABC be an equilateral triangle and consider the origin $O = B$. We denote by z_A, $z_B = 0$, z_C the affixes of A, B and C. Then*

$$\begin{aligned} z_C + \omega z_A &= 0, \\ \omega z_C + \omega^2 z_A &= 0, \textit{ and} \\ \omega^2 z_C + z_A &= 0, \end{aligned}$$

where $\omega = e^{i\frac{2\pi}{3}}$ and we assumed the triangle oriented such that $z_A = z_C e^{i\frac{\pi}{3}}$.

Proof: We write $1 + e^{i\pi} = 0$ as $1 + e^{i\frac{\pi}{3}}e^{i\frac{2\pi}{3}} = 0$. Multiplying by z_C we obtain $z_C + z_C e^{i\frac{\pi}{3}}e^{i\frac{2\pi}{3}} = 0$. Using $z_A = z_C e^{i\frac{\pi}{3}}$ yields $z_C + \omega z_A = 0$. The other relations are obtained by multiplication with ω and ω^2 and using $\omega^3 = 1$. ■

Solved Exercise 4.3.18 *Let ΔOAB, ΔOCD and ΔOEF be three (not necessarily congruent) equilateral triangles. Prove that the midpoints of the segments BC, DE and FA are the vertices of an equilateral triangle.*

Proof: By Exercise 4.3.17 we have

$$z_A + \omega z_B = 0, \quad z_C + \omega z_D = 0, \quad z_E + \omega z_F = 0.$$

Multiplying the second and the third equation by ω and ω^2, respectively, and using $\omega^3 = 1$ we obtain the relations

$$z_A + \omega z_B = 0, \quad \omega z_C + \omega^2 z_D = 0, \quad \omega^2 z_E + z_F = 0.$$

Adding these relations and dividing by 2 yields

$$\frac{z_A + z_F}{2} + \frac{z_B + z_C}{2}\omega + \frac{z_D + z_E}{2}\omega^2 = 0.$$

By Exercise 4.3.16, part (*ii*), we obtain that $\{\frac{z_A+z_F}{2}, \frac{z_B+z_C}{2}, \frac{z_D+z_E}{2}\}$ are the affixes of an equilateral triangle. Since these are the affixes of the midpoints of AF, BC and DE, we obtain the desired result. ■

4.4 The quotient of three complex numbers

Definition 4.4.1 *Let A, B, C be three points in the plane, with $B \neq C$. Let z_A, z_B, z_C be their affixes. We define their quotient by*

$$(A, B; C) = \frac{z_A - z_C}{z_B - z_C}.$$

Exercise 4.4.2 *Show the following properties:*

(i) $(A, B; C) = 0 \Longleftrightarrow A = C$,

(ii) $(A, B; C)(B, A; C) = 1$, *and*

(iii) $(A, B; C) + (A, C; B) = 1$.

Exercise 4.4.3 *If $z = (A, B; C)$, there are six complex numbers obtained by the permutation of the points A, B and C:*

$$\begin{aligned}
(A, B; C) &= z, \\
(B, A; C) &= \frac{1}{z}, \\
(A, C; B) &= 1 - z, \\
(C, A; B) &= \frac{1}{1 - z}, \\
(B, C; A) &= 1 - \frac{1}{z} = \frac{z - 1}{z}, \text{ and} \\
(C, B; A) &= \frac{z}{z - 1}.
\end{aligned}$$

Exercise 4.4.4 *Draw the above six complex numbers for $z = e^{i\frac{\pi}{3}}$.*

Solved Exercise 4.4.5 *Let A, B, C, A', B', C' be six points in the plane. Show that*

$$(B, C; A')(C, A; B')(A, B; C') = (A', C'; B)(B', A'; C)(C', B'; A).$$

Proof: Using the definition of the quotient yields

$$\begin{aligned}(B,C;A')(C,A;B')(A,B;C') &= \frac{z_B - z_{A'}}{z_C - z_{A'}} \cdot \frac{z_C - z_{B'}}{z_A - z_{B'}} \cdot \frac{z_A - z_{C'}}{z_B - z_{C'}} \\ &= \frac{z_B - z_{A'}}{z_B - z_{C'}} \cdot \frac{z_C - z_{B'}}{z_C - z_{A'}} \cdot \frac{z_A - z_{C'}}{z_A - z_{B'}} \\ &= \frac{z_{A'} - z_B}{z_{C'} - z_B} \cdot \frac{z_{B'} - z_C}{z_{A'} - z_C} \cdot \frac{z_{C'} - z_A}{z_{B'} - z_A} \\ &= (A',C';B)(B',A';C)(C',B';A).\end{aligned}$$

■

Based on Exercise 4.4.3, it makes sense to consider the following two transformations

$$z \to \frac{1}{z}, \qquad z \to 1 - z. \tag{4.4.18}$$

It can be easily checked that the set $\{-1, \frac{1}{2}, 2\}$ is closed with respect to the above operations. One can show that there are also other sets with this property, see Exercise 4.4.7. In the following we shall discuss the geometrical significance when the quotient of three complex numbers is equal to one of the above values.

If $(A,B;C) = -1$, then

$$\frac{z_A - z_C}{z_B - z_C} = -1.$$

Solving for z_C yields $z_c = \frac{1}{2}(z_A + z_B)$. We arrive at the following result:

Proposition 4.4.6 $(A,B;C) = -1$ *if and only if* C *is the midpoint of line segment* AB.

The other cases are left as an exercise to the reader, see Exercise 4.4.8.

Exercise 4.4.7 *(i) Let* $S = \{-1, \frac{1}{2}, 2\}$. *Show that the set* S *is invariant under the transformations* (4.4.18). *Is this the only triplet of real numbers with this property?*
(ii) Show that $\{e^{i\pi/3}, e^{-i\pi/3}\}$ *is the only set with two elements which is invariant by the transformations* (4.4.18).

Exercise 4.4.8 *Show that*

$$(A,B;C) = \frac{1}{2} \Longleftrightarrow A \textit{ midpoint of } BC, \textit{and}$$

$$(A,B;C) = 2 \Longleftrightarrow B \textit{ midpoint of } AC.$$

Proposition 4.4.9 *Three points A, B, C form an equilateral triangle if and only if*

$$(A, B; C) \in \{e^{i\pi/3}, e^{-i\pi/3}\}.$$

Proof: Let $(A, B; C) = e^{i\pi/3}$. Then

$$\frac{z_A - z_C}{z_B - z_C} = e^{i\pi/3} \Longrightarrow \arg \frac{z_A - z_C}{z_B - z_C} = \frac{\pi}{3},$$

so $\widehat{ACB} = \dfrac{\pi}{3}$. On the other side

$$(A, C; B) = 1 - (A, B; C) = 1 - e^{i\pi/3} = 2 \sin \frac{\pi}{6} e^{i\frac{\pi/3-\pi}{2}} = e^{-i\pi/3},$$

so $\widehat{ABC} = |\arg(A, C; B)| = \frac{\pi}{3}$. Since ΔABC is a triangle with two angles equal to $\pi/3$, it follows that the triangle has all the angles equal to $\pi/3$ (Why?). Hence ΔABC is equilateral.

A similar proof applies to the case $(A, B; C) = e^{-i\pi/3}$. The reciprocal is left as an exercise for the reader. ■

4.5 Collinearity and intersection

The quotient of three numbers can be used to describe collinearity of points and intersection of lines.

Proposition 4.5.1 *Three distinct points A, B, C in the complex plane are collinear if and only if $(A, B; C) \in \mathbb{R}$.*

Proof: We have

$$(A, B; C) \in \mathbb{R} \Longleftrightarrow \frac{z_A - z_C}{z_B - z_C} \in \mathbb{R} \Longleftrightarrow \arg \frac{z_A - z_C}{z_B - z_C} \in \{0, \pi\}.$$

This means that either $\widehat{ACB} = 0$, or $\widehat{ACB} = \pi$, *i.e.*, the points A, B, C are collinear. ■

Proposition 4.5.2 *Let r be a positive real number. If the point M divides the segment AB in the quotient r, then*

$$(A, B; M) = \begin{cases} -r, & \textit{if } M \textit{ is inside the segment } [AB]; \\ r, & \textit{if } M \textit{ is outside the segment } [AB]. \end{cases}$$

Proof: Let M be between A and B, with $|AM| = r\,|BM|$. Since $\widehat{AMB} = \pi$,

$$\arg \frac{z_A - z_M}{z_B - z_M} = \pi,$$

and then

$$\begin{aligned}(A, B; M) &= \frac{z_A - z_M}{z_B - z_M} = \left|\frac{z_A - z_M}{z_B - z_M}\right| e^{i\pi} = -\left|\frac{z_A - z_M}{z_B - z_M}\right| \\ &= -\frac{|z_A - z_M|}{|z_B - z_M|} = -\frac{|AM|}{|BM|} = -r.\end{aligned}$$

The case when M belongs to the exterior of the segment AB is similar and is left as an exercise to the reader. ■

Corollary 4.5.3 *If the point M divides the segment AB in the quotient r, then*

$$z_M = \frac{1}{r-1}(rz_B - z_A). \tag{4.5.19}$$

Proof: If the point M divides the segment AB in the quotient r, by Proposition 4.5.2 we have $(A, B; M) = r$, *i.e.*, $\dfrac{z_A - z_M}{z_B - z_M} = r$. Solving for z_M yields relation (4.5.19). ■

Proposition 4.5.4 *The lines AB and CD intersect at the point M if and only if*

$$(A, B; M) \in \mathbb{R} \quad \textit{and} \quad (C, M; D) \in \mathbb{R}. \tag{4.5.20}$$

Proof: The lines AB and CD intersect at the point M if and only if A, B, M and C, D, M are collinear. Applying Proposition 4.5.1 yields (4.5.20). ■

Corollary 4.5.5 *The lines A_iB_i pass through the point M if and only if*

$$(A_i, B_i; M) \in \mathbb{R}, \qquad \forall i = 1, \ldots, n.$$

In the following we shall present an important application regarding the centroid of a triangle.

Solved Exercise 4.5.6 *The medians of a triangle ΔABC pass through the same point G, which has the affix*

$$z_G = \frac{1}{3}(z_A + z_B + z_C).$$

Proof: Let A' be the midpoint of the segment BC with affix

$$z_{A'} = \frac{1}{2}(z_B + z_C).$$

Using Proposition 4.5.1, we shall show that the points A, G, A' are collinear. We have

$$\begin{aligned}(A, A'; G) &= \frac{z_A - z_G}{z_{A'} - z_G} = \frac{z_A - \frac{1}{3}(z_A + z_B + z_C)}{\frac{1}{2}(z_B + z_C) - \frac{1}{3}(z_A + z_B + z_C)} \\ &= \frac{\frac{2}{3}z_A - \frac{1}{3}(z_B + z_C)}{\frac{1}{6}(z_B + z_C) - \frac{1}{3}z_A} = \frac{4z_A - 2z_B - 2z_C}{z_B + z_C - 2z_A} \\ &= -2 \in \mathbb{R}.\end{aligned}$$

Since the quotient is real, the points A, G, A' are collinear. Moreover, $\dfrac{|AG|}{|GA'|} = 2$. Similarly, we can show that B, G, B' and C, G, C' are collinear and

$$\frac{|BG|}{|GB'|} = 2, \qquad \frac{|CG|}{|GC'|} = 2.$$

The point G with this property is called the *centroid* of triangle ΔABC. ■

In general, the **centroid** (center of mass) of a convex polygon $A_0A_1 \cdots A_{n-1}$ has the affix equal to the mean of the vertices' affixes

$$z_G = \frac{1}{n}(z_{A_0} + z_{A_1} + \cdots + z_{z_{n-1}}).$$

Exercise 4.5.7 *Show that the centroid of a convex quadrilateral is the point where the bimedians intersect. (Recall that a bimedian is the line segment joining the midpoints of the opposite sides).*

Exercise 4.5.8 *(i) Let $A_1A_2 \cdots A_n$ be a polygon with the symmetry axis ℓ. Show that the centroid G belongs to ℓ.*

(ii) If ℓ_1 and ℓ_2 are two symmetry axes of the polygon $A_1A_2 \cdots A_n$, show that the centroid G is situated at the intersection of ℓ_1 and ℓ_2.

Another application is Pappus' theorem.

Theorem 4.5.9 (Pappus) *Let A', B', C' be three points which divide the sides BC, CA, AB of triangle ΔABC in the same quotient r. Then triangles ΔABC and $\Delta A'B'C'$ have the same centroid.*

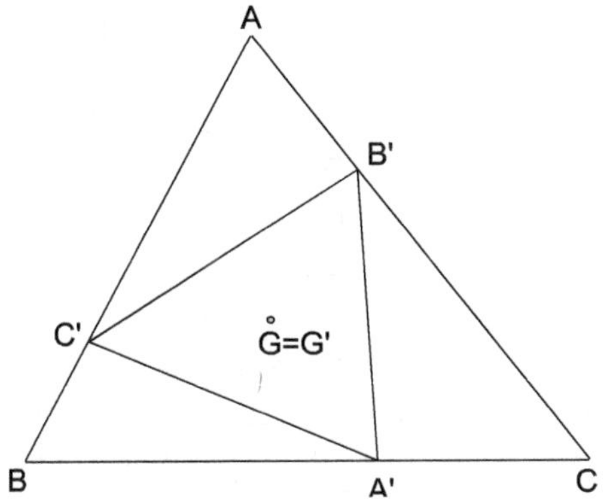

Figure 4.8: Triangles ΔABC and $\Delta A'B'C'$ have the same centroid.

Proof: Let $A' \in BC$, $B' \in AC$ and $C' \in AB$ such that

$$(B, C; A') = (C, A; B') = (A, B; C') = r.$$

By Corollary 4.5.3 we obtain

$$\begin{aligned} z_{A'} &= \frac{r}{r-1} z_C + \frac{1}{r-1} z_B, \\ z_{B'} &= \frac{r}{r-1} z_A + \frac{1}{r-1} z_C, \text{ and} \\ z_{C'} &= \frac{r}{r-1} z_B + \frac{1}{r-1} z_A. \end{aligned}$$

Let G, G' be the centroids of triangles ΔABC and $\Delta A'B'C'$, respectively. Since

$$z_{A'} + z_{B'} + z_{C'} = \frac{r}{r-1}(z_A + z_B + z_C) + \frac{1}{r-1}(z_A + z_B + z_C) = z_A + z_B + z_C,$$

it follows that $z_{G'} = z_G$, *i.e.*, $G' = G$, see Figure 4.8. ■

It is interesting that the converse of the above theorem also holds true. In order to prove this, we need the following lemma.

Lemma 4.5.10 *Let ΔABC be a triangle and α, β, $\gamma \in \mathbb{R}$, such that $\alpha + \beta + \gamma = 0$. If*

$$\alpha z_A + \beta z_B + \gamma z_C = 0, \tag{4.5.21}$$

then $\alpha = \beta = \gamma = 0$.

Proof: Substituting $\alpha = -(\beta + \gamma)$ in (4.5.21) yields

$$\begin{aligned} -(\beta + \gamma)z_A + \beta z_B + \gamma z_C &= 0 \Longleftrightarrow \\ \beta(z_B - z_A) + \gamma(z_C - z_A) &= 0 \Longleftrightarrow \end{aligned}$$

$$\frac{z_B - z_A}{z_C - z_A} = -\frac{\gamma}{\beta},$$

where we assumed $\beta \neq 0$. Then $(B, C; A) \in \mathbb{R}$, and by Proposition 4.5.1, it follows that A, B, C are collinear points, which contradicts that ΔABC is a triangle. Hence $\beta = 0$. In a similar way, starting from other linear combinations, we can show that $\alpha = 0$ and $\gamma = 0$. ■

We also have the following reciprocal of Pappus' theorem.

Theorem 4.5.11 *Let A', B', C' be three points on the sides BC, CA, AB of triangle ΔABC. If triangles ΔABC and $\Delta A'B'C'$ have the same centroids, then $(B, C; A') = (C, A; B') = (A, B; C')$.*

Proof: Let $(B, C; A') = r_1$, $(C, A; B') = r_2$, $(A, B; C') = r_3$, and $t_j = r_j/(r_j - 1)$. Therefore

$$\begin{aligned} z_{A'} &= t_1 z_C + (1 - t_1) z_B, \\ z_{B'} &= t_2 z_A + (1 - t_2) z_C, \text{and} \\ z_{C'} &= t_3 z_B + (1 - t_3) z_A. \end{aligned}$$

Taking the sum yields

$$z_{A'} + z_{B'} + z_{C'} = (t_2 + 1 - t_3) z_A + (t_3 + 1 - t_1) z_B + (t_1 + 1 - t_2) z_C. \quad (4.5.22)$$

Since triangles ΔABC and $\Delta A'B'C'$ have the same centroid

$$z_{A'} + z_{B'} + z_{C'} = z_A + z_B + z_C. \quad (4.5.23)$$

From (4.5.22) and (4.5.23) yields

$$\begin{aligned} (t_2 + 1 - t_3) z_A + (t_3 + 1 - t_1) z_B + (t_1 + 1 - t_2) z_C &= z_A + z_B + z_C \Longleftrightarrow \\ (t_2 - t_3) z_A + (t_3 - t_1) z_B + (t_1 - t_2) z_C &= 0 \Longleftrightarrow \\ \alpha z_A + \beta z_B + \gamma z_C &= 0, \end{aligned}$$

with $\alpha = t_2 - t_3$, $\beta = t_3 - t_1$, $\gamma = t_1 - t_2$. Since $\alpha + \beta + \gamma = 0$, Lemma 4.5.10 yields $\alpha = \beta = \gamma = 0$, *i.e.*, $t_1 = t_2 = t_3$. Hence $r_1 = r_2 = r_3$ and the theorem is proved. ■

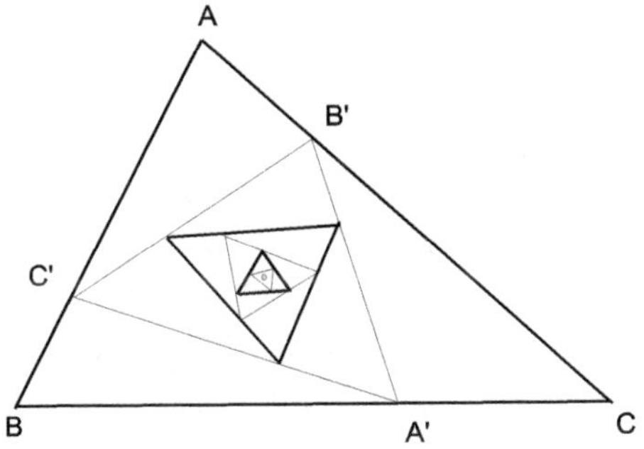

Figure 4.9: Iterating Pappus theorem yields the centroid.

Pappus theorem can now be applied to obtain the triangle $\Delta A''B''C''$ with the same centroid as triangle $\Delta A'B'C'$. Iterating the procedure, we obtain a sequence of inscribed triangles all having the same centroid. In the limit, the triangles will approach the centroid of the initial triangle ΔABC, see Figure 4.9.

This works in general for any polygon. In the next exercise we are asking to show this in the case of a quadrilateral.

Exercise 4.5.12 *Let $ABCD$ be a convex quadrilateral*[5] *and consider the points $A_1 \in AB$, $B_1 \in BC$, $C_1 \in CD$, $D_1 \in DA$ such that*

$$(A, B; A_1) = (B, C; B_1) = (C, D; C_1) = (D, A; D_1).$$

Show that $ABCD$ and $A_1B_1C_1D_1$ have the same centroids.

Theorem 4.5.13 (Napoleon) *Let ΔABC be an arbitrary triangle and consider the points A', B', $C' \notin Int(\Delta ABC)$ such that $\Delta ABC'$, $\Delta ACB'$ and $\Delta BCA'$ are equilateral triangles. Let G_c, G_b, G_a be the centroids of the above triangles. Then*

1) Triangles ΔABC and $\Delta G_aG_bG_c$ have the same centroid.

2) The triangle $\Delta G_aG_bG_c$ is equilateral.

Proof: (i) Since B' is obtained from C by a $60°$ counter-clockwise rotation centered at A, see Figure 4.10, we have

$$z_{B'} = z_A + (z_C - z_A)e^{i\frac{\pi}{3}}.$$

[5]The result works for any quadrilateral. Convexity is assumed for the sake of simplicity.

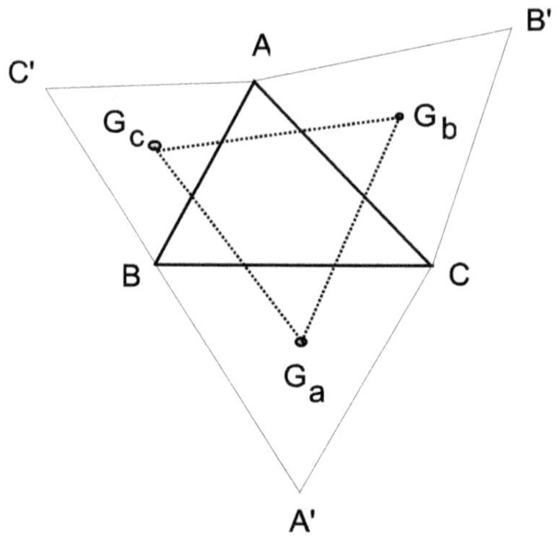

Figure 4.10: $\Delta G_aG_bG_c$ is equilateral.

Using similar reasons, we get

$$\begin{aligned} z_{A'} &= z_C + (z_B - z_C)e^{i\frac{\pi}{3}}, \text{and} \\ z_{C'} &= z_B + (z_A - z_B)e^{i\frac{\pi}{3}}. \end{aligned}$$

Using the property of the centroid, see Example 4.5.6, we can write

$$\begin{aligned} z_{G_a} &= \frac{1}{3}(z_B + z_C + z_{A'}) = \frac{1}{3}\left(z_B + 2z_C + (z_B - z_C)e^{i\frac{\pi}{3}}\right), \\ z_{G_b} &= \frac{1}{3}(z_A + z_C + z_{B'}) = \frac{1}{3}\left(2z_A + z_C + (z_C - z_A)e^{i\frac{\pi}{3}}\right), \text{and} \\ z_{G_c} &= \frac{1}{3}(z_A + z_B + z_{C'}) = \frac{1}{3}\left(z_A + 2z_B + (z_A - z_B)e^{i\frac{\pi}{3}}\right). \end{aligned}$$

Let G and G' be the centroids of triangles ΔABC and $\Delta G_aG_bG_c$. Using the above relations, we have

$$\begin{aligned} z_{G'} &= \frac{z_{G_a} + z_{G_b} + z_{G_c}}{3} \\ &= \frac{1}{9}\left(3z_A + 3z_B + 3z_C\right) + \frac{1}{9}e^{i\frac{\pi}{3}}\left(z_B - z_C + z_A - z_B + z_C - z_A\right) \\ &= \frac{1}{3}\left(z_A + z_B + z_C\right) = z_G, \end{aligned}$$

and hence $G = G'$.

(*ii*) It suffices to show that $(G_b, G_c; G_a) = e^{-i\frac{\pi}{3}}$, see Proposition 4.4.9. Using the above formulas, we have

$$\begin{aligned}\frac{z_{G_b} - z_{G_a}}{z_{G_c} - z_{G_a}} &= \frac{(2z_A - z_C - z_B) + e^{i\frac{\pi}{3}}(2z_C - z_A - z_B)}{(z_A + z_B - 2z_C) + e^{i\frac{\pi}{3}}(z_A + z_C - 2z_B)} \\ &= e^{-i\frac{\pi}{3}} \frac{(2z_A - z_C - z_B) + e^{i\frac{\pi}{3}}(2z_C - z_A - z_B)}{e^{-i\frac{\pi}{3}}(z_A + z_B - 2z_C) + (z_A + z_C - 2z_B)} \\ &= e^{-i\frac{\pi}{3}},\end{aligned}$$

since

$$\begin{aligned}(2z_A - z_C - z_B) + e^{i\frac{\pi}{3}}(2z_C - z_A - z_B) &= e^{-i\frac{\pi}{3}}(z_A + z_B - 2z_C) \\ &\quad + (z_A + z_C - 2z_B) \iff \\ 2z_A - z_C - z_B - z_A - z_C + 2z_B &= (z_A + z_B - 2z_C)(e^{i\frac{\pi}{3}} + e^{-i\frac{\pi}{3}}) \\ &\iff z_A + z_B - 2z_C = z_A + z_B - 2z_C,\end{aligned}$$

which holds true. We used that $e^{i\frac{\pi}{3}} + e^{-i\frac{\pi}{3}} = 2Re(e^{i\frac{\pi}{3}}) = 2 \cdot \frac{1}{2} = 1$. ■

Exercise 4.5.14 *Replace the equilateral triangles in Theorem 4.5.13 by squares with centers O_1, O_2, O_3. Show that triangles ΔABC and $\Delta O_1O_2O_3$ have the same centroid.*

Exercise 4.5.15 *Let $ABCD$ be a convex quadrilateral and consider the exterior points A', B', C', D' such that $\Delta ABA'$, $\Delta BCB'$, $\Delta CDC'$, and $\Delta DAD'$ are equilateral triangles. Let G_c, G_b, G_a, G_d be the centroids of the above triangles. Show that quadrilaterals $ABCD$ and $G_aG_bG_cG_d$ have the same centroid.*

Solved Exercise 4.5.16 *Let ΔABC be a triangle and let $AA'B'C$ and $AA''B''B$ be squares constructed in the exterior of triangle ΔABC, with centers O', and O'' respectively. Let M be the midpoint of the side BC. Show that*

(i) $O'M$ and $O''M$ are perpendicular lines.

(ii) $|O'M| = |O''M|$.

Proof: (*i*) Since B' is obtained from A by a 90° clockwise rotation centered at C, we have

$$z_{B'} = z_C + (z_A - z_C)e^{-i\frac{\pi}{2}} = z_C + i(z_C - z_A).$$

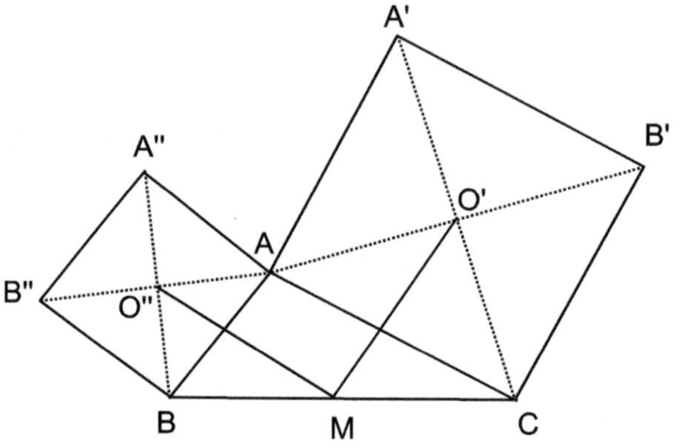

Figure 4.11: The lines $O'M$ and $O''M$ are perpendicular.

Using that O' is the midpoint of segment AB', we get

$$\begin{aligned} z_{O'} &= \frac{1}{2}(z_A + z_{B'}) = \frac{1}{2}(z_A + z_C + i(z_C - z_A)) \\ &= \frac{1}{2}(z_A + z_C) + \frac{i}{2}(z_C - z_A). \end{aligned} \tag{4.5.24}$$

Since A'' is obtained from B by a 90° clockwise rotation centered at A, we have

$$z_{A''} = z_A + (z_B - z_A)e^{-i\frac{\pi}{2}} = z_A + i(z_A - z_B).$$

Using that O'' is the midpoint of segment $A''B$ yields

$$\begin{aligned} z_{O''} &= \frac{1}{2}(z_{A''} + z_B) = \frac{1}{2}(z_A + i(z_A - z_B) + z_B) \\ &= \frac{1}{2}(z_A + z_B) + \frac{i}{2}(z_A - z_B). \end{aligned} \tag{4.5.25}$$

Using (4.5.24) and (4.5.25) we have

$$\begin{aligned} \frac{z_{O'} - z_M}{z_{O''} - z_M} &= \frac{\frac{1}{2}(z_A + z_C - z_B - z_C) + \frac{i}{2}(z_C - z_A)}{\frac{1}{2}(z_A + z_B - z_B - z_C) + \frac{i}{2}(z_A - z_B)} \\ &= \frac{(z_A - z_B) + i(z_C - z_A)}{(z_A - z_C) + i(z_A - z_B)} = (-i)\frac{(z_A - z_B) + i(z_C - z_A)}{(z_A - z_B) + i(z_C - z_A)} \\ &= -i = e^{-i\frac{\pi}{2}}. \end{aligned}$$

Then

$$\widehat{O'MO''} = \arg\frac{z_{O'} - z_M}{z_{O''} - z_M} = \arg e^{-i\frac{\pi}{2}} = -\frac{\pi}{2},$$

and hence $O'M$ and $O''M$ are perpendicular lines.

(*ii*) Taking the modulus in the above expression yields

$$\frac{|z_{O'} - z_M|}{|z_{O''} - z_M|} = |-i|,$$

which is equivalent with $\dfrac{|O'M|}{|O''M|} = 1$, which leads to $|O'M| = |O''M|$. ■

Exercise 4.5.17 *Let $ABCD$ be a parallelogram and denote by O_1, O_2, O_3, O_4 the centers of the squares constructed on the sides AB, BC, CD, and DA, respectively. Show the following*

(i) $|O_1O_3| = |O_2O_4|$.

(ii) O_1O_3 *is perpendicular on* O_2O_4.

(iii) $O_1O_2O_3O_4$ *is a square.*

Exercise 4.5.18 *Let $ABCD$ be a convex quadrilateral and let O_1, O_2, O_3, O_4 be the centers of the squares constructed in the exterior of $ABCD$ on its sides. Show that the lines O_1O_3 and O_2O_4 are perpendicular.*

4.6 Two important theorems

The law of sines and the law of cosines are two important theorems of metrical geometry. They provide relations between the trigonometric functions of the angles and the length of the sides of an arbitrary triangle. In the following we shall prove these theorems using complex numbers.

Theorem 4.6.1 (The law of sines) *Let ΔABC be a triangle with the circumradius R. Then*

$$\frac{|BC|}{\sin \widehat{A}} = \frac{|CA|}{\sin \widehat{B}} = \frac{|AB|}{\sin \widehat{C}} = 2R. \tag{4.6.26}$$

Proof: We shall denote by O the center of the circumcircle and consider the counter-clockwise oriented angles α, β, γ, made by the rays $(OA$, $(OB$, $(OC$ with the x-axis, see Figure 4.12. Choosing the origin at the circumcenter O, then the affixes of the vertices can be written as

$$z_A = Re^{i\alpha}, \qquad z_B = Re^{i\beta}, \qquad z_C = Re^{i\gamma}.$$

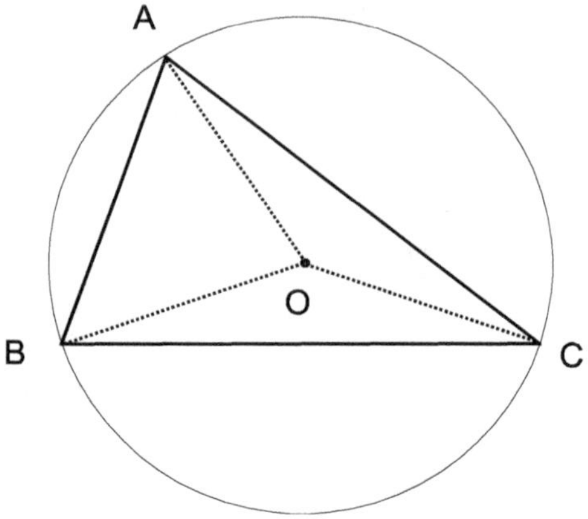

Figure 4.12: Triangle ΔABC inscribed in a circle.

Since a rotation centered at O does not change the distances and angles, we may assume that $(OC$ is the x-axis, *i.e.*, $\gamma = 0$. Then $\alpha = \widehat{AOC} = 2\widehat{B}$, and $\beta = 2\pi - \widehat{BOC} = 2\pi - 2\widehat{A} = 2(\pi - \widehat{A})$. The length of the side AC is

$$\begin{aligned} |AC| &= |z_C - z_A| = |Re^0 - Re^{i\alpha}| = R|1 - e^{i\alpha}| \\ &= 2R \sin\frac{\alpha}{2} = 2R\sin\widehat{B}, \end{aligned}$$

see Exercise 4.3.8. Hence $\dfrac{|AC|}{\sin\widehat{B}} = 2R$. In a similar way we can show $\dfrac{|BC|}{\sin\widehat{A}} = 2R$ and $\dfrac{|AB|}{\sin\widehat{C}} = 2R$. ∎

Theorem 4.6.2 (The law of cosines) *Let ΔABC be a triangle with side lengths $a = |BC|$, $b = |AC|$, and $c = |AB|$. Then*

$$a^2 = b^2 + c^2 - 2bc\cos\widehat{A}. \qquad (4.6.27)$$

Proof: Let α and γ be the angles made by the rays $(AB$ and $(AC$ with the x-axis, see Figure 4.14. Then

$$z_B = z_A + ce^{i\alpha}, \quad z_C = z_A + be^{i\gamma}.$$

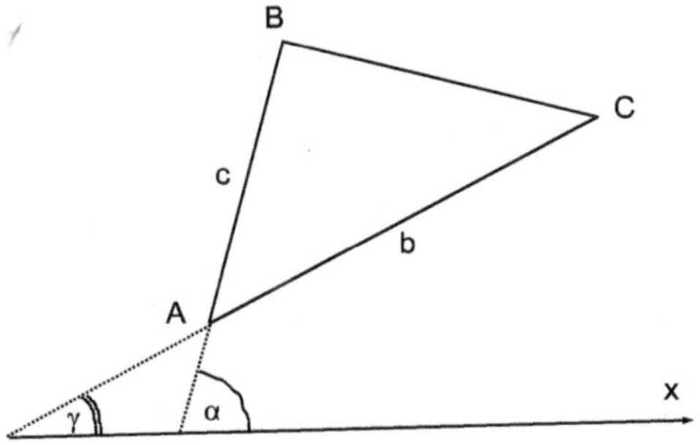

Figure 4.13: The angles α and γ.

We have

$$\begin{aligned}
|BC|^2 &= |z_B - z_C|^2 = |(z_A + ce^{i\alpha}) - (z_A + be^{i\gamma})|^2 = |ce^{i\alpha} - be^{i\gamma}|^2 \\
&= |c\cos\alpha + ic\sin\alpha - b\cos\gamma - ib\sin\gamma|^2 \\
&= |(c\cos\alpha - b\cos\gamma) + i(c\sin\alpha - b\sin\gamma)|^2 \\
&= (c\cos\alpha - b\cos\gamma)^2 + (c\sin\alpha - b\sin\gamma)^2 \\
&= c^2 + b^2 - 2bc\cos\alpha\cos\gamma - 2bc\sin\alpha\sin\gamma \\
&= c^2 + b^2 - 2bc\big(\cos\alpha\cos\gamma + \sin\alpha\sin\gamma\big) \\
&= c^2 + b^2 - 2bc\cos(\alpha - \gamma) \\
&= c^2 + b^2 - 2bc\cos\widehat{A},
\end{aligned}$$

since $\widehat{A} = \alpha - \gamma$. ∎

4.7 Applications

The distance $|OG|$

Consider a triangle ΔABC inscribed in a circle centered at O and denote its centroid by G. In the following we shall express the distance OG in terms of the angles $\widehat{A}$, $\widehat{B}$, $\widehat{C}$ and circumradius R of triangle ΔABC.

Proposition 4.7.1 *The distance between O and G is given by*

$$|OG| = \frac{R}{3}\sqrt{1 - 8\cos\widehat{A}\cos\widehat{B}\cos\widehat{C}}.$$

Proof: Let α, β, γ be the angles formed by the radii $(OA$, $(OB$, $(OC$ with the x-axis. Then

$$z_A = Re^{i\alpha}, \qquad z_B = Re^{i\beta}, \qquad z_C = Re^{i\gamma}.$$

The centroid has the affix

$$z_G = \frac{1}{3}(z_A + z_B + z_C) = \frac{R}{3}(e^{i\alpha} + e^{i\beta} + e^{i\gamma}).$$

Then

$$|OG|^2 = |z_G - z_O|^2 = |z_G|^2 = \frac{R^2}{9}|e^{i\alpha} + e^{i\beta} + e^{i\gamma}|^2. \quad (4.7.28)$$

The right side is obtained by the following trigonometric computation

$$\begin{aligned}
|e^{i\alpha} + e^{i\beta} + e^{i\gamma}|^2 &= (\cos\alpha + \cos\beta + \cos\gamma)^2 + (\sin\alpha + \sin\beta + \sin\gamma)^2 \\
&= \cos^2\alpha + \cos^2\beta + \cos^2\gamma \\
&\quad +2\cos\alpha\cos\beta + 2\cos\alpha\cos\gamma + 2\cos\beta\cos\gamma \\
&\quad +\sin^2\alpha + \sin^2\beta + \sin^2\gamma \\
&\quad +2\sin\alpha\sin\beta + 2\sin\alpha\sin\gamma + 2\sin\beta\sin\gamma \\
&= 3 + 2(\cos\alpha\cos\beta + \sin\alpha\sin\beta) \\
&\quad +2(\cos\alpha\cos\gamma + \sin\alpha\sin\gamma) \\
&\quad +2(\cos\beta\cos\gamma + \sin\beta\sin\gamma) \\
&= 3 + 2\cos(\beta - \alpha) + 2\cos(\alpha - \gamma) + 2\cos(\beta - \gamma) \\
&= 3 + 2\cos\widehat{AOB} + 2\cos\widehat{AOC} + 2\cos(2\pi - \widehat{BOC}) \\
&= 3 + 2\cos(2\widehat{C}) + 2\cos(2\widehat{B}) + 2\cos(2\widehat{A}) \\
&= 3 + 2\Big(\cos(2\widehat{C}) + \cos(2\widehat{B}) + \cos(2\widehat{A})\Big) \\
&= 3 + 2(-1 - 4\cos\widehat{A}\cos\widehat{B}\cos\widehat{C}) \\
&= 1 - 8\cos\widehat{A}\cos\widehat{B}\cos\widehat{C}.
\end{aligned}$$

Substituting in (4.7.28) yields the desired relation. ∎

Corollary 4.7.2 *In any triangle ΔABC we have the inequality*

$$\cos\widehat{A}\cos\widehat{B}\cos\widehat{C} \le \frac{1}{8},$$

with identity for the equilateral triangle.

Exercise 4.7.3 *(i) Let $ABCD$ be a quadrilateral of centroid G, inscribed in a circle of center O and radius R. Find a formula for $|OG|^2$ using a similar method applied in Proposition 4.7.1.*
(ii) Prove a generalization to a polygon inscribed in a circle.

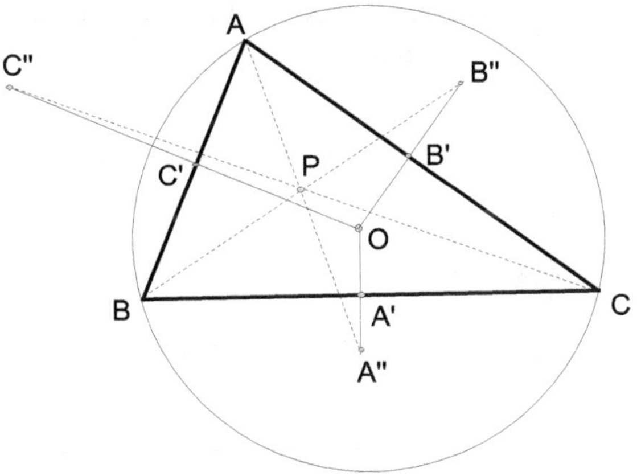

Figure 4.14: AA'', BB'', CC'' intersect at a point collinear with O and G.

Solved Exercise 4.7.4 *Let ΔABC be a triangle and A'', B'', C'' be the symmetric points of the circumcenter O with respect to the midpoints A', B', C' of the sides BC, AC and AB. Show the following.*

(i) The lines AA'', BB'' and CC'' intersect at a point P.

(ii) The point P is the midpoint of AA'', BB'' and CC''.

(iii) Let G, G'' be the centroids of triangles ΔABC and $\Delta A''B''C''$. Then the points O, G, G'', P are collinear.

Proof: (i) We shall choose the origin at the circumcenter O. The midpoints of the sides BC, AC and AB have the affixes

$$z_{A'} = \frac{1}{2}(z_B + z_C), \qquad z_{B'} = \frac{1}{2}(z_A + z_C), \qquad z_{C'} = \frac{1}{2}(z_A + z_B).$$

Using the symmetry equation (4.3.13) yields

$$\begin{aligned} z_{A''} &= 2z_{A'} - z_O = z_B + z_C \\ z_{B''} &= 2z_{B'} - z_O = z_A + z_C \\ z_{C''} &= 2z_{C'} - z_O = z_A + z_B. \end{aligned}$$

The midpoint of the segment AA'' has the affix

$$\frac{z_A + z_{A''}}{2} = \frac{1}{2}z_A + \frac{1}{2}(z_B + z_C) = \frac{1}{2}(z_A + z_B + z_C).$$

Similarly, the midpoints of BB'' and CC'' have affixes

$$\frac{z_B + z_{B''}}{2} = \frac{1}{2}(z_A + z_B + z_C), \qquad \frac{z_C + z_{C''}}{2} = \frac{1}{2}(z_A + z_B + z_C).$$

It follows that the segments AA'', BB'' and CC'' have the same midpoint P with affix $\frac{1}{2}(z_A + z_B + z_C)$ and hence the lines intersect at P.

(ii) It follows from (i).

(iii) The points O, G, G'', P have the following affixes

$$\begin{aligned}
z_O &= 0 \\
z_G &= \frac{1}{3}(z_A + z_B + z_C) \\
z_{G''} &= \frac{1}{3}(z_{A''} + z_{B''} + z_{C''}) = \frac{2}{3}(z_A + z_B + z_C) = 2z_G \\
z_P &= \frac{1}{2}(z_A + z_B + z_C) = \frac{3}{2}z_G.
\end{aligned}$$

Since

$$\begin{aligned}
(G, G''; O) &= \frac{z_G - z_O}{z_{G''} - z_O} = \frac{z_G}{z_{G''}} \in \mathbb{R} \\
(G, P; O) &= \frac{z_G - z_O}{z_P - z_O} = \frac{z_G}{z_P} = \frac{2}{3} \in \mathbb{R},
\end{aligned}$$

by Proposition 4.5.1 we obtain the desired collinearity. ■

Exercise 4.7.5 *Let ΔABC be a triangle inscribed in a circle of center O. Let A', B' and C' be the midpoints of the sides BC, AC and AB respectively. Let $A'' \in (OA'$, $B'' \in (OB'$, and $C'' \in (OC'$ such that $OA'' = \lambda|OA'|$, $OB'' = \lambda|OB'|$ and $OC'' = \lambda|OC'|$, with $\lambda > 0$. See Figure 4.14.*

(i) Show that AA'', BB'' and CC'' intersect at a point P.

(ii) Show that the point P belongs to the Euler line of triangle ΔABC, i.e., $P \in OG$, where G is the centroid.

4.8 The biquotient of four points

For any four distinct points A, B, C, D in the plane, their *biquotient* is defined by

$$(A, B; C, D) = \frac{(A, B; C)}{(A, B; D)}.$$

Proposition 4.8.1 *(i) The biquotient is invariant if the first pair (A, B) is swapped with the second pair (C, D):*

$$(A, B; C, D) = (C, D; A, B).$$

(ii) Swapping the points of the first pair (A, B) or the second pair (C, D) flips the biquotient:

$$(A, B; C, D) = \frac{1}{(B, A; C, D)} = \frac{1}{(A, B; D, C)}.$$

(iii) Flipping the inner points B, C yields to a complementary biquotient:

$$(A, B; C, D) = 1 - (A, C; B, D).$$

Proof: It is left as an exercise to the reader. ■

Corollary 4.8.2 *Combining the above transformations (i)–(iii) yields six different values for the biquotient. More precisely, if*

$$z = (A, B; C, D),$$

then

$$\begin{aligned}
(A, B; C, D) &= (B, A; D, C) = (C, D; A, B) = (D, C; B, A) = z, \\
(A, B; D, C) &= (B, A; C, D) = (C, D; B, A) = (D, C; A, B) = \frac{1}{z}, \\
(A, C; B, D) &= (B, D; A, C) = (C, A; D, B) = (D, B; C, A) = 1 - z, \\
(A, C; D, B) &= (B, D; C, A) = (C, A; B, D) = (D, B; A, C) = \frac{1}{1 - z}, \\
(A, D; C, B) &= (B, C; D, A) = (C, B; A, D) = (D, A; B, C) = \frac{z}{z - 1}, \\
(A, D; B, C) &= (B, C; A, D) = (C, B; D, A) = (D, A; C, B) = \frac{z - 1}{z}.
\end{aligned}$$

The next results show that the biquotient is intimately related with collinearity and concyclity of four distinct points in the plane.

Proposition 4.8.3 *If A, B, C, D are four collinear points, then*

$$(A, B; C, D) \in \mathbb{R}.$$

Proof: Since A, B, C and A, B, D are collinear, by Proposition 4.5.1 we get

$$(A, B; C) \in \mathbb{R}, \qquad (A, B; D) \in \mathbb{R}.$$

Hence $(A, B; C, D) = (A, B; C)/(A, B; D) \in \mathbb{R}$. ■

The converse does not hold true; in fact we have the following result.

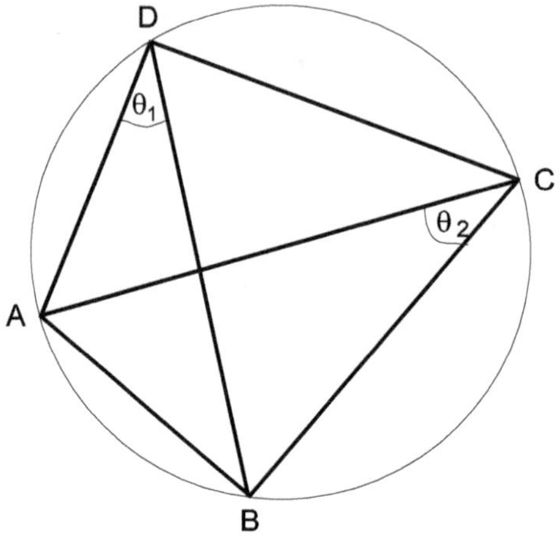

Figure 4.15: The angles α and γ.

Proposition 4.8.4 *Let A, B, C, D be four non-collinear points. The points belong to the same circle if and only if*

$$(A, B; C, D) \in \mathbb{R}.$$

Proof: Four distinct points A, B, C, D belong to the same circle if and only if

$$\widehat{ADB} = \widehat{ACB}, \tag{4.8.29}$$

see Figure 4.15. Let

$$\begin{aligned} \theta_1 &= \widehat{ADB} = \arg\frac{z_A - z_D}{z_B - z_D}, \text{ and} \\ \theta_2 &= \widehat{ACB} = \arg\frac{z_A - z_C}{z_B - z_C}. \end{aligned}$$

Then

$$\frac{z_A - z_D}{z_B - z_D} = \rho_1 e^{i\theta_1}, \qquad \frac{z_A - z_C}{z_B - z_C} = \rho_2 e^{i\theta_2}.$$

The biquotient of A, B, C, D becomes

$$\begin{aligned} (A, B; C, D) &= \frac{(A, B; C)}{(A, B; D)} = \frac{z_A - z_C}{z_B - z_C} : \frac{z_A - z_D}{z_B - z_D} \\ &= \frac{\rho_2 e^{i\theta_2}}{\rho_1 e^{i\theta_1}} = \frac{\rho_2}{\rho_1} e^{i(\theta_2 - \theta_1)}. \end{aligned}$$

Therefore $(A, B; C, D) \in \mathbb{R}$ if and only if $\theta_1 = \theta_2$, *i.e.*, relation (4.8.29) holds. ■

Remark 4.8.5 If the points are distinct, the above biquotient is non-zero (Why?).

Exercise 4.8.6 *Let ABCD be a convex quadrilateral like in Figure 4.15. Show that*

$$|(A, B; C, D)| > 1.$$

Exercise 4.8.7 *Let A_1, A_2, A_3, A_4 be four points on the unit circle with the corresponding arguments θ_1, θ_2, θ_3, θ_4. Show that*

$$(A_1, A_2; A_3, A_4) = \frac{\sin\frac{\theta_1-\theta_3}{2}\sin\frac{\theta_2-\theta_4}{2}}{\sin\frac{\theta_2-\theta_3}{2}\sin\frac{\theta_1-\theta_4}{2}}.$$

In the following we shall give a few applications of the previous result.

Theorem 4.8.8 (Ptolemy) *Let ABCD be a cyclic quadrilateral. Then the product of the diagonals is equal with the sum of the products of the opposite sides, i.e.,*

$$|AC| \cdot |BD| = |AD| \cdot |BC| + |AB| \cdot |DC|. \tag{4.8.30}$$

Proof: Dividing by $|AC| \cdot |BD|$, equation (4.8.30) becomes

$$\begin{aligned}
1 &= \frac{|AD| \cdot |BC|}{|AC| \cdot |BD|} + \frac{|AB| \cdot |DC|}{|AC| \cdot |BD|} \Longleftrightarrow \\
1 &= \frac{|z_A - z_D|}{|z_B - z_D|} \cdot \frac{|z_B - z_C|}{|z_A - z_C|} + \frac{|z_B - z_A|}{|z_C - z_A|} \cdot \frac{|z_C - z_D|}{|z_B - z_D|} \Longleftrightarrow \\
1 &= \frac{|(A, B; D)|}{|(A, B; C)|} + \frac{|(B, C; A)|}{|(B, C; D)|} \Longleftrightarrow \\
1 &= |(A, B; D, C)| + |(B, C; A, D)|. \qquad (4.8.31)
\end{aligned}$$

Since A, B, C, D are the vertices of a cyclic quadrilateral, by Proposition 4.8.4 we have

$$(A, B; C, D) = t \in \mathbb{R}\backslash\{0\}.$$

From Corollary 4.8.2,

$$(A, B; D, C) = \frac{1}{t}, \qquad (B, C; A, D) = \frac{t-1}{t}.$$

If $t > 0$, relation (4.8.31) becomes

$$\begin{aligned} 1 &= (A, B; D, C) + (B, C; A, D) \Longleftrightarrow \\ 1 &= \frac{1}{t} + \frac{t-1}{t}, \end{aligned}$$

which holds true.

If $t < 0$, relation (4.8.31) becomes

$$\begin{aligned} 1 &= -(A, B; D, C) - (B, C; A, D) \Longleftrightarrow \\ 1 &= \frac{1}{-t} + \frac{-t-1}{-t}, \end{aligned}$$

which also holds true. Hence relation (4.8.30) holds. ■

Proposition 4.8.9 (Ptolemy's inequality) *For any four points, A, B, C and D in the plane we have*

$$|AC| \cdot |BD| \leq |AD| \cdot |BC| + |AB| \cdot |DC|.$$

Proof: Let z_A, z_B, z_C and z_D be the affixes of the points A, B, C and D. It can be checked by a direct computation that

$$(z_A - z_B)(z_C - z_D) + (z_A - z_D)(z_B - z_C) = (z_A - z_C)(z_B - z_D).$$

Applying the modulus and the triangle inequality as well as the multiplicative property of the modulus, yields

$$\begin{aligned} |AC| \cdot |BD| &= |z_A - z_C| \cdot |z_B - z_D| = |(z_A - z_C)(z_B - z_D)| \\ &= |(z_A - z_B)(z_C - z_D)| + |(z_A - z_D)(z_B - z_C)| \\ &\leq |z_A - z_B||z_C - z_D| + |z_A - z_D||z_B - z_C)| \\ &= |AB| \cdot |DC| + |AD| \cdot |BC|. \end{aligned}$$

■

Theorem 4.8.10 (Van Schouten) *Let M be a point on the circumcircle of the equilateral triangle ΔABC, between the points A and C. Then*

$$|AM| + |CM| = |BM|. \tag{4.8.32}$$

Proof: *Variant I*: Applying Ptolemy's theorem to the cyclic quadrilateral $ABCM$ yields

$$|BM| \cdot |AC| = |AM| \cdot |BC| + |MC| \cdot |AB|.$$

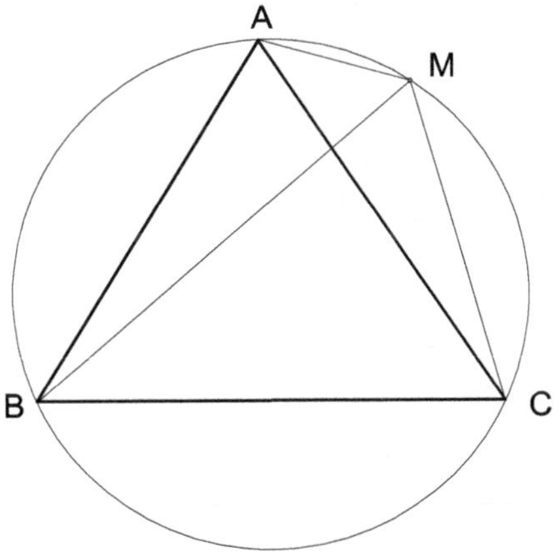

Figure 4.16: $|AM| + |CM| = |BM|$.

Since $|AC| = |BC| = |AB|$ the above relation becomes after simplification

$$|AM| + |CM| = |BM|.$$

Variant II: Since triangle ΔABC is equilateral, we have $(A, B; C) = \epsilon$, where ϵ is either $e^{i\frac{\pi}{3}}$ or $e^{-i\frac{\pi}{3}}$, see Proposition 4.4.9. Since A, B, C, M are on the same circle, by Proposition 4.8.4 we have the following real biquotients

$$(A, B; C, M) = t \in \mathbb{R}\backslash\{0\}, \tag{4.8.33}$$

$$(C, B; A, M) = \frac{t}{t-1} \in \mathbb{R}\backslash\{0\}. \tag{4.8.34}$$

We shall assume $t > 0$. By Exercise 4.8.6, we have $t > 1$. Relation (4.8.33) becomes

$$\begin{aligned} \frac{(A, B; C)}{(A, B; M)} &= t \Longleftrightarrow \frac{\epsilon}{(A, B; M)} = t \Longleftrightarrow (A, B; M) = \frac{\epsilon}{t} \\ \frac{z_A - z_M}{z_B - z_M} &= \frac{\epsilon}{t} \Longrightarrow \frac{|z_A - z_M|}{|z_B - z_M|} = \frac{|\epsilon|}{t} \Longleftrightarrow \\ \frac{|AM|}{|BM|} &= \frac{1}{t} \Longleftrightarrow |AM| = \frac{1}{t}|BM|. \end{aligned} \tag{4.8.35}$$

On the other side, relation (4.8.34) becomes

$$\begin{aligned}
\frac{(C,B;A)}{(C,B;M)} &= \frac{t}{t-1} \Longleftrightarrow \frac{\frac{\epsilon}{\epsilon-1}}{(C,B;M)} = \frac{t}{t-1} \Longleftrightarrow \\
(C,B;M) &= \frac{\epsilon}{\epsilon-1}\cdot\frac{t-1}{t} \Longleftrightarrow \frac{z_C - z_M}{z_B - z_M} = \frac{\epsilon}{\epsilon-1}\cdot\frac{t-1}{t}.
\end{aligned}$$

Taking the modulus yields

$$\frac{|z_C - z_M|}{|z_B - z_M|} = \frac{|\epsilon|}{|\epsilon-1|}\cdot\frac{|t-1|}{|t|}.$$

Since $t > 1$, $|\epsilon| = |\epsilon - 1| = 1$, see Exercise 4.3.9, we obtain

$$\frac{|CM|}{|BM|} = \frac{t-1}{t} \Longrightarrow |CM| = \frac{t-1}{t}|BM|. \tag{4.8.36}$$

Adding relations (4.8.35) and (4.8.36) yields

$$|AM| + |CM| = \frac{1}{t}|BM| + \frac{t-1}{t}|BM| = |BM|,$$

which ends the proof.

The case $t < 0$ is left as an exercise for the reader. ■

The converse of the above theorem also holds true.

Theorem 4.8.11 (Van Schouten, converse) *Let ΔABC be an equilateral triangle, and M be a point such that*

$$|AM| + |CM| = |BM|. \tag{4.8.37}$$

Then M belongs to the circumcircle of triangle ΔABC, between A and C.

Proof: Let $(A,B;C,M) = z \in \mathbb{C}$. We need to show that $z \in \mathbb{R}$. Since A, B, C are non-collinear, applying Proposition 4.8.4, yields that A, B, C, M are concyclic points.

In a way similar with the previous proof, we can obtain

$$\begin{aligned}
|AM| &= \frac{1}{|z|}|BM| \\
|CM| &= \frac{|z-1|}{|z|}|BM|.
\end{aligned}$$

Then (4.8.37) becomes the following equation for z

$$\frac{1}{|z|} + \frac{|z-1|}{|z|} = 1.$$

Multiplying by $|z|$ yields

$$1 + |z - 1| = |z|.$$

By Exercise 4.1.2, we have $|z| - 1 \le |z - 1|$, or $|z - 1| + 1 \ge |z|$. The identity is reached when $z - 1$ and 1 are proportional, *i.e.*, when z is real. Hence M belongs to the circumcircle of triangle ΔABC. Then we have the following three disjoint cases:

(1) M between A and C;

(2) M between A and B;

(3) M between B and C.

Applying Theorem 4.8.10 in each of the above cases yields

(1) $|AM| + |CM| = |BM|$;

(2) $|AM| + |BM| = |BC|$;

(3) $|BM| + |CM| = |BA|$.

Since only the case (1) occurs, the point M is between A and C. The theorem is completely proved. ■

Exercise 4.8.12 *Let AB be a chord on the circle centered at O with radius R. Let M be a variable point on the arc AB. Find the position of the point M such that the sum $|AM| + |MB|$ is minimum.*

4.9 The distance from a point to a line

Consider a point M and a line AB in the plane, see Figure 4.17. Let

$$z_M = a_M + ib_M, \quad z_A = a_A + ib_A, \quad z_B = a_B + ib_B$$

be the affixes of the above points. The equation of the line AB is

$$z_X = tz_A + (1 - t)z_B, \qquad t \in \mathbb{R},$$

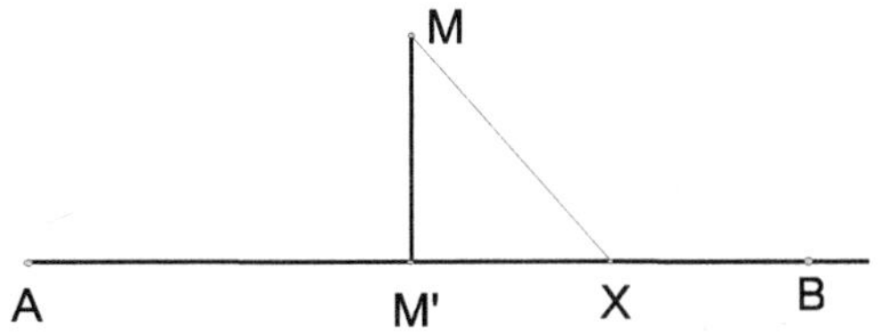

Figure 4.17: MM' perpendicular on AB.

where X is an arbitrary point on AB. The square of the distance between M and X is a quadratic function of the parameter t, which can be expressed as

$$\begin{aligned}
|MX|^2 &= |z_M - z_X|^2 = |z_M - tz_A - (1-t)z_B|^2 \\
&= \Big(a_M - ta_A - (1-t)a_B\Big)^2 + \Big(b_M - tb_A - (1-t)b_B\Big)^2 \\
&= \Big(t(a_B - a_A) + a_M - a_B\Big)^2 + \Big(t(b_B - b_A) + b_M - b_B\Big)^2 \\
&= \alpha t^2 + \beta t + \gamma,
\end{aligned}$$

with

$$\begin{aligned}
\alpha &= (a_B - a_A)^2 + (b_B - b_A)^2 = |z_B - z_A|^2 = |AB|^2 > 0, \\
\beta &= 2(a_B - a_A)(a_M - a_B) + 2(b_B - b_A)(b_M - b_B) \\
&= 2Re(z_B - z_A)\, Re(z_M - z_B) + 2Im(z_B - z_A)\, Im(z_M - z_B) \\
&= 2Re[(z_M - z_B)\overline{(z_B - z_A)}] = 2Re[\overline{(z_M - z_B)}(z_B - z_A)], \\
\gamma &= (a_M - a_B)^2 + (b_M - b_B)^2 = |z_M - z_B|^2 = |MB|^2.
\end{aligned}$$

see Exercise 4.3.7. Since $\alpha > 0$, the function

$$f(t) = \alpha t^2 + \beta t + \gamma$$

has a minimum equal to

$$\begin{aligned}
\min f &= -\frac{\Delta}{4a} = \frac{-\beta^2 - 4\alpha\gamma}{4\alpha} = -\frac{\beta^2}{4\alpha} + \gamma \\
&= \gamma - \frac{\beta^2}{4\alpha}.
\end{aligned}$$

We arrive at the following result which provides a formula for the distance from a point to a line.

Proposition 4.9.1 *The distance from the point M to the line AB is given by*

$$dist(M, AB) = \min f = \sqrt{|MB|^2 - \frac{[Re(z_M - z_B)\overline{(z_B - z_A)}]^2}{|AB|^2}}.$$

Denote by M' the foot of the perpendicular from M to line AB. In the following we shall find the affix of M'. The minimum of the parabolic function $f(t)$ is reached for $t_v = -\frac{\beta}{2\alpha}$. Then

$$\begin{aligned}
z_{M'} &= f(t_v) = t_v z_A + (1 - t_v)z_B = t(z_A - z_B) + z_B \\
&= -\frac{\beta}{2\alpha}(z_A - z_B) + z_B.
\end{aligned}$$

This can be written as

$$z_{M'} = z_B - \frac{Re[\overline{(z_M - z_B)}(z_B - z_A)]}{|AB|^2}(z_A - z_B).$$

Exercise 4.9.2 *Find the distance from the origin to the line passing through the points with affixes i and 1.*

In the following we shall discuss two applications of Proposition 4.9.1 to computing areas and proving perpendicularity.

4.10 The area of a triangle

Let A, B and M be three points with affixes z_A, z_B and z_M. We shall evaluate the area of triangle ΔABM in terms of these affixes.

Proposition 4.10.1 *The area of triangle ΔABC is*

$$\mathcal{A}_{ABM} = \frac{1}{2}|Im(\overline{z_M - z_B})(z_A - z_B)|. \qquad (4.10.38)$$

Proof: We shall provide two proof variants. The first is a consequence of the distance formula given by Proposition 4.9.1, while the latter follows from a classical area formula.

Variant I: Using Proposition 4.9.1 we have

$$\begin{aligned}
\mathcal{A}_{ABM} &= \frac{1}{2}|AB|\, dist(M, AB) \\
&= \frac{1}{2}|AB| \frac{\sqrt{|MB|^2 \cdot |AB|^2 - [Re(z_M - z_B)(\overline{z_B - z_A})]^2}}{|AB|} \\
&= \frac{1}{2}\big[(z_M - z_B)(\overline{z_M - z_B})(z_B - z_A)(\overline{z_B - z_A}) \\
&\quad - [Re(z_M - z_B)(\overline{z_B - z_A})]^2\big]^{1/2}.
\end{aligned}$$

For the sake of simplicity we denote $z_1 = z_M - z_B$ and $z_2 = z_B - z_A$. Then using formula (4.3.16) the previous relation becomes

$$\begin{aligned}
\mathcal{A}_{ABM} &= \frac{1}{2}\big[z_1 \cdot \overline{z_1} \cdot z_2 \cdot \overline{z_2} - Re[z_1\overline{z_2}]Re[\overline{z_1}z_2]\big]^{1/2} \\
&= \frac{1}{2}\big[Re(z_1\overline{z_2} \cdot \overline{z_1}z_2) - Re[z_1\overline{z_2}]Re[\overline{z_1}z_2]\big]^{1/2} \\
&= \frac{1}{2}\big[Re(z_1\overline{z_2})Re(\overline{z_1}z_2) - Im(z_1\overline{z_2})Im(\overline{z_1}z_2) - Re[z_1\overline{z_2}]Re[\overline{z_1}z_2]\big]^{1/2} \\
&= \frac{1}{2}|Im(z_1\overline{z_2})| = \frac{1}{2}|Im(\overline{z_M - z_B})(z_A - z_B)|,
\end{aligned}$$

where we canceled the product of the real parts and used that $Im(\overline{z}) = -Im(z)$ for $z = z_1\overline{z_2}$.
Variant II: We start from the formula $\mathcal{A}_{ABM} = \frac{1}{2}|AB| \cdot |MB| \sin\theta$, where $\theta = \widehat{MBA}$. Let $A' \in (BA$ such that $|MB| = |A'B|$. Then

$$z_{A'} - z_B = (z_M - z_B)e^{i\theta}. \tag{4.10.39}$$

Since from proportionality we have

$$z_A - z_B = (z_{A'} - z_B) \cdot \frac{|AB|}{|A'B|},$$

then (4.10.39) becomes

$$z_A - z_B = (z_M - z_B)e^{i\theta} \cdot \frac{|AB|}{|A'B|}.$$

Then multiplying by $\overline{z_M - z_B}$ yields

$$\begin{aligned}(z_A - z_B)\overline{(z_M - z_B)} &= |z_M - z_B|^2 e^{i\theta} \cdot \frac{|AB|}{|A'B|} = |MB|^2 e^{i\theta} \cdot \frac{|AB|}{|MB|} \\ &= |AB| \cdot |MB| e^{i\theta} = |AB| \cdot |MB|(\cos\theta + i\sin\theta).\end{aligned}$$

Hence, taking the imaginary part we obtain

$$Im(z_A - z_B)\overline{(z_M - z_B)} = |AB| \cdot |MB| \sin\theta = 2\mathcal{A}_{ABM}.$$

■

There are more ways to asses the area of a triangle in determinant form as follows.

Corollary 4.10.2 *Let $z_A = x_A + iy_A$, $z_B = x_B + iy_B$ and $z_M = x_M + iy_M$ be the affixes of the points A, B and M. Then the area of triangle ΔABM can be written as*

$$\mathcal{A}_{ABM} = \frac{1}{2}\epsilon \begin{vmatrix} x_M - x_B & y_M - y_B \\ x_A - x_B & y_A - y_B \end{vmatrix} = \frac{1}{2}\epsilon \begin{vmatrix} x_M & y_M & 1 \\ x_A & y_A & 1 \\ x_B & x_B & 1 \end{vmatrix},$$

where $\epsilon = 1$ if the determinant is positive and $\epsilon = -1$ otherwise.

Proof: An algebraic computation provides

$$\begin{aligned}(z_A - z_B)\overline{(z_M - z_B)} &= [(x_A - x_B) + i(y_A - y_B)] \cdot [(x_M - x_B) - i(y_M - y_B)] \\ &= (x_A - x_B)(x_M - x_B) + (y_A - y_B)(y_M - y_B) \\ &\quad + i[(x_M - x_B)(y_A - y_B) - (x_A - x_B)(y_M - y_B)],\end{aligned}$$

and taking the imaginary part we have

$$Im(z_A - z_B)(\overline{z_M - z_B}) = (x_M - x_B)(y_A - y_B) - (x_A - x_B)(y_M - y_B)$$
$$= \begin{vmatrix} x_M - x_B & y_M - y_B \\ x_A - x_B & y_A - y_B \end{vmatrix}.$$

Using now Proposition 4.10.1 we obtain the first formula for the area. To obtain the second determinant formula it suffices to show that the two determinants are equal. To this end, we subtract the last row from the first two rows and expand the determinant over the last column as follows

$$\begin{vmatrix} x_M & y_M & 1 \\ x_A & y_A & 1 \\ x_B & x_B & 1 \end{vmatrix} = \begin{vmatrix} x_M - x_B & y_M - y_B & 0 \\ x_A - x_B & y_A - y_B & 0 \\ x_B & x_B & 1 \end{vmatrix} = \begin{vmatrix} x_M - x_B & y_M - y_B \\ x_A - x_B & y_A - y_B \end{vmatrix}.$$

The role of the coefficient ϵ is to keep the area positive. ■

Corollary 4.10.3 *Let A, B and M be three distinct points. The following statements are equivalent.*

(i) The points A, B and M are collinear.

(ii)

$$\begin{vmatrix} x_M - x_B & y_M - y_B \\ x_A - x_B & y_A - y_B \end{vmatrix} = 0.$$

(iii)

$$\begin{vmatrix} x_M & y_M & 1 \\ x_A & y_A & 1 \\ x_B & x_B & 1 \end{vmatrix} = 0.$$

Proof: The collinearity of the points is equivalent to a degenerate triangle ΔABM, which means that its area vanishes. The result follows now from Corollary 4.10.2. ■

Exercise 4.10.4 *Let $\omega_0 = 1$, ω_1 and ω_2 be the 3rd roots of unity and denote by A_0, A_1 and A_2 their images. Use Corollary 4.10.2 to find the area of triangle $\Delta A_0A_1A_2$.*

4.11 Perpendicularity

We note that if in Proposition 4.9.1 we have $MB \perp AB$, then $\text{dist}(M, AB) = |MB|$, fact that can be written as

$$\sqrt{|MB|^2 - \frac{[Re(z_M - z_B)\overline{(z_B - z_A)}]^2}{|AB|^2}} = |MB|.$$

This is equivalent to $Re(z_M - z_B)\overline{(z_B - z_A)} = 0$. We arrived at the following result.

Proposition 4.11.1 *We have $MB \perp AB$ if and only if*

$$Re(z_M - z_B)\overline{(z_B - z_A)} = 0.$$

In the following we shall provide another variant of proof for this relation. We shall use the angle argument to express the perpendicularity as

$$MB \perp AB \Leftrightarrow \widehat{ABM} = 90^\circ \Leftrightarrow arg\frac{z_A - z_B}{z_M - z_B} = \frac{\pi}{2}. \qquad (4.11.40)$$

On the other side, the angle argument does not change by multiplication with a scalar, so we have

$$\begin{aligned} arg\frac{z_A - z_B}{z_M - z_B} &= arg\Big(\frac{z_A - z_B}{z_M - z_B}|z_M - z_B|^2\Big) \\ &= arg\Big(\frac{z_A - z_B}{z_M - z_B}(z_M - z_B)\overline{(z_M - z_B)}\Big) \\ &= arg(z_A - z_B)\overline{(z_M - z_B)}. \end{aligned}$$

Using (4.11.40) yields $arg(z_A - z_B)\overline{(z_M - z_B)} = \frac{\pi}{2}$, namely $(z_A - z_B)\overline{(z_M - z_B)}$ is a pure imaginary complex number. Hence, $Re(z_M - z_B)\overline{(z_B - z_A)} = 0$.

Part III

Vectorial Approach

Chapter 5

Vectors and their Basic Properties

The last thing that we find in making a book
is to know what we must put first.
Blaise Pascal

The goal of this section is to introduce the basic set theory notions and notations, which will be used in the rest of the book. Set theory language is handy not only for algebra but also for geometry in its modern approach.

5.1 Algebra of sets

A set is a collection of objects. These objects are called *elements*. A set will be denoted by a capital letter such as A, B, C, X, $Y, \ldots$. A set can be defined in several ways. We can define a set enumerating its elements, like $A = \{x_1, x_2, x_3, \ldots\}$. Another way to define a set is to characterize its elements by a specific property P:

$$A = \{x; P(x)\},$$

which reads as "*the set of elements x for which the proposition $P(x)$ is satisfied*".
If x is an element of the set A, we write $x \in A$ and read "*x belongs to A*", or "*x is an element of the set A.*" For instance, if ℓ is a line and M is a point on the line ℓ, we shall write $M \in \ell$. If x is not an element of the set A, we write $x \notin A$. It is useful to consider the set with no elements, called the *empty set*, which is denoted by $\emptyset$.

Let A and B be two sets. The set A is called a *subset* of B if

$$\forall x \in A \Longrightarrow x \in B,$$

which reads as *"for any element x of the set A, x is an element of the set B."* "$\forall$" is the logic quantifier *"for any"* and the arrow " $\Longrightarrow$ " is the *"logic implication."* We shall write $A \subset B$ and read A is *included* in B. Two sets A and B are called *equal* if $A \subset B$ and $B \subset A$. We write in this case $A = B$. If the sets are not equal we write $A \neq B$. Sometimes, when we need to prove an iff statement, we shall use the double inclusion method based on the above definition. Statements of the form *"$P(x)$ holds if and only if $Q(x)$ holds"* should be interpreted as an identity between two sets A and B, where $A = \{x; P(x)\}$ and $B = \{x; Q(x)\}$. The proofs of those kinds of theorems will be done in two steps:

Step 1. $P(x) \Rightarrow Q(x)$, which means $A \subset B$, and
Step 2. $Q(x) \Rightarrow P(x)$, which means $B \subset A$.

The inclusion $\subset$ has the following properties:

1) Reflexivity: $A \subset A$;
2) Transitivity: $A \subset B,\ B \subset C \Longrightarrow A \subset C$;
3) Antisymmetry: $A \subset B,\ B \subset A \Longrightarrow A = B$.

Antisymmetry expresses the fact that a set is uniquely determined by its elements, regardless of the order. For instance $\{a, b, c, d\} = \{b, c, d, a\}$.

Given two sets A and B, the set $A \cup B = \{x; x \in A \text{ or } x \in B\}$ is called the *union* of the sets A and B.

The set $A \cap B = \{x; x \in A \text{ and } x \in B\}$ is called the *intersection* of the sets A and B.

The above definitions can be easily extended to any number (finite or not) of sets.

Exercise 5.1.1 (*i*) *Let ℓ, m be two lines in the plane. What are the possible elements of the set $\ell \cap m$?*
(*ii*) *If Π is a plane and ℓ is a line, find the possible elements of the sets $\ell \cap \Pi$ and $\ell \cup \Pi$.*

We note that the union and intersection have the following properties:

1. $A \cup B = B \cup A, A \cap B = B \cap A$ (commutativity);
2. $A \cup (B \cup C) = (A \cup B) \cup C$ (associativity of union);
3. $A \cap (B \cap C) = (A \cap B) \cap C$ (associativity of intersection);
4. $A \subset A \cup B, A \cap B \subset A$;
5. $A \subset B \Longrightarrow A \cup B = B, A \subset B \Longrightarrow A \cap B = A$;
6. $A \cup \emptyset = A, A \cap \emptyset = \emptyset$;
7. $A \cup A = A, A \cap A = A$;
8. $A \cup (B \cap C) = (A \cup B) \cap (A \cup C)$ (distributivity);
9. $A \cap (B \cup C) = (A \cap B) \cup (A \cap C)$ (distributivity).

Exercise 5.1.2 *Prove the above properties, using double inclusion, when necessary.*

The set $A \backslash B = \{x; x \in A, x \notin B\}$ is called the *difference* of the sets A and B.

Example 5.1.1 (Separation axioms) *(i) Let M be a point on the line ℓ. Then $\ell \backslash M$ is the union of two half-lines.*
(ii) Let ℓ be a line included in the plane Π. Then $\Pi \backslash \ell$ is the union of two half-planes.
(iii) If Π is a plane and let $\mathbf{E}$ denote the three dimensional Euclidean space. Then $\mathbf{E} \backslash \Pi$ is a union of two half-spaces.

If $\mathbf{E}$ denotes the three dimensional Euclidean space and A, $B \subset \mathbf{E}$ are two subsets, we shall denote by $\overline{A} = \mathbf{E} \backslash A$, and call it *the complementary set* of A.

Exercise 5.1.3 *Show that*
(i) $A \subset B \Longrightarrow \overline{B} \subset \overline{A}$,
(ii) $\overline{\overline{A}} = A$,
(iii) $\overline{A \cap B} = \overline{A} \cup \overline{B}$, *and*
(iv) $\overline{A \cup B} = \overline{A} \cap \overline{B}$.

The last two relations are called *De Morgan's laws.*

The sets A and B are called *disjoint* if $A \cap B = \emptyset$. If ℓ, m are two lines in a plane, which are disjoint sets, then ℓ and m are called *parallel lines*, and we shall write $\ell || m$. For reflexivity purposes we consider that each line is parallel to itself. Hence the definition of two parallel lines can be stated as follows:

Definition 5.1.4 *Two lines ℓ and m are parallel if either $\ell = m$, or $\ell \cap m = \emptyset$.*

5.2 Relations

A man is a bundle of relations, a knot of roots, whose flower and fruitage is the world.
Ralph Waldo Emerson (1803–1882)

The set $\{x, y\}$ has two elements. Since it is the same as the set $\{y, x\}$, we cannot specify which element is the first and which is the second. To introduce an order between the elements we shall define the *ordered pair* $(x, y) = \{\{x, y\}, x\}$ and read *"x is the first element and y is the second"*.

Let X and Y be two sets and define *the cartesian product* as

$$X \times Y = \{(x, y); x \in X, y \in Y\}.$$

Definition 5.2.1 *A binary relation $\mathcal{R}$ is a triplet $\mathcal{R} = (X, Y, G)$, where X and Y are sets and $G \subset X \times Y$. The set G is called the graph of the relation $\mathcal{R}$. We shall write $x\mathcal{R}y$ and read x is in the relation $\mathcal{R}$ with y iff $(x, y) \in G$.*

In the particular case when $X = Y$, the binary relation $\mathcal{R}$ is called *homogeneous.* In this case $\mathcal{R} = (X, X, G)$, with $G \subset X \times X$. The equivalence and the order relations are among the most important homogeneous binary relations. We shall define these relations in the following.

Definition 5.2.2 *The binary relation $\mathcal{R} = (X, X, G)$ is an equivalence relation on the set X if the following three conditions are satisfied:*

(i) $x \in X \Longrightarrow x\mathcal{R}x$ (reflexivity);
(ii) $x\mathcal{R}y \Longrightarrow y\mathcal{R}x$ (symmetry);
(iii) $x\mathcal{R}y$ and $y\mathcal{R}z \Longrightarrow x\mathcal{R}z$ (transitivity).

All the elements which are in relation $\mathcal{R}$ with a given element x is a set called the *equivalence class* of x. We shall write $\mathcal{R}_x = \{y; y\mathcal{R}x\}$. If $x \neq z$, their equivalence classes $\mathcal{R}_x$ and $\mathcal{R}_y$ are disjoint or coincide. Since $x \in \mathcal{R}_x$, it follows that $\bigcup_{x \in X} \mathcal{R}_x = X$. We say that the equivalence classes form a partition of the set X.

The set of the equivalence classes $X/\mathcal{R} = \{\mathcal{R}_x; x \in X\}$ is called the *factor set.* The application $p : X \to X/\mathcal{R}$ defined by $p(x) = \mathcal{R}_x$ is called the *canonical projection* of the equivalence $\mathcal{R}$.

Example 5.2.1 (*i*) *The relation "$\|$" introduced by Definition 5.1.4 is an equivalence relation between lines;*
(*ii*) *The relation "is a friend of" is an not an equivalence relation between people, because it is not transitive.*
(*iii*) *The relation "to live in the same country" is an equivalence relation between people. The equivalence classes in this case coincide with the countries populations.*

Exercise 5.2.3 *Let $d \subset \Pi$ be a line included in the plane. Consider the binary relation $\mathcal{R} \subset (\Pi \backslash d) \times (\Pi \backslash d)$ defined by $X\mathcal{R}Y \iff [XY] \cap d = \emptyset$. Show that $\mathcal{R}$ is an equivalence relation. ($[XY]$ denotes the line segment XY.)*

The equivalence relation will be used to define the vector notion in the plane or space. In a later section we shall define a vector as an equivalence class of a certain binary equivalence relation.

In the following we shall define the order relations.

Definition 5.2.4 *The binary relation $\mathcal{R}$ on the set X is called a strict order relation if the following properties are satisfied:*

1) *there is no $x \in X$ such that $x\mathcal{R}x$ (non-reflexive),*
2) *$x\mathcal{R}y$ and $y\mathcal{R}x$ cannot hold simultaneously (non-symmetric), and*
3) *$x\mathcal{R}y, y\mathcal{R}z \Longrightarrow x\mathcal{R}z$ (transitive).*

If the *non-reflexivity* property is replaced by *reflexivity, i.e.,* $x\mathcal{R}x, \forall x \in X$, the relation $\mathcal{R}$ is called a *(non-strict) order relation.* The order relation $\mathcal{R}$ on the set X is called *total* if for any two elements $x, y \in X$ we have $x\mathcal{R}y$ or $y\mathcal{R}x$.

An element $m \in X$ is called *last element* of the order relation $\mathcal{R}$ if

$$m\mathcal{R}x \Longrightarrow m = x.$$

Let $\mathcal{R}$ and $\mathcal{R}'$ be two total order relation such that

$$x\mathcal{R}y \iff y\mathcal{R}'x.$$

Then $\mathcal{R}$ and $\mathcal{R}'$ are called *opposite* relations. The order relations are used by the order axioms of a line. For instance, the order relations on a line, "$\leq$" and "$\geq$", are opposite, since "$x \leq y$" is equivalent to "$y \geq x$", for x, y reals.

The next axiom postulates the existence of two opposite orientations of a line. It is compatible with the human cortex, which has neurons that perceive the concepts of "right" and "left" directions.

Line Orientations Axiom. *For any line d there is a set* $\Omega = \{\omega_-, \omega_+\}$ *of two opposite total order relations on the set d, called the orientations of d.*

The next axiom states that the line is infinite in both directions:

Incompleteness Axiom on a Line. *For any line d, the relations* $\{\omega_-, \omega_+\}$ *do not have a last element.*

The concept that one point is situated "between" other two points can now be defined using orientations.

Definition 5.2.5 *A line endowed with an orientation is called an oriented line. The point C is between the points A and B if:*

1) $A \neq C, B \neq C$;

2) A, B, C *are collinear (belong to a line);*

3) *we have either* $A\omega B$ *and* $B\omega C$, *or* $C\omega B$ *and* $B\omega A$, *where* $\omega \in \Omega$ *is one of the orientations of d.*

The orientation ω_+ will be called the *positive orientation* of d, while ω_- will be called the *negative orientation* of d.

Given two distinct points A and B, we define the line segment $[AB]$ between A and B as the set of the points M, which are situated between A and B. If M is between A and B we also write $M \in [AB]$.

Exercise 5.2.6 *Let A, B, C be three distinct points. Show that at most one point can be situated between the other two points.*

Exercise 5.2.7 *Given two distinct points A and B, show that there is a point C between A and B.*

Let $O \in d$ be a point on a line. The sets

$$d_+ = \{M; M \in d, O\omega_+ M\},$$

$$d_- = \{M; M \in d, O\omega_- M\}$$

are called the *half-lines* defined by the point O on the line d. $d_+(d_-)$ is called the half-line with the positive (negative) orientation starting at O, see Figure 5.1.

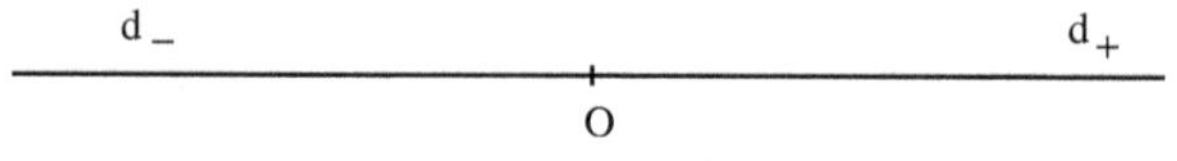

Figure 5.1: The two half-lines defined by the positive and negative orientations.

Exercise 5.2.8 *Show that* $d_+ \cup d_- = d$.

Definition 5.2.9 *Given three non-collinear points* A, B, C, *the set*

$$\Delta ABC = \{A, B, C\} \cup [BC] \cup [CA] \cup [AB]$$

is called a triangle with vertices A, B, C *and sides* $[BC]$, $[CA]$, $[AB]$.

5.3 Oriented segments

Every orientation presupposes a disorientation.
Hans Magnus Enzensberger (b. 1929)

An ordered pair of points (A, B) is called an *oriented segment* and it will be denoted by $\overline{AB}$. If $A = B$, the segment $\overline{AA}$ will be called the *null segment.* If $\overline{AB}$ is an oriented segment, the points A and B are called the *origin* and the *endpoint*, respectively.

Let A, $B \in d$. The segment $\overline{AB}$ is *positive (negative) oriented* if the half-line which starts at A and contains the point B is positive (negative) oriented.

The oriented segments $\overline{AB}$ and $\overline{CD}$ are called *equal* if $A = C$ and $B = D$. If $A = D$ and $B = C$, they are called *opposite segments* and we shall write $\overline{AB} = -\overline{BA}$ (the minus sign denoting the opposite orientations of the segments).

Let $A \neq B$. The line defined by the points A and B is called the *support line* of $\overline{AB}$. Two oriented segments are called *collinear* if their support lines are the same. The segments are called *parallel* if their support lines are parallel.

An oriented segment is characterized by *direction*, *orientation* and *length.* In the following we shall define each of these notions.

Direction

Two lines ℓ, ℓ' in the plane are called *equal or parallel* if $\ell = \ell'$ or $\ell \cap \ell' = \emptyset$, and we shall write $\ell || \ell'$. Let $\mathcal{L}$ be the set of the lines in the plane. Then "$||$" is an equivalence relation on $\mathcal{L}$.

Exercise 5.3.1 *Verify that the relation* "$||$" *is a reflexive, symmetric and transitive relation on* $\mathcal{L}$.

The equivalence classes of the relation "$||$" are called *directions.* In other words, a direction is a family of parallel lines.

The direction of an oriented segment $\overline{AB}$ is the direction of its support line AB. Hence two nonzero oriented segments have the same direction iff their support lines coincide or are parallel.

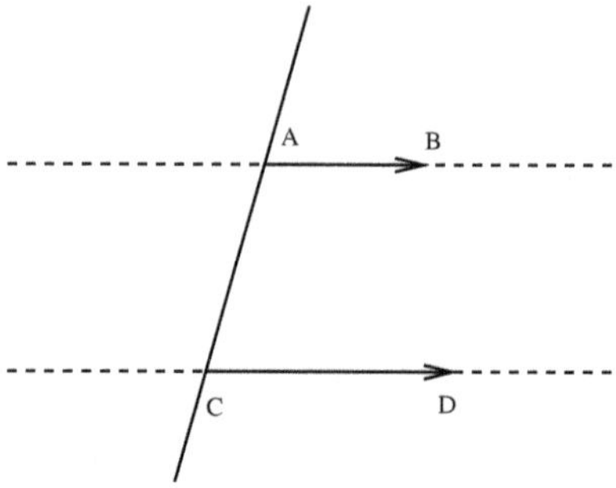

Figure 5.2: The half-plane determined by the line AC.

Exercise 5.3.2 *Define the relation δ on the set of oriented segments as $\overline{AB}\,\delta\,\overline{CD}$ if and only if the oriented segments $\overline{AB}$ and $\overline{CD}$ have the same direction. Prove that δ is an equivalence relation. What is the factor set?*

Orientation

Two collinear oriented segments have *the same orientation* if the orientations of the support lines of each of the segments are the same. If the oriented segments $\overline{AB}$ and $\overline{CD}$ are parallel, they have *the same orientation* if the endpoints B, D belong to the same half-plane determined by the line AC, see Figure 5.2.

Exercise 5.3.3 *Let Ω be a relation defined on the set of oriented segments by:*

$$\overline{AB}\,\Omega\,\overline{CD} \Longleftrightarrow \overline{AB} \text{ and } \overline{CD} \text{ have the same orientation.}$$

Show that Ω is an equivalence relation.

Length

The length of the oriented segment $\overline{AB}$ is the Euclidean distance between the points A and B, *i.e.*, $dist(A, B)$, which will be denoted by $|AB|$. Two oriented segments of the same length are called *congruent*.

Exercise 5.3.4 *Consider the following relation defined on the set of oriented segments: $\overline{AB} \equiv \overline{CD} \Longleftrightarrow |AB| = |CD|$. Show that " $\equiv$ " is an equivalence relation.*

Exercise 5.3.5 *Let A, B, C, D be four collinear points such that the segments $\overline{AB}$ and $\overline{CD}$ have the same length. Show that the segments $\overline{AC}$ and $\overline{BD}$ have the same length.*

5.4 Vectors

I shot an arrow into the air,
It fell to earth, I knew not where;
Henry W. Longfellow (1809–1882)

Let " $\wedge$ " be the logic connective *"AND"*. Consider the relation $\mathcal{E}$ defined on the set of oriented segments by

$$\overline{AB}\,\mathcal{E}\,\overline{CD} \Longleftrightarrow (\overline{AB}\,\delta\,\overline{CD}) \wedge (\overline{AB}\,\Omega\,\overline{CD}) \wedge (\overline{AB}{\equiv}\overline{CD}),$$

i.e., two segments are in relation $\mathcal{E}$ iff they have the same direction, the same orientation and they are congruent. $\mathcal{E}$ is an equivalence relation on the set of oriented vectors, since any of the relations δ, Ω, $\equiv$ are reflexive, symmetric and transitive.

Exercise 5.4.1 *Show that* $\overline{AB}\,\mathcal{E}\,\overline{CD}$ *if and only if* $\overline{BD}\,\mathcal{E}\,\overline{AC}$. *Consider both cases of collinear and non-collinear oriented segments.*

Exercise 5.4.2 *Show that if* $\overline{AB}\,\mathcal{E}\,\overline{A'B'}$ *and* $\overline{BC}\,\mathcal{E}\,\overline{B'C'}$, *then* $\overline{AC}\,\mathcal{E}\,\overline{A'C'}$.

Exercise 5.4.3 *Let* $\overline{AB}$ *and* $\overline{CD}$ *be two oriented segments. Show that* $\overline{AB}\,\mathcal{E}\,\overline{CD}$ *if and only if* $[AD]$ *and* $[BC]$ *have the same midpoint.*

Definition 5.4.4 *The equivalence classes of the relation* $\mathcal{E}$ *are called vectors.*

A vector can be regarded as an oriented segment, which moves keeping its direction and orientation. If $\overline{AB}$ is an oriented segment, its equivalence class is denoted by $\overrightarrow{AB}$. Usually we shall work with one element of the class, which is an oriented segment. For instance, if $ABCD$ is a parallelogram, the vectors $\overrightarrow{AB}$ and $\overrightarrow{DC}$ are equal, *i.e.*, they have the same direction, orientation and length, while the oriented segments $\overline{AB}$, $\overline{DC}$ are not equal, since they have different origin points and endpoints.

Sometimes we shall denote vectors by small letters with an arrow, like $\overrightarrow{a}, \overrightarrow{b}, \overrightarrow{c}, \overrightarrow{u}, \overrightarrow{v}, \ldots$. Let $\overrightarrow{v}$ be a vector and $\overline{AB}$ be an element of the class. We shall write $\overrightarrow{v} = \overrightarrow{AB}$. The length of the vector $\overrightarrow{v}$ is $dist(A, B)$. The direction and the orientation of the vector $\overrightarrow{v}$ are the same as the direction and orientation of $\overline{AB}$. The equivalence class of the oriented segment with the opposite orientation $\overline{BA}$ is the opposite vector $-\overrightarrow{v}$. See Figure 5.3(a).

Let $\mathcal{P}$ be the plane and let $O \in \mathcal{P}$ be a fixed point, called the origin. For any point $M \in \mathcal{P}$ we associate the oriented segment $\overline{OM}$, called the *position vector* of the point M. Its equivalence class is the vector $\overrightarrow{OM}$. Let $\mathcal{V}$ be the set of the vectors in the plane $\mathcal{P}$. Consider the function $p : \mathcal{P} \to \mathcal{V}$, defined by $p(M) = \overrightarrow{OM}$, see Figure 5.3(b).

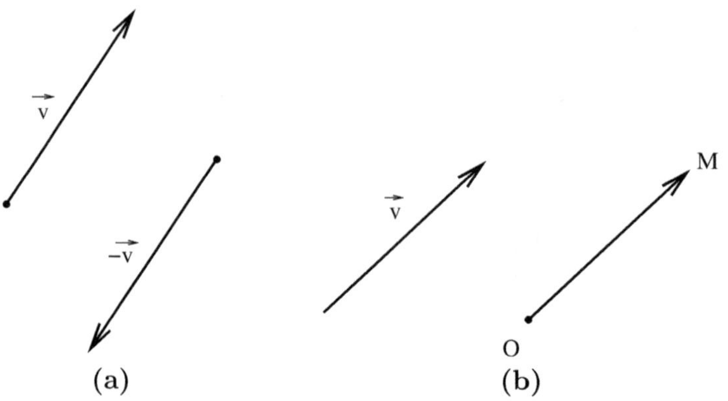

Figure 5.3: (a) Opposite vectors; (b) The position vector of the point M.

Exercise 5.4.5 *The function p is one-to-one, i.e., if $p(M) = p(M')$, then $M = M'$.*

Exercise 5.4.6 *The function p is onto (surjective), i.e., for any vector $\overrightarrow{v} \in \mathcal{V}$, there is a unique point $M \in \mathcal{P}$ such that $p(M) = \overrightarrow{v}$.*

From the previous two exercises, it follows that the plane $\mathcal{P}$ can be identified with the set of the vectors $\mathcal{V}$ by the bijection p, once an origin has been fixed in $\mathcal{P}$.

Similar properties hold for the set of the vectors in the Euclidean space.

5.5 Operations with vectors

"There is no other royal path which leads to geometry,"
said Euclid to Ptolemy I.

Vector addition

We shall present two rules of adding vectors: the triangle rule and the parallelogram rule.

The triangle rule

Let $\overrightarrow{a}$ and $\overrightarrow{b}$ be two vectors. Let $O \in \mathcal{P}$ be the origin in the plane. By Exercise 5.4.6, there are unique points A, $B \in \mathcal{P}$ such that $\overrightarrow{a} = \overrightarrow{OA}$ and $\overrightarrow{b} = \overrightarrow{AB}$. We shall define the sum $\overrightarrow{a} + \overrightarrow{b} = \overrightarrow{c}$, with $\overrightarrow{c} = \overrightarrow{OB}$. We can also write as

$$\overrightarrow{OA} + \overrightarrow{AB} = \overrightarrow{OB},$$

which is called *the triangle identity*. The method of finding the vector $\overrightarrow{OB}$ as the sum of $\overrightarrow{OA}$ and $\overrightarrow{AB}$ is called the *triangle rule*.

The above identity involves equivalence classes. In order to have a correct definition of the sum, we need to show that the vector $\overrightarrow{c}$ does not depend on the origin O. Let $O' \in \mathcal{P}$ be another point. By Exercise 5.4.6, there are unique points A', $B' \in \mathcal{P}$ such that $\overrightarrow{a} = \overrightarrow{O'A'}$ and $\overrightarrow{b} = \overrightarrow{A'B'}$. We need to show that $\overrightarrow{c} = \overrightarrow{O'B'}$, *i.e.*, $\overrightarrow{OB}$ and $\overrightarrow{O'B'}$ are the same vectors, *i.e.*, $\overline{OB}\,\mathcal{E}\,\overline{O'B'}$. There are two cases to investigate:

The case 1: The points O, A, B are non-collinear.
Since $\overrightarrow{OA} = \overrightarrow{O'A'}$, then $\overline{OA}\,\mathcal{E}\,\overline{O'A'}$, and by Exercise 5.4.1 we have $\overline{OO'}\,\mathcal{E}\,\overline{AA'}$. In a similar way, since $\overrightarrow{AB} = \overrightarrow{A'B'}$, it follows that $\overline{AA'}\,\mathcal{E}\,\overline{BB'}$. Hence $\overline{OO'}\,\mathcal{E}\,\overline{AA'}$ and $\overline{AA'}\,\mathcal{E}\,\overline{BB'}$. By transitivity we obtain $\overline{OO'}\mathcal{E}\overline{BB'}$. Using Exercise 5.4.1 it follows that $\overline{OB}\,\mathcal{E}\,\overline{O'B'}$, see Figure 5.4.

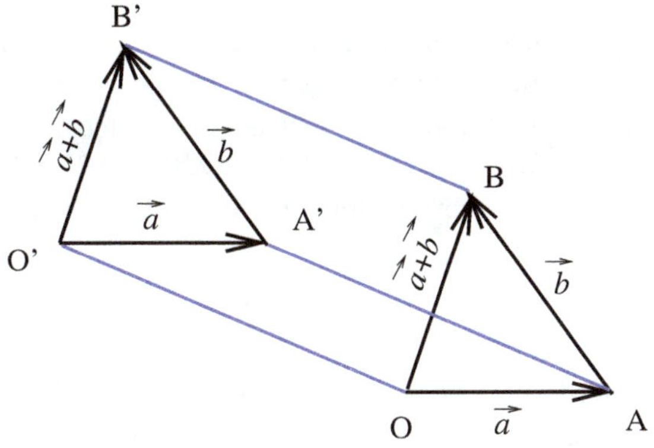

Figure 5.4: The case when O, A and B are non-collinear.

The case 2: The points O, A, B are collinear.
The points O', A', B' are also collinear. Then the vectors $\overrightarrow{OA}$, $\overrightarrow{AB}$, $\overrightarrow{O'A'}$, $\overrightarrow{A'B'}$ have the same direction and orientation (why?). Then $\overrightarrow{OB}$ and $\overrightarrow{O'B'}$ will have the same direction and orientation, too. Since $|OA| + |AB| = |O'A'| + |A'B'|$, it follows that $|OB| = |O'B'|$, *i.e.*, the vectors $\overrightarrow{OB}$ and $\overrightarrow{O'B'}$ have the same length. Hence they are equal vectors, and we proved that $\overrightarrow{OA} + \overrightarrow{AB} = \overrightarrow{O'A'} + \overrightarrow{A'B'}$. See Figure 5.5.

Exercise 5.5.1 *Let A, B, C be three points. Show that*

(i) $\overrightarrow{AB} + \overrightarrow{BC} + \overrightarrow{CA} = 0$ *(triangle identity),*

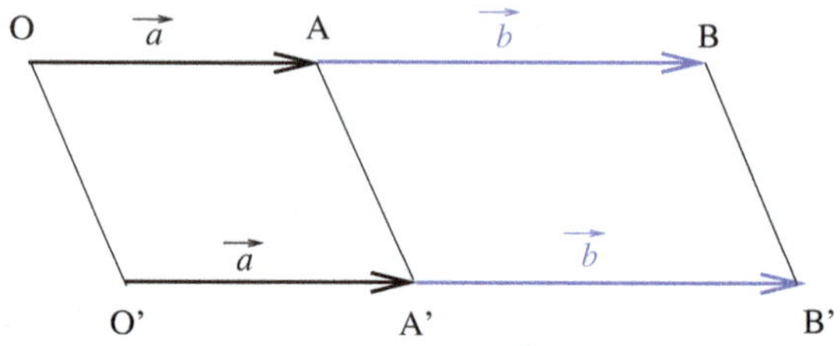

Figure 5.5: The case when O, A and B are collinear.

(ii) $|\overrightarrow{AB} + \overrightarrow{BC}| \leq |\overrightarrow{AB}| + |\overrightarrow{BC}|$ *(triangle inequality), and*

(iii) $|\overrightarrow{AB} - \overrightarrow{BC}| \geq \Big||\overrightarrow{AB}| - |\overrightarrow{BC}|\Big|$.

The parallelogram rule

This rule is a consequence of the triangle rule. Let $\vec{a}$ and $\vec{b}$ be two vectors. If $O \in \mathcal{P}$ is a fixed point in the plane, let A, $B \in \mathcal{P}$ be such that $\vec{a} = \overrightarrow{OA}$ and $\vec{b} = \overrightarrow{OB}$. Let M be a point in the plane such that $OAMB$ is a parallelogram. Since $AM||OB$ and $|AM| = |OB|$ and B, M belong to the same half-plane determined by the line OA, it follows that $\overrightarrow{AM} = \overrightarrow{OB}$. The triangle identity yields

$$\vec{a} + \vec{b} = \overrightarrow{OA} + \overrightarrow{AM} = \overrightarrow{OM},$$

which is called *the parallelogram rule.* Hence the vector defined by the parallelogram diagonal $\overline{OM}$ is the sum of $\overrightarrow{OA}$ and $\overrightarrow{OB}$. See Figure 5.6.

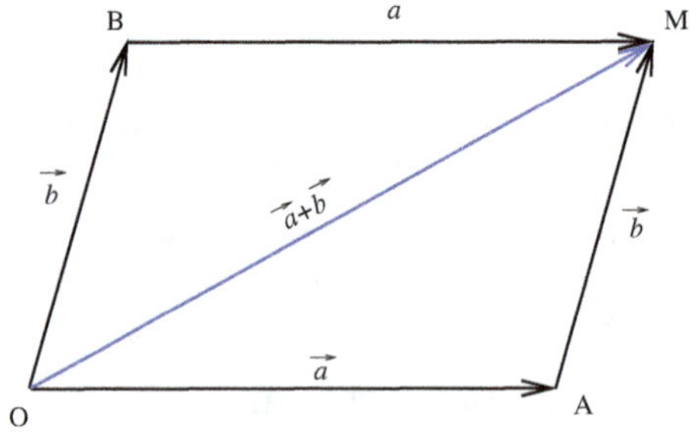

Figure 5.6: The parallelogram rule.

The following proposition states the properties of the vector addition:

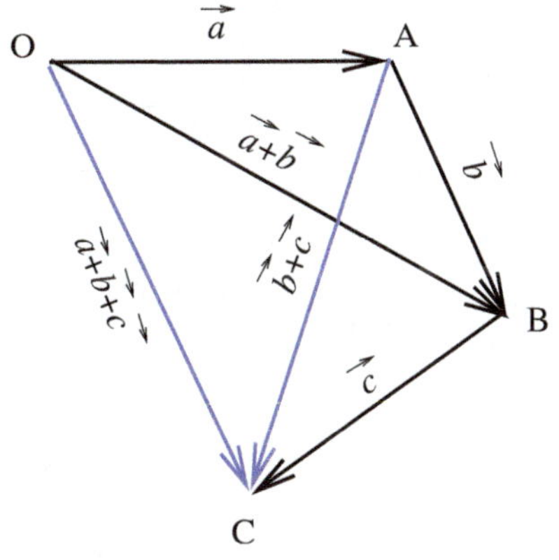

Figure 5.7: The associativity of vector addition.

Proposition 5.5.2 *Let $\mathcal{V}$ be the set of the vectors in the plane. The vector addition $+ : \mathcal{V}\times\mathcal{V} \to \mathcal{V}$, with $(\vec{a}, \vec{b}) \to \vec{a}+\vec{b}$ has the following properties:*
(i) $\vec{a} + (\vec{b} + \vec{c}) = (\vec{a} + \vec{b}) + \vec{c}$, $\forall \vec{a}, \vec{b}, \vec{c} \in \mathcal{V}$ (associativity);
(ii) $\vec{a} + \vec{b} = \vec{b} + \vec{a}$, $\forall \vec{a}, \vec{b} \in \mathcal{V}$ (commutativity).

Proof: (*i*) Let O be a fixed point in the plane $\mathcal{P}$. By Exercise 5.4.6, there are A, B, $C \in \mathcal{P}$ such that $\vec{a} = \overrightarrow{OA}$, $\vec{b} = \overrightarrow{AB}$ and $\vec{c} = \overrightarrow{BC}$. See Figure 5.7. Using the triangle identity twice, yields

$$\begin{aligned} \vec{a} + \vec{b} &= \overrightarrow{OA} + \overrightarrow{AB} = \overrightarrow{OB} \\ (\vec{a} + \vec{b}) + \vec{c} &= \overrightarrow{OB} + \overrightarrow{BC} = \overrightarrow{OC}. \end{aligned} \tag{5.5.1}$$

$$\begin{aligned} \vec{b} + \vec{c} &= \overrightarrow{AB} + \overrightarrow{BC} = \overrightarrow{AC} \\ \vec{a} + (\vec{b} + \vec{c}) &= \overrightarrow{OA} + \overrightarrow{AC} = \overrightarrow{OC}. \end{aligned} \tag{5.5.2}$$

From (5.5.1) and (5.5.2) we obtain $\vec{a} + (\vec{b} + \vec{c}) = (\vec{a} + \vec{b}) + \vec{c}$. This enables us to write from now on $\vec{a} + \vec{b} + \vec{c}$ instead of $\vec{a} + (\vec{b} + \vec{c})$ or $(\vec{a} + \vec{b}) + \vec{c}$, see Figure 1.8.

The cases when some of the points O, A, B, C are collinear are left as an exercise for the reader.
(*ii*) Following the notations on the Figure 5.6 and using the parallelogram rule, yields

$$\begin{aligned} \vec{a} + \vec{b} &= \overrightarrow{OA} + \overrightarrow{AM} = \overrightarrow{OM}; \\ \vec{b} + \vec{a} &= \overrightarrow{OB} + \overrightarrow{BM} = \overrightarrow{OM}. \end{aligned}$$

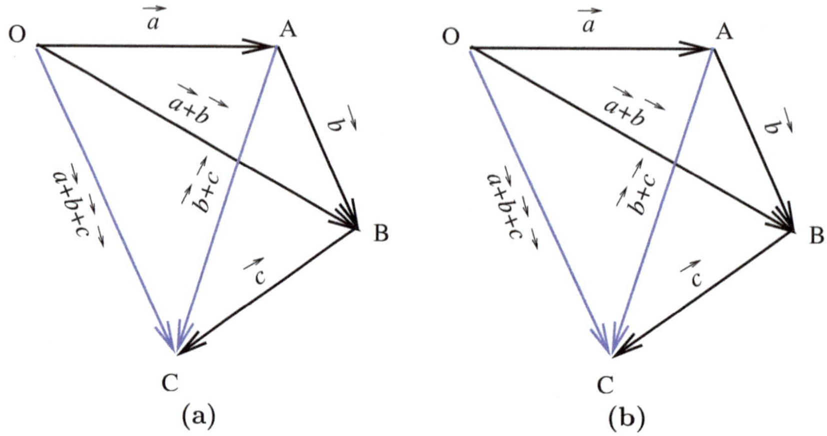

Figure 5.8: (a) Vector subtraction; (b) Parallelogram rule for vector subtraction.

Hence $\vec{a} + \vec{b} = \vec{b} + \vec{a}$, which is called the commutativity of the vector addition. ■

If $\vec{a} = \overrightarrow{OA}$ is a vector, then the vector $\overrightarrow{AO}$ has the same length and direction but opposite orientation. The vector $\overrightarrow{AO}$ is called the opposite vector of $\vec{a}$ and it will be denoted by $-\vec{a}$.

The vector $\overrightarrow{AA}$ is the zero vector and it will be denoted by $\vec{0}$. The zero vector has zero length and undeterminate direction and orientation.

The difference of two vectors

To subtract two vectors $\vec{a}$ and $\vec{b}$ means to add the opposite vector of $\vec{b}$ to $\vec{a}$:

$$\vec{a} - \vec{b} = \vec{a} + (-\vec{b}).$$

Exercise 5.5.3 *Show that we have*

$$\begin{aligned} \vec{a} - \vec{a} &= \vec{0} \text{ and} \\ \vec{a} + \vec{0} &= \vec{0} + \vec{a} = \vec{a}. \end{aligned}$$

The difference $\vec{a} - \vec{b}$ can be computed using a variant of the triangle rule:

Let $\vec{a} = \overrightarrow{OA}$ and $\vec{b} = \overrightarrow{OB}$. Then $\vec{a} - \vec{b} = \overrightarrow{BA}$. See Figure 5.8(a).
Let M be a point such that $OBAM$ is a parallelogram, see Figure 5.8(b). Then $\overrightarrow{AM} = \overrightarrow{BO} = -\overrightarrow{OB} = \vec{b}$. Then

$$\vec{a} - \vec{b} = \vec{a} + (-\vec{b}) = \overrightarrow{OA} + \overrightarrow{AM} = \overrightarrow{OM} = \overrightarrow{BA}.$$

Exercise 5.5.4 *Prove the following identities.*
(i) $(\overrightarrow{a} + \overrightarrow{b}) - (\overrightarrow{c} + \overrightarrow{b}) = \overrightarrow{a} - \overrightarrow{c}$, *and*
(ii) $(\overrightarrow{a} - \overrightarrow{b}) + (\overrightarrow{b} - \overrightarrow{c}) = \overrightarrow{a} - \overrightarrow{c}$.

Scalar multiplication

Let $\overrightarrow{v}$ be a vector and $s \in \mathbb{R}$ be a real number, called *scalar*. The product between s and $\overrightarrow{v}$ is a new vector $s\overrightarrow{v}$ defined as follows:
1) if $\overrightarrow{v} = \overrightarrow{0}$ or $s = 0$, then $s\overrightarrow{v} = \overrightarrow{0}$,
2) if $\overrightarrow{v} \neq \overrightarrow{0}$ and $s \neq 0$, then $s\overrightarrow{v}$ is a vector with
- the same direction as $\overrightarrow{v}$,
- the same orientation as $\overrightarrow{v}$ if $s > 0$,
- the opposite orientation of $\overrightarrow{v}$ if $s < 0$, and
- the length equal to $|s| \cdot |\overrightarrow{v}|$.

Remark 5.5.5 The vectors $s\overrightarrow{v}$ and $\overrightarrow{v}$ are collinear. In the case when $s = -1$, we obtain the opposite vector $(-1)\overrightarrow{v} = -\overrightarrow{v}$.

Proposition 5.5.6 *Let $\mathcal{V}$ be the set of the vectors in the plane (or space). The multiplication by scalars is a binary law defined on $\mathbb{R} \times \mathcal{V}$ with values in $\mathcal{V}$, given by $(s, \overrightarrow{v}) \to s\overrightarrow{v}$, and has the following properties.*
(i) $1\overrightarrow{v} = \overrightarrow{v}$, $\forall \overrightarrow{v} \in \mathcal{V}$,
(ii) $s(t\overrightarrow{v}) = (st)\overrightarrow{v}$, $\forall s, t \in \mathbb{R}$, $\forall \overrightarrow{v} \in \mathcal{V}$,
(iii) $(s + t)\overrightarrow{v} = s\overrightarrow{v} + t\overrightarrow{v}$, $\forall s, t \in \mathbb{R}$, $\forall \overrightarrow{v} \in \mathcal{V}$, *and*
(iv) $s(\overrightarrow{v} + \overrightarrow{w}) = s\overrightarrow{v} + s\overrightarrow{w}$, $\forall s \in \mathbb{R}$, $\forall v, w \in \mathcal{V}$.

Proof: Relations *(i)*, *(ii)* and *(iii)* are obvious and are left as an exercise for the reader.

(iv) Let $\overrightarrow{v}, \overrightarrow{w} \in \mathcal{V}$ be two vectors, and let $s \in \mathbb{R}$ be a fixed scalar. Choose an origin O in the plane. There are points A, B, in the plane such that $\overrightarrow{v} = \overrightarrow{OA}$, $\overrightarrow{w} = \overrightarrow{AB}$. We shall assume $s > 0$, see Figure 5.9.

By the triangle rule we have $\overrightarrow{OB} = \overrightarrow{OA} + \overrightarrow{AB} = \overrightarrow{v} + \overrightarrow{w}$. Let B' be a point on the line OB such that $s(\overrightarrow{v} + \overrightarrow{w}) = \overrightarrow{OB'}$. Since $\widehat{AOB} = \widehat{A'OB'}$ and

$$\frac{|OA|}{|OA'|} = \frac{|OB|}{|OB'|} = \frac{1}{s},$$

it follows that the triangles OAB and $OA'B'$ are similar. Then $|A'B'| = s|AB|$ and $\widehat{OAB} = \widehat{OA'B'}$ (Why?). It follows that $AB||A'B'$ (Why?) and hence $\overrightarrow{AB} = s\overrightarrow{A'B'}$, since they have the same direction and orientation.

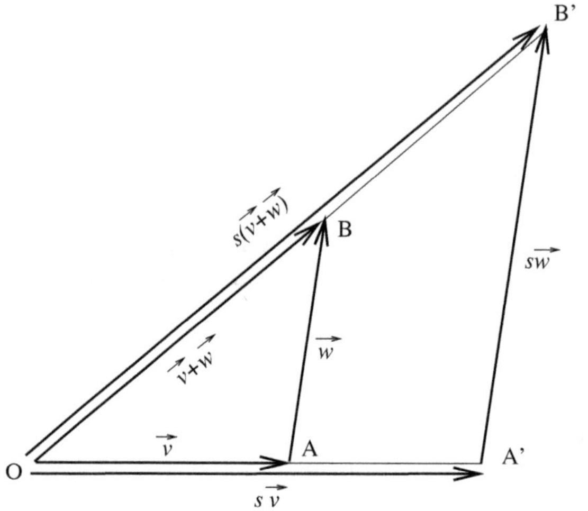

Figure 5.9: For part (iv) of Proposition 5.5.6.

Hence $\overrightarrow{A'B'} = s\vec{w}$. Using the triangle rule again, yields $\overrightarrow{OA'} + \overrightarrow{A'B'} = \overrightarrow{OB'}$, which can be written also as $s\vec{v} + s\vec{w} = s(\vec{v} + \vec{w})$, and the distributivity is proved. ■

Remark 5.5.7 The set of vectors $\mathcal{V}$ endowed with the vector addition and multiplication with scalars form a vectorial space. The definition can be extended to any set $\mathcal{V}$ endowed with two operations $+ : \mathcal{V} \times \mathcal{V} \to \mathcal{V}$ and $\cdot : \mathbb{R} \times \mathcal{V} \to \mathcal{V}$, which satisfy the properties of Propositions 5.5.2–5.5.6.

Solved Exercise 5.5.8 *Let $ABCD$ be a parallelogram and M be the midpoint of $[AD]$. Express the sum $\overrightarrow{MD} + \overrightarrow{MC} + \overrightarrow{MB}$ in terms of $\overrightarrow{AB}$ and $\overrightarrow{AD}$. See Figure 5.10.*

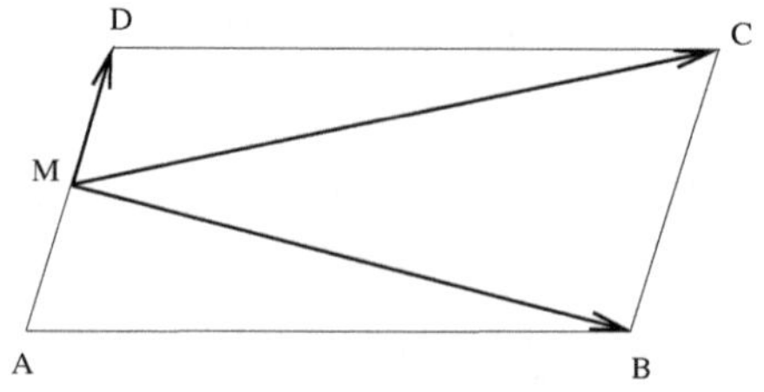

Figure 5.10: For Exercise 5.5.8.

Proof: Using the vectorial triangle identity in ΔMDC and ΔMAB, yields

$$\begin{aligned}\overrightarrow{MD}+\overrightarrow{MC}+\overrightarrow{MB} &= \overrightarrow{MD}+(\overrightarrow{MD}+\overrightarrow{DC})+(\overrightarrow{MA}+\overrightarrow{AB})\\ &= \underbrace{(\overrightarrow{MD}+\overrightarrow{MA})}_{=0}+\overrightarrow{MD}+2\overrightarrow{AB}\\ &= \overrightarrow{MD}+2\overrightarrow{AB}=\frac{1}{2}\overrightarrow{AD}+2\overrightarrow{AB},\end{aligned}$$

where we used that $\overrightarrow{AB}=\overrightarrow{DC}$. ■

The next exercise deals with the converse of Exercise 5.5.8.

Exercise 5.5.9 *Let M be a point and $ABCD$ be a parallelogram. Show that if M satisfies*

$$\overrightarrow{MD}+\overrightarrow{MC}+\overrightarrow{MB}=\frac{1}{2}\overrightarrow{AD}+2\overrightarrow{AB},$$

then M is the midpoint of $[AD]$.

Exercise 5.5.10 *Let $A_1A_2A_3A_4A_5A_6$ be a regular hexagon. Show that*
(i) $\overrightarrow{A_1A_4}=2(\overrightarrow{A_1A_3}-\overrightarrow{A_1A_2})$,
(ii) $\overrightarrow{A_1A_5}=2\overrightarrow{A_1A_3}-3\overrightarrow{A_1A_2}$, *and*
(iii) $\overrightarrow{A_1A_6}=\overrightarrow{A_1A_3}-2\overrightarrow{A_1A_2}$.

Exercise 5.5.11 *Solve the following vectorial system*

$$\begin{aligned}2\vec{x}-\vec{y} &= \vec{a}\\ \vec{x}+\vec{y} &= \vec{b},\end{aligned}$$

with the vectors $\vec{a}$ and $\vec{b}$ given.

Exercise 5.5.12 *Let A, B, $C \in \mathcal{P}$ be three points in the plane.*
(i) Show that there is a point $D \in \mathcal{P}$ such that

$$\overrightarrow{DA}+\overrightarrow{DB}=\overrightarrow{DC}.$$

(ii) Show that the point D with the property (i) is unique.
(iii) Prove that for any $M \in \mathcal{P}$ we have

$$\overrightarrow{MA}+\overrightarrow{MB}-\overrightarrow{MC}=\overrightarrow{MD}.$$

(iv) Show that the expression $\overrightarrow{MA}+\overrightarrow{MB}-2\overrightarrow{MC}$ does not depend on the point M.
(v) Find a point $M \in \mathcal{P}$ such that

$$\overrightarrow{MA}+\overrightarrow{MB}-\overrightarrow{MC}=\overrightarrow{AB}.$$

5.6 Linear independence

Let $\overrightarrow{v_1}$, $\overrightarrow{v_2}, \ldots, \overrightarrow{v_n}$ be n vectors in the plane, and let $a_1, a_2, \ldots, a_n \in \mathbb{R}$ be n real numbers. The vector

$$\overrightarrow{v} = a_1\overrightarrow{v_1} + a_2\overrightarrow{v_2} + \cdots + a_n\overrightarrow{v_n}$$

is called the *linear combination* of the vectors $\overrightarrow{v_i}$ with coefficients a_i, with $i \in \{1, \ldots, n\}$.

Definition 5.6.1 *The vectors $\overrightarrow{v_1}$, $\overrightarrow{v_2}, \ldots, \overrightarrow{v_n}$ are called linearly dependent if there is a vanishing linear combination with not all zero coefficients, i.e.,*

$$\sum_{i=1}^{n} a_i\overrightarrow{v_i} = 0, \qquad \textit{with} \qquad \sum_{i=1}^{n} a_i^2 \neq 0.$$

The vectors $\overrightarrow{v_1}$, $\overrightarrow{v_2}, \ldots, \overrightarrow{v_n}$ are called linearly independent if they are not linearly dependent.

Exercise 5.6.2 *Show that the vectors $\overrightarrow{v_1}$, $\overrightarrow{v_2}, \ldots, \overrightarrow{v_n}$ are linearly independent if and only if*

$$\sum_{i=1}^{n} a_i\overrightarrow{v_i} = 0 \Longrightarrow a_1 = a_2 = \cdots = a_n = 0.$$

Exercise 5.6.3 *If the vectors $\overrightarrow{v_1}$, $\overrightarrow{v_2}, \ldots, \overrightarrow{v_n}$ are linearly dependent, show that the vectors $\overrightarrow{v_1}$, $\overrightarrow{v_2}, \ldots, \overrightarrow{v_n}$, $\overrightarrow{v_{n+1}}, \ldots, \overrightarrow{v_{n+p}}$ are also linearly dependent.*

Exercise 5.6.4 *Let $\overrightarrow{v_1}$, $\overrightarrow{v_2}, \ldots, \overrightarrow{v_n}$ be linearly independent vectors. Show that for any $1 \leq p \leq n$, the vectors $\overrightarrow{v_1}$, $\overrightarrow{v_2}, \ldots, \overrightarrow{v_p}$ are linearly independent.*

Exercise 5.6.5 *Show that the vectors $\overrightarrow{v_1}$, $\overrightarrow{v_2}, \ldots, \overrightarrow{v_n}$ are linearly dependent if and only if there is a $p \in \{1, \ldots, n\}$ such that $\overrightarrow{v_p}$ is a linear combination of the other $p-1$ vectors.*

Proposition 5.6.6 *Let $\overrightarrow{v}$, $\overrightarrow{w}$ be two vectors in the plane. The vectors $\overrightarrow{v}$ and $\overrightarrow{w}$ are linearly dependent if and only if they are collinear.*

Proof: " $\Longrightarrow$ " Let $\overrightarrow{v}$ and $\overrightarrow{w}$ be linearly dependent vectors. Then there are a, $b \in \mathbb{R}$, not both equal to zero, such that $a\overrightarrow{v} + b\overrightarrow{w} = 0$, or $a\overrightarrow{v} = -b\overrightarrow{w}$.

If $a \neq 0$, then dividing by a, yields $\overrightarrow{v} = -\dfrac{b}{a}\overrightarrow{w} = \lambda\overrightarrow{w}$, which means that the vectors $\overrightarrow{v}$ and $\overrightarrow{w}$ are collinear.

If $a = 0$, using that $b \neq 0$, it follows that $\overrightarrow{w} = \overrightarrow{0}$ (Why?). Hence $\overrightarrow{w} = 0 \cdot \overrightarrow{v}$ and then $\overrightarrow{v}, \overrightarrow{w}$ are collinear.

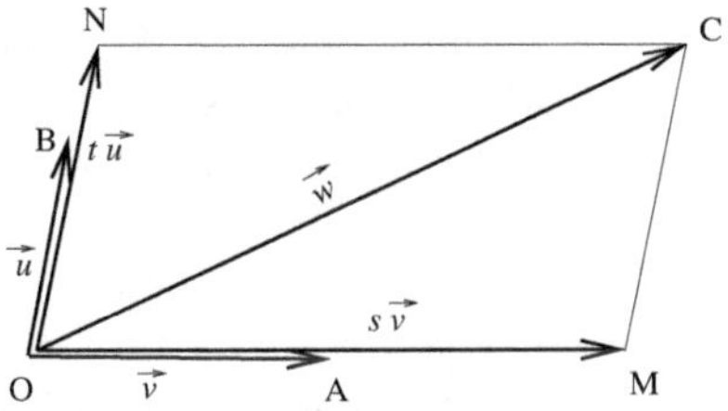

Figure 5.11: For the proof of Proposition 5.6.8.

" $\Longleftarrow$ " Let $\vec{v} = \lambda\vec{w}$. Then $1\vec{v} + (-\lambda)\vec{w} = 0$, *i.e.*, the vectors $\vec{v}$ and $\vec{w}$ are linearly dependent. ■

From the previous proposition, by negation, we obtain the following result:

Corollary 5.6.7 *Any two non-collinear vectors are linearly independent.*

The following result deals with the linear dependence of three vectors in the plane.

Proposition 5.6.8 *Any three vectors in the plane are linearly dependent.*

Proof: Let $\vec{u}$, $\vec{v}$, $\vec{w}$ be three vectors in the plane. If the vectors $\vec{u}$ and $\vec{v}$ are collinear, then they are linearly dependent. By Exercise 5.6.3 it follows that $\vec{u}$, $\vec{v}$, $\vec{w}$ are linearly dependent.

If the vectors $\vec{u}$ and $\vec{v}$ are not collinear, by Corollary 5.6.7, they are linearly independent. Let $O \in \mathcal{P}$ be a fixed point in the plane. There are points A, B, $C \in \mathcal{P}$ such that

$$\vec{v} = \overrightarrow{OA}, \qquad \vec{u} = \overrightarrow{OB}, \qquad \vec{w} = \overrightarrow{OC},$$

see Figure 5.11.

By the parallelogram rule, $\overrightarrow{OC} = \overrightarrow{OM} + \overrightarrow{ON}$, where $(\overrightarrow{OM}, \overrightarrow{OA})$ and $(\overrightarrow{ON}, \overrightarrow{OB})$ are pairs of collinear vectors. Hence there are $s, t \in \mathbb{R}$ such that $\overrightarrow{OM} = s\overrightarrow{OA}$ and $\overrightarrow{ON} = t\overrightarrow{OB}$. Then $\overrightarrow{OC} = s\overrightarrow{OA} + t\overrightarrow{OB}$, which becomes $\vec{w} = s\vec{v} + t\vec{u}$, which can be written as a vanishing linear combination: $s\vec{v} + t\vec{u} - \vec{w} = 0$. It follows that the vectors $\vec{u}, \vec{v}, \vec{w}$ are linear dependent. ■

Corollary 5.6.9 *The maximum number of linearly independent vectors in the plane is* 2.

Definition 5.6.10 *The maximum number of the linearly independent vectors in a vectorial space $\mathcal{V}$ is called the dimension of the space. Any maximal set of linearly independent vectors is called a basis of the space.*

In our case, the set of the vectors in the plane form a vectorial space of dimension 2. Any two non-collinear vectors form a basis of the space.

The next result deals with the decomposition of a vector with respect to a basis.

Proposition 5.6.11 *Let $\{\overrightarrow{u}, \overrightarrow{v}\}$ be a basis in the plane. Then for any vector $\overrightarrow{w}$, there are unique numbers $a, b \in \mathbb{R}$ such that $\overrightarrow{w} = a\overrightarrow{u} + b\overrightarrow{v}$.*

Proof: We shall prove the existence and the uniqueness of the numbers a, b.
Existence: From Corollary 5.6.9, the vectors $\overrightarrow{u}$, $\overrightarrow{v}$, $\overrightarrow{w}$ are linearly dependent, *i.e.*, we can write

$$\alpha\overrightarrow{u} + \beta\overrightarrow{v} + \gamma\overrightarrow{w} = 0, \qquad \alpha, \beta, \gamma \in \mathbb{R}.$$

If $\gamma = 0$, then $\alpha\overrightarrow{u} + \beta\overrightarrow{v} = 0$, *i.e.*, $\overrightarrow{u}$ and $\overrightarrow{v}$ are linearly dependent, which is a contradiction, because $\{\overrightarrow{u}, \overrightarrow{v}\}$ is a basis. Hence $\gamma \neq 0$. Dividing by γ, yields

$$\begin{aligned} \frac{\alpha}{\gamma}\overrightarrow{u} + \frac{\beta}{\gamma}\overrightarrow{v} + \overrightarrow{w} &= 0 \Longleftrightarrow \\ \overrightarrow{w} = -\frac{\alpha}{\gamma}\overrightarrow{u} - \frac{\beta}{\gamma}\overrightarrow{v} &= a\overrightarrow{u} + b\overrightarrow{v}, \end{aligned}$$

with $a = -\alpha/\gamma$, $b = -\beta/\gamma$.
Uniqueness: If there are two decompositions

$$\overrightarrow{w} = a\overrightarrow{u} + b\overrightarrow{v}, \qquad \overrightarrow{w} = a'\overrightarrow{u} + b'\overrightarrow{v},$$

subtracting, yields

$$(a - a')\overrightarrow{u} + (b - b')\overrightarrow{v} \;=\; 0.$$

Since $\overrightarrow{u}$ and $\overrightarrow{v}$ are linearly dependent, it follows that $a = a'$ and $b = b'$. Hence the decomposition is unique. ■

Remark 5.6.12 If $\{\overrightarrow{u}, \overrightarrow{v}\}$ is a basis in the plane, the numbers a and b are called the components of the vector $\overrightarrow{w}$ in the basis $\{\overrightarrow{u}, \overrightarrow{v}\}$.

Exercise 5.6.13 *Let $\{\overrightarrow{u}, \overrightarrow{v}\}$ be a basis and let $\overrightarrow{w_1} = a_1\overrightarrow{u} + b_1\overrightarrow{v}$ and $\overrightarrow{w_2} = a_2\overrightarrow{u} + b_2\overrightarrow{v}$ be two vectors in the plane. Show that:*
(i) $\overrightarrow{w_1} + \overrightarrow{w_2} = (a_1 + a_2)\overrightarrow{u} + (b_1 + b_2)\overrightarrow{v}$;
(ii) $\lambda\overrightarrow{w_1} = (\lambda a_1)\overrightarrow{u} + (\lambda a_2)\overrightarrow{v}$;
(iii) $\overrightarrow{w_1}$ and $\overrightarrow{w_2}$ are collinear if and only if $a_1b_2 = a_2b_1$.

Exercise 5.6.14 *Let $\{\overrightarrow{u}, \overrightarrow{v}\}$, $\{\overrightarrow{u'}, \overrightarrow{v'}\}$ be two bases in the plane related by*

$$\begin{aligned}\overrightarrow{u'} &= a_{11}\overrightarrow{u} + a_{12}\overrightarrow{v}, \text{ and}\\ \overrightarrow{v'} &= a_{21}\overrightarrow{u} + a_{22}\overrightarrow{v}.\end{aligned}$$

Let $\overrightarrow{w} = \alpha\overrightarrow{u} + \beta\overrightarrow{v}$ be a vector. Write the vector $\overrightarrow{w}$ in terms of $\overrightarrow{u'}$ and $\overrightarrow{v'}$.

The results of this section can be extended to the vectorial space of the vectors in the three dimensional space. In this case, the dimension is equal to 3 and the maximal number of linear independent vectors is 3.

In this case, Propositions 5.6.6, 5.6.8 and 5.6.11 will become respectively:

Proposition 5.6.15 *Let $\overrightarrow{u}$, $\overrightarrow{v}$, $\overrightarrow{w}$ be three vectors in the space. The vectors are linearly dependent if and only if they are coplanar.*

Proposition 5.6.16 *Any 4 vectors in the space are linearly dependent.*

Proposition 5.6.17 *Let $\{\overrightarrow{u}, \overrightarrow{v}, \overrightarrow{w}\}$ be a basis in the space. Then for any vector $\overrightarrow{m}$, there are three unique numbers a, b, $c \in \mathbb{R}$ such that $\overrightarrow{m} = a\overrightarrow{u} + b\overrightarrow{v} + c\overrightarrow{w}$.*

The proofs of the previous three statements is similar with the proof of Propositions 5.6.6, 5.6.8 and 5.6.11.

Solved Exercise 5.6.18 *Let ABC be a triangle. Find $\lambda \in \mathbb{R}$ such that $\overrightarrow{MA} + \overrightarrow{MB} + \lambda\overrightarrow{MC}$ does not depend on the point M.*

Proof: Let M and M' be two distinct points. By contradiction, we assume that

$$\overrightarrow{MA} + \overrightarrow{MB} + \lambda\overrightarrow{MC} = \overrightarrow{M'A} + \overrightarrow{M'B} + \lambda\overrightarrow{M'C}.$$

Subtracting, we obtain

$$\begin{aligned}0 &= (\overrightarrow{MA} + \overrightarrow{MB} + \lambda\overrightarrow{MC}) - (\overrightarrow{M'A} + \overrightarrow{M'B} + \lambda\overrightarrow{M'C})\\ &= (\overrightarrow{MA} - \overrightarrow{M'A}) + (\overrightarrow{MB} - \overrightarrow{M'B}) + \lambda(\overrightarrow{MC} - \overrightarrow{M'C})\\ &= \overrightarrow{MM'} + \overrightarrow{MM'} + \lambda\overrightarrow{MM'}\\ &= (2 + \lambda)\overrightarrow{MM'}.\end{aligned}$$

Hence $\lambda = -2$. ∎

Exercise 5.6.19 *Show that the linear combination*

$$\alpha\overrightarrow{MA} + \beta\overrightarrow{MB} + \gamma\overrightarrow{MC}$$

does not depend on the point M if and only if

$$\alpha + \beta + \gamma = 0.$$

Exercise 5.6.20 *Let $A_1A_2\cdots A_n$ be a polygon. Show that the linear combination $\sum_{i=1}^{n} a_i\overrightarrow{MA_i}$ is independent of M if and only if $\sum_{i=1}^{n} a_i = 0$.*

Chapter 6

Elementary Properties of Vectors

6.1 Medians

And Lucy, dear child, mind your arithmetic... What would life be without arithmetic, but a scene of horrors?

Syndey Smith, 1835

We shall start this section defining the midpoint of a line segment and the medians of a triangle.

Definition 6.1.1 *Let* $[AB]$ *be a line segment. The point* M *is called the middle of* $[AB]$ *if* $\overrightarrow{MA} + \overrightarrow{MB} = 0$, *see Figure 6.1.*

Figure 6.1: The midpoint M of the segment AB.

Exercise 6.1.2 *Show that if* M *and* M' *are two points such that*

$$\overrightarrow{MA} + \overrightarrow{MB} = \overrightarrow{M'A} + \overrightarrow{M'B} = 0,$$

then $M = M'$. *In other words, the midpoint of a segment is unique.*

Definition 6.1.3 *Let ABC be a triangle and A', B', C' be the midpoints of the sides BC, AC and AB. The line segments $[AA']$, $[BB']$ and $[CC']$ are called the medians of the triangle ABC.*

The following result is very useful and it is a consequence of the parallelogram rule:

Lemma 6.1.4 *Let $[AA']$ be the median of triangle ABC. Then*

$$\overrightarrow{AA'} = \frac{1}{2}\left(\overrightarrow{AB} + \overrightarrow{AC}\right).$$

Proof: Let D be a point in the plane such that $ABDC$ (in this order) is a parallelogram. The parallelogram rule yields

$$\overrightarrow{AB} + \overrightarrow{AC} = \overrightarrow{AD} = 2\overrightarrow{AA'},$$

since A' is the midpoint of diagonals $[AD]$ and $[BC]$. See Figure 6.2. ∎

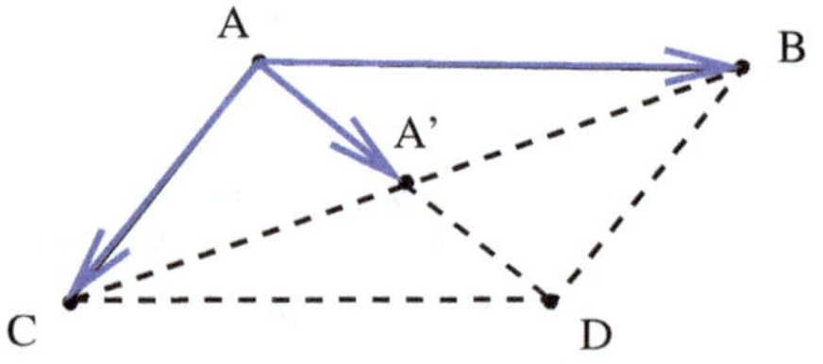

Figure 6.2: For Lemma 6.1.4.

The following result is called the *polygon rule*:

Lemma 6.1.5 *Let $A_1, A_2, A_3, \cdots, A_n$ be n points in the plane (space). Then*

$$\overrightarrow{A_1A_n} = \overrightarrow{A_1A_2} + \overrightarrow{A_2A_3} + \overrightarrow{A_3A_4} + \cdots + \overrightarrow{A_{n-1}A_n}.$$

Proof: Iterating the triangle rule, yields

$$\begin{aligned}
\overrightarrow{A_1A_2} + \overrightarrow{A_2A_3} &= \overrightarrow{A_1A_3}\\
\overrightarrow{A_1A_3} + \overrightarrow{A_3A_4} &= \overrightarrow{A_1A_4}\\
\overrightarrow{A_1A_4} + \overrightarrow{A_4A_5} &= \overrightarrow{A_1A_5}\\
\cdots\cdots\cdots\cdots &= \cdots\cdots\\
\overrightarrow{A_1A_{n-2}} + \overrightarrow{A_{n-2}A_{n-1}} &= \overrightarrow{A_1A_{n-1}}\\
\overrightarrow{A_1A_{n-1}} + \overrightarrow{A_{n-1}A_n} &= \overrightarrow{A_1A_n}.
\end{aligned}$$

Adding and cancelling in pairs yields the desired result.

Another shorter proof variant is as follows. We note that

$$\overrightarrow{A_1A_2} + \overrightarrow{A_2A_3} + \overrightarrow{A_3A_4} + \cdots + \overrightarrow{A_nA_1} = 0,$$

see Figure 6.3. Then solving for $\overrightarrow{A_1A_n} = -\overrightarrow{A_nA_1}$ we also obtain the result. ■

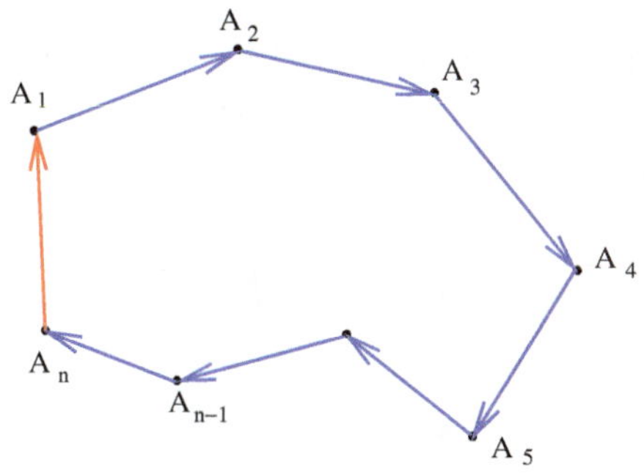

Figure 6.3: Polygonal line.

We say the vectors $\vec{u}$, $\vec{v}$, $\vec{w}$ form a triangle if there is a triangle ABC such that $\{\vec{u}, \vec{v}, \vec{w}\} = \{\overrightarrow{AB}, \overrightarrow{BC}, \overrightarrow{CA}\}$.

Proposition 6.1.6 *The vectors $\vec{u}, \vec{v}, \vec{w}$ form a triangle if and only if $\vec{u} + \vec{v} + \vec{w} = 0$.*

Proof: "$\Longrightarrow$" Let $\vec{u} = \overrightarrow{AB}$, $\vec{v} = \overrightarrow{BC}$, $\vec{w} = \overrightarrow{CA}$. It is obvious that

$$\overrightarrow{AB} + \overrightarrow{BC} + \overrightarrow{CA} = \overrightarrow{AC} + \overrightarrow{CA} = \overrightarrow{AC} - \overrightarrow{AC} = 0,$$

see Figure 6.4.

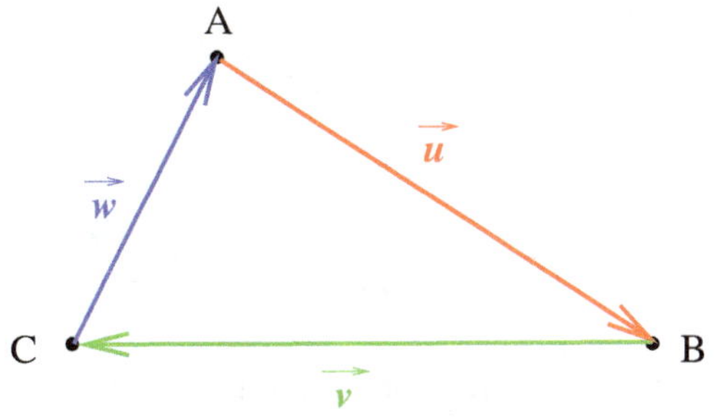

Figure 6.4: Vectors forming a triangle.

" $\Longleftarrow$ " Let A be a fixed point in the plane $\mathcal{P}$. Given the vector $\overrightarrow{u}$, there is a unique point $B \in \mathcal{P}$ such that $\overrightarrow{u} = \overrightarrow{AB}$. Choose the points C, $D \in \mathcal{P}$ such that $\overrightarrow{v} = \overrightarrow{BC}$ and $\overrightarrow{w} = \overrightarrow{CD}$. We need to show that $A = D$. The polygon rule yields

$$\overrightarrow{AD} = \overrightarrow{AB} + \overrightarrow{BC} + \overrightarrow{CD} = \overrightarrow{u} + \overrightarrow{v} + \overrightarrow{w} = 0,$$

and hence $A = D$. See Figure 6.5. ■

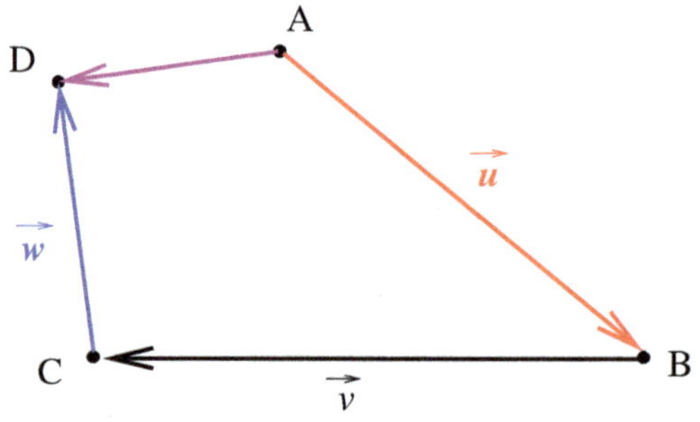

Figure 6.5: For Proposition 6.1.6.

Solved Exercise 6.1.7 *Let ABC be a triangle and let A', B', C' be the midpoints of the sides $[BC]$, $[CA]$ and $[AB]$, see Figure 6.6. Show that with the line segments $[AA']$, $[BB']$, $[CC']$ we can construct a triangle.*

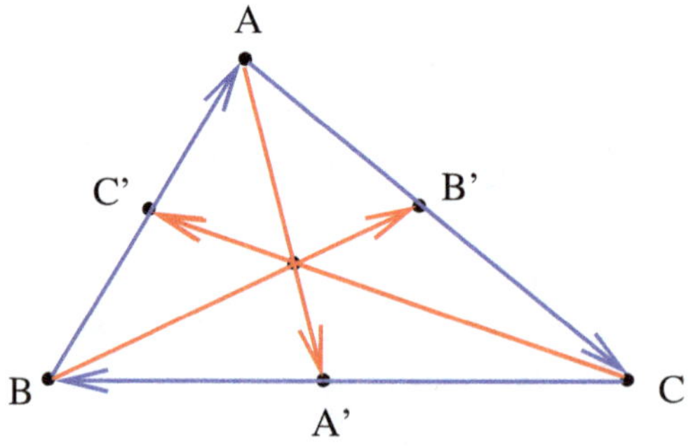

Figure 6.6: For Proposition 6.1.6.

Proof: We need to show that $\overrightarrow{AA'} + \overrightarrow{BB'} + \overrightarrow{CC'} = 0$. Using Lemma 6.1.4

we have

$$\begin{aligned}\overrightarrow{AA'} &= \frac{1}{2}\left(\overrightarrow{AB}+\overrightarrow{AC}\right),\\ \overrightarrow{BB'} &= \frac{1}{2}\left(\overrightarrow{BC}+\overrightarrow{BA}\right), \text{ and}\\ \overrightarrow{CC'} &= \frac{1}{2}\left(\overrightarrow{CA}+\overrightarrow{CB}\right).\end{aligned}$$

Adding, yields

$$\overrightarrow{AA'}+\overrightarrow{BB'}+\overrightarrow{CC'} = \frac{1}{2}\left(\overrightarrow{AB}+\overrightarrow{BA}+\overrightarrow{AC}+\overrightarrow{CA}+\overrightarrow{BC}+\overrightarrow{CB}\right) = 0.$$

Applying Proposition 6.1.6, we obtain the desired result. ■

Remark 6.1.8 (*i*) The previous result implies that with the medians of any triangle we can always construct another triangle.

(*ii*) The reciprocal is in general false, in the sense that there are three segments in the plane which can be considered as sides of a triangle, but for which the vectorial equation does not hold. To this end, we shall consider the following counterexample. Let A, B, C be the vertices of an equilateral triangle and M a point in the same plane, which is not the center of ΔABC. Then the sum $\overrightarrow{MA}+\overrightarrow{MB}+\overrightarrow{MC} \neq 0$, but the line segments MA, MB and MC are the length of a triangle by Pompeiu's theorem.

The following result is a generalization of Lemma 6.1.4:

Lemma 6.1.9 *Let ABC be a triangle and $M \in BC$ such that $\overrightarrow{BM} = k\overrightarrow{MC}$. Then*

$$\overrightarrow{AM} = \frac{\overrightarrow{AB}+k\overrightarrow{AC}}{k+1}.$$

Proof: We have the following equivalences

$$\begin{aligned}\overrightarrow{BM} &= k\overrightarrow{MC} \Longleftrightarrow\\ -\overrightarrow{AB}+\overrightarrow{AM} &= k\left(\overrightarrow{AC}-\overrightarrow{AM}\right) \Longleftrightarrow\\ (k+1)\overrightarrow{AM} &= k\overrightarrow{AC}+\overrightarrow{AB} \Longleftrightarrow\\ \overrightarrow{AM} &= \frac{\overrightarrow{AB}+k\overrightarrow{AC}}{k+1},\end{aligned}$$

where we have used the triangle rule. See Figure 6.7. ■

The previous result can be stated by saying that $\overrightarrow{AM}$ is a convex combination of $\overrightarrow{AB}$ and $\overrightarrow{AC}$, with weights $\frac{1}{k+1}$ and $1-\frac{1}{k+1}$, respectively.

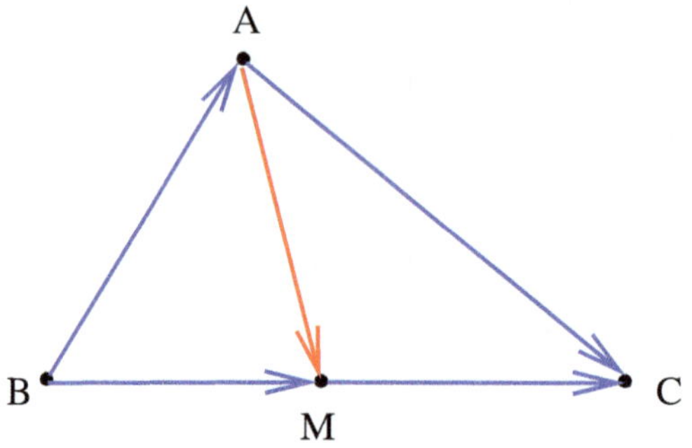

Figure 6.7: For Lemma 6.1.9.

Solved Exercise 6.1.10 *Consider an equilateral triangle* ΔABC. *Choose* $E \in AB$ *and* $D \in AC$ *such that* $|AE| = 2|BE|$ *and* $|DC| = 2|AD|$. *Let* $F = CE \cap BD$. *Show that* $\widehat{AFC} = 90°$. *See Figure 6.8.*

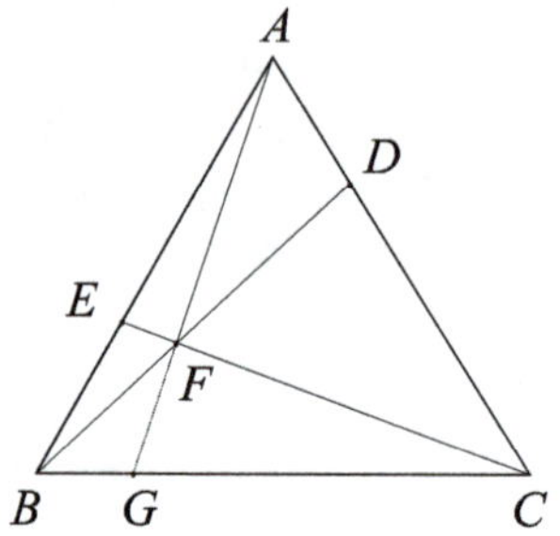

Figure 6.8: AF is perpendicular on FC.

Proof: Let $G = AF \cap BC$. Since AG, BD and CE intersect at F, by Ceva's theorem, Theorem 2.32.28, we have

$$\frac{|AE|}{|EB|}\,\frac{|BG|}{|GC|}\,\frac{|CD|}{|DA|} = 1,$$

which implies $\dfrac{|BG|}{|GC|} = \dfrac{1}{4}$. Therefore we have

$$\overrightarrow{BG} = \frac{1}{4}\overrightarrow{GC}, \quad \overrightarrow{CD} = 2\overrightarrow{DA}, \quad \overrightarrow{AE} = 2\overrightarrow{EB}.$$

Then by Lemma 6.1.9 we have

$$\overrightarrow{AG} = \frac{\overrightarrow{AB} + \frac{1}{4}\overrightarrow{AC}}{5/4}, \quad \overrightarrow{BD} = \frac{\overrightarrow{BC} + 2\overrightarrow{BA}}{3}, \quad \overrightarrow{CE} = \frac{\overrightarrow{CA} + 2\overrightarrow{CB}}{3}.$$

In order to prove $AF \perp FC$ it suffices to show $\overrightarrow{AG} \cdot \overrightarrow{CE} = 0$. Using the aforementioned relations we have

$$\begin{aligned}
\overrightarrow{AG} \cdot \overrightarrow{CE} &= \frac{4}{15}\left(\overrightarrow{AB} + \frac{1}{4}\overrightarrow{AC}\right) \cdot \left(\overrightarrow{CA} + 2\overrightarrow{CB}\right) \\
&= \frac{4}{15}\left[\overrightarrow{AB} \cdot \overrightarrow{CA} + 2\overrightarrow{AB} \cdot \overrightarrow{CB} + \frac{1}{4}\overrightarrow{AC} \cdot \overrightarrow{CA} + \frac{1}{2}\overrightarrow{AC} \cdot \overrightarrow{CB}\right] \\
&= \frac{4}{15}\left[-a^2 \cos 60^\circ + 2a^2 \cos 60^\circ - \frac{1}{4}a^2 - \frac{1}{2}a^2 \cos 60^\circ\right] \\
&= \frac{4a^2}{15}\left(-\frac{1}{2} + 1 - \frac{1}{4} - \frac{1}{4}\right) = 0.
\end{aligned}$$

Hence, $AF \perp CF$, i.e., $\widehat{AFC} = 90^\circ$. ■

6.2 Thales' theorem

Thales said there was no difference between life and death. "Why, then," said some one to him, "do not you die?" "Because," said he, "it does make no difference."

A simple application of the triangle identity is the well-known Thales' theorem:

Theorem 6.2.1 (Thales' theorem) *Let ABC be a triangle and let $B' \in AB$ and $C' \in AC$ such that $\overrightarrow{AB'} = \lambda\overrightarrow{AB}$, $\overrightarrow{AC'} = \lambda\overrightarrow{AC}$, with $\lambda \in \mathbb{R}$. Then $\overrightarrow{B'C'} = \lambda\overrightarrow{BC}$. In particular, we have $\dfrac{|B'C'|}{|BC|} = \lambda$ and $B'C' \| BC$.*

Proof: Using the triangle identity, we write

$$\begin{aligned}
\overrightarrow{B'C'} &= \overrightarrow{B'A} + \overrightarrow{AC'} = -\overrightarrow{AB'} + \overrightarrow{AC'} \\
&= -\lambda\overrightarrow{AB} + \lambda\overrightarrow{AC} \\
&= \lambda(\overrightarrow{AC} - \overrightarrow{AB}) = \lambda\overrightarrow{BC},
\end{aligned}$$

see Figure 6.9. ■

The following corollary is called the midpoint connector theorem:

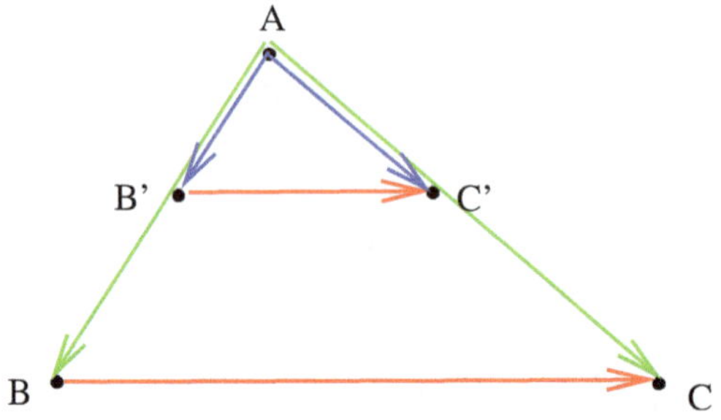

Figure 6.9: Thales' theorem.

Corollary 6.2.2 *Let ABC be a triangle and let B', C' be the midpoints of the segments $[AB]$ and $[AC]$, respectively. Then $B'C'||BC$ and $|B'C'| = \frac{1}{2}|BC|$.*

6.3 The centroid

> *Archimedes had stated, that given the force, any given weight might be moved; and even boasted that if there were another earth, by going into it he could remove this.*
>
> *Plutarch*

In this section, we shall define and characterize the *centroid* of a triangle. Vectorial calculus is very useful when dealing with centroids of systems of three or more points. The reason is the fact that the vectorial calculus comes from Physics, where forces are characterized by vectors, and centroids are centers of gravity.

Theorem 6.3.1 *The medians of a triangle intersect at a point G, called the centroid, situated at $2/3$ from the vertex and $1/3$ from the base on each median.*

Proof: Let $AA' \cap BB' = \{G\}$, where A' and B' are the midpoints of the segments $[BC]$ and $[AC]$, see Figure 6.10. Using that collinear vectors are proportional, there are two real numbers λ_A, $\lambda_B \in \mathbb{R}$ such that

$$\overrightarrow{AG} = \lambda_A \overrightarrow{AA'}, \qquad \overrightarrow{BG} = \lambda_B \overrightarrow{BB'}, \qquad \overrightarrow{GB'} = (1 - \lambda_B)\overrightarrow{BB'}.$$

Then the vectorial identity

$$\overrightarrow{GA} + \overrightarrow{AB'} = \overrightarrow{GB'}$$

can be written as

$$-\lambda_A\overrightarrow{AA'} + \frac{1}{2}\overrightarrow{AC} = (1-\lambda_B)\overrightarrow{BB'}.$$

Using Lemma 6.1.4 yields

$$\begin{aligned}
-\frac{\lambda_A}{2}\left(\overrightarrow{AB}+\overrightarrow{AC}\right) + \frac{1}{2}\overrightarrow{AC} = \frac{1}{2}(1-\lambda_B)\left(\overrightarrow{BA}+\overrightarrow{BC}\right) &\Longleftrightarrow \\
-\lambda_A\overrightarrow{AB} - \lambda_A\overrightarrow{AC} + \overrightarrow{AC} = (1-\lambda_B)\overrightarrow{BA} + (1-\lambda_B)\overrightarrow{BC} &\Longleftrightarrow \\
-\lambda_A\overrightarrow{AB} + (1-\lambda_A)\overrightarrow{AC} = -(1-\lambda_B)\overrightarrow{AB} + (1-\lambda_B)\left(\overrightarrow{AC}-\overrightarrow{AB}\right) &\Longleftrightarrow \\
-\lambda_A\overrightarrow{AB} + (1-\lambda_A)\overrightarrow{AC} = -2(1-\lambda_B)\overrightarrow{AB} + (1-\lambda_B)\overrightarrow{AC} &\Longleftrightarrow \\
0 = \left(2(1-\lambda_B)-\lambda_A\right)\overrightarrow{AB} + (1-\lambda_A-1+\lambda_B)\overrightarrow{AC}.&
\end{aligned}$$

Since the vectors $\overrightarrow{AB}$ and $\overrightarrow{AC}$ are linear independent, their coefficients must vanish. It follows that λ_A and λ_B satisfy the linear system

$$\begin{aligned}
\lambda_A &= 2(1-\lambda_B) \\
\lambda_A &= \lambda_B.
\end{aligned}$$

The solution is $\lambda_A = \lambda_B = 2/3$, and hence

$$\frac{|AG|}{|AA'|} = \frac{|BG|}{|BB'|} = \frac{2}{3}.$$

Let C' be the midpoint of the segment $[AB]$ and let $\{G'\} = AA' \cap CC'$. In a similar way we can show that

$$\frac{|AG'|}{|AA'|} = \frac{|CG'|}{|CC'|} = \frac{2}{3}.$$

Hence $G = G'$ (why?), and hence AA', BB', CC' intersect at G. ■

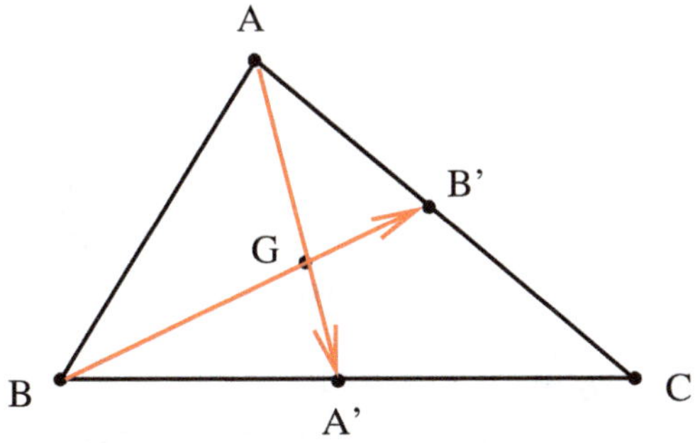

Figure 6.10: The centroid G of ΔABC.

In the following we shall deal with one of the most important vectorial properties of the centroid. The generalization of the next theorem will be taken as a definition for the centroid of systems with $n \geq 2$ points in space.

Theorem 6.3.2 *Let ABC be a triangle and let G be its centroid. Then*
1) $\overrightarrow{GA} + \overrightarrow{GB} + \overrightarrow{GC} = 0$.
2) *For any point M in the plane (or space) we have*

$$\overrightarrow{MA} + \overrightarrow{MB} + \overrightarrow{MC} = 3\overrightarrow{MG}.$$

3) *The point M is the centroid of triangle ABC if and only if*

$$\overrightarrow{MA} + \overrightarrow{MB} + \overrightarrow{MC} = 0.$$

Proof: (i) Using Theorem 6.3.1 and Lemma 6.1.4 yields

$$\begin{aligned}
\overrightarrow{GA} + \overrightarrow{GB} + \overrightarrow{GC} &= \frac{2}{3}\overrightarrow{A'A} + \frac{2}{3}\overrightarrow{B'B} + \frac{2}{3}\overrightarrow{C'C} = -\frac{2}{3}\left(\overrightarrow{AA'} + \overrightarrow{BB'} + \overrightarrow{CC'}\right) \\
&= -\frac{2}{3}\left[\frac{1}{2}\left(\overrightarrow{AB} + \overrightarrow{AC}\right) + \frac{1}{2}\left(\overrightarrow{BA} + \overrightarrow{BC}\right) + \frac{1}{2}\left(\overrightarrow{CA} + \overrightarrow{CB}\right)\right] \\
&= -\frac{1}{3}\left(\overrightarrow{AB} + \overrightarrow{AC} + \overrightarrow{BA} + \overrightarrow{BC} + \overrightarrow{CA} + \overrightarrow{CB}\right) \\
&= 0,
\end{aligned}$$

since the vectors cancel in pairs.

(ii) The triangle rule yields

$$\begin{aligned}
\overrightarrow{MA} &= \overrightarrow{MG} + \overrightarrow{GA} \\
\overrightarrow{MB} &= \overrightarrow{MG} + \overrightarrow{GB} \\
\overrightarrow{MC} &= \overrightarrow{MG} + \overrightarrow{GC}.
\end{aligned}$$

Adding, we obtain

$$\begin{aligned}
\overrightarrow{MA} + \overrightarrow{MB} + \overrightarrow{MC} &= 3\overrightarrow{MG} + \underbrace{(\overrightarrow{GA} + \overrightarrow{GB} + \overrightarrow{GC})}_{=0} \\
&= 3\overrightarrow{MG},
\end{aligned}$$

where we have used (i), see Figure 6.11.

(iii) It is a consequence of (ii). If $\overrightarrow{MA} + \overrightarrow{MB} + \overrightarrow{MC} = 0$, then $\overrightarrow{MG} = 0$ and hence $M = G$. ∎

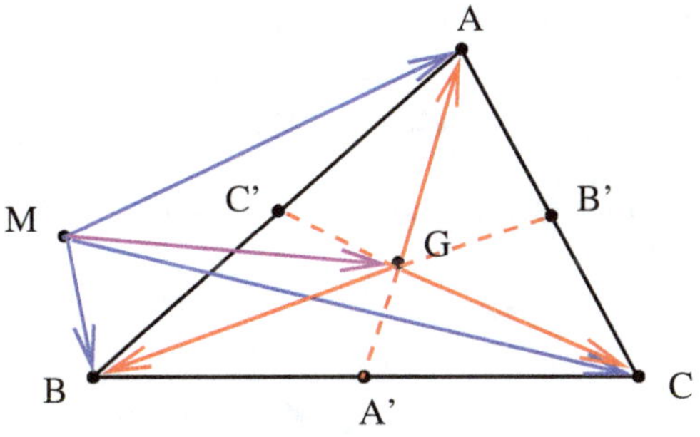

Figure 6.11: For the proof of Theorem 6.3.2.

Exercise 6.3.3 *Let AA', BB', CC' be the medians of triangle ABC and let M be a point in the plane. Show that*

$$3\overrightarrow{MA} + 2\overrightarrow{AA'} = 3\overrightarrow{MB} + 2\overrightarrow{BB'} = 3\overrightarrow{MC} + 2\overrightarrow{CC'} = 3\overrightarrow{MG}.$$

As an application, we have the following result:

Proposition 6.3.4 *Let ABC and $A'B'C'$ be two triangles with centroids G and G'.*
(i) Then we have

$$\overrightarrow{AA'} + \overrightarrow{BB'} + \overrightarrow{CC'} = 3\overrightarrow{GG'}.$$

(ii) Triangles ABC and $A'B'C'$ have the same centroid if and only if

$$\overrightarrow{AA'} + \overrightarrow{BB'} + \overrightarrow{CC'} = 0.$$

Proof: (i) Applying part (ii) of Theorem 6.3.2 yields

$$\begin{aligned} 3\overrightarrow{GG'} &= \overrightarrow{GA'} + \overrightarrow{GB'} + \overrightarrow{GC'} \\ &= \left(\overrightarrow{GA} + \overrightarrow{AA'}\right) + \left(\overrightarrow{GB} + \overrightarrow{BB'}\right) + \left(\overrightarrow{GC} + \overrightarrow{CC'}\right) \\ &= \left(\overrightarrow{GA} + \overrightarrow{GB} + \overrightarrow{GC}\right) + \left(\overrightarrow{AA'} + \overrightarrow{BB'} + \overrightarrow{CC'}\right). \end{aligned}$$

Using part (i) of Theorem 6.3.2 we obtain

$$3\overrightarrow{GG'} = \overrightarrow{AA'} + \overrightarrow{BB'} + \overrightarrow{CC'}.$$

(ii) It is a consequence of (i). ∎

As an application of the above proposition, we have:

Proposition 6.3.5 (Pappus' theorem) *Let A', B', C' be points on the sides BC, AC and AB of triangle ABC such that*

$$\overrightarrow{BA'} = \lambda\overrightarrow{A'C}, \qquad \overrightarrow{CB'} = \lambda\overrightarrow{B'A}, \qquad \overrightarrow{AC'} = \lambda\overrightarrow{C'B},$$

for some $\lambda \in \mathbb{R}$, see Figure 6.12. Then the triangles ABC and $A'B'C'$ have the same centroid.

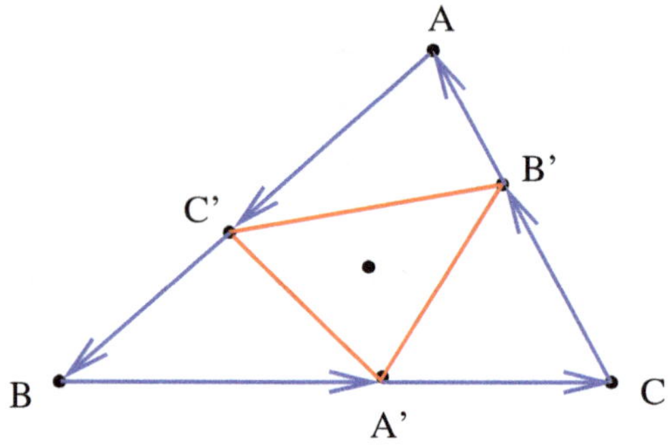

Figure 6.12: For Pappus' theorem.

Proof: Using Lemma 6.1.9 we have

$$\begin{aligned}
\overrightarrow{AA'} &= \frac{1}{\lambda+1}\left(\overrightarrow{AB}+\lambda\overrightarrow{AC}\right)\\
\overrightarrow{BB'} &= \frac{1}{\lambda+1}\left(\overrightarrow{BC}+\lambda\overrightarrow{BA}\right)\\
\overrightarrow{CC'} &= \frac{1}{\lambda+1}\left(\overrightarrow{CA}+\lambda\overrightarrow{CB}\right).
\end{aligned}$$

Adding and using Exercise 5.5.1, we obtain

$$\overrightarrow{AA'}+\overrightarrow{BB'}+\overrightarrow{CC'} = \frac{1}{\lambda+1}\left[\left(\overrightarrow{AB}+\overrightarrow{BC}+\overrightarrow{CA}\right)+\lambda\left(\overrightarrow{AC}+\overrightarrow{CB}+\overrightarrow{BA}\right)\right] = 0.$$

Using Proposition 6.3.4, part (*ii*), yields that triangles ABC and $A'B'C'$ have the same centroid. ■

It is worth to note that the converse is also true. In order to prove the converse, we need first the following lemma:

Lemma 6.3.6 *Given two distinct points A and B, the point M belongs to the line AB if and only if there are (unique) numbers x, $y \in \mathbb{R}$, with $x + y = 1$, such that for any point O (in the plane or space) we have*

$$\overrightarrow{OM} = x\overrightarrow{OA} + y\overrightarrow{OB}.$$

The numbers x and y do not depend on the point O. See Figure 6.13.

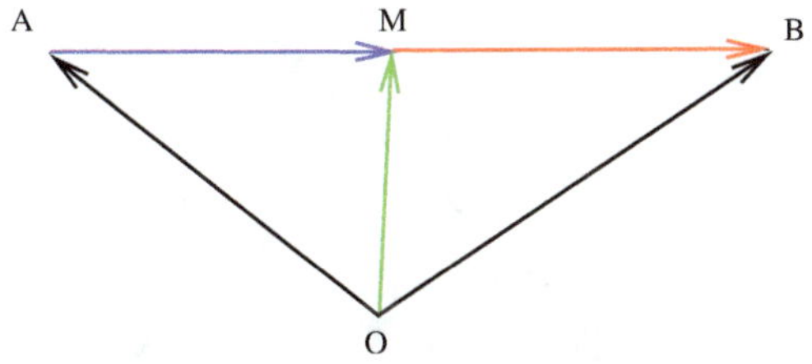

Figure 6.13: For Lemma 6.3.6.

Proof: " $\Longrightarrow$ " If $M \in AB$, then the vectors $\overrightarrow{AM}$ and $\overrightarrow{MB}$ are collinear, so there is a $k \in \mathbb{R}$ such that $\overrightarrow{AM} = k\overrightarrow{MB}$. For any point O the identity becomes

$$\overrightarrow{OM} - \overrightarrow{OA} = k\left(\overrightarrow{OB} - \overrightarrow{OM}\right)$$

and solving for $\overrightarrow{OM}$ yields

$$\overrightarrow{OM} = (1-k)\overrightarrow{OA} + k\overrightarrow{OB}.$$

Choose $x = 1 - k$ and $y = k$. Since k does not depend on the point O, the same will be true for the numbers x and y.

" $\Longleftarrow$ " Let O be a point such that $\overrightarrow{OM} = x\overrightarrow{OA} + y\overrightarrow{OB}$ with $x + y = 1$. Then

$$\begin{aligned} \overrightarrow{OM} &= x\overrightarrow{OA} + (1-x)\overrightarrow{OB} = x\left(\overrightarrow{OA} - \overrightarrow{OB}\right) + \overrightarrow{OB} \\ &= x\overrightarrow{BA} + \overrightarrow{OB} \Rightarrow \overrightarrow{OM} - \overrightarrow{OB} = x\overrightarrow{BA}, \end{aligned}$$

which becomes $\overrightarrow{BM} = x\overrightarrow{BA}$, *i.e.*, the vectors $\overrightarrow{MB}$ and $\overrightarrow{AB}$ are collinear, and hence M is collinear with A and B, *i.e.*, $M \in AB$. ■

The converse of Proposition 6.3.5 is given in the following:

Proposition 6.3.7 *Let A', B', C' be points on the sides (or on their extensions) BC, CA, AB of triangle ABC such that triangles $A'B'C'$ and ABC have the same centroid. Then there is a $\lambda \in \mathbb{R}$ such that*

$$\overrightarrow{BA'} = \lambda\overrightarrow{A'C}, \quad \overrightarrow{CB'} = \lambda\overrightarrow{B'A}, \quad \overrightarrow{AC'} = \lambda\overrightarrow{C'B}.$$

Proof: Let M be an arbitrary fixed point in the plane, see Figure 6.14. Set $\overrightarrow{a} = \overrightarrow{MA}$, $\overrightarrow{b} = \overrightarrow{MB}$, $\overrightarrow{c} = \overrightarrow{MC}$, $\overrightarrow{a'} = \overrightarrow{MA'}$, $\overrightarrow{b'} = \overrightarrow{MB'}$, $\overrightarrow{c'} = \overrightarrow{MC'}$. Since (B, A', C), (C, B', A), (A, C', B) are collinear, using Lemma 6.3.6, there are

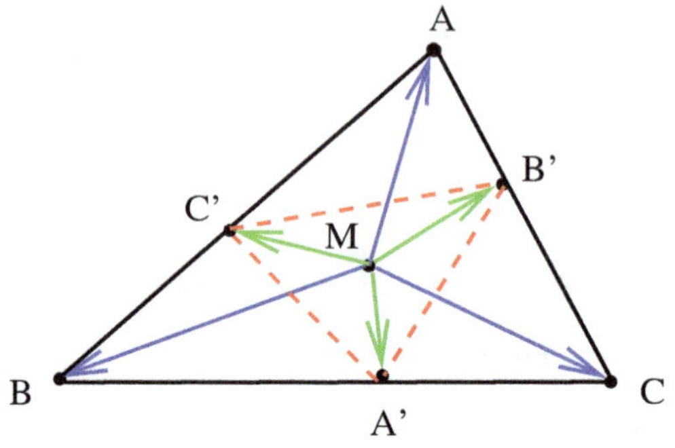

Figure 6.14: For Proposition 6.3.7.

$k_1, k_2, k_3 \in \mathbb{R}$ such that

$$\begin{aligned}\overrightarrow{a'} &= k_1\overrightarrow{b} + (1-k_1)\overrightarrow{c},\\ \overrightarrow{b'} &= k_2\overrightarrow{c} + (1-k_2)\overrightarrow{a}, \text{and}\\ \overrightarrow{c'} &= k_3\overrightarrow{a} + (1-k_3)\overrightarrow{b}.\end{aligned}$$

Summing, we obtain

$$\overrightarrow{a'} + \overrightarrow{b'} + \overrightarrow{c'} = (1+k_3-k_2)\overrightarrow{a} + (1+k_1-k_3)\overrightarrow{b} + (1+k_2-k_1)\overrightarrow{c}.$$

Applying Proposition 6.3.4, part (i), triangles ABC and $A'B'C'$ have the same centroid if and only if

$$\overrightarrow{a} + \overrightarrow{b} + \overrightarrow{c} = \overrightarrow{a'} + \overrightarrow{b'} + \overrightarrow{c'},$$

which becomes

$$(k_3-k_2)\overrightarrow{a} + (k_1-k_3)\overrightarrow{b} + (k_2-k_1)\overrightarrow{c} = 0.$$

Since the vectors $\overrightarrow{a}$, $\overrightarrow{b}$, $\overrightarrow{c}$ are linearly independent, it follows that $k_1 = k_2 = k_3 = k$. Choosing λ such that $\dfrac{1}{\lambda+1} = k$ finishes the proof. ■

We recall the following concept.

Definition 6.3.8 *The triangle obtained by joining the midpoints of the sides of a given triangle is called the median triangle, see Figure 6.15.*

Corollary 6.3.9 *A triangle and its median triangle have the same centroid.*

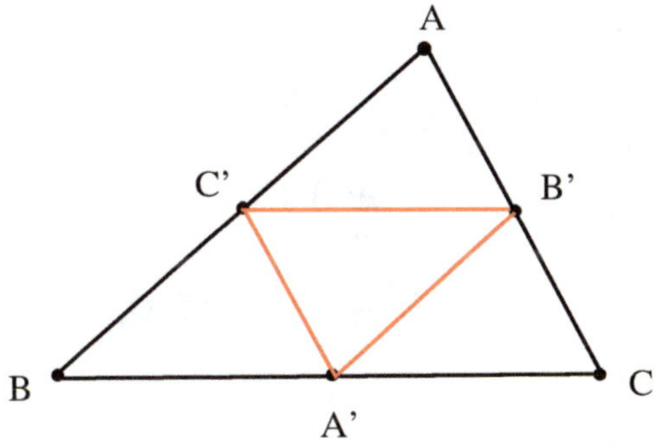

Figure 6.15: For Proposition 6.3.7.

Proof: Taking $\lambda = 1$ in Proposition 6.3.7, we obtain the desired result. ■

Solved Exercise 6.3.10 *Let ABC be a triangle and M be an arbitrary point.*
(i) Find the length of $\overrightarrow{MA} + \overrightarrow{MB} - 2\overrightarrow{MC}$.
(ii) Show that

$$m_c < \frac{|MA| + |MB|}{2} + |MC|,$$

where m_c is the length of the median corresponding to the vertex C.
(iii) Prove that

$$m_a + m_b + m_c < 2(|MA| + |MB| + |MC|),$$

where m_a, m_b, m_c are the lengths of the medians of triangle ABC.

Proof: (*i*) Since $\overrightarrow{MA} + \overrightarrow{MB} - 2\overrightarrow{MC}$ does not depend on M (see Exercise 5.6.18), we shall choose $M = C$. Then using Lemma 6.1.4 yields

$$\overrightarrow{MA} + \overrightarrow{MB} - 2\overrightarrow{MC} = \overrightarrow{CA} + \overrightarrow{CB} = 2\overrightarrow{CC'},$$

where C' is the midpoint of the line segment $[AB]$. Using $|\overrightarrow{CC'}| = m_c$, we obtain the desired length $|\overrightarrow{MA} + \overrightarrow{MB} - 2\overrightarrow{MC}| = m_c$.
(*ii*) The triangle inequality (see Exercise 5.5.1) yields

$$2m_c = 2|\overrightarrow{CC'}| = |\overrightarrow{MA} + \overrightarrow{MB} - 2\overrightarrow{MC}| < |\overrightarrow{MA}| + |\overrightarrow{MB}| + 2|\overrightarrow{MC}|.$$

Dividing by 2 leads to the desired result.

(*iii*) By cyclic permutations, we have

$$\begin{aligned} m_a &< \frac{|MB|+|MC|}{2} + |MA|, \\ m_b &< \frac{|MA|+|MC|}{2} + |MB|, \text{and} \\ m_c &< \frac{|MA|+|MB|}{2} + |MC|. \end{aligned}$$

Adding, yields

$$m_a + m_b + m_c < 2(|MA| + |MB| + |MC|).$$

■

Exercise 6.3.11 *Show that*

$$m_a + m_b + m_c < 6R,$$

where m_a, m_b, m_c are the length of the medians of triangle ABC and R is its circumradius.

6.4 Bimedians in quadrilaterals

Definition 6.4.1 *The line segments connecting the midpoints of opposite sides of a convex quadrilateral are called bimedians.*

The following result provides an expression of the bimedian in terms of the sides vectors.

Proposition 6.4.2 *Let $ABCD$ be a convex quadrilateral and let M, N be the midpoints of the sides $[AB]$ and $[DC]$. Then*

$$\overrightarrow{MN} = \frac{1}{2}\left(\overrightarrow{AD} + \overrightarrow{BC}\right) = \frac{1}{2}\left(\overrightarrow{AC} + \overrightarrow{BD}\right).$$

Proof: We shall refer to Figure 6.16. The polygon rule yields

$$\begin{aligned} \overrightarrow{MN} &= \overrightarrow{MA} + \overrightarrow{AD} + \overrightarrow{DN} = -\frac{1}{2}\overrightarrow{AB} + \overrightarrow{AD} + \frac{1}{2}\overrightarrow{DC} \\ &= \frac{1}{2}\left(2\overrightarrow{AD} + \overrightarrow{DC} - \overrightarrow{AB}\right) = \frac{1}{2}\left(\overrightarrow{AD} + \overrightarrow{AD} + \overrightarrow{DC} - \overrightarrow{AB}\right) \\ &= \frac{1}{2}\left(\overrightarrow{AD} + \overrightarrow{AC} - \overrightarrow{AB}\right) = \frac{1}{2}\left(\overrightarrow{AD} + \overrightarrow{BC}\right), \end{aligned}$$

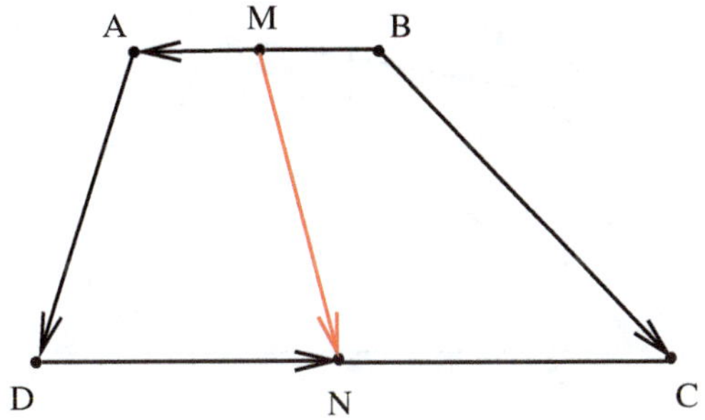

Figure 6.16: For Proposition 6.4.2.

which is the first identity. For the second identity, we apply the polygon rule twice

$$\begin{aligned}\overrightarrow{MN} &= \overrightarrow{MB}+\overrightarrow{BD}+\overrightarrow{DN}\\ \overrightarrow{MN} &= \overrightarrow{MA}+\overrightarrow{AC}+\overrightarrow{CN}\end{aligned}$$

and then add to obtain

$$2\overrightarrow{MN} = \left(\overrightarrow{MB}+\overrightarrow{MA}\right)+\overrightarrow{BD}+\overrightarrow{AC}+\left(\overrightarrow{DN}+\overrightarrow{CN}\right).$$

Using that M and N are midpoints, the first and the last parentheses vanish. Dividing by 2 yields the desired result. ∎

Remark 6.4.3 The above result will be used for finding the length of the bimedians of a quadrilateral in terms of its sides and the angle of two opposite sides.

We shall prove next a similar result regarding the line segment connecting the midpoints of the diagonals:[1]

Proposition 6.4.4 *Let $ABCD$ be a convex quadrilateral. Denote by O_1 and O_2 the midpoints of the diagonals $[AC]$ and $[BD]$, respectively. Then*

$$\overrightarrow{O_1O_2} = \frac{1}{2}\left(\overrightarrow{AB}-\overrightarrow{DC}\right).$$

[1]Sometimes this is considered as the third bimedian of the quadrilateral.

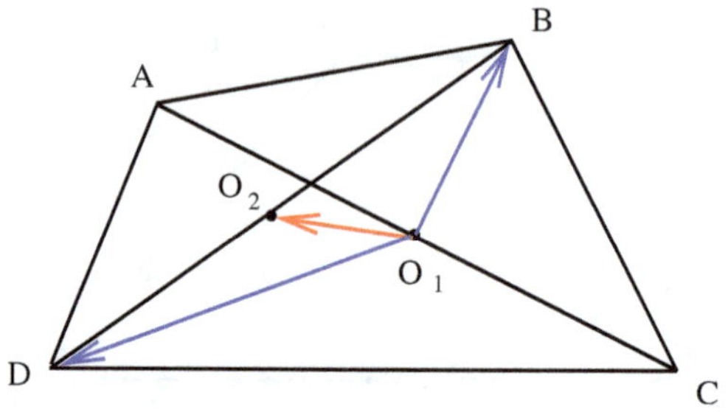

Figure 6.17: For Proposition 6.4.4.

Proof: We refer to Figure 6.17. Lemma 6.1.4 applied in triangle DBO_1 yields

$$\begin{aligned}\overrightarrow{O_1O_2} &= \frac{1}{2}\left(\overrightarrow{O_1B}+\overrightarrow{O_1D}\right) = -\frac{1}{2}\left(\overrightarrow{BO_1}+\overrightarrow{DO_1}\right)\\ &= -\frac{1}{2}\left[\frac{1}{2}\left(\overrightarrow{BA}+\overrightarrow{BC}\right)+\frac{1}{2}\left(\overrightarrow{DA}+\overrightarrow{DC}\right)\right]\\ &= -\frac{1}{4}\left[\overrightarrow{BA}+\overrightarrow{BC}+\overrightarrow{DA}+\overrightarrow{DC}\right]\\ &= \frac{1}{4}\left[\overrightarrow{AB}-\overrightarrow{BC}+\overrightarrow{AD}-\overrightarrow{DC}\right].\end{aligned}$$

Adding $\frac{1}{4}[\overrightarrow{AB}+\overrightarrow{BC}+\overrightarrow{CD}+\overrightarrow{DA}]=0$, yields

$$\begin{aligned}\overrightarrow{O_1O_2} &= \frac{1}{4}\left[\overrightarrow{AB}-\overrightarrow{BC}+\overrightarrow{AD}-\overrightarrow{DC}+\overrightarrow{AB}+\overrightarrow{BC}+\overrightarrow{CD}+\overrightarrow{DA}\right]\\ &= \frac{1}{4}\left[\overrightarrow{AB}+\overrightarrow{AD}+\overrightarrow{CD}+\overrightarrow{AB}+\overrightarrow{CD}-\overrightarrow{AD}\right]\\ &= \frac{1}{4}\left[2\left(\overrightarrow{AB}+\overrightarrow{CD}\right)\right] = \frac{1}{2}\left(\overrightarrow{AB}+\overrightarrow{CD}\right)\\ &= \frac{1}{2}\left(\overrightarrow{AB}-\overrightarrow{DC}\right).\end{aligned}$$

■

Consider a quadrilateral with unit masses concentrated at vertices and the sides considered as massless rods. Its center of mass can be defined vectorially as follows.

Definition 6.4.5 *The point G is called the centroid of the quadrilateral $ABCD$ if*

$$\overrightarrow{GA}+\overrightarrow{GB}+\overrightarrow{GC}+\overrightarrow{GD}=0. \tag{6.4.1}$$

The following result proves the uniqueness of the centroid:

Proposition 6.4.6 *If G is a point such that (6.4.1) holds, then G is unique.*

Proof: Suppose there are two points G and G' which verify relation (6.4.1). Subtracting the relations

$$\begin{aligned} \overrightarrow{GA} + \overrightarrow{GB} + \overrightarrow{GC} + \overrightarrow{GD} &= 0 \\ \overrightarrow{G'A} + \overrightarrow{G'B} + \overrightarrow{G'C} + \overrightarrow{G'D} &= 0, \end{aligned}$$

and using the triangle identity, yields

$$\begin{aligned} 0 &= \left(\overrightarrow{GA} - \overrightarrow{G'A}\right) + \left(\overrightarrow{GB} - \overrightarrow{G'B}\right) + \left(\overrightarrow{GC} - \overrightarrow{G'C}\right) + \left(\overrightarrow{GD} - \overrightarrow{G'D}\right) \\ &= \left(\overrightarrow{GA} + \overrightarrow{AG'}\right) + \left(\overrightarrow{GB} + \overrightarrow{BG'}\right) + \left(\overrightarrow{GC} + \overrightarrow{CG'}\right) + \left(\overrightarrow{GD} + \overrightarrow{DG'}\right) \\ &= 4\overrightarrow{GG'}, \end{aligned}$$

and hence $G = G'$. ■

The next theorem proves the existence of the centroid:

Theorem 6.4.7 *Let M, N be the midpoints of the sides $[AB]$, $[DC]$ of a convex quadrilateral $ABCD$. Then the midpoint of the bimedian $[MN]$ is the centroid of $ABCD$.*

Proof: Let X be the midpoint of the bimedian $[MN]$. Lemma 6.1.4 yields

$$\begin{aligned} 0 &= \overrightarrow{XM} + \overrightarrow{XN} \\ &= \frac{1}{2}\left(\overrightarrow{XA} + \overrightarrow{XB}\right) + \frac{1}{2}\left(\overrightarrow{XD} + \overrightarrow{XC}\right) \\ &= \frac{1}{2}\left(\overrightarrow{XA} + \overrightarrow{XB} + \overrightarrow{XC} + \overrightarrow{XD}\right), \end{aligned}$$

and hence X verifies (6.4.1) and then $X = G$. See Figure 6.18. ■

Corollary 6.4.8 *The bimedians of a convex quadrilateral intersect at a point G, which is the centroid of $ABCD$. Furthermore, G is the midpoint of each bimedian.*

Proof: We have shown that the midpoint of the bimedian $[MN]$ is the centroid G. In a similar way we can show that the midpoint of the other bimedian is the centroid. Since the centroid is unique, the bimedians bisect. ■

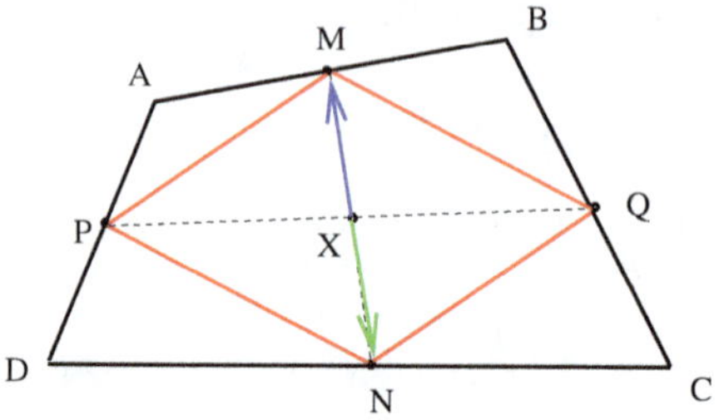

Figure 6.18: For Theorem 6.4.7.

Exercise 6.4.9 *Let $ABCD$ be a quadrilateral with centroid G. Show that for any point Y we have*

$$\overrightarrow{YA} + \overrightarrow{YB} + \overrightarrow{YC} + \overrightarrow{YD} = 4\overrightarrow{YG}.$$

Exercise 6.4.10 *Show that the line defined by the midpoints of the diagonals of a convex quadrilateral passes through its centroid.*

Chapter 7

The Inner Product and its Applications

There is no great genius without a tincture of madness.
Seneca

This chapter deals with the inner product of two vectors and its applications to geometry. The main application of the inner product deals with segment lengths and perpendicular lines in a very quick and elegant manner.

7.1 Orthogonal projections

Let $\overrightarrow{a}$ and $\overrightarrow{b}$ be two vectors in the plane. Define the *orthogonal projection* of the vector $\overrightarrow{a}$ on the vector $\overrightarrow{b}$ as

$$pr_{\overrightarrow{b}}\overrightarrow{a} = |\overrightarrow{a}|\cos(\widehat{\overrightarrow{a}, \overrightarrow{b}}),$$

see Figure 7.1.

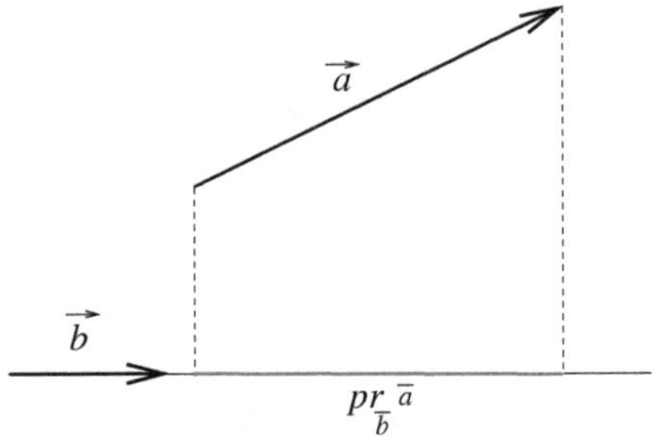

Figure 7.1: The orthogonal projection of a vector.

Proposition 7.1.1 *The function* $pr_{\overrightarrow{b}}$ *is linear, i.e., for any* $\overrightarrow{a}$, $\overrightarrow{c}$ *vectors in the plane, we have*

1) $pr_{\overrightarrow{b}}(\overrightarrow{a}+\overrightarrow{c}) = pr_{\overrightarrow{b}}\overrightarrow{a} + pr_{\overrightarrow{b}}\overrightarrow{c}$;

2) $pr_{\overrightarrow{b}}(t\overrightarrow{a}) = t\, pr_{\overrightarrow{b}}\overrightarrow{a}$, $\forall t \in \mathbb{R}$.

Proof: 1) Let $O \in \mathcal{P}$ be the origin, see Figure 7.2(a). Then there are unique points A, $B \in \mathcal{P}$ such that $\overrightarrow{a} = \overrightarrow{OA}$, $\overrightarrow{b} = \overrightarrow{OB}$. Let A', $B' \in OB$ such that $AA' \perp OB$ and $CC' \perp OB$. Then

$$|OA'| = |OA| \cos \widehat{AOA'} = |\overrightarrow{a}| \cos(\widehat{\overrightarrow{a}, \overrightarrow{b}}) = pr_{\overrightarrow{b}}\overrightarrow{a}.$$

In a similar way we can show that $|A'C'| = pr_{\overrightarrow{b}}\overrightarrow{c}$. Using the triangle rule, the obvious relation $|OC'| = |OA'| + |A'C'|$ becomes

$$pr_{\overrightarrow{b}}(\overrightarrow{a}+\overrightarrow{c}) = pr_{\overrightarrow{b}}\overrightarrow{a} + pr_{\overrightarrow{b}}\overrightarrow{c}.$$

2) Let A', B' be the orthogonal projections of A, B onto the support line of the vector $\overrightarrow{b}$, which passes through O. Using the similar triangles $OAA' \sim OBB'$ (why?) yields

$$|OB'| = t|OA'| \qquad \text{or} \qquad pr_{\overrightarrow{b}}(t\overrightarrow{a}) = t pr_{\overrightarrow{b}}\overrightarrow{a},$$

see Figure 7.2(b). ■

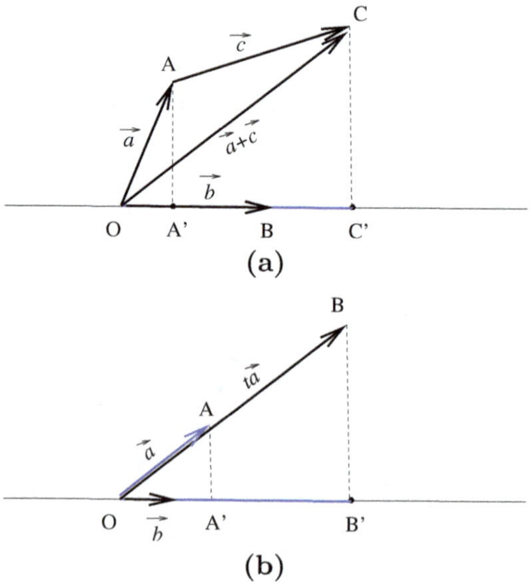

Figure 7.2: For Proposition 7.1.1.

7.2 Definitions and basic properties

In this section we shall introduce a *metric structure* on the set of vectors. This assigns to each pair of vectors a real number. In particular, when the vectors in the pair coincide, the number is the length of the vector. This metric structure has to be compatible with the Euclidean metric, *i.e.*, the length of a vector given by the inner product has to be the same as the length given in the Euclidean geometry. In order to accomplish this, we need to define the *inner product* or the *scalar product* of two vectors.

Definition 7.2.1 *Let* $\overrightarrow{u}$ *and* $\overrightarrow{v}$ *be two vectors. If* $\overrightarrow{u} \neq 0$ *and* $\overrightarrow{v} \neq 0$, *the real number*

$$\overrightarrow{u} \cdot \overrightarrow{v} = |\overrightarrow{u}| \cdot |\overrightarrow{v}| \cos(\overrightarrow{u}, \overrightarrow{v}),$$

is called the inner product of the vectors $\overrightarrow{u}$ *and* $\overrightarrow{v}$. *If* $\overrightarrow{u} = 0$ *or* $\overrightarrow{v} = 0$ *we define* $\overrightarrow{u} \cdot \overrightarrow{v} = 0$.

Exercise 7.2.2 *Let B be the orthogonal projection of A on OC, see Figure 7.3. Show that*

$$\overrightarrow{OA} \cdot \overrightarrow{OC} = |OC| \cdot |OB|.$$

The following result deals with the properties of the inner product:

Proposition 7.2.3 *For any vectors* $\overrightarrow{u}$, $\overrightarrow{v}$ *and* $\overrightarrow{w}$ *we have:*
1) $\overrightarrow{u} \cdot \overrightarrow{v} = \overrightarrow{v} \cdot \overrightarrow{u}$, *(commutativity)*
2) $\lambda(\overrightarrow{u} \cdot \overrightarrow{v}) = (\lambda\overrightarrow{u}) \cdot \overrightarrow{v} = \overrightarrow{u} \cdot (\lambda\overrightarrow{v})$, $\forall\lambda \in \mathbb{R}$, *(homogeneity)*
3) $\overrightarrow{u} \cdot (\overrightarrow{v} + \overrightarrow{w}) = \overrightarrow{u} \cdot \overrightarrow{v} + \overrightarrow{u} \cdot \overrightarrow{w}$, *(distributivity)*
4) $\overrightarrow{u} \cdot \overrightarrow{u} = |u|^2 \geq 0$, *(positive definite)*
5) $\overrightarrow{u} \perp \overrightarrow{v}$ *if and only if* $\overrightarrow{u} \cdot \overrightarrow{v} = 0$ *for nonzero vectors* $\overrightarrow{u}$, $\overrightarrow{v}$.

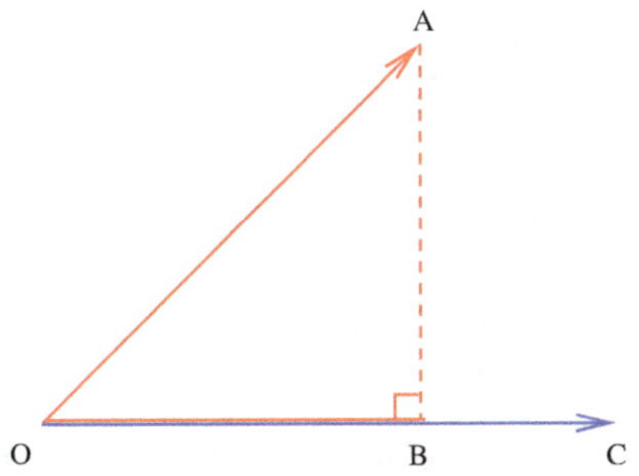

Figure 7.3: Orthogonal projection.

Proof: 1) If $\overrightarrow{u} = 0$ or $\overrightarrow{v} = 0$, the identity is obvious. If $\overrightarrow{u} \neq 0$, $\overrightarrow{v} \neq 0$, then

$$\overrightarrow{u} \cdot \overrightarrow{v} = |\overrightarrow{u}|\,|\overrightarrow{v}| \cos(\widehat{\overrightarrow{u}, \overrightarrow{v}}) = |\overrightarrow{v}|\,|\overrightarrow{u}| \cos(\widehat{\overrightarrow{v}, \overrightarrow{u}}) = \overrightarrow{v} \cdot \overrightarrow{u}.$$

2) Using that

$$(\text{sgn}\lambda)\cos(\widehat{\overrightarrow{u}, \overrightarrow{v}}) = \cos(\widehat{\lambda\overrightarrow{u}, \overrightarrow{v}}), \qquad \lambda = (\text{sgn}\lambda)|\lambda|, \qquad \forall\lambda \in \mathbb{R},$$

we have

$$\begin{aligned} \lambda(\overrightarrow{u} \cdot \overrightarrow{v}) &= \lambda|\overrightarrow{u}|\,|\overrightarrow{v}| \cos(\widehat{\overrightarrow{u}, \overrightarrow{v}}) \\ &= |\lambda\overrightarrow{u}|\,|\overrightarrow{v}| \cos(\widehat{\lambda\overrightarrow{u}, \overrightarrow{v}}) \\ &= (\lambda\overrightarrow{u}) \cdot \overrightarrow{v}. \end{aligned}$$

3) Let $\overrightarrow{b} = \dfrac{\overrightarrow{u}}{|\overrightarrow{u}|}$. Using the orthogonal projection's definition, we obtain

$$\begin{aligned} pr_{\overrightarrow{b}}\overrightarrow{v} &= |\overrightarrow{v}| \cos(\widehat{\overrightarrow{b}, \overrightarrow{v}}) = \overrightarrow{b} \cdot \overrightarrow{v} \\ pr_{\overrightarrow{b}}\overrightarrow{w} &= |\overrightarrow{w}| \cos(\widehat{\overrightarrow{b}, \overrightarrow{w}}) = \overrightarrow{b} \cdot \overrightarrow{w}. \end{aligned}$$

Adding,

$$pr_{\overrightarrow{b}}\overrightarrow{v} + pr_{\overrightarrow{b}}\overrightarrow{w} = \overrightarrow{b} \cdot \overrightarrow{v} + \overrightarrow{b} \cdot \overrightarrow{w}. \tag{7.2.1}$$

Using that $pr_{\overrightarrow{b}}$ is a linear function, (7.2.1) yields

$$\overrightarrow{b} \cdot \overrightarrow{v} + \overrightarrow{b} \cdot \overrightarrow{w} = pr_{\overrightarrow{b}}(\overrightarrow{v} + \overrightarrow{w}) = \overrightarrow{b} \cdot (\overrightarrow{v} + \overrightarrow{w}).$$

Multiplying by $|\overrightarrow{u}|$, and using $|\overrightarrow{u}|\overrightarrow{b} = \overrightarrow{u}$, yields

$$\overrightarrow{u} \cdot \overrightarrow{v} + \overrightarrow{u} \cdot \overrightarrow{w} = \overrightarrow{u} \cdot (\overrightarrow{v} + \overrightarrow{w}),$$

and the distributivity is proved.

4) From the definition of the inner product we have

$$\overrightarrow{u} \cdot \overrightarrow{u} = |\overrightarrow{u}|\,|\overrightarrow{u}| \cos 0 = |\overrightarrow{u}|^2 \geq 0.$$

5) Let $\overrightarrow{u}$, $\overrightarrow{v}$ be two non-zero vectors. Then $\overrightarrow{u} \perp \overrightarrow{v} \Longleftrightarrow \widehat{\overrightarrow{u}, \overrightarrow{v}} = 90^\circ \Longleftrightarrow \cos(\widehat{\overrightarrow{u}, \overrightarrow{v}}) = 0 \Longleftrightarrow \overrightarrow{u} \cdot \overrightarrow{v} = 0$. ■

Exercise 7.2.4 (Euler's relation, 1747) *If A, B, C, D are four collinear points, show that*

$$\overrightarrow{AB} \cdot \overrightarrow{CD} + \overrightarrow{BC} \cdot \overrightarrow{AD} + \overrightarrow{CA} \cdot \overrightarrow{BD} = 0.$$

Solved Exercise 7.2.5 *Let $\overrightarrow{u}$, $\overrightarrow{v}$, $\overrightarrow{w}$ be three vectors such that $\overrightarrow{u} + \overrightarrow{v} + \overrightarrow{w} = 0$. Show that*

$$\overrightarrow{u} \cdot \overrightarrow{v} + \overrightarrow{v} \cdot \overrightarrow{w} + \overrightarrow{w} \cdot \overrightarrow{u} = 0$$

if and only if $\overrightarrow{u} = \overrightarrow{v} = \overrightarrow{w} = 0$.

Proof: Using the properties of the inner product, we have

$$\begin{aligned} 0 &= |\overrightarrow{u} + \overrightarrow{v} + \overrightarrow{w}|^2 = (\overrightarrow{u} + \overrightarrow{v} + \overrightarrow{w}) \cdot (\overrightarrow{u} + \overrightarrow{v} + \overrightarrow{w}) \\ &= |\overrightarrow{u}|^2 + |\overrightarrow{v}|^2 + |\overrightarrow{w}|^2 + 2(\overrightarrow{u} \cdot \overrightarrow{v} + \overrightarrow{v} \cdot \overrightarrow{w} + \overrightarrow{w} \cdot \overrightarrow{u}). \end{aligned}$$

Then

$$\overrightarrow{u} \cdot \overrightarrow{v} + \overrightarrow{v} \cdot \overrightarrow{w} + \overrightarrow{w} \cdot \overrightarrow{u} = 0$$

if and only if $|\overrightarrow{u}|^2 + |\overrightarrow{v}|^2 + |\overrightarrow{w}|^2 = 0$, *i.e.*, $\overrightarrow{u} = \overrightarrow{v} = \overrightarrow{w} = 0$, where we used that $|\overrightarrow{u}| = 0 \iff \overrightarrow{u} = 0$. ■

Solved Exercise 7.2.6 *Let $ABCD$ be a quadrilateral. Show that*

$$\overrightarrow{AB} = \overrightarrow{DC} \quad \textit{if and only if} \quad \overrightarrow{AD} = \overrightarrow{BC}. \tag{7.2.2}$$

Proof: Relation (7.2.2) can be written as

$$\overrightarrow{AB} + \overrightarrow{CD} = 0 \quad \text{if and only if} \quad \overrightarrow{DA} + \overrightarrow{BC} = 0. \tag{7.2.3}$$

Using the polygon rule, we have

$$\begin{aligned} 0 &= \overrightarrow{AB} + \overrightarrow{BC} + \overrightarrow{CD} + \overrightarrow{DA} \\ &= (\overrightarrow{AB} + \overrightarrow{CD}) + (\overrightarrow{DA} + \overrightarrow{BC}), \end{aligned}$$

which yields (7.2.3), see Figure 7.4. ■

A parallelogram can be defined via vectors as follows:

Definition 7.2.7 *The quadrilateral $ABCD$ is called a parallelogram if $\overrightarrow{AB} = \overrightarrow{DC}$ or if $\overrightarrow{AB} + \overrightarrow{CD} = 0$.*

Exercise 7.2.8 *Show that in a parallelogram $ABCD$, we have*

$$|AB| = |DC|, \qquad |BC| = |AD|, \qquad AB \| DC, \qquad BC \| AD.$$

Figure 7.4: A parallelogram defined by vectors.

Definition 7.2.9 *A convex quadrilateral ABCD is called a rhombus if*

$$|AB| = |BC| = |CD| = |DA|.$$

In particular, every square is a rhombus.

Solved Exercise 7.2.10 *Show that any parallelogram with perpendicular diagonals is a rhombus.*

Proof: Let $ABCD$ be a parallelogram with $AC \perp DB$, see Figure 7.5. Then we have the following equivalencies

$$\begin{aligned}
\overrightarrow{AC} \cdot \overrightarrow{DB} &= 0 \iff \\
(\overrightarrow{AD} + \overrightarrow{DC}) \cdot (\overrightarrow{DC} + \overrightarrow{CB}) &= 0 \iff \\
\overrightarrow{AD} \cdot \overrightarrow{DC} + \overrightarrow{AD} \cdot \overrightarrow{CB} + \overrightarrow{DC} \cdot \overrightarrow{DC} + \overrightarrow{DC} \cdot \overrightarrow{CB} &= 0 \iff \\
\overrightarrow{AD} \cdot \overrightarrow{DC} + \overrightarrow{DC} \cdot \overrightarrow{CB} + \overrightarrow{AD} \cdot \overrightarrow{CB} + |DC|^2 &= 0 \iff \\
\overrightarrow{DC} \cdot \underbrace{(\overrightarrow{AD} + \overrightarrow{CB})}_{=0} - \overrightarrow{AD} \cdot \overrightarrow{BC} + |DC|^2 &= 0 \iff \\
|DC|^2 - |AD|^2 &= 0 \iff \\
(|DC| - |AD|)(|DC| + |AD|) &= 0 \iff \\
|AD| &= |DC|.
\end{aligned}$$

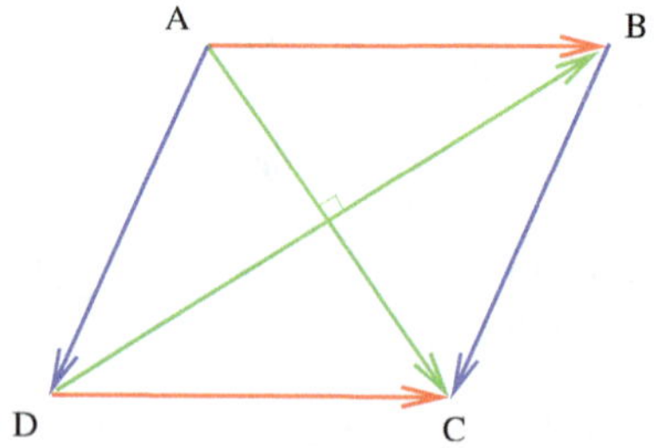

Figure 7.5: A rhombus.

Since the opposite sides in a parallelogram are equal, if follows that all sides are equal and hence $ABCD$ is a rhombus. ■

Solved Exercise 7.2.11 *Let $ABCD$ be a parallelogram and let E and F be the orthogonal projections of C on the lines AB and AD, respectively. Show that*

$$|AB|\cdot|AE|+|AD|\cdot|AF|=|AC|^2.$$

Proof: We have

$$\begin{aligned}
|AB|\cdot|AE|+|AD|\cdot|AF| &= \overrightarrow{AB}\cdot\overrightarrow{AE}+\overrightarrow{AD}\cdot\overrightarrow{AF}\\
&= \overrightarrow{AB}\cdot(\overrightarrow{AC}+\overrightarrow{CE})+\overrightarrow{AD}\cdot(\overrightarrow{AC}+\overrightarrow{CF})\\
&= \overrightarrow{AB}\cdot\overrightarrow{AC}+\underbrace{\overrightarrow{AB}\cdot\overrightarrow{CE}}_{=0}+\overrightarrow{AD}\cdot\overrightarrow{AC}+\underbrace{\overrightarrow{AD}\cdot\overrightarrow{CF}}_{=0}\\
&= \overrightarrow{AB}\cdot\overrightarrow{AC}+\overrightarrow{AD}\cdot\overrightarrow{AC}=(\overrightarrow{AB}+\overrightarrow{AD})\cdot\overrightarrow{AC}\\
&= \overrightarrow{AC}\cdot\overrightarrow{AC}=|AC|^2,
\end{aligned}$$

where we used that $AD\perp CF$, $AB\perp CE$ and Proposition 7.2.3, part 5). See Figure 7.6. ■

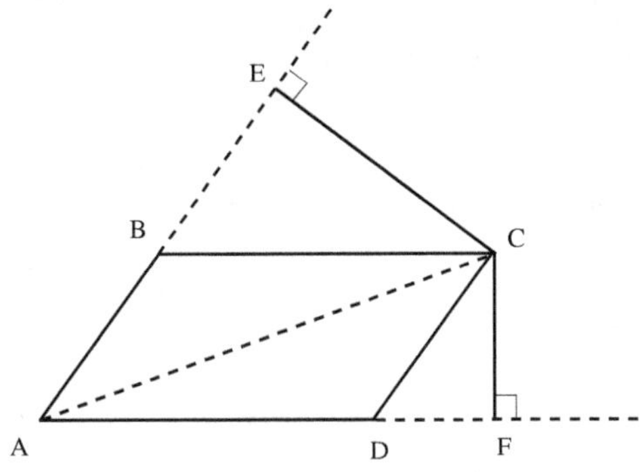

Figure 7.6: For Exercise 7.2.11.

7.3 Rotations

For any $\phi\in\mathbb{R}$ let R_ϕ be a transform, which maps vectors into vectors. For any vector $\overrightarrow{v}$ we define the vector $\overrightarrow{v'}=R_\phi\overrightarrow{v}$ such that $|\overrightarrow{v}|=|\overrightarrow{v'}|$, and the angle between the directions of $\overrightarrow{v}$ and $\overrightarrow{v'}$ is $\widehat{\delta_{\overrightarrow{v}},\delta_{\overrightarrow{v'}}}=\phi$, see Figure 7.7. The vector $R_\phi\overrightarrow{v}$ is called the *rotation* of the vector $\overrightarrow{v}$ with the angle ϕ.

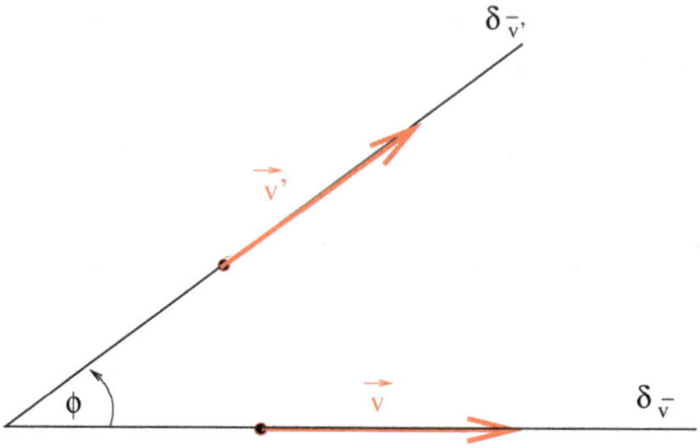

Figure 7.7: A vector rotation.

Exercise 7.3.1 *Show that for any two vectors $\overrightarrow{u}$, $\overrightarrow{v}$ we have $\widehat{R_\phi \overrightarrow{u}, R_\phi \overrightarrow{v}} = \widehat{\overrightarrow{u}, \overrightarrow{v}}$, i.e., the rotation preserves the angle.*

The rotational invariance of the inner product is given by the following result:

Lemma 7.3.2 *For any two given vectors $\overrightarrow{u}$ and $\overrightarrow{v}$ let $\overrightarrow{v'} = R_\phi \overrightarrow{v}$ and $\overrightarrow{u'} = R_\phi \overrightarrow{u}$ for some $\phi \in \mathbb{R}$. Then $\overrightarrow{u} \cdot \overrightarrow{v} = \overrightarrow{u'} \cdot \overrightarrow{v'}$.*

Proof: Using Exercise 7.3.1 and the definition of the rotation, we have

$$\overrightarrow{u} \cdot \overrightarrow{v} = |\overrightarrow{u}| \cdot |\overrightarrow{v}| \cos(\widehat{\overrightarrow{u}, \overrightarrow{v}}) = |\overrightarrow{u'}| \cdot |\overrightarrow{v'}| \cos(\widehat{\overrightarrow{u'}, \overrightarrow{v'}}) = \overrightarrow{u'} \cdot \overrightarrow{v'}.$$

■

Exercise 7.3.3 *Let $\overrightarrow{u}$, $\overrightarrow{v}$ be two vectors such that $|\overrightarrow{u}| = |\overrightarrow{v}|$. Show that there is $\phi \in \mathbb{R}$ such that $\overrightarrow{v} = R_\phi \overrightarrow{u}$.*

Exercise 7.3.4 *The rotations form a group, i.e., for any ϕ, $\phi_i \in \mathbb{R}$, with $i \in \{1, 2, 3\}$, we have*

i) $R_{\phi_1} \circ (R_{\phi_2} \circ R_{\phi_3}) = (R_{\phi_1} \circ R_{\phi_2}) \circ R_{\phi_3}$,

ii) $R_{\phi_1} \circ R_{\phi_2} = R_{\phi_1 + \phi_2}$,

iii) $R_\phi \circ R_{-\phi} = Id$, $R_0 = Id$,

iv) $R_{\phi_1} \circ R_{\phi_2} = R_{\phi_2} \circ R_{\phi_1}$,

where Id is the identical map defined by $Id(v) = v$, for any vector v.

7.4 Orthonormal systems

The following notion is fundamental for the rest of the vectorial approach of geometry.

Definition 7.4.1 *A basis $\{e_1, e_2, e_3\}$ in $\mathbb{R}^3$ is called orthonormal if $e_i \cdot e_j = \delta_{ij}$, where $\delta_{ij} = 0$ for $i \neq j$ and $\delta_{ii} = 1$. This means that $|e_1| = |e_2| = |e_3|$ and $e_1 \perp e_2$, $e_2 \perp e_3$, $e_1 \perp e_3$, see Figure 7.8.*

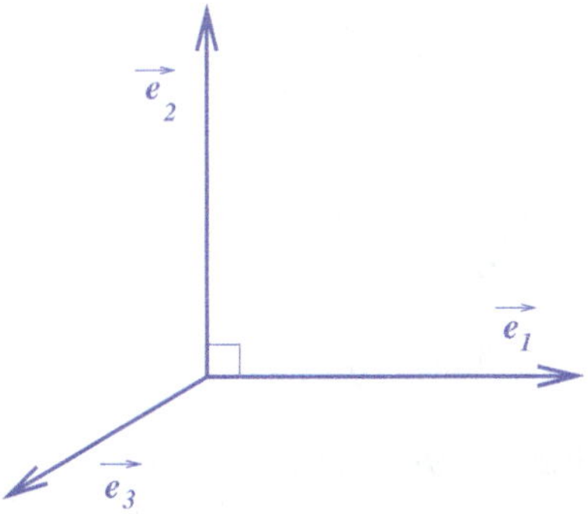

Figure 7.8: An orthonormal system.

Exercise 7.4.2 *Show that $e_1 = (1, 0, 0)$, $e_2 = (0, 1, 0)$, $e_3 = (0, 0, 1)$ is an orthonormal basis in $\mathbb{R}^3$.*

There are infinitely many orthonormal bases in $\mathbb{R}^3$. The next exercise deals with this issue.

Exercise 7.4.3 *Let $\{e_1, e_2, e_3\}$ be an orthonormal basis in $\mathbb{R}^3$. For any $\phi \in \mathbb{R}$, let $e_i' = R_\phi e_i$. Show that $\{e_1', e_2', e_3'\}$ is an orthonormal basis of $\mathbb{R}^3$.*

The next result expresses the inner product in terms of the vector components in an orthonormal basis.

Theorem 7.4.4 *Let $\{e_1, e_2, e_3\}$ be an orthonormal basis in $\mathbb{R}^3$, and consider two vectors $\vec{u} = \sum_{i=1}^{3} u^i e_i$, $\vec{v} = \sum_{i=1}^{3} v^i e_i$. Then their inner product is*

$$\vec{u} \cdot \vec{v} = u^1 v^1 + u^2 v^2 + u^3 v^3.$$

Proof: Using the distributivity of the inner product and the definition of the orthonormal basis, we have

$$\begin{aligned}
\vec{u}\cdot\vec{v} &= \vec{u}\cdot(v^1 e_1 + v^2 e_2 + v^3 e_3)\\
&= v^1(\vec{u}\cdot e_1) + v^2(\vec{u}\cdot e_2) + v^3(\vec{u}\cdot e_3)\\
&= \sum_{i=1}^{3} v^i(\vec{u}\cdot e_i) = \sum_{i=1}^{3} v^i(u^1 e_1\cdot e_i + u^2 e_2\cdot e_i + u^3 e_3\cdot e_i)\\
&= \sum_{i=1}^{3} v^i\left(\sum_{j=1}^{3} u^j\delta_{ij}\right) = u^1v^1 + u^2v^2 + u^3v^3.
\end{aligned}$$

■

Exercise 7.4.5 (Pythagorean relation) *Let* $\{e_1, e_2, e_3\}$ *be an orthonormal basis. If* $\vec{u} = \sum_{i=1}^{3} u^i e_i$*, show that*

$$|\vec{u}|^2 = (u^1)^2 + (u^2)^2 + (u^3)^2.$$

Exercise 7.4.6 (Polar decomposition) *For any two vectors* $\vec{u}$ *and* $\vec{v}$*, show that*

$$\vec{u}\cdot\vec{v} = \frac{1}{2}\Big(|\vec{u}+\vec{v}|^2 - |\vec{u}|^2 - |\vec{v}|^2\Big).$$

Exercise 7.4.7 (The parallelogram rule) *For any two vectors* $\vec{u}$ *and* $\vec{v}$*, prove the following identity*

$$|\vec{u}|^2 + |\vec{v}|^2 = \frac{1}{2}\Big(|\vec{u}+\vec{v}|^2 + |\vec{u}-\vec{v}|^2\Big).$$

Exercise 7.4.8 *Show that the sum of the squares of the opposite sides of a parallelogram is equal to the sum of the squares of the diagonals.*

7.5 The law of cosines

The law of cosines provides a formula for the cosine of the angles of a triangle in terms of the sides length.

Theorem 7.5.1 *Let* ABC *be a triangle with* $a = |BC|$, $b = |AC|$, $c = |AB|$*. Then*

$$\cos\widehat{A} = \frac{b^2 + c^2 - a^2}{2bc}.$$

Proof: Since $\overrightarrow{BC} = \overrightarrow{BA} + \overrightarrow{AC}$, we obtain

$$\begin{aligned}
\overrightarrow{BC} \cdot \overrightarrow{BC} &= (\overrightarrow{BA} + \overrightarrow{AC}) \cdot (\overrightarrow{BA} + \overrightarrow{AC}) \Longleftrightarrow \\
|BC|^2 &= |BA|^2 + |AC|^2 + 2\overrightarrow{BA} \cdot \overrightarrow{AC} \Longleftrightarrow \\
a^2 &= c^2 + b^2 - c \cdot b \cdot \cos \widehat{A} \Longleftrightarrow \\
\cos \widehat{A} &= \frac{b^2 + c^2 - a^2}{2bc}.
\end{aligned}$$

■

Corollary 7.5.2 (Pythagorean theorem) *In triangle ABC we have*

$$\widehat{A} = 90^\circ \Longleftrightarrow |BC|^2 = |AC|^2 + |AB|^2.$$

Proof: It follows from the law of cosines,

$$\widehat{A} = 90^\circ \Longleftrightarrow \cos \widehat{A} = 0 \Longleftrightarrow b^2 + c^2 = a^2.$$

■

Corollary 7.5.3 *We have*

$$\overrightarrow{AC} \cdot \overrightarrow{AB} = \frac{1}{2}[|AC|^2 + |AB|^2 - |BC|^2].$$

Proof: It follows from an application of the law of cosines

$$\begin{aligned}
\overrightarrow{AC} \cdot \overrightarrow{AB} &= |AC|\,|AB| \cos \widehat{A} \\
&= |AC|\,|AB| \frac{|AC|^2 + |AB|^2 - |BC|^2}{2|AC|\,|AB|} \\
&= \frac{1}{2}[|AC|^2 + |AB|^2 - |BC|^2].
\end{aligned}$$

■

7.6 The orthocenter

The perpendicular from one vertex to the opposite side of a triangle is called an *altitude*. A triangle has three altitudes. Next we shall prove the existence of the orthocenter of a triangle.

Theorem 7.6.1 *The altitudes of a triangle intersect at a point, called the orthocenter.*

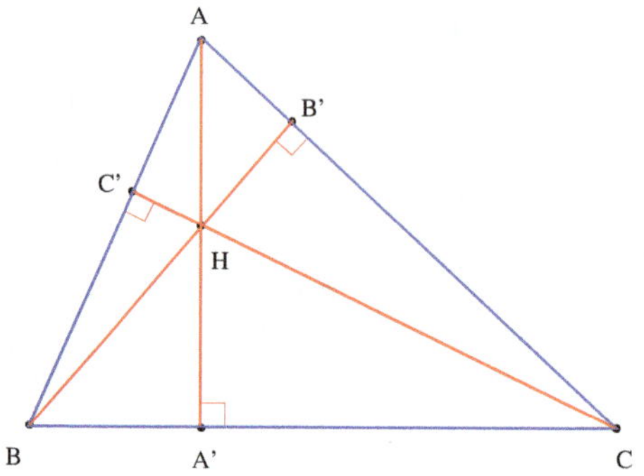

Figure 7.9: The orthocenter of a triangle.

Proof: Let AA' and CC' be the altitudes of triangle ABC from A and C. Let $\{H\} = AA' \cap CC'$. See Figure 7.9. We shall show that $BH \perp AC$. This is equivalent to $\overrightarrow{HB} \cdot \overrightarrow{CA} = 0$. Using $CB \perp HA$ and $AB \perp HC$, we have

$$\begin{aligned}
\overrightarrow{HB} \cdot \overrightarrow{CA} &= \overrightarrow{HB} \cdot (\overrightarrow{HA} - \overrightarrow{HC}) = \overrightarrow{HB} \cdot \overrightarrow{HA} - \overrightarrow{HB} \cdot \overrightarrow{HC} \\
&= (\overrightarrow{HC} + \overrightarrow{CB}) \cdot \overrightarrow{HA} - (\overrightarrow{HA} + \overrightarrow{AB}) \cdot \overrightarrow{HC} \\
&= \overrightarrow{HC} \cdot \overrightarrow{HA} + \overrightarrow{CB} \cdot \overrightarrow{HA} - \overrightarrow{HA} \cdot \overrightarrow{HC} - \overrightarrow{AB} \cdot \overrightarrow{HC} \\
&= \underbrace{\overrightarrow{CB} \cdot \overrightarrow{HA}}_{=0} - \underbrace{\overrightarrow{AB} \cdot \overrightarrow{HC}}_{=0} = 0,
\end{aligned}$$

i.e., $HB \perp CA$ and hence BH is an altitude. It follows that the altitudes pass through the same point H. ■

7.7 The midpoint connector theorem for trapezoid

A quadrilateral $ABCD$ with two opposite parallel sides is called a *trapezoid*. The following result deals with the midpoint connector of a trapezoid:

Proposition 7.7.1 *Let $ABCD$ be a trapezoid with $AB \| DC$ and let E, F be the midpoints of the sides $[AD]$ and $[BC]$, respectively. Then*

1) $EF \| AB$,

2) $|EF| = \frac{1}{2}(|AB| + |DC|)$.

Proof: We shall express $\overrightarrow{EF}$ in two ways using the polygon rule, see Figure 7.10

$$\overrightarrow{EF} = \overrightarrow{EA} + \overrightarrow{AB} + \overrightarrow{BF} = \frac{1}{2}\overrightarrow{DA} + \overrightarrow{AB} + \frac{1}{2}\overrightarrow{BC},$$

and

$$\overrightarrow{EF} = \overrightarrow{ED} + \overrightarrow{DC} + \overrightarrow{CF} = \frac{1}{2}\overrightarrow{AD} + \overrightarrow{DC} + \frac{1}{2}\overrightarrow{CB}.$$

Adding, yields

$$\begin{aligned} 2\overrightarrow{EF} &= \frac{1}{2}\underbrace{(\overrightarrow{DA} + \overrightarrow{AD})}_{=0} + \overrightarrow{AB} + \overrightarrow{DC} + \frac{1}{2}\underbrace{(\overrightarrow{BC} + \overrightarrow{CB})}_{=0} \\ &= \overrightarrow{AB} + \overrightarrow{DC}. \end{aligned}$$

Therefore

$$\overrightarrow{EF} = \frac{1}{2}(\overrightarrow{AB} + \overrightarrow{DC}). \tag{7.7.4}$$

Since the vectors $\overrightarrow{AB}$ and $\overrightarrow{DC}$ are parallel, their sum will be a vector parallel with $\overrightarrow{AB}$, and hence $EF \| AB$.

Using (7.7.4) we have

$$\begin{aligned} |EF|^2 &= \overrightarrow{EF} \cdot \overrightarrow{EF} = \frac{1}{4}(|AB|^2 + |DC|^2 + 2\overrightarrow{AB} \cdot \overrightarrow{DC}) \\ &= \frac{1}{4}(|AB|^2 + |DC|^2 + 2|AB| \cdot |DC|) \\ &= \frac{1}{4}(|AB| + |DC|)^2, \end{aligned}$$

where we have used Exercise 7.7.2. Hence $|EF| = (|AB|+|DC|)/2$. ■

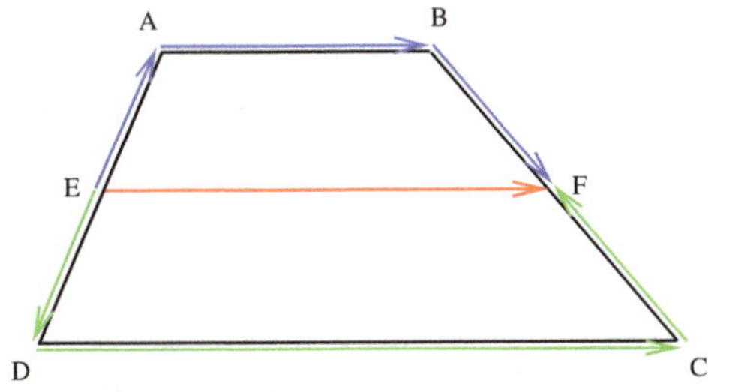

Figure 7.10: The midpoint connector in a trapezoid.

Exercise 7.7.2 *If the vectors* $\overrightarrow{v}$ *and* $\overrightarrow{w}$ *are collinear and have the same orientation, then show that*

$$\overrightarrow{v} \cdot \overrightarrow{w} = |\overrightarrow{v}| \cdot |\overrightarrow{w}|.$$

Exercise 7.7.3 *Show that for any two vectors* $\overrightarrow{u}, \overrightarrow{v}$ *we have*

$$|\overrightarrow{u}|^2 - |\overrightarrow{v}|^2 = (\overrightarrow{u} - \overrightarrow{v}) \cdot (\overrightarrow{u} + \overrightarrow{v}).$$

Exercise 7.7.4 *Let* ABC *be a triangle. Show that*

(i) $\overrightarrow{BA} \cdot \overrightarrow{BC} = \frac{1}{2}(|BA|^2 + |BC|^2 - |AC|^2)$

(ii) $\overrightarrow{AC} \cdot \overrightarrow{AB} + \overrightarrow{BA} \cdot \overrightarrow{BC} + \overrightarrow{CB} \cdot \overrightarrow{CA} = \frac{1}{2}(|AB|^2 + |BC|^2 + |AC|^2)$.

(iii) Let $[AA']$, $[BB']$, $[CC']$ *be the altitude segments of triangle* ABC, *see Figure 3.8. Show that*

$$|BA'| \cdot |BC| + |CB'| \cdot |CA| + |AC'| \cdot |AB| = \frac{1}{2}(|AB|^2 + |BC|^2 + |AC|^2).$$

7.8 Geometric locus problems

In this section we shall approach a geometric locus problem using vectorial calculus. These problems have also a longer traditional approach.

Lemma 7.8.1 *Let* A, B *be two points and let* D *be the midpoint of the segment* $[AB]$. *Then for any point* M *we have*

$$|MA|^2 - |MB|^2 = 2\overrightarrow{MD} \cdot \overrightarrow{BA}.$$

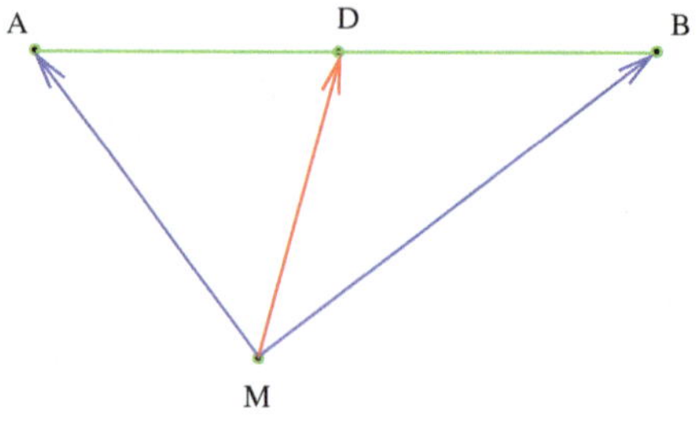

Figure 7.11: For Lemma 7.8.1.

Proof: Using Exercise 7.7.3 and Lemma 6.1.4 we have

$$\begin{aligned}|MA|^2 - |MB|^2 &= (\overrightarrow{MA} - \overrightarrow{MB}) \cdot (\overrightarrow{MA} + \overrightarrow{MB}) \\ &= \overrightarrow{BA} \cdot 2\overrightarrow{MD},\end{aligned}$$

see Figure 7.11. ■

Proposition 7.8.2 *Let A and B be two distinct points of the plane and let $k \in \mathbb{R}$. Then the set*

$$\{M; |MA|^2 - |MB|^2 = k\}$$

is a line perpendicular to the line AB.

Proof: Let $P \in AB$ with $MP \perp AB$. Then by Lemma 7.8.1 we have

$$\begin{aligned}|MA|^2 - |MB|^2 = k &\iff \overrightarrow{MD} \cdot \overrightarrow{BA} = \frac{k}{2} \iff \\ (\overrightarrow{MP} + \overrightarrow{PD}) \cdot \overrightarrow{BA} = \frac{k}{2} &\iff \underbrace{\overrightarrow{MP} \cdot \overrightarrow{BA}}_{=0} + \overrightarrow{PD} \cdot \overrightarrow{BA} = \frac{k}{2},\end{aligned}$$

so $\overrightarrow{PD} \cdot \overrightarrow{BA} = \frac{k}{2}$. We encounter three cases:

(*a*) The case $k < 0$: The vectors $\overrightarrow{PD}$ and $\overrightarrow{BA}$ have opposite orientation. This means that P is on the left side of D, see Figure 7.12(a). In this case we have

$$|PD| \cdot |AB| = -\frac{k}{2} \iff |PD| = -\frac{k}{2a},$$

where $a = |AB|$. It follows that M belongs to a line perpendicular to AB at P.

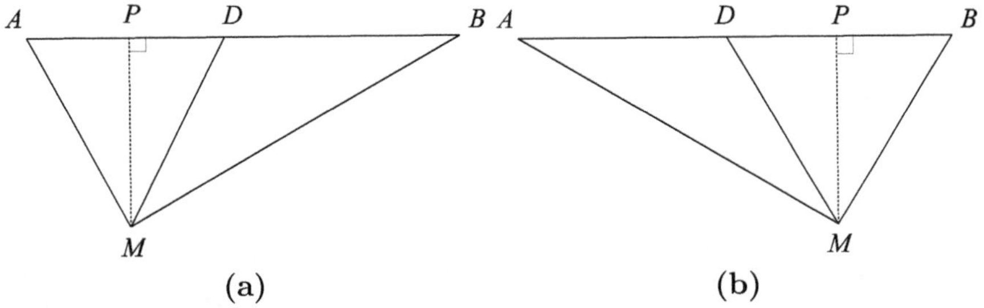

Figure 7.12: (a) The case $k < 0$; (b) The case $k > 0$.

(b) The case $k > 0$: The vectors $\overrightarrow{PD}$ and $\overrightarrow{BA}$ the same orientation. This means that P is on the right side of D, see Figure 7.12(b). We have

$$|PD| \cdot |AB| = \frac{k}{2} \Longleftrightarrow |PD = \frac{k}{2a},$$

where $a = |AB|$. Again, it follows that M belongs to a line perpendicular to AB at P.

(c) The case $k = 0$: From $|PD| \cdot |AB| = 0$ it follows that $|PD| = 0$, so $P = D$. In this case the line on which M belongs to is the perpendicular line on AB that passes through the midpoint of AB. Hence, the locus in this case is the perpendicular bisector of the line segment $[AB]$. ■

7.9 The median theorem

In this section we shall compute the length of a median of a triangle in terms of the sides length, and provide a few applications. The following theorem is a consequence of Lemma 6.1.4.

Theorem 7.9.1 (The median theorem) *Let ABC be a triangle with $a = |BC|$, $b = |AC|$, $c = |AB|$. Let m_a denote the length of the median from A. Then*

$$m_a^2 = \frac{b^2 + c^2}{2} - \frac{a^2}{4}.$$

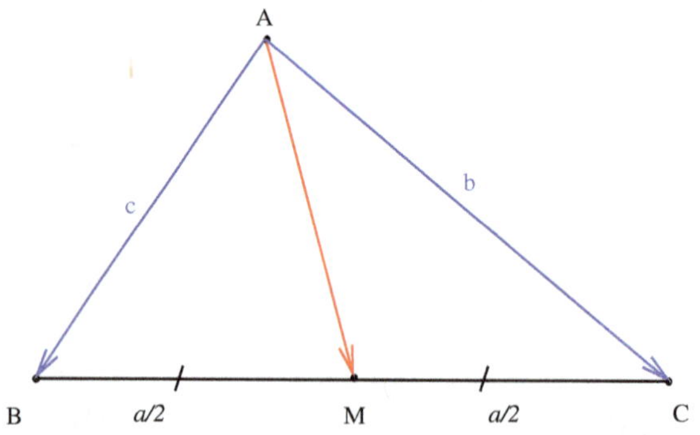

Figure 7.13: For the median theorem.

Proof: Let M be the midpoint of the segment $[BC]$, see Figure 7.13. Then

$$\begin{aligned}
m_a^2 &= |AM|^2 = \overrightarrow{AM}\cdot\overrightarrow{AM}\\
&= \frac{1}{2}(\overrightarrow{AB}+\overrightarrow{AC})\cdot\frac{1}{2}(\overrightarrow{AB}+\overrightarrow{AC})\\
&= \frac{1}{4}(\overrightarrow{AB}\cdot\overrightarrow{AB}+\overrightarrow{AB}\cdot\overrightarrow{AC}+\overrightarrow{AC}\cdot\overrightarrow{AB}+\overrightarrow{AC}\cdot\overrightarrow{AC})\\
&= \frac{1}{4}(|AB|^2+2\overrightarrow{AB}\cdot\overrightarrow{AC}+|AC|^2)\\
&= \frac{1}{4}(c^2+2|AB|\cdot|AC|\cos A+b^2)\\
&= \frac{1}{4}(c^2+2bc\cdot\frac{b^2+c^2-a^2}{2bc}+b^2)\\
&= \frac{1}{4}(c^2+b^2+c^2-a^2+b^2)\\
&= \frac{1}{4}(2(b^2+c^2)-a^2)\\
&= \frac{b^2+c^2}{2}-\frac{a^2}{4}.
\end{aligned}$$

■

Exercise 7.9.2 *Find the length of the medians in a right triangle.*

Exercise 7.9.3 *Find the length of the medians in an equilateral triangle.*

The next result is similar to Proposition 7.8.2, involving the sum instead of the difference.

Proposition 7.9.4 *Let A and B be two distinct points of the plane and $k \in \mathbb{R}$ such that $k > \frac{1}{2}|AB|^2$. Then the set*

$$\{M; |MA|^2 + |MB|^2 = k\}$$

is a circle centered at the midpoint of AB.

Proof: Let D be the midpoint of AB. Then the median theorem applied to triangle MAB yields

$$|MD|^2 = \frac{|MA|^2+|MB|^2}{2} - \frac{|AB|^2}{4} = \frac{1}{2}\left(k - \frac{|AB|^2}{2}\right) = \text{constant},$$

so M belongs to the circle $\mathcal{C}(D, r)$, centered at D and having radius $r = \sqrt{\left(\frac{k}{2} - \frac{|AB|^2}{4}\right)}$. We note that if $k = \frac{1}{2}|AB|^2$, the locus of M is just the point D and there are no solutions if $k < \frac{1}{2}|AB|^2$. ■

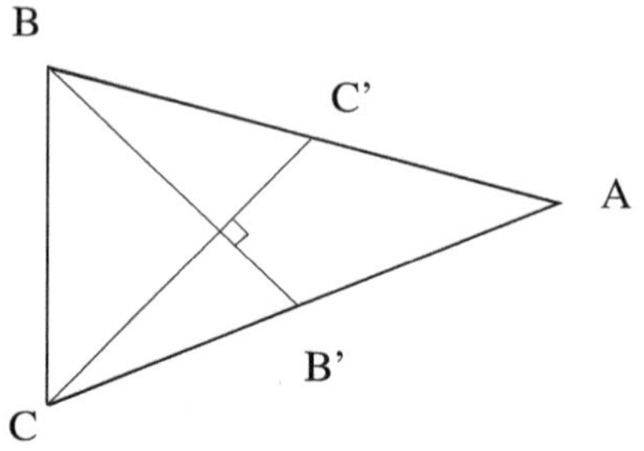

Figure 7.14: For Exercise 7.9.5.

The following exercise deals with triangles, which have two perpendicular medians.

Solved Exercise 7.9.5 *Let CC', BB' be the medians of triangle ABC, which correspond to the sides AB, AC respectively. Show that if $BB' \perp CC'$ then $5|BC|^2 = |AB|^2 + |AC|^2$.*

Proof: We shall refer to Figure 7.14. By Proposition 7.2.3, we have that $BB' \perp CC'$ implies $\overrightarrow{BB'} \cdot \overrightarrow{CC'} = 0$. Since $2\overrightarrow{BB'} = \overrightarrow{BA} + \overrightarrow{BC}$, $2\overrightarrow{CC'} = \overrightarrow{CA} + \overrightarrow{CB}$, we can write

$$\begin{aligned}
0 &= 4\overrightarrow{BB'} \cdot \overrightarrow{CC'} = (\overrightarrow{BA} + \overrightarrow{BC}) \cdot (\overrightarrow{CA} + \overrightarrow{CB}) \\
&= \overrightarrow{BA} \cdot \overrightarrow{CA} + \overrightarrow{BA} \cdot \overrightarrow{CB} + \overrightarrow{BC} \cdot \overrightarrow{CA} + \overrightarrow{BC} \cdot \overrightarrow{CB} \\
&= (\overrightarrow{BC} + \overrightarrow{CA}) \cdot \overrightarrow{CA} + (\overrightarrow{BC} + \overrightarrow{CA}) \cdot \overrightarrow{CB} + \overrightarrow{BC} \cdot \overrightarrow{CA} - |BC|^2 \\
&= \overrightarrow{BC} \cdot \overrightarrow{CA} + \overrightarrow{CA} \cdot \overrightarrow{CA} + \overrightarrow{BC} \cdot \overrightarrow{CB} + \overrightarrow{CA} \cdot \overrightarrow{CB} + \overrightarrow{BC} \cdot \overrightarrow{CA} - |BC|^2 \\
&= \overrightarrow{BC} \cdot \overrightarrow{CA} + |CA|^2 - |BC|^2 + \overrightarrow{CB} \cdot \overrightarrow{BA} + \overrightarrow{BC} \cdot \overrightarrow{CA} - |BC|^2 \\
&= (\overrightarrow{BC} + \underbrace{\overrightarrow{CB} + \overrightarrow{BC})}_{=0} \cdot \overrightarrow{CA} + |CA|^2 - 2|BC|^2 \\
&= \overrightarrow{BC} \cdot \overrightarrow{CA} + |CA|^2 - 2|BC|^2,
\end{aligned}$$

Hence, if $BB' \perp CC'$, then

$$\overrightarrow{BC} \cdot \overrightarrow{CA} + |CA|^2 - 2|BC|^2 = 0. \tag{7.9.5}$$

In a similar way we can obtain the symmetric relation (obtained by swapping C and B)

$$\overrightarrow{CB} \cdot \overrightarrow{BA} + |BA|^2 - 2|BC|^2 = 0. \tag{7.9.6}$$

Adding the equations (7.9.5) and (7.9.6) yields

$$\begin{aligned}
\overrightarrow{BC}\cdot\overrightarrow{CA}-\overrightarrow{BC}\cdot\overrightarrow{BA}+|CA|^2+|BA|^2-4|BC|^2 &= 0 \Longleftrightarrow \\
\overrightarrow{BC}\cdot(\overrightarrow{CA}-\overrightarrow{BA})+|CA|^2+|BA|^2-4|BC|^2 &= 0 \Longleftrightarrow \\
-|BC|^2+|CA|^2+|BA|^2-4|BC|^2 &= 0 \Longleftrightarrow \\
5|BC|^2=|CA|^2+|BA|^2.
\end{aligned}$$

■

Exercise 7.9.6 *Show that if in a triangle ABC the relation $5|BC|^2 = |CA|^2 + |BA|^2$ is satisfied, then the medians from B and C are perpendicular.*

Exercise 7.9.7 *Let ABC be an equilateral triangle and $E \in [AB]$ such that $\overrightarrow{AE} = 2\overrightarrow{EB}$. Let $F \in [BC]$ be a point such that $AF \perp CE$. Find the ratio $\dfrac{|BF|}{|FC|}$. See Figure 7.15.*

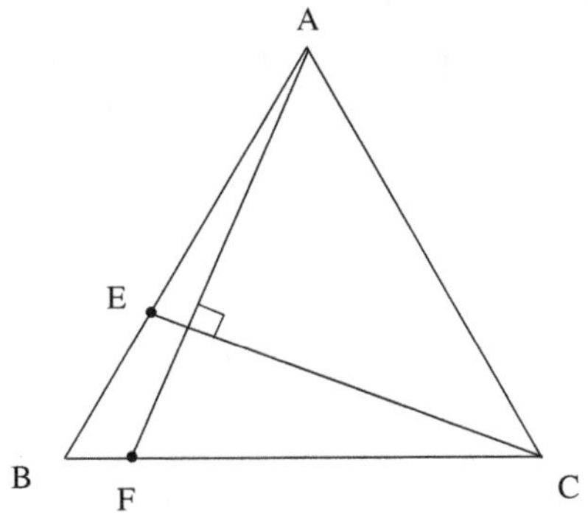

Figure 7.15: For Exercise 7.9.7.

7.10 Leibnitz's formula

We have shown that the medians of a triangle intersect at the centroid G, situated at $2/3$ from each vertex and $1/3$ from each side. In this section, we shall prove Leibnitz's formula, which enables us to compute the distance between any point and the centroid G.

We first need the following result dealing with the cyclic sum of inner products of the side vectors:

Lemma 7.10.1 *Let G be the centroid of the triangle ABC. Then*

$$\overrightarrow{GA}\cdot\overrightarrow{GB}+\overrightarrow{GA}\cdot\overrightarrow{GC}+\overrightarrow{GB}\cdot\overrightarrow{GC}=-\frac{1}{6}(|AB|^2+|BC|^2+|CA|^2).$$

Proof: Let $[AA']$, $[BB']$, $[CC']$ be the medians of the triangle ABC as in Figure 7.16. Using the property of the centroid, we have

$$\begin{aligned}\overrightarrow{GA}\cdot\overrightarrow{GB} &= \overrightarrow{AG}\cdot\overrightarrow{BG}=\left(\frac{2}{3}\overrightarrow{AA'}\right)\cdot\left(\frac{2}{3}\overrightarrow{BB'}\right)=\frac{4}{9}\overrightarrow{AA'}\cdot\overrightarrow{BB'}\\ &= \frac{4}{9}\cdot\frac{1}{2}(\overrightarrow{AB}+\overrightarrow{AC})\cdot\frac{1}{2}(\overrightarrow{BA}+\overrightarrow{BC})\\ &= \frac{1}{9}(\overrightarrow{AB}\cdot\overrightarrow{BA}+\overrightarrow{AB}\cdot\overrightarrow{BC}+\overrightarrow{AC}\cdot\overrightarrow{BA}+\overrightarrow{AC}\cdot\overrightarrow{BC}).\end{aligned}$$

Hence

$$\overrightarrow{GA}\cdot\overrightarrow{GB}=\frac{1}{9}(-|AB|^2+\overrightarrow{AB}\cdot\overrightarrow{BC}-\overrightarrow{AC}\cdot\overrightarrow{AB}+\overrightarrow{AC}\cdot\overrightarrow{BC}). \quad (7.10.7)$$

In a similar way we can show that

$$\overrightarrow{GA}\cdot\overrightarrow{GC}=\frac{1}{9}(-\overrightarrow{AB}\cdot\overrightarrow{AC}-\overrightarrow{AB}\cdot\overrightarrow{BC}-|AC|^2-\overrightarrow{AC}\cdot\overrightarrow{BC}), \quad (7.10.8)$$

$$\overrightarrow{GB}\cdot\overrightarrow{GC}=\frac{1}{9}(-\overrightarrow{BC}\cdot\overrightarrow{AC}-|BC|^2+\overrightarrow{AB}\cdot\overrightarrow{AC}+\overrightarrow{AB}\cdot\overrightarrow{BC}). \quad (7.10.9)$$

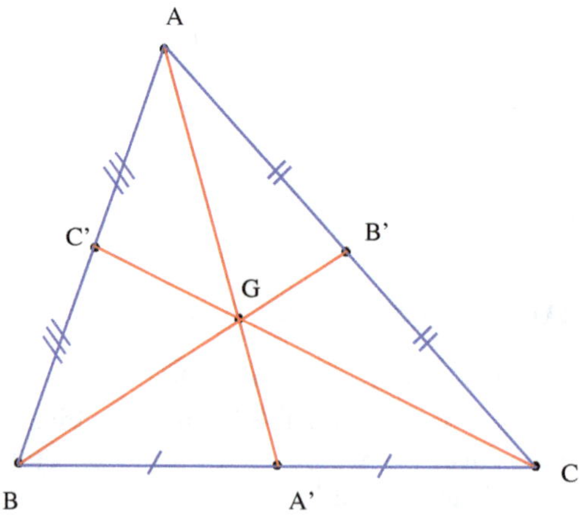

Figure 7.16: For Lemma 7.10.1.

Adding (7.10.7), (7.10.8) and (7.10.9), after cancellations we obtain

$$\overrightarrow{GA}\cdot\overrightarrow{GB}+\overrightarrow{GA}\cdot\overrightarrow{GC}+\overrightarrow{GB}\cdot\overrightarrow{GC}=-\frac{1}{9}\Big(|AB|^2+|AC|^2+|BC|^2 + \overrightarrow{BA}\cdot\overrightarrow{BC}+\overrightarrow{AC}\cdot\overrightarrow{AB}+\overrightarrow{CB}\cdot\overrightarrow{CA}\Big). \tag{7.10.10}$$

From the law of cosines, we have

$$\overrightarrow{BA}\cdot\overrightarrow{BC}=|BA|\cdot|BC|\cdot\frac{|BA|^2+|BC|^2-|AC|^2}{2|BA|\cdot|BC|}=\frac{1}{2}(|BA|^2+|BC|^2-|AC|^2)$$

$$\overrightarrow{AC}\cdot\overrightarrow{AB}=|AC|\cdot|AB|\cdot\frac{|AC|^2+|AB|^2-|BC|^2}{2|AC|\cdot|AB|}=\frac{1}{2}(|AC|^2+|AB|^2-|BC|^2)$$

$$\overrightarrow{CB}\cdot\overrightarrow{CA}=|CB|\cdot|CA|\cdot\frac{|CB|^2+|CA|^2-|AB|^2}{2|CB|\cdot|CA|}=\frac{1}{2}(|CB|^2+|CA|^2-|AB|^2).$$

Substituting in (7.10.10) yields

$$\begin{aligned}
&\overrightarrow{GA}\cdot\overrightarrow{GB}+\overrightarrow{GA}\cdot\overrightarrow{GC}+\overrightarrow{GB}\cdot\overrightarrow{GC}\\
&\quad=-\frac{1}{9}\Big(|AB|^2+|AC|^2+|BC|^2+\frac{1}{2}\Big(|BA|^2+|BC|^2-|AC|^2\\
&\qquad+|AC|^2+|AB|^2-|BC|^2+|CB|^2+|CA|^2-|AB|^2\Big)\Big)\\
&\quad=-\frac{1}{9}\Big(|AB|^2+|AC|^2+|BC|^2+\frac{1}{2}(|AB|^2+|AC|^2+|BC|^2)\Big)\\
&\quad=-\frac{1}{6}(|AB|^2+|AC|^2+|BC|^2),
\end{aligned}$$

which is the desired result. ∎

Now we are ready to prove one of the most interesting property of the centroid.

Theorem 7.10.2 (Leibnitz) *Let M be a point and ABC be a triangle with the centroid G. Then*

$$|MA|^2+|MB|^2+|MC|^2=3|MG|^2+\frac{1}{3}(|AB|^2+|BC|^2+|AC|^2).$$

Proof: From part (*ii*) of Theorem 6.3.2

$$\overrightarrow{MA} + \overrightarrow{MB} + \overrightarrow{MC} = 3\overrightarrow{MG}.$$

Then

$$\begin{aligned} 9|MG|^2 &= (\overrightarrow{MA} + \overrightarrow{MB} + \overrightarrow{MC}) \cdot (\overrightarrow{MA} + \overrightarrow{MB} + \overrightarrow{MC}) \iff \\ 9|MG|^2 &= |MA|^2 + |MB|^2 + |MC|^2 \\ &\quad +2(\overrightarrow{MA} \cdot \overrightarrow{MB} + \overrightarrow{MA} \cdot \overrightarrow{MC} + \overrightarrow{MB} \cdot \overrightarrow{MC}). \end{aligned} \tag{7.10.11}$$

The inner product terms are computed as follows

$$\begin{aligned} \overrightarrow{MA} \cdot \overrightarrow{MB} &= (\overrightarrow{MG} + \overrightarrow{GA}) \cdot (\overrightarrow{MG} + \overrightarrow{GB}) \\ &= |MG|^2 + \overrightarrow{MG} \cdot \overrightarrow{GB} + \overrightarrow{GA} \cdot \overrightarrow{MG} + \overrightarrow{GA} \cdot \overrightarrow{GB}, \\ \overrightarrow{MA} \cdot \overrightarrow{MC} &= (\overrightarrow{MG} + \overrightarrow{GA}) \cdot (\overrightarrow{MG} + \overrightarrow{GC}) \\ &= |MG|^2 + \overrightarrow{MG} \cdot \overrightarrow{GC} + \overrightarrow{GA} \cdot \overrightarrow{MG} + \overrightarrow{GA} \cdot \overrightarrow{GC}, \\ \overrightarrow{MB} \cdot \overrightarrow{MC} &= (\overrightarrow{MG} + \overrightarrow{GB}) \cdot (\overrightarrow{MG} + \overrightarrow{GC}) \\ &= |MG|^2 + \overrightarrow{MG} \cdot \overrightarrow{GC} + \overrightarrow{GB} \cdot \overrightarrow{MG} + \overrightarrow{GB} \cdot \overrightarrow{GC}. \end{aligned}$$

Adding the last three identities we obtain

$$\begin{aligned} & \overrightarrow{MA} \cdot \overrightarrow{MB} + \overrightarrow{MA} \cdot \overrightarrow{MC} + \overrightarrow{MB} \cdot \overrightarrow{MC} \\ = \; & 3|MG|^2 + \overrightarrow{MG} \cdot (\overrightarrow{GB} + \overrightarrow{GC} + \overrightarrow{GC} + \overrightarrow{GA} + \overrightarrow{GA} + \overrightarrow{GB}) \\ & +\overrightarrow{GA} \cdot \overrightarrow{GB} + \overrightarrow{GA} \cdot \overrightarrow{GC} + \overrightarrow{GB} \cdot \overrightarrow{GC} \\ = \; & 3|MG|^2 + 2\overrightarrow{MG} \cdot (\overrightarrow{GA} + \overrightarrow{GB} + \overrightarrow{GC}) \\ & +\overrightarrow{GA} \cdot \overrightarrow{GB} + \overrightarrow{GA} \cdot \overrightarrow{GC} + \overrightarrow{GB} \cdot \overrightarrow{GC}. \end{aligned}$$

Using $\overrightarrow{GA} + \overrightarrow{GB} + \overrightarrow{GC} = 0$ and Lemma 7.10.1, we have

$$\overrightarrow{MA} \cdot \overrightarrow{MB} + \overrightarrow{MA} \cdot \overrightarrow{MC} + \overrightarrow{MB} \cdot \overrightarrow{MC} = 3|MG|^2 - \frac{1}{6}(|AB|^2 + |BC|^2 + |CA|^2). \tag{7.10.12}$$

Substituting (7.10.12) into (7.10.11) yields

$$\begin{aligned} |MA|^2 + |MB|^2 + |MC|^2 + |MC|^2 + 6|MG|^2 - \frac{1}{3}(|AB|^2 + |BC|^2 + |CA|^2) \\ = 9|MG|^2, \end{aligned}$$

which leads to the desired result. ∎

Corollary 7.10.3 *Let O and G be the circumcenter and the centroid of triangle ABC with the side lengths a, b, c and circumradius R. Then*

$$|OG|^2 = R^2 - \frac{1}{9}(a^2 + b^2 + c^2).$$

Proof: Choosing $M = O$ we have $|OA| = |OB| = |OC| = R$ and apply the previous theorem. ■

Corollary 7.10.4 *Let ABC be a triangle and M be a point. The sum*

$$|MA|^2 + |MB|^2 + |MC|^2$$

is minimum if and only if $M = G$, the centroid of triangle ABC. The minimum is equal to

$$\frac{1}{3}(|AB|^2 + |BC|^2 + |CA|^2).$$

Exercise 7.10.5 *Let A and B be two points. Show that the sum $|MA|^2 + |MB|^2$ is minimum if and only if M is the midpoint of the line segment $[AB]$.*

Exercise 7.10.6 *Let ΔABC be a triangle of centroid G. Let $k > 0$ be a constant such that $3k > a^2 + b^2 + c^2$. Show that the locus of the point M such that*

$$|MA|^2 + |MB|^2 + |MC|^2 = k$$

is a circle centered at G.

Exercise 7.10.7 *Let $ABCD$ be a square and let M be a point. The sum*

$$|MA|^2 + |MB|^2 + |MC|^2 + |MD|^2$$

is minimum if and only if M is the center of the square. State and prove a generalization!

Solved Exercise 7.10.8 *Let ABC be an arbitrary triangle and let N, M, P be points on the line segments $[BC]$, $[AB]$ and $[AC]$ such that*

$$|BN| = |CP| = |AM| = x > 0.$$

(i) Let G, G' be the centroids of triangles ABC and NMP. Show that

$$x^2\left(3 - 2(\cos \widehat{A} + \cos \widehat{B} + \cos \widehat{C})\right) = 9|GG'|^2.$$

(ii) Show that triangles ABC and NMP have the same centroid if and only if ABC is equilateral.

(iii) Show that in any triangle

$$\cos\widehat{A} + \cos\widehat{B} + \cos\widehat{C} \leq \frac{3}{2},$$

with identity for the equilateral triangle. See Figure 7.17.

Proof: (*i*) From Proposition 6.3.4

$$\overrightarrow{AN} + \overrightarrow{BP} + \overrightarrow{CM} = 3\overrightarrow{GG'}. \tag{7.10.13}$$

Substituting

$$\overrightarrow{AN} = \overrightarrow{AB} + \overrightarrow{BN}, \qquad \overrightarrow{BP} = \overrightarrow{BC} + \overrightarrow{CP}, \qquad \overrightarrow{CM} = \overrightarrow{CA} + \overrightarrow{AM}$$

into (7.10.13) yields

$$\begin{aligned} 3\overrightarrow{GG'} &= \overrightarrow{AB} + \overrightarrow{BN} + \overrightarrow{BC} + \overrightarrow{CP} + \overrightarrow{CA} + \overrightarrow{AM} \iff \\ 3\overrightarrow{GG'} &= (\underbrace{\overrightarrow{AB} + \overrightarrow{BC} + \overrightarrow{CA}}_{=0}) + (\overrightarrow{BN} + \overrightarrow{CP} + \overrightarrow{AM}) \iff \\ 3\overrightarrow{GG'} &= \overrightarrow{BN} + \overrightarrow{CP} + \overrightarrow{AM}. \end{aligned}$$

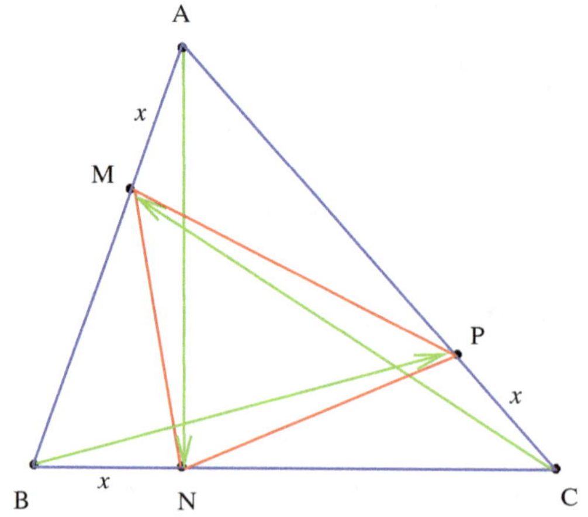

Figure 7.17: For Exercise 7.10.8.

Using Proposition 7.2.3, part 4), yields

$$9|GG'|^2 = |\overrightarrow{BN} + \overrightarrow{CP} + \overrightarrow{AM}|^2 \iff$$
$$9|GG'|^2 = |\overrightarrow{BN}|^2 + |\overrightarrow{CP}|^2 + |\overrightarrow{AM}|^2 + 2(\overrightarrow{BN} \cdot \overrightarrow{CP} + \overrightarrow{CP} \cdot \overrightarrow{AM} + \overrightarrow{AM} \cdot \overrightarrow{BN})$$
$$9|GG'|^2 = 3x^2 + 2\Big(x^2 \cos(180 - \widehat{C}) + x^2 \cos(180 - \widehat{A}) + x^2 \cos(180 - \widehat{B})\Big)$$
$$9|GG'|^2 = x^2\Big(3 - 2(\cos \widehat{A} + \cos \widehat{B} + \cos \widehat{C})\Big).$$

(*ii*) It is left as an exercise for the reader.
(*iii*) It follows from the fact that $|GG'|^2 \geq 0$ and part (*i*). ■

7.11 The distance $|IG|$

Let I and G denote the incenter and the centroid of a triangle ΔABC. We note that both points belong to the interior of the triangle. We shall compute their distance applying Leibnitz' formula from page 407

$$|IG|^2 = \frac{1}{3}(|AI|^2 + |BI|^2 + |CI|^2) - \frac{1}{9}(a^2 + b^2 + c^2). \tag{7.11.14}$$

This distance will be computed in two ways.

1. *The first variant* Considering the identities

$$|AI| = \frac{r}{\sin\frac{\widehat{A}}{2}}, \quad |BI| = \frac{r}{\sin\frac{\widehat{B}}{2}}, \quad |CI| = \frac{r}{\sin\frac{\widehat{C}}{2}},$$

the sum of their squares can be computed as

$$\begin{aligned}
\sum_{circ.perm.} |AI|^2 &= r^2 \sum_{circ.perm.} \frac{1}{\sin^2\frac{\widehat{A}}{2}} = r^2 \sum_{circ.perm.} \frac{bc}{(p-b)(p-c)} \\
&= r^2 \frac{bc(p-a) + ac(p-b) + ab(p-c)}{(p-a)(p-b)(p-c)} \\
&= \frac{r^2}{(p-a)(p-b)(p-c)}[p(bc + ca + ab) - 3abc] \\
&= \frac{(pr)^2}{p^2(p-a)(p-b)(p-c)}[p(bc + ca + ab) - 12R\mathcal{A}] \\
&= \frac{\mathcal{A}^2}{p\mathcal{A}^2}[p(bc + ca + ab) - 12Rpr] \\
&= bc + ca + ab - 12Rr = p^2 + r^2 - 8Rr,
\end{aligned}$$

where we used the formula for $bc + ca + ab$ from page 117. Substituting into (7.11.14) we obtain

$$\begin{aligned}|IG|^2 &= \frac{1}{3}(|AI|^2 + |BI|^2 + |CI|^2) - \frac{1}{9}(a^2 + b^2 + c^2)\\ &= \frac{1}{3}(p^2 + r^2 - 8Rr) - \frac{1}{9}(2p^2 - 2r^2 - 8Rr)\\ &= \frac{1}{9}(p^2 + 5r^2 - 16Rr),\end{aligned} \tag{7.11.15}$$

where we used the formula for $a^2 + b^2 + c^2$ from page 117.

2. The second variant Considering the identities

$$|AI| = \sqrt{r^2 + (p-a)^2}, \quad |BI| = \sqrt{r^2 + (p-b)^2}, \quad |CI| = \sqrt{r^2 + (p-c)^2}$$

the sum of the squares becomes

$$\begin{aligned}\sum_{circ.perm.} |AI|^2 &= 3r^2 + \sum (p-a)^2\\ &= 3r^2 + 3p^2 + \sum a^2 - 2p \sum a\\ &= 3r^2 + 3p^2 + \sum a^2 - 4p^2 = 3r^2 - p^2 + \sum a^2.\end{aligned}$$

Substituting into (7.11.14) yields

$$\begin{aligned}|IG|^2 &= \frac{1}{3}(|AI|^2 + |BI|^2 + |CI|^2) - \frac{1}{9}\sum a^2\\ &= r^2 - \frac{1}{3}p^2 + \frac{2}{9}\sum a^2.\end{aligned} \tag{7.11.16}$$

Exercise 7.11.1 *Show that formulas (7.11.16) and (7.11.16) are equal.*

7.12 The Feuerbach point

Theorem 7.12.1 *The Euler circle and the inner circle of any triangle are tangent at a point.*

Proof: To prove that the Euler circle and the inner circle are tangent it suffices to show that the distance between their centers is equal to the difference of their radii, see Figure 7.18(a). More specifically, if O_9 and I denote the centers of the Euler and inner circles, and R_0 and r denote respectively their radii, we need to show the relation

$$|IO_9| = R_0 - r. \tag{7.12.17}$$

This will be shown by computation. First, we recall that Euler's circle has the radius equal to half the circumradius, namely $R_0 = R/2$. We also use that the points O, G, H and O_9 are collinear. Since O_9 is the midpoint of OH and G is located at $1/3$ from O and $2/3$ from H, denoting $x = |GO|$, we have $|HO| = 3x$, $|OO_9| = \frac{3}{2}x$ and $|O_9G| = \frac{x}{2}$, see Figure 7.18(b). Stewart's relation applied to triangle ΔIO_9O yields

$$|IG|^2\,|OO_9| = |IO_9|^2|GO| + |OI|^2|O_9G| - |O_9G|\,|GO|\,|O_9O| \iff$$
$$|IG|^2\frac{3}{2}x = |IO_9|^2x + |OI|^2\frac{x}{2} - \frac{x}{2}\cdot x\cdot\frac{3}{2}x.$$

Simplifying by x and solving for $|IO_9|^2$ and then using the formula for $|IG|^2$ given by (7.11.16) as well as Euler's formula $|OI|^2 = R(R-2r)$, and the formula for $|GO|$ given on page 409, we have

$$\begin{aligned}
|IO_9|^2 &= \frac{3}{2}|IG|^2 - \frac{1}{2}|OI|^2 + \frac{3}{4}x^2\\
&= \frac{3}{2}\left(r^2 - \frac{1}{3}p^2 + \frac{2}{9}\sum a^2\right) - \frac{1}{2}R(R-2r) + \frac{3}{4}\cdot\frac{1}{9}\left(9R^2 - \sum a^2\right)\\
&= \frac{3}{2}r^2 - \frac{1}{2}p^2 + \frac{1}{3}\sum a^2 - \frac{1}{2}R^2 + Rr + \frac{3}{4}R^2 - \frac{1}{12}\sum a^2\\
&= \frac{1}{4}R^2 + \frac{3}{2}r^2 + Rr + \frac{1}{4}\sum a^2 - \frac{1}{2}p^2\\
&= \frac{1}{4}R^2 + \frac{3}{2}r^2 + Rr + \frac{1}{4}(2p^2 - 2r^2 - 8Rr) - \frac{1}{2}p^2\\
&= \left(\frac{R}{2}\right)^2 + r^2 - Rr = \left(\frac{R}{2} - r\right)^2 = (R_0 - r)^2,
\end{aligned}$$

and hence relation (7.12.17) is proved. ∎

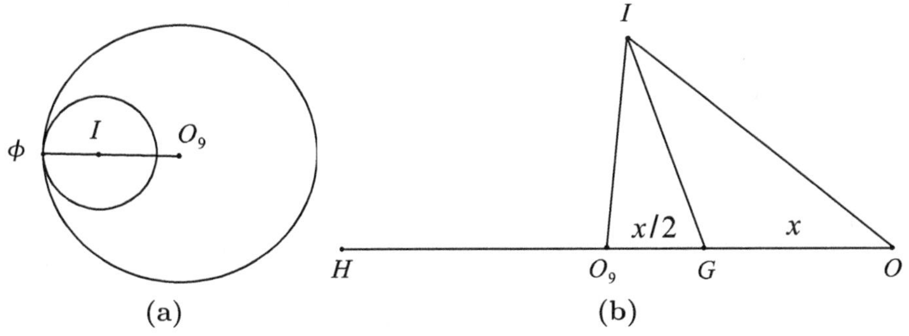

Figure 7.18: (a) The Euler and inner circles are tangent; (b) $|IG|$ can be computed by Stewart's relation from ΔIO_9O.

The point where the Euler circle and the inner circle are tangent is called the *Feuerbach point* of the triangle and it is denoted by Φ.

Remark 7.12.2 Let I_a, I_b and I_c be the centers of the circles with radii r_a, r_b and r_c, respectively. A similar computation can show that $|I_aO_9| = R_0 + r_a$, $|I_bO_9| = R_0 + r_b$, and $|I_cO_9| = R_0 + r_c$, which means that the excircles are tangent to the incircle, at three points, Φ_a, Φ_b and Φ_c, called the *adjoint Feuerbach points* of the triangle.

7.13 The length of a cevian

Let ABC be a triangle and let $M \in [BC]$ such that $\overrightarrow{BM} = k\overrightarrow{MC}$, with $k \in \mathbb{R}$. The segment $[AM]$ is called a *cevian*. It is named after the Italian mathematician Giovanni Ceva.

In this section, we shall compute the length of the cevian $[AM]$ in terms of the elements of triangle ABC. By Lemma 6.1.9, we have

$$\overrightarrow{AM} = \frac{\overrightarrow{AB} + k\overrightarrow{AC}}{k+1},$$

see Figure 7.19. Then

$$\begin{aligned}
|AM|^2 = \overrightarrow{AM} \cdot \overrightarrow{AM} &= \frac{1}{(k+1)^2}(\overrightarrow{AB} + k\overrightarrow{AC}) \cdot (\overrightarrow{AB} + k\overrightarrow{AC}) \\
&= \frac{1}{(k+1)^2}(\overrightarrow{AB} \cdot \overrightarrow{AB} + k\overrightarrow{AB} \cdot \overrightarrow{AC} + k\overrightarrow{AC} \cdot \overrightarrow{AB} + k^2\overrightarrow{AC} \cdot \overrightarrow{AC}) \\
&= \frac{1}{(k+1)^2}(|AB|^2 + 2k|AB| \cdot |AC| \cos A + k^2|AC|^2) \\
&= \frac{1}{(k+1)^2}\left(|AB|^2 + 2k|AB| \cdot |AC| \frac{|AB|^2 + |AC|^2 - |BC|^2}{2|AB| \cdot |AC|}\right. \\
&\quad \left. + k^2|AC|^2\right) \\
&= \frac{1}{(k+1)^2}\left(|AB|^2 + k(|AB|^2 + |AC|^2 - |BC|^2) + k^2|AC|^2\right) \\
&= \frac{1}{(k+1)^2}\left((k+1)|AB|^2 + k(k+1)|AC|^2 - k|BC|^2\right) \\
&= \frac{|AB|^2 + k|AC|^2}{k+1} - \frac{k}{(k+1)^2}|BC|^2.
\end{aligned}$$

We arrived at the following result:

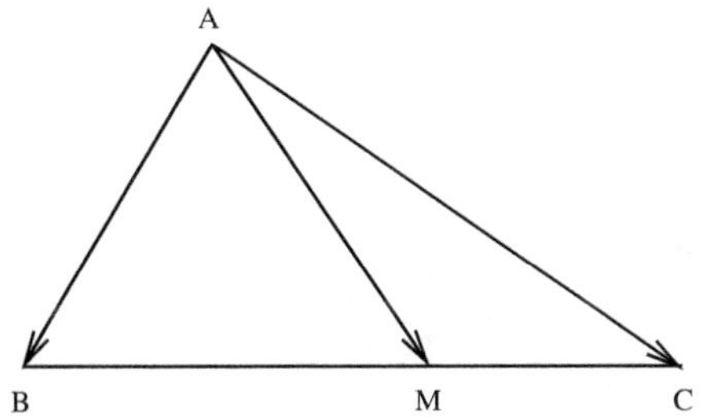

Figure 7.19: A cevian AM.

Proposition 7.13.1 *Let ABC be a triangle. If $M \in [BC]$ such that $\frac{|BM|}{|MC|} = k$, the length of the cevian $[AM]$ is given by*

$$|AM|^2 = \frac{c^2 + kb^2}{k+1} - \frac{k}{(k+1)^2}a^2, \tag{7.13.18}$$

where $a = |BC|$, $b = |AC|$, $c = |AB|$.

Solved Exercise 7.13.2 *Let $A' \in [BC]$, $B' \in [AC]$ and $C' \in [AB]$ such that*

$$\frac{|BA'|}{|A'C|} = \frac{|CB'|}{|B'A|} = \frac{|AC'|}{|C'B|} = k.$$

(i) Show that

$$|AA'|^2 + |BB'|^2 + |CC'|^2 = \frac{k^2 + k + 1}{(k+1)^2}(a^2 + b^2 + c^2).$$

(ii) Show that the sum $|AA'|^2 + |BB'|^2 + |CC'|^2$ is minimum if and only if A', B' and C' are the side midpoints.

Proof: (*i*) By Proposition 7.13.1 we have

$$\begin{aligned}
|AA'|^2 &= \frac{1}{k+1}c^2 + \frac{k}{k+1}b^2 - \frac{k}{(k+1)^2}a^2 \\
|BB'|^2 &= \frac{1}{k+1}a^2 + \frac{k}{k+1}c^2 - \frac{k}{(k+1)^2}b^2 \\
|CC'|^2 &= \frac{1}{k+1}b^2 + \frac{k}{k+1}a^2 - \frac{k}{(k+1^2}c^2.
\end{aligned}$$

Taking the sum yields

$$\begin{aligned}|AA'|^2+|BB'|^2+|CC'|^2 &= (a^2+b^2+c^2)\left(\frac{1}{k+1}+\frac{k}{k+1}-\frac{k}{(k+1)^2}\right)\\ &= \frac{k^2+k+1}{(k+1)^2}(a^2+b^2+c^2).\end{aligned}$$

(*ii*) We shall write

$$\frac{k^2+k+1}{(k+1)^2}=\frac{1}{(k+1)^2}-\frac{1}{k+1}+1.$$

Denoting $x=\frac{1}{k+1}$, the problem reduces to finding the value of x for which x^2-x+1 is minimum. Since $x^2-x+1=(x-\frac{1}{2})^2+\frac{3}{4}$, it follows that the minimum is reached for $x=\frac{1}{2}$, which corresponds to the value $k=1$. This means that A', B' and C' are the side midpoints. ■

Particular cases

In the case $k=1$ the point M is the midpoint of $[BC]$ and hence $[AM]$ is a median. The above theorem recovers the median theorem:

$$|AM|^2=\frac{b^2+c^2}{2}-\frac{a^2}{4}.$$

Exercise 7.13.3 *Show that the cevian AM is the angle bisector if and only if*

$$\frac{|BM|}{|MC|}=\frac{|AB|}{|AC|}.$$

In the case of the angle bisector, $k=\dfrac{c}{b}$, see Exercise 7.13.3. Then the square of the length of the angle bisector AM is

$$\begin{aligned}
|AM|^2 &= \frac{c^2 + \frac{c}{b}b^2}{\frac{c}{b}+1} - \frac{\frac{c}{b}}{(\frac{c}{b}+1)^2}a^2 \\
&= \frac{(c^2+bc)b}{b+c} - \frac{a^2bc}{(b+c)^2} \\
&= \frac{bc(b+c)}{b+c} - \frac{a^2bc}{(b+c)^2} \\
&= bc\Big(1 - \frac{a^2}{(b+c)^2}\Big) = bc\frac{(b+c)^2 - a^2}{(b+c)^2} \\
&= bc\frac{(b+c-a)(b+c+a)}{(b+c)^2} \\
&= \frac{4bc}{(b+c)^2}\cdot\frac{b+c-a}{2}\cdot\frac{b+c+a}{2} \\
&= \frac{4bc}{(b+c)^2}(p-a)p,
\end{aligned}$$

where $p = \dfrac{1}{2}(a+b+c)$ is the *semiperimeter* of triangle ABC. Hence we arrived at the following formula for the length of the angle bisector:

$$|AM|^2 = \frac{2\sqrt{bcp(p-a)}}{b+c}.$$

Exercise 7.13.4 *The cevian $[AM]$ with $\dfrac{|BM|}{|MC|} = \left(\dfrac{c}{b}\right)^2$ is called a symmedian. Find the length of the symmedian $[AM]$.*

7.14 The geometry of quadrilaterals

In this section, we shall take advantage of vectorial calculus to prove results regarding diagonals, sides and bimedians of quadrilaterals. The next result deals with *orthodiagonal* quadrilaterals:

Proposition 7.14.1 *Let $ABCD$ be a convex quadrilateral. Then the diagonals AC, BD are perpendicular if and only if*

$$|AB|^2 + |CD|^2 = |BC|^2 + |AD|^2.$$

Proof: Let $\vec{a} = \overrightarrow{AB}$, $\vec{b} = \overrightarrow{AC}$, $\vec{c} = \overrightarrow{AD}$, see Figure 7.20. We have $AC \perp BD$ if and only if $\overrightarrow{AC}\cdot\overrightarrow{BD} = 0$, which becomes

$$\vec{b}\cdot(\vec{c}-\vec{a}) = 0 \iff \vec{b}\cdot\vec{c} = \vec{a}\cdot\vec{b}. \tag{7.14.19}$$

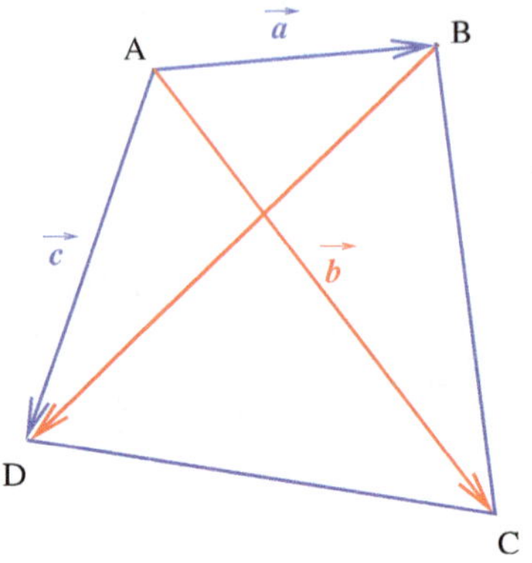

Figure 7.20: For Proposition 7.14.1.

Using (7.14.19), we write

$$\begin{aligned}
|CD|^2 + |AB|^2 &= (\vec{b} - \vec{c}) \cdot (\vec{b} - \vec{c}) + \vec{a} \cdot \vec{a} \\
&= \vec{b} \cdot \vec{b} - 2\vec{b} \cdot \vec{c} + \vec{c} \cdot \vec{c} + \vec{a} \cdot \vec{a} \\
&= \vec{b} \cdot \vec{b} - 2\vec{a} \cdot \vec{b} + \vec{a} \cdot \vec{a} + \vec{c} \cdot \vec{c} \\
&= (\vec{a} - \vec{b}) \cdot (\vec{a} - \vec{b}) + \vec{c} \cdot \vec{c} \\
&= \overrightarrow{CB} \cdot \overrightarrow{CB} + \overrightarrow{AD} \cdot \overrightarrow{AD} \\
&= |CB|^2 + |AD|^2.
\end{aligned}$$

Hence, if $AC \perp BD$, then $|CD|^2 + |AB|^2 = |CB|^2 + |AD|^2$.
The proof can be easily inverted to prove the converse implication. ■

The next result deals with a necessary and sufficient condition for the bimedians to be perpendicular.

Proposition 7.14.2 *Let $ABCD$ be a convex quadrilateral. Then the bimedians are perpendicular if and only if the diagonals $[AC]$ and $[BD]$ are equal.*

Proof: Let M, N, E, and F be the midpoints of the sides $[AD]$, $[BC]$, $[AB]$, and $[DC]$, respectively, see Figure 7.21. The bimedians $[MN]$ and $[EF]$ are perpendicular if and only if $\overrightarrow{MN} \cdot \overrightarrow{EF} = 0$. It suffices to show that

$$\overrightarrow{MN} \cdot \overrightarrow{EF} = \frac{1}{4}(|AC|^2 - |BD|^2).$$

Using Proposition 6.4.2 we express the bimedians vectors in terms of

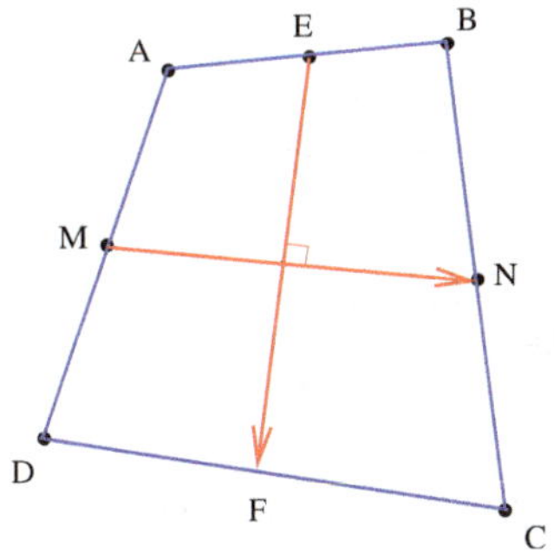

Figure 7.21: For Proposition 7.14.2.

diagonals

$$\begin{aligned}\overrightarrow{MN} &= \frac{1}{2}(\overrightarrow{AC} + \overrightarrow{DB})\\ \overrightarrow{EF} &= \frac{1}{2}(\overrightarrow{AC} + \overrightarrow{BD}) = \frac{1}{2}(\overrightarrow{AC} - \overrightarrow{DB}).\end{aligned}$$

Then by Exercise 7.7.4 we have

$$\overrightarrow{MN} \cdot \overrightarrow{EF} = \frac{1}{4}(\overrightarrow{AC} + \overrightarrow{DB}) \cdot (\overrightarrow{AC} - \overrightarrow{DB}) = \frac{1}{4}(|AC|^2 - |BD|^2).$$

Hence, $\overrightarrow{MN} \cdot \overrightarrow{EF} = 0$ if and only if $(|AC| - |BD|)(|AC| + |BD|) = 0$, *i.e.*, $|AC| = |BD|$, which ends the proof. ∎

Remark 7.14.3 We can also provide a proof using an expression of the bimedian vectors in terms of the side vectors as follows. By Proposition 6.4.2 we have

$$\begin{aligned}\overrightarrow{MN} &= \frac{1}{2}(\overrightarrow{AB} + \overrightarrow{DC})\\ \overrightarrow{EF} &= \frac{1}{2}(\overrightarrow{AD} + \overrightarrow{BC}).\end{aligned}$$

Then

$$\begin{aligned}4\overrightarrow{MN} \cdot \overrightarrow{EF} &= (\overrightarrow{AB} + \overrightarrow{DC}) \cdot (\overrightarrow{AD} + \overrightarrow{BC})\\ &= \overrightarrow{AB} \cdot \overrightarrow{AD} + \overrightarrow{AB} \cdot \overrightarrow{BC} + \overrightarrow{DC} \cdot \overrightarrow{AD} + \overrightarrow{DC} \cdot \overrightarrow{BC}\\ &= \overrightarrow{AB} \cdot \overrightarrow{AD} - \overrightarrow{BA} \cdot \overrightarrow{BC} - \overrightarrow{DC} \cdot \overrightarrow{DA} + \overrightarrow{CD} \cdot \overrightarrow{CB}\\ &= |AB| \cdot |AD| \cos\widehat{A} - |BA| \cdot |BC| \cos\widehat{B}\\ &\quad -|DC| \cdot |DA| \cos\widehat{D} + |CD| \cdot |CB| \cos\widehat{C}. \qquad (7.14.20)\end{aligned}$$

The law of cosines yields

$$\cos\widehat{A} = \frac{|AB|^2+|AD|^2-|BD|^2}{2|AB|\cdot|AD|}, \qquad \cos\widehat{B} = \frac{|AB|^2+|BC|^2-|AC|^2}{2|AB|\cdot|BC|},$$

$$\cos\widehat{C} = \frac{|DC|^2+|BC|^2-|BD|^2}{2|DC|\cdot|BC|}, \qquad \cos\widehat{D} = \frac{|DC|^2+|DA|^2-|AC|^2}{2|DC|\cdot|DA|}.$$

Substituting in (7.14.20) we obtain

$$\begin{aligned} 4\overrightarrow{MN}\cdot\overrightarrow{EF} &= \frac{1}{2}\Big(|AB|^2+|AD|^2-|BD|^2-|AB|^2-|BC|^2+|AC|^2 \\ &\quad -|DC|^2-|DA|^2+|AC|^2+|DC|^2+|BC|^2-|BD|^2\Big) \\ &= |AC|^2-|BD|^2. \end{aligned}$$

Solved Exercise 7.14.4 *Let ABC be a triangle. Let $AB'B''B$, $AC'C''C$ be squares constructed on the sides $[AB]$, $[AC]$ in the exterior of triangle ABC. Let O be the midpoint of $[BC]$ and let M, N be the centers of the squares $AB'B''B$ and $AC'C''C$, see Figure 7.22.*

(i) Show that $|B'C| = |C'B|$.

(ii) Prove that $B'C \perp C'B$.

(iii) Show that $|OM| = |ON|$ and $OM \perp ON$.

(iv) Show that

$$|BC|^2+|B'C'|^2 = 2(|AB|^2+|AC|^2).$$

(v) Let Q be the midpoint of $[B'C']$. Show that $|OQ| = |MN|$.

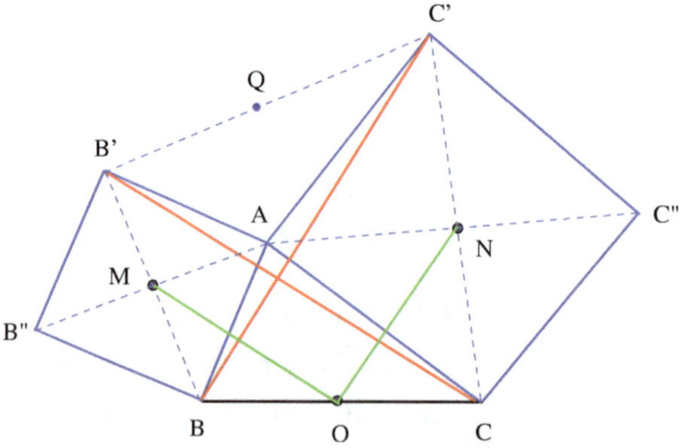

Figure 7.22: For Exercise 7.14.4.

Proof: (*i*) Since $\overrightarrow{BC'} = \overrightarrow{BA} + \overrightarrow{AC'}$, we have

$$\begin{aligned}|BC'|^2 &= |BA|^2 + |AC'|^2 + 2\overrightarrow{BA}\cdot\overrightarrow{AC'}\\ &= |BA|^2 + |AC'|^2 - 2|BA|\cdot|AC'|\cos\widehat{BAC'}. \end{aligned} \tag{7.14.21}$$

Using $\overrightarrow{CB'} = \overrightarrow{CA} + \overrightarrow{AB'}$ we obtain

$$\begin{aligned}|CB'|^2 &= |CA|^2 + |AB'|^2 + 2\overrightarrow{CA}\cdot\overrightarrow{AB'}\\ &= |CA|^2 + |AB'|^2 - 2|CA|^2\cdot|AB'|\cos\widehat{CAB'}. \end{aligned} \tag{7.14.22}$$

Since $\widehat{BAC'} = \widehat{A} + 90^\circ = \widehat{CAB'}$, $|BA| = |AB'|$ and $|AC| = |AC'|$ from (7.14.21) and (7.14.22) we obtain $|BC'| = |CB'|$.

(*ii*) Using the properties of the inner product, we obtain

$$\begin{aligned}\overrightarrow{BC'}\cdot\overrightarrow{CB'} &= (\overrightarrow{BA}+\overrightarrow{AC'})\cdot(\overrightarrow{CA}+\overrightarrow{AB'})\\ &= \overrightarrow{BA}\cdot\overrightarrow{CA} + \underbrace{\overrightarrow{BA}\cdot\overrightarrow{AB'}}_{=0} + \underbrace{\overrightarrow{AC'}\cdot\overrightarrow{CA}}_{=0} + \overrightarrow{AC'}\cdot\overrightarrow{AB'}\\ &= \overrightarrow{AB}\cdot\overrightarrow{AC} + \overrightarrow{AC'}\cdot\overrightarrow{AB'}\\ &= |AB|\cdot|AC|\cos\widehat{A} + |AC'|\cdot|AB'|\cos(360^\circ - 90^\circ - 90^\circ - \widehat{A})\\ &= |AB|\cdot|AC|\cos\widehat{A} + |AC|\cdot|AB|\cos(180^\circ - \widehat{A})\\ &= 0,\end{aligned}$$

since $\cos(180^\circ - \widehat{A}) = -\cos\widehat{A}$. Hence $\overrightarrow{BC'}\cdot\overrightarrow{CB'} = 0$, *i.e.*, $BC' \perp CB'$.

(*iii*) The midpoint connector theorem yields

$$\overrightarrow{MO} = \frac{1}{2}\overrightarrow{B'C}, \qquad \overrightarrow{NO} = \frac{1}{2}\overrightarrow{C'B},$$

and apply parts (*i*) and (*ii*).

(*iv*) The diagonals $[BC']$, $[B'C]$ in the quadrilateral $BCC'B'$ are perpendicular. Using Proposition 7.14.1 yields

$$|BC|^2 + |B'C'|^2 = |BB'|^2 + |CC'|^2. \tag{7.14.23}$$

The Pythagorean theorem applied to triangles ABB' and ACC' yields

$$\begin{aligned} |BB'|^2 &= |AB|^2 + |AB'|^2 = 2|AB|^2, \\ |CC'|^2 &= |AC|^2 + |AC'|^2 = 2|AC|^2. \end{aligned}$$

Substituting in (7.14.23) yields the desired identity.

(*v*) $[OQ]$ and $[MN]$ are bimedians in the quadrilateral $BCC'B'$. By Proposition 7.14.2 we obtain $|OQ| = |MN|$ if and only if the diagonals BC' and $B'C$ are equal, then apply part (*i*). ∎

Exercise 7.14.5 *Let M, N be the midpoints of the sides $[AB]$ and $[DC]$ of the quadrilateral $ABCD$. Show that*

$$|MN|^2 = \frac{1}{4}\Big(|AD|^2 + |BC|^2 + 2\overrightarrow{AD} \cdot \overrightarrow{BC}\Big).$$

Proposition 7.14.6 (Euler's relation) *Let $ABCD$ be a convex quadrilateral. Denote by M, N, E, F, O_1, O_2 the midpoints of the sides $[AB]$, $[DC]$, $[AD]$, $[BC]$ and the diagonals $[AC]$, $[BD]$, respectively. Then*

$$4|O_1O_2|^2 + 2(|MN|^2 + |EF|^2) = |AB|^2 + |BC|^2 + |CD|^2 + |DA|^2. \tag{7.14.24}$$

Proof: We shall refer to Figure 7.23. Using Exercise 7.14.5 and Proposition 6.4.4, yields

$$\begin{aligned} 4|MN|^2 &= |AD|^2 + |BC|^2 + 2\overrightarrow{AD} \cdot \overrightarrow{BC} \\ 4|O_1O_2|^2 &= |AD|^2 + |BC|^2 - 2\overrightarrow{AD} \cdot \overrightarrow{BC}. \end{aligned}$$

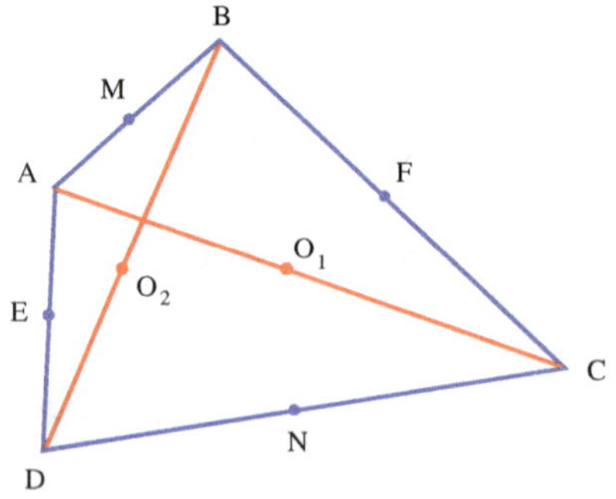

Figure 7.23: For Euler's relation.

Adding the previous two relations we obtain

$$\begin{aligned} 4|MN|^2 + 4|O_1O_2|^2 &= 2(|AD|^2 + |BC|^2) \Longleftrightarrow \\ 2|MN|^2 + 2|O_1O_2|^2 &= |AD|^2 + |BC|^2. \end{aligned} \tag{7.14.25}$$

In a similar way

$$2|EF|^2 + 2|O_1O_2|^2 = |AB|^2 + |DC|^2. \tag{7.14.26}$$

Adding relations (7.14.25) and (7.14.26) yields (7.14.24). ■

The rest of this section will deal with vectorial relations among the elements of a quadrilateral. The quadrilateral $ABCD$ will be considered convex and let $\{O\} = AC \cap BD$. Let $\overrightarrow{a} = \overrightarrow{AB}$, $\overrightarrow{b} = \overrightarrow{BC}$, $\overrightarrow{c} = \overrightarrow{CD}$, $\overrightarrow{d} = \overrightarrow{DA}$, $\overrightarrow{d_1} = \overrightarrow{AC}$, $\overrightarrow{d_2} = \overrightarrow{BD}$, $\overrightarrow{\gamma} = \overrightarrow{AO}$, $\overrightarrow{\delta} = \overrightarrow{OC}$, $\overrightarrow{\alpha} = \overrightarrow{BO}$, $\overrightarrow{\beta} = \overrightarrow{OD}$, see Figure 7.24.

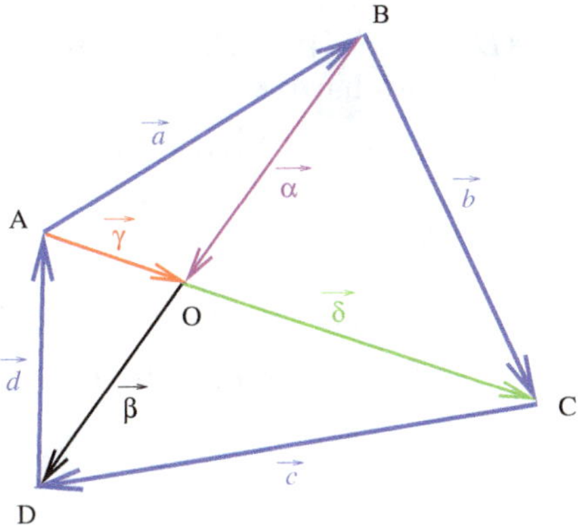

Figure 7.24: Vectors in a convex quadrilateral.

Lemma 7.14.7 *With the above notations we have*

$$\overrightarrow{d_1} \cdot \overrightarrow{d_2} = \overrightarrow{a} \cdot \overrightarrow{c} - \overrightarrow{b} \cdot \overrightarrow{d}.$$

Proof: We shall compute the inner products in both sides. The left side becomes

$$\overrightarrow{d_1} \cdot \overrightarrow{d_2} = (\overrightarrow{\gamma} + \overrightarrow{\delta}) \cdot (\overrightarrow{\alpha} + \overrightarrow{\beta}) = \overrightarrow{\gamma} \cdot \overrightarrow{\alpha} + \overrightarrow{\gamma} \cdot \overrightarrow{\beta} + \overrightarrow{\delta} \cdot \overrightarrow{\alpha} + \overrightarrow{\delta} \cdot \overrightarrow{\beta}. \tag{7.14.27}$$

The right side can be computed as

$$\begin{aligned}\overrightarrow{a}\cdot\overrightarrow{c}-\overrightarrow{b}\cdot\overrightarrow{d} &= (\overrightarrow{\gamma}-\overrightarrow{\alpha})\cdot(\overrightarrow{\beta}-\overrightarrow{\delta})-(\overrightarrow{\alpha}+\overrightarrow{\delta})\cdot(-\overrightarrow{\gamma}-\overrightarrow{\beta})\\ &= \overrightarrow{\gamma}\cdot\overrightarrow{\beta}-\overrightarrow{\gamma}\cdot\overrightarrow{\delta}-\overrightarrow{\alpha}\cdot\overrightarrow{\beta}+\overrightarrow{\alpha}\cdot\overrightarrow{\delta}+\overrightarrow{\alpha}\cdot\overrightarrow{\gamma}\\ &\quad+\overrightarrow{\alpha}\cdot\overrightarrow{\beta}+\overrightarrow{\delta}\cdot\overrightarrow{\gamma}+\overrightarrow{\delta}\cdot\overrightarrow{\beta}\\ &= \overrightarrow{\gamma}\cdot\overrightarrow{\beta}+\overrightarrow{\alpha}\cdot\overrightarrow{\delta}+\overrightarrow{\alpha}\cdot\overrightarrow{\gamma}+\overrightarrow{\delta}\cdot\overrightarrow{\beta}. \qquad (7.14.28)\end{aligned}$$

Comparing (7.14.27) and (7.14.28) yields $\overrightarrow{d_1}\cdot\overrightarrow{d_2}=\overrightarrow{a}\cdot\overrightarrow{c}-\overrightarrow{b}\cdot\overrightarrow{d}$. ■

Solved Exercise 7.14.8 *With the above notations show that*
(*i*) $\overrightarrow{a}\cdot\overrightarrow{c}=a\,c\cos(\widehat{A}+\widehat{D})$, $\overrightarrow{b}\cdot\overrightarrow{d}=b\,d\cos(\widehat{C}+\widehat{D})$

(*ii*) $d_1\,d_2=\dfrac{ac\cos(\widehat{A}+\widehat{D})-bd\cos(\widehat{C}+\widehat{D})}{\cos\varphi}$,

where φ *is the angle between the diagonals* $\overrightarrow{d_1}$ *and* $\overrightarrow{d_2}$,

(*iii*) $d_1d_2\le\dfrac{ac+bd}{|\cos\phi|}$.

Proof: (*i*) Let $\{V\}=AD\cap BC$, $\{U\}=AB\cap DC$ and $\widehat{u}=\widehat{AUD}$, $\widehat{v}=\widehat{AVB}$. In triangles VDC and UAD we have

$$180^\circ-\widehat{u}=\widehat{A}+\widehat{D},\qquad 180^\circ-\widehat{v}=\widehat{D}+\widehat{C}.$$

Since $(\widehat{\overrightarrow{a},\overrightarrow{c}})=180^\circ-\widehat{u}$, $(\widehat{\overrightarrow{b},\overrightarrow{d}})=180^\circ-\widehat{v}$, we obtain

$$\begin{aligned}\overrightarrow{a}\cdot\overrightarrow{c} &= a\,c\cos(180-\widehat{u})=a\,c\cos(\widehat{A}+\widehat{D})\\ \overrightarrow{b}\cdot\overrightarrow{d} &= b\,d\cos(180-\widehat{v})=b\,d\cos(\widehat{D}+\widehat{C}).\end{aligned}$$

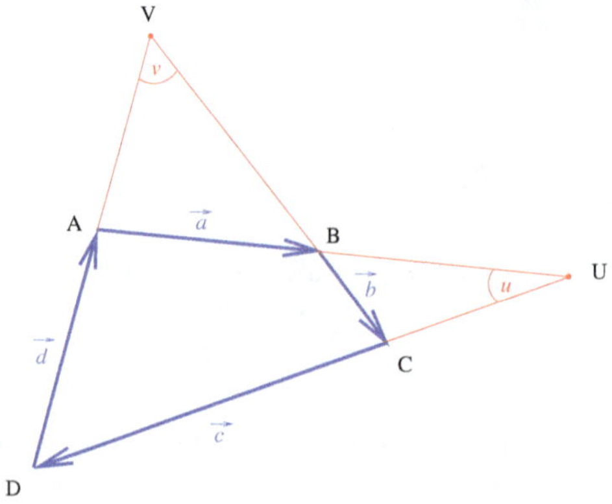

Figure 7.25: For Exercise 7.14.8.

(*ii*) We shall express $\overrightarrow{d_1} \cdot \overrightarrow{d_2}$ in two ways. From the definition of the inner product

$$\overrightarrow{d_1} \cdot \overrightarrow{d_2} = d_1\, d_2 \cos\varphi. \tag{7.14.29}$$

From Lemma 7.14.7

$$\overrightarrow{d_1} \cdot \overrightarrow{d_2} = \overrightarrow{a} \cdot \overrightarrow{c} - \overrightarrow{b} \cdot \overrightarrow{d} = ac\cos(\widehat{A} + \widehat{D}) - bd\cos(\widehat{D} + \widehat{C}). \tag{7.14.30}$$

Comparing (7.14.29) and (7.14.30) easily leads to the desired result.

(*iii*) It follows from (*ii*). ■

Proposition 7.14.9 *Let $ABCD$ be a convex quadrilateral and let $\varphi \in (0, 90°]$ be the acute angle between the diagonals AC and BD. With the previous notations we have*

$$\cos\varphi = \frac{|b^2 + d^2 - a^2 - c^2|}{2d_1 d_2}.$$

Proof: From Figure 3.24 we have

$$2\overrightarrow{d_1} = \overrightarrow{a} + \overrightarrow{b} - \overrightarrow{c} - \overrightarrow{d}, \qquad 2\overrightarrow{d_2} = \overrightarrow{b} + \overrightarrow{c} - \overrightarrow{a} - \overrightarrow{d}.$$

Then

$$\begin{aligned}4\overrightarrow{d_1} \cdot \overrightarrow{d_2} &= (\overrightarrow{a} + \overrightarrow{b} - \overrightarrow{c} - \overrightarrow{d}) \cdot (\overrightarrow{b} + \overrightarrow{c} - \overrightarrow{a} - \overrightarrow{d})\\
&= \overrightarrow{a} \cdot \overrightarrow{b} + \overrightarrow{a} \cdot \overrightarrow{c} - a^2 - \overrightarrow{a} \cdot \overrightarrow{d} + b^2 + \overrightarrow{b} \cdot \overrightarrow{c} - \overrightarrow{b} \cdot \overrightarrow{a} - \overrightarrow{b} \cdot \overrightarrow{d}\\
&\quad - \overrightarrow{c} \cdot \overrightarrow{b} - c^2 + \overrightarrow{c} \cdot \overrightarrow{a} + \overrightarrow{c} \cdot \overrightarrow{d} - \overrightarrow{d} \cdot \overrightarrow{b} - \overrightarrow{d} \cdot \overrightarrow{c} + \overrightarrow{d} \cdot \overrightarrow{a} + d^2\\
&= b^2 + d^2 - a^2 - c^2 + 2(\overrightarrow{a} \cdot \overrightarrow{c} - \overrightarrow{b} \cdot \overrightarrow{d}).\end{aligned}$$

Using Lemma 7.14.7 yields

$$\begin{aligned}4\overrightarrow{d_1} \cdot \overrightarrow{d_2} &= b^2 + d^2 - a^2 - c^2 + 2\overrightarrow{d_1} \cdot \overrightarrow{d_2} \Longleftrightarrow\\
2\overrightarrow{d_1} \cdot \overrightarrow{d_2} &= b^2 + d^2 - a^2 - c^2 \Longleftrightarrow\\
2d_1 d_2 \cos\varphi &= b^2 + d^2 - a^2 - c^2 \Longleftrightarrow\\
\cos\varphi &= \frac{b^2 + d^2 - a^2 - c^2}{2d_1 d_2},\end{aligned}$$

which is the desired result. ■

Corollary 7.14.10 *The diagonals AC and BD are perpendicular if and only if the sum of the squares of the opposite sides are equal:*

$$b^2 + d^2 = a^2 + c^2.$$

Proof: The diagonals are perpendicular if and only if $\cos\varphi = 0$. Then apply Proposition 7.14.9. ■

Theorem 7.14.11 *In a convex quadrilateral the sums of the squares of the opposite sides are equal if and only if the bimedians are equal.*

Proof: Let M, N, U, and V be the midpoints of segment lines $[AB]$, $[DC]$, $[BC]$, and $[AD]$, see Figure 7.26. We need to show

$$|MN| = |UV| \iff |AB|^2 + |DC|^2 = |AD|^2 + |BC|^2.$$

With the above notations, Lemma 7.14.7 can be written as

$$\overrightarrow{AC}\cdot\overrightarrow{BD} = \overrightarrow{AB}\cdot\overrightarrow{CD} - \overrightarrow{BC}\cdot\overrightarrow{DA}. \qquad (7.14.31)$$

Using Proposition 6.4.2 we have

$$\begin{aligned} 2\overrightarrow{NM} &= \overrightarrow{DA} + \overrightarrow{CB} \Longrightarrow 4|NM|^2 = |DA|^2 + |CB|^2 - 2\overrightarrow{BC}\cdot\overrightarrow{DA} \\ 2\overrightarrow{VU} &= \overrightarrow{AB} + \overrightarrow{DC} \Longrightarrow 4|VU|^2 = |AB|^2 + |DC|^2 - 2\overrightarrow{AB}\cdot\overrightarrow{CD}. \end{aligned}$$

Subtracting,

$$4(|NM|^2 - |VU|^2) = |DA|^2 + |CB|^2 - |AB|^2 - |DC|^2 + 2(\overrightarrow{AB}\cdot\overrightarrow{CD} - \overrightarrow{BC}\cdot\overrightarrow{DA})$$

and using (7.14.31) yields

$$\begin{aligned} 4(|NM|^2 - |VU|^2) &= |DA|^2 + |CB|^2 - |AB|^2 - |DC|^2 + 2\overrightarrow{AC}\cdot\overrightarrow{BD} \\ &= d^2 + b^2 - a^2 - c^2 + 2\overrightarrow{d_1}\cdot\overrightarrow{d_2} \\ &= d^2 + b^2 - a^2 - c^2 + 2d_1 d_2 \cos\varphi. \end{aligned}$$

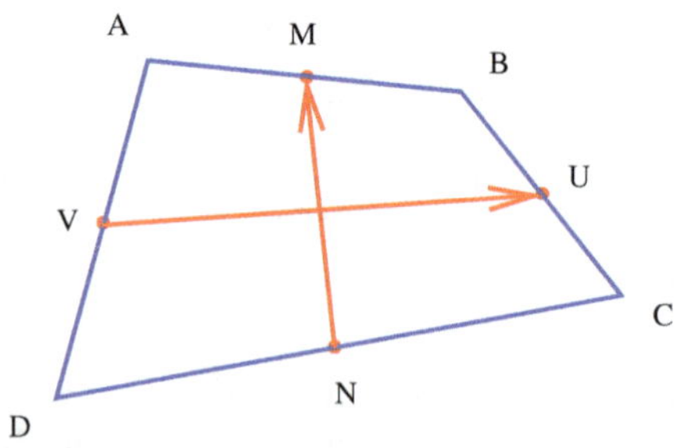

Figure 7.26: Perpendicular bimedians in a convex quadrilateral.

Substituting $\cos\varphi$ from Proposition 7.14.9 yields

$$\begin{aligned} 4(|NM|^2 - |VU|^2) &= d^2 + b^2 - a^2 - c^2 + 2d_1d_2\frac{b^2 + d^2 - a^2 - c^2}{2d_1d_2} \\ &= 2(d^2 + b^2 - a^2 - c^2). \end{aligned}$$

Hence

$$|NM|^2 - |VU|^2 = \frac{1}{2}(d^2 + b^2 - a^2 - c^2)$$

and $|NM| = |VU|$ if and only if $a^2 + c^2 = d^2 + b^2$, which means $\cos\varphi = 0$, or $\varphi = 90°$, and hence then $AC \perp BD$. The reasoning can be easily inverted. ■

Corollary 7.14.12 *In a convex quadrilateral the following statements are equivalent:*

(i) The diagonals are perpendicular;

(ii) The sums of the squares of the opposite sides are equal;

(iii) The bimedians have equal lengths.

Remark 7.14.13 The equivalence between (i) and (iii) can be shown directly as follows. By Proposition 6.4.2 we express the bimedian vectors in terms of diagonals as

$$\overrightarrow{NM} = \frac{1}{2}(\overrightarrow{CA} + \overrightarrow{DB}), \qquad \overrightarrow{VU} = \frac{1}{2}(\overrightarrow{AC} + \overrightarrow{DB}).$$

Then we have the following equivalencies

$$\begin{aligned} &|NM|^2 = |VU|^2 \Longleftrightarrow \overrightarrow{NM} \cdot \overrightarrow{NM} = \overrightarrow{VU} \cdot \overrightarrow{VU} \Longleftrightarrow \\ &|CA|^2 + |DB|^2 + 2\overrightarrow{CA} \cdot \overrightarrow{DB} = |AC|^2 + |DB|^2 + 2\overrightarrow{AC} \cdot \overrightarrow{DB} \Longleftrightarrow \\ &\overrightarrow{CA} \cdot \overrightarrow{DB} - \overrightarrow{AC} \cdot \overrightarrow{DB} = 0 \Longleftrightarrow \\ &2\overrightarrow{CA} \cdot \overrightarrow{DB} = 0 \Longleftrightarrow CA \perp DB. \end{aligned}$$

In the following we shall supply a consequence of Proposition 7.14.9 in computing the area of a quadrilateral in terms of its sides and diagonals. We shall recall first a formula for the area $\mathcal{A}$ of a convex quadrilateral with diagonals lengths d_1, d_2 and angle φ between diagonals

$$\mathcal{A} = \frac{1}{2}d_1d_2\sin\varphi.$$

Corollary 7.14.14 *Let $\mathcal{A}$ be the area of a convex quadrilateral with the side length a, b, c, d (in this order) and diagonals lengths d_1, d_2. Then*

$$16\mathcal{A}^2 = 4d_1^2d_2^2 - (b^2 + d^2 - a^2 - c^2)^2. \tag{7.14.32}$$

Proof: We have

$$\begin{aligned} 4\mathcal{A}^2 &= d_1^2d_2^2 \sin^2\varphi = d_1^2d_2^2(1 - \cos^2\varphi) \\ &= d_1^2d_2^2 - d_1^2d_2^2\frac{(b^2 + d^2 - a^2 - c^2)^2}{4d_1^2d_2^2} \\ &= d_1^2d_2^2 - \frac{1}{4}(b^2 + d^2 - a^2 - c^2)^2. \end{aligned}$$

Multiplying by 4 we obtain (7.14.32). ■

7.15 Solved exercises

Solved Exercise 7.15.1 *In any convex quadrilateral*

$$d_1d_2 > \frac{|b^2 + d^2 - a^2 - c^2|}{2}.$$

Proof: We may use that $\mathcal{A} > 0$ or $|\cos\varphi| < 1$. ■

Solved Exercise 7.15.2 *Let $ABCD$ be a rectangle and let E be the point where the perpendicular line through A intersects the line BC, see Figure 7.27(a). Show that $|AB|^2 = |EB| \cdot |CB|$.*

Proof: Since $EB \perp AB$ we have

$$0 = \overrightarrow{EB} \cdot \overrightarrow{AB} = \overrightarrow{EB} \cdot (\overrightarrow{AD} + \overrightarrow{DB}) = \overrightarrow{EB} \cdot \overrightarrow{AD} + \overrightarrow{EB} \cdot \overrightarrow{DB} \tag{7.15.33}$$

Using the triangle inequality we can write

$$\begin{aligned} |AB|^2 &= \overrightarrow{AB} \cdot \overrightarrow{AB} = (\overrightarrow{AE} + \overrightarrow{EB}) \cdot (\overrightarrow{AD} + \overrightarrow{DB}) \\ &= \overrightarrow{AE} \cdot \overrightarrow{AD} + \underbrace{\overrightarrow{AE} \cdot \overrightarrow{DB}}_{=0} + \underbrace{\overrightarrow{EB} \cdot \overrightarrow{AD} + \overrightarrow{EB} \cdot \overrightarrow{DB}}_{=0} \\ &= \overrightarrow{AE} \cdot \overrightarrow{AD}, \end{aligned}$$

where we used $AE \perp DB$ and equation (7.15.33). Hence

$$\begin{aligned} |AB|^2 &= \overrightarrow{AE} \cdot \overrightarrow{AD} = \overrightarrow{AE} \cdot \overrightarrow{BC} = \overrightarrow{EA} \cdot \overrightarrow{CB} \\ &= |EA| \cos\widehat{AEB} \cdot |CB| = |EB| \cdot |CB|. \end{aligned}$$

■

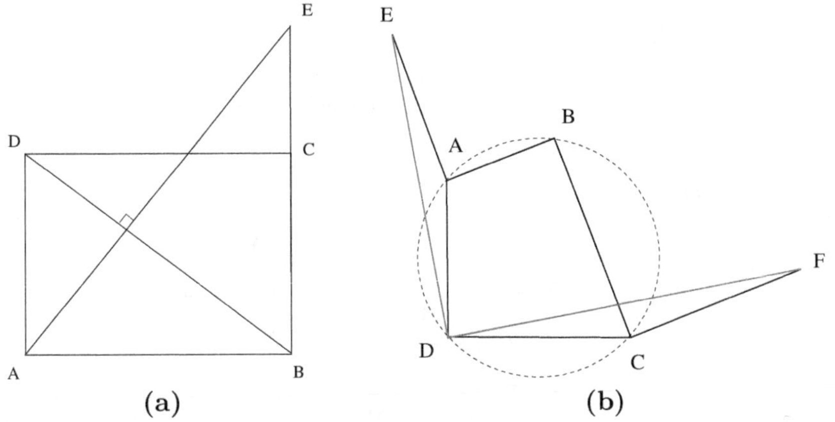

Figure 7.27: (a) For Exercise 7.15.2; (b) For Exercise 7.15.3.

Solved Exercise 7.15.3 *Let $ABCD$ be a quadrilateral. Construct points E and F such that $AE||BC$, $|AE| = |AD|$, $CF||AB$, $|CF| = |CD|$. Show that $DF \perp DE$ if and only if $\widehat{A} + \widehat{C} = 180°$, i.e., $ABCD$ is cyclic. See Figure 7.27(b).*

Proof: We have $\overrightarrow{AE} = \frac{|AD|}{|BC|}\overrightarrow{CB}$ and $\overrightarrow{CF} = \frac{|DC|}{|AB|}\overrightarrow{AB}$. Using $\overrightarrow{DE} = \overrightarrow{DA} + \overrightarrow{AE}$ and $\overrightarrow{DF} = \overrightarrow{DC} + \overrightarrow{CF}$, we have

$$\begin{aligned}
\overrightarrow{DE} \cdot \overrightarrow{DF} &= (\overrightarrow{DA} + \overrightarrow{AE}) \cdot (\overrightarrow{DC} + \overrightarrow{CF}) \\
&= \overrightarrow{DA} \cdot \overrightarrow{DC} + \overrightarrow{DA} \cdot \overrightarrow{AB}\,\frac{|DC|}{|AB|} + \overrightarrow{CB} \cdot \overrightarrow{DC}\,\frac{|AD|}{|BC|} \\
&\quad + \overrightarrow{CB} \cdot \overrightarrow{AB}\,\frac{|AD|}{|BC|}\,\frac{|DC|}{|AB|} \\
&= |DA| \cdot |DC| \cos \widehat{D} - |DA| \cdot |AB| \cos \widehat{A} \frac{|DC|}{|AB|}
\end{aligned}$$

$$\begin{aligned}
&\quad -|CB| \cdot |DC| \cos \widehat{C}\,\frac{|AD|}{|BC|} + |CB| \cdot |AB| \cos \widehat{B}\,\frac{|AD|}{|BC|}\,\frac{|DC|}{|AB|} \\
&= |DA| \cdot |DC| \cos \widehat{D} - |DA| \cdot |DC| \cos \widehat{A} \\
&\quad -|DC| \cdot |AD| \cos \widehat{C} + |AD| \cdot |DC| \cos \widehat{B} \\
&= |DA| \cdot |DC| (\cos \widehat{D} + \cos \widehat{B} - \cos \widehat{A} - \cos \widehat{C}) \\
&= 2|DA| \cdot |DC| \left(\cos \frac{\widehat{B} + \widehat{D}}{2} \cos \frac{\widehat{B} - \widehat{D}}{2} - \cos \frac{\widehat{A} + \widehat{C}}{2} \cos \frac{\widehat{A} - \widehat{C}}{2} \right).
\end{aligned}$$

Since $\widehat{A}+\widehat{B}+\widehat{C}+\widehat{D}=360^\circ$, we have $\dfrac{\widehat{A}+\widehat{C}}{2}=180^\circ-\dfrac{\widehat{B}+\widehat{D}}{2}$ and $\cos\dfrac{\widehat{A}+\widehat{C}}{2}=-\cos\dfrac{\widehat{B}+\widehat{D}}{2}$. Transforming the difference into product, we obtain

$$\begin{aligned}
&\overrightarrow{DE}\cdot\overrightarrow{DF}\\
&= 2|DA|\cdot|DC|\cos\frac{\widehat{B}+\widehat{D}}{2}\left[\cos\frac{\widehat{B}-\widehat{D}}{2}-\cos\frac{\widehat{A}-\widehat{C}}{2}\right]\\
&= 2|DA|\cdot|DC|\cos\frac{\widehat{B}+\widehat{D}}{4}\cos\frac{\widehat{B}-\widehat{D}+\widehat{A}-\widehat{C}}{4}\cos\frac{\widehat{B}-\widehat{D}-\widehat{A}+\widehat{C}}{4}.
\end{aligned}$$

Hence $\overrightarrow{DE}\cdot\overrightarrow{DF}=0$ if and only if at least one of the above trigonometric functions vanish. We have three cases to investigate:

1) $\cos\dfrac{\widehat{B}+\widehat{D}}{2}=0\iff\dfrac{\widehat{B}+\widehat{D}}{2}=90^\circ\iff\widehat{B}+\widehat{D}=180^\circ\iff\widehat{A}+\widehat{C}=180^\circ$, *i.e.*, $ABCD$ is cyclic.

2) $\cos\dfrac{\widehat{B}-\widehat{D}+\widehat{A}-\widehat{C}}{4}=0\iff\widehat{A}+\widehat{B}-(\widehat{C}+\widehat{D})=360^\circ$, *i.e.*, $\widehat{A}+\widehat{B}=360^\circ+(\widehat{C}+\widehat{D})>360^\circ$, which contradicts the obvious inequality $\widehat{A}+\widehat{B}<\widehat{A}+\widehat{B}+\widehat{C}+\widehat{D}=360^\circ$.

3) $\cos\dfrac{\widehat{B}-\widehat{D}-\widehat{A}+\widehat{C}}{4}=0\iff\widehat{B}+\widehat{C}-\widehat{D}-\widehat{A}=360^\circ$ and hence $\widehat{B}+\widehat{C}=360^\circ+\widehat{A}+\widehat{D}>360^\circ$, which is again a contradiction.

Hence only case 1) may take place.

■

Exercise 7.15.4 *Let $ABCD$ be a quadrilateral such that $\widehat{A}+\widehat{C}=90^\circ$ (or $\widehat{B}+\widehat{D}=90^\circ$). Show that*

$$|AC|^2\cdot|DB|^2=|AB|^2\cdot|CD|^2+|BC|^2\cdot|AD|^2.$$

Exercise 7.15.5 *Let p,q be the length of the diagonals of a rhombus with side length ℓ. Show that $p^2+q^2=4\ell^2$.*

Exercise 7.15.6 *Let $ABCD$ be a quadrilateral such that $|AB|=|CD|$. Show that the set of points M such that*

$$|MA|^2+|MB|^2=|MC|^2+|MD|^2$$

is a line perpendicular to the bimedian which joins the sides AB and CD.

Exercise 7.15.7 *Let $ABCD$ be a trapezoid. Show that*

$$|AC|^2 + |BD|^2 = |AD|^2 + |BC|^2 + 2|AB| \cdot |DC|.$$

Exercise 7.15.8 *Let $ABCD$ be a rectangle and $M \in [BD]$ be an arbitrary point. Show that*
(i) $|MB|^2 \cdot |AD|^2 + |MD|^2 \cdot |AB|^2 = |MA|^2 \cdot |BD|^2$
(ii) $|MB|^2 \cdot |CD|^2 + |MD|^2 \cdot |CB|^2 = |MC|^2 \cdot |BD|^2$
(iii) $|MB|^2 + |MD|^2 = |MA|^2 + |MC|^2$.

Exercise 7.15.9 *Let $ABCD$ be a square and let M, N, P, Q be the midpoints of its sides. Show that for any point O in the plane, the expression*

$$|OA|^2 + |OB|^2 + |OC|^2 + |OD|^2 - |OM|^2 - |ON|^2 - |OP|^2 - |OQ|^2$$

is constant. Find the value of the constant in terms of the side of the square. See Figure 7.28(b).

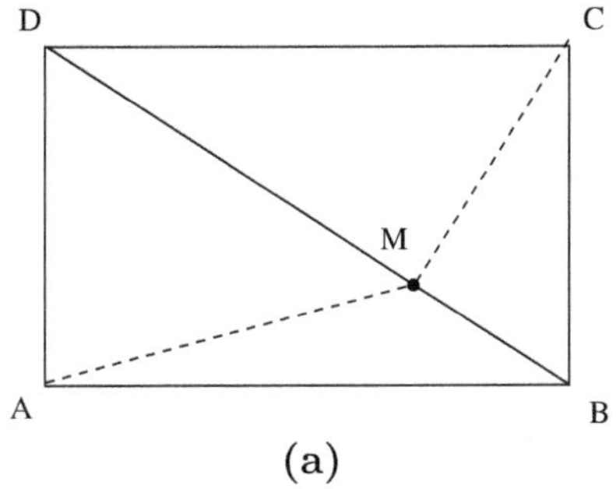

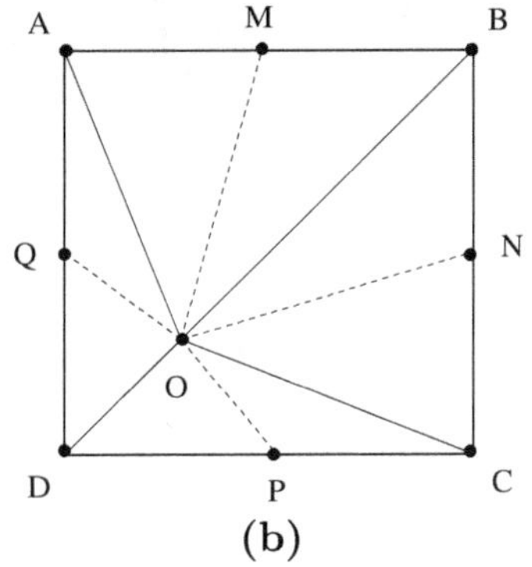

Figure 7.28: (a) For Exercise 7.15.8; (b) For Exercise 7.15.9.

Chapter 8

The Cross Product and its Applications

Nature and Nature's laws lay hid in night:
God said, "Let Newton be!" and all was light.
Alexander Pope (1688–1744)

This chapter presents the cross product of two vectors, which is also called the vectorial product, or the exterior product, since its outcome is a vector. It will be mainly used to problems involving area.

8.1 Positive oriented basis

Let $\{e_1, e_2\}$ be the canonical basis of $\mathbb{R}^2$ given by $e_1 = (1, 0)$ and $e_2 = (0, 1)$. A basis $\{\overrightarrow{u}, \overrightarrow{v}\}$ in $\mathbb{R}^2$ is called *positive oriented* if it has the same orientation as $\{e_1, e_2\}$, namely, if

$$\det \begin{pmatrix} u^1 & u^2 \\ v^1 & v^2 \end{pmatrix} > 0,$$

where

$$\overrightarrow{u} = u^1 e_1 + u^2 e_2, \qquad \overrightarrow{v} = v^1 e_1 + v^2 e_2.$$

If the above determinant is negative, the basis $\{\overrightarrow{u}, \overrightarrow{v}\}$ is called *negative oriented.*

Exercise 8.1.1 *If $\{\overrightarrow{u}, \overrightarrow{v}\}$ is a positive oriented basis, then show that $\{-\overrightarrow{u}, \overrightarrow{v}\}$ and $\{\overrightarrow{u}, -\overrightarrow{v}\}$ are negative oriented bases.*

In a similar way, we can define a positive oriented basis in $\mathbb{R}^3$. To this end, let $e_1 = (1,0,0)$, $e_2 = (0,1,0)$ and $e_3 = (0,0,1)$ be the canonical basis in $\mathbb{R}^3$, and let $\{\vec{u}, \vec{v}, \vec{w}\}$ be another base defined by

$$\begin{aligned} \vec{u} &= u^1 e_1 + u^2 e_2 + u^3 e_3, \\ \vec{v} &= v^1 e_1 + v^2 e_2 + v^3 e_3, \text{and} \\ \vec{w} &= w^1 e_1 + w^2 e_2 + w^3 e_3. \end{aligned}$$

The basis $\{\vec{u}, \vec{v}, \vec{w}\}$ is called *positive oriented* if

$$\det \begin{pmatrix} u^1 & u^2 & u^3 \\ v^1 & v^2 & v^3 \\ w^1 & w^2 & w^3 \end{pmatrix} > 0.$$

Exercise 8.1.2 *Determine whether the following triplets are positive oriented bases:*

$$\begin{aligned} (i)\ \vec{u} &= (1,3,0), & \vec{v} &= (0,1,4), & \vec{w} &= (3,0,2), \\ (ii)\ \vec{u} &= (2,0,2), & \vec{v} &= (0,1,0), & \vec{w} &= (5,0,3), \\ (iii)\ \vec{u} &= (3,2,1), & \vec{v} &= (1,0,4), & \vec{w} &= (7,4,6). \end{aligned}$$

Exercise 8.1.3 *Find all the values of x for which the vectors*

$$\vec{u} = (x,0,1), \quad \vec{v} = (1,0,x), \quad \vec{w} = (1,x,0)$$

form a positive oriented basis in $\mathbb{R}^3$.

8.2 Definition and properties

Definition 8.2.1 *Let $\vec{u}$ and $\vec{v}$ be two vectors in the plane $\mathcal{P}$. The cross product (vector product) of $\vec{u}$ and $\vec{v}$ is a vector denoted by $\vec{u} \times \vec{v}$ defined by the following properties, see Figure 8.1:*

1) *direction: $\vec{u} \times \vec{v}$ is a vector perpendicular to $\mathcal{P}$,*

2) *orientation: the triplet $\{\vec{u}, \vec{v}, \vec{u} \times \vec{v}\}$ is positive oriented, and*

3) *length: $|\vec{u} \times \vec{v}| = |\vec{u}||\vec{v}| \sin\varphi$, where $\varphi \in [0, 180°]$ is the angle formed by the vectors $\vec{u}$ and $\vec{v}$.*

Remark 8.2.2 The vector $\vec{u} \times \vec{v}$ is perpendicular to both vectors $\vec{u}$ and $\vec{v}$.

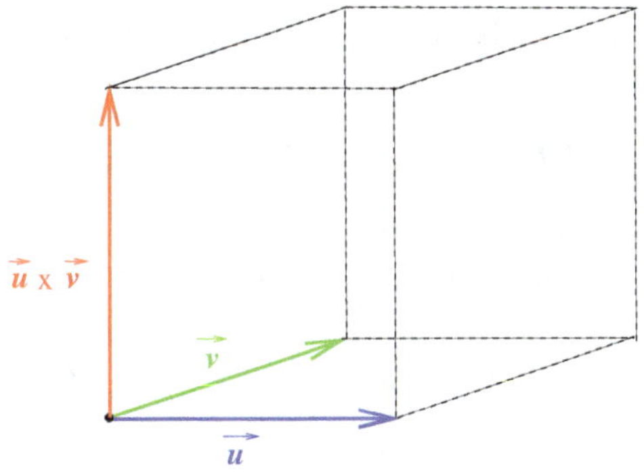

Figure 8.1: The cross product of two vectors.

Exercise 8.2.3 *Let $\{e_1, e_2\}$ be a basis in the plane $\mathcal{P}$. If*

$$u = u^1 e_1 + u^2 e_2, \quad v = v^1 e_1 + v^2 e_2$$

show that

$$\begin{aligned} e_1 \times (u^1 e_1 + u^2 e_2) &= u^2 e_1 \times e_2 \\ e_2 \times (u^1 e_1 + u^2 e_2) &= u^1 e_2 \times e_1 \\ e_1 \times e_2 &= -e_2 \times e_1 \\ e_1 \times e_2 &= e_3. \end{aligned}$$

Solved Exercise 8.2.4 *Consider the notations from Exercise 8.2.3. Show that*

$$\vec{u} \times \vec{v} = (u^1 v^2 - u^2 v^1) e_3.$$

Proof: Using Exercise 8.2.3 we obtain

$$\begin{aligned} \vec{u} \times \vec{v} &= (u^1 e_1 + u^2 e_2) \times (v^1 e_1 + v^2 e_2) \\ &= u^1 v^2 e_1 \times e_2 + u^2 v^1 e_2 \times e_1 \\ &= u^1 v^2 e_3 - u^2 v^1 e_3 = (u^1 v^2 - u^2 v^1) e_3. \end{aligned}$$

■

Exercise 8.2.5 *Let $\vec{u} = u^1 e_1 + u^2 e_2 + u^3 e_3$ and $\vec{v} = v^1 e_1 + v^2 e_2 + v^3 e_3$. Show that*

$$\vec{u} \times \vec{v} = (u^2 v^3 - v^2 u^3) e_1 - (u^1 v^3 - v^1 u^3) e_2 + (u^1 v^2 - v^1 u^2) e_3.$$

By abuse of notation, the relation given by Exercise 8.2.5 can be written as

$$\vec{u} \times \vec{v} = \det \begin{pmatrix} u^1 & u^2 & u^3 \\ v^1 & v^2 & v^3 \\ e_1 & e_2 & e_3 \end{pmatrix}.$$

Next we shall present a few basic properties of the cross product.

Proposition 8.2.6 *For any vectors $\vec{u}, \vec{v}$ and $\vec{w}$ we have:*

(i) $\vec{u} \times \vec{v} = -\vec{v} \times \vec{u}$ *(anticommutativity),*
(ii) $\vec{u} \times (\vec{v} + \vec{w}) = \vec{u} \times \vec{v} + \vec{u} \times \vec{w}$ *(distributivity),*
(iii) $\vec{u} \times \vec{v} = 0$ *if and only if there is* $\lambda \in \mathbb{R}$ *such that* $\vec{v} = \lambda \vec{u}$*, and*
(iv) The cross product is not associative, i.e., there are $\vec{u}, \vec{v}, \vec{w}$ *such that* $\vec{u} \times (\vec{v} \times \vec{w}) \neq (\vec{u} \times \vec{v}) \times \vec{w}$.

Proof: (*i*) Using the properties of the determinants we have

$$\vec{u} \times \vec{v} = \det \begin{pmatrix} u^1 & u^2 & u^3 \\ v^1 & v^2 & v^3 \\ e_1 & e_2 & e_3 \end{pmatrix} = -\det \begin{pmatrix} v^1 & v^2 & v^3 \\ u^1 & u^2 & u^3 \\ e_1 & e_2 & e_3 \end{pmatrix} = -\vec{v} \times \vec{u}.$$

(*ii*) From the additivity property of the determinants

$$\begin{aligned} \vec{u} \times (\vec{v} + \vec{w}) &= \det \begin{pmatrix} u^1 & u^2 & u^3 \\ v^1 + w^1 & v^2 + w^2 & v^3 + w^3 \\ e_1 & e_2 & e_3 \end{pmatrix} \\ &= \det \begin{pmatrix} u^1 & u^2 & u^3 \\ v^1 & v^2 & v^3 \\ e_1 & e_2 & e_3 \end{pmatrix} + \det \begin{pmatrix} u^1 & u^2 & u^3 \\ w^1 & w^2 & w^3 \\ e_1 & e_2 & e_3 \end{pmatrix} \\ &= \vec{u} \times \vec{v} + \vec{u} \times \vec{w}. \end{aligned}$$

(*iii*) Assume $\vec{u}, \vec{v} \neq 0$. From the definition of the cross product

$$\vec{u} \times \vec{v} = 0 \Longleftrightarrow |\vec{u} \times \vec{v}| = 0 \Longleftrightarrow \sin \varphi = 0.$$

Hence $\varphi \in \{0, 180°\}$, *i.e.*, $\vec{u}$ and $\vec{v}$ have the same direction and hence they are proportional. In particular, we have $\vec{u} \times \vec{u} = 0$.

(*iv*) Choose $0 \neq u \neq v = w \neq 0$. The left side is zero, see (*iii*), while the right side is not equal to zero. ∎

8.3 Applications to area problems

No problem can stand the assault of sustained thinking.

Voltaire

One of the important properties of the cross product is its relationship with the area.

Proposition 8.3.1 *Let ABC be a triangle with the area $\mathcal{A}_{ABC}$. Then*

$$\mathcal{A}_{ABC} = \frac{1}{2}|\overrightarrow{AB} \times \overrightarrow{AC}| = \frac{1}{2}|\overrightarrow{BC} \times \overrightarrow{BA}| = \frac{1}{2}|\overrightarrow{CA} \times \overrightarrow{CB}|.$$

Proof: Let $[BD]$ be the altitude corresponding to the vertex B. Then

$$\begin{aligned} \frac{1}{2}|\overrightarrow{AB} \times \overrightarrow{AC}| &= \frac{1}{2}|AB| \cdot |AC| \sin \hat{A} \\ &= \frac{1}{2}|BD| \cdot |AC| = \mathcal{A}_{ABC}, \end{aligned}$$

where we have used the area triangle formula given by half the product between an altitude and its corresponding base, see Figure 8.2. In a similar way we obtain the other two identities. ■

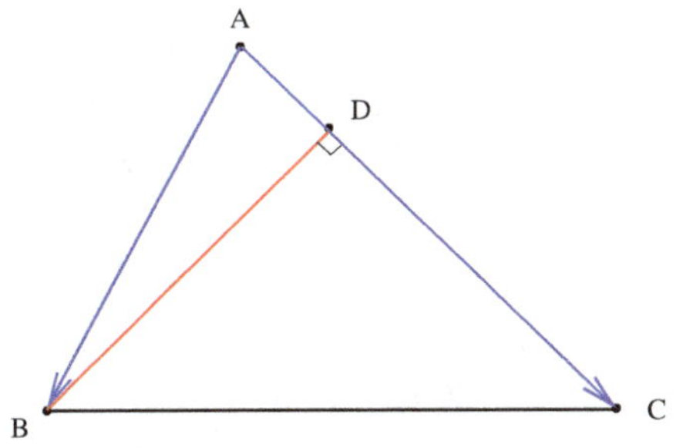

Figure 8.2: For the area of a triangle ΔABC.

Exercise 8.3.2 *Show that the area of the median triangle is 1/4 of the area of the initial triangle.*

An important application is the law of sines:

Theorem 8.3.3 (Law of sines) *Let ABC be a triangle. Then*

$$\frac{|AB|}{\sin \widehat{C}} = \frac{|AC|}{\sin \widehat{B}} = \frac{|BC|}{\sin \widehat{A}}. \tag{8.3.1}$$

Proof: From $|\overrightarrow{BC} \times \overrightarrow{BA}| = |\overrightarrow{AB} \times \overrightarrow{AC}|$ we have

$$|BC| \cdot |BA| \sin \widehat{B} = |AB| \cdot |AC| \sin \widehat{A},$$

which after dividing by $|AB|$ yields

$$|BC| \sin \widehat{B} = |AC| \sin \widehat{A} \Longleftrightarrow \frac{|BC|}{\sin \widehat{A}} = \frac{|AC|}{\sin \widehat{B}}.$$

Similarly, we obtain the relation

$$\frac{|AB|}{\sin \widehat{C}} = \frac{|BC|}{\sin \widehat{A}},$$

and hence relation (8.3.1) holds. ■

In the following we shall investigate properties of points which satisfy certain relations involving areas. The next result refers to Figure 8.3.

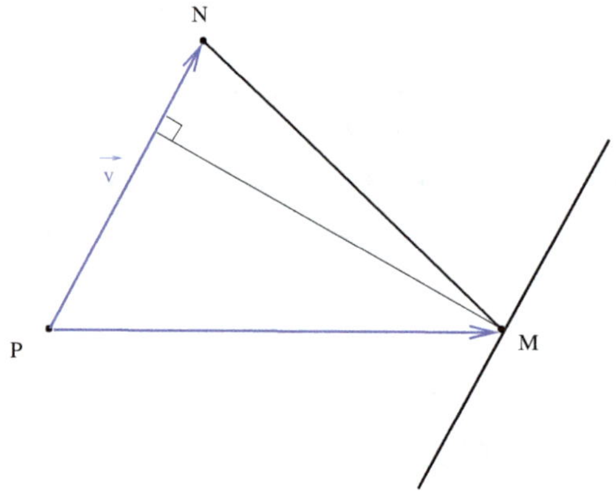

Figure 8.3: For Lemma 8.3.4.

Lemma 8.3.4 *Let P be a point and $\overrightarrow{v}$ be a vector in the plane. Then the set*

$$\{M; |\overrightarrow{PM} \times \overrightarrow{v}| = constant\}$$

is a line parallel to the support line of $\overrightarrow{v}$.

Proof: Let N be a point such that $\overrightarrow{PN} = \overrightarrow{v}$. Then by Proposition 8.3.1 we have

$$\begin{aligned} |\overrightarrow{PM} \times \overrightarrow{v}| &= |\overrightarrow{PM} \times \overrightarrow{PN}| = 2\mathcal{A}_{PMN} \\ &= |PN|\, dist(M, PN) = |\overrightarrow{v}|\, dist(M, PN). \end{aligned}$$

Then $dist(M, PN) =$ constant, and hence M will belong to a line parallel to the support line of $\overrightarrow{v}$. The distance between these two lines is $\dfrac{|\overrightarrow{PM} \times \overrightarrow{v}|}{|\overrightarrow{v}|}$.

■

Corollary 8.3.5 *Let P be a point in the plane and let $\overrightarrow{v}$ be a given vector. Then the set*

$$\{M; \overrightarrow{PM} \times \overrightarrow{v} = 0\}$$

is a line which passes through P and it is parallel to the support line of $\overrightarrow{v}$.

Proof: It follows from Lemma 8.3.4. In the following we shall provide a direct proof from the definition of the cross product.

$$\begin{aligned} \overrightarrow{PM} \times \overrightarrow{v} = 0 \quad &\Longleftrightarrow \quad |PM| \cdot |v| \sin(\overrightarrow{PM}, \overrightarrow{v}) \Longleftrightarrow \\ (\overrightarrow{PM}, \overrightarrow{v}) \in \{0, 180^\circ\} \quad &\Longleftrightarrow \quad \overrightarrow{PM} \| \overrightarrow{v}, \end{aligned}$$

and hence M belongs to a line parallel to $\overrightarrow{v}$, which contains P. ■

Solved Exercise 8.3.6 *Let δ_1, δ_2 be two lines such that $\delta_1 \cap \delta_2 = \{P\}$. Let A, $B \in \delta_1$, C, $D \in \delta_2$ be four given points. Show that the set*

$$\{M; \mathcal{A}_{MAB} = \mathcal{A}_{MCD}\}$$

is a line which passes through the point P.

Proof: From the cross product definition

$$\mathcal{A}_{MAB} = \mathcal{A}_{MCD} \Longleftrightarrow \overrightarrow{AB} \times \overrightarrow{AM} = \overrightarrow{CM} \times \overrightarrow{CD}. \tag{8.3.2}$$

Using $\overrightarrow{AM} = \overrightarrow{AP} + \overrightarrow{PM}$, $\overrightarrow{CM} = \overrightarrow{CP} + \overrightarrow{PM}$, relation (8.3.2) yields

$$\begin{aligned} \overrightarrow{AB} \times (\overrightarrow{AP} + \overrightarrow{PM}) &= (\overrightarrow{CP} + \overrightarrow{PM}) \times \overrightarrow{CD} \Longleftrightarrow \\ \underbrace{\overrightarrow{AB} \times \overrightarrow{AP}}_{=0} + \overrightarrow{AB} \times \overrightarrow{PM} &= \underbrace{\overrightarrow{CP} \times \overrightarrow{CD}}_{=0} + \overrightarrow{PM} \times \overrightarrow{CD} \Longleftrightarrow \\ \overrightarrow{AB} \times \overrightarrow{PM} &= \overrightarrow{PM} \times \overrightarrow{CD} \Longleftrightarrow \\ \overrightarrow{AB} \times \overrightarrow{PM} - \overrightarrow{PM} \times \overrightarrow{CD} &= 0 \Longleftrightarrow \\ -\overrightarrow{PM} \times \overrightarrow{AB} - \overrightarrow{PM} \times \overrightarrow{CD} &= 0 \Longleftrightarrow \\ \overrightarrow{PM} \times (\overrightarrow{AB} + \overrightarrow{CD}) &= 0. \end{aligned}$$

Let $\overrightarrow{v} = \overrightarrow{AB} + \overrightarrow{CD}$. Then M satisfies

$$\overrightarrow{PM} \times \overrightarrow{v} = 0.$$

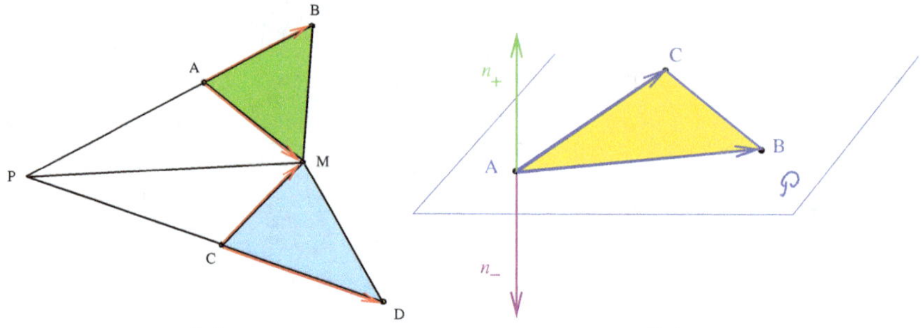

Figure 8.4: (a) For Exercise 8.3.6; (b) For Definition 8.3.7.

Applying Corollary 8.3.5 we obtain the desired result. The line passes through P since for $M = P$ both areas are vanishing, so they are equal. See Figure 8.4(a). ■

Definition 8.3.7 *Let ABC be a triangle in the plane $\mathcal{P}$. Let n_+, n_- be the upward and the downward normals of the plane $\mathcal{P}$. There are two oriented areas of the triangle ABC:*

(i) positive oriented area: $\mathcal{A}_{ABC} \cdot n_+$, and

(ii) negative oriented area: $\mathcal{A}_{ABC} \cdot n_-$, see Figure 8.4(b).

We shall denote by $\overrightarrow{\mathcal{A}}_{ABC}$ either of the above orientations. We shall understand from the context if it is the positive or the negative orientation:

$$\overrightarrow{\mathcal{A}}_{ABC} = \frac{1}{2}(\overrightarrow{AB} \times \overrightarrow{AC}) = \mathcal{A}_{ABC} \cdot n_+,$$

or

$$\overrightarrow{\mathcal{A}}_{ABC} = \frac{1}{2}(\overrightarrow{BA} \times \overrightarrow{BC}) = \mathcal{A}_{ABC} \cdot n_-.$$

In this way the oriented area becomes a vector perpendicular to the plane $\mathcal{P}$. When solving a problem, which involves areas, the areas have to be considered with the same orientation. The next exercise is using this fact.

Solved Exercise 8.3.8 *Let ℓ_1, ℓ_2 be two non-parallel lines and consider the points A, $B \in \ell_1$, and C, $D \in \ell_2$ fixed. Show that the set*

$$\{M; \mathcal{A}_{ABM} + \mathcal{A}_{CDM} = constant\}$$

is a line.

Proof: Let P be the intersection point of ℓ_1 and ℓ_2. Considering the positive oriented areas, we have

$$\begin{aligned}
\vec{\mathcal{A}}_{MAB} + \vec{\mathcal{A}}_{MCD} &= \frac{1}{2}\overrightarrow{AM} \times \overrightarrow{AB} + \frac{1}{2}\overrightarrow{CD} \times \overrightarrow{CM} \\
&= \frac{1}{2}\left[(\overrightarrow{AP} + \overrightarrow{PM}) \times \overrightarrow{AB} + \overrightarrow{CD} \times (\overrightarrow{CP} + \overrightarrow{PM})\right] \\
&= \frac{1}{2}\left[\overrightarrow{AP} \times \overrightarrow{AB} + \overrightarrow{PM} \times \overrightarrow{AB} + \overrightarrow{CD} \times \overrightarrow{CP} + \overrightarrow{CD} \times \overrightarrow{PM}\right] \\
&= \frac{1}{2}\left[\overrightarrow{PM} \times \overrightarrow{AB} - \overrightarrow{PM} \times \overrightarrow{CD}\right] \\
&= \frac{1}{2}\overrightarrow{PM} \times (\overrightarrow{AB} - \overrightarrow{CD}) = \frac{1}{2}\overrightarrow{PM} \times \vec{v},
\end{aligned}$$

where $\vec{v} = \overrightarrow{AB} - \overrightarrow{CD}$ does not depend on M. Lemma 8.3.4 leads to the desired result. ■

Exercise 8.3.9 *Solve Exercise 8.3.8 in the case when the lines ℓ_1 and ℓ_2 are parallel.*

8.4 Solved problems

In this section we shall solve a few problems using the cross product.

Solved Exercise 8.4.1 *Let ABC be a triangle and let $M \in [AB]$, $N \in [BC]$, $P \in [AC]$ such that $\overrightarrow{AM} = \lambda\overrightarrow{AB}$, $\overrightarrow{BN} = \lambda\overrightarrow{BC}$, $\overrightarrow{CP} = \lambda\overrightarrow{CA}$, where $\lambda \in \mathbb{R}$, see Figure 8.5. Find the value of λ such that the area of triangle MNP is minimum.*

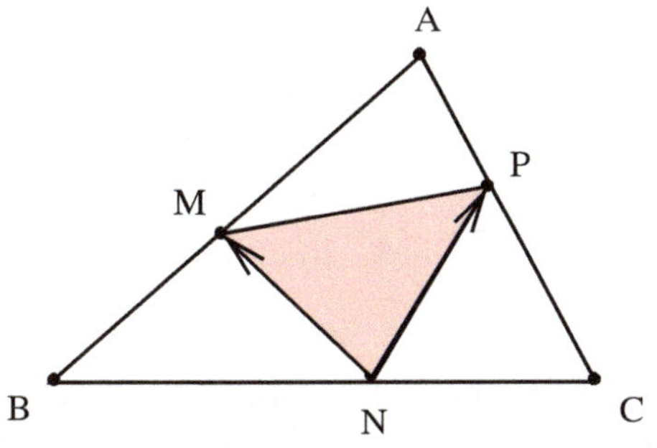

Figure 8.5: For Exercise 8.4.1.

Proof: Considering the areas oriented, we have

$$\begin{aligned}
-2\overrightarrow{\mathcal{A}}_{NMP} &= \overrightarrow{NM} \times \overrightarrow{NP} = \left(\overrightarrow{NB} + \overrightarrow{BM}\right) \times \left(\overrightarrow{NC} + \overrightarrow{CP}\right) \\
&= \left(-\overrightarrow{BN} + \overrightarrow{BM}\right) \times \left(\overrightarrow{NC} + \overrightarrow{CP}\right) \\
&= \left(-\lambda\overrightarrow{BC} - \overrightarrow{AB} + \overrightarrow{AM}\right) \times \left(\overrightarrow{BC} - \overrightarrow{BN} + \overrightarrow{CP}\right) \\
&= \left(-\lambda\overrightarrow{BC} - \overrightarrow{AB} + \lambda\overrightarrow{AB}\right) \times \left(\overrightarrow{BC} - \lambda\overrightarrow{BC} + \lambda\overrightarrow{CA}\right) \\
&= \left(-\lambda\overrightarrow{BC} + (\lambda - 1)\overrightarrow{AB}\right) \times \left((1 - \lambda)\overrightarrow{BC} + \lambda\overrightarrow{CA}\right) \\
&= -\lambda^2\overrightarrow{BC} \times \overrightarrow{CA} + (\lambda - 1)(1 - \lambda)\overrightarrow{AB} \times \overrightarrow{BC} \\
&\quad + \lambda(\lambda - 1)\overrightarrow{AB} \times \overrightarrow{CA} \\
&= \lambda^2\overrightarrow{CB} \times \overrightarrow{CA} + (1 - \lambda)^2\overrightarrow{BA} \times \overrightarrow{BC} - \lambda(\lambda - 1)\overrightarrow{AB} \times \overrightarrow{AC} \\
&= -2\lambda^2\overrightarrow{\mathcal{A}}_{ABC} - 2(1 - \lambda)^2\overrightarrow{\mathcal{A}}_{ABC} - 2\lambda(\lambda - 1)^2\overrightarrow{\mathcal{A}}_{ABC} \\
&= -2\overrightarrow{\mathcal{A}}_{ABC}\left(\lambda^2 + (1 - \lambda)^2 + \lambda(\lambda - 1)\right) \\
&= -2\overrightarrow{\mathcal{A}}_{ABC}\left(3\lambda^2 - 3\lambda + 1\right).
\end{aligned}$$

Hence

$$\overrightarrow{\mathcal{A}}_{NMP} = (3\lambda^2 - 3\lambda + 1)\overrightarrow{\mathcal{A}}_{ABC}.$$

Completing the square in the quadratic polynomial, we have

$$\begin{aligned}
3\lambda^2 - 3\lambda + 1 &= 3\left(\lambda^2 - \lambda + \frac{1}{3}\right) \\
&= 3\left(\lambda^2 - 2\lambda \cdot \frac{1}{2} + \frac{1}{4} + \frac{1}{3} - \frac{1}{4}\right) \\
&= 3\left(\lambda - \frac{1}{2}\right)^2 + \frac{1}{4}.
\end{aligned}$$

Hence the minimum value is equal to $\frac{1}{4}$ and it is reached for $\lambda = \frac{1}{2}$. This means that the points M, N, P are midpoints, *i.e.*, MNP is the median triangle of triangle ABC. ∎

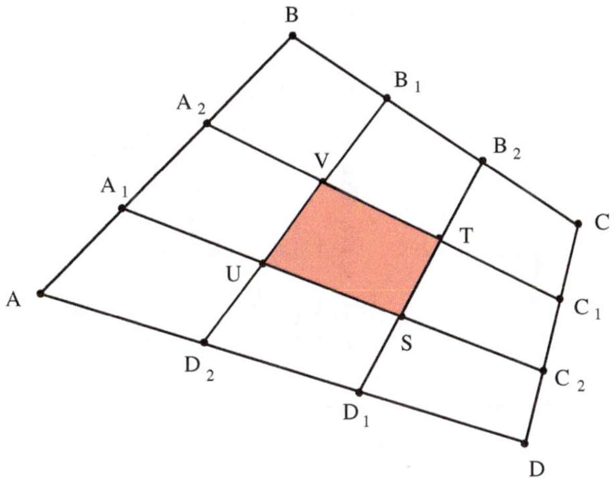

Figure 8.6: For Proposition 8.4.2.

Proposition 8.4.2 *Let $ABCD$ be a convex quadrilateral and let A_1, $A_2 \in [AB]$, B_1, $B_2 \in [BC]$, C_1, $C_2 \in [CD]$, D_1, $D_2 \in [AD]$ such that*

$$|AA_1| = |A_1A_2| = |A_2B|, \qquad |BB_1| = |B_1B_2| = |B_2C|,$$

$$|CC_1| = |C_1C_2| = |C_2D|, \qquad |DD_1| = |D_1D_2| = |D_2A|.$$

Then

$$\overrightarrow{D_2B_1} = \frac{2\overrightarrow{AB} + \overrightarrow{DC}}{3}, \qquad \overrightarrow{D_1B_2} = \frac{2\overrightarrow{DC} + \overrightarrow{AB}}{3},$$

$$\overrightarrow{A_1C_2} = \frac{2\overrightarrow{AD} + \overrightarrow{BC}}{3}, \qquad \overrightarrow{A_2C_1} = \frac{2\overrightarrow{BC} + \overrightarrow{AD}}{3}.$$

Proof: We shall refer to Figure 8.6. Applying Proposition 6.4.2 in quadrilaterals ABB_2D_1 and D_2B_1CD yields

$$\overrightarrow{D_2B_1} = \frac{1}{2}\left(\overrightarrow{AB} + \overrightarrow{D_1B_2}\right), \tag{8.4.3}$$

$$\overrightarrow{D_1B_2} = \frac{1}{2}\left(\overrightarrow{D_2B_1} + \overrightarrow{DC}\right). \tag{8.4.4}$$

Substituting (8.4.3) into (8.4.4) we have

$$\begin{aligned}
\overrightarrow{D_2B_1} &= \frac{1}{2}\left(\overrightarrow{AB} + \frac{1}{2}(\overrightarrow{D_2B_1} + \overrightarrow{DC})\right) \\
&= \frac{1}{2}\overrightarrow{AB} + \frac{1}{4}\overrightarrow{D_2B_1} + \frac{1}{4}\overrightarrow{DC} \Longleftrightarrow \\
\left(1 - \frac{1}{4}\right)\overrightarrow{D_2B_1} &= \frac{1}{2}\overrightarrow{AB} + \frac{1}{4}\overrightarrow{DC} \Longleftrightarrow \\
\frac{3}{4}\overrightarrow{D_2B_1} &= \frac{1}{2}\overrightarrow{AB} + \frac{1}{4}\overrightarrow{DC} \Longleftrightarrow \\
3\overrightarrow{D_2B_1} &= 2\overrightarrow{AB} + \overrightarrow{DC} \Longleftrightarrow \\
\overrightarrow{D_2B_1} &= \frac{2\overrightarrow{AB} + \overrightarrow{DC}}{3}.
\end{aligned}$$

In a similar way, we can obtain the other identities. ■

Exercise 8.4.3 *With the previous notations, show that*

$$9|D_2B_1|^2 = 4|AB|^2 + |DC|^2 + 4\overrightarrow{AB} \cdot \overrightarrow{DC}.$$

Exercise 8.4.4 *Show that* $|\overrightarrow{u} + \overrightarrow{v}| = |\overrightarrow{u}| + |\overrightarrow{v}|$ *if and only if* $\overrightarrow{u} = \lambda\overrightarrow{v}$, *with* $\lambda > 0$.

Solved Exercise 8.4.5 *Show that the area of a quadrilateral* $ABCD$ *is given by*

$$\mathcal{A}_{ABCD} = \frac{1}{2}|\overrightarrow{BC} \times \overrightarrow{BA} + \overrightarrow{DA} \times \overrightarrow{DC}| = \frac{1}{2}|\overrightarrow{BA} \times \overrightarrow{BC} + \overrightarrow{DC} \times \overrightarrow{DA}|.$$

Proof: We consider Figure 8.7. The vectors $\overrightarrow{BC} \times \overrightarrow{BA}$ and $\overrightarrow{DA} \times \overrightarrow{DC}$ are collinear and they have the same direction. Then

$$\begin{aligned}
\mathcal{A}_{ABCD} &= \mathcal{A}_{ABC} + \mathcal{A}_{ACD} \\
&= \frac{1}{2}|\overrightarrow{BC} \times \overrightarrow{BA}| + \frac{1}{2}|\overrightarrow{DA} \times \overrightarrow{DC}| \\
&= \frac{1}{2}|\overrightarrow{BC} \times \overrightarrow{BA} + \overrightarrow{DA} \times \overrightarrow{DC}|.
\end{aligned}$$

■

Solved Exercise 8.4.6 *Consider the notations from Figure 8.6. Show that*

$$\mathcal{A}_{UVTS} = \frac{1}{9}\mathcal{A}_{ABCD}.$$

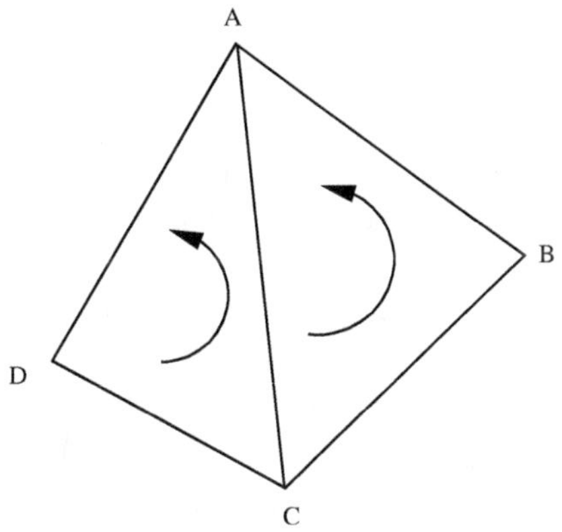

Figure 8.7: For Exercise 8.4.5.

Proof: Using Exercise 8.4.5 and oriented areas we have

$$\begin{aligned}
2\overrightarrow{\mathcal{A}}_{UVTS} &= \overrightarrow{US} \times \overrightarrow{UV} + \overrightarrow{TV} \times \overrightarrow{TS} \\
&= \frac{1}{9}[\overrightarrow{A_1C_2} \times \overrightarrow{D_2B_1} + \overrightarrow{C_1A_2} \times \overrightarrow{B_2D_1}] \\
&= \frac{1}{9}[\overrightarrow{A_1C_2} \times \overrightarrow{D_2B_1} + \overrightarrow{A_2C_1} \times \overrightarrow{D_1B_2}] \\
&= \frac{1}{9}\left[\frac{2\overrightarrow{AD} + \overrightarrow{BC}}{3} \times \frac{2\overrightarrow{AB} + \overrightarrow{DC}}{3} + \frac{2\overrightarrow{BC} + \overrightarrow{AD}}{3} \times \frac{2\overrightarrow{DC} + \overrightarrow{AB}}{3}\right] \\
&= \frac{1}{81}\Big[4\overrightarrow{AD} \times \overrightarrow{AB} + 2\overrightarrow{AD} \times \overrightarrow{DC} + 2\overrightarrow{BC} \times \overrightarrow{AB} + \overrightarrow{BC} \times \overrightarrow{DC} \\
&\quad +4\overrightarrow{BC} \times \overrightarrow{DC} + 2\overrightarrow{BC} \times \overrightarrow{AB} + 2\overrightarrow{AD} \times \overrightarrow{DC} + \overrightarrow{AD} \times \overrightarrow{AB}\Big] \\
&= \frac{1}{81}\left[(5\overrightarrow{AD} \times \overrightarrow{AB} + 5\overrightarrow{CB} \times \overrightarrow{CD}) + 4(\overrightarrow{DC} \times \overrightarrow{DA} + \overrightarrow{BA} \times \overrightarrow{BC})\right] \\
&= \frac{1}{81}[5 \cdot 2\overrightarrow{\mathcal{A}}_{ABCD} + 4 \cdot 2\overrightarrow{\mathcal{A}}_{ABCD}] \\
&= \frac{18}{81}\overrightarrow{\mathcal{A}}_{ABCD} = \frac{2}{9}\overrightarrow{\mathcal{A}}_{ABCD}.
\end{aligned}$$

Hence $\overrightarrow{\mathcal{A}}_{UVTS} = \frac{1}{9}\overrightarrow{\mathcal{A}}_{ABCD}$ and then $\mathcal{A}_{UVTS} = \frac{1}{9}\mathcal{A}_{ABCD}$, since $|\overrightarrow{\mathcal{A}}| = \mathcal{A}$. ■

Exercise 8.4.7 *Let ABCD be a convex quadrilateral. Show that its area* *is given by*

$$\mathcal{A}_{ABCD} = \frac{1}{2}|\overrightarrow{AC} \times \overrightarrow{BD}|.$$

Exercise 8.4.8 *Let $ABCD$ be a convex quadrilateral and denote by M, N, P and Q the midpoints of its sides. Show that*

$$\mathcal{A}_{MNPQ} = \frac{1}{2}\mathcal{A}_{ABCD}.$$

Exercise 8.4.9 *Let $ABCD$ be a convex quadrilateral and let $M \in [AB]$, $N \in [BC]$, $P \in [CD]$ and $Q \in [AD]$ such that $\overrightarrow{AM} = \lambda\overrightarrow{AB}$, $\overrightarrow{BN} = \lambda\overrightarrow{BC}$, $\overrightarrow{CP} = \lambda\overrightarrow{CD}$, $\overrightarrow{DQ} = \lambda\overrightarrow{DA}$. Find the value of the real parameter λ such that the area $\mathcal{A}_{MNPQ}$ is minimum.*

Exercise 8.4.10 *Let $ABCD$ be a square with side length a. Consider the points $M \in [AD]$, $N \in [AB]$, and $P \in [BC]$ such that $|AN| = |NB|$, $\overrightarrow{AM} = 2\overrightarrow{MD}$ and $\overrightarrow{BP} = \frac{1}{2}\overrightarrow{PC}$. Find the area of triangle MNP.*

Chapter 9

The Mixed Scalar Product

Never express yourself more clearly than you are able to think.
Niels Bohr

This chapter introduces the mixed scalar product, which is a ternary operation on vectors, involving both the inner and the cross products of vectors. It will be used to compute volumes of parallelepipeds and tetrahedrons.

9.1 Definitions and properties

This section introduces the concept of a new operation among vectors, which is applied to three vectors and will be useful in computing volumes.

Lemma 9.1.1 *For any vectors $\overrightarrow{u}$, $\overrightarrow{v}$, $\overrightarrow{w}$ in $\mathbb{R}^3$ we have*

$$(\overrightarrow{u} \times \overrightarrow{v}) \cdot \overrightarrow{w} = \overrightarrow{u} \cdot (\overrightarrow{v} \times \overrightarrow{w}).$$

Proof: Let $\{e_1, e_2, e_3\}$ be the canonical base of $\mathbb{R}^3$ and write

$$\begin{aligned}
\overrightarrow{u} &= u^1 e_1 + u^2 e_2 + u^3 e_3 \\
\overrightarrow{v} &= v^1 e_1 + v^2 e_2 + v^3 e_3 \\
\overrightarrow{w} &= w^1 e_1 + w^2 e_2 + w^3 e_3.
\end{aligned}$$

Using the definition of the cross product in coordinates, we have

$$\overrightarrow{u} \times \overrightarrow{v} = (u^2 v^3 - u^3 v^2) e_1 + (u^3 v^1 - u^1 v^3) e_2 + (u^1 v^2 - u^2 v^1) e_3,$$

and using the distributivity of the inner product and the properties of the orthonormal basis, we obtain

$$\begin{aligned}(\overrightarrow{u} \times \overrightarrow{v}) \cdot \overrightarrow{w} &= (u^2v^3 - u^3v^2)w^1 + (u^3v^1 - u^1v^3)w^2 + (u^1v^2 - u^2v^1)w^3 \\ &= u^2v^3w^1 + u^3v^1w^2 + u^1v^2w^3 - u^3v^2w^1 - u^1v^3w^2 - u^2v^1w^3 \\ &= \det \begin{pmatrix} u^1 & u^2 & u^3 \\ v^1 & v^2 & v^3 \\ w^1 & w^2 & w^3 \end{pmatrix}. \end{aligned} \tag{9.1.1}$$

In a similar way, we can show that

$$\overrightarrow{u} \cdot (\overrightarrow{v} \times \overrightarrow{w}) = (\overrightarrow{v} \times \overrightarrow{w}) \cdot \overrightarrow{u} = \det \begin{pmatrix} v^1 & v^2 & v^3 \\ w^1 & w^2 & w^3 \\ u^1 & u^2 & u^3 \end{pmatrix}. \tag{9.1.2}$$

Since swapping two rows in a determinant changes its sign, we have

$$\begin{aligned}\det \begin{pmatrix} u^1 & u^2 & u^3 \\ v^1 & v^2 & v^3 \\ w^1 & w^2 & w^3 \end{pmatrix} &\overset{R_1 \leftrightarrow R_2}{=} -\det \begin{pmatrix} v^1 & v^2 & v^3 \\ u^1 & u^2 & u^3 \\ w^1 & w^2 & w^3 \end{pmatrix} \\ &\overset{R_2 \leftrightarrow R_3}{=} \det \begin{pmatrix} v^1 & v^2 & v^3 \\ w^1 & w^2 & w^3 \\ u^1 & u^2 & u^3 \end{pmatrix},\end{aligned}$$

and from (9.1.1) and (9.1.2) yields $(\overrightarrow{u} \times \overrightarrow{v}) \cdot \overrightarrow{w} = \overrightarrow{u} \cdot (\overrightarrow{v} \times \overrightarrow{w})$. ∎

Definition 9.1.2 *Let $\overrightarrow{u}$, $\overrightarrow{v}$, $\overrightarrow{w}$ be three vectors. The mixed scalar product of $\overrightarrow{u}$, $\overrightarrow{v}$, $\overrightarrow{w}$ is the number*

$$[\overrightarrow{u}, \overrightarrow{v}, \overrightarrow{w}] = (\overrightarrow{u} \times \overrightarrow{v}) \cdot \overrightarrow{w}.$$

Using vectors components, we have

$$[\overrightarrow{u}, \overrightarrow{v}, \overrightarrow{w}] = \det \begin{pmatrix} u^1 & u^2 & u^3 \\ v^1 & v^2 & v^3 \\ w^1 & w^2 & w^3 \end{pmatrix}.$$

Exercise 9.1.3 *Prove the following relations:*

(i) $[\overrightarrow{u}, \overrightarrow{v}, \overrightarrow{w}] = [\overrightarrow{v}, \overrightarrow{w}, \overrightarrow{u}] = [\overrightarrow{w}, \overrightarrow{u}, \overrightarrow{v}]$,

(ii) $[\overrightarrow{u}, \overrightarrow{u}, \overrightarrow{w}] = [\overrightarrow{u}, \overrightarrow{w}, \overrightarrow{u}] = [\overrightarrow{w}, \overrightarrow{u}, \overrightarrow{u}] = 0$,

(iii) $[r_1\overrightarrow{u}, r_2\overrightarrow{v}, r_3\overrightarrow{w}] = r_1r_2r_3[\overrightarrow{u}, \overrightarrow{v}, \overrightarrow{w}]$, $\forall r_1, r_2, r_3 \in \mathbb{R}$.

Exercise 9.1.4 *Show that the following statements are equivalent:*
(i) $\overrightarrow{u}$, $\overrightarrow{v}$, $\overrightarrow{w}$ *are linear dependent,*
(ii) $\overrightarrow{w} = \alpha\overrightarrow{u} + \beta\overrightarrow{v}$, *with* α, $\beta \in \mathbb{R}$,
(iii) $[\overrightarrow{u}, \overrightarrow{v}, \overrightarrow{w}] = 0$,
(iv) $\det\begin{pmatrix} u^1 & u^2 & u^3 \\ v^1 & v^2 & v^3 \\ w^1 & w^2 & w^3 \end{pmatrix} = 0.$

Exercise 9.1.5 *The mixed scalar product is additive in each of the arguments, i.e., the following identities hold:*
(i) $[\overrightarrow{u} + \overrightarrow{t}, \overrightarrow{v}, \overrightarrow{w}] = [\overrightarrow{u}, \overrightarrow{v}, \overrightarrow{w}] + [\overrightarrow{t}, \overrightarrow{v}, \overrightarrow{w}]$,
(ii) $[\overrightarrow{u}, \overrightarrow{v} + \overrightarrow{t}, \overrightarrow{w}] = [\overrightarrow{u}, \overrightarrow{v}, \overrightarrow{w}] + [\overrightarrow{u}, \overrightarrow{t}, \overrightarrow{w}]$,
(iii) $[\overrightarrow{u}, \overrightarrow{v}, \overrightarrow{w} + \overrightarrow{t}] = [\overrightarrow{u}, \overrightarrow{v}, \overrightarrow{w}] + [\overrightarrow{u}, \overrightarrow{v}, \overrightarrow{t}]$, *for any vectors* $\overrightarrow{u}, \overrightarrow{v}, \overrightarrow{w}, \overrightarrow{t}$.

The following result deals with the geometric meaning of the mixed scalar product of three vectors.

Lemma 9.1.6 *The absolute value of the mixed scalar product* $|[\overrightarrow{u}, \overrightarrow{v}, \overrightarrow{w}]|$ *is the volume of the parallelopiped spanned by the vectors* $\overrightarrow{u}$, $\overrightarrow{v}$, $\overrightarrow{w}$.

Proof: We shall refer to Figure 9.1. Let $\mathcal{A}(\overrightarrow{u}, \overrightarrow{v})$ be the area of the parallelogram spanned by the vectors $\overrightarrow{u}$ and $\overrightarrow{v}$. Then $\mathcal{A}(\overrightarrow{u}, \overrightarrow{v}) = |\overrightarrow{u} \times \overrightarrow{v}|$. Let ϕ be the angle between $\overrightarrow{w}$ and the vertical direction. The quantity $h = |\overrightarrow{w}| \cdot |\cos\phi|$ is the altitude of the parallelopiped perpendicular to the plane of the parallelogram spanned by $\overrightarrow{u}$ and $\overrightarrow{v}$. Then the parallelopiped has the volume

$$\begin{aligned} vol &= \mathcal{A}(\overrightarrow{u}, \overrightarrow{v}) \cdot h = |\overrightarrow{u} \times \overrightarrow{v}| \cdot |\overrightarrow{w}| \cdot |\cos\phi| \\ &= |(\overrightarrow{u} \times \overrightarrow{v}) \cdot \overrightarrow{w}| = |[\overrightarrow{u}, \overrightarrow{v}, \overrightarrow{w}]|. \end{aligned}$$

■

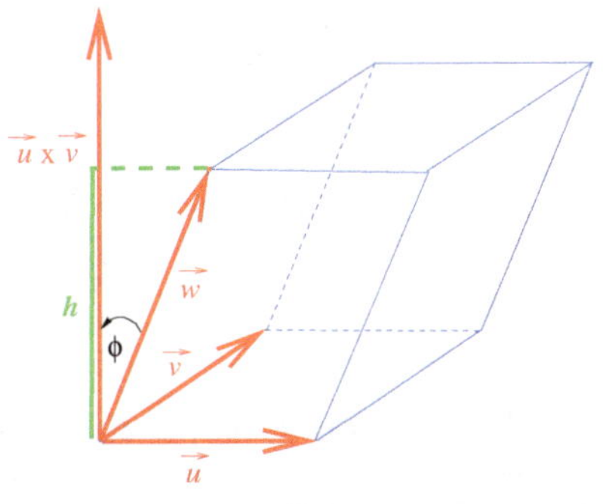

Figure 9.1: The volume of a parallelopiped spanned by three vectors.

9.2 Gibbs' formula

The goal of this section is to prove a formula, which deals with the non-associativity of the cross product of three vectors, called Gibbs' formula. We need a few preliminary results, some of which are left to the reader as exercises.

Solved Exercise 9.2.1 *Let $\overrightarrow{b}$ and $\overrightarrow{c}$ be two linear independent vectors with $\overrightarrow{b} \neq 0$. Find a scalar $s \in \mathbb{R}$ such that the vector $\overrightarrow{c} + s\overrightarrow{b}$ is perpendicular to $\overrightarrow{b}$.*

Proof: From $(\overrightarrow{c} + s\overrightarrow{b}) \cdot \overrightarrow{b} = 0$ we obtain $\overrightarrow{c} \cdot \overrightarrow{b} + s|\overrightarrow{b}|^2 = 0$, or $s = -\overrightarrow{c} \cdot \overrightarrow{b}/|\overrightarrow{b}|^2$. ■

Exercise 9.2.2 *Let $\overrightarrow{a}, \overrightarrow{b}, \overrightarrow{c}$ be three vectors in the space and let s be given as in Exercise 9.2.1. Let $\overrightarrow{c'} = \overrightarrow{c} + s\overrightarrow{b}$. Show that*

(i) $\overrightarrow{a} \times (\overrightarrow{b} \times \overrightarrow{c}) = \overrightarrow{a} \times (\overrightarrow{b} \times \overrightarrow{c'})$, *and*

(ii) $(\overrightarrow{a} \cdot \overrightarrow{c})\overrightarrow{b} - (\overrightarrow{a} \cdot \overrightarrow{b})\overrightarrow{c} = (\overrightarrow{a} \cdot \overrightarrow{c'})\overrightarrow{b} - (\overrightarrow{a} \cdot \overrightarrow{b})\overrightarrow{c'}$.

Exercise 9.2.3 *Let $\overrightarrow{c'} = \lambda\overrightarrow{c}$, $\lambda \in \mathbb{R}$. Show that*

(i) $\overrightarrow{a} \times (\overrightarrow{b} \times \overrightarrow{c'}) = \lambda\overrightarrow{a} \times (\overrightarrow{b} \times \overrightarrow{c})$,

(ii) $(\overrightarrow{a} \cdot \overrightarrow{c'})\overrightarrow{b} - (\overrightarrow{a} \cdot \overrightarrow{b})\overrightarrow{c'} = \lambda\Big((\overrightarrow{a} \cdot \overrightarrow{c})\overrightarrow{b} - (\overrightarrow{a} \cdot \overrightarrow{b})\overrightarrow{c}\Big)$.

Solved Exercise 9.2.4 *Show that*

$$\Big(\overrightarrow{a} \times (\overrightarrow{b} \times \overrightarrow{c})\Big) \cdot \overrightarrow{c} = \Big((\overrightarrow{b} \times \overrightarrow{c}) \times \overrightarrow{c}\Big) \cdot \overrightarrow{a}.$$

Proof: Using the invariance property of the mixed scalar product under a cyclic permutation of terms, we have

$$\begin{aligned}\Big(\overrightarrow{a} \times (\overrightarrow{b} \times \overrightarrow{c})\Big) \cdot \overrightarrow{c} &= [\overrightarrow{a}, \overrightarrow{b} \times \overrightarrow{c}, \overrightarrow{c}] = [\overrightarrow{b} \times \overrightarrow{c}, \overrightarrow{c}, \overrightarrow{a}] \\ &= \Big((\overrightarrow{b} \times \overrightarrow{c}) \times \overrightarrow{c}\Big) \cdot \overrightarrow{a}.\end{aligned}$$

■

Solved Exercise 9.2.5 *Let $\overrightarrow{b}$ and $\overrightarrow{c}$ be two perpendicular vectors with $|\overrightarrow{c}| = 1$. Show that*

$$(\overrightarrow{b} \times \overrightarrow{c}) \times \overrightarrow{c} = -\overrightarrow{b}.$$

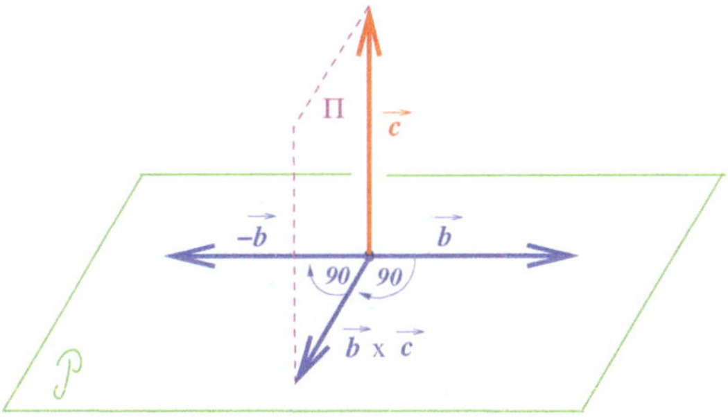

Figure 9.2: For Exercise 9.2.5.

Proof: Let $\mathcal{P}$ be the plane perpendicular to $\vec{c}$. Then $\vec{b}$ and $\vec{b} \times \vec{c}$ are parallel to $\mathcal{P}$ (Why?). Then $\{\vec{b}, \vec{c}, \vec{b} \times \vec{c}\}$ is a positive oriented basis, with $\vec{b} \perp \vec{c}$, $\vec{b} \perp \vec{b} \times \vec{c}$.

Let Π be the plane defined by the vectors $\vec{c}$ and $\vec{b} \times \vec{c}$. Then $\vec{b}$ and $(\vec{b} \times \vec{c}) \times \vec{c}$ are perpendicular to Π (why?), and hence the vectors are collinear, see Figure 9.2.

Since $\vec{b}$ and $(\vec{b} \times \vec{c}) \times \vec{c}$ have the same length (why?) and opposite directions (why?), it follows that

$$(\vec{b} \times \vec{c}) \times \vec{c} = -\vec{b}.$$

■

Exercise 9.2.6 *Let $\vec{b} = \lambda \vec{c}$, with $\lambda \in \mathbb{R}$. Show that*

(i) $\vec{a} \times (\vec{b} \times \vec{c}) = 0$,

(ii) $(\vec{a} \cdot \vec{c})\vec{b} = (\vec{a} \cdot \vec{b})\vec{c}$.

Now we are ready to prove a result regarding a double cross product.

Lemma 9.2.7 (Gibbs' formula) *Given three vectors $\vec{a}$, $\vec{b}$, $\vec{c}$ in space, we have*

$$\vec{a} \times (\vec{b} \times \vec{c}) = (\vec{a} \cdot \vec{c})\vec{b} - (\vec{a} \cdot \vec{b})\vec{c}. \tag{9.2.3}$$

Proof: If the vectors $\vec{b}$ and $\vec{c}$ are linear dependent, applying Exercise 9.2.6, both terms are equal to zero and the identity is obviously satisfied.

Assume $\vec{b}$ and $\vec{c}$ are linear independent. Let $\vec{u} = \vec{a} \times (\vec{b} \times \vec{c})$. Then $\vec{u}$ is perpendicular to both $\vec{a}$ and $\vec{b} \times \vec{c}$. Let $\mathcal{P}$ be the plane

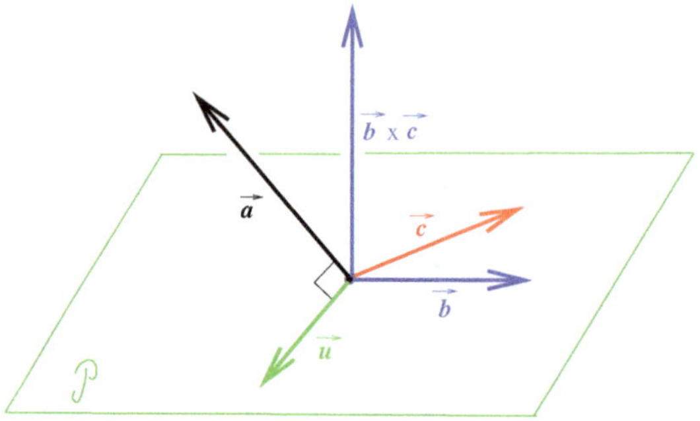

Figure 9.3: For Gibbs' formula.

defined by the vectors $\vec{b}$ and $\vec{c}$. Since $\vec{b} \times \vec{c}$ is perpendicular to both $\vec{c}$ and $\vec{b}$, then $\vec{b} \times \vec{c}$ will be perpendicular to $\mathcal{P}$. Hence $\vec{u}$ is parallel to $\mathcal{P}$.

Since $\vec{b}$ and $\vec{c}$ are linearly independent, they form a basis in the plane $\mathcal{P}$. Then there are scalars α, $\beta \in \mathbb{R}$ such that

$$\vec{u} = \alpha \vec{b} + \beta \vec{c}. \tag{9.2.4}$$

On the other side, $\vec{u}$ is perpendicular to $\vec{a}$, which leads to

$$0 = \vec{a} \cdot \vec{u} = \vec{a} \cdot (\alpha \vec{b} + \beta \vec{c}) = \alpha \vec{a} \cdot \vec{b} + \beta \vec{a} \cdot \vec{c}.$$

Let

$$\rho = \frac{\alpha}{\vec{a} \cdot \vec{c}} = -\frac{\beta}{\vec{a} \cdot \vec{b}}$$

and write

$$\alpha = \rho \vec{a} \cdot \vec{c}, \qquad \beta = -\rho \vec{a} \cdot \vec{b}.$$

Then (9.2.4) becomes

$$\vec{u} = \rho \Big((\vec{a} \cdot \vec{c}) \vec{b} - (\vec{a} \cdot \vec{b}) \vec{c} \Big). \tag{9.2.5}$$

The final step of the proof is to show that $\rho = 1$. We shall write equation (9.2.5) as

$$\vec{a} \times (\vec{b} \times \vec{c}) = \rho \Big((\vec{a} \cdot \vec{c}) \vec{b} - (\vec{a} \cdot \vec{b}) \vec{c} \Big). \tag{9.2.6}$$

Let s be the scalar given by Exercise 9.2.1. Then replacing $\vec{c}$ by $\vec{c'} = \vec{c} + s \vec{b}$, using Exercise 9.2.2, both left and right sides of (9.2.6) do not change, so ρ will do the same.

Hence we have reduced our problem to the case when the vectors $\overrightarrow{b}$ and $\overrightarrow{c}$ are perpendicular.

Replacing $\overrightarrow{c}$ by $\overrightarrow{c'} = \lambda\overrightarrow{c}$ and using Exercise 9.2.3, it follows that the coefficient ρ is not affected. Assuming $\lambda = 1/|\overrightarrow{c}|$, we may consider $|\overrightarrow{c}| = 1$. Using Exercise 9.2.4 yields

$$\overrightarrow{u} \cdot \overrightarrow{c} = \left(\overrightarrow{a} \times (\overrightarrow{b} \times \overrightarrow{c})\right) \cdot \overrightarrow{c} = \left((\overrightarrow{b} \times \overrightarrow{c}) \times \overrightarrow{c}\right) \cdot \overrightarrow{a}. \tag{9.2.7}$$

Using (9.2.5), equation (9.2.7) becomes

$$\left((\overrightarrow{b} \times \overrightarrow{c}) \times \overrightarrow{c}\right) \cdot \overrightarrow{a} = \rho\left((\overrightarrow{a} \cdot \overrightarrow{c})(\overrightarrow{b} \cdot \overrightarrow{c}) - (\overrightarrow{a} \cdot \overrightarrow{b})(\overrightarrow{c} \cdot \overrightarrow{c})\right). \tag{9.2.8}$$

Since $\overrightarrow{b} \perp \overrightarrow{c}$, $\overrightarrow{c} \cdot \overrightarrow{c} = 1$, equation (9.2.8) becomes

$$\left((\overrightarrow{b} \times \overrightarrow{c}) \times \overrightarrow{c}\right) \cdot \overrightarrow{a} = -\rho(\overrightarrow{a} \cdot \overrightarrow{b}). \tag{9.2.9}$$

Replacing the left term by the expression provided by Exercise 9.2.5, we obtain

$$-\overrightarrow{b} \cdot \overrightarrow{a} = -\rho(\overrightarrow{a} \cdot \overrightarrow{b}),$$

and hence $\rho = 1$, since the inner product is commutative. Hence (9.2.6) becomes (9.2.3). ∎

We shall provide another proof for Gibbs' formula involving explicit computations in coordinates. We shall write the vectors with respect to the canonical orthonormal basis in $\mathbb{R}^3$ as

$$\overrightarrow{a} = \sum_{i=1}^{3} a^i e_i, \quad \overrightarrow{b} = \sum_{i=1}^{3} b^i e_i, \quad \overrightarrow{c} = \sum_{i=1}^{3} c^i e_i.$$

Then

$$\overrightarrow{b} \times \overrightarrow{c} = (b^2c^3 - b^3c^2)e_1 + (b^3c^1 - b^1c^3)e_2 + (b^1c^2 - b^2c^1)e_3, \tag{9.2.10}$$

and applying the cross product formula again we have

$$\begin{aligned}\overrightarrow{a} \times (\overrightarrow{b} \times \overrightarrow{c}) &= [a^2(b^1c^2 - b^2c^1) - a^3(b^3c^1 - b^1c^3)]e_1 \\ &\quad + [a^3(b^2c^3 - b^3c^2) - a^1(b^1c^2 - b^2c^1)]e_2 \\ &\quad + [a^1(b^3c^1 - b^1c^3) - a^2(b^2c^3 - b^3c^2)]e_3.\end{aligned} \tag{9.2.11}$$

Using Theorem 7.4.4 we have

$$\overrightarrow{a} \cdot \overrightarrow{c} = a^1c^1 + a^2c^2 + a^3c^3, \qquad \overrightarrow{a} \cdot \overrightarrow{b} = a^1b^1 + a^2b^2 + a^3b^3,$$

and hence

$$\begin{aligned}(\vec{a}\cdot\vec{c})\vec{b}-(\vec{a}\cdot\vec{b})\vec{c} &= (a^1c^1+a^2c^2+a^3c^3)(b^1e_1+b^2e_2+b^3e_3)\\ &\quad -(a^1b^1+a^2b^2+a^3b^3)(c^1e_1+c^2e_2+c^3e_3)\\ &= [a^2(b^1c^2-b^2c^1)-a^3(b^3c^1-b^1c^3)]e_1\\ &\quad +[a^3(b^2c^3-b^3c^2)-a^1(b^1c^2-b^2c^1)]e_2\\ &\quad +[a^1(b^3c^1-b^1c^3)-a^2(b^2c^3-b^3c^2)]e_3,\end{aligned}$$

which is the same as formula (9.2.11).

Remark 9.2.8 Gibbs' formula can be written symbolically as

$$\vec{a}\times(\vec{b}\times\vec{c})=\det\begin{pmatrix}\vec{b} & \vec{c}\\ \vec{a}\cdot\vec{b} & \vec{a}\cdot\vec{c}\end{pmatrix}.$$

The following consequence deals with the non-associativity of the cross product.

Corollary 9.2.9 *Let $\vec{a}$ and $\vec{c}$ be two linear independent vectors. Then for any nonzero vector $\vec{b}$, which is not perpendicular on both vectors $\vec{a}$ and $\vec{c}$, we have*

$$\vec{a}\times(\vec{b}\times\vec{c})\neq(\vec{a}\times\vec{b})\times\vec{c}.$$

Proof: We shall present a proof by contradiction. Assume that $\vec{a}\times(\vec{b}\times\vec{c})=(\vec{a}\times\vec{b})\times\vec{c}$. Then we have the following sequence of equivalent relations

$$\begin{aligned}\vec{a}\times(\vec{b}\times\vec{c}) &= -\vec{c}\times(\vec{a}\times\vec{b})\Longleftrightarrow\\ \det\begin{pmatrix}\vec{b} & \vec{c}\\ \vec{a}\cdot\vec{b} & \vec{a}\cdot\vec{c}\end{pmatrix} &= -\det\begin{pmatrix}\vec{a} & \vec{b}\\ \vec{c}\cdot\vec{a} & \vec{c}\cdot\vec{b}\end{pmatrix}\Longleftrightarrow\\ (\vec{a}\cdot\vec{c})\vec{b}-(\vec{a}\cdot\vec{b})\vec{c} &= (\vec{c}\cdot\vec{a})\vec{b}-(\vec{c}\cdot\vec{b})\vec{a}\Longleftrightarrow\\ \underbrace{(\vec{a}\cdot\vec{b})}_{\neq 0}\vec{c} &= \underbrace{(\vec{c}\cdot\vec{b})}_{\neq 0}\vec{a},\end{aligned}$$

which contradicts the fact that $\vec{a}$ and $\vec{c}$ are linear independent. ■

9.3 Applications

A consequence of Gibbs' formula is the Lie's identity. The cross product is not associative. However, the following Lie algebra identity holds:

Corollary 9.3.1 (Lie's identity) *For any three vectors* $\vec{a}$, $\vec{b}$, $\vec{c}$ *we have*

$$\vec{a} \times (\vec{b} \times \vec{c}) + \vec{b} \times (\vec{c} \times \vec{a}) + \vec{c} \times (\vec{a} \times \vec{b}) = 0.$$

Proof: Gibbs' formula provides

$$\begin{aligned}
\vec{a} \times (\vec{b} \times \vec{c}) &= (\vec{a} \cdot \vec{c})\vec{b} - (\vec{a} \cdot \vec{b})\vec{c}, \\
\vec{b} \times (\vec{c} \times \vec{a}) &= -(\vec{b} \cdot \vec{c})\vec{a} + (\vec{b} \cdot \vec{a})\vec{c}, \\
\vec{c} \times (\vec{a} \times \vec{b}) &= (\vec{c} \cdot \vec{b})\vec{a} - (\vec{c} \cdot \vec{a})\vec{b}.
\end{aligned}$$

Summing, and using the commutativity of the inner product, yields

$$\begin{aligned}
& \vec{a} \times (\vec{b} \times \vec{c}) + \vec{b} \times (\vec{c} \times \vec{a}) + \vec{c} \times (\vec{a} \times \vec{b}) \\
= & (\vec{c} \cdot \vec{b} - \vec{b} \cdot \vec{c})\vec{a} + (\vec{a} \cdot \vec{c} - \vec{c} \cdot \vec{a})\vec{b} + (\vec{b} \cdot \vec{a} - \vec{a} \cdot \vec{b})\vec{c} \\
= & 0.
\end{aligned}$$

■

Solved Exercise 9.3.2 (Lagrange's identity) *Show that*

$$(\vec{a} \times \vec{b}) \cdot (\vec{c} \times \vec{d}) = \det \begin{pmatrix} \vec{a} \cdot \vec{c} & \vec{a} \cdot \vec{d} \\ \vec{b} \cdot \vec{c} & \vec{b} \cdot \vec{d} \end{pmatrix}.$$

Proof: Using the definition and the cyclic invariance of the mixed scalar product and Gibbs' formula, we have

$$\begin{aligned}
(\vec{a} \times \vec{b}) \cdot (\vec{c} \times \vec{d}) &= [\vec{a}, \vec{b}, \vec{c} \times \vec{d}] = [\vec{b}, \vec{c} \times \vec{d}, \vec{a}] \\
&= \left(\vec{b} \times (\vec{c} \times \vec{d})\right) \cdot \vec{a} \\
&= \left((\vec{b} \cdot \vec{d})\vec{c} - (\vec{b} \cdot \vec{c})\vec{d}\right) \cdot \vec{a} \\
&= (\vec{b} \cdot \vec{d})(\vec{c} \cdot \vec{a}) - (\vec{b} \cdot \vec{c})(\vec{d} \cdot \vec{a}) \\
&= \det \begin{pmatrix} \vec{a} \cdot \vec{c} & \vec{a} \cdot \vec{d} \\ \vec{b} \cdot \vec{c} & \vec{b} \cdot \vec{d} \end{pmatrix}.
\end{aligned}$$

■

Exercise 9.3.3 *Show that*
(i) $|\vec{a} \times \vec{b}|^2 = |\vec{a}|^2 \cdot |\vec{b}|^2 - (\vec{a} \cdot \vec{b})^2$.
(ii) $|\vec{a}| \cdot |\vec{b}| \geq |\vec{a} \cdot \vec{b}|$ *(Cauchy's inequality).*
For (ii), when is the identity reached? Why?

Solved Exercise 9.3.4 *Prove the formula*

$$(\vec{a}\times\vec{b})\times(\vec{c}\times\vec{d}) = \det\begin{pmatrix}\vec{c} & \vec{d}\\ [\vec{a},\vec{b},\vec{c}] & [\vec{a},\vec{b},\vec{d}]\end{pmatrix}.$$

Proof: Let $\vec{v}=\vec{a}\times\vec{b}$. Gibbs' formula yields

$$\begin{aligned}\vec{v}\times(\vec{c}\times\vec{d}) &= \det\begin{pmatrix}\vec{c} & \vec{d}\\ \vec{v}\cdot\vec{c} & \vec{v}\cdot\vec{d}\end{pmatrix} = (\vec{v}\cdot\vec{d})\vec{c}-(\vec{v}\cdot\vec{c})\vec{d}\\ &= \left((\vec{a}\times\vec{b})\cdot\vec{d}\right)\vec{c}-\left((\vec{a}\times\vec{b})\cdot\vec{c}\right)\vec{d}\\ &= [\vec{a},\vec{b},\vec{d}]\vec{c}-[\vec{a},\vec{b},\vec{c}]\vec{d}\\ &= \det\begin{pmatrix}\vec{c} & \vec{d}\\ [\vec{a},\vec{b},\vec{c}] & [\vec{a},\vec{b},\vec{d}]\end{pmatrix}.\end{aligned}$$

■

Exercise 9.3.5 *Show that*

$$(\vec{a}\times\vec{b})\times(\vec{a}\times\vec{c}) = [\vec{a},\vec{b},\vec{c}]\vec{a}.$$

Exercise 9.3.6 *Show that*

$$(\vec{a}\times\vec{b})\times(\vec{c}\times\vec{d}) = \det\begin{pmatrix}[\vec{a},\vec{b},\vec{d}] & [\vec{b},\vec{c},\vec{d}]\\ \vec{a} & \vec{b}\end{pmatrix}.$$

Exercise 9.3.7 *Show that*

$$\det\begin{pmatrix}\vec{c} & \vec{d}\\ [\vec{a},\vec{b},\vec{c}] & [\vec{a},\vec{b},\vec{d}]\end{pmatrix} = -\det\begin{pmatrix}\vec{a} & \vec{b}\\ [\vec{a},\vec{c},\vec{d}] & [\vec{b},\vec{c},\vec{d}]\end{pmatrix}.$$

Exercise 9.3.8 *Let* $\vec{a_1}$, $\vec{a_2}$, $\vec{a_3}$, $\vec{b_1}$, $\vec{b_2}$, $\vec{b_3}$ *be six vectors in space. Let* $c_{ij}=\vec{a_i}\cdot\vec{b_j}$, *with* $i,j\in\{1,2,3\}$.
(i) Show that $\det(c_{ij}) = [\vec{a_1},\vec{a_2},\vec{a_3}]\cdot[\vec{b_1},\vec{b_2},\vec{b_3}]$.
(ii) Find a formula for $[\vec{a_1},\vec{a_2},\vec{a_3}]^2$ *using part (i).*

Solved Exercise 9.3.9 *Prove the identity*

$$[\vec{a}\times\vec{b},\vec{c}\times\vec{d},\vec{e}\times\vec{f}] = \det\begin{pmatrix}[\vec{a},\vec{b},\vec{e}] & [\vec{a},\vec{b},\vec{f}]\\ [\vec{c},\vec{d},\vec{e}] & [\vec{c},\vec{d},\vec{f}]\end{pmatrix}.$$

Proof: Set $\vec{v} = \vec{a} \times \vec{b}$, $\vec{w} = \vec{c} \times \vec{d}$. Using the definition and the properties of the mixed scalar product, Gibbs' formula yields

$$\begin{aligned}
& [\vec{a} \times \vec{b}, \vec{c} \times \vec{d}, \vec{e} \times \vec{f}] \\
= \; & [\vec{v}, \vec{w}, \vec{e} \times \vec{f}] = [\vec{w}, \vec{e} \times \vec{f}, \vec{v}] \\
= \; & \left(\vec{w} \times (\vec{e} \times \vec{f})\right) \cdot \vec{v} = \det \begin{pmatrix} \vec{e} & \vec{f} \\ \vec{w} \cdot \vec{e} & \vec{w} \cdot \vec{f} \end{pmatrix} \cdot \vec{v} \\
= \; & \left((\vec{w} \cdot \vec{f})\vec{e} - (\vec{w} \cdot \vec{e})\vec{f}\right) \cdot \vec{v} \\
= \; & (\vec{w} \cdot \vec{f})(\vec{e} \cdot \vec{v}) - (\vec{w} \cdot \vec{e})(\vec{f} \cdot \vec{v}) \\
= \; & \det \begin{pmatrix} \vec{e} \cdot \vec{v} & \vec{f} \cdot \vec{v} \\ \vec{w} \cdot \vec{e} & \vec{w} \cdot \vec{f} \end{pmatrix} = \det \begin{pmatrix} \vec{v} \cdot \vec{e} & \vec{v} \cdot \vec{f} \\ \vec{w} \cdot \vec{e} & \vec{w} \cdot \vec{f} \end{pmatrix} \\
= \; & \det \begin{pmatrix} (\vec{a} \times \vec{b}) \cdot \vec{e} & (\vec{a} \times \vec{b}) \cdot \vec{f} \\ (\vec{c} \times \vec{d}) \cdot \vec{e} & (\vec{c} \times \vec{d}) \cdot \vec{f} \end{pmatrix} \\
= \; & \det \begin{pmatrix} [\vec{a}, \vec{b}, \vec{e}] & [\vec{a}, \vec{b}, \vec{f}] \\ [\vec{c}, \vec{d}, \vec{e}] & [\vec{c}, \vec{d}, \vec{f}] \end{pmatrix}.
\end{aligned}$$

■

Let $\{\vec{a}, \vec{b}, \vec{c}\}$ be three linear independent (and hence non-coplanar) vectors. Define $\sigma = 1/[\vec{a}, \vec{b}, \vec{c}]$ and let

$$\vec{a'} = \sigma(\vec{b} \times \vec{c}), \quad \vec{b'} = \sigma(\vec{c} \times \vec{a}), \quad \vec{c'} = \sigma(\vec{a} \times \vec{b}).$$

Solved Exercise 9.3.10 *Show that*

$$[\vec{a'}, \vec{b'}, \vec{c'}] = \frac{1}{[\vec{a}, \vec{b}, \vec{c}]}.$$

Proof: Using the properties of the mixed scalar product and Exercise 9.3.9 yields

$$\begin{aligned}
[\vec{a'}, \vec{b'}, \vec{c'}] &= \sigma^3[\vec{b} \times \vec{c}, \vec{c} \times \vec{a}, \vec{a} \times \vec{b}] \\
&= \sigma^3 \det \begin{pmatrix} [\vec{b}, \vec{c}, \vec{a}] & [\vec{b}, \vec{c}, \vec{b}] \\ [\vec{c}, \vec{a}, \vec{a}] & [\vec{c}, \vec{a}, \vec{b}] \end{pmatrix} \\
&= \sigma^3[\vec{b}, \vec{c}, \vec{a}] \cdot [\vec{c}, \vec{a}, \vec{b}] = \sigma^3[\vec{a}, \vec{b}, \vec{c}]^2 \\
&= \sigma = \frac{1}{[\vec{a}, \vec{b}, \vec{c}]}.
\end{aligned}$$

■

Exercise 9.3.11 *Show that*

$$\begin{array}{lll} \overrightarrow{a} \cdot \overrightarrow{a'} = 1, & \overrightarrow{a} \cdot \overrightarrow{b'} = 0, & \overrightarrow{a} \cdot \overrightarrow{c'} = 0, \\ \overrightarrow{b} \cdot \overrightarrow{a'} = 0, & \overrightarrow{b} \cdot \overrightarrow{b'} = 1, & \overrightarrow{b} \cdot \overrightarrow{c'} = 0, \\ \overrightarrow{c} \cdot \overrightarrow{a'} = 0, & \overrightarrow{c} \cdot \overrightarrow{b'} = 0, & \overrightarrow{c} \cdot \overrightarrow{c'} = 1. \end{array}$$

This explains why $\{\overrightarrow{a'}, \overrightarrow{b'}, \overrightarrow{c'}\}$ is sometimes called the dual system of $\{\overrightarrow{a}, \overrightarrow{b}, \overrightarrow{c}\}$.

Solved Exercise 9.3.12 *Let $\overrightarrow{a}$, $\overrightarrow{b}$, $\overrightarrow{c}$ be three non-coplanar vectors. Show that for any vector $\overrightarrow{v}$ the following decomposition holds*

$$\overrightarrow{v} = (\overrightarrow{v} \cdot \overrightarrow{a})\overrightarrow{a'} + (\overrightarrow{v} \cdot \overrightarrow{b})\overrightarrow{b'} + (\overrightarrow{v} \cdot \overrightarrow{c})\overrightarrow{c'}. \tag{9.3.12}$$

Proof: Making $\overrightarrow{d} = \overrightarrow{v}$ in Exercise 9.3.7 yields

$$[\overrightarrow{a}, \overrightarrow{b}, \overrightarrow{v}]\overrightarrow{c} - [\overrightarrow{a}, \overrightarrow{b}, \overrightarrow{c}]\overrightarrow{v} = [\overrightarrow{a}, \overrightarrow{c}, \overrightarrow{v}]\overrightarrow{b} - [\overrightarrow{b}, \overrightarrow{c}, \overrightarrow{v}]\overrightarrow{a},$$

which becomes after separating terms and using the definition of the mixed scalar product becomes

$$\begin{aligned} [\overrightarrow{a}, \overrightarrow{b}, \overrightarrow{c}]\overrightarrow{v} &= [\overrightarrow{b}, \overrightarrow{c}, \overrightarrow{v}]\overrightarrow{a} - [\overrightarrow{a}, \overrightarrow{c}, \overrightarrow{v}]\overrightarrow{b} + [\overrightarrow{a}, \overrightarrow{b}, \overrightarrow{v}]\overrightarrow{c} \\ &= \left((\overrightarrow{b} \times \overrightarrow{c}) \cdot \overrightarrow{v}\right)\overrightarrow{a} - ((\overrightarrow{a} \times \overrightarrow{c}) \cdot \overrightarrow{v})\,\overrightarrow{b} \\ &\quad + \left((\overrightarrow{a} \times \overrightarrow{b}) \cdot \overrightarrow{v}\right)\overrightarrow{c}. \end{aligned}$$

Dividing by $[\overrightarrow{a}, \overrightarrow{b}, \overrightarrow{c}]$ yields (9.3.12). ■

Exercise 9.3.13 *With the previous notations, show that for any vector $\overrightarrow{v}$ we have*

$$\overrightarrow{v} = (\overrightarrow{v} \cdot \overrightarrow{a'})\overrightarrow{a} + (\overrightarrow{v} \cdot \overrightarrow{b'})\overrightarrow{b} + (\overrightarrow{v} \cdot \overrightarrow{c'})\overrightarrow{c}.$$

Exercise 9.3.14 *Consider the vectors $\overrightarrow{u} = (1, 0, 3)$, $\overrightarrow{v} = (3, 1, 2)$ and $\overrightarrow{w} = (0, 1, 5)$. Find the volume of the parallelopiped generated by the vectors $\overrightarrow{u}$, $\overrightarrow{v}$, $\overrightarrow{w}$.*

Exercise 9.3.15 *Let $\overrightarrow{u} = (1, 0, 3)$, $\overrightarrow{v} = (2, 1, 0)$ and $\overrightarrow{w} = (0, 4, 6)$. Compute $[2\overrightarrow{u} + \overrightarrow{v}, \overrightarrow{v} + \overrightarrow{w}, 3\overrightarrow{w} + \overrightarrow{u}]$.*

Exercise 9.3.16 *Let $\overrightarrow{a} = (2, 0, 2)$, $\overrightarrow{b} = (1, 1, 0)$ and $\overrightarrow{c} = (3, 3, 2)$. Compute $\overrightarrow{a} \times (\overrightarrow{b} \times \overrightarrow{c})$ and $(\overrightarrow{a} \times \overrightarrow{b}) \times \overrightarrow{c}$. Are they equal? Why?*

Exercise 9.3.17 *Let* $\overrightarrow{a} = (2,0,0)$, $\overrightarrow{b} = (3,0,8)$, $\overrightarrow{c} = (1,0,1)$ *and* $\overrightarrow{d} = (0,0,3)$. *Compute* $(\overrightarrow{a} \times \overrightarrow{b}) \cdot (\overrightarrow{c} \times \overrightarrow{d})$.

Exercise 9.3.18 *Let* $\overrightarrow{a}, \overrightarrow{b}$ *be two vectors such that* $|\overrightarrow{a}| = 2$, $|\overrightarrow{b}| = 5$ *and* $\overrightarrow{a} \cdot \overrightarrow{b} = 6$. *Find* $|\overrightarrow{a} \times \overrightarrow{b}|$.

Chapter 10

Applications to Plane Geometry

The most dangerous moment comes with victory.
Napoleon Bonaparte

This chapter provides various applications of vectorial tools introduced in the previous chapters to Euclidean geometry.

10.1 Napoleon-type theorems

Napoleon Bonaparte (1769–1821), the emperor of France, manifested even as a child, interest in geometry. He stated and demonstrated the following problem:

Theorem 10.1.1 (Napoleon) *Consider a general triangle and construct outward on its sides three equilateral triangles. Then their centers are the corners of an equilateral triangle.*

Proof: We shall refer to the construction given in Figure 10.1. In order to prove that the triangle $\Delta G_A G_B G_C$ is equilateral it suffices to show that $|G_A G_B| = |G_B G_C| = |G_C G_A|$. We shall do this by computing one of these sides and showing that it can be expressed as a symmetric expression of the sides a, b, c of ΔABC. Using Proposition 6.3.4 (from page 377) applied to triangles $\Delta AB'C$ and $\Delta BCA'$, we obtain

$$3\overrightarrow{G_A G_B} = \overrightarrow{BA} + \overrightarrow{CB'} + \overrightarrow{A'C}.$$

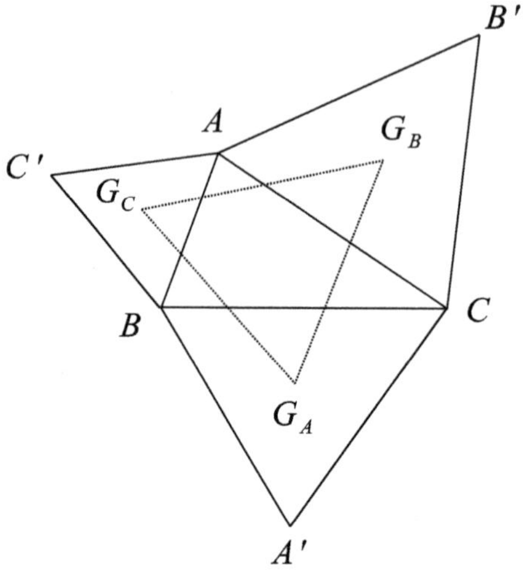

Figure 10.1: Napoleon's problem.

In the following we take the inner products and use Exercise 10.1.3 we have

$$\begin{aligned}
9|G_AG_B|^2 &= |BA|^2 + |CB'|^2 + |A'C|^2 + 2\overrightarrow{BA}\cdot\overrightarrow{CB'} + 2\overrightarrow{BA}\cdot\overrightarrow{A'C}\\
&\quad + 2\overrightarrow{CB'}\cdot\overrightarrow{A'C}\\
&= c^2 + b^2 + a^2 + 2cb\cos(\widehat{\overrightarrow{BA},\overrightarrow{CB'}}) + 2ca\cos(\widehat{\overrightarrow{BA},\overrightarrow{A'C}})\\
&\quad + 2ba\cos(\widehat{\overrightarrow{CB'},\overrightarrow{A'C}})\\
&= a^2 + b^2 + c^2 + 2bc\cos(\widehat{A} - 60^\circ) + 2ac\cos(\widehat{B} - 60^\circ)\\
&\quad + 2ab\cos(\widehat{C} - 60^\circ)\\
&= a^2 + b^2 + c^2 + 2bc\left(\frac{1}{2}\cos\widehat{A} - \frac{\sqrt{3}}{2}\sin\widehat{A}\right)\\
&\quad + 2ac\left(\frac{1}{2}\cos\widehat{B} - \frac{\sqrt{3}}{2}\sin\widehat{B}\right) + 2ab\left(\frac{1}{2}\cos\widehat{C} - \frac{\sqrt{3}}{2}\sin\widehat{C}\right).
\end{aligned}$$

Now, using the law of cosines and the law of sines we obtain

$$\begin{aligned}
9|G_AG_B|^2 = a^2 + b^2 + c^2 + 2bc\frac{b^2+c^2-a^2}{2bc}\frac{1}{2} - \sqrt{3}bc\sin\widehat{A}\\
+ 2ac\frac{a^2+c^2-b^2}{2ac}\frac{1}{2} - \sqrt{3}ac\sin\widehat{B}
\end{aligned}$$

$$+ 2ab\frac{a^2+b^2-c^2}{2ab}\frac{1}{2} - \sqrt{3}ab\sin\widehat{C}$$
$$= \frac{3}{2}(a^2+b^2+c^2) - \sqrt{3}\left(bc\frac{a}{2R} + ac\frac{b}{2R} + ab\frac{c}{2R}\right)$$
$$= \frac{3}{2}(a^2+b^2+c^2) - \frac{3\sqrt{3}}{2R}abc.$$

Hence

$$|G_AG_B|^2 = \frac{1}{6}\left(a^2+b^2+c^2 - \frac{\sqrt{3}}{R}abc\right),$$

which is a symmetric expression in a, b and c. By a similar computation we arrive to the same expression for $|G_BG_C|^2$ and $|G_CG_A|^2$. We conclude that $\Delta G_AG_BG_C$ is an equilateral triangle. ■

Remark 10.1.2 The areas of triangles $\Delta G_AG_BG_C$ and ΔABC are related by the relation

$$\mathcal{A}_{G_AG_BG_C} = \frac{\sqrt{3}}{4}|G_AG_B|^2 = \frac{\sqrt{3}}{4\cdot 6}\left(a^2+b^2+c^2 - \frac{\sqrt{3}}{R}abc\right)$$
$$= \frac{\sqrt{3}}{24}(a^2+b^2+c^2) - \frac{1}{2}\mathcal{A}_{ABC}.$$

Solved Exercise 10.1.3 *Let ℓ and ℓ' be two lines in the plane. A third line, m, cuts the previous lines under alternate interior angles $\alpha = \widehat{\ell, m}$ and $\beta = \widehat{\ell', m}$. Show that the angle made by the lines is $\widehat{\ell, \ell'} = |\alpha - \beta|$.*

Proof: If $\ell \| \ell'$ then the lines are parallel, so $\widehat{\ell, \ell'} = 0$. In this case the alternate interior angles are equal, $\alpha = \beta$, so $\widehat{\ell, \ell'} = |\alpha - \beta|$ is verified. If $\ell \not\| \ell'$ then the lines intersect at a point P. Let $\theta = \widehat{\ell, \ell'}$. If $\{A\} = \ell \cap m$ and $\{B\} = \ell' \cap m$, applying the exterior angle theorem in triangle ABP we obtain $\alpha = \theta + \beta$ or $\beta = \alpha + \theta$. ■

In the rest of the section we shall continue to investigate properties dealing with exterior constructions on the sides of a triangle or a quadrilateral. In order to do this, we need a few preliminary results.

Theorem 10.1.4 (Cauchy's inequality) *Let $\overrightarrow{a}$ and $\overrightarrow{b}$ be two vectors. Then*

$$|\overrightarrow{a} \cdot \overrightarrow{b}| \le |\overrightarrow{a}| \cdot |\overrightarrow{b}|.$$

The identity holds if and only if the vectors $\overrightarrow{a}$ and $\overrightarrow{b}$ are proportional.

Proof: Using the definition on the inner product, we have

$$|\overrightarrow{a} \cdot \overrightarrow{b}| = |\overrightarrow{a}| \cdot |\overrightarrow{b}| \cdot |\cos(\widehat{\overrightarrow{a}, \overrightarrow{b}})| \leq |\overrightarrow{a}| \cdot |\overrightarrow{b}|,$$

since cosine function is bounded above by 1 and below by -1. The identity holds for the case $|\cos(\widehat{\overrightarrow{a}, \overrightarrow{b}})| = 1$, *i.e.*, when the angle $\widehat{\overrightarrow{a}, \overrightarrow{b}} \in \{0, 180°\}$. In this case the vectors have the same support line and hence they are proportional. ■

Solved Exercise 10.1.5 *Let $\overrightarrow{a}, \overrightarrow{b}$ be two vectors. Show that the matrix*

$$\begin{pmatrix} \overrightarrow{a} \cdot \overrightarrow{a} & \overrightarrow{a} \cdot \overrightarrow{b} \\ \overrightarrow{b} \cdot \overrightarrow{a} & \overrightarrow{b} \cdot \overrightarrow{b} \end{pmatrix} \tag{10.1.1}$$

is non-singular if and only if the vectors $\overrightarrow{a}$ and $\overrightarrow{b}$ are linear independent.

Proof: We shall prove the complementary statement:

The matrix (10.1.1) *is singular if and only if $\overrightarrow{a}$ and $\overrightarrow{b}$ are linearly dependent.*

The matrix (10.1.1) is singular if and only if its determinant vanishes

$$0 = \det \begin{pmatrix} \overrightarrow{a} \cdot \overrightarrow{a} & \overrightarrow{a} \cdot \overrightarrow{b} \\ \overrightarrow{b} \cdot \overrightarrow{a} & \overrightarrow{b} \cdot \overrightarrow{b} \end{pmatrix} = |\overrightarrow{a}|^2 \cdot |\overrightarrow{b}|^2 - (\overrightarrow{a} \cdot \overrightarrow{b})^2.$$

This can be written as $|\overrightarrow{a} \cdot \overrightarrow{b}| = |\overrightarrow{a}| \cdot |\overrightarrow{b}|$ and hence we have identity in the Cauchy's inequality. Using Theorem 10.1.4 yields $\overrightarrow{a} = \lambda \overrightarrow{b}$, with $\lambda \in \mathbb{R}$, *i.e.*, the vectors are linear dependent. ■

Solved Exercise 10.1.6 *Let $\overrightarrow{a}, \overrightarrow{b}$ be two linear independent vectors in the plane. If $\overrightarrow{c}$ is a vector in the same plane such that $\overrightarrow{a} \cdot \overrightarrow{c} = \overrightarrow{b} \cdot \overrightarrow{c} = 0$, then $\overrightarrow{c} = 0$. In other words, if the vector $\overrightarrow{c}$ is perpendicular to two linear independent vectors $\overrightarrow{a}$ and $\overrightarrow{b}$, then it is the zero vector.*

Proof: Since $\{\overrightarrow{a}, \overrightarrow{b}\}$ is a basis in the plane, we can write

$$\overrightarrow{c} = x\overrightarrow{a} + y\overrightarrow{b}, \qquad \text{with} \quad x, y \in \mathbb{R}. \tag{10.1.2}$$

Taking the inner product with $\overrightarrow{a}$ and then with $\overrightarrow{b}$, yields

$$\begin{aligned} 0 &= \overrightarrow{a} \cdot \overrightarrow{c} = x\overrightarrow{a} \cdot \overrightarrow{a} + y\overrightarrow{a} \cdot \overrightarrow{b} \\ 0 &= \overrightarrow{b} \cdot \overrightarrow{c} = x\overrightarrow{b} \cdot \overrightarrow{a} + y\overrightarrow{b} \cdot \overrightarrow{b}, \end{aligned}$$

which can be written as

$$\begin{pmatrix} \overrightarrow{a}\cdot\overrightarrow{a} & \overrightarrow{a}\cdot\overrightarrow{b} \\ \overrightarrow{b}\cdot\overrightarrow{a} & \overrightarrow{b}\cdot\overrightarrow{b} \end{pmatrix}\begin{pmatrix} x \\ y \end{pmatrix} = \begin{pmatrix} 0 \\ 0 \end{pmatrix}.$$

Since $\overrightarrow{a}$ and $\overrightarrow{b}$ are linear independent, the matrix (10.1.1) is nonsingular, see Exercise 10.1.5. Then the solution of the above linear system is

$$\begin{pmatrix} x \\ y \end{pmatrix} = \begin{pmatrix} \overrightarrow{a}\cdot\overrightarrow{a} & \overrightarrow{a}\cdot\overrightarrow{b} \\ \overrightarrow{b}\cdot\overrightarrow{a} & \overrightarrow{b}\cdot\overrightarrow{b} \end{pmatrix}^{-1}\begin{pmatrix} 0 \\ 0 \end{pmatrix} = \begin{pmatrix} 0 \\ 0 \end{pmatrix}.$$

Hence $x = y = 0$ and (10.1.2) becomes $\overrightarrow{c} = 0\overrightarrow{a} + 0\overrightarrow{b} = 0$. ■

The following theorem deals with a Napoleon type construction:

Theorem 10.1.7 *If squares $ACC''A'$, $CBB''C'$, $BAA''B'$ are constructed externally on the sides of any triangle ABC, then their centers form a triangle with the same centroid as triangle ABC.*

Proof: Let M be an arbitrary point in the plane. Denote by O_1, O_2, O_3 the centers of the external squares, see Figure 10.2. From Lemma 6.1.4 we have

$$\overrightarrow{MO_2} = \frac{1}{2}(\overrightarrow{MA'} + \overrightarrow{MC}) = \frac{1}{2}(\overrightarrow{MA} + \overrightarrow{AA'} + \overrightarrow{MC}). \qquad (10.1.3)$$

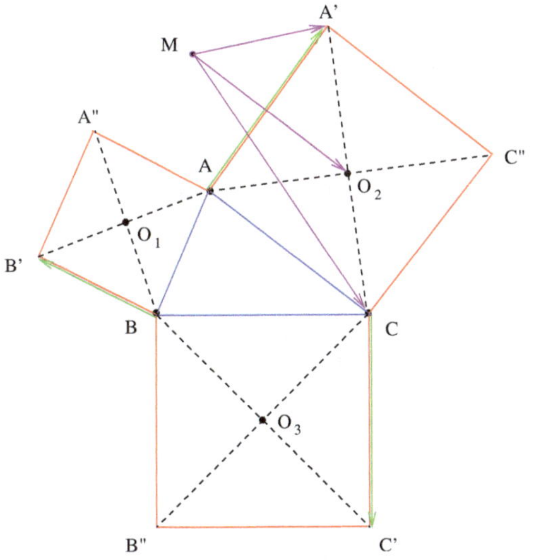

Figure 10.2: Triangles ΔABC and $\Delta O_1O_2O_3$ share the same centroid.

In a similar way we get

$$\overrightarrow{MO_1} = \frac{1}{2}(\overrightarrow{MB} + \overrightarrow{BB'} + \overrightarrow{MA}), \tag{10.1.4}$$

$$\overrightarrow{MO_3} = \frac{1}{2}(\overrightarrow{MC} + \overrightarrow{CC'} + \overrightarrow{MB}). \tag{10.1.5}$$

Summing (10.1.3), (10.1.4) and (10.1.5) yields

$$\overrightarrow{MO_1} + \overrightarrow{MO_2} + \overrightarrow{MO_3} = \overrightarrow{MA} + \overrightarrow{MB} + \overrightarrow{MC} + \frac{1}{2}(\overrightarrow{AA'} + \overrightarrow{BB'} + \overrightarrow{CC'}),$$

which can be written using Theorem 6.3.2 as

$$3\overrightarrow{MG'} = 3\overrightarrow{MG} + \frac{1}{2}(\overrightarrow{AA'} + \overrightarrow{BB'} + \overrightarrow{CC'}), \tag{10.1.6}$$

where G and G' are the centroids of ΔABC and $\Delta O_1O_2O_3$, respectively. It suffices to show that $\overrightarrow{AA'} + \overrightarrow{BB'} + \overrightarrow{CC'} = 0$. To this end, we compute

$$\begin{aligned}
& \overrightarrow{BC} \cdot (\overrightarrow{AA'} + \overrightarrow{BB'} + \overrightarrow{CC'}) \\
= \; & \overrightarrow{BC} \cdot (\overrightarrow{AA'} + \overrightarrow{BB'}) + \underbrace{\overrightarrow{BC} \cdot \overrightarrow{CC'}}_{=0} \\
= \; & \overrightarrow{BC} \cdot \overrightarrow{AA'} + \overrightarrow{BC} \cdot \overrightarrow{BB'} \\
= \; & (\overrightarrow{BA} + \overrightarrow{AC}) \cdot \overrightarrow{AA'} + (\overrightarrow{BA} + \overrightarrow{AC}) \cdot \overrightarrow{BB'} \\
= \; & \overrightarrow{BA} \cdot \overrightarrow{AA'} + \underbrace{\overrightarrow{AC} \cdot \overrightarrow{AA'}}_{=0} + \underbrace{\overrightarrow{BA} \cdot \overrightarrow{BB'}}_{=0} + \overrightarrow{AC} \cdot \overrightarrow{BB'} \\
= \; & \overrightarrow{BA} \cdot \overrightarrow{AA'} + \overrightarrow{AC} \cdot \overrightarrow{BB'} \\
= \; & -\overrightarrow{AB} \cdot \overrightarrow{AA'} + \overrightarrow{AC} \cdot \overrightarrow{AA''} \\
= \; & -|AB| \cdot |AA'| \cos(\widehat{A} + 90^\circ) + |AC| \cdot |AA''| \cos(90^\circ - \widehat{A}) \\
= \; & 0,
\end{aligned}$$

since $|AB| = |AA''|$, $|AA'| = |AC|$. Hence

$$\overrightarrow{BC} \cdot (\overrightarrow{AA'} + \overrightarrow{BB'} + \overrightarrow{CC'}) = 0.$$

In a similar way we obtain

$$\overrightarrow{BA} \cdot (\overrightarrow{AA'} + \overrightarrow{BB'} + \overrightarrow{CC'}) = 0.$$

Since the vectors $\overrightarrow{BA}$ and $\overrightarrow{BC}$ are linear independent, applying Exercise 10.1.6 yields

$$\overrightarrow{AA'} + \overrightarrow{BB'} + \overrightarrow{CC'} = 0.$$

Hence (10.1.6) becomes

$$3\overrightarrow{MG'} = 3\overrightarrow{MG},$$

which implies $G = G'$. ∎

Remark 10.1.8 By constructing squares in the exterior of the sides of triangle ΔABC is like changing the weights of the triangle sides, such that the centroid of the new construction remains the same as the centroid of the original triangle.

Theorem 10.1.9 *Let O be the circumcenter of triangle ABC. On the perpendicular lines from O to AB, BC, and CA we choose points C', A' and B' such that*

$$|OA'| = \lambda|BC|, \quad |OB'| = \lambda|AC|, \quad |OC'| = \lambda|AB|,$$

for a given $\lambda > 0$. Then O is the centroid of triangle $A'B'C'$.

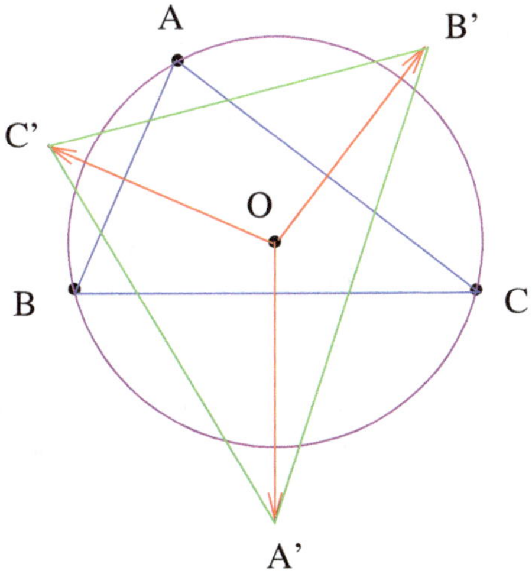

Figure 10.3: O is the centroid of $\Delta A'B'C'$.

Proof: We shall refer to Figure 10.3. Let $\overrightarrow{v} = \overrightarrow{OA'} + \overrightarrow{OB'} + \overrightarrow{OC'}$. It suffices to show that $\overrightarrow{v} = 0$. Then Theorem 6.3.2, part 3) yields the desired result.

We shall show that $\overrightarrow{v}$ is perpendicular to $\overrightarrow{BC}$, $\overrightarrow{CA}$ and $\overrightarrow{AB}$. Applying Exercise 10.1.6 will yield $\overrightarrow{v} = 0$. In order to do this we shall use

Proposition 7.2.3, part 5).

$$\begin{aligned}
\vec{v}\cdot\overrightarrow{BC} &= (\overrightarrow{OA'}+\overrightarrow{OB'}+\overrightarrow{OC'})\cdot\overrightarrow{BC} = \underbrace{\overrightarrow{OA'}\cdot\overrightarrow{BC}}_{=0}+(\overrightarrow{OB'}+\overrightarrow{OC'})\cdot\overrightarrow{BC}\\
&= (\overrightarrow{OB'}+\overrightarrow{OC'})\cdot\overrightarrow{BC} = (\overrightarrow{OB'}+\overrightarrow{OC'})\cdot(\overrightarrow{BA}+\overrightarrow{AC})\\
&= \overrightarrow{OB'}\cdot\overrightarrow{BA}+\underbrace{\overrightarrow{OB'}\cdot\overrightarrow{AC}}_{=0}+\underbrace{\overrightarrow{OC'}\cdot\overrightarrow{BA}}_{=0}+\overrightarrow{OC'}\cdot\overrightarrow{AC}\\
&= \overrightarrow{OB'}\cdot\overrightarrow{BA}+\overrightarrow{OC'}\cdot\overrightarrow{AC} = \overrightarrow{OB'}\cdot\overrightarrow{BA}-\overrightarrow{OC'}\cdot\overrightarrow{CA}\\
&= |OB'|\cdot|BA|\cos\varphi-|OC'|\cdot|CA|\cos\varphi\\
&= \lambda|AC|\cdot|BA|\cos\varphi-\lambda|AB|\cdot|CA|\cos\varphi\\
&= 0,
\end{aligned}$$

where $\varphi = |90^\circ - \hat{A}|$ by Exercise 10.1.3. In a similar way, $\vec{v}\cdot\overrightarrow{CA} = 0$ and $\vec{v}\cdot\overrightarrow{AB} = 0$, and the proof is finished. ■

Remark 10.1.10 It can be shown that the perpendicular lines through A, B, C to $B'C'$, $A'C'$, $A'B'$ respectively, intersect at a point. The proof is not appropriate to vectorial techniques and will be done later by other methods. We also note that triangles ABC and $A'B'C'$ are orthologic, see the next section.

In the following we shall present another Napoleon-type theorem, where the initial triangle is replaced by a convex quadrilateral and the exterior triangles are replaced by exterior squares.

Theorem 10.1.11 *Given a convex quadrilateral $ABCD$, place a square outwardly on each side and denote by O_i, $i \in \{1,\ldots,4\}$ their centers. Then the quadrilaterals $ABCD$ and $O_1O_2O_3O_4$ have the same centroid, see Figure 10.4.*

Before starting the proof we shall prepare the ground with a few useful results.

Lemma 10.1.12 *The following relations take place.*

(i) $\overrightarrow{AC}$ is perpendicular to the vectors $\overrightarrow{BB'}+\overrightarrow{BB''}$ and $\overrightarrow{DD'}+\overrightarrow{DD''}$.

(ii) $\overrightarrow{BD}$ is perpendicular to the vectors $\overrightarrow{CC'}+\overrightarrow{CC''}$ and $\overrightarrow{AA'}+\overrightarrow{AA''}$.

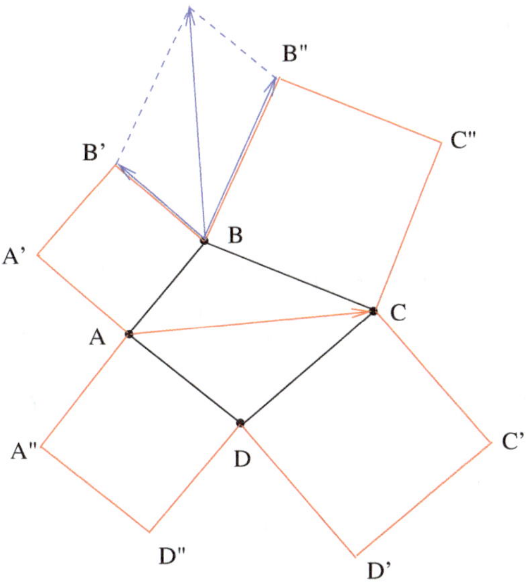

Figure 10.4: For Theorem 10.1.11.

Proof: 1) We shall consider the notations of Figure 10.4. Using the properties of the inner product we have

$$\begin{aligned}
\overrightarrow{AC}\cdot(\overrightarrow{BB'}+\overrightarrow{BB''}) &= \overrightarrow{AC}\cdot\overrightarrow{BB'}+\overrightarrow{AC}\cdot\overrightarrow{BB''}\\
&= (\overrightarrow{AB}+\overrightarrow{BC})\cdot\overrightarrow{BB'}+(\overrightarrow{AB}+\overrightarrow{BC})\cdot\overrightarrow{BB''}\\
&= \underbrace{\overrightarrow{AB}\cdot\overrightarrow{BB'}}_{=0}+\overrightarrow{BC}\cdot\overrightarrow{BB'}+\overrightarrow{AB}\cdot\overrightarrow{BB''}+\underbrace{\overrightarrow{BC}\cdot\overrightarrow{BB''}}_{=0}\\
&= \overrightarrow{BC}\cdot\overrightarrow{BB'}-\overrightarrow{BA}\cdot\overrightarrow{BB''}\\
&= |BC|\cdot|BB'|\cos(90+\widehat{B'BB''})\\
&\quad -|BA|\cdot|BB''|\cos(90+\widehat{B'BB''})\\
&= 0,
\end{aligned}$$

since $|BC| = |BB''|$, $|BA| = |BB'|$. Hence $\overrightarrow{AC}$ is perpendicular to the vector $\overrightarrow{BB'}+\overrightarrow{BB''}$. In a similar way we prove the other relations. ■

Lemma 10.1.13 *We have*

$$\overrightarrow{BB'}+\overrightarrow{BB''}+\overrightarrow{DD'}+\overrightarrow{DD''}=0. \tag{10.1.7}$$

Proof: Lemma 10.1.12 yields

$$\overrightarrow{AC} \cdot (\overrightarrow{BB'} + \overrightarrow{BB''}) = 0$$

$$\overrightarrow{AC} \cdot (\overrightarrow{DD'} + \overrightarrow{DD''}) = 0.$$

Adding, we obtain

$$\overrightarrow{AC} \cdot (\overrightarrow{BB'} + \overrightarrow{BB''} + \overrightarrow{DD'} + \overrightarrow{DD''}) = 0. \tag{10.1.8}$$

Using that

$$\overrightarrow{BB'} = \overrightarrow{AA'}, \quad \overrightarrow{BB''} = \overrightarrow{CC''}, \quad \overrightarrow{DD'} = \overrightarrow{CC'}, \quad \overrightarrow{DD''} = \overrightarrow{AA''},$$

it follows that

$$\overrightarrow{BB'} + \overrightarrow{BB''} + \overrightarrow{DD'} + \overrightarrow{DD''} = \overrightarrow{AA'} + \overrightarrow{AA''} + \overrightarrow{CC'} + \overrightarrow{CC''}. \tag{10.1.9}$$

Using again Lemma 10.1.12, yields

$$\overrightarrow{BD} \cdot (\overrightarrow{AA'} + \overrightarrow{AA''}) = 0,$$

$$\overrightarrow{BD} \cdot (\overrightarrow{CC'} + \overrightarrow{CC''}) = 0,$$

and hence by addition we obtain

$$\overrightarrow{BD} \cdot (\overrightarrow{AA'} + \overrightarrow{AA''} + \overrightarrow{CC'} + \overrightarrow{CC''}),$$

or, using (10.1.9) yields

$$\overrightarrow{BD} \cdot (\overrightarrow{BB'} + \overrightarrow{BB''} + \overrightarrow{DD'} + \overrightarrow{DD''}) = 0. \tag{10.1.10}$$

From (10.1.8) and (10.1.10), using Exercise 10.1.6 yields (10.1.7), because $\overrightarrow{AC}$ and $\overrightarrow{BD}$ are linear independent. ∎

In the following we shall present the proof of Theorem 10.1.11.

Proof: Let M be an arbitrary point in the plane. Let G and G' be the centroids of quadrilaterals $ABCD$ and $O_1O_2O_3O_4$, respectively. Then

$$4\overrightarrow{MG} = \overrightarrow{MA} + \overrightarrow{MB} + \overrightarrow{MC} + \overrightarrow{MD}, \tag{10.1.11}$$

$$4\overrightarrow{MG'} = \overrightarrow{MO_1} + \overrightarrow{MO_2} + \overrightarrow{MO_3} + \overrightarrow{MO_4}. \tag{10.1.12}$$

We shall compute the vectors $\overrightarrow{MO_i}$, $i \in \{1, \ldots, 4\}$, see Figure 10.5.

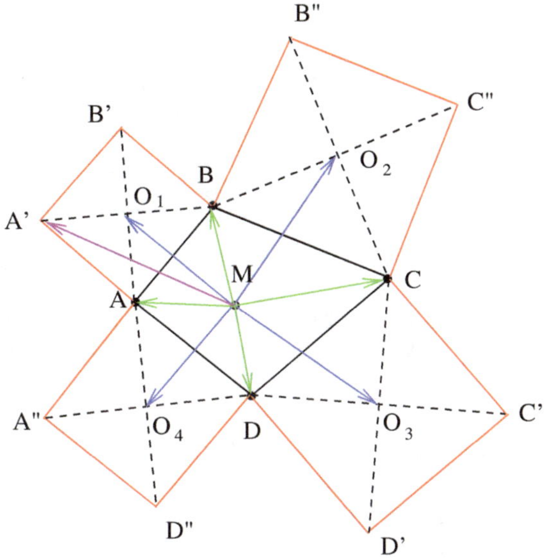

Figure 10.5: For Theorem 10.1.11.

Using Lemma 6.1.4, yields

$$\begin{aligned}\overrightarrow{MO_1} &= \frac{1}{2}(\overrightarrow{MA'} + \overrightarrow{MB}) = \frac{1}{2}(\overrightarrow{MA} + \overrightarrow{AA'} + \overrightarrow{MB}) \\ &= \frac{1}{2}\overrightarrow{MA} + \frac{1}{2}\overrightarrow{AA'} + \frac{1}{2}\overrightarrow{MB}.\end{aligned}$$

Writing similar formulas for each center O_i we have

$$\begin{aligned}\overrightarrow{MO_1} &= \frac{1}{2}\overrightarrow{MA} + \frac{1}{2}\overrightarrow{MB} + \frac{1}{2}\overrightarrow{AA'} \\ \overrightarrow{MO_2} &= \frac{1}{2}\overrightarrow{MB} + \frac{1}{2}\overrightarrow{MC} + \frac{1}{2}\overrightarrow{BB''} \\ \overrightarrow{MO_3} &= \frac{1}{2}\overrightarrow{MC} + \frac{1}{2}\overrightarrow{MD} + \frac{1}{2}\overrightarrow{CC'} \\ \overrightarrow{MO_4} &= \frac{1}{2}\overrightarrow{MD} + \frac{1}{2}\overrightarrow{MA} + \frac{1}{2}\overrightarrow{DD''}.\end{aligned}$$

Summing,

$$
\begin{aligned}
& \overrightarrow{MO_1} + \overrightarrow{MO_2} + \overrightarrow{MO_3} + \overrightarrow{MO_4} \\
= \; & \overrightarrow{MA} + \overrightarrow{MB} + \overrightarrow{MC} + \overrightarrow{MD} + \frac{1}{2}(\overrightarrow{AA'} + \overrightarrow{BB''} + \overrightarrow{CC'} + \overrightarrow{DD''}) \\
= \; & \overrightarrow{MA} + \overrightarrow{MB} + \overrightarrow{MC} + \overrightarrow{MD} + \frac{1}{2}\underbrace{(\overrightarrow{BB'} + \overrightarrow{BB''} + \overrightarrow{DD'} + \overrightarrow{DD''})}_{=0} \\
= \; & \overrightarrow{MA} + \overrightarrow{MB} + \overrightarrow{MC} + \overrightarrow{MD},
\end{aligned}
$$

where we used Lemma 10.1.13. Hence, the terms on the left-hand side of relations (10.1.11) and (10.1.12) are equal. It follows that $\overrightarrow{MG} = \overrightarrow{MG'}$ and hence $G = G'$. ∎

Exercise 10.1.14 (Yaglom) *Let $ABCD$ be a parallelogram. Construct in the exterior, on the sides, the squares $ABB'A''$, $BCC'B''$, $CDD'C''$ and $DAA'D''$ with centers A^*, B^*, C^* and D^*, respectively, see Figure 10.6.*

(i) Show that $A^*B^*C^*D^*$ is a square.

(ii) Prove that the lines AC, BD, A^*C^*, $A''C''$, $B'D'$, $B''D''$, $A'C'$ and B^*D^* intersect at a point. What is this point?

(iii) Show that $|A'A''| = |C'C''| = |AC|$ and $|B'B''| = |D'D''| = |BD|$.

(iv) Prove that $A'A'' \perp AC$ and $B'B'' \perp BD$.

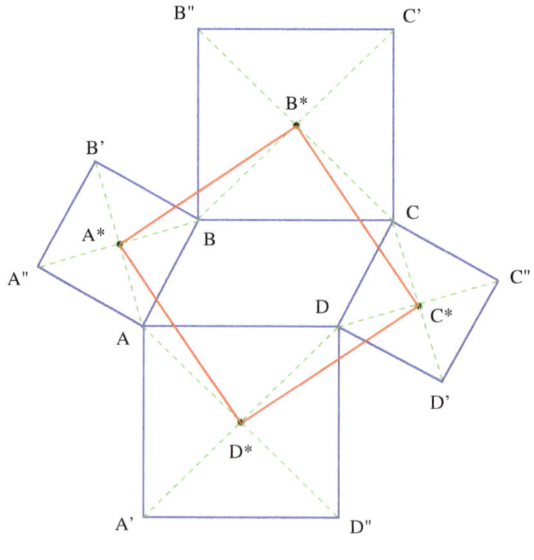

Figure 10.6: For Exercise 10.1.14.

Exercise 10.1.15 *Let $ABDE$, $BCGF$, $CAIJ$ be squares constructed externally on the sides of the triangle ABC with centers Z, X, Y. Let A' be the midpoint of $[BC]$ and let L be a point such that $AELI$ is a parallelogram.*

(i) Show that $AA' \perp EI$ and $|EI| = 2|AA'|$.

(ii) Prove that $AL \perp BC$, $CL \perp BJ$, $BL \perp CD$, $BI \perp CE$.

(iii) Show that the following segments are congruent:

$$|AL| = |BC|, \quad |CL| = |BJ|, \quad |BL| = |CD|, \quad |BI| = |CE|.$$

(iv) Prove that the lines AX, BY, CZ intersect at a point.

(v) Show that the lines CD, BJ, AL intersect at a point.

(vi) Show that $AX \perp YZ$ and $|AX| = |YZ|$.

Exercise 10.1.16 (Aubel) *Given a convex quadrilateral, construct outwardly squares on each side, and connect the centers of opposite squares. Then the segments are congruent and cross at a right angle. See Figure 10.7.*

Solved Exercise 10.1.17 *Let A', B' and C' be the midpoints of the sides BC, AC and AB of triangle ΔABC and consider an arbitrary point M*

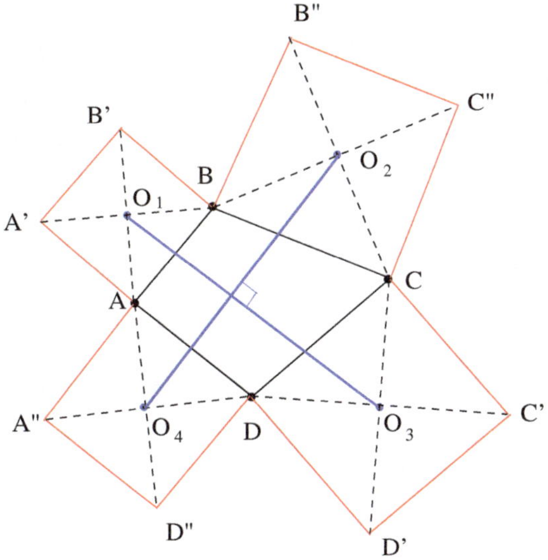

Figure 10.7: For Exercise 10.1.16.

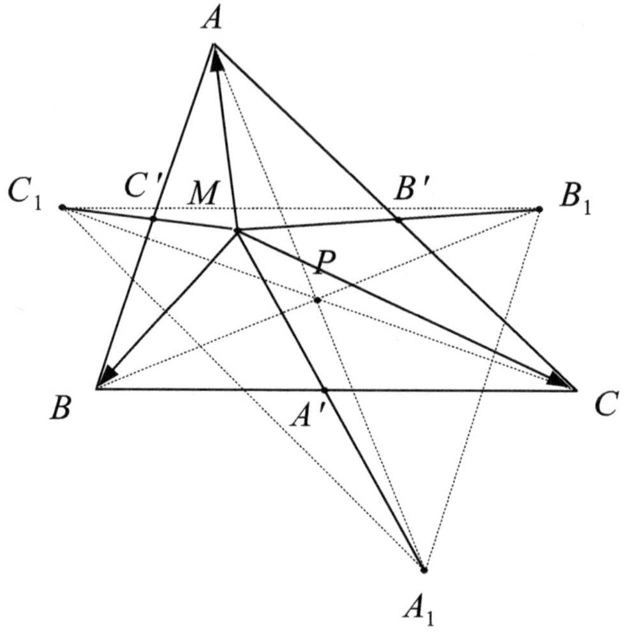

Figure 10.8: For Exercise 10.1.17.

in the same plane. Denote by A_1, B_1 and C_1 the symmetric points of M with respect to A, B' and C', respectively. Let G and G_1 be the centroids of triangles ΔABC and $\Delta A_1B_1C_1$, respectively. Show the following.

(i) The points M, G and G_1 are collinear, with G midpoint of MG_1.

(ii) $\Delta ABC \equiv \Delta A_1B_1C_1$.

(iii) The lines AA_1, BB_1 and CC_1 are concurrent at a point P. This point is the common midpoint of segments AA_1, BB_1, CC_1 and GG_1.

Proof: We shall refer to Figure 10.8.

(*i*) We note that MBA_1C, MCB_1A and MAC_1B are parallelograms (as quadrilaterals with bisecting diagonals). Then by the parallelogram rule we have

$$\begin{aligned}
\overrightarrow{MB} + \overrightarrow{MC} &= \overrightarrow{MA_1}\\
\overrightarrow{MC} + \overrightarrow{MA} &= \overrightarrow{MB_1}\\
\overrightarrow{MA} + \overrightarrow{MB} &= \overrightarrow{MC_1}.
\end{aligned}$$

Adding we obtain

$$2(\overrightarrow{MA} + \overrightarrow{MB} + \overrightarrow{MC}) = \overrightarrow{MA_1} + \overrightarrow{MB_1} + \overrightarrow{MC_1},$$

which is equivalent to $6\overrightarrow{MG} = 3\overrightarrow{MG_1}$, which implies that M, G and G_1 are collinear and G is the midpoint of the segment MG_1.

(*ii*) Using the triangle rule and the aforementioned relations, we have

$$\begin{aligned}\overrightarrow{A_1B_1} &= \overrightarrow{A_1M} + \overrightarrow{MB_1} = \overrightarrow{MB_1} - \overrightarrow{MA_1} = (\overrightarrow{MA} + \overrightarrow{MC}) - (\overrightarrow{MB} + \overrightarrow{MC}) \\ &= \overrightarrow{MA} - \overrightarrow{MB} = \overrightarrow{BM} + \overrightarrow{MA} = \overrightarrow{BA}.\end{aligned}$$

Taking the lengths yields $|A_1B_1| = |AB|$. In a similar way we obtain $|A_1C_1| = |AC|$ and $|B_1C_1| = |BC|$ and hence $\Delta ABC \equiv \Delta A_1B_1C_1$ by the (SSS) congruence case.

(*iii*) From part (*ii*) we have $\overrightarrow{A_1B_1} = \overrightarrow{BA}$, so A_1B_1AB is a parallelogram. Then its diagonals, AA_1 and BB_1 intersect at a point P, which is the common midpoint of AA_1 and BB_1. In a similar way we can show that P is the common midpoint of AA_1 and CC_1. Then it follows that the lines AA_1, BB_1 and CC_1 are concurrent at the point P. Since P is the midpoint of the aforementioned segments, we can write

$$\overrightarrow{PA} + \overrightarrow{PA_1} = 0, \quad \overrightarrow{PB} + \overrightarrow{PB_1} = 0, \quad \overrightarrow{PC} + \overrightarrow{PC_1} = 0.$$

Adding and grouping, yields

$$(\overrightarrow{PA} + \overrightarrow{PB} + \overrightarrow{PC}) + (\overrightarrow{PA_1} + \overrightarrow{PB_1} + \overrightarrow{PC_1}) = 0,$$

which is equivalent to

$$3\overrightarrow{PG} + 3\overrightarrow{PG_1} = 0,$$

which implies that P is the midpoint of GG_1. ■

10.2 Menelaos' theorem

If we are not ashamed to think it,
we should not be ashamed to say it.
Marcus T. Cicero

An important tool in proving collinearity of three points is Menelaos' theorem. In the following we shall use the following convention. If $\overrightarrow{u}$ and $\overrightarrow{v}$ are two collinear vectors with $\overrightarrow{u} = \lambda\overrightarrow{v}$ then we shall write $\dfrac{\overrightarrow{u}}{\overrightarrow{v}} = \lambda$. If the vectors are not collinear the division of vectors does not make sense.

Theorem 10.2.1 (Menelaos) *Let ABC be a triangle and let $P \in BC$ and $Q \in AC$, $R \in AB$. The points R, Q, P are collinear if and only if*

$$\frac{\overrightarrow{PB}}{\overrightarrow{PC}} \cdot \frac{\overrightarrow{QC}}{\overrightarrow{QA}} \cdot \frac{\overrightarrow{RA}}{\overrightarrow{RB}} = 1. \tag{10.2.13}$$

Proof: We note that in the left side of (10.2.13) only two quotients are negative or all of them are positive. This corresponds to the following two cases:

(a) Two of the points P, Q, R lie inside of the sides $[BC]$, $[AC]$, $[AB]$ and one point is outside. See Figure 10.9(a).

(b) All three points P, Q, R lie outside of the sides of the triangle. See Figure 10.9(b).

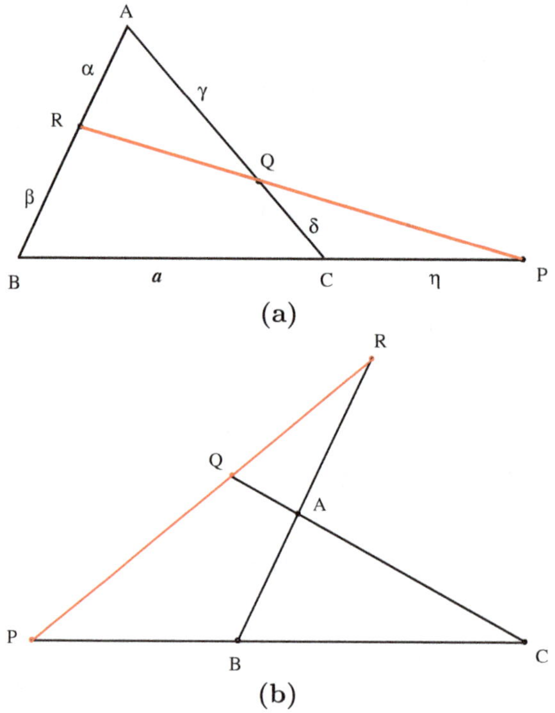

Figure 10.9: (a) Q between A and C; (b) A between Q and C.

In the following we shall consider the case (a) and leave the case (b) as an exercise for the reader.

In order to simplify the computations we shall make the following notations: $\alpha = |AR|$, $\beta = |BR|$, $a = |BC|$, $\eta = |CP|$, $\gamma = |AQ|$. Relation (10.2.13) can be written as

$$\frac{a+\eta}{\eta}\frac{\delta}{\gamma}\frac{\alpha}{\beta} = 1. \tag{10.2.14}$$

We note that the points R, Q, P are collinear if and only if the vectors $\overrightarrow{RQ}$ and $\overrightarrow{QP}$ are collinear. These vectors will be expressed using the base $\{\overrightarrow{AB}, \overrightarrow{AC}\}$. We have

$$\begin{aligned}\overrightarrow{RQ} &= \overrightarrow{RA} + \overrightarrow{AQ} = -\overrightarrow{AR} + \overrightarrow{AQ} \\ &= -\frac{\alpha}{\alpha+\beta}\overrightarrow{AB} + \frac{\gamma}{\gamma+\delta}\overrightarrow{AC}.\end{aligned} \tag{10.2.15}$$

Since $\overrightarrow{BC}$ and $\overrightarrow{CP}$ are collinear and $|\overrightarrow{BC}| = a$ and $|\overrightarrow{CP}| = \eta$,

$$\overrightarrow{CP} = \frac{\eta}{a}\overrightarrow{BC}.$$

Then

$$\begin{aligned}\overrightarrow{QP} &= \overrightarrow{QC} + \overrightarrow{CP} = \frac{\delta}{\gamma+\delta}\overrightarrow{AC} + \frac{\eta}{a}\overrightarrow{BC} \\ &= \frac{\delta}{\gamma+\delta}\overrightarrow{AC} + \frac{\eta}{a}(\overrightarrow{AC} - \overrightarrow{AB}) \\ &= \left(\frac{\delta}{\gamma+\delta} + \frac{\eta}{a}\right)\overrightarrow{AC} - \frac{\eta}{a}\overrightarrow{AB}.\end{aligned} \tag{10.2.16}$$

From (10.2.15) and (10.2.16) the vectors $\overrightarrow{RQ}$ and $\overrightarrow{QP}$ are collinear if and only if the coefficients of $\overrightarrow{AB}$ and $\overrightarrow{AC}$ respectively, are proportional:

$$\begin{aligned}\frac{\delta}{\gamma+\delta} + \frac{\eta}{a} &= \lambda\frac{\gamma}{\gamma+\delta} \\ \frac{\eta}{a} &= \lambda\frac{\alpha}{\alpha+\beta},\end{aligned}$$

which can be written as

$$\begin{aligned}\frac{\delta a + \eta(\gamma+\delta)}{a(\gamma+\delta)} &= \lambda\frac{\gamma}{\gamma+\delta} \\ \frac{\eta}{a} &= \lambda\frac{\alpha}{\alpha+\beta}.\end{aligned}$$

Dividing the above equations, yields

$$\begin{aligned}
\frac{\delta a+\eta(\gamma+\delta)}{a}\frac{a}{\eta} &= \gamma\frac{\alpha+\beta}{\alpha} \Longleftrightarrow \\
\frac{\delta a}{\eta}+\gamma+\delta &= \gamma+\frac{\gamma\beta}{\alpha} \Longleftrightarrow \\
\delta\left(\frac{a}{\eta}+1\right) &= \frac{\gamma\beta}{\alpha} \Longleftrightarrow \\
\frac{a+\eta}{\eta}\frac{\delta}{\gamma}\frac{\alpha}{\beta} &= 1,
\end{aligned}$$

which is relation (10.2.14). ■

Exercise 10.2.2 *Let $P \in BC$, $Q \in AC$, $R \in AB$ such that*

$$|BC| = |CP|, \qquad \overrightarrow{AQ} = \frac{1}{2}\overrightarrow{AC}, \qquad \overrightarrow{AR} = \frac{1}{3}\overrightarrow{AB}.$$

Show that the vectors $\overrightarrow{RQ}$ and $\overrightarrow{RP}$ are collinear.

10.3 Euler's line

As far as the laws of mathematics refer to reality, they are not certain; and as far as they are certain, they do not refer to reality.

Albert Einstein (1879–1955)

In this section we shall show a relationship between three remarkable points of a triangle. Recall that the altitudes of a triangle intersect at the orthocenter, the medians intersect at the centroid and the perpendicular bisectors of the triangle sides intersect at the circumcenter.

Theorem 10.3.1 (Euler's line) *Let O, G, H be the circumcenter, centroid and orthocenter of a triangle ABC. Then the points O, G, H are collinear and*

$$\overrightarrow{OH} = 3\overrightarrow{OG}.$$

Proof: We shall compute $\overrightarrow{OA}+\overrightarrow{OB}+\overrightarrow{OC}$ in two ways, see Figure 10.10.

From part (*ii*) of Theorem 6.3.2 we have

$$\overrightarrow{OA}+\overrightarrow{OB}+\overrightarrow{OC} = 3\overrightarrow{OG}. \tag{10.3.17}$$

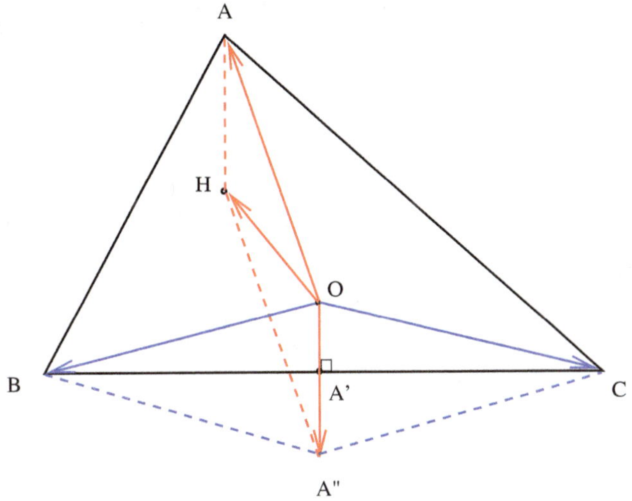

Figure 10.10: For Theorem 10.3.1.

Let A' be the midpoint of the segment $[BC]$ and A'' be the reflection of O in the line BC. Then $BOCA''$ is a parallelogram (because the diagonals bisect). Hence

$$\overrightarrow{OB} + \overrightarrow{OC} = \overrightarrow{OA''}. \tag{10.3.18}$$

Since $|HA| = 2|OA'|$ (see Exercise 10.3.2), and $AH \perp BC$, $OA' \perp BC$, then (10.3.18) implies that $OAHA''$ is a parallelogram, and hence

$$\begin{aligned} \overrightarrow{OA} + \overrightarrow{OB} + \overrightarrow{OC} &= \overrightarrow{OA} + \overrightarrow{OA''} \\ &= \overrightarrow{OA} + \overrightarrow{AH} = \overrightarrow{OH}. \end{aligned} \tag{10.3.19}$$

From (10.3.17) and (10.3.19) we obtain

$$\overrightarrow{OH} = 3\overrightarrow{OG},$$

which implies that O, G, H are collinear. The line that contains these points is called the *Euler line* of triangle ABC. ■

Exercise 10.3.2 *Show that* $|HA| = 2|OA'| = 2R\cos A$, *where* R *is the circumradius of triangle* ABC.

10.4 Paralelogic and orthologic triangles

What a marvel that so simple a figure as the triangle is so inexhaustible in its properties!
A. L. Crelle, 1821

This section presents a few interesting applications of the inner and cross products to line intersection problems. The following two theorems have a symmetric character.

Theorem 10.4.1 *Let ABC and $A'B'C'$ be two triangles. Assume the parallel lines through A, B, C to $B'C'$, $A'C'$, $A'B'$ respectively, intersect at a point K. Then the property is symmetric, i.e., the parallel lines through A', B', C' to BC, AC, AB respectively, intersect at a point L (refer to Figure 10.11).*

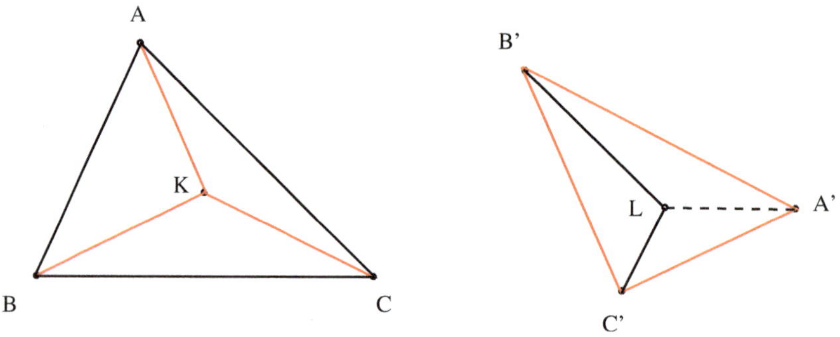

Figure 10.11: For Theorem 10.4.1.

Proof: Let L be the intersection of the parallel lines through B' and C' to AC and AB, respectively. (Why these lines intersect?) We shall prove that $A'L \| BC$. It suffice to show that $\overrightarrow{A'L} \times \overrightarrow{BC} = 0$. Using that $AK \| B'C'$, $BK \| C'A'$, $CK \| A'B'$, $C'L \| BA$, $B'L \| AC$ we have the corresponding cross products equal to zero. Then the triangle rule yields

$$\begin{aligned}
\overrightarrow{A'L} \times \overrightarrow{BC} &= \overrightarrow{A'L} \times (\overrightarrow{BK} + \overrightarrow{KC}) = \overrightarrow{A'L} \times \overrightarrow{BK} + \overrightarrow{A'L} \times \overrightarrow{KC} \\
&= (\overrightarrow{A'C'} + \overrightarrow{C'L}) \times \overrightarrow{BK} + (\overrightarrow{A'B'} + \overrightarrow{B'L}) \times \overrightarrow{KC} \\
&= \underbrace{\overrightarrow{A'C'} \times \overrightarrow{BK}}_{=0} + \overrightarrow{C'L} \times \overrightarrow{BK} + \underbrace{\overrightarrow{A'B'} \times \overrightarrow{KC}}_{=0} + \overrightarrow{B'L} \times \overrightarrow{KC} \\
&= \overrightarrow{C'L} \times \overrightarrow{BK} + \overrightarrow{B'L} \times \overrightarrow{KC}
\end{aligned}$$

$$
\begin{aligned}
&= \overrightarrow{C'L} \times (\overrightarrow{BA} + \overrightarrow{AK}) + \overrightarrow{B'L} \times (\overrightarrow{KA} + \overrightarrow{AC}) \\
&= \underbrace{\overrightarrow{C'L} \times \overrightarrow{BA}}_{=0} + \overrightarrow{C'L} \times \overrightarrow{AK} + \overrightarrow{B'L} \times \overrightarrow{KA} + \underbrace{\overrightarrow{B'L} \times \overrightarrow{AC}}_{=0} \\
&= \overrightarrow{C'L} \times \overrightarrow{AK} + \overrightarrow{B'L} \times \overrightarrow{KA} = \overrightarrow{C'L} \times \overrightarrow{AK} - \overrightarrow{B'L} \times \overrightarrow{AK} \\
&= (\overrightarrow{C'L} - \overrightarrow{B'L}) \times \overrightarrow{AK} = (\overrightarrow{C'L} + \overrightarrow{LB'}) \times \overrightarrow{AK} \\
&= \overrightarrow{C'B'} \times \overrightarrow{AK} = 0.
\end{aligned}
$$

Hence $A'L \| BC$ and the proof is finished. ∎

The triangles ABC and $A'B'C'$ are called *parallelogic*. The points K and L are called the parallelogy centers.

It is interesting to note that if in Theorem 10.4.1 the word *parallel* is replaced by the word *perpendicular* the property still holds. In this case the triangles are called *orthologic*. The next theorem states the existence of orthologic triangles. Its proof is very similar with the proof of Theorem 10.4.1, and it uses inner products instead of cross products.

Theorem 10.4.2 *Let ABC and $A'B'C'$ be two triangles such that the perpendicular lines through A, B, C to $B'C'$, $A'C'$, $A'B'$ respectively, intersect at a point K. Then the property is symmetric, i.e., the perpendicular lines through A', B', C' to BC, AC, AB respectively, intersect at a point L.*

Proof: Let L be the intersection of the perpendicular lines through B' and C' to AC and AB, respectively. We need to show that $A'L \perp BC$. For this it suffices to show that $\overrightarrow{A'L} \cdot \overrightarrow{BC} = 0$. See Figure 10.12.

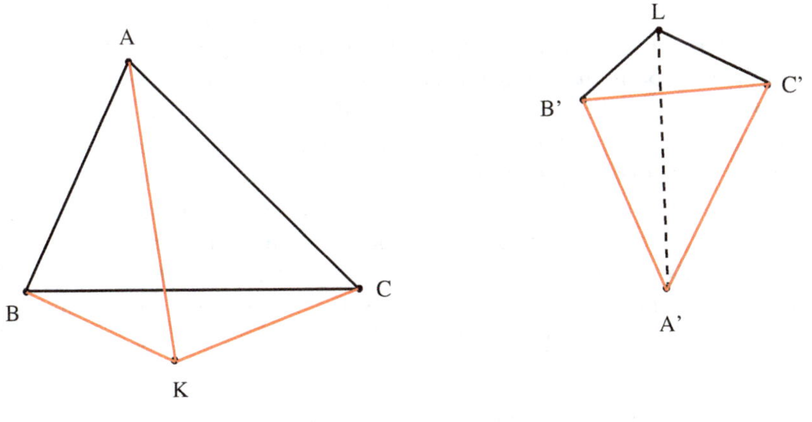

Figure 10.12: For Theorem 10.4.2.

Using the triangle rule and the distributivity of the inner product, we have

$$\begin{aligned}
\overrightarrow{A'L}\cdot\overrightarrow{BC} &= \overrightarrow{A'L}\cdot(\overrightarrow{BK}+\overrightarrow{KC}) = \overrightarrow{A'L}\cdot\overrightarrow{BK}+\overrightarrow{A'L}\cdot\overrightarrow{KC}\\
&= (\overrightarrow{A'C'}+\overrightarrow{C'L})\cdot\overrightarrow{BK}+(\overrightarrow{A'B'}+\overrightarrow{B'L})\cdot\overrightarrow{KC}\\
&= \underbrace{\overrightarrow{A'C'}\cdot\overrightarrow{BK}}_{=0}+\overrightarrow{C'L}\cdot\overrightarrow{BK}+\underbrace{\overrightarrow{A'B'}\cdot\overrightarrow{KC}}_{=0}+\overrightarrow{B'L}\cdot\overrightarrow{KC}\\
&= \overrightarrow{C'L}\cdot\overrightarrow{BK}+\overrightarrow{B'L}\cdot\overrightarrow{KC}\\
&= \overrightarrow{C'L}\cdot(\overrightarrow{BA}+\overrightarrow{AK})+\overrightarrow{B'L}\cdot(\overrightarrow{KA}+\overrightarrow{AC})\\
&= \underbrace{\overrightarrow{C'L}\cdot\overrightarrow{BA}}_{=0}+\overrightarrow{C'L}\cdot\overrightarrow{AK}+\overrightarrow{B'L}\cdot\overrightarrow{KA}+\underbrace{\overrightarrow{B'L}\cdot\overrightarrow{AC}}_{=0}\\
&= \overrightarrow{C'L}\cdot\overrightarrow{AK}+\overrightarrow{B'L}\cdot\overrightarrow{KA} = \overrightarrow{C'L}\cdot\overrightarrow{AK}-\overrightarrow{B'L}\cdot\overrightarrow{AK}\\
&= (\overrightarrow{C'L}-\overrightarrow{B'L})\cdot\overrightarrow{AK} = (\overrightarrow{C'L}+\overrightarrow{LB'})\cdot\overrightarrow{AK}\\
&= \overrightarrow{C'B'}\cdot\overrightarrow{AK} = 0.
\end{aligned}$$

Hence $A'L \perp BC$, which finishes the proof. ∎

The points L and K are called the *orthology centers.*

Exercise 10.4.3 *Show that any triangle and its median triangle are orthologic. The orthology centers are the orthocenter H and the circumcenter O of the initial triangle.*

Let ABC be a triangle and A', B', C' be the midpoints of the sides $[BC]$, $[AC]$, $[AB]$, see Figure 10.13. The median triangle $A'B'C'$ has the sides parallel to the sides of triangle ABC (because of the midpoint connector theorem). Hence the perpendicular line through A to $B'C'$ is also perpendicular to BC and hence it is an altitude in triangle ABC. In a similar way the perpendicular lines from B and C to AC and AB are altitudes in triangle ABC and hence all of them intersect at the point H, which is the orthocenter of triangle ABC.

By Theorem 10.4.2 the property is symmetric. We shall check this directly for this particular case. The perpendicular lines through A', B', C' to BC, AC, AB are the segment bisectors of the sides $[BC]$, $[AC]$, $[AB]$ respectively. It is a known fact that the segment bisectors of the sides of a triangle intersect at a point O, called the *circumcenter* (the center of the circumcircle).

It is worth noting that Theorems 10.4.1 and 10.4.2 have the following common generalization:

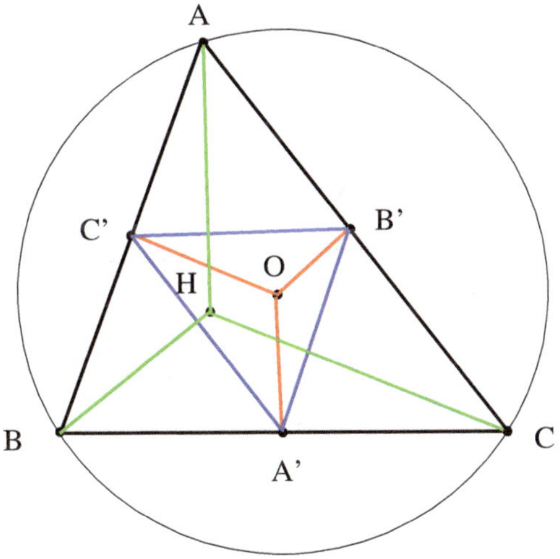

Figure 10.13: For Exercise 10.4.3.

Theorem 10.4.4 *Let ABC and $A'B'C'$ be two triangles. Assume the lines through A, B, C, which make with the lines $B'C'$, $A'C'$, $A'B'$ respectively, the same angle φ, intersect at a point. Then the lines through A', B', C', which make with the lines BC, AC, AB respectively, the same angle $180° - \varphi$ intersect at a point.*

Two triangles with the above property are called *isologic*. The intersection points are called *isology centers*.

10.5 Neuberg's theorem

Mathematics may be defined as the subject in which we never know what we are talking about, nor whether what we are saying is true.
Bertrand Russell,
Mysticism and Logic (1917) ch. 4.

In the following we shall present a property discovered by Neuberg in 1895.

Theorem 10.5.1 (Neuberg) *Let ABC be a triangle and let ℓ be a line in the same plane. Let A', B', C' be the orthogonal projections of A, B, C onto the line ℓ. Let A'', B'', C'' be the orthogonal projections of A', B', C' onto the sides BC, CA, AB, respectively. Then the lines $A'A''$, $B'B''$, $C'C''$ intersect at a point.*

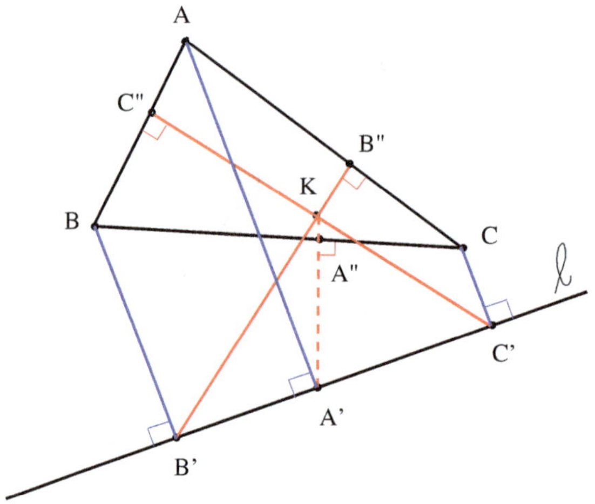

Figure 10.14: For Neuberg's theorem.

Proof: We shall refer to Figure 10.14. Let $\{K\} = B'B'' \cap C'C''$. We shall show that $A'K \perp BC$. It suffices to prove that $\overrightarrow{A'K} \cdot \overrightarrow{BC} = 0$. Using that $C'K \perp AB$, $B'K \perp AC$, yields

$$\begin{aligned}
\overrightarrow{A'K} \cdot \overrightarrow{BC} &= \overrightarrow{A'K} \cdot (\overrightarrow{BA} + \overrightarrow{AC}) \\
&= \overrightarrow{A'K} \cdot \overrightarrow{BA} + \overrightarrow{A'K} \cdot \overrightarrow{AC} \\
&= (\overrightarrow{A'C'} + \overrightarrow{C'K}) \cdot \overrightarrow{BA} + (\overrightarrow{A'B'} + \overrightarrow{B'K}) \cdot \overrightarrow{AC} \\
&= \overrightarrow{A'C'} \cdot \overrightarrow{BA} + \overrightarrow{A'B'} \cdot \overrightarrow{AC} \\
&= |A'C'| \cdot |B'A'| - |A'B'| \cdot |A'C'| = 0,
\end{aligned}$$

where we have used Exercise 7.2.2.
Hence $A'K \perp BC$, see Proposition 7.2.3, part 5), and then $A'A''$ will pass through K. ∎

The point K is called the *orthopol center* of triangle ABC with respect to the line ℓ.

Remark 10.5.2 There is a generalization of the Neuberg theorem, which replaces the orthogonal projections by projections of angle $0 < \varphi \leq 90°$.

Exercise 10.5.3 *Let ABC, $A'B'C'$ be two triangles symmetric with respect to a line, see Figure 10.15.*

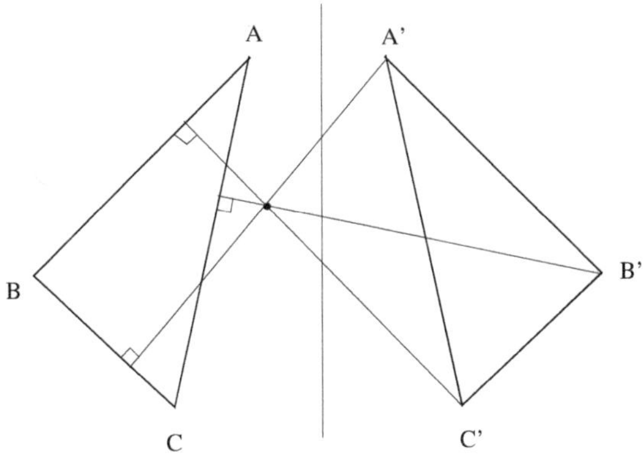

Figure 10.15: For Exercise 10.5.3.

(i) Show that the perpendicular lines from A', B', C' to BC, AC, AB, respectively, intersect at a point.

(ii) Triangles ABC and $A'B'C'$ are orthologic.

Exercise 10.5.4 (Oppel) *Let ABC be a triangle and $A_1 \in BC$, $B_1 \in AC$, $C_1 \in AB$. Show that the perpendicular lines through the points A_1, B_1, C_1 to BC, CA, AB, respectively, intersect at a point if and only if*

$$|BA_1|^2 + |CB_1|^2 + |AC_1|^2 = |A_1C|^2 + |B_1A|^2 + |C_1B|^2.$$

10.6 Desargues' theorem

In mathematics you don't understand things. You just get used to them.
Johann von Neumann (1903–1957)

When dealing with collinearity of three points the following lemma is useful:

Lemma 10.6.1 *Let $O \notin AB$. Then the point $M \in [AB]$ if and only if there are α, $\beta \in (0, 1)$ with $\alpha + \beta = 1$ such that*

$$\overrightarrow{OM} = \alpha\overrightarrow{OA} + \beta\overrightarrow{OB}. \tag{10.6.20}$$

Proof: We consider the Figure 10.16. Since the vectors $\overrightarrow{OA}$, and $\overrightarrow{OB}$ are linear independent, they form a basis in the plane. There are $\alpha, \beta \in \mathbb{R}$ such that

$$\overrightarrow{OM} = \alpha\overrightarrow{OA} + \beta\overrightarrow{OB}. \tag{10.6.21}$$

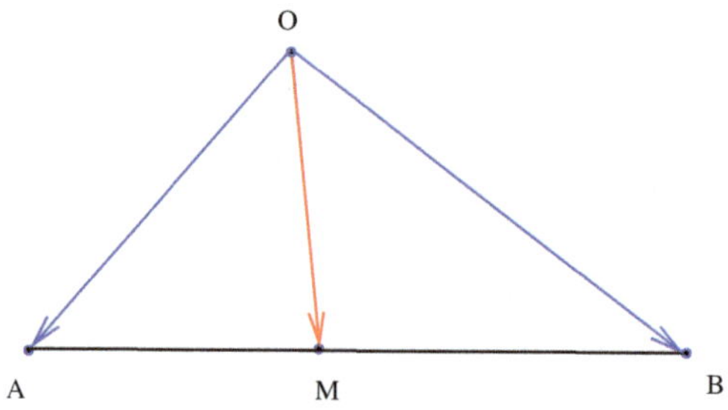

Figure 10.16: For Lemma 10.6.1.

Using

$$\overrightarrow{OM} = \overrightarrow{OA} + \overrightarrow{AM}, \qquad \overrightarrow{OB} = \overrightarrow{OA} + \overrightarrow{AB},$$

equation (10.6.21) becomes

$$\begin{aligned} \overrightarrow{OA} + \overrightarrow{AM} &= \alpha\overrightarrow{OA} + \beta\overrightarrow{OA} + \beta\overrightarrow{AB} \Longleftrightarrow \\ \overrightarrow{AM} &= (\alpha + \beta - 1)\overrightarrow{OA} + \beta\overrightarrow{AB}. \end{aligned}$$

Hence $\alpha + \beta = 1$ if and only if $\overrightarrow{AM} = \beta\overrightarrow{AB}$, *i.e.*, the points A, M, B are collinear. We have $\beta \in (0, 1)$, and hence $\alpha \in (0, 1)$, if and only if M is between A and B. ∎

The set

$$\{M; \overrightarrow{OM} = \alpha\overrightarrow{OA} + (1 - \alpha)\overrightarrow{OB}, \forall\alpha \in \mathbb{R}\}$$

is the line AB, while the set

$$\{M; \overrightarrow{OM} = \alpha\overrightarrow{OA} + (1 - \alpha)\overrightarrow{OB}, \alpha \in [0, 1]\}$$

is the closed segment $[AB]$. We say that the *convex combination* of $\overrightarrow{OA}$ and $\overrightarrow{OB}$ is the segment $[AB]$.

A useful consequence in applications is the following result:

Corollary 10.6.2 *Let $O \notin AB$. Then the points M, A, B are collinear if and only if there are a, b, $c \in \mathbb{R}$ with $a + b + c = 0$, such that*

$$a\,\overrightarrow{OM} + b\,\overrightarrow{OA} + c\,\overrightarrow{OB} = 0.$$

Theorem 10.6.3 (Desargues) *Let ℓ_1, ℓ_2, ℓ_3 be three lines in the plane (or space) passing through the point O. Let A_1, $B_1 \in \ell_1$, A_2, $B_2 \in \ell_2$, A_3, $B_3 \in \ell_3$. Suppose the following lines intersect and consider the intersection points $\{L_3\} = A_1A_2 \cap B_1B_2$, $\{L_1\} = A_2A_3 \cap B_2B_3$ and $\{L_2\} = A_1A_3 \cap B_1B_3$. Then the points L_1, L_2, L_3 are collinear.*

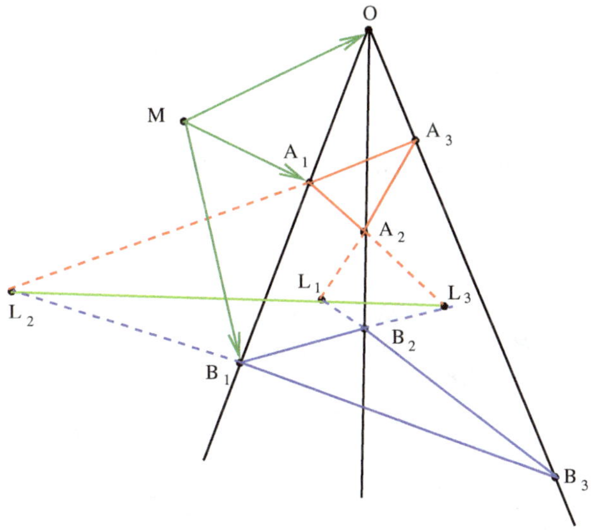

Figure 10.17: For Desargues' theorem.

Proof: We refer to Figure 10.17. Let M be an arbitrary point in the plane or space. Since (O, A_1, B_1), (O, A_2, B_2), and (O, A_3, B_3) are collinear, applying Lemma 10.6.1, there are $\lambda_i \in \mathbb{R}$, $i \in \{1, 2, 3\}$ such that we have

$$\begin{aligned} \overrightarrow{MO} &= \lambda_1 \overrightarrow{MA_1} + (1-\lambda_1)\overrightarrow{MB_1} \\ &= \lambda_2 \overrightarrow{MA_2} + (1-\lambda_2)\overrightarrow{MB_2} \\ &= \lambda_3 \overrightarrow{MA_3} + (1-\lambda_3)\overrightarrow{MB_3}. \end{aligned} \tag{10.6.22}$$

If $\lambda_1 = \lambda_2 = \lambda$, subtracting we obtain

$$\begin{aligned} 0 &= \overrightarrow{MO} - \overrightarrow{MO} = \lambda(\overrightarrow{MA_1} - \overrightarrow{MA_2}) + (1-\lambda)(\overrightarrow{MB_1} - \overrightarrow{MB_2}) \\ &= \lambda\overrightarrow{A_2A_1} + (1-\lambda)\overrightarrow{B_2B_1}, \end{aligned}$$

from where

$$\overrightarrow{A_2A_1} = \frac{\lambda - 1}{\lambda}\overrightarrow{B_2B_1},$$

and hence $A_1A_2 \| B_1B_2$, which is a contradiction, because the lines A_1A_2 and B_1B_2 intersect at L_3.
It follows that $\lambda_1 \neq \lambda_2$. In a similar way, we can show that $\lambda_2 \neq \lambda_3$ and $\lambda_1 \neq \lambda_3$.
From the first identity of (10.6.22) we have by subtracting

$$\lambda_1\overrightarrow{MA_1} - \lambda_2\overrightarrow{MA_2} = (\lambda_1 - 1)\overrightarrow{MB_1} + (1-\lambda_2)\overrightarrow{MB_2},$$

and dividing by $\lambda_1 - \lambda_2 \neq 0$, yields

$$\frac{\lambda_1}{\lambda_1-\lambda_2}\overrightarrow{MA_1} - \frac{\lambda_2}{\lambda_1-\lambda_2}\overrightarrow{MA_2} = \frac{\lambda_1-1}{\lambda_1-\lambda_2}\overrightarrow{MB_1} - \frac{\lambda_2-1}{\lambda_1-\lambda_2}\overrightarrow{MB_2}. \quad (10.6.23)$$

Both left and right terms are convex combinations, since

$$\frac{\lambda_1}{\lambda_1-\lambda_2} - \frac{\lambda_2}{\lambda_1-\lambda_2} = 1, \qquad \frac{\lambda_1-1}{\lambda_1-\lambda_2} - \frac{\lambda_2-1}{\lambda_1-\lambda_2} = 1.$$

Hence both terms of (10.6.23) are equal to the same vector $\overrightarrow{MP}$, with (P, A_1, A_2) and (P, B_1, B_2) collinear, see Lemma 10.6.1. This means that $\{P\} = A_1A_2 \cap B_1B_2$, *i.e.*, $P = L_3$. It follows that

$$\begin{aligned} \overrightarrow{ML_3} &= \frac{\lambda_1}{\lambda_1-\lambda_2}\overrightarrow{MA_1} - \frac{\lambda_2}{\lambda_1-\lambda_2}\overrightarrow{MA_2} \Longleftrightarrow \\ (\lambda_1-\lambda_2)\overrightarrow{ML_3} &= \lambda_1\overrightarrow{MA_1} - \lambda_2\overrightarrow{MA_2}. \end{aligned} \quad (10.6.24)$$

In a similar way, we can show that

$$(\lambda_2-\lambda_3)\overrightarrow{ML_1} = \lambda_2\overrightarrow{MA_2} - \lambda_3\overrightarrow{MA_3}, \quad (10.6.25)$$

$$(\lambda_3-\lambda_1)\overrightarrow{ML_2} = \lambda_3\overrightarrow{MA_3} - \lambda_1\overrightarrow{MA_1}. \quad (10.6.26)$$

Summing (10.6.24), (10.6.25) and (10.6.26), we obtain

$$(\lambda_1-\lambda_2)\overrightarrow{ML_3} + (\lambda_2-\lambda_3)\overrightarrow{ML_1} + (\lambda_3-\lambda_1)\overrightarrow{ML_2} = 0.$$

Since the sum of the coefficients of the above linear combination is zero, Corollary 10.6.2 implies that L_1, L_2 and L_3 are three collinear points. ■

10.7 Stewart's relation

Proof is the idol before whom the
pure mathematician tortures himself.
Sir Arthur Eddington (1882–1944),
The Nature of the Physical World

Theorem 10.7.1 (Stewart) *Let $D \in [BC]$ be a point on the side BC of a triangle ABC. Then*

$$|AD|^2\overrightarrow{BC} = |AC|^2\overrightarrow{BD} + |AB|^2\overrightarrow{DC} - |BD|\cdot|DC|\overrightarrow{BC}. \quad (10.7.27)$$

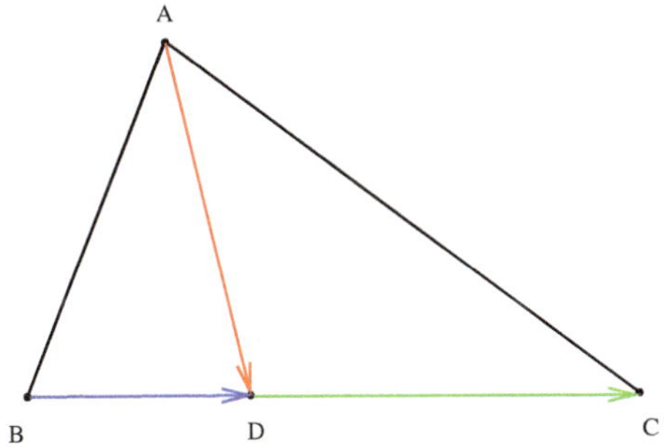

Figure 10.18: For Stewart's relation.

Proof: We consider Figure 10.18. The triangle identity applied to triangles ABD and ABC yields

$$\begin{aligned} \overrightarrow{AD} = \overrightarrow{AB} + \overrightarrow{BD} &\implies |AD|^2 = |AB|^2 + |BD|^2 + \overrightarrow{AB} \cdot \overrightarrow{BD} \\ \overrightarrow{AC} = \overrightarrow{AB} + \overrightarrow{BC} &\implies |AC|^2 = |AB|^2 + |BC|^2 + \overrightarrow{AB} \cdot \overrightarrow{BC}. \end{aligned}$$

Multiplying the first relation by $\overrightarrow{BC}$ and the second by $\overrightarrow{DB}$ yields

$$\begin{aligned} |AD|^2\overrightarrow{BC} &= |AB|^2\overrightarrow{BC} + |BD|^2\overrightarrow{BC} + (\overrightarrow{AB} \cdot \overrightarrow{BD})\overrightarrow{BC} \\ |AC|^2\overrightarrow{DB} &= |AB|^2\overrightarrow{DB} + |BC|^2\overrightarrow{DB} + (\overrightarrow{AB} \cdot \overrightarrow{BC})\overrightarrow{DB}. \end{aligned}$$

Adding and using Exercise 10.7.2, yields

$$\begin{aligned} |AD|^2\overrightarrow{BC} + |AC|^2\overrightarrow{DB} &= |AB|^2(\overrightarrow{BC} + \overrightarrow{DB}) + |BD|^2\overrightarrow{BC} + |BC|^2\overrightarrow{DB} \\ &= |AB|^2\overrightarrow{DC} + |BD|^2\overrightarrow{BC} + |BC|^2\overrightarrow{DB}. \end{aligned} \tag{10.7.28}$$

Using Exercise 10.7.3 yields

$$\begin{aligned} |BD|^2\overrightarrow{BC} + |BC|^2\overrightarrow{DB} &= |BD|^2\overrightarrow{BC} - |BC|^2\overrightarrow{BD} \\ &= |BD|^2\overrightarrow{BC} - |BC|^2 \cdot \frac{|BD|}{|BC|}\overrightarrow{BC} \\ &= (|BD|^2 - |BC| \cdot |BD|)\overrightarrow{BC} \\ &= |BD|(|BD| - |BC|)\overrightarrow{BC} \\ &= -|BD| \cdot |DC|\overrightarrow{BC}. \end{aligned}$$

Substituting in (10.7.28) yields (10.7.27). ■

Exercise 10.7.2 *Let $D \in [BC]$. Show that*

$$(\overrightarrow{AB} \cdot \overrightarrow{BD})\overrightarrow{BC} + (\overrightarrow{AB} \cdot \overrightarrow{BC})\overrightarrow{DB} = 0.$$

Exercise 10.7.3 *Let $D \in [BC]$. Show that*

$$\overrightarrow{BD} = \frac{|BD|}{|BC|}\overrightarrow{BC}.$$

Exercise 10.7.4 *Prove or disprove:*
Given a triangle ABC and D a point such that relation (10.7.27) *holds, then the points C, D, B are collinear.*

Exercise 10.7.5 *Let A', B', C' be the midpoints of the sides $[BC]$, $[CA]$ and $[AB]$ of triangle ABC. Let D be a point in the plane such that $AD \cap B'C' = \{P\}$, $BD \cap A'C' = \{N\}$, $CD \cap A'B' = \{M\}$. Prove that the lines $A'P$, $B'N$ and $C'M$ intersect at a point or they are parallel. See Figure 10.19.*

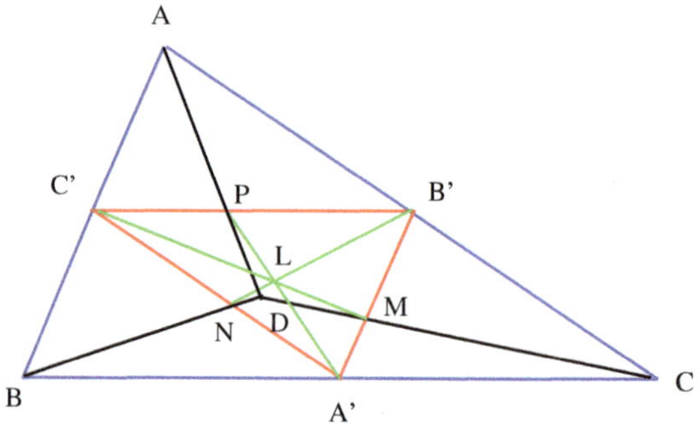

Figure 10.19: For Exercise 10.7.5.

10.8 Homothety

Do not worry about your difficulties in Mathematics.
I can assure you mine are still greater.
Albert Einstein

Definition 10.8.1 *Let $O \in \mathcal{P}$ be a point in the plane and let $H_k : \mathcal{P}\backslash\{O\} \to \mathcal{P}$ be an application defined by $H_k(M) = M'$, where $\overrightarrow{OM'} = k\overrightarrow{OM}$ with $k \in \mathbb{R}\backslash\{0\}$.*

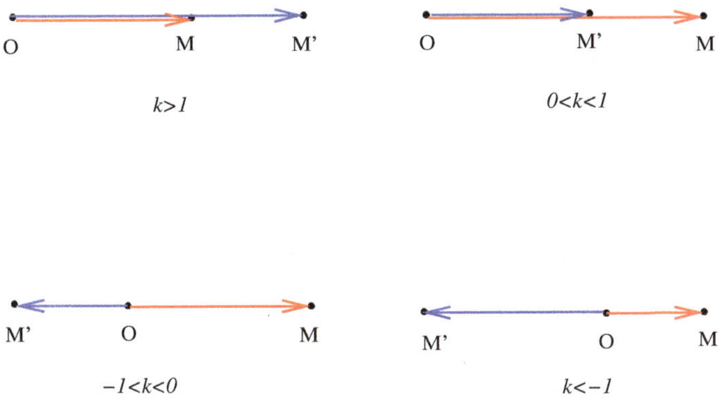

Figure 10.20: Homotheties.

H_k is called the homothety of center O and power k. The cases $k > 1$, $0 < k < 1$, $-1 < k < 0$, and $k < -1$ are represented in Figure 10.20.

Exercise 10.8.2 *Show the following properties:*

(i) $H_k \circ H_{k'} = H_{kk'}$;

(ii) $H_1 = Id$;

(iii) $(H_k)^{-1} = H_{1/k}$;

(iv) $H_k \circ (H_{k'} \circ H_{k''}) = (H_k \circ H_{k'}) \circ H_{k''}$.

Remark 10.8.3 The above properties state that the set of homotheties H_k form a group, which is isomorphic with the multiplicative group $(\mathbb{R}^*, \cdot)$ under the correspondence $\mathbb{R} \ni k \to H_k$.

Exercise 10.8.4 *Let ℓ be a line an $O \notin \ell$ be a point. Show that for any $k \neq 0$ the set $H_k(\ell) = \{H_k(P); P \in \ell\}$ is a line parallel to ℓ.*

Exercise 10.8.5 *Let ℓ be a line and $O \in \ell$ be a point. Show that for any $k \neq 0$ we have $H_k(\ell \backslash \{O\}) = \ell \backslash \{O\}$.*

Exercise 10.8.6 *Let H_k be a homothety of center O and power k. Let M, N be two points in the plane and let $M' = H_k(M)$ and $N' = H_k(N)$. Show that $|M'N'| = k|MN|$.*

The following theorem deals with the homothety of a circle.

Theorem 10.8.7 *Let $\mathcal{C}(A, r)$ be the circle of center A and radius r. If O is a given point in the plane and k is a nonzero real number, then the homothety H_k of center O and power k transforms the circle $\mathcal{C}(A, r)$ into the circle centered at $H_k(A)$ with the radius $|k|r$.*

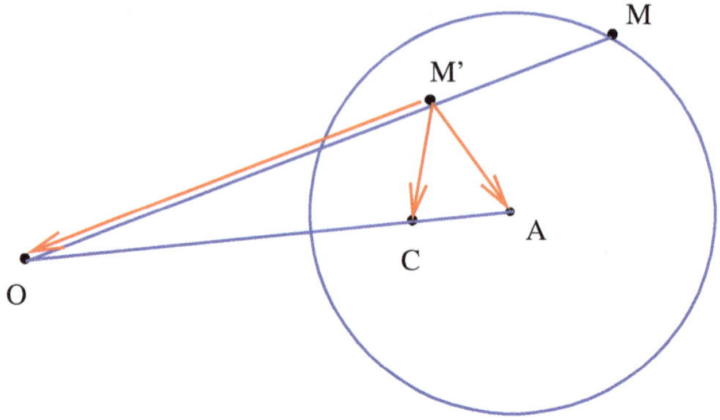

Figure 10.21: The homothethy of a circle.

Proof: Consider Figure 10.21, which represents the case $O \in Ext\,\mathcal{C}(A,r)$ and $0 < k < 1$. Let $M \in \mathcal{C}(A,r)$ be an arbitrary point and $M' = H_k(M)$ be its image. Let $C = H_k(A)$. Then $C \in [OA]$, since $0 < k < 1$. We shall compute the distance $|M'C|$ and show that it is independent of the point M.
Since $\overrightarrow{OC} = k\overrightarrow{OA}$, we have

$$\begin{aligned}
\overrightarrow{OC} &= k(\overrightarrow{OC} + \overrightarrow{CA}) \Longleftrightarrow (1-k)\overrightarrow{OC} = k\overrightarrow{CA} \\
(1-k)(\overrightarrow{OM'} + \overrightarrow{M'C}) &= k(\overrightarrow{CM'} + \overrightarrow{M'A}) \\
(1-k)\overrightarrow{OM'} + (1-k)\overrightarrow{M'C} &= k\overrightarrow{CM'} + k\overrightarrow{M'A} \\
(1-k)\overrightarrow{M'C} + k\overrightarrow{M'C} &= k\overrightarrow{M'A} + (1-k)\overrightarrow{M'O} \\
\overrightarrow{M'C} &= k\overrightarrow{M'A} + (1-k)\overrightarrow{M'O}. \qquad (10.8.29)
\end{aligned}$$

Using the triangle rule, we write $\overrightarrow{M'A} = \overrightarrow{M'O} + \overrightarrow{OA}$, and then (10.8.29) becomes

$$\begin{aligned}
\overrightarrow{M'C} &= k\overrightarrow{M'O} + k\overrightarrow{OA} + (1-k)\overrightarrow{M'O} \\
&= k\overrightarrow{M'O} + k\overrightarrow{OA} + \overrightarrow{M'O} - k\overrightarrow{M'O} \\
&= k\overrightarrow{OA} + \overrightarrow{M'O} = k\overrightarrow{OA} - k\overrightarrow{OM} \\
&= k(\overrightarrow{OA} - \overrightarrow{OM}) = k(\overrightarrow{OA} + \overrightarrow{MO}) \\
&= k(\overrightarrow{MO} + \overrightarrow{OA}) = k\overrightarrow{MA}.
\end{aligned}$$

Hence

$$\overrightarrow{M'C} = k\overrightarrow{MA}. \qquad (10.8.30)$$

Then we have

$$|M'C| = |\overrightarrow{M'C}| = |k| \cdot |\overrightarrow{MA}| = |k|r,$$

which means that $M' \in \mathcal{C}(C, |k|r)$.
This proves the inclusion $H_k(\mathcal{C}(A,r)) \subset \mathcal{C}(C, |k|r)$. The inverse inclusion is left as an exercise for the reader. ■

Exercise 10.8.8 *With the notations as in Theorem 10.8.7, show the inclusion* $\mathcal{C}(C, |k|r) \subset H_k(\mathcal{C}(A,r))$.

Exercise 10.8.9 *If in Theorem 10.8.7* $O \in \mathcal{C}(A,r)$*, show that* $H_k(\mathcal{C}(A,r))$ *is a circle tangent to the circle* $\mathcal{C}(A,r)$.

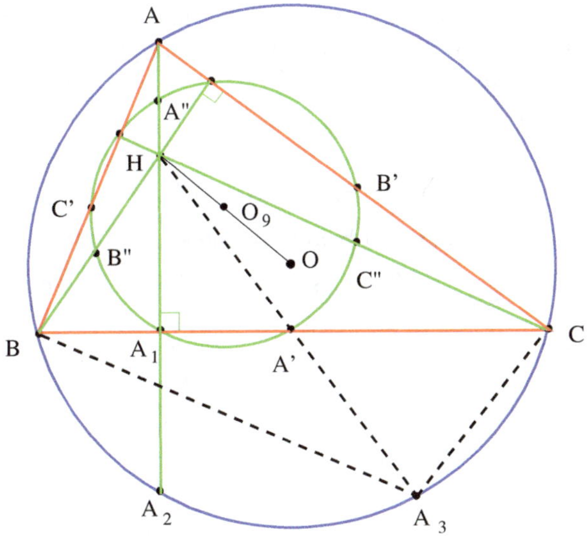

Figure 10.22: Euler's circle.

10.9 Euler's circle

The beginning is the most important part of the work.
Plato (427 BC–347 BC),
The Republic

Let ABC be a triangle with the orthocenter H and the circumcenter O. Let AA_1, BB_1, CC_1 be the altitudes of triangle ABC. Let A', B', C' be the midpoints of the sides $[BC]$, $[AC]$ and $[AB]$ and A'', B'', C'' be the midpoints of the segments $[AH]$, $[BH]$ and $[CH]$. Let A_2, A_3 be points on the circumcircle of triangle ABC such that A, A_1, A_2 and H, A', A_3 are collinear, see Figure 10.22.

Exercise 10.9.1 *Show that* $|HA_1| = |A_1A_2|$.

Exercise 10.9.2 *Prove that* A' *is the midpoint of the segment* $[HA_3]$.

Theorem 10.9.3 (Euler's circle) *The points* A', B', C', A_1, B_1, C_1, A'', B'', C'' *belong to the same circle.*

Proof: Consider the homothety of center H (the orthocenter) and power $k = 1/2$. Using Theorem 10.8.7 the circumcircle $\mathcal{C}(O, R)$ of triangle ABC will be transformed into a circle centered at the midpoint of $[HO]$ with the radius $R_0 = R/2$. The new circle will pass through the midpoints between H and points on the circumcircle. The homothety sends the points as follows:

$$A \to A'', \quad B \to B'' \quad C \to C''$$

$$A_2 \to A_1, \qquad A_3 \to A'.$$

In a similar way B_1, C_1, B', C' belong to the same circle. ■

This circle is called the *9-points circle* or the *Euler's circle* of triangle ΔABC.

Chapter 11

Baricentric Coordinates

The mathematics is not there till we put it there.
Sir Arthur Eddington (1882–1944),
The Philosophy of Physical Science

11.1 Baricentric coordinates using oriented areas

In this section all the triangle areas will be considered with a sign orientation. By convention we shall suppose that areas $\mathcal{A}_{XYZ}$ with X, Y, Z followed counter-clockwise are positive. With this convention we have

$$\mathcal{A}_{XYZ} = \mathcal{A}_{YZX} = \mathcal{A}_{ZXY} = -\mathcal{A}_{XZY} = -\mathcal{A}_{ZYX} = -\mathcal{A}_{YXZ}.$$

Definition 11.1.1 *Let ABC be a triangle and M be a point in the plane. The baricentric coordinates of M is a triplet (α, β, γ) of numbers proportional with the signed oriented areas $\mathcal{A}_{MBC}$, $\mathcal{A}_{MCA}$, $\mathcal{A}_{MAB}$:*

$$\frac{\alpha}{\mathcal{A}_{MBC}} = \frac{\beta}{\mathcal{A}_{MCA}} = \frac{\gamma}{\mathcal{A}_{MAB}},$$

which satisfy the condition

$$\alpha + \beta + \gamma = 1.$$

Using the sign of the oriented areas, it follows that in Figure 11.1(a) all the baricentric coordinates are positive, while in Figure 11.1(b) we have $\alpha < 0$ and $\beta, \gamma > 0$.

Proposition 11.1.2 *The baricentric coordinates are unique, i.e., if (α, β, γ) and $(\alpha', \beta', \gamma')$ are baricentric coordinates of the same point M, then $\alpha = \alpha'$, $\beta = \beta'$ and $\gamma = \gamma'$.*

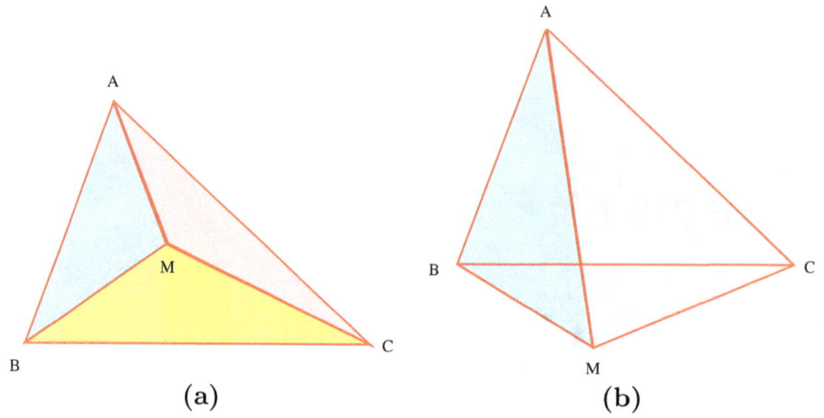

Figure 11.1: (a) $M \in Int(ABC)$; (b) $M \in Ext(ABC)$.

Proof: Let

$$\lambda = \frac{\alpha}{\mathcal{A}_{MBC}} = \frac{\beta}{\mathcal{A}_{MCA}} = \frac{\gamma}{\mathcal{A}_{MAB}} \Longleftrightarrow$$

$$\alpha = \lambda \mathcal{A}_{MBC}, \quad \beta = \lambda \mathcal{A}_{MCA}, \quad \gamma = \lambda \mathcal{A}_{MAB}, \tag{11.1.1}$$

and

$$\lambda' = \frac{\alpha'}{\mathcal{A}_{MBC}} = \frac{\beta'}{\mathcal{A}_{MCA}} = \frac{\gamma'}{\mathcal{A}_{MAB}} \Longleftrightarrow$$

$$\alpha' = \lambda' \mathcal{A}_{MBC}, \quad \beta' = \lambda' \mathcal{A}_{MCA}, \quad \gamma' = \lambda' \mathcal{A}_{MAB}. \tag{11.1.2}$$

We have $\lambda \neq 0, \lambda' \neq 0$, because otherwise, $\alpha = \beta = \gamma = 0$ and $\alpha' = \beta' = \gamma' = 0$, which contradicts the conditions $\alpha + \beta + \gamma = 1$ and $\alpha' + \beta' + \gamma' = 1$. Dividing equations (11.1.1) and (11.1.2) yields

$$\alpha = k\alpha', \qquad \beta = k\beta', \qquad \gamma = k\gamma', \tag{11.1.3}$$

where $k = \lambda/\lambda'$. Adding the relations in (11.1.3) we obtain

$$1 = \alpha + \beta + \gamma = k \underbrace{(\alpha' + \beta' + \gamma')}_{=1} = k,$$

and hence $k = 1$. Substituting back into (11.1.3) we arrive at

$$\alpha = \alpha', \qquad \beta = \beta', \qquad \gamma = \gamma',$$

which finishes the proof. ∎

Exercise 11.1.3 *Let M be a point in the plane of triangle ABC. Show that the baricentric coordinates are given by*

$$\alpha = \frac{\mathcal{A}_{MBC}}{\mathcal{A}_{ABC}}, \qquad \beta = \frac{\mathcal{A}_{MCA}}{\mathcal{A}_{ABC}}, \qquad \gamma = \frac{\mathcal{A}_{MAB}}{\mathcal{A}_{ABC}},$$

where the areas are considered with the sign orientation.

Exercise 11.1.4 *Let G be the centroid of triangle ABC, see Figure 11.2. Show that*

$$\overrightarrow{GA} \times \overrightarrow{GB} = \overrightarrow{GB} \times \overrightarrow{GC} = \overrightarrow{GC} \times \overrightarrow{GA}.$$

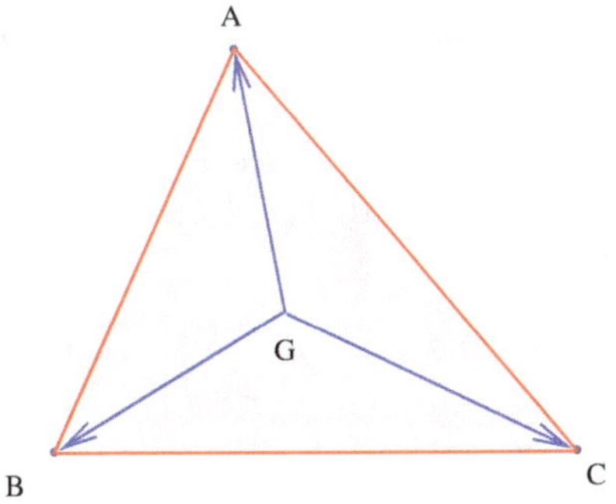

Figure 11.2: For Exercise 11.1.4.

In the following we are concerned with the baricentric coordinates of a few remarkable points in a triangle.

Proposition 11.1.5 *The baricentric coordinates of the centroid G of triangle ABC are $(\frac{1}{3}, \frac{1}{3}, \frac{1}{3})$.*

Proof: Using the interpretation of the cross product as an area and Exercise 11.1.4 we obtain $\mathcal{A}_{GBC} = \mathcal{A}_{GCA} = \mathcal{A}_{GAB}$. Using Exercise 11.1.3 we obtain $\alpha = \beta = \gamma$. Since $\alpha + \beta + \gamma = 1$, it follows that $\alpha = \beta = \gamma = \frac{1}{3}$. ∎

Proposition 11.1.6 *Let I be the incenter of triangle ABC. The baricentric coordinates of I are*

$$\alpha = \frac{a}{2p}, \qquad \beta = \frac{b}{2p}, \qquad \gamma = \frac{c}{2p},$$

where $2p = a + b + c$ is the perimeter and $a = |BC|$, $b = |AC|$, $c = |AB|$.

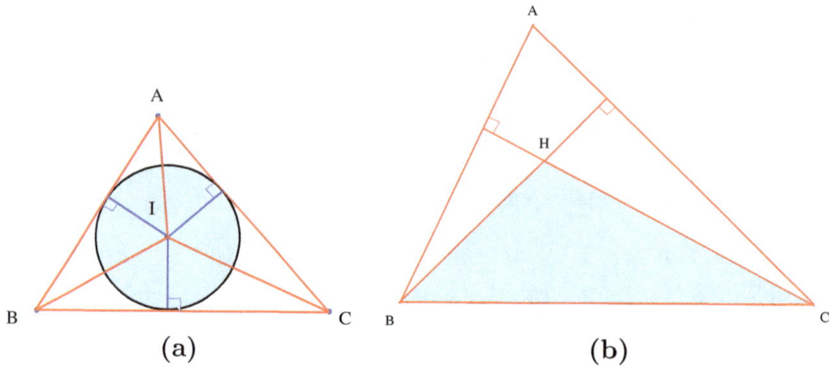

Figure 11.3: (a) The incenter I; (b) The orthocenter H.

Proof: We refer to Figure 11.3(a). Since $I \in IntABC$, all the baricentric coordinates are positive. Let $r = dist(I, BC) = dist(I, AC) = dist(I, AB)$ be the inradius. Then

$$\mathcal{A}_{IBC} = \frac{1}{2}ra, \qquad \mathcal{A}_{ICA} = \frac{1}{2}rb, \qquad \mathcal{A}_{IAB} = \frac{1}{2}rc.$$

Exercise 11.1.3 yields

$$\alpha = \frac{\mathcal{A}_{IBC}}{\mathcal{A}_{ABC}} = \frac{\frac{1}{2}ra}{pr} = \frac{a}{2p},$$

where we used the formula for the triangle area $\mathcal{A}_{ABC} = pr$. In a similar way we obtain $\beta = \dfrac{b}{2p}$ and $\gamma = \dfrac{c}{2p}$. ■

Exercise 11.1.7 *Show that the vertices of triangle* ABC *have the following baricentric coordinates:* $A(1, 0, 0)$, $B(0, 1, 0)$, $C(0, 0, 1)$.

Exercise 11.1.8 *Show that the midpoints of the sides* $[AB]$, $[BC]$, $[CA]$ *have the baricentric coordinates* $\left(\frac{1}{2}, \frac{1}{2}, 0\right)$, $\left(0, \frac{1}{2}, \frac{1}{2}\right)$ *and* $\left(\frac{1}{2}, 0, \frac{1}{2}\right)$.

We recall two sets of formulas that will be useful in the proof of the next proposition.

1. If H denotes the orthocenter of triangle ABC, then

$$|AH| = 2R|\cos\widehat{A}|, \quad |BH| = 2R|\cos\widehat{B}|, \quad |CH| = 2R|\cos\widehat{C}|, \qquad (11.1.4)$$

where R denotes the circumradius of triangle ABC.

2. The area of any triangle ABC can be written as

$$\mathcal{A}_{ABC} = 2R^2 \sin \widehat{A} \sin \widehat{B} \sin \widehat{C}. \tag{11.1.5}$$

Proposition 11.1.9 *The orthocenter H of an acute triangle ABC has the following baricentric coordinates*

$$\alpha = \cot \widehat{B} \cdot \cot \widehat{C}, \qquad \beta = \cot \widehat{A} \cdot \cot \widehat{C}, \qquad \gamma = \cot \widehat{A} \cdot \cot \widehat{B}.$$

Proof: Since ABC is an acute triangle, then $H \in IntABC$. Using relations (11.1.4)

$$\begin{aligned}
\mathcal{A}_{HBC} &= \frac{1}{2}|BH| \cdot |CH| \sin \widehat{BHC} \\
&= \frac{1}{2} 2R \cos \widehat{B} \cdot 2R \cos \widehat{C} \cdot \sin \widehat{A} \\
&= 2R^2 \cos \widehat{B} \cos \widehat{C} \sin \widehat{A},
\end{aligned}$$

where we have used that $\widehat{BHC} = 180° - \widehat{BAC}$ (why?). See Figure 11.3(b). Using area formula (11.1.5), Exercise 11.1.3 yields

$$\alpha = \frac{\mathcal{A}_{HBC}}{\mathcal{A}_{ABC}} = \frac{2R^2 \cos \widehat{B} \cos \widehat{C} \sin \widehat{A}}{2R^2 \sin \widehat{A} \sin \widehat{B} \sin \widehat{C}} = \cot \widehat{B} \cot \widehat{C}.$$

In a similar way, we obtain the formulas for β and γ. ■

Proposition 11.1.10 *Let O be the circumcenter of the triangle ABC. Then the baricentric coordinates of O are*

$$\alpha = \frac{\cos \widehat{A}}{2 \sin \widehat{B} \sin \widehat{C}}, \qquad \beta = \frac{\cos \widehat{B}}{2 \sin \widehat{A} \sin \widehat{C}}, \qquad \gamma = \frac{\cos \widehat{C}}{2 \sin \widehat{A} \sin \widehat{B}}.$$

Proof: Suppose that ABC is an acute triangle, see Figure 11.4(a). Then $O \in IntABC$ (Why?). As $\widehat{BOC}$ is a central angle cutting the same arc on the circumcircle as the inscribed angle $\widehat{BAC}$, we have $\widehat{BOC} = 2\widehat{BAC} = 2\widehat{A}$. Using $|BO| = |CO| = R$, the circumradius, we can write

$$\mathcal{A}_{BOC} = \frac{1}{2}|BO| \cdot |CO| \sin \widehat{BOC} = \frac{1}{2} R^2 \sin 2\widehat{A}, \tag{11.1.6}$$

Using Exercise 11.1.5 we have

$$\begin{aligned}
\alpha &= \frac{\mathcal{A}_{OBC}}{\mathcal{A}_{ABC}} = \frac{\frac{1}{2} R^2 \sin 2\widehat{A}}{2R^2 \sin \widehat{A} \sin \widehat{B} \sin \widehat{C}} \\
&= \frac{\sin \widehat{A} \cos \widehat{A}}{2 \sin \widehat{A} \sin \widehat{B} \sin \widehat{C}} = \frac{\cos \widehat{A}}{2 \sin \widehat{B} \sin \widehat{C}}.
\end{aligned}$$

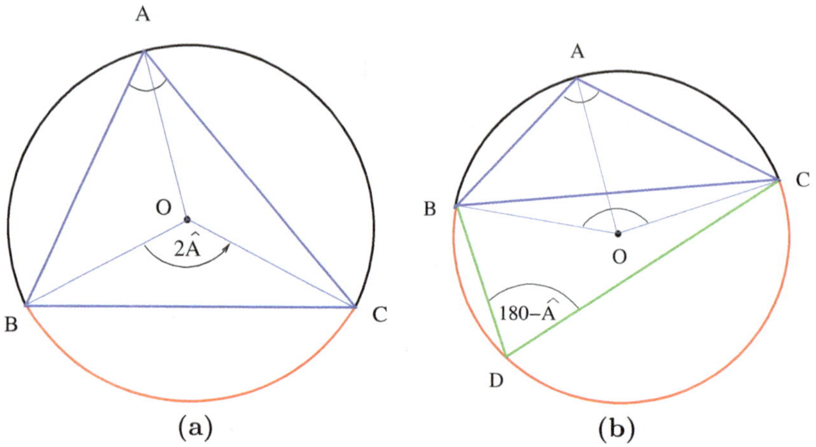

Figure 11.4: (a) $O \in Int(ABC)$; (b) $O \in Ext(ABC)$.

In a similar way we can obtain the formulas for the other two baricentric coordinates. ■

Exercise 11.1.11 *Prove Proposition 11.1.10 in the case when* $90° \leq \widehat{A} <$ $180°$, *see Figure 11.4(b).*

11.2 The equation of a line in the baricentric coordinates

If scientific reasoning were limited to the logical
processes of arithmetic, we should not get very
far in our understanding of the physical word
One might as well attempt to grasp the game
of poker entirely by the use of the mathematics
of probability.
Vannevar Bush (1890–1974)

The following result shows that a vanishing linear combination of the baricentric coordinates defines a line in the plane.

Theorem 11.2.1 *Let* M *be a point in the plane with the baricentric coordinates* (α, β, γ). *Given the numbers* k_1, k_2, k_3, *the set*

$$\{M; \alpha k_1 + \beta k_2 + \gamma k_3 = 0\}$$

is a line in the plane.

Proof: Let $\lambda = \dfrac{\alpha}{\mathcal{A}_{MBC}} = \dfrac{\beta}{\mathcal{A}_{MCA}} = \dfrac{\gamma}{\mathcal{A}_{MAB}}$. Then we have

$$\alpha = \lambda\mathcal{A}_{MBC}, \qquad \beta = \lambda\mathcal{A}_{MCA}, \qquad \gamma = \lambda\mathcal{A}_{MAB},$$

and hence the equation

$$\alpha k_1 + \beta k_2 + \gamma k_3 = 0$$

becomes

$$k_1\mathcal{A}_{MBC} + k_2\mathcal{A}_{MCA} + k_3\mathcal{A}_{MAB} = 0. \tag{11.2.7}$$

The areas are considered positive if the vertices of the triangle are followed counter-clockwise. Multiplying (11.2.7) by the normal n_+ of the plane, we obtain

$$k_1\overrightarrow{\mathcal{A}}_{MBC} + k_2\overrightarrow{\mathcal{A}}_{MCA} + k_3\overrightarrow{\mathcal{A}}_{MAB} = 0, \tag{11.2.8}$$

which can be written as

$$\begin{aligned}
k_1\overrightarrow{BC} \times \overrightarrow{BM} + k_2\overrightarrow{CA} \times \overrightarrow{CM} + k_3\overrightarrow{AB} \times \overrightarrow{AM} &= 0 \Longleftrightarrow \\
k_1\overrightarrow{BC} \times \overrightarrow{BM} + k_2\overrightarrow{CA} \times (\overrightarrow{CB} + \overrightarrow{BM}) + k_3\overrightarrow{AB} \times (\overrightarrow{AB} + \overrightarrow{BM}) &= 0 \Longleftrightarrow \\
k_1\overrightarrow{BC} \times \overrightarrow{BM} + k_2\overrightarrow{CA} \times \overrightarrow{CB} + k_2\overrightarrow{CA} \times \overrightarrow{BM} + k_3\overrightarrow{AB} \times \overrightarrow{BM} &= 0 \Longleftrightarrow \\
(k_1\overrightarrow{BC} + k_2\overrightarrow{CA} + k_3\overrightarrow{AB}) \times \overrightarrow{BM} = -k_2\overrightarrow{CA} \times \overrightarrow{CB}.&
\end{aligned}$$

Let $\overrightarrow{v} = k_1\overrightarrow{BC} + k_2\overrightarrow{CA} + k_3\overrightarrow{AB}$. Then

$$|\overrightarrow{BM} \times \overrightarrow{v}| = 2|k_2|\,|\mathcal{A}_{ABC}| = \text{constant}.$$

By Lemma 8.3.4 the point M describes a line parallel to the support line of $\overrightarrow{v}$, which finishes the proof. ■

Corollary 11.2.2 *The set $\{M; \alpha k_1 + \beta k_2 = 0\}$ is a line parallel to the support line of the vector $k_1\overrightarrow{BC} + k_3\overrightarrow{AB}$.*

Another consequence is the determinant formula for the equation of a line.

Proposition 11.2.3 *Let M_1, M_2, M_3 be three points in the plane of triangle ABC with the baricentric coordinates $(\alpha_i, \beta_i, \gamma_i)$, $i \in \{1, 2, 3\}$. Then the points M_1, M_2, M_3 are collinear if and only if*

$$\det\begin{pmatrix} \alpha_1 & \beta_1 & \gamma_1 \\ \alpha_2 & \beta_2 & \gamma_2 \\ \alpha_3 & \beta_3 & \gamma_3 \end{pmatrix} = 0.$$

Proof: Using Theorem 11.2.1 there are three numbers k_1, k_2, k_3 such that

$$\begin{aligned} k_1\alpha_1 + k_2\beta_1 + k_3\gamma_1 &= 0 \\ k_1\alpha_2 + k_2\beta_2 + k_3\gamma_2 &= 0 \\ k_1\alpha_3 + k_2\beta_3 + k_3\gamma_3 &= 0, \end{aligned}$$

which can be written as a linear system

$$\begin{pmatrix} \alpha_1 & \beta_1 & \gamma_1 \\ \alpha_2 & \beta_2 & \gamma_2 \\ \alpha_3 & \beta_3 & \gamma_3 \end{pmatrix} \cdot \begin{pmatrix} k_1 \\ k_2 \\ k_3 \end{pmatrix} = \begin{pmatrix} 0 \\ 0 \\ 0 \end{pmatrix}.$$

Since k_1, k_2, k_3 are not all zero, it follows that the determinant of the matrix of the baricentric coordinates vanishes. ■

Corollary 11.2.4 *Let $M(\alpha,\beta,\gamma)$ and $N(\alpha',\beta',\gamma')$ be two points in the plane (A,B,C). A point X belongs to the line MN if and only if its baricentric coordinates (x,y,z) satisfy the equation*

$$\det \begin{pmatrix} x & y & z \\ \alpha & \beta & \gamma \\ \alpha' & \beta' & \gamma' \end{pmatrix} = 0.$$

We can retrieve the Euler line by means of baricentric coordinates as follows.

Theorem 11.2.5 *The points O, G and H are collinear.*

Proof: We start be recomputing the baricentric coordinates of the circumcenter. A trigonometric computation shows

$$\frac{\cos \widehat{A}}{2\sin \widehat{B} \sin \widehat{C}} = \frac{\cos(180^\circ - \widehat{B} - \widehat{C})}{2\sin \widehat{B} \sin \widehat{C}} = -\frac{\cos(\widehat{B} + \widehat{C})}{2\sin \widehat{B} \sin \widehat{C}}$$

$$= \frac{-\cos \widehat{B} \cos \widehat{C} + \sin \widehat{B} \sin \widehat{C}}{2\sin \widehat{B} \sin \widehat{C}} = \frac{1}{2}(\cot \widehat{B} \cot \widehat{C} + 1).$$

Therefore, the baricentric coordinates of O are

$$O\left(\frac{1}{2}(\cot \widehat{B} \cot \widehat{C} + 1), \frac{1}{2}(\cot \widehat{C} \cot \widehat{A} + 1), \frac{1}{2}(\cot \widehat{A} \cot \widehat{B} + 1)\right).$$

Using $G(\frac{1}{3},\frac{1}{3},\frac{1}{3})$ and $H(\cot\widehat{B}\cot\widehat{C},\cot\widehat{A}\cot\widehat{C},\cot\widehat{A}\cot\widehat{B})$, it follows that O, G and H are collinear if and only if

$$\det\begin{pmatrix}\frac{1}{2}(\cot\widehat{B}\cot\widehat{C}+1) & \frac{1}{2}(\cot\widehat{C}\cot\widehat{A}+1) & \frac{1}{2}(\cot\widehat{A}\cot\widehat{B}+1)\\ \frac{1}{3} & \frac{1}{3} & \frac{1}{3}\\ \cot\widehat{B}\cot\widehat{C} & \cot\widehat{A}\cot\widehat{C} & \cot\widehat{A}\cot\widehat{B}\end{pmatrix}=0.$$

This determinant vanishes because the first row is a linear combination of the second and the third row, namely $R_1=\frac{3}{2}R_2+\frac{1}{2}R_3$. ■

Solved Exercise 11.2.6 *The incenter I of triangle ABC belongs to a median if and only if the triangle is isosceles.*

Proof: Let A' be the midpoint of $[BC]$. Then A' has the baricentric coordinates $(0,\frac{1}{2},\frac{1}{2})$. By Proposition 11.2.3, $I\in AA'$ if and only if

$$\det\begin{pmatrix}1 & 0 & 0\\ \frac{a}{2p} & \frac{b}{2p} & \frac{c}{2p}\\ 0 & \frac{1}{2} & \frac{1}{2}\end{pmatrix}=0\iff\det\begin{pmatrix}1 & 0 & 0\\ a & b & c\\ 0 & 1 & 1\end{pmatrix}=0,$$

which becomes $b=c$, *i.e.*, the triangle is isosceles. We arrive at a similar conclusion if I belongs to any other median. ■

Solved Exercise 11.2.7 *Let I,G be the incenter and the centroid of triangle ABC. Show that the line IG passes through the vertex A if and only if the triangle ABC is isosceles, with $|AB|=|AC|$.*

Proof: Using Proposition 11.2.3 and the baricentric coordinates of A, I and G, we have

$$\det\begin{pmatrix}1 & 0 & 0\\ \dfrac{a}{2p} & \dfrac{b}{2p} & \dfrac{c}{2p}\\ \dfrac{1}{3} & \dfrac{1}{3} & \dfrac{1}{3}\end{pmatrix}=0\iff\det\begin{pmatrix}1 & 0 & 0\\ a & b & c\\ 1 & 1 & 1\end{pmatrix}=0,$$

which becomes $b=c$, and hence the triangle is isosceles. ■

Exercise 11.2.8 *The line OI passes through a vertex of triangle ABC if and only if the triangle is isosceles.*

In general, the incenter, I, is not collinear with O, G and H. The next exercise provides an equivalent condition for these points to be collinear.

Solved Exercise 11.2.9 *The incenter I of triangle ABC belongs to the Euler line OG if and only if*

$$\sum_{circ.\ perm.} \sin 2\widehat{A}\sin(\widehat{B}-\widehat{C}) = 0.$$

Proof: As in the previous exercises we set the determinant of the baricentric coordinates of I, O, G equal to zero. The rest of the proof is a trigonometric computation. We have I, O, G collinear if and only if

$$0 = \det\begin{pmatrix} \dfrac{a}{2p} & \dfrac{b}{2p} & \dfrac{c}{2p} \\ \dfrac{\cos\widehat{A}}{2\sin\widehat{B}\sin\widehat{C}} & \dfrac{\cos\widehat{B}}{2\sin\widehat{A}\sin\widehat{C}} & \dfrac{\cos\widehat{C}}{2\sin\widehat{A}\sin\widehat{B}} \\ \dfrac{1}{3} & \dfrac{1}{3} & \dfrac{1}{3} \end{pmatrix}$$

$$= \frac{1}{12p\sin\widehat{A}\sin\widehat{B}\sin\widehat{C}}\cdot\det\begin{pmatrix} a & b & c \\ \cos\widehat{A}\sin\widehat{A} & \cos\widehat{B}\sin\widehat{B} & \cos\widehat{C}\sin\widehat{C} \\ 1 & 1 & 1 \end{pmatrix}$$

$$\Longleftrightarrow 0 = \det\begin{pmatrix} a & b & c \\ \sin 2\widehat{A} & \sin 2\widehat{B} & \sin 2\widehat{C} \\ 1 & 1 & 1 \end{pmatrix}$$
$$= a\sin 2\widehat{B} + b\sin 2\widehat{C} + c\sin 2\widehat{A} - c\sin 2\widehat{B} - b\sin 2\widehat{A} - a\sin 2\widehat{C}$$

$$\begin{aligned}
\Longleftrightarrow 0 &= a(\sin 2\widehat{B} - \sin 2\widehat{C}) + b(\sin 2\widehat{C} - \sin 2\widehat{A}) + c(\sin 2\widehat{A} - \sin 2\widehat{B}) \\
&= a\cos(\widehat{B}+\widehat{C})\sin(\widehat{B}-\widehat{C}) + b\cos(\widehat{C}+\widehat{A})\sin(\widehat{C}-\widehat{A}) \\
&\quad + c\cos(\widehat{A}+\widehat{B})\sin(\widehat{A}-\widehat{B}) \\
&= a(-\cos\widehat{A})\sin(\widehat{B}-\widehat{C}) + b(-\cos\widehat{B})\sin(\widehat{C}-\widehat{A}) \\
&\quad + c(-\cos\widehat{C})\sin(\widehat{A}-\widehat{B}) \\
&= -2R\sin\widehat{A}\cos\widehat{A}\sin(\widehat{B}-\widehat{C}) - 2R\sin\widehat{B}\cos\widehat{B}\sin(\widehat{C}-\widehat{A}) \\
&\quad - 2R\sin\widehat{C}\cos\widehat{C}\sin(\widehat{A}-\widehat{B}),
\end{aligned}$$

which becomes after dividing by $-R$

$$\sin 2\widehat{A}\sin(\widehat{B}-\widehat{C}) + \sin 2\widehat{B}\sin(\widehat{C}-\widehat{A}) + \sin 2\widehat{C}\sin(\widehat{A}-\widehat{B}) = 0,$$

where we have used the trigonometric formulas

$$\sin 2x = 2\sin x \cos x, \qquad \cos(180^\circ - x) = -\cos x,$$

$$\sin x - \sin y = 2\cos\frac{x+y}{2}\sin\frac{x-y}{2},$$

and the law of sines: $a = 2R\sin\widehat{A}$, $b = 2R\sin\widehat{B}$, $c = 2R\sin\widehat{C}$, see Theorem 8.3.3. ■

11.3 Vectorial properties of baricentric coordinates

Science is organized knowledge.
Wisdom is organized life.
Immanuel Kant (1724–1804)

In the following we shall use the following result proved in Chapter 1, see Lemma 6.3.6:

Lemma 11.3.1 *Given two distinct points A and B, the point M belongs to the line AB if and only if there are (unique) numbers x, $y \in \mathbb{R}$, with $x + y = 1$, such that for any point O (in the plane or the space) we have*

$$\overrightarrow{OM} = x\overrightarrow{OA} + y\overrightarrow{OB}.$$

The numbers x and y do not depend on the point O. See Figure 11.5.

Proof: We shall prove this result now using the cross product and then make the connection to baricentric coordinates. We consider the cross product of the relation $\overrightarrow{OM} = x\overrightarrow{OA} + y\overrightarrow{OB}$ with $\overrightarrow{OA}$ and $\overrightarrow{OB}$ and obtain respectively

$$\overrightarrow{OM} \times \overrightarrow{OA} = y\overrightarrow{OB} \times \overrightarrow{OA}$$
$$\overrightarrow{OM} \times \overrightarrow{OB} = x\overrightarrow{OA} \times \overrightarrow{OB}.$$

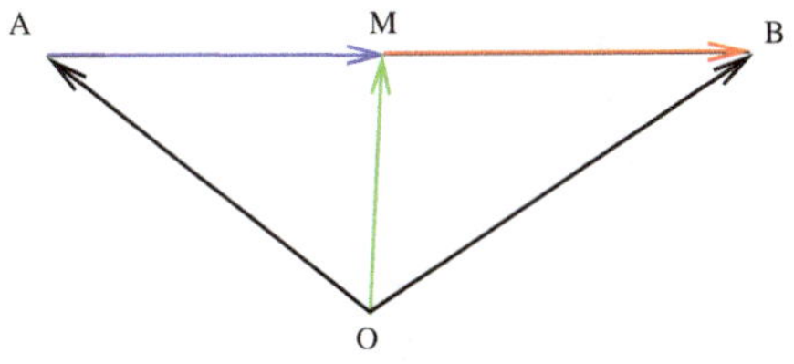

Figure 11.5: For Lemma 11.3.1.

Considering the orientations vectors $n_{\pm}$, the aforementioned identities become

$$2n_{+}\mathcal{A}_{OAM} = 2y\, n_{+}\mathcal{A}_{OAB}$$
$$2n_{-}\mathcal{A}_{OMM} = 2x\, n_{-}\mathcal{A}_{OAB}.$$

Therefore

$$x = \frac{\mathcal{A}_{OAM}}{\mathcal{A}_{OAB}} = \frac{|MB|}{|AB|}, \quad y = \frac{\mathcal{A}_{OAM}}{\mathcal{A}_{OAB}} = \frac{|AM|}{|AB|}.$$

Hence, the coefficients x and y are independent of O. They play the role of baricentric coordinates for the segment AB. ■

We shall extend this result to three vectors, case in which we obtain the usual baricentric coordinates. The following two results will emphasize the relationship between vectors and baricentric coordinates.

Theorem 11.3.2 *Let ABC be a triangle and M be an arbitrary point. The point M belongs to the plane (ABC) if and only if there are (unique) numbers x, y, $z \in \mathbb{R}$, with $x + y + z = 1$ such that for any point O (in the plane or space) we have*

$$\overrightarrow{OM} = x\overrightarrow{OA} + y\overrightarrow{OB} + z\overrightarrow{OC}.$$

Proof: We consider Figure 11.6.

" $\Longrightarrow$ " Let M be a point in the plane (ABC). Assume there are points $\{M_a\} = AM \cap BC$, $\{M_b\} = BM \cap AC$, $\{M_c\} = CM \cap AB$. Let O be an arbitrary point in the plane or space. By Lemma 11.3.1 there are numbers

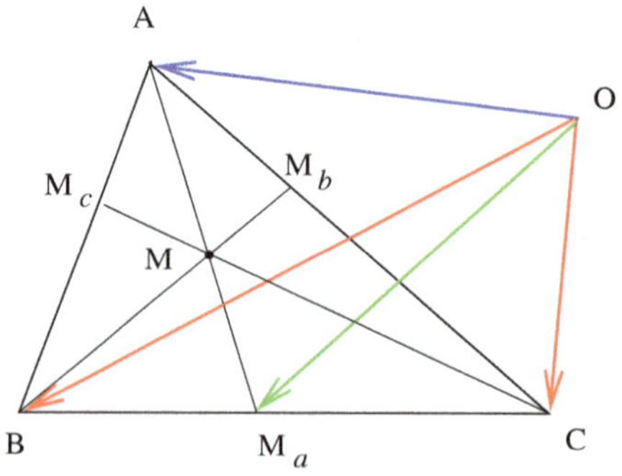

Figure 11.6: For Theorem 11.3.2.

ρ, $\xi \in \mathbb{R}$ such that

$$\begin{aligned}\overrightarrow{OM} &= \rho\overrightarrow{OA} + (1-\rho)\overrightarrow{OM_a}\\ \overrightarrow{OM_a} &= \xi\overrightarrow{OB} + (1-\xi)\overrightarrow{OC}.\end{aligned}$$

Then

$$\begin{aligned}\overrightarrow{OM} &= \rho\overrightarrow{OA} + (1-\rho)\left(\xi\overrightarrow{OB} + (1-\xi)\overrightarrow{OC}\right)\\ &= \rho\overrightarrow{OA} + (1-\rho)\xi\overrightarrow{OB} + (1-\rho)(1-\xi)\overrightarrow{OC}\\ &= x\overrightarrow{OA} + y\overrightarrow{OB} + z\overrightarrow{OC},\end{aligned}$$

with $x = \rho$, $y = (1-\rho)\xi$, $z = (1-\rho)(1-\xi)$. A simple computation yields

$$\begin{aligned}x + y + z &= \rho + (1-\rho)\xi + (1-\rho)(1-\xi)\\ &= \rho + \xi - \rho\xi + 1 - \xi - \rho + \rho\xi\\ &= 1.\end{aligned}$$

By Lemma 11.3.1 the factors ρ, ξ are defined independently of O, and hence the same property will hold for x, y and z.

In order to prove uniqueness, we shall consider another triplet (x', y', z') with the same property

$$\overrightarrow{OM} = x\overrightarrow{OA} + y\overrightarrow{OB} + z\overrightarrow{OC} = x'\overrightarrow{OA} + y'\overrightarrow{OB} + z'\overrightarrow{OC}.$$

Subtracting, we obtain

$$(x-x')\overrightarrow{OA} + (y-y')\overrightarrow{OB} + (z-z')\overrightarrow{OC} = 0,$$

with

$$(x-x')+(y-y')+(z-z') = (x+y+z)-(x'+y'+z') = 1-1 = 0. \quad (11.3.9)$$

Assume for instance that $z \neq z'$. Since the numbers x, y, z, x', y', z' do not depend on the point O, we shall choose $O = C$ and have

$$(x-x')\overrightarrow{CA} + (y-y')\overrightarrow{CB} = 0. \quad (11.3.10)$$

Since the vectors $\overrightarrow{CA}$ and $\overrightarrow{CB}$ are linear independent in the plane, the linear combination (11.3.10) should have zero coefficients, *i.e.*, $x = x'$ and $y = y'$, and hence using (11.3.9) yields $z = z'$, which is a contradiction.

It follows that $x = x', y = y'$, and $z = z'$.

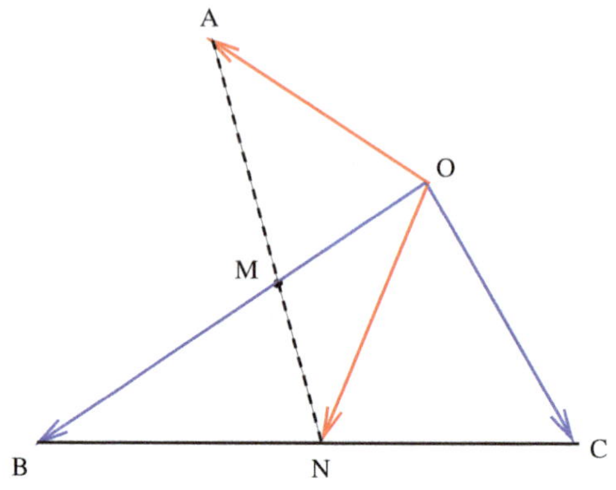

Figure 11.7: For Theorem 11.3.2.

" $\Longleftarrow$ " Assume that $\overrightarrow{OM} = x\overrightarrow{OA} + y\overrightarrow{OB} + z\overrightarrow{OC}$. Since $x + y + z = 1$, it follows that the identities $x = 1$, $y = 1$, $z = 1$ cannot hold simultaneously. Supposing $x \neq 1$, let

$$u = \frac{y}{1-x}, \qquad v = \frac{z}{1-x}$$

and obviously $u + v = 1$. By Lemma 11.3.1 there is a point $N \in BC$ such that $\overrightarrow{ON} = u\overrightarrow{OB} + v\overrightarrow{OC}$. Since

$$\begin{aligned} x\overrightarrow{OA} + (1-x)\overrightarrow{ON} &= x\overrightarrow{OA} + (1-x)u\overrightarrow{OB} + (1-x)v\overrightarrow{OC} \\ &= x\overrightarrow{OA} + y\overrightarrow{OB} + z\overrightarrow{OC} = \overrightarrow{OM}, \end{aligned}$$

applying Lemma 11.3.1 yields that M belongs to the line AN which is included in the plane (ABC), see Figure 11.7. Hence M belongs to the plane (ABC). ■

Theorem 11.3.3 *The numbers x, y, z determined by Theorem 11.3.2 are the baricentric coordinates of the point M.*

Proof: From Theorem 11.3.2 for any point O we have

$$\overrightarrow{OM} = x\overrightarrow{OA} + y\overrightarrow{OB} + z\overrightarrow{OC}.$$

Choosing $O \in \{A, B, C\}$, we obtain the following relations

$$\begin{aligned} \overrightarrow{AM} &= y\overrightarrow{AB} + z\overrightarrow{AC}, \\ \overrightarrow{BM} &= x\overrightarrow{BA} + z\overrightarrow{BC}, \text{and} \\ \overrightarrow{CM} &= x\overrightarrow{CA} + y\overrightarrow{CB}. \end{aligned}$$

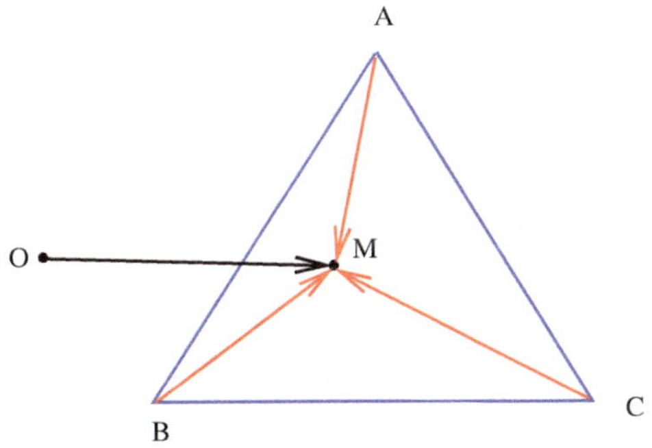

Figure 11.8: For Theorem 11.3.3.

Taking cross products and considering oriented areas, we have

$$\begin{aligned} 2\overrightarrow{\mathcal{A}}_{MBC} &= \overrightarrow{BC} \times \overrightarrow{BM} = \overrightarrow{BC} \times (x\overrightarrow{BA} + z\overrightarrow{BC}) \\ &= x\overrightarrow{BC} \times \overrightarrow{BA} + z\underbrace{\overrightarrow{BC} \times \overrightarrow{BC}}_{=0} \\ &= x\overrightarrow{BC} \times \overrightarrow{BA} = 2x\overrightarrow{\mathcal{A}}_{ABC}. \end{aligned}$$

Hence

$$x = \frac{\mathcal{A}_{MBC}}{\mathcal{A}_{ABC}} = \alpha,$$

the first baricentric coordinate. In a similar way, we obtain

$$y = \frac{\mathcal{A}_{MCA}}{\mathcal{A}_{ABC}} = \beta, \qquad z = \frac{\mathcal{A}_{MAB}}{\mathcal{A}_{ABC}} = \gamma.$$

■

Remark 11.3.4 If M is a point in the plane with baricentric coordinates (α, β, γ) then for any point O we have

$$\overrightarrow{OM} = \alpha\overrightarrow{OA} + \beta\overrightarrow{OB} + \gamma\overrightarrow{OC}.$$

Physical interpretation

Consider a triangle ABC in the plane Π and let $\overrightarrow{F}_A = \alpha\overrightarrow{n}_-$, $\overrightarrow{F}_B = \beta\overrightarrow{n}_-$, $\overrightarrow{F}_C = \gamma\overrightarrow{n}_-$ be forces applied at the points A, B, C perpendicular to the plane Π, see Figure 11.9. Then the sum $\overrightarrow{F}_A + \overrightarrow{F}_B + \overrightarrow{F}_C$ is a force with the origin at the point M with the baricentric coordinates (α, β, γ).

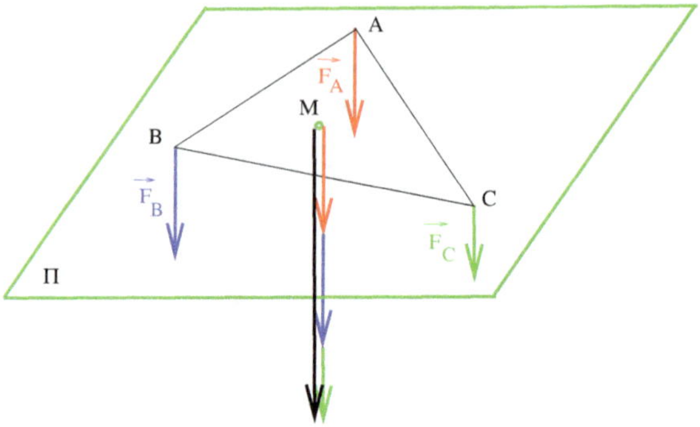

Figure 11.9: Physical interpretation of baricentric coordinates.

11.4 Metric properties in a triangle

In this section we shall apply baricentric coordinates techniques to find metric properties in a triangle.

The next result contains a formula for the distance between two points in terms of the tripolar coordinates (distances to the vertices) of one point and the baricentric coordinates of the other.

Theorem 11.4.1 *Let M be a point in the plane of a triangle ABC. Then for any point O in the plane (or space), we have*

$$|OM|^2 = \alpha|OA|^2 + \beta|OB|^2 + \gamma|OC|^2 - \alpha\beta c^2 - \alpha\gamma b^2 - \beta\gamma a^2, \quad (11.4.11)$$

where (α, β, γ) are the baricentric coordinates of M and $a = |BC|$, $b = |AC|$, and $c = |AB|$.

Proof: We have $\overrightarrow{OM} = \alpha\overrightarrow{OA} + \beta\overrightarrow{OB} + \gamma\overrightarrow{OC}$ and hence

$$\begin{aligned}
|OM|^2 &= \overrightarrow{OM} \cdot \overrightarrow{OM} = (\alpha\overrightarrow{OA} + \beta\overrightarrow{OB} + \gamma\overrightarrow{OC}) \cdot (\alpha\overrightarrow{OA} + \beta\overrightarrow{OB} + \gamma\overrightarrow{OC}) \\
&= \alpha^2|OA|^2 + \beta^2|OB|^2 + \gamma^2|OC|^2 \\
&\quad +2(\alpha\beta^2\overrightarrow{OA} \cdot \overrightarrow{OB} + \alpha\gamma\overrightarrow{OA} \cdot \overrightarrow{OC} + \beta\gamma\overrightarrow{OB} \cdot \overrightarrow{OC}) \\
&= \alpha^2|OA|^2 + \beta|OB|^2 + \gamma^2|OC|^2 + \alpha\beta(|OA|^2 + |OB|^2 - |AB|^2) \\
&\quad +\alpha\gamma(|OA|^2 + |OC|^2 - |AC|^2) + \beta\gamma(|OB|^2 + |OC|^2 - |BC|^2) \\
&= |OA|^2(\alpha^2 + \alpha\beta + \alpha\gamma) + |OB|^2(\beta^2 + \alpha\beta + \beta\gamma) \\
&\quad +|OC|^2(\gamma^2 + \alpha\gamma + \beta\gamma) - \alpha\beta|AB|^2 - \alpha\gamma|AC|^2 - \beta\gamma|BC|^2 \\
&= \alpha|OA|^2 + \beta|OB|^2 + \gamma|OC|^2 - \alpha\beta|AB|^2 - \alpha\gamma|AC|^2 - \beta\gamma|BC|^2,
\end{aligned}$$

where we have used that $\alpha+\beta+\gamma=1$, and $2\overrightarrow{OA}\cdot\overrightarrow{OB}=|OA|^2+|OB|^2-|AB|^2$, and two more similar relations. ∎

This formula is useful when the distances to the triangle vertices is known for one of the points. As an application of Theorem 11.4.1 we shall compute the distance between the circumcenter O and the incenter I of a triangle ABC.

Theorem 11.4.2 (Euler's relation) *Let R, r denote the circumradius and the inradius of the triangle ABC. Then $|OI|^2=R(R-2r)$.*

Proof: We shall apply Theorem 11.4.1 for $M=I$, the incenter and for O, the circumcenter. In this case $|OA|=|OB|=|OC|=R$, the circumradius. Using that the baricentric coordinates of I are $\alpha=\dfrac{a}{2p}$, $\beta=\dfrac{b}{2p}$, $\gamma=\dfrac{c}{2p}$, formula (11.4.11) becomes

$$\begin{aligned}
|OI|^2 &= \frac{a}{2p}R^2+\frac{b}{2p}R^2+\frac{c}{2p}R^2-\frac{abc^2}{4p^2}-\frac{ab^2c}{4p^2}-\frac{a^2bc}{4p^2}\\
&= \frac{R^2}{2p}(a+b+c)-\frac{1}{4p^2}abc(a+b+c)\\
&= \frac{R^2}{2p}2p-\frac{1}{4p^2}abc\cdot 2p=R^2-\frac{abc}{2p}\\
&= R^2-\frac{abc}{4R}\cdot\frac{4R}{2p}=R^2-\mathcal{A}_{ABC}\cdot\frac{2R}{p}\\
&= R^2-pr\frac{2R}{p}=R^2-2Rr=R(R-2r),
\end{aligned}$$

where we have used the following formulas for the area of a triangle:

$$\mathcal{A}_{ABC}=pr=\frac{abc}{4R}.$$

∎

Another application of Theorem 11.4.1 is the Leibnitz relation:

Corollary 11.4.3 (Leibnitz) *Let G be the centroid of the triangle ABC and let O be any point (in the plane or space). Then*

$$|OG|^2=\frac{1}{3}(|OA|^2+|OB|^2+|OC|^2)-\frac{1}{9}(a^2+b^2+c^2). \tag{11.4.12}$$

In the case when O is the circumcenter, we obtain

$$|OG|^2=R^2-\frac{1}{9}(a^2+b^2+c^2). \tag{11.4.13}$$

Proof: The baricentric coordinates of G are $\alpha = \beta = \gamma = 1/3$. Substituting in (11.4.11) yields formula (11.4.12). When $|OA| = |OB| = |OC| = R$, relation (11.4.12) becomes (11.4.13). ■

Exercise 11.4.4 *Let H and O be the orthocenter and the circumcenter of triangle ABC. Find the distance $|OH|$ in terms of the side length and the triangle circumradius.*

Exercise 11.4.5 *Let I_a, I_b, I_c be the centers of the excircles of triangle ABC, see Figure 11.10.*

(i) Show that the baricentric coordinates for the above points are

$$I_a\left(\frac{-a}{2(p-a)}, \frac{b}{2(p-a)}, \frac{c}{2(p-c)}\right), \quad I_b\left(\frac{a}{2(p-a)}, \frac{-b}{2(p-a)}, \frac{c}{2(p-c)}\right),$$

$$I_c\left(\frac{a}{2(p-a)}, \frac{b}{2(p-a)}, \frac{-c}{2(p-c)}\right).$$

(ii) Show that $|OI_a|^2 = R(R+2r_a)$, $|OI_b|^2 = R(R+2r_b)$, and $|OI_c|^2 = R(R+2r_c)$, where O is the circumcenter and r_a, r_b, r_c are the exradii of triangle ABC.

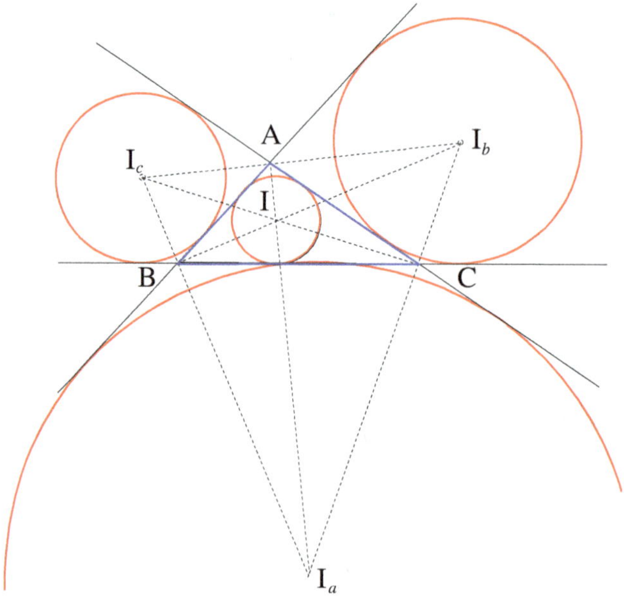

Figure 11.10: Circumcenters and circumcircles.

The next result contains a formula for the distance between two points in terms of their baricentric coordinates. It can replace the use of Theorem 11.4.1 when the distances to vertices are difficult to compute.

In the following we shall consider two points $M(\alpha, \beta, \gamma)$ and $M'(\alpha', \beta', \gamma')$ in the plane of triangle ABC, with the baricentric coordinates indicated in the parentheses.

Theorem 11.4.6 *The distance between the points M and M' in terms of their baricentric coordinates and the side lengths of triangle ABC is*

$$|MM'|^2 = (\alpha - \alpha')(\beta' - \beta)c^2 + (\gamma - \gamma')(\alpha' - \alpha)b^2 + (\beta - \beta')(\gamma' - \gamma)a^2. \tag{11.4.14}$$

Proof: For any point O we have

$$\begin{aligned} \overrightarrow{OM} &= \alpha\overrightarrow{OA} + \beta\overrightarrow{OB} + \gamma\overrightarrow{OC} \\ \overrightarrow{OM'} &= \alpha'\overrightarrow{OA} + \beta'\overrightarrow{OB} + \gamma'\overrightarrow{OC}. \end{aligned}$$

Subtracting, yields

$$\overrightarrow{M'M} = \overrightarrow{OM} - \overrightarrow{OM'} = (\alpha - \alpha')\overrightarrow{OA} + (\beta - \beta')\overrightarrow{OB} + (\gamma - \gamma')\overrightarrow{OC}.$$

Then taking the inner product of $\overrightarrow{M'M}$ with itself we obtain the sum of squares of $|OA|$, $|OB|$ and $|OC|$. Using also that

$$\alpha + \beta + \gamma = \alpha' + \beta' + \gamma' = 1$$

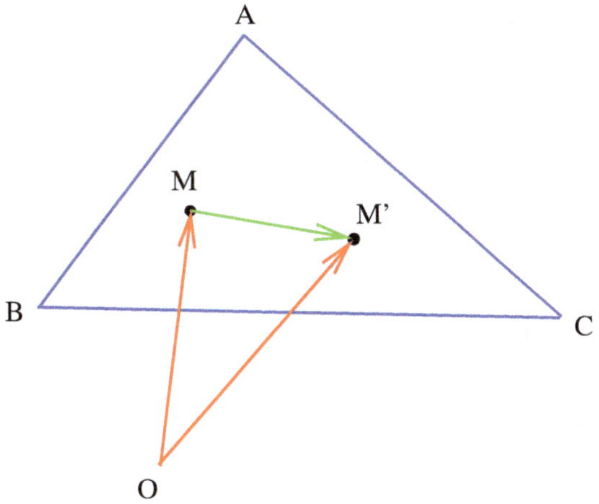

Figure 11.11: For Theorem 11.4.6.

we obtain

$$\begin{aligned}
|M'M|^2 &= \overrightarrow{M'M} \cdot \overrightarrow{M'M} \\
&= (\alpha-\alpha')^2|OA|^2 + (\beta-\beta')^2|OB|^2 + (\gamma-\gamma')^2|OC|^2 \\
&\quad +2(\alpha-\alpha')(\beta-\beta')\overrightarrow{OA}\cdot\overrightarrow{OB} + 2(\alpha-\alpha')(\gamma-\gamma')\overrightarrow{OA}\cdot\overrightarrow{OC} \\
&\quad +2(\beta-\beta')(\gamma-\gamma')\overrightarrow{OB}\cdot\overrightarrow{OC} \\
&= (\alpha-\alpha')^2|OA|^2 + (\beta-\beta')^2|OB|^2 + (\gamma-\gamma')^2|OC|^2 \\
&\quad +(\alpha-\alpha')(\beta-\beta')(|OA|^2+|OB|^2-|AB|^2) \\
&\quad +(\alpha-\alpha')(\gamma-\gamma')(|OA|^2+|OC|^2-|AC|^2) \\
&\quad +(\beta-\beta')(\gamma-\gamma')(|OB|^2+|OC|^2-|BC|^2) \\
&= [(\alpha-\alpha')^2 + (\alpha-\alpha')(\beta-\beta') + (\alpha-\alpha')(\gamma-\gamma')]\cdot|OA|^2 \\
&\quad +[(\beta-\beta')^2 + (\alpha-\alpha')(\beta-\beta') + (\beta-\beta')(\gamma-\gamma')]\cdot|OB|^2 \\
&\quad +[(\gamma-\gamma')^2 + (\alpha-\alpha')(\gamma-\gamma') + (\beta-\beta')(\gamma-\gamma')]\cdot|OC|^2 \\
&\quad -(\alpha-\alpha')(\beta-\beta')|AB|^2 - (\alpha-\alpha')(\gamma-\gamma')|AC|^2 \\
&\quad -(\beta-\beta')(\gamma-\gamma')|BC|^2 \\
&= (\alpha-\alpha')\underbrace{(\alpha+\beta+\gamma-\alpha'-\beta'-\gamma')}_{1-1=0}|OA|^2 \\
&\quad +(\beta-\beta')\underbrace{(\beta+\alpha+\gamma-\beta'-\alpha'-\gamma')}_{1-1=0}|OB|^2 \\
&\quad +(\gamma-\gamma')\underbrace{(\gamma+\alpha+\beta-\gamma'-\alpha'-\beta')}_{1-1=0}|OB|^2 \\
&\quad -(\alpha-\alpha')(\beta-\beta')|AB|^2 - (\alpha-\alpha')(\gamma-\gamma')|AC|^2 \\
&\quad -(\beta-\beta')(\gamma-\gamma')|BC|^2 \\
&= (\alpha-\alpha')(\beta'-\beta)|AB|^2 \\
&\quad +(\gamma-\gamma')(\alpha'-\alpha)|AC|^2 \\
&\quad +(\beta-\beta')(\gamma'-\gamma)|BC|^2.
\end{aligned}$$

■

Theorem 11.4.6 is a very powerful tool in computing distances between points in the plane. In the following we shall include a few applications.

The median theorem

Let A' be the midpoint of the side $[BC]$ of triangle ABC. Using that $A(1,0,0)$, $A'\left(0,\frac{1}{2},\frac{1}{2}\right)$, formula (11.4.14) yields the length of the median

$$\begin{aligned}|AA'|^2 &= (1-0)\left(\frac{1}{2}-0\right)c^2+\left(0-\frac{1}{2}\right)(0-1)b^2+\left(0-\frac{1}{2}\right)\left(\frac{1}{2}-0\right)a^2\\ &= \frac{1}{2}(b^2+c^2)-\frac{1}{4}a^2.\end{aligned}$$

The distance between M and G

Let $M(\alpha,\beta,\gamma)$ be a point in the plane and let $G\left(\frac{1}{3},\frac{1}{3},\frac{1}{3}\right)$ be its centroid. Then

$$|MG|^2=\left(\alpha-\frac{1}{3}\right)\left(\frac{1}{3}-\beta\right)c^2+\left(\gamma-\frac{1}{3}\right)\left(\frac{1}{3}-\alpha\right)b^2+\left(\beta-\frac{1}{3}\right)\left(\frac{1}{3}-\gamma\right)a^2.$$

We note that the distance $|MG|$ can be also evaluated by Leibnitz formula (11.4.12) when the distances $|MA|$, $|MB|$ and $|MC|$ are known.

Exercise 11.4.7 *Let H be the orthocenter of triangle ABC. Find the distance $|HG|$ in terms of the triangle sides length.*

Exercise 11.4.8 *Let I be the orthocenter of triangle ABC. Find the distance $|IG|$ in terms of the triangle sides length.*

The distance between the incenter I and the excenter I_a

We have $I(\alpha,\beta,\gamma)$ and $I_a(\alpha',\beta',\gamma')$, where

$$\alpha=\frac{a}{2p},\qquad \beta=\frac{b}{2p},\qquad \gamma=\frac{c}{2p}$$

$$\alpha'=\frac{-a}{2(p-a)},\qquad \beta'=\frac{b}{2(p-a)},\qquad \gamma'=\frac{c}{2(p-a)}.$$

Then formula (11.4.14) yields

$$\begin{aligned}
|II_a|^2 &= \left(\frac{a}{2p}+\frac{a}{2(p-a)}\right)\left(\frac{b}{2(p-a)}-\frac{b}{2p}\right)c^2\\
&\quad+\left(\frac{c}{2p}-\frac{c}{2(p-a)}\right)\left(\frac{-a}{2(p-a)}-\frac{a}{2p}\right)b^2\\
&\quad+\left(\frac{b}{2p}-\frac{b}{2(p-a)}\right)\left(\frac{c}{2(p-a)}-\frac{c}{2p}\right)a^2 \Longleftrightarrow\\
4|II_a|^2 &= abc^2\left(\frac{1}{p}+\frac{1}{p-a}\right)\left(\frac{1}{p-a}-\frac{1}{p}\right)\\
&\quad+ab^2c\left(\frac{1}{p}-\frac{1}{p-a}\right)\left(-\frac{1}{p-a}-\frac{1}{p}\right)\\
&\quad+a^2bc\left(\frac{1}{p}-\frac{1}{p-a}\right)\left(\frac{1}{p-a}-\frac{1}{p}\right)\Longleftrightarrow\\
\frac{4|II_a|^2}{abc} &= \frac{c(b+c)a-ab(-b-c)-a^3}{p^2(p-a)^2}=\frac{(b+c)(ac+ab)-a^3}{p^2(p-a)^2}\\
&= \frac{a}{p^2(p-a)^2}[(b+c)^2-a^2]=\frac{a(b+c-a)(b+c+a)}{p^2(p-a)^2}\\
&= \frac{a\cdot 2(p-a)\cdot 2p}{p^2(p-a)^2}=\frac{4a}{p(p-a)}.
\end{aligned}$$

It follows that

$$|II_a|^2=\frac{a^2bc}{p(p-a)}.$$

Remark 11.4.9 Using the trigonometric formula $\cos\frac{\widehat{A}}{2}=\sqrt{\frac{p(p-a)}{bc}}$, we can also write

$$|II_a|=\frac{a}{\cos\frac{\widehat{A}}{2}}=\frac{2R\sin\widehat{A}}{\cos\frac{\widehat{A}}{2}}=\frac{4R\sin\frac{\widehat{A}}{2}\cos\frac{\widehat{A}}{2}}{\cos\frac{\widehat{A}}{2}}=4R\sin\frac{\widehat{A}}{2}.$$

Exercise 11.4.10 *Find the distances $|I_aI_b|$, $|I_aI_c|$, $|I_bI_c|$ in terms of the sides length of triangle ABC.*

The equation of a circle in baricentric coordinates

Consider the circle of radius $\rho>0$ centered at $P(\alpha,\beta,\gamma)$ defined by

$$\mathcal{C}(P,\rho)=\{M;|MP|^2=\rho^2\}.$$

Using Theorem 11.4.6 it follows that the baricentric coordinates (x, y, z) of a point $M \in \mathcal{C}(P, \rho)$ satisfy the relation

$$(\alpha - x)(y - \beta)c^2 + (\gamma - z)(x - \alpha)b^2 + (\beta - y)(z - \gamma)a^2 = \rho^2.$$

Exercise 11.4.11 *Write the equation using baricentric coordinates for the circumcircle $\mathcal{C}(O, R)$.*

Exercise 11.4.12 *Write the equations using baricentric coordinates for the incircle $\mathcal{C}(I, r)$ and for the excircles $\mathcal{C}(I_a, r_a)$, $\mathcal{C}(I_b, r_b)$, $\mathcal{C}(I_c, r_c)$.*

Exercise 11.4.13 *Let $M(\alpha, \beta, \gamma)$ and $M'(\alpha', \beta', \gamma')$ be two points in the plane with the baricentric coordinates indicated in the parenthesis. Let $X(x, y, z)$ be the midpoint of the line segment $[MM']$. Show that*

$$x = \frac{\alpha + \alpha'}{2}, \qquad y = \frac{\beta + \beta'}{2}, \qquad z = \frac{\gamma + \gamma'}{2}.$$

Exercise 11.4.14 *Let $N(\alpha, \beta, \gamma)$ and $N'(\alpha', \beta', \gamma')$ be two points in the plane with the baricentric coordinates indicated in the parenthesis. Let $M(x, y, z)$ be a point on the line NN' such that $\overrightarrow{NM} = \lambda \overrightarrow{MN'}$, for $\lambda \in \mathbb{R}$. Find the baricentric coordinates of M.*

Exercise 11.4.15 (Feuerbach's point) (i) *Find the baricentric coordinates of the center of the Euler circle.*

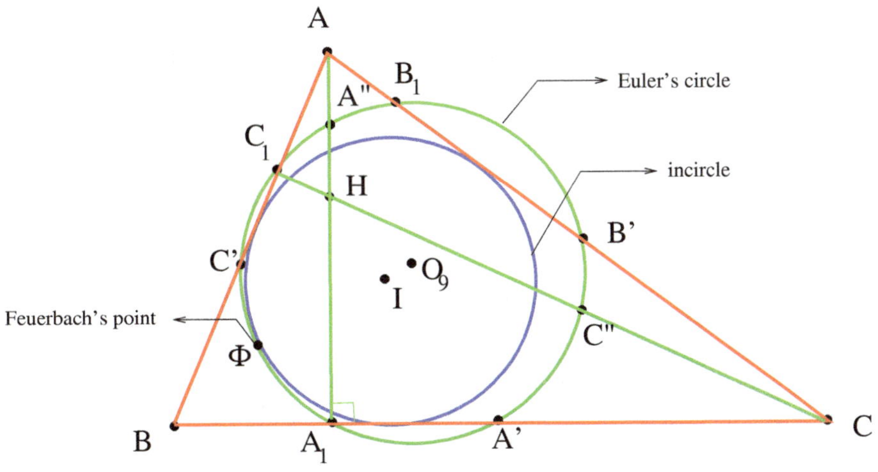

Figure 11.12: Feuerbach's point.

(ii) Write Euler's circle equation using baricentric coordinates.

(iii) Show that the incircle and the Euler circle have a point Φ *in common, see Figure 11.12.*

Chapter 12

Momenta and Applications

The concepts introduced and explored in this chapter are deeply rooted in physical principles and have numerous applications in mechanics, particularly in describing the rotational motion of various solid shapes.

12.1 Angular momentum

Let a point O and a vector $\overrightarrow{v}$ be given. The capacity of rotation of the vector $\overrightarrow{v}$ about the point O is described by a vectorial measure called *angular momentum*, which is denoted by $L(O, \overrightarrow{v})$ and is defined by

$$L(O, \overrightarrow{v}) = \overrightarrow{OA} \times \overrightarrow{v}, \tag{12.1.1}$$

where A is the origin of $\overrightarrow{v}$. The angular momentum $L(O, \overrightarrow{v})$ is a vector perpendicular to the plane determined by $(O, \overrightarrow{v})$ and has the magnitude $|L(O, \overrightarrow{v})| = |\overrightarrow{v}||OA| \sin(\widehat{\overrightarrow{OA}, \overrightarrow{v}})$.

Physical significance We consider $\overrightarrow{v} = \overrightarrow{AA'}$. The vector $\overrightarrow{v}$ can be decomposed over two orthogonal directions

$$\overrightarrow{v} = \overrightarrow{v}^{tan} + \overrightarrow{v}^{nor},$$

where the normal component, $\overrightarrow{v}^{nor}$, is collinear to the position vector $\overrightarrow{OA}$, and the tangent component, $\overrightarrow{v}^{tan}$, has a perpendicular direction to OA, see Figure 12.1. Its magnitude is given by

$$|\overrightarrow{v}^{tan}| = |\overrightarrow{v}| \cos\theta = |AA'| \sin(\widehat{\overrightarrow{OA}, \overrightarrow{v}}) = \frac{1}{|OA|}|\overrightarrow{OA} \times \overrightarrow{v}| = \frac{1}{|OA|}|L(O, \overrightarrow{v})|,$$

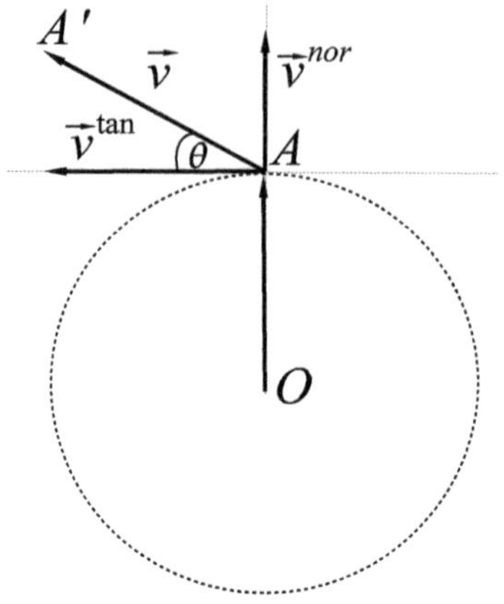

Figure 12.1: Angular momentum of a vector with respect to a point.

where θ is the angle made by $\overrightarrow{v}$ and the perpendicular on OA. We note that $\overrightarrow{v}^{tan}$ and $\overrightarrow{v}^{nor}$ are responsible respectively for rotation and elongation. Now, considering O fixed and the point A constrained to move along a circle centered at O of fixed radius $|OA|$, it follows that the magnitude of the angular momentum, $|L(O, \overrightarrow{v})|$, represents the rotational tendency about O due to the tangent component, $\overrightarrow{v}^{tan}$.

We may also think of $\overrightarrow{v}$ as a force acting at the point A on a rigid arm OA, with O fixed.

From the geometric point of view, the angular momentum represents an oriented area, as it follows from the next computation

$$\begin{aligned} L(O, \overrightarrow{v}) &= \overrightarrow{OA} \times \overrightarrow{v} = \overrightarrow{OA} \times \overrightarrow{AA'} = \overrightarrow{OA} \times (\overrightarrow{OA'} - \overrightarrow{OA}) \\ &= \overrightarrow{OA} \times \overrightarrow{OA'} - \underbrace{\overrightarrow{OA} \times \overrightarrow{OA}}_{=0} \\ &= 2\mathcal{A}_{OAA'}\, n_+, \end{aligned}$$

where n_+ represents the unit normal to the plane.
The angular momentum can be extended to an ensemble of points as follows.

12.1.1 Angular momentum of a triangle

Next, we shall study some aspects of the rotational kinematics of a rigid triangle ΔABC with respect to a fixed point. To this end, we consider three vectors, $\overrightarrow{v_A} = \overrightarrow{AA'}$, $\overrightarrow{v_B} = \overrightarrow{BB'}$, $\overrightarrow{v_C} = \overrightarrow{CC'}$, applied at the vertices of triangle ΔABC. The *total angular momentum* of these vectors with respect to a given point O in the plane is given by the sum

$$L(O, \overrightarrow{v_A}) + L(O, \overrightarrow{v_B}) + L(O, \overrightarrow{v_C}) = \overrightarrow{OA} \times \overrightarrow{AA'} + \overrightarrow{OB} \times \overrightarrow{BB'} + \overrightarrow{OC} \times \overrightarrow{CC'}.$$

This represents the cumulative rotational effect about the point O due to forces given by the vectors $\overrightarrow{v_A}$, $\overrightarrow{v_B}$, and $\overrightarrow{v_C}$, which act on the triangle vertices.

The next result investigates how this total rotational tendency depends on the pivot point.

Proposition 12.1.1 *Let O' be another point in the plane. Then*

$$L(O, \overrightarrow{v_A}) + L(O, \overrightarrow{v_B}) + L(O, \overrightarrow{v_C}) = \overrightarrow{OO'} \times \overrightarrow{v} + L(O', \overrightarrow{v_A}) + L(O', \overrightarrow{v_B}) + L(O', \overrightarrow{v_C}),$$

where $\overrightarrow{v} = \overrightarrow{v_A} + \overrightarrow{v_B} + \overrightarrow{v_C}$.

Proof: Using the properties of distributivity of the cross product we have

$$\begin{aligned} L(O, \overrightarrow{v_A}) &= \overrightarrow{OA} \times \overrightarrow{v_A} = (\overrightarrow{OO'} + \overrightarrow{O'A}) \times \overrightarrow{v_A} \\ &= \overrightarrow{OO'} \times \overrightarrow{v_A} + L(O', \overrightarrow{v_A}). \end{aligned}$$

Writing two more relations, for $\overrightarrow{v_B}$ and $\overrightarrow{v_C}$, and then taking the sum we arrive at the desired result. ■

This result has several interesting applications.

Application 12.1.2 Let O denote the circumcenter of triangle ΔABC and consider the concurrent vector forces $\overrightarrow{v_A} = \overrightarrow{AO}$, $\overrightarrow{v_B} = \overrightarrow{BO}$, and $\overrightarrow{v_C} = \overrightarrow{CO}$. For this choice of vectors, the following momenta vanish

$$L(O, \overrightarrow{v_A}) = L(O, \overrightarrow{v_B}) = L(O, \overrightarrow{v_C}) = 0.$$

Also, the total force is

$$\overrightarrow{v} = \overrightarrow{v_A} + \overrightarrow{v_B} + \overrightarrow{v_C} = -(\overrightarrow{OA} + \overrightarrow{OB} + \overrightarrow{OC}) = -3\overrightarrow{OG},$$

where G denotes the triangle centroid. Then Proposition 12.1.1 yields

$$3\overrightarrow{OO'} \times \overrightarrow{OG} = L(O', \overrightarrow{v_A}) + L(O', \overrightarrow{v_B}) + L(O', \overrightarrow{v_C}).$$

Since $O' \in OG$ if and only if $\overrightarrow{OO'} \times \overrightarrow{OG} = 0$, we arrive at the following result:

Proposition 12.1.3 *A point O' belongs to the Euler's line of triangle ΔABC if and only if*

$$\overrightarrow{O'A} \times \overrightarrow{AO} + \overrightarrow{O'B} \times \overrightarrow{BO} + \overrightarrow{O'C} \times \overrightarrow{CO} = 0.$$

Corollary 12.1.4 *Let O and H denote the circumcenter and the orthocenter of triangle ΔABC. Then we have*

$$\overrightarrow{HA} \times \overrightarrow{AO} + \overrightarrow{HB} \times \overrightarrow{BO} + \overrightarrow{HC} \times \overrightarrow{CO} = 0. \qquad (12.1.2)$$

Proof: It follows from Proposition 12.1.3 by considering $O' = H$ and using that O, H and G are collinear. ∎

Proof variant: The relation can be verified directly, using brute force computation as follows. We use the fact that AH and AO are isogonals in angle $\widehat{BAC}$, so we have $\widehat{BAH} = \widehat{OAC}$. Since $\widehat{BAH} = 90° - \widehat{B}$, then

$$\widehat{HAO} = \widehat{A} - 2\widehat{BAH} = \widehat{A} - 180° + 2\widehat{B} = \widehat{B} - \widehat{C}.$$

Therefore, we can compute the first cross product term as

$$\begin{aligned}\overrightarrow{HB} \times \overrightarrow{OB} &= 2R^2 \cos\widehat{B}\sin(\widehat{C} - \widehat{A})\overrightarrow{e}_3,\\ \overrightarrow{HC} \times \overrightarrow{OC} &= 2R^2 \cos\widehat{C}\sin(\widehat{A} - \widehat{B})\overrightarrow{e}_3.\end{aligned}$$

Substituting in (12.1.2) and dividing by $2R^2$ yields

$$\cos\widehat{A}\sin(\widehat{B} - \widehat{C}) + \cos\widehat{B}\sin(\widehat{C} - \widehat{A}) + \cos\widehat{C}\sin(\widehat{A} - \widehat{B}) = 0. \qquad (12.1.3)$$

To show this relation we compute each of its terms

$$\begin{aligned}\overrightarrow{HA} \times \overrightarrow{OA} &= |HA|\,|OA|\sin(\overrightarrow{HA}, \overrightarrow{OA})\overrightarrow{e}_3\\ &= 2R^2 \cos\widehat{A}\sin(\widehat{B} - \widehat{C})\overrightarrow{e}_3,\end{aligned}$$

where we used the relations $|HA| = 2R\cos\widehat{A}$ and $|OA| = R$. Similarly, we obtain two more relations

$$\begin{aligned}\cos\widehat{A}\sin(\widehat{B} - \widehat{C}) &= \cos\widehat{A}\sin\widehat{B}\cos\widehat{C} - \cos\widehat{A}\cos\widehat{B}\sin\widehat{C},\\ \cos\widehat{B}\sin(\widehat{C} - \widehat{A}) &= \cos\widehat{B}\sin\widehat{C}\cos\widehat{A} - \cos\widehat{B}\cos\widehat{C}\sin\widehat{A}, \text{ and}\\ \cos\widehat{C}\sin(\widehat{A} - \widehat{B}) &= \cos\widehat{C}\sin\widehat{A}\cos\widehat{B} - \cos\widehat{C}\cos\widehat{A}\sin\widehat{B}.\end{aligned}$$

Adding and canceling the terms in pairs we obtain (12.1.3).

Application 12.1.5 We consider the notations I, O and H for the incenter, circumcenter and orthocenter of a triangle ΔABC. We choose the concurrent vector forces $\overrightarrow{v_A} = \overrightarrow{AH}$, $\overrightarrow{v_B} = \overrightarrow{BH}$, and $\overrightarrow{v_C} = \overrightarrow{CH}$. Then the total force vector becomes

$$\overrightarrow{v} = -(\overrightarrow{HA} + \overrightarrow{HB} + \overrightarrow{HC}) = -3\overrightarrow{HG} = 3\overrightarrow{GH} = 2\overrightarrow{OH}.$$

Let $O' = I$. Then Proposition 12.1.1 yields

$$L(O, \overrightarrow{v_A}) + L(O, \overrightarrow{v_B}) + L(O, \overrightarrow{v_C}) = 2\overrightarrow{OI} \times \overrightarrow{OH} + L(I, \overrightarrow{v_A}) + L(I, \overrightarrow{v_B}) + L(I, \overrightarrow{v_C}). \tag{12.1.4}$$

The total sum of angular momenta with respect to O vanishes by Corollary 12.1.4

$$\begin{aligned} L(O, \overrightarrow{v_A}) + L(O, \overrightarrow{v_B}) + L(O, \overrightarrow{v_C}) &= \overrightarrow{OA} \times \overrightarrow{AH} + \overrightarrow{OB} \times \overrightarrow{BH} + \overrightarrow{OC} \times \overrightarrow{CH} \\ &= 0. \end{aligned}$$

Noting that

$$\widehat{HAI} = |\widehat{BAI} - \widehat{BAH}| = \left| \frac{\widehat{A}}{2} - (90^\circ - \widehat{B}) \right| = \frac{|\widehat{B} - \widehat{C}|}{2},$$

we compute the angular momenta with respect to the incenter as

$$\begin{aligned} L(I, \overrightarrow{v_A}) &= \overrightarrow{IA} \times \overrightarrow{AH} = \overrightarrow{AH} \times \overrightarrow{AI} \\ &= |AH|\,|AI| \sin \widehat{HAI}\, \overrightarrow{e}_3 \\ &= 2R \cos \widehat{A}\, \frac{r}{\sin \frac{\widehat{A}}{2}} \sin \frac{\widehat{B} - \widehat{C}}{2}\, \overrightarrow{e}_3. \end{aligned}$$

Computing two more similar relations and adding them and using (12.1.4) yields

$$-2\overrightarrow{OI} \times \overrightarrow{OH} = 2Rr \sum_{circ.perm.} \frac{\cos \widehat{A}}{\sin \frac{\widehat{A}}{2}} \sin \frac{\widehat{B} - \widehat{C}}{2}\, \overrightarrow{e}_3.$$

This can be also stated as:

Proposition 12.1.6 *Let ΔABC be an acute-angled triangle. Then the area of triangle ΔOIH is given by*

$$\mathcal{A}_{OIH} = \frac{1}{2} Rr \sum_{circ.perm.} \frac{\cos \widehat{A}}{\sin \frac{\widehat{A}}{2}} \sin \frac{\widehat{B} - \widehat{C}}{2}.$$

Application 12.1.7 In this case we consider $O' = G$, the centroid, and $O = M$, an arbitrary point in the plane. We also take the vectors $\overrightarrow{v_A} = \overrightarrow{AG}$, $\overrightarrow{v_B} = \overrightarrow{BG}$, and $\overrightarrow{v_C} = \overrightarrow{CG}$. From the centroid property, we have

$$\overrightarrow{v} = \overrightarrow{v_A} + \overrightarrow{v_B} + \overrightarrow{v_C} = -(\overrightarrow{GA} + \overrightarrow{GB} + \overrightarrow{GC}) = 0.$$

Then by Proposition 12.1.1 we obtain

$$L(M, \overrightarrow{v_A}) + L(M, \overrightarrow{v_B}) + L(M, \overrightarrow{v_C}) = L(G, \overrightarrow{v_A}) + L(G, \overrightarrow{v_B}) + L(G, \overrightarrow{v_C}).$$

Using that the cross product of collinear vectors vanishes,

$$L(G, \overrightarrow{v_A}) = \overrightarrow{GA} \times \overrightarrow{AG} = 0,$$

$$L(G, \overrightarrow{v_B}) = \overrightarrow{GB} \times \overrightarrow{BG} = 0,$$

$$L(G, \overrightarrow{v_B}) = \overrightarrow{GB} \times \overrightarrow{BG} = 0,$$

the aforementioned relation simplifies to

$$\begin{aligned} L(M, \overrightarrow{v_A}) + L(M, \overrightarrow{v_B}) + L(M, \overrightarrow{v_C}) &= 0 \Longleftrightarrow \\ \overrightarrow{MA} \times \overrightarrow{AG} + \overrightarrow{MB} \times \overrightarrow{BG} + \overrightarrow{MC} \times \overrightarrow{CG} &= 0 \Longleftrightarrow \\ \overrightarrow{GA} \times \overrightarrow{AM} + \overrightarrow{GB} \times \overrightarrow{BM} + \overrightarrow{GC} \times \overrightarrow{CM} &= 0 \Longleftrightarrow \\ L(G, \overrightarrow{AM}) + L(G, \overrightarrow{BM}) + L(G, \overrightarrow{CM}) &= 0. \end{aligned}$$

We arrived at the following result regarding a system of three concurrent vector forces:

Proposition 12.1.8 *Given an arbitrary point M in the plane of the triangle ΔABC, we consider the vector forces $\overrightarrow{w_A} = \overrightarrow{AM}$, $\overrightarrow{w_B} = \overrightarrow{BM}$, $\overrightarrow{w_C} = \overrightarrow{CM}$. Then the total angular momentum of these forces with respect to the centroid G is equal to zero.*

The kinematic representation of this result is as follows. Consider a triangular-shaped slab pinned at its centroid, G, namely, it can only rotate freely about G, see Figure 12.2. Then consider a point M in the plane and anchor the triangle vertices to M by some elastic strings, such that the tension forces are proportional to the distances from vertices to M. Then the triangle does not rotate, i.e., it is in an equilibrium state.

Remark 12.1.9 Some of the results proved for the angular momentum of triangles, such as Propositions 12.1.1 and 12.1.8, can easily be generalized to polygons, see Exercises 12.1.10 and 12.1.11.

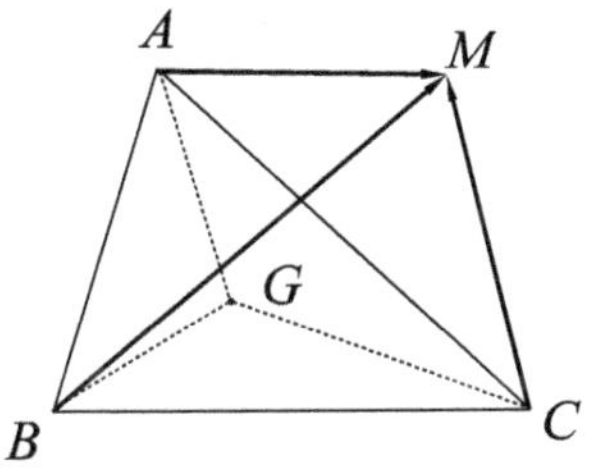

Figure 12.2: Total angular momentum about G vanishes.

Exercise 12.1.10 *Let O and O' be two points in the plane of the polygon $A_0A_1 \cdots A_{n-1}$. Then for any system of vectors, $\{v_1, \ldots, v_{n-1}\}$, we have*

$$\sum_{j=1}^{n-1} L(O, \overrightarrow{v_j}) = \overrightarrow{OO'} \times \sum_{j=1}^{n-1} \overrightarrow{v_j} + \sum_{j=1}^{n-1} L(O', \overrightarrow{v_j}).$$

Exercise 12.1.11 *Let G be the centroid of the polygon $A_0A_1 \cdots A_{n-1}$. Then for any point M in the plane we have*

$$\sum_{j=1}^{n-1} L(G, \overrightarrow{A_jM}) = 0.$$

12.2 Moment of inertia

Roughly speaking, the capacity of rotation of a planar shape with respect to a point in the plane is assessed by the *moment of inertia* with respect to that point. Similarly, the difficulty of rotation of a shape with respect to an axis represents the moment of inertia with respect to that axis.

Since these momenta reach their extrema along the symmetry axes, the moments of inertia provide a useful tool to investigate the rotational symmetry of a triangle, or any other polygonal line.

12.3 Moment of inertia with respect to a point

We shall introduce the moment of inertia with respect to a point using the total angular moment concept as follows. We consider a point A subject to rotation about a fixed point, O, along a circle of radius r, with a constant angular speed, $\omega = \frac{d\theta}{dt}$, see Figure 12.3(a). Here, θ denotes the central angle of rotation and t is time. The speed magnitude of the point A is

$$|\overrightarrow{v_A}| = \frac{ds}{dt} = \frac{d(r\theta)}{dt} = r\omega,$$

where $s = r\theta$ denotes the arc along the circle corresponding to a central angle of rotation θ. The direction of velocity vector of A is perpendicular to the radius OA, so $\overrightarrow{v_A} \perp \overrightarrow{OA}$. Then the angular momentum of the velocity of point A with respect to O is

$$L(O, \overrightarrow{v_A}) = \overrightarrow{OA} \times \overrightarrow{v_A} = |OA||\overrightarrow{v_A}|\overrightarrow{e}_3 = r^2\omega\overrightarrow{e}_3.$$

This is a vector normal to the plane, with the magnitude proportional to the square of the rotation radius. The idea is to consider a system of rotating points and to add their angular momenta, which is given by the previous formula. We shall do this in the case of three points.

We consider a rigid triangle ΔABC, which can rotate freely about a fixed point, O, see Figure 12.3(b). The total angular momentum of its vertices is given by

$$\begin{aligned} L_{tot} &= L(O, \overrightarrow{v_A}) + L(O, \overrightarrow{v_B}) + L(O, \overrightarrow{v_C}) \\ &= (|OA|^2\omega_A + |OB|^2\omega_B + |OC|^2\omega_C)\overrightarrow{e}_3 \\ &= (|OA|^2 + |OB|^2 + |OC|^2)\omega\overrightarrow{e}_3, \end{aligned}$$

where use used that $\omega_A = \omega_B = \omega_C = \omega$, namely, all vertices rotate at the same angular speed. The amount $|L_{tot}|$ is called the moment of inertia of ΔABC with respect to the point O. Without losing generality, we assume $\omega = 1$ and consider the moment of inertia given by the sum of squares as

$$\mathcal{I}(O; A, B, C) = |OA|^2 + |OB|^2 + |OC|^2. \qquad (12.3.5)$$

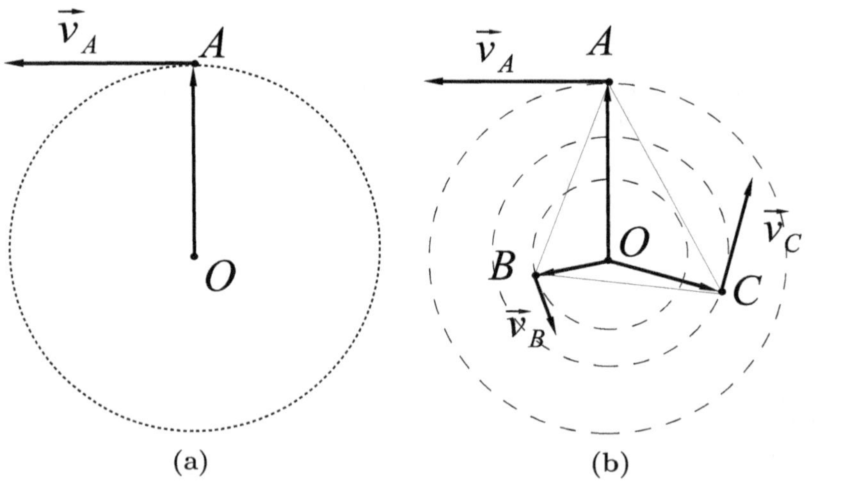

Figure 12.3: Moment of inertia: (a) Of the point A; (b) Of the triangle ΔABC.

We shall encounter first a few examples of moments of inertia with respect to some remarkable centers of a triangle:

(*i*) If O is the circumcenter of ΔABC, then $\mathcal{I}(O; A, B, C) = 3R^2$, where R denotes the circumradius.

(*ii*) If O is the centroid, G, of ΔABC, then

$$\begin{aligned}\mathcal{I}(G; A, B, C) &= |GA|^2 + |GB|^2 + |GC|^2 = \left(\frac{2}{3}m_a\right)^2 + \left(\frac{2}{3}m_b\right)^2 + \left(\frac{2}{3}m_c\right)^2 \\ &= \frac{4}{9}(m_a^2 + m_b^2 + m_c^2) = \frac{1}{3}(a^2 + b^2 + c^2),\end{aligned}$$

where m_a, m_b, m_c represent the lengths of medians corresponding to the triangle sides of lengths a, b and c. The last identity had used Exercise 2.21.5.

(*iii*) If O is the orthocenter, H, of the acute-angled triangle ΔABC, then

$$\begin{aligned}\mathcal{I}(H; A, B, C) &= |HA|^2+|HB|^2+|HC|^2 = 4R^2\left(\cos^2 \widehat{A} + \cos^2 \widehat{B} + \cos^2 \widehat{C}\right) \\ &= 4R^2\left(1 - 2\cos\widehat{A}\cos\widehat{B}\cos\widehat{C}\right) = 4R^2\left(1 - \frac{r_0}{R}\right),\end{aligned}$$

where we used the formula for the inradius of the orthic triangle, $r_0 = 2R\cos\widehat{A}\cos\widehat{B}\cos\widehat{C}$ proved on page 108. We also used Proposition 2.13.6 and the trigonometric formula (2.1.2) from page 55.

(*iv*) If I denotes the incenter of triangle ΔABC, we have

$$\begin{aligned}\mathcal{I}(I; A, B, C) &= |AI|^2 + |BI|^2 + |CI|^2 = \frac{r^2}{\sin^2 \frac{\widehat{A}}{2}} + \frac{r^2}{\sin^2 \frac{\widehat{B}}{2}} + \frac{r^2}{\sin^2 \frac{\widehat{C}}{2}} \\ &= r^2 \sum_{circ.perm} \frac{bc}{(p-b)(p-c)} \\ &= r^2 \frac{(p-a)bc + (p-b)ac + (p-c)ab}{(p-a)(p-b)(p-c)} \\ &= r^2 p \frac{p(bc + ca + ab) - 3abc}{\mathcal{A}^2} \\ &= bc + ca + ab - \frac{3abc}{p} \\ &= p^2 + r^2 + 4Rr - \frac{12R\mathcal{A}}{p} = p^2 + r^2 + 4Rr - 12Rrp \\ &= p^2 + r^2 - 8Rr.\end{aligned}$$

We used in the computation the area formulas

$$\mathcal{A} = pr = \frac{abc}{4R} = \sqrt{p(p-a)(p-b)(p-c)}$$

as well as the formula $bc + ca + ab = p^2 + r^2 + 4Rr$, see page 117.

A problem of physical significance is the following: *With respect to which point M in the plane is the moment of inertia of triangle ΔABC minimum?*

Proposition 12.3.1 *Let M be a point in the plane of triangle ΔABC with centroid G. Then we have*

$$\mathcal{I}(M; A, B, C) = 3|MG|^2 + \mathcal{I}(G; A, B, C). \tag{12.3.6}$$

Proof: By Leibnitz' theorem, see page 407, we have

$$|MA|^2 + |MB|^2 + |MC|^2 = 3|MG|^2 + \frac{1}{3}(|AB|^2 + |BC|^2 + |AC|^2).$$

For $M = G$ this becomes

$$|GA|^2 + |GB|^2 + |GC|^2 = \frac{1}{3}(|AB|^2 + |BC|^2 + |AC|^2).$$

Then the last two relations can be written as

$$|MA|^2 + |MB|^2 + |MC|^2 = 3|MG|^2 + |GA|^2 + |GB|^2 + |GC|^2,$$

which is equivalent to (12.3.6). ■

Consequently,

$$\mathcal{I}(M; A, B, C) \geq \mathcal{I}(G; A, B, C), \tag{12.3.7}$$

so there is a unique point M in the plane for which the moment of inertia of triangle ΔABC is minimum, and this point is the centroid, $M = G$.

It is worth noting that relation (12.3.7) is a source of inequalities among the elements of a triangle. For instance, choosing $M = O$, the circumcenter, the inequality (12.3.7) becomes $9R^2 \geq a^2 + b^2 + c^2$.

12.4 Moment of inertia with respect to a line

Let $\mathcal{S} = \{A_1, \ldots, A_n\}$ be system of n points in the plane. The moment of inertia of the system $\mathcal{S}$ with respect to a line ℓ in the plane is given by

$$\mathcal{I}(\ell; \mathcal{S}) = \sum_{j=1}^{n} \text{dist}^2(A_j, \ell),$$

where $\text{dist}(A_j, \ell)$ denotes the distance from the point A_j to the line ℓ.

We shall deal in this section only with systems of 3 points, but similar relations can be proved in the general case.

The next result shows that the centroid is preserved by orthogonal projections on a line.

Lemma 12.4.1 *Let A_1, B_1, C_1, G_1 be the orthogonal projections of the vertices of triangle ΔABC and its centroid, G, onto a line ℓ in the plane, see Figure 12.4. Then*

$$\overrightarrow{G_1A_1} + \overrightarrow{G_1B_1} + \overrightarrow{G_1C_1} = 0, \tag{12.4.8}$$

namely, G_1 is the centroid of the system $\{A_1, B_1, C_1\}$.

Proof: Let B' be the midpoint of AC. Then $\overrightarrow{BG} = 2\overrightarrow{GB'}$ and by projection, $\overrightarrow{B_1G_1} = 2\overrightarrow{G_1B_2}$, where B_2 is the projection of B' on ℓ. Since B' is the midpoint of AC, then also B_2 is the midpoint of A_1C_1 (note that $B'B_2$ is a midpoint connector in the trapezoid AA_1C_1C), so we have $\overrightarrow{G_1B_2} = \frac{1}{2}(\overrightarrow{G_1A_1} + \overrightarrow{G_1C_1})$. Substituting in the previous expression yields

$$\overrightarrow{B_1G_1} = \overrightarrow{G_1A_1} + \overrightarrow{G_1C_1}.$$

Similarly, we obtain two more relations

$$\overrightarrow{A_1G_1} = \overrightarrow{G_1B_1} + \overrightarrow{G_1C_1}.$$

$$\overrightarrow{C_1G_1} = \overrightarrow{G_1A_1} + \overrightarrow{G_1B_1}.$$

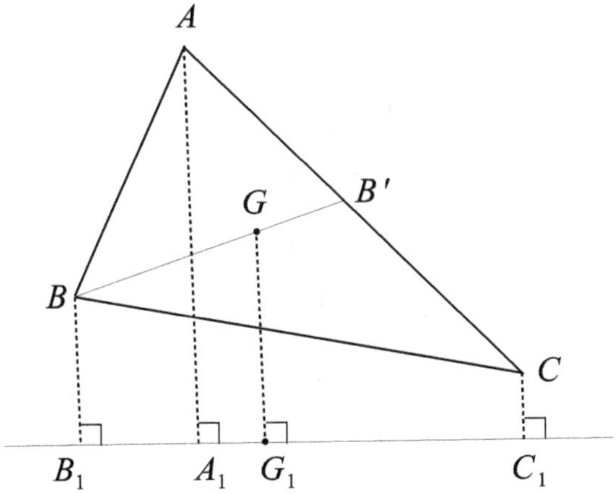

Figure 12.4: G_1 is the centroid of $\{A_1, B_1, C_1\}$.

We add the last three relations to obtain

$$\overrightarrow{A_1G_1} + \overrightarrow{B_1G_1} + \overrightarrow{C_1G_1} = 2(\overrightarrow{G_1A_1} + \overrightarrow{G_1B_1} + \overrightarrow{G_1C_1}) \Longleftrightarrow$$

$$3(\overrightarrow{A_1G_1} + \overrightarrow{B_1G_1} + \overrightarrow{C_1G_1}) = 0,$$

which implies identity (12.4.8). ■

Proposition 12.4.2 *Let A_1, B_1, C_1, G_1 be the orthogonal projections of the vertices of triangle ΔABC and its centroid, G, onto a line ℓ in the plane. Then*

$$\overrightarrow{AA_1} + \overrightarrow{BB_1} + \overrightarrow{CC_1} = 3\overrightarrow{GG_1}. \qquad (12.4.9)$$

Proof: By vector addition we obtain

$$\begin{aligned}\overrightarrow{AA_1} &= \overrightarrow{AG} + \overrightarrow{GG_1} + \overrightarrow{G_1A_1}\\ \overrightarrow{BB_1} &= \overrightarrow{BG} + \overrightarrow{GG_1} + \overrightarrow{G_1B_1}\\ \overrightarrow{CC_1} &= \overrightarrow{CG} + \overrightarrow{GG_1} + \overrightarrow{G_1C_1},\end{aligned}$$

see Figure 12.5. Adding, yields

$$\begin{aligned}\overrightarrow{AA_1}+\overrightarrow{BB_1}+\overrightarrow{CC_1} &= \underbrace{(\overrightarrow{AG} + \overrightarrow{BG} + \overrightarrow{CG})}_{=0} + 3\overrightarrow{GG_1} + \underbrace{(\overrightarrow{G_1A_1} + \overrightarrow{G_1B_1} + \overrightarrow{G_1C_1})}_{=0}\\ &= 3\overrightarrow{GG_1},\end{aligned}$$

where the first parenthesis vanishes due to the property of the centroid and the second parenthesis vanishes by Lemma 12.4.1. ■

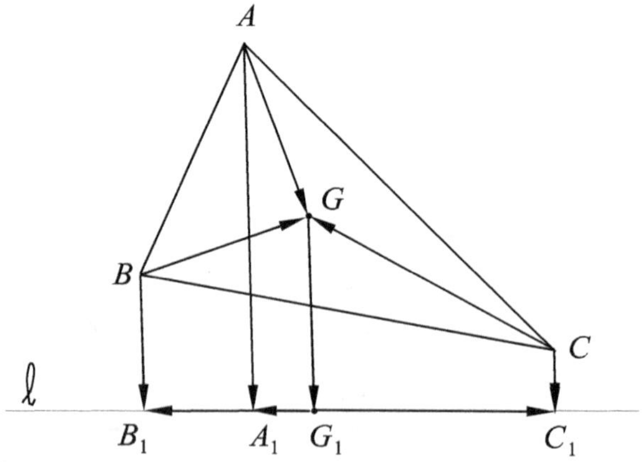

Figure 12.5: For Proposition 12.4.2.

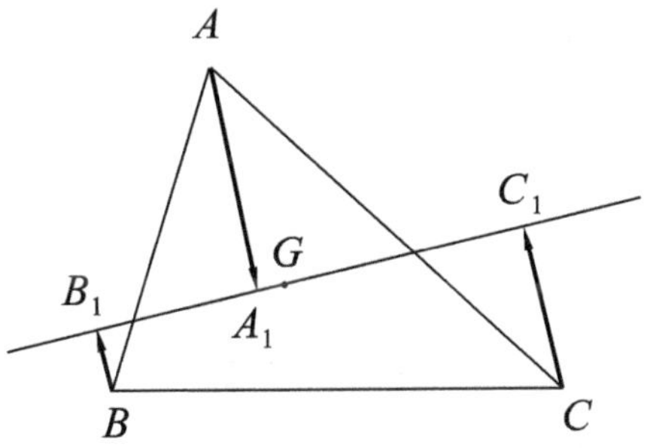

Figure 12.6: For Corollary 12.4.3.

Corollary 12.4.3 *A line ℓ passes through the centroid of triangle ΔABC if and only if*

$$\overrightarrow{AA_1} + \overrightarrow{BB_1} + \overrightarrow{CC_1} = 0. \tag{12.4.10}$$

Proof: If follows from Proposition 12.4.2 by choosing $G = G_1$. See Figure 12.6. ■

Remark 12.4.4 Assume ℓ does not pass through any of the vertices. Since G is inside of the triangle, the line ℓ separates the vertices into one vertex on one side and two on the other. Since the vectors are parallel, two of them have the same sense, and the other has the opposite sense. Then equation (12.4.10) is equivalent to one of the following segment identities:

$$|AA_1| = |BB_1| + |CC_1|, \quad |BB_1| = |AA_1| + |CC_1|, \quad |CC_1| = |BB_1| + |AA_1|.$$

If the line ℓ passes through a vertex, let's say the vertex A, then the aforementioned relation becomes $|BB_1| = |CC_1|$.

Exercise 12.4.5 *A line passing through the centroid of ΔABC intersects sides $[AB]$ and $[AC]$ at points N and M. Prove that*

$$\frac{|BN|}{|NA|} + \frac{|CM|}{|MA|} = 1.$$

Moment of inertia of ΔABC

The moment of inertia of a triangle ΔABC with respect to a line ℓ in the plane is given by the sum of squares of distances from vertices to the line ℓ

$$\mathcal{I}(\ell; A, B, C) = \text{dist}^2(A, \ell) + \text{dist}^2(B, \ell) + \text{dist}^2(C, \ell).$$

This represents a physical measure that assesses the capacity of rotation of the triangle with respect to the axis ℓ. The triangle is considered as three unit masses concentrated at the vertices, neglecting the weights of the triangle sides.

We shall compute a few moments of inertia with respect to some particular lines.

(*i*) If $\ell = BC$ then $\mathcal{I}(BC; A, B, C) = \text{dist}^2(A, BC) = h_a^2$, where h_a the length of the altitude from A.

(*ii*) If ℓ is the altitude from vertex A of triangle ΔABC then an application on the law of cosines yields

$$\begin{aligned}\mathcal{I}(\ell; A, B, C) &= |BA_1|^2 + |A_1C|^2 = c^2 \cos^2 \widehat{B} + b^2 \cos^2 \widehat{C}\\ &= \frac{1}{4a^2}[(a^2 + c^2 - b^2)^2 + (a^2 + b^2 - c^2)^2]\\ &= \frac{1}{4a^2}[2a^4 + 2(b^2 - c^2)^2] = \frac{1}{2}a^2 + \frac{1}{2}\frac{(b^2 - c^2)^2}{a^2}.\end{aligned}$$

We note that in the case of an isosceles triangle, with $a = b$, the momentum is minimum, and is equal to $\frac{1}{2}a^2$.

(*iii*) We assume now that ℓ is the angle bisector from vertex A of triangle ΔABC. Then

$$\begin{aligned}\mathcal{I}(\ell; A, B, C) &= |BB_1|^2 + |CC_1|^2 = c^2 \sin^2 \frac{\widehat{A}}{2} + b^2 \sin^2 \frac{\widehat{A}}{2}\\ &= (b^2 + c^2) \sin^2 \frac{\widehat{A}}{2} = \frac{b^2 + c^2}{bc}(p - b)(p - c)\\ &= \frac{b^2 + c^2}{bc}\frac{(-b + a + c)(-c + a + b)}{4} = \frac{b^2 + c^2}{4bc}(a^2 - (b - c)^2).\end{aligned}$$

(*iv*) Let ℓ be the median from the vertex A. We denote by M the midpoint of segment BC and let B_1 and C_1 be the orthogonal projections of B and C on AM, see Figure 12.7. By (ASA) congruence case we have $\Delta MBB_1 \equiv \Delta MCC_1$ (since $|MB| = |MC|$, $\widehat{BMB_1} = \widehat{CMC_1}$, and $\widehat{B_1BM} = \widehat{C_1CM}$ as alternate interior angles for $BB_1 \| CC_1$). Therefore, $|BB_1| = |CC_1|$. We shall compute this length by expressing the area of triangle ΔABM as $\mathcal{A}_{ABM} = \frac{1}{2}|BB_1|\,|AM| = \frac{1}{2}|BM|h_a = \frac{1}{2}\mathcal{A}_{ABC}$. Then $|BB_1| = \dfrac{\mathcal{A}_{ABC}}{m_a}$, where $m_a = |AM|$ is the length of the median from A. Now we can compute the moment of inertia as follows

$$\mathcal{I}(\ell; A, B, C) = |BB_1|^2 + |CC_1|^2 = 2|BB_1|^2 = \frac{2\mathcal{A}_{ABC}^2}{m_a^2}.$$

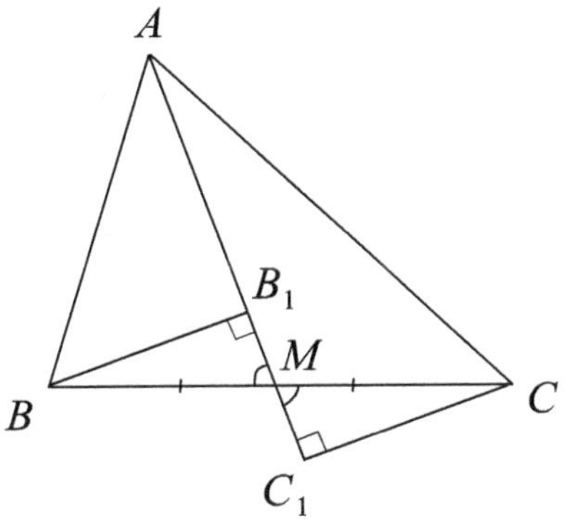

Figure 12.7: For the case (*iv*).

A problem of physical significance is the following: *With respect to which line ℓ in the plane is the moment of inertia of triangle ΔABC minimum?*

A partial answer to this question is provided next. This is an analog of the *parallel axis theorem* from mechanics:

Theorem 12.4.6 *Let (δ) be a given direction in the plane of triangle ΔABC. The line ℓ of direction (δ) having the smallest inertia moment $I(\ell; A, B, C)$ is the one passing through the centroid of ΔABC.*

Proof: Let ℓ be a line of direction (δ) and A_1, B_1 and C_1 be the projections of A, B and C on the line ℓ. Let ℓ' be another line of direction (δ) passing through the centroid G of ΔABC, see Figure 12.8. Let A_2, B_2 and C_2 be the projections of A, B and C on the line ℓ'. We shall denote by x the distance between the parallel lines ℓ and ℓ'. Consequently, $x = |A_1A_2| = |B_1B_2| = |C_1C_2|$. We also denote by $\overrightarrow{v} = \overrightarrow{A_2A_1} = \overrightarrow{B_2B_1} = \overrightarrow{C_2C_1}$. Using inner product we evaluate

$$\begin{aligned}|AA_1|^2 &= \overrightarrow{AA_1} \cdot \overrightarrow{AA_1} = (\overrightarrow{AA_2} + \overrightarrow{A_2A_1}) \cdot (\overrightarrow{AA_2} + \overrightarrow{A_2A_1}) \\ &= |AA_2|^2 + |A_2A_1|^2 + 2\overrightarrow{AA_2} \cdot \overrightarrow{A_2A_1} \\ &= |AA_2|^2 + x^2 + 2\overrightarrow{AA_2} \cdot \overrightarrow{v}.\end{aligned}$$

Using similar evaluations for $|BB_1|^2$ and $|CC_1|^2$ by addition we get

$$\begin{aligned}|AA_1|^2 + |BB_1|^2 + |CC_1|^2 = {} & |AA_2|^2 + |BB_2|^2 + |CC_2|^2 + 3x^2 \\ & + 2\overrightarrow{v} \cdot (\overrightarrow{AA_2} + \overrightarrow{BB_2} + \overrightarrow{CC_2}).\end{aligned}$$

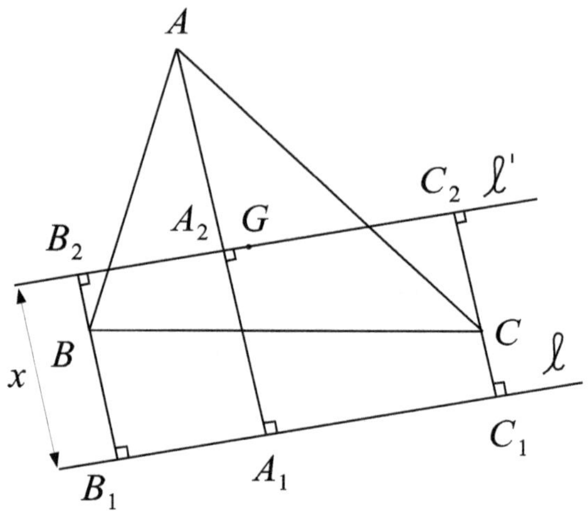

Figure 12.8: For Theorem 12.4.6.

By Corollary 12.4.3 the last sum vanishes. Therefore

$$|AA_1|^2 + |BB_1|^2 + |CC_1|^2 = |AA_2|^2 + |BB_2|^2 + |CC_2|^2 + 3x^2,$$

which can be written in terms of momenta as

$$\mathcal{I}(\ell; A, B, C) = \mathcal{I}(\ell'; A, B, C) + 3x^2.$$

Consequently, $\mathcal{I}(\ell; A, B, C)$ is minimum for $x = 0$, which corresponds to the case when ℓ passes through G. ■

Remark 12.4.7 The previous result reduces the problem of looking for an axis with the minimum moment of inertia to the case when the axis passes through the centroid. These directions can be parameterized by an angle on the unit circle, $\theta \in \mathbf{S}^1$. Among all lines, there is (at least) one of a distinguished direction (δ) for which $\mathcal{I}(\ell; A, B, C)$ is minimum. The existence of this direction follows from the fact that a continuous function on the unit circle $\mathbf{S}^1$ always reaches its global minimum.

To find the exact direction corresponding to the moment of inertia minimum is an intricate computational problem, which will be treated next. We note that this direction does not correspond in general to any a priori studied line in a triangle.

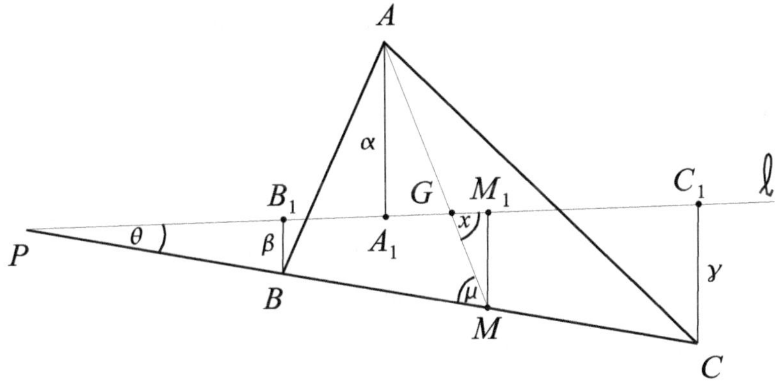

Figure 12.9: The axis ℓ and the base BC form the angle θ.

The minimum moment of inertia line

We consider a line ℓ passing through the centroid G of triangle ΔABC. We also assume that the side BC is the longest. Heuristic reasons indicate that the minimum inertia axis should align (more or less) to the direction of BC, so we are in the situation presented in Figure 12.9. We shall denote by θ the acute angle made by the lines ℓ and BC. Let A_1, B_1 and C_1 be the projections of A, B and C on the line ℓ and denote $\alpha = |AA_1|$, $\beta = |BB_1|$ and $\gamma = |CC_1|$. By Remark 12.4.4 from page 531 we have $\alpha = \beta + \gamma$. Therefore, the momentum can be written only in terms of variables β and γ as

$$\mathcal{I}(\ell; A, B, C) = \alpha^2 + \beta^2 + \gamma^2 = 2(\beta^2 + \gamma^2 + \beta\gamma). \tag{12.4.11}$$

The goal is to write this expression only as a function of the angle θ.

We proceed by computing $|B_1C_1|$ in two ways. By projection we have

$$|B_1C_1| = |BC| \cos\theta = a\cos\theta.$$

Also, applying the Pythagorean relation in the trapezoid BB_1C_1C, we obtain

$$|B_1C_1|^2 = |BC|^2 - (|CC_1| - |BB_1|)^2 = a^2 - (\gamma - \beta)^2.$$

Equating the last two relations and simplify yields

$$\beta^2 + \gamma^2 = a^2 \sin^2\theta + 2\beta\gamma. \tag{12.4.12}$$

Substituting in (12.4.11) we obtain

$$\mathcal{I}(\ell; A, B, C) = 2(a^2 \sin^2\theta + 3\beta\gamma). \tag{12.4.13}$$

To be able to minimize the momentum with respect to θ, it suffices to write the product $\beta\gamma$ in terms of θ. To this end, let M be the midpoint of AB and M_1 be the projection of M on ℓ. Since MM_1 is a midpoint connector in the trapezoid BB_1C_1C, we have

$$|MM_1| = \frac{1}{2}(\beta + \gamma).$$

Let $\mu = \widehat{AMB}$ be the angle made by the median AM and the base BC. We note that $\sin\mu = \dfrac{h_a}{m_a}$, where h_a and m_a are the lengths of the altitude and median from A. We also denote by $x = \widehat{M_1GM}$ the angle made by ℓ and the medial AM. By the exterior angle identity in triangle ΔGPM, we have $x = \theta + \mu$. In the right triangle ΔMM_1G we have

$$\sin x = \frac{|MM_1|}{|GM|} = \frac{3}{2}\frac{\beta + \gamma}{m_a},$$

so $\beta + \gamma = \frac{2}{3}m_a \sin x$. Taking the square yields

$$\frac{4}{9}m_a^2 \sin^2 x = \beta^2 + \gamma^2 + 2\beta\gamma.$$

Subtracting the equation (12.4.12)

$$a^2 \sin^2\theta = \beta^2 + \gamma^2 - 2\beta\gamma$$

we obtain the product of β and γ in terms of θ as

$$\beta\gamma = \frac{1}{9}m_a^2 \sin^2 x - \frac{1}{4}a^2 \sin^2\theta.$$

We substitute this relation in (12.4.13) to obtain

$$\begin{aligned}
\mathcal{I}(\ell; A, B, C) &= 2(a^2 \sin^2\theta + 3\beta\gamma) \\
&= 2\left(a^2 \sin^2\theta + \frac{1}{3}m_a^2 \sin^2 x - \frac{3}{4}a^2 \sin^2\theta\right) \\
&= 2\left(\frac{a^2}{4}\sin^2\theta + \frac{1}{3}m_a^2 \sin^2 x\right) \\
&= 2\left(\frac{a^2}{4}\sin^2\theta + \frac{1}{3}m_a^2 \sin^2(\theta + \mu)\right),
\end{aligned}$$

which is an expression depending only on the variable θ. In order to find the minimum of this function we associate the function

$$f(\theta) = \frac{a^2}{4}\sin^2\theta + \frac{1}{3}m_a^2 \sin^2(\theta + \mu)$$

and differentiate

$$\begin{aligned} f'(\theta) &= \frac{a^2}{4} 2\sin\theta\cos\theta + \frac{1}{3} m_a^2 2\sin(\theta+\mu)\cos(\theta+\mu) \\ &= \frac{a^2}{4}\sin(2\theta) + \frac{1}{3} m_a^2 \sin(2\theta+2\mu). \end{aligned}$$

Solving the trigonometric equation $f'(\theta) = 0$, we obtain

$$\begin{aligned} 3a^2\sin(2\theta) &= -4m_a^2\sin(2\theta+2\mu) \iff \\ 3a^2\sin(2\theta) &= -4m_a^2\sin(2\theta)\cos(2\mu) - 4m_a^2\cos(2\mu)\sin(2\mu) \iff \\ \tan(2\mu) &= \frac{-4m_a^2\sin(2\mu)}{3a^2 + 4m_a^2\cos(2\mu)}. \end{aligned}$$

Both the nominator and the denominator of the right side have geometric interpretations as follows. Using $h_a = m_a \sin\mu$, we have

$$\begin{aligned} m_a^2\cos(2\mu) &= m_a^2(1 - 2\sin^2\mu) = m_a^2 - 2h_a^2 \\ m_a^2\sin(2\mu) &= 2m_a\sin\mu\, m_a\cos\mu = 2h_a m_a\cos\mu. \end{aligned}$$

Let A' denote the leg of the altitude from the vertex A, see Figure 12.10. Then in the right triangle $\Delta AA'M$ we have $|A'M| = m_a\cos\mu$. Therefore, $2h_a m_a\cos\mu = 2h_a|A'M| = 4\mathcal{A}_{AA'M}$. Substituting in the previous equation

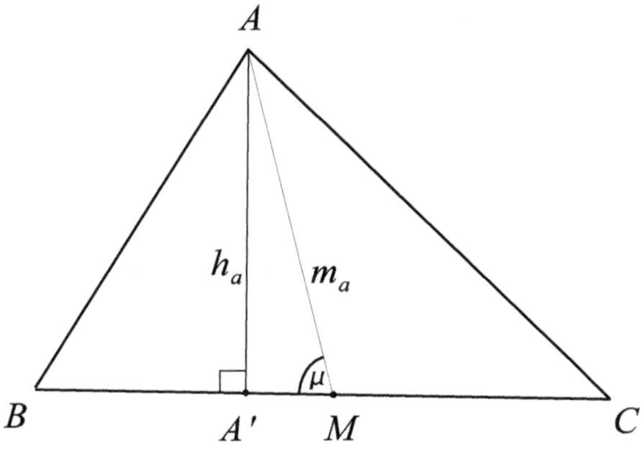

Figure 12.10: The angle θ depends on the area of triangle $\Delta AA'M$.

and solving for θ we obtain

$$\begin{aligned}\theta &= \frac{1}{2}\tan^{-1}\left(\frac{-16\mathcal{A}_{AA'M}}{3a^2+4m_a^2-8h_a^2}\right) = \frac{1}{2}\tan^{-1}\left(\frac{-16\mathcal{A}_{AA'M}}{2(a^2+b^2+c^2)-8h_a^2}\right) \\ &= \frac{1}{2}\tan^{-1}\left(\frac{8\mathcal{A}_{AA'M}}{4h_a^2-(a^2+b^2+c^2)}\right). \qquad (12.4.14)\end{aligned}$$

We note that in the denominator we used the following computation involving the median theorem

$$3a^2+4m_a^2 = 3a^2+4\left(\frac{b^2+b^2}{2}-\frac{a^2}{4}\right) = 2(a^2+b^2+c^2).$$

It is worth noting that in the case of an isosceles triangle (with $|BC| > |AB| = |AC|$) the median is perpendicular to the base, so $\mu = 90°$. It follows that $\theta = 0$, and hence the line ℓ is parallel to the base. The converse also holds true, namely, if the minimum moment of inertia line ℓ is parallel to the base BC, then the triangle is isosceles, with $|AB| = |AC|$.

In the case of a non-isosceles triangle the altitude and median from the vertex A and the base BC form the triangle $\Delta AA'M$. Its area plays an important role in determining the angle θ. Since $\tan^{-1}(x)$ is an increasing function, larger values of the area $\mathcal{A}_{AA'M}$ correspond to larger values of the angle θ. We may regard in a certain sense, that the area $\mathcal{A}_{AA'M}$ assess how non-isosceles the triangle ΔABC is.

Solved Exercise 12.4.8 *Let AB be a fixed line segment and ℓ a line in the plane of a given direction. Show that the moment of inertia $\mathcal{I}(\ell; A, B)$ is minimum if and only if the line ℓ passes through the midpoint of AB.*

Proof: Let $\{M\} = \ell \cap AB$ and denote $d = |AB|$, $x = |AM|$, so $|MB| = d - x$, see Figure 12.11. Let θ be the angle between the lines ℓ and AB. If

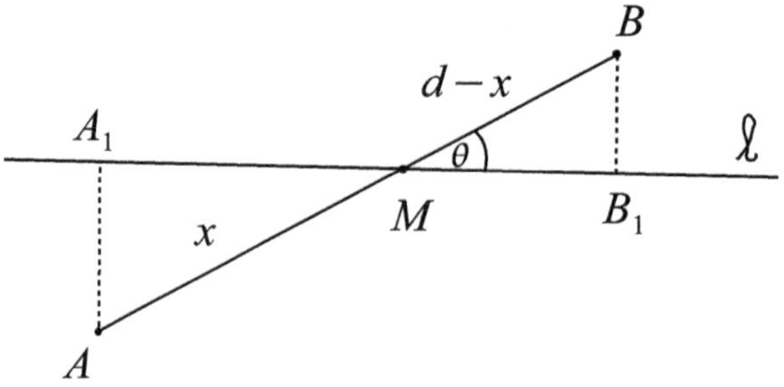

Figure 12.11: For Exercise 12.4.8.

A_1 and B_1 are the projections of A and B on ℓ, we have $|AA_1| = x\sin\theta$ and $|BB_1| = (d-x)\sin x$. Then

$$|AA_1|^2 + |BB_1|^2 = \sin^2\theta(x^2 + (d-x)^2) = \sin^2\theta\left(x^2 - dx + \frac{1}{2}d^2\right).$$

The quadratic function in x reaches its minimum for $x = \frac{d}{2}$, which means that M is the midpoint of AB. We note that if $\theta = 0$ then the moment is zero, regardless of the position of the point M. ■

Chapter 13

Applications to Solid Geometry

Force without wisdom falls of its own weight.
Horace (65 BC–8 BC), Odes

The main object of study in this chapter is the tetrahedron. This can be thought of as a generalization to 3 dimensions of a triangle, or a quadrilateral from plane geometry.
We start the presentation with some preliminary properties of lines and planes.

13.1 Lines perpendicular to planes

Definition 13.1.1 *A line d is perpendicular to a plane Π if d is perpendicular to any line contained in Π.*

A practical way to show that a line is perpendicular to a plane is as follows.

Theorem 13.1.2 *A line d is perpendicular to a plane Π if and only if there are two non-parallel lines a, $b \subset \Pi$ such that $d \perp a$ and $d \perp b$.*

Proof: " $\Longleftarrow$ " We shall refer to Figure 13.1(a). Let $\overrightarrow{u}$ and $\overrightarrow{v}$ be vectors with directions a and b. Since a and b are not parallel, the vectors $\overrightarrow{u}$ and $\overrightarrow{v}$ are linear independent and hence they form a basis in the plane Π. Then any vector $\overrightarrow{w}$ in the plane Π can be written as a linear combination of the basis vectors: $\overrightarrow{w} = \alpha\overrightarrow{u} + \beta\overrightarrow{v}$, with $\alpha, \beta \in \mathbb{R}$. Let $\overrightarrow{t}$ be a vector with the support line d. Since $d \perp a$ and $d \perp b$ it follows that $\overrightarrow{t} \cdot \overrightarrow{u} = 0$ and

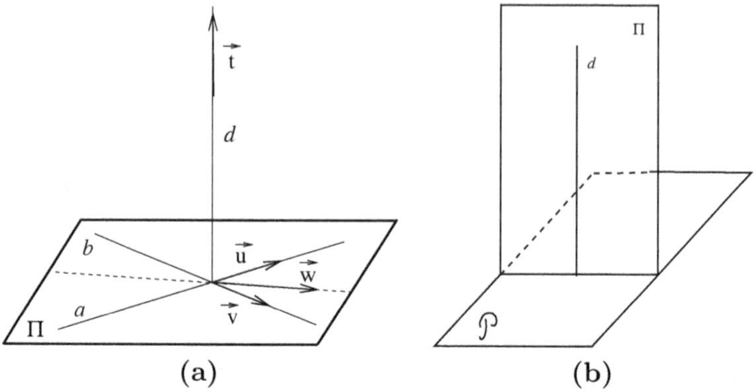

Figure 13.1: (a) A line d perpendicular on two lines, a and b, in the plane Π; (b) Two perpendicular planes, Π and $\mathcal{P}$.

$\overrightarrow{t} \cdot \overrightarrow{v} = 0$ and hence $\overrightarrow{t} \cdot \overrightarrow{w} = \alpha \overrightarrow{t} \cdot \overrightarrow{u} + \beta \overrightarrow{t} \cdot \overrightarrow{v} = 0$, *i.e.*, d and the line with the direction $\overrightarrow{w}$ are perpendicular.
The reverse implication " $\Longrightarrow$ " is obviously satisfied. ■

Exercise 13.1.3 *Let A be a point, which does not belong to the plane Π. Show that there is a unique perpendicular line through A to Π.*

Definition 13.1.4 *A plane Π is perpendicular to a plane $\mathcal{P}$ if there is a line $d \subset \Pi$ with d perpendicular to $\mathcal{P}$. We shall write $\Pi \perp \mathcal{P}$. See Figure 13.1(b).*

Exercise 13.1.5 *Prove that the perpendicularity relation " $\perp$ " between planes is symmetric, i.e., if $\Pi \perp \mathcal{P}$ then $\mathcal{P} \perp \Pi$.*

Exercise 13.1.6 *Prove that the perpendicularity relation " $\perp$ " between planes is not transitive. However, if Π_1, Π_2 and Π_3 are planes such that $\Pi_1 \perp \Pi_2$ and $\Pi_2 \perp \Pi_3$, then $\Pi_1 \cap \Pi_3 = \emptyset$ or $\Pi_1 = \Pi_3$. See Figure 13.2(a).*

Exercise 13.1.7 *Let $\Pi \perp \mathcal{P}$ be two perpendicular planes and let $A \in \Pi$. Show that the perpendicular line from A to $\mathcal{P}$ is contained in the plane Π. See Figure 13.2(b).*

13.2 The common perpendicular of two non-coplanar lines

Let d and m be two non-coplanar lines in space. Let $\overrightarrow{v}$ and $\overrightarrow{w}$ be vectors collinear with the directions of d and m, respectively. If $\{e_1, e_2, e_3\}$ is the

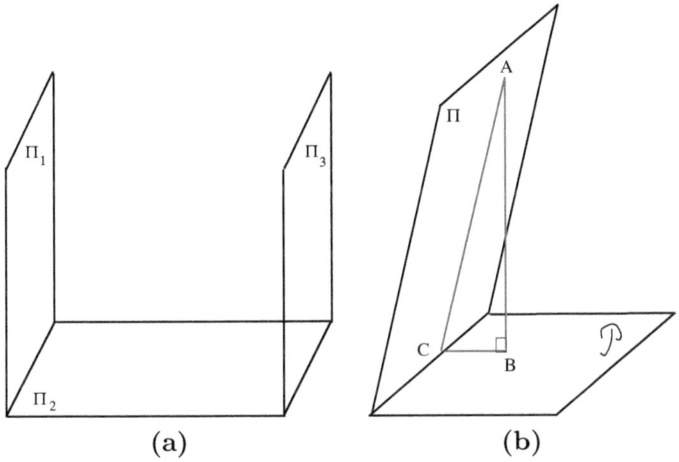

Figure 13.2: (a) Two planes perpendicular on the third; (b) For Exercise 13.1.7.

canonical base of $\mathbb{R}^3$, we can write

$$\overrightarrow{v} = v^1 e_1 + v^2 e_2 + v^3 e_3$$
$$\overrightarrow{w} = w^1 e_1 + w^2 e_2 + w^3 e_3.$$

We are interested in finding a direction perpendicular on both directions of d and m. Let $\overrightarrow{u} = u^1 e_1 + u^2 e_2 + u^3 e_3$ be a vector collinear with the desired perpendicular direction. Then $\overrightarrow{u}$ is perpendicular on both vectors $\overrightarrow{v}$ and $\overrightarrow{w}$. Then $\overrightarrow{u} \cdot \overrightarrow{v} = 0$ and $\overrightarrow{u} \cdot \overrightarrow{w} = 0$, and using Theorem 7.4.4 yields

$$\begin{aligned} u^1 v^1 + u^2 v^2 + u^3 v^3 &= 0 \\ u^1 w^1 + u^2 w^2 + u^3 w^3 &= 0, \end{aligned} \tag{13.2.1}$$

which is a linear system with three unknowns u^1, u^2, u^3 and two equations. Let $u^3 = -\lambda$. Then the system can be written in the matrix form

$$\begin{pmatrix} v^1 & v^2 \\ w^1 & w^2 \end{pmatrix} \begin{pmatrix} u^1 \\ u^2 \end{pmatrix} = \lambda \begin{pmatrix} v^3 \\ w^3 \end{pmatrix}.$$

Since the vectors $\overrightarrow{v}, \overrightarrow{w}$ are linear independent (why?), we can write the solution of the system (13.2.1) as

$$\begin{pmatrix} u^1 \\ u^2 \end{pmatrix} = \lambda \begin{pmatrix} v^1 & v^2 \\ w^1 & w^2 \end{pmatrix}^{-1} \begin{pmatrix} v^3 \\ w^3 \end{pmatrix}, \qquad u^3 = -\lambda.$$

Hence, the direction of the vector $\vec{u}$ is perpendicular to both lines d and m. We have shown that this direction exists and it is unique.[1] However, we still need to show that there is a line with this direction, which intersects the given lines d and m.

Definition 13.2.1 *Let d and m be two non-coplanar lines. A line perpendicular on both lines d and m, which intersects them, is called the common perpendicular of the lines d and m.*

Theorem 13.2.2 *The common perpendicular of two non-coplanar lines exists and is unique.*

Proof: The proof is based on a construction. Let d and m be two non-coplanar lines. See Figure 13.3(a).
Step 1: Let $A \in d$ be a point and consider the line m' parallel to m passing through A. Let Π be the plane defined by d and m'. See Figure 13.3(b).

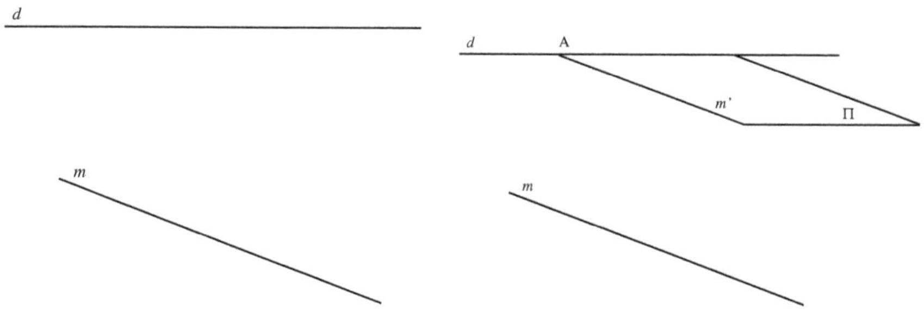

(a) (b)

Figure 13.3: For the common perpendicular of two lines.

Step 2: Let $\mathcal{P}$ be the plane perpendicular to the plane Π, which contains the line d. Let $\mathcal{P} \cap m = \{B\}$. See Figure 13.4(a).
Step 3: The perpendicular line through B to d is the common perpendicular of the lines d and m. See Figure 13.4(b). This proves the existence. The uniqueness comes from the previous discussion. The details are left as an exercise to the reader. ■

13.3 The three perpendicular lines theorem

The first application is called the *three perpendicular lines theorem*, see Figure 13.5.

[1]The vector $\vec{u}$ is not unique, since it depends on the parameter λ; however, its direction is unique.

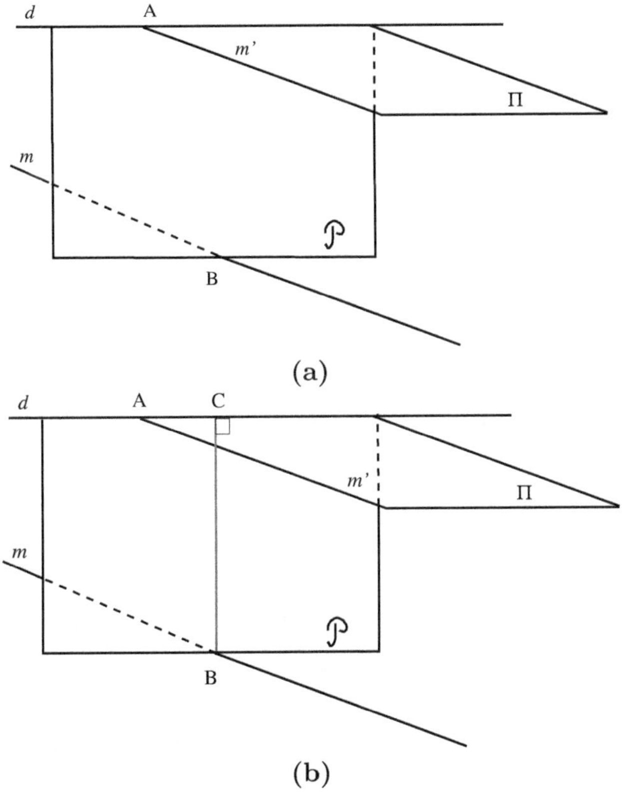

Figure 13.4: Construction used in the proof of Theorem 13.2.2: (a) For Step 2; (b) For Step 3.

Theorem 13.3.1 *Let $\mathcal{P}$ be a plane and A, B, $C \in \mathcal{P}$ be three non-collinear points. Let $M \notin \mathcal{P}$ such that $MA \perp \mathcal{P}$ and $AB \perp BC$. Then $MB \perp BC$.*

Proof: We shall use that $MB \perp BC$ if and only if $\overrightarrow{MB} \cdot \overrightarrow{BC} = 0$. We have

$$\begin{aligned}\overrightarrow{MB} \cdot \overrightarrow{BC} &= (\overrightarrow{MA} + \overrightarrow{AB}) \cdot \overrightarrow{BC} \\ &= \overrightarrow{MA} \cdot \overrightarrow{BC} + \overrightarrow{AB} \cdot \overrightarrow{BC} \\ &= 0,\end{aligned}$$

since $MA \perp BC$ (because $MA \perp \mathcal{P}$ and $BC \subset \mathcal{P}$) and $AB \perp BC$. ■

The next exercise states the converses of Theorem 13.3.1.

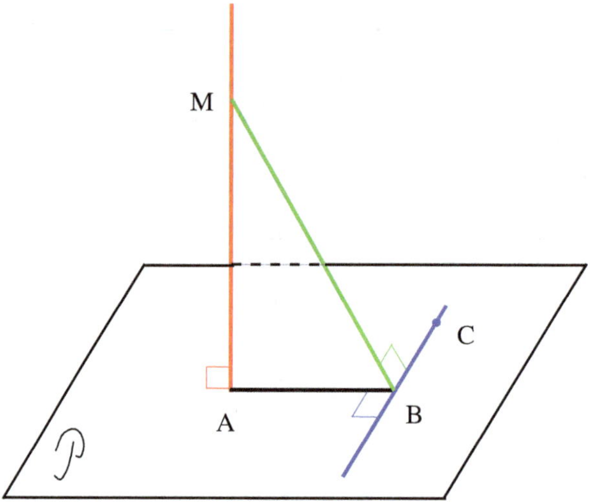

Figure 13.5: The three perpendicular lines theorem.

Exercise 13.3.2 *Let A, B, $C \in \mathcal{P}$ be three non-collinear points and let $M \notin \mathcal{P}$.*

(i) Show that if $MA \perp \mathcal{P}$ and $MB \perp BC$, then $AB \perp BC$.

(ii) Show that if $MA \perp AB$, $MB \perp BC$ and $AB \perp BC$, then $MA \perp \mathcal{P}$.

Solved Exercise 13.3.3 *Let a and b be two non-parallel lines in the plane $\mathcal{P}$. Let $\{O\} = a \cap b$ and let $P \in \mathcal{P}$, $P \neq O$. Consider the line d perpendicular to $\mathcal{P}$ through P. Let $M \in d$ be an arbitrary point. Denote by A, B the orthogonal projections of M to the lines a and b. Show that the quadrilateral $PAOB$ is cyclic (there is a circle passing through its vertices). See Figure 13.6.*

Proof: By the three perpendicular lines Theorem 13.3.1 we have $PA \perp AO$, $PB \perp BO$, and hence $\hat{A} + \hat{B} = 180$. Using a well-known theorem about cyclic quadrilaterals yields that $PAOB$ is cyclic. ■

13.4 Lines parallel to planes

Definition 13.4.1 *A line d is parallel to a plane Π if $d \cap \Pi = \emptyset$, i.e., the lines does not have any points in common with the plane. In this case we shall use the notation $d||\Pi$.*

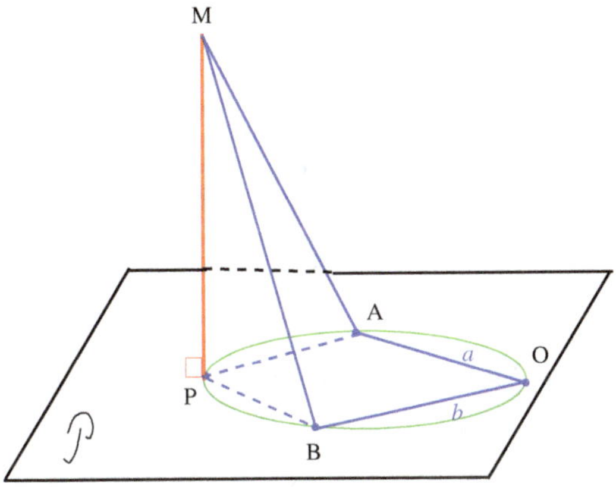

Figure 13.6: For Exercise 13.3.3.

This definition is usually difficult to apply to practical problems. The next result is much more useful.

Theorem 13.4.2 *A line d is parallel to a plane Π if there is a line m, which is included in Π, such that $d||m$.*

Proof: " $\Longrightarrow$ " Assume $d||\Pi$, so $d \cap \Pi = \emptyset$. Choose an arbitrary point P in Π and consider the plane $\mathcal{P} = (d, P)$. Since the planes $\mathcal{P}$ and Π have a point in common, P, then their intersection is a line, $\mathcal{P} \cap \Pi = m$, passing through P. Then the lines m and d, being contained in the plane $\mathcal{P}$, are coplanar. Since $m \cap d \subset \Pi \cap d = \emptyset$, then $d||m$.

" $\Longleftarrow$ " Consider a line m, with $m \subset \Pi$ and $m||d$. By contradiction, we assume that d is not parallel to the plane Π, so there is a point $A \in d \cap \Pi$. Let $\mathcal{P} = (d, m)$ be the plane defined by the lines d and m. Since the intersection between the (nonparallel) planes Π and $\mathcal{P}$ is a line, and we have $m \subset \Pi \cap \mathcal{P}$ and $A \in \Pi \cap \mathcal{P}$, it follows that $A \in m$. Therefore, $A \in d \cap m$, which contradicts the fact that d and m are parallel. ∎

Tetrahedron definition

Let A, B, C, D be four non-coplanar points. Then $ABCD$ is called a *tetrahedron* with vertices A, B, C, D. $[AB]$, $[BC]$, $[AC]$, $[AD]$, $[BD]$ and $[CD]$ are called *edges* and (ABC), (ACD), (ABD) and (BCD) are called *faces*, see Figure 13.7.
The next result deals with the opposite perpendicular sides of a tetrahedron.

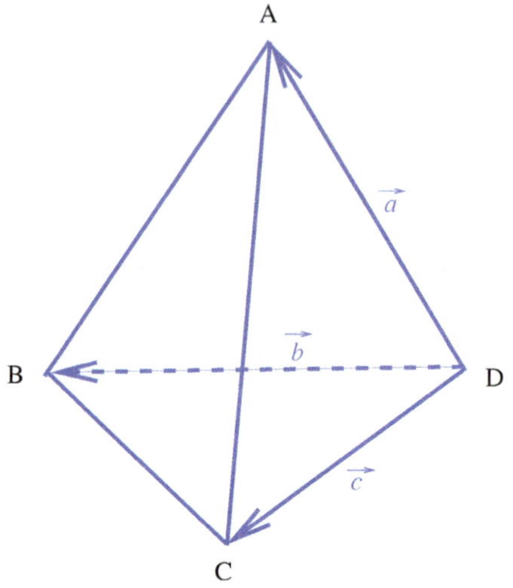

Figure 13.7: A tetrahedron $ABCD$.

Proposition 13.4.3 *Let $ABCD$ be a tetrahedron. Show that if $AB \perp CD$ and $AC \perp BD$ then $AD \perp BC$.*

Proof: Let $\vec{a} = \overrightarrow{DA}$, $\vec{b} = \overrightarrow{DB}$, $\vec{c} = \overrightarrow{DC}$. We have $AB \perp CD$ if and only if $\overrightarrow{AB} \cdot \overrightarrow{CD} = 0$, which can be written as

$$(\vec{b} - \vec{a}) \cdot \vec{c} = 0 \Longleftrightarrow \vec{b} \cdot \vec{c} = \vec{a} \cdot \vec{c}. \tag{13.4.2}$$

On the other hand, $AC \perp BD$ if and only if $\overrightarrow{AC} \cdot \overrightarrow{BD} = 0$, which becomes

$$(\vec{c} - \vec{a}) \cdot \vec{b} = 0 \Longleftrightarrow \vec{b} \cdot \vec{c} = \vec{a} \cdot \vec{b}. \tag{13.4.3}$$

Subtracting (13.4.2) and (13.4.3) yields

$$\begin{aligned} 0 &= \vec{b} \cdot \vec{c} - \vec{b} \cdot \vec{c} = \vec{a} \cdot \vec{b} - \vec{a} \cdot \vec{c} \\ &= \vec{a} \cdot (\vec{b} - \vec{c}) = \overrightarrow{DA} \cdot \overrightarrow{BC}. \end{aligned}$$

Hence $AD \perp BC$. ∎

13.5 Computing volumes

The main tool in computing volumes will be the mixed scalar product.

Lemma 13.5.1 *Let $ABCD$ be a tetrahedron. Then the absolute value of the mixed scalar products of the vectors starting at the same vertex is the same, i.e.,*

$$|[\overrightarrow{AB}, \overrightarrow{AC}, \overrightarrow{AD}]| = |[\overrightarrow{BA}, \overrightarrow{BC}, \overrightarrow{BD}]| = |[\overrightarrow{CA}, \overrightarrow{CB}, \overrightarrow{CD}]| = |[\overrightarrow{DA}, \overrightarrow{DB}, \overrightarrow{DC}]|.$$

Proof: Using the triangle rule and the additivity of the mixed scalar product, we have

$$\begin{aligned}
[\overrightarrow{BA}, \overrightarrow{BC}, \overrightarrow{BD}] &= [\overrightarrow{BA}, \overrightarrow{BA} + \overrightarrow{AC}, \overrightarrow{BD}] \\
&= \underbrace{[\overrightarrow{BA}, \overrightarrow{BA}, \overrightarrow{BD}]}_{=0} + [\overrightarrow{BA}, \overrightarrow{AC}, \overrightarrow{BD}] \\
&= [\overrightarrow{BA}, \overrightarrow{AC}, \overrightarrow{BA} + \overrightarrow{AD}] \\
&= \underbrace{[\overrightarrow{BA}, \overrightarrow{AC}, \overrightarrow{BA}]}_{=0} + [\overrightarrow{BA}, \overrightarrow{AC}, \overrightarrow{AD}] \\
&= -[\overrightarrow{AB}, \overrightarrow{AC}, \overrightarrow{AD}],
\end{aligned}$$

and hence we proved the first identity. The other identities have a similar proof. ∎

The geometric significance of Lemma 13.5.1 is given by the following definition.

Definition 13.5.2 *The volume of the tetrahedron $ABCD$ is the positive number*

$$\mathcal{V}_{ABCD} = \frac{1}{6}|[\overrightarrow{AB}, \overrightarrow{AC}, \overrightarrow{AD}]|.$$

The reason for the coefficient $\frac{1}{6}$ will be explained later.

An useful formula for computing the volume of a tetrahedron involves the area of one of the faces and the distance between the opposite vertex to that face.

Proposition 13.5.3 *Show that the volume of the tetrahedron $ABCD$ is*

$$\mathcal{V}_{ABCD} = \frac{1}{3}h_A \cdot \mathcal{A}_{BCD},$$

where $h_A = dist(A, (BCD))$ and $\mathcal{A}_{BCD}$ is the area of the triangle BCD.

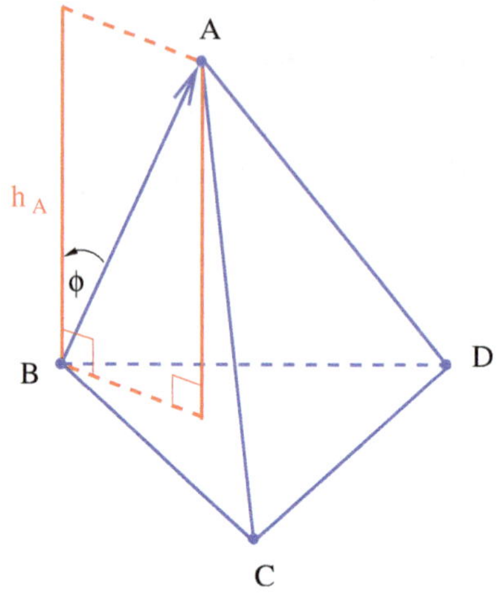

Figure 13.8: The volume of a tetrahedron $ABCD$.

Proof: We have

$$\begin{aligned}\mathcal{V}_{ABCD} &= \frac{1}{6}|[\overrightarrow{BA}, \overrightarrow{BC}, \overrightarrow{BD}]| \\ &= \frac{1}{6}|[\overrightarrow{BC}, \overrightarrow{BD}, \overrightarrow{BA}]| = \frac{1}{6}|(\overrightarrow{BC} \times \overrightarrow{BD}) \cdot \overrightarrow{BA}| \\ &= \frac{1}{6}|\overrightarrow{BC} \times \overrightarrow{BD}| \cdot h_A = \frac{1}{3} \cdot \frac{1}{2}|\overrightarrow{BC} \times \overrightarrow{BD}| \cdot h_A \\ &= \frac{1}{3}\mathcal{A}_{BCD} \cdot h_A,\end{aligned}$$

where we used that the distance between the point A and the plane BCD is $h_A = |BA| \cos\phi$, where ϕ is the angle between $\overrightarrow{AB}$ and the vertical direction, see Figure 13.8. ■

Remark 13.5.4 We also have

$$h_A \cdot \mathcal{A}_{BCD} = h_B \cdot \mathcal{A}_{ACD} = h_C \cdot \mathcal{A}_{ABD}.$$

Exercise 13.5.5 *Find the volume of the tetrahedron $ABCD$, where $\overrightarrow{AB} = (1, 0, 3)$, $\overrightarrow{AC} = (0, 3, 2)$, $\overrightarrow{AD} = (2, 6, 1)$.*

Triangular prismatoid

In Figure 13.9 $ABB'A'$, $ACC'A'$ and $BCC'B'$ are parallelograms and the points A, B, C belong to a plane parallel to the plane passing through A', B', C'. Then $ABCA'B'C'$ is called a *triangular prismatoid.*

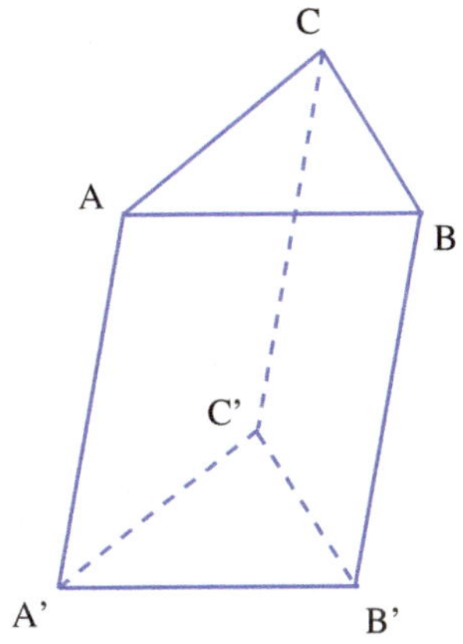

Figure 13.9: A triangular prismatoid.

Exercise 13.5.6 *Show that the bases ABC and $A'B'C'$ are congruent triangles.*

The following result divides a prismatoid into tetrahedrons.[2]

Lemma 13.5.7 (i) *Any triangular prismatoid can be decomposed into three tetrahedrons of the same volume.*
(ii) *Every given tetrahedron can be completed to a triangular prismatoid.*

Proof: (i) Let $ABCA'B'C'$ be a triangular prismatoid, see Figure 13.10. Joining the points A and B', A and C', B and C', the following three tetrahedrons are formed: $AA'B'C'$, $ABCC'$ and $ABB'C'$, with the volumes $\mathcal{V}_1$, $\mathcal{V}_2$ and $\mathcal{V}_3$, respectively. We shall show that $\mathcal{V}_1 = \mathcal{V}_2 = \mathcal{V}_3$.

Since $ABB'A'$, $ACC'A'$ are parallelograms, we have $\overrightarrow{A'B'} = \overrightarrow{AB}$, $\overrightarrow{A'A} = \overrightarrow{C'C}$ and $\overrightarrow{A'C'} = \overrightarrow{AC}$. Hence

$$\begin{aligned}
\mathcal{V}_1 &= \frac{1}{6}|[\overrightarrow{A'C'}, \overrightarrow{A'B'}, \overrightarrow{A'A}]| = \frac{1}{6}|[\overrightarrow{AC}, \overrightarrow{AB}, \overrightarrow{C'C}]| \\
&= \frac{1}{6}|[\overrightarrow{AC}, \overrightarrow{AB}, \overrightarrow{C'A} + \overrightarrow{AC}]| \\
&= \frac{1}{6}|[\overrightarrow{AC} + \overrightarrow{AB} + \overrightarrow{C'A}] + \underbrace{[\overrightarrow{AC} + \overrightarrow{AB} + \overrightarrow{AC}]}_{=0} \\
&= \frac{1}{6}|[\overrightarrow{AC}, \overrightarrow{AB}, \overrightarrow{AC}]| = \mathcal{V}_2.
\end{aligned}$$

Hence $\mathcal{V}_1 = \mathcal{V}_2$.

[2]This is the 3-dimensional analog of the result stating that any parallelogram can be decomposed into two congruent triangles.

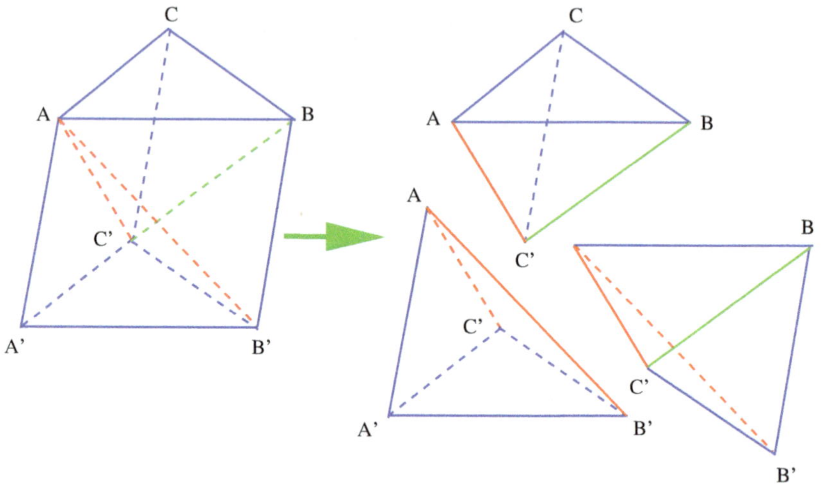

Figure 13.10: The decomposition of a tetrahedron into three triangular prismatoids.

Since $BCC'B'$ is a parallelogram, $\overrightarrow{BC} = \overrightarrow{B'C'}$. Then

$$\begin{aligned}
\mathcal{V}_2 &= \frac{1}{6}|[\overrightarrow{BA}, \overrightarrow{BC}, \overrightarrow{BC'}]| = \frac{1}{6}|[\overrightarrow{BA}, \overrightarrow{B'C'}, \overrightarrow{BC'}]| \\
&= \frac{1}{6}|[\overrightarrow{BA}, \overrightarrow{B'B} + \overrightarrow{BC'}, \overrightarrow{BC'}]| \\
&= \frac{1}{6}|[\overrightarrow{BA}, \overrightarrow{B'B}, \overrightarrow{BC'}] + \underbrace{[\overrightarrow{BA}, \overrightarrow{BC'}, \overrightarrow{BC'}]}_{=0}| \\
&= \frac{1}{6}|[\overrightarrow{BA}, \overrightarrow{BB'}, \overrightarrow{BC'}]| = \mathcal{V}_3,
\end{aligned}$$

and hence $\mathcal{V}_2 = \mathcal{V}_3$.

(ii) Consider the tetrahedron $AA'B'C'$, see Figure 13.11. Choose the points C and B such that $\overrightarrow{C'C} = \overrightarrow{A'A}$ and $\overrightarrow{B'B} = \overrightarrow{A'A}$. Then $ABB'A'$, $ACC'A'$, $BCC'B'$ are parallelograms and hence $ABCA'B'C'$ becomes a triangular prismatoid. ∎

Proposition 13.5.8 *Let $ABCA'B'C'$ be a triangular prismatoid. Then its volume is given by*

$$\mathcal{V}_{ABCA'B'C'} = \frac{1}{2}|[\overrightarrow{AC}, \overrightarrow{AB}, \overrightarrow{AA'}]|.$$

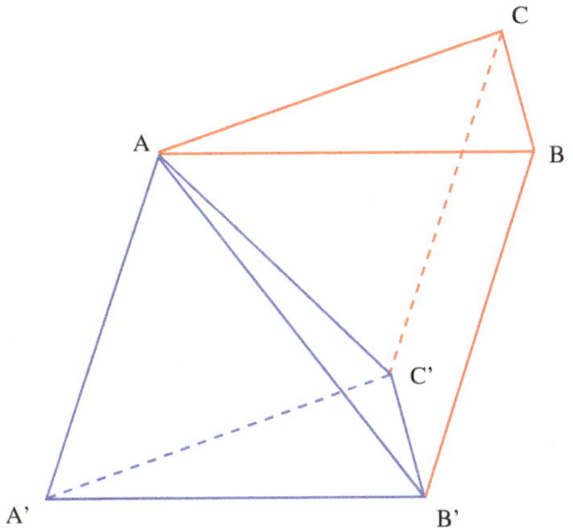

Figure 13.11: For the proof of Lemma 13.5.7, part (*ii*).

Proof: Using Lemma 13.5.7 the prismatoid can be decomposed into three tetrahedrons of the same volume. Then

$$\begin{aligned}\mathcal{V}_{ABCA'B'C'} &= \mathcal{V}_1 + \mathcal{V}_2 + \mathcal{V}_3 = 3\mathcal{V}_1 = 3 \cdot \frac{1}{6}|[\overrightarrow{A'A}, \overrightarrow{A'B'}, \overrightarrow{A'C'}]| \\ &= \frac{1}{2}|[\overrightarrow{AA'}, \overrightarrow{AB}, \overrightarrow{AC}]| = \frac{1}{2}|[\overrightarrow{AC}, \overrightarrow{AB}, \overrightarrow{AA'}]|,\end{aligned}$$

where we used that $ABB'A'$ and $ACC'A'$ are parallelograms. ■

Corollary 13.5.9 *The volume of a triangular prismatoid is given by*

$$\mathcal{V}_{ABCA'B'C'} = h\mathcal{A}_{ABC},$$

where h denotes the altitude of the prismatoid.

Proof: By Proposition 13.5.8 the prismatoid volume is three times the volume of one of the tetrahedrons. Applying Remark 13.5.4 we obtain the desired relation. ■

Exercise 13.5.10 *Show that every triangular prismatoid can be extended to a parallelepiped of double volume, see Figure 13.12.*

Remark 13.5.11 Using Proposition 13.5.8, the parallelopiped volume becomes

$$\mathcal{V}_{ABCDA'B'C'D'} = 2\mathcal{V}_{ABCA'B'C'} = |[\overrightarrow{AA'}, \overrightarrow{AB}, \overrightarrow{AC}]|,$$

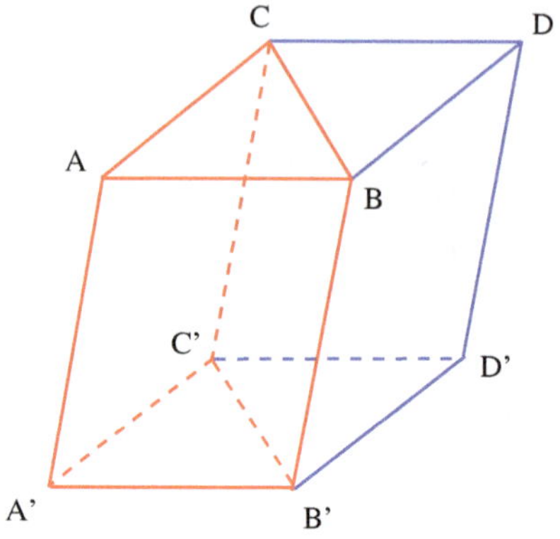

Figure 13.12: For Exercise 13.5.10.

which recovers the formula for the volume of a parallelopiped used in Chapter 5. Hence, the coefficient $1/6$ in the definition of the tetrahedron's volume makes sense.

Solved Exercise 13.5.12 *Let $ABCA'B'C'$ be a triangular prismatoid. Show that*

$$dist(AA', B'C') = dist(A, \mathcal{P}),$$

where $\mathcal{P}$ is the plane of the parallelogram $BB'C'C$.

Proof: By Theorem 13.2.2, there are $X \in AA'$ and $Y \in B'C'$ such that

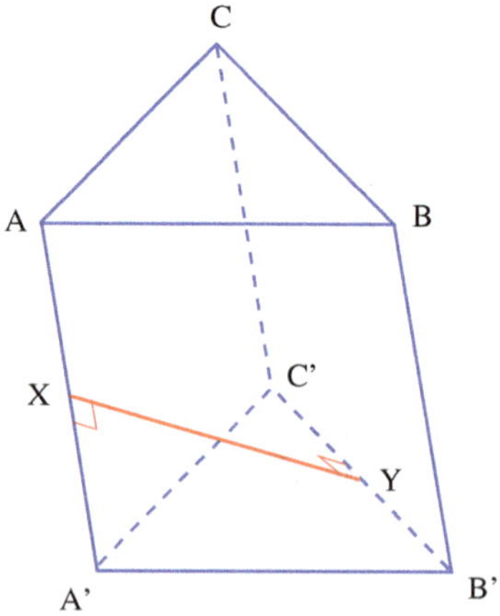

Figure 13.13: For Exercise 13.5.12.

$XY \perp AA'$ and $XY \perp B'C'$. Since $\overrightarrow{AA'} = \overrightarrow{BB'}$ (why?) and $\overrightarrow{XY} \cdot \overrightarrow{AA'} = 0$, we have $\overrightarrow{XY} \cdot \overrightarrow{BB'} = 0$, *i.e.*, $XY \perp BB'$. Since XY is perpendicular on two lines in the plane $\mathcal{P}$, using Theorem 13.1.2 it follows that $XY \perp \mathcal{P}$. Since $AA' || \mathcal{P}$ (why?), it follows that

$$dist(AA', B'C') = |XY| = dist(AA', \mathcal{P}) = dist(A, \mathcal{P}).$$

■

The next result provides a formula for the volume of a tetrahedron in terms of the information provides by two opposite sides.

Proposition 13.5.13 *Consider a tetrahedron and let ϕ and d be the angle and the distance between two opposite sides of lengths m and n. Then the tetrahedron's volume is given by*

$$\mathcal{V} = \frac{1}{6} m \cdot n \cdot d \cdot \sin \phi.$$

Proof: By Lemma 13.5.7 the tetrahedron $ABCD$ can be completed to the triangular prismatoid $AD'C'CDB$, see Figure 13.14. Let $|AB| = m$, $|DC| = n$, $\widehat{AB, DC} = \phi$ and $d = dist(AB, DC) = dist(A, \mathcal{P})$, see also

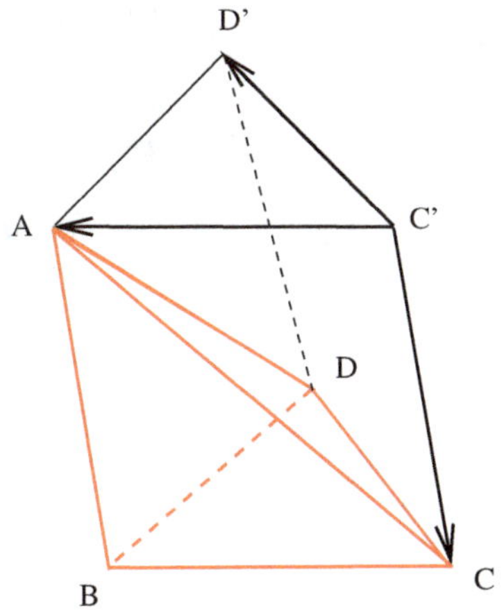

Figure 13.14: For Exercise 13.5.13.

Exercise 13.5.12. Using Lemma 13.5.7 and Proposition 13.5.8, we obtain

$$\begin{aligned}
\mathcal{V}_{ABCD} &= \frac{1}{3}\mathcal{V}_{AD'C'BDC} = \frac{1}{3}\cdot\frac{1}{2}|[\overrightarrow{C'D'},\overrightarrow{C'C},\overrightarrow{C'A}]| \\
&= \frac{1}{6}|[\overrightarrow{CD},\overrightarrow{AB},\overrightarrow{C'A}]| = \frac{1}{6}|(\overrightarrow{CD}\times\overrightarrow{AB})\cdot\overrightarrow{C'A}| \\
&= \frac{1}{6}|\overrightarrow{CD}\times\overrightarrow{AB}|dist(A,\mathcal{P}) = \frac{1}{6}|CD|\cdot|AB|\cdot\sin\phi\cdot d \\
&= \frac{1}{6}m\cdot n\cdot d\cdot\sin\phi.
\end{aligned}$$

■

Exercise 13.5.14 *Let $ABCD$ be an orthogonal tetrahedron (opposite sides are perpendicular). Show that*

$$\mathcal{V}_{ABCD} = \frac{1}{6}|AB|\cdot|CD|\cdot dist(AB,CD).$$

Exercise 13.5.15 *Let $ABCD$ be a tetrahedron with the opposite edges equal. Let M, N be the midpoints of $[AB]$, $[CD]$. Show that the volume of the tetrahedron is*

$$\mathcal{V}_{ABCD} = \frac{1}{6}|AB|^2\cdot|MN|\sin(\widehat{AB,CD}).$$

In the following we shall find a formula for the volume of a tetrahedron involving lengths of the edges only. This will be the analog of Heron's formula from the triangle geometry. To this end, we shall prepare the ground with a few results.

Lemma 13.5.16 *Let $\overrightarrow{u}$, $\overrightarrow{v}$, $\overrightarrow{w}$ be three vectors in the space. Then we have*

$$[\overrightarrow{u},\overrightarrow{v},\overrightarrow{w}]^2 = \det\begin{pmatrix}\overrightarrow{u}\cdot\overrightarrow{u} & \overrightarrow{u}\cdot\overrightarrow{v} & \overrightarrow{u}\cdot\overrightarrow{w}\\ \overrightarrow{v}\cdot\overrightarrow{u} & \overrightarrow{v}\cdot\overrightarrow{v} & \overrightarrow{v}\cdot\overrightarrow{w}\\ \overrightarrow{w}\cdot\overrightarrow{u} & \overrightarrow{w}\cdot\overrightarrow{v} & \overrightarrow{w}\cdot\overrightarrow{w}\end{pmatrix}.$$

Proof: Let $\{e_1,e_2,e_3\}$ be the canonical base of $\mathbb{R}^3$. Considering $\overrightarrow{u} = \sum_{i=1}^{3} u^i e_i$, $\overrightarrow{v} = \sum_{i=1}^{3} v^i e_i$, and $\overrightarrow{w} = \sum_{i=1}^{3} w^i e_i$, we have

$$[\overrightarrow{u},\overrightarrow{v},\overrightarrow{w}] = \det\begin{pmatrix}u^1 & u^2 & u^3\\ v^1 & v^2 & v^3\\ w^1 & w^2 & w^3\end{pmatrix}.$$

Using that $\det(AB) = \det(A)\det(B)$ and $\det(A^T) = \det A$, yields

$$\begin{aligned}
&[\vec{u}, \vec{v}, \vec{w}]^2 \\
&= \det \begin{pmatrix} u^1 & u^2 & u^3 \\ v^1 & v^2 & v^3 \\ w^1 & w^2 & w^3 \end{pmatrix} \cdot \det \begin{pmatrix} u^1 & v^1 & w^1 \\ u^2 & v^2 & w^2 \\ u^3 & v^3 & w^3 \end{pmatrix} \\
&= \det \begin{pmatrix} u^1u^1+u^2u^2+u^3u^3 & u^1v^1+u^2v^2+u^3v^3 & u^1w^1+u^2w^2+u^3w^3 \\ v^1u^1+v^2u^2+v^3u^3 & v^1v^1+v^2v^2+v^3v^3 & v^1w^1+v^2w^2+v^3w^3 \\ w^1u^1+w^2u^2+w^3u^3 & w^1v^1+w^2v^2+w^3v^3 & w^1w^1+w^2w^2+w^3w^3 \end{pmatrix} \\
&= \det \begin{pmatrix} \vec{u}\cdot\vec{u} & \vec{u}\cdot\vec{v} & \vec{u}\cdot\vec{w} \\ \vec{v}\cdot\vec{u} & \vec{v}\cdot\vec{v} & \vec{v}\cdot\vec{w} \\ \vec{w}\cdot\vec{u} & \vec{w}\cdot\vec{v} & \vec{w}\cdot\vec{w} \end{pmatrix}.
\end{aligned}$$

■

Solved Exercise 13.5.17 *Let $A_1A_2A_3$ be a triangle and denote by $a_1 = |A_2A_3|$, $a_2 = |A_1A_3|$, $a_3 = |A_1A_2|$ the side lengths. Show that*

$$\overrightarrow{A_1A_2} \cdot \overrightarrow{A_1A_3} = \frac{1}{2}(a_2^2 + a_3^2 - a_1^2).$$

Proof: The definition of the inner product and the law of cosines yield

$$\begin{aligned}
\overrightarrow{A_1A_2} \cdot \overrightarrow{A_1A_3} &= |A_1A_2| \cdot |A_1A_3| \cos \widehat{A_2A_1A_3} \\
&= a_3a_2 \frac{a_2^2 + a_3^2 - a_1^2}{2a_3a_2} = \frac{1}{2}(a_2^2 + a_3^2 - a_1^2).
\end{aligned}$$

■

Let $\Delta A_1A_2A_3A_4$ be a tetrahedron and denote by $d_{ij} = |A_iA_j|$, with $i, j \in \{1, \ldots, 4\}$ be the distances between vertices, *i.e.*, the length of the vertices, see Figure 13.15. Let $\mathcal{V} = \mathcal{V}_{A_1A_2A_3A_4}$ be its volume.

Proposition 13.5.18 *With the above notations, the tetrahedron's volume $\mathcal{V}$ satisfies*

$$288\mathcal{V}^2 = \det \begin{pmatrix} 2d_{12}^2 & d_{12}^2 + d_{13}^2 - d_{23}^2 & d_{12}^2 + d_{14}^2 - d_{24}^2 \\ d_{13}^2 + d_{12}^2 - d_{23}^2 & 2d_{13}^2 & d_{13}^2 + d_{14}^2 - d_{34}^2 \\ d_{12}^2 + d_{14}^2 - d_{24}^2 & d_{14}^2 + d_{13}^2 - d_{34}^2 & 2d_{14}^2 \end{pmatrix}. \tag{13.5.4}$$

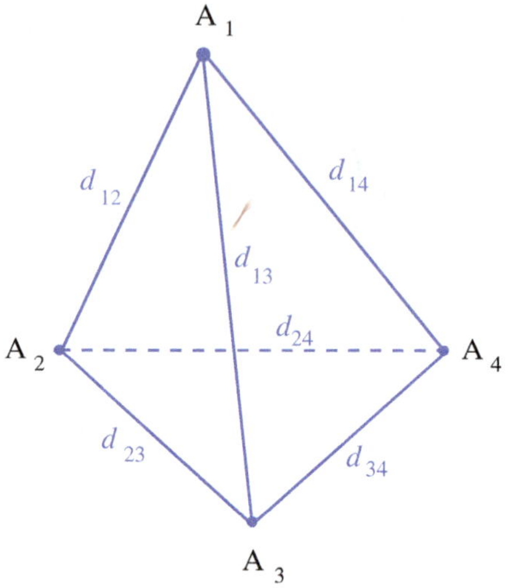

Figure 13.15: The side lengths d_{ij} in a tetrahedron.

Proof: The volume is given by $\mathcal{V} = \frac{1}{6}|[\overrightarrow{A_1A_2}, \overrightarrow{A_1A_3}, \overrightarrow{A_1A_4}]|$. Using Lemma 13.5.16 and Exercise 13.5.17 yields

$$
\begin{aligned}
36\mathcal{V}^2 &= [\overrightarrow{A_1A_2}, \overrightarrow{A_1A_3}, \overrightarrow{A_1A_4}]^2 \\
&= \det\begin{pmatrix} \overrightarrow{A_1A_2}\cdot\overrightarrow{A_1A_2} & \overrightarrow{A_1A_2}\cdot\overrightarrow{A_1A_3} & \overrightarrow{A_1A_2}\cdot\overrightarrow{A_1A_4} \\ \overrightarrow{A_1A_3}\cdot\overrightarrow{A_1A_2} & \overrightarrow{A_1A_3}\cdot\overrightarrow{A_1A_3} & \overrightarrow{A_1A_3}\cdot\overrightarrow{A_1A_4} \\ \overrightarrow{A_1A_4}\cdot\overrightarrow{A_1A_2} & \overrightarrow{A_1A_4}\cdot\overrightarrow{A_1A_3} & \overrightarrow{A_1A_4}\cdot\overrightarrow{A_1A_4} \end{pmatrix} \\
&= \det\begin{pmatrix} d_{12}^2 & \frac{1}{2}(d_{12}^2+d_{13}^2-d_{23}^2) & \frac{1}{2}(d_{12}^2+d_{14}^2-d_{24}^2) \\ \frac{1}{2}(d_{13}^2+d_{12}^2-d_{23}^2) & d_{13}^2 & \frac{1}{2}(d_{13}^2+d_{14}^2-d_{34}^2) \\ \frac{1}{2}(d_{12}^2+d_{14}^2-d_{24}^2) & \frac{1}{2}(d_{14}^2+d_{13}^2-d_{34}^2) & d_{14}^2 \end{pmatrix}.
\end{aligned}
$$

Multiplying by 8 leads to formula (13.5.4). ■

The above determinant can be written in different ways. See the following two exercises.

Exercise 13.5.19 *Consider the following notations regarding the sums and products of the squares of opposite sides of a tetrahedron*

$$s_1 = d_{14}^2 + d_{23}^2, \qquad s_2 = d_{13}^2 + a_{24}^2, \qquad s_3 = d_{12}^2 + d_{34}^2,$$

$$p_1 = d_{14}^2 d_{23}^2, \qquad p_2 = d_{13}^2 d_{24}^2, \qquad p_3 = d_{12}^2 d_{34}^2.$$

For the sides situated on the same face we set

$$q_1 = (d_{23}d_{24}d_{34})^2,\ q_2 = (d_{13}d_{14}d_{34})^2,\ q_3 = (d_{12}d_{14}d_{24})^2,\ q_4 = (d_{12}d_{13}d_{23})^2.$$

Let $s = \frac{1}{2}(s_1 + s_2 + s_3)$, $p = \frac{1}{2}(p_1 + p_2 + p_3)$, $r = \frac{1}{2}(p_1 s_1 + p_2 s_2 + p_3 s_3)$, *and* $q = \frac{1}{4}(q_1 + q_2 + q_3 + q_4)$. *Show that*

$$36\mathcal{V}^2 = s \cdot p - r - q.$$

Exercise 13.5.20 (Euler, 1752) *With the previous notations, the volume* $\mathcal{V}$ *of a tetrahedron is given by*

$$288\mathcal{V}^2 = \det \begin{pmatrix} 0 & 1 & 1 & 1 & 1 \\ 1 & 0 & d_{12}^2 & d_{13}^2 & d_{14}^2 \\ 1 & d_{12}^2 & 0 & d_{23}^2 & d_{24}^2 \\ 1 & d_{13}^2 & d_{23}^2 & 0 & d_{34}^2 \\ 1 & d_{14}^2 & d_{24}^2 & d_{34}^2 & 0 \end{pmatrix}.$$

The above formula corresponds to Heron's formula from triangle geometry. The area $\mathcal{A}$ of a triangle ABC with side lengths a, b, c and semiperimeter $p = (a+b+c)/2$, is given by $\mathcal{A}^2 = p(p-a)(p-b)(p-c)$. The vertices A, B, C are collinear if and only if the area $\mathcal{A} = 0$, which leads to one of the following identities:

$$p - a = 0 \quad \text{or} \quad p - b = 0 \quad \text{or} \quad p - c = 0,$$

which become a scalar triangle identity

$$b + c = a \quad \text{or} \quad a + c = b \quad \text{or} \quad a + b = c. \tag{13.5.5}$$

Relation (13.5.5) is a necessary and sufficient condition for three points with distances a, b, c among them, to be collinear. The following exercise provides a generalization to four coplanar points.

Exercise 13.5.21 *Let* A_1, A_2, A_3, A_4 *be four points in the space, with* $d_{ij} = |A_iA_j|$. *Show that the points* A_i, $i \in \{1, \ldots, 4\}$ *are coplanar if and only if*

$$\begin{aligned} & d_{13}^2 d_{24}^2 (d_{13}^2 + d_{24}^2 - d_{12}^2 - d_{23}^2 - d_{34}^2 - d_{14}^2) + d_{13}^2 (d_{12}^2 - d_{14}^2)(d_{23}^2 - d_{34}^2) \\ & + d_{24}^2 (d_{12}^2 - d_{23}^2)(d_{14}^2 - d_{34}^2) + (d_{12}^2 - d_{23}^2 + d_{34}^2 - d_{14}^2)(d_{12}^2 \cdot d_{34}^2 - d_{23}^2 \cdot d_{14}^2) \\ & = 0. \end{aligned}$$

The above condition can be written also as $s \cdot p = r + q$.

13.6 Orthogonal tetrahedrons

Science has proof without any certainty.
Creationists have certainty without any proof.
Ashley Montague

When considering the altitude of a tetrahedron, we expect the altitude to intersect the opposite face exactly at its orthocenter. However, this happens only for certain tetrahedrons, called *orthocentric* or *orthogonal.* A precise definition is stated in the following:

Definition 13.6.1 *A tetrahedron $ABCD$ is called orthogonal if the opposite edges are perpendicular, i.e., $AB \perp CD$, $AD \perp BC$, $BD \perp AC$.*

By Proposition 13.4.3 it suffices just two pairs of opposite sides to be perpendicular for a tetrahedron to be orthogonal.

Lemma 13.6.2 *Let $ABCD$ be an orthogonal tetrahedron and let H_A be the orthogonal projection of the vertex A onto the face (BCD). Then H_A is the orthocenter of triangle BCD.*

Proof: We shall refer to Figure 13.16. Let $H_A \in (BCD)$ such that $AH_A \perp (BCD)$. Then, in particular, $AH_A \perp DC$. We have

$$\begin{aligned}\overrightarrow{BH_A} \cdot \overrightarrow{CD} &= (\overrightarrow{BA} + \overrightarrow{AH_A}) \cdot \overrightarrow{CD} \\ &= \underbrace{\overrightarrow{BA} \cdot \overrightarrow{CD}}_{=0} + \underbrace{\overrightarrow{AH_A} \cdot \overrightarrow{CD}}_{=0} = 0,\end{aligned}$$

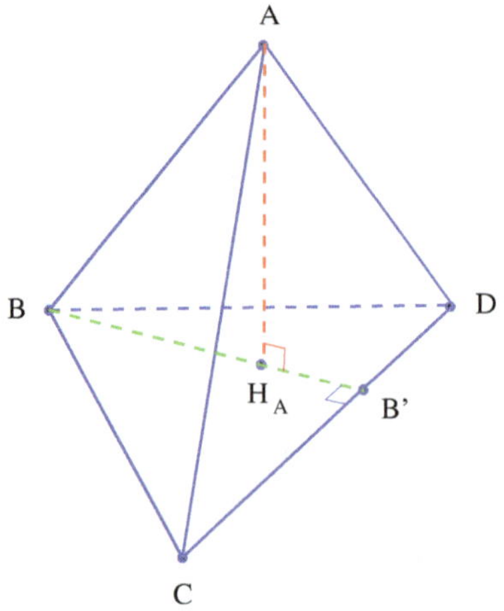

Figure 13.16: For Lemma 13.6.2.

since $BA \perp CD$ and $AH_A \perp CD$. Hence $BH_A \perp CD$, *i.e.*, H_A belongs to the altitude from B of triangle BCD.

In a similar way, we can show that $\overrightarrow{CH_A} \cdot \overrightarrow{BD} = 0$ and $\overrightarrow{DH_A} \cdot \overrightarrow{BC} = 0$, and hence H_A will belong to the other two altitudes of triangle BCD and hence H_A will be the orthocenter of triangle BCD. ■

The reverse of the above result is also true, as demonstrated in the following.

Lemma 13.6.3 *Let $ABCD$ be a tetrahedron, and let H_A be the orthocenter of triangle ΔBCD. If $AH_A \perp (BCD)$ then the tetrahedron is orthogonal.*

Proof: We refer to Figure 13.17. Since H_A is the orthocenter of triangle BCD, then $BH_A \perp DC$, $CH_A \perp BD$ and $DH_A \perp BC$. Then

$$\begin{aligned} \overrightarrow{AB} \cdot \overrightarrow{CD} &= (\overrightarrow{AH_A} + \overrightarrow{H_AB}) \cdot \overrightarrow{CD} \\ &= \underbrace{\overrightarrow{AH_A} \cdot \overrightarrow{CD}}_{=0} + \underbrace{\overrightarrow{H_AB} \cdot \overrightarrow{CD}}_{=0} = 0, \end{aligned}$$

since $AH_A \perp CD$ and $BH_A \perp CD$. Hence $AB \perp CD$.

In a similar way,

$$\overrightarrow{AD} \cdot \overrightarrow{BC} = \overrightarrow{DH_A} \cdot \overrightarrow{BC} + \overrightarrow{H_AA} \cdot \overrightarrow{BC} = 0,$$

$$\overrightarrow{AC} \cdot \overrightarrow{BD} = \overrightarrow{AH_A} \cdot \overrightarrow{BD} + \overrightarrow{H_AC} \cdot \overrightarrow{BD} = 0,$$

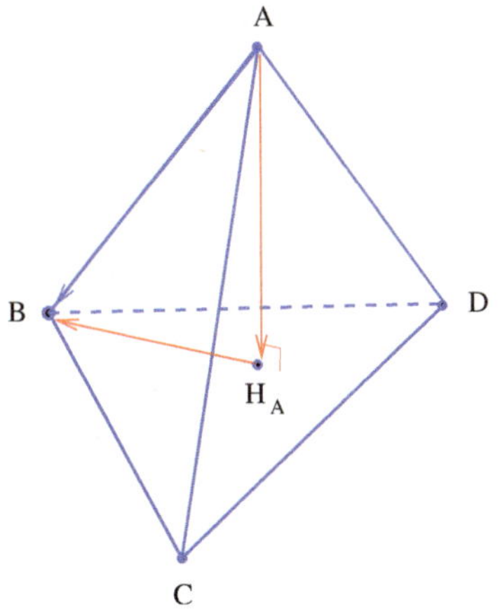

Figure 13.17: For Lemma 13.6.3.

and hence $AD \perp BC$ and $AC \perp BD$. Hence $ABCD$ is an orthogonal tetrahedron. ■

The following result was proved by Vecten in 1817, and provides necessary and sufficient conditions for a tetrahedron to be orthogonal.

Theorem 13.6.4 (Vecten) *Let $ABCD$ be a tetrahedron. The following statements are equivalent:*

1) $ABCD$ is orthogonal;

2) The line which joins the vertex A and the orthocenter of triangle BCD is perpendicular to the plane (BCD);

3) The line which joins the vertex B and the orthocenter of triangle ACD is perpendicular to the plane (ACD);

4) The line which joins the vertex C and the orthocenter of triangle ABD is perpendicular to the plane (ABD);

5) The line which joins the vertex D and the orthocenter of triangle ABC is perpendicular to the plane (ABC);

6) $AB \perp CD$ and $AD \perp BC$.

Proof: 1) $\Longrightarrow$ 2) It follows from Lemma 13.6.2.
2) $\Longrightarrow$ 1) Apply Lemma 13.6.3.
In a similar way we can prove that the statement 1) is equivalent with statements 3), 4) and 5).

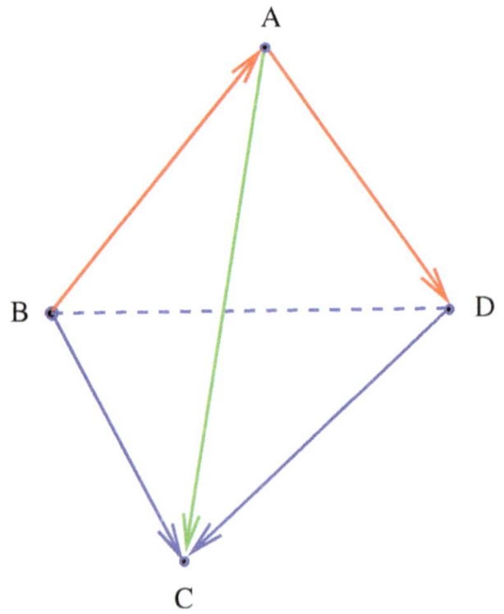

Figure 13.18: For Theorem 13.6.4.

Obviously we have 1) $\Longrightarrow$ 6). Next we shall prove that 6) $\Longrightarrow$ 1), *i.e.*, $AC \perp BD$. Using the triangle rule, we have

$$\begin{aligned}
\overrightarrow{AC} \cdot \overrightarrow{BD} &= (\overrightarrow{AD} + \overrightarrow{DC}) \cdot \overrightarrow{BD} \\
&= \overrightarrow{AD} \cdot \overrightarrow{BD} + \overrightarrow{DC} \cdot \overrightarrow{BD} \\
&= \overrightarrow{AD} \cdot (\overrightarrow{BC} + \overrightarrow{CD}) + \overrightarrow{DC} \cdot (\overrightarrow{BA} + \overrightarrow{AD}) \\
&= \underbrace{\overrightarrow{AD} \cdot \overrightarrow{BC}}_{=0} + \overrightarrow{AD} \cdot \overrightarrow{CD} + \underbrace{\overrightarrow{DC} \cdot \overrightarrow{BA}}_{=0} + \overrightarrow{DC} \cdot \overrightarrow{AD} \\
&= \overrightarrow{AD} \cdot \overrightarrow{CD} + \overrightarrow{AD} \cdot DC = \overrightarrow{AD} \cdot \underbrace{(\overrightarrow{CD} + \overrightarrow{DC})}_{=0} \\
&= 0,
\end{aligned}$$

where we used that $DC \perp BA$ and $AD \perp BC$. Hence $AC \perp BD$ *i.e.*, part 1) holds. See Figure 13.18. ∎

Definition 13.6.5 *The line passing through a vertex of a tetrahedron that is perpendicular to the opposite face is called an altitude of the tetrahedron.*

A tetrahedron has four altitudes. Vecten's theorem has the following consequence:

Corollary 13.6.6 *If one altitude of a tetrahedron intersects the opposite face at its orthocenter, then the same property holds for all other three altitudes.*

Exercise 13.6.7 *Prove Corollary 13.6.6.*

Exercise 13.6.8 *Show that in a tetrahedron having all faces of equal areas the bimedians are perpendicular on the opposite sides.*

Exercise 13.6.9 *Let H_A, H_B, H_C, H_D be the orthocenters of faces (BCD), (ACD), (ABD), and (ABC) of the orthogonal tetrahedron $ABCD$. Show that*

$$|AH_A| = |BH_B| = |CH_C| = |DH_D|$$

if and only if

$$|AC| = |BD|, \qquad |AB| = |CD|, \qquad |BC| = |AD|.$$

13.7 The orthocenter of a tetrahedron

Everything you can imagine is real.
Pablo Picasso (1881–1973)

The altitudes of a tetrahedron do not always intersect. In this section we search for necessary and sufficient conditions for the intersection of the altitudes of a tetrahedron. The following lemma is a useful tool in proving the intersection of three lines in space.

Lemma 13.7.1 *Let α, β, γ be three distinct, non-coplanar lines in the space such that any two of them intersect. Then α, β, γ intersect at a point.*

Proof: Let $\{C\} = \alpha \cap \beta$, $\{B\} = \alpha \cap \gamma$, and $\{A\} = \beta \cap \gamma$.
First, we assume that $\{A, B, C\}$ are three distinct points and let $\Pi = (A, B, C)$ be the plane determined by these points. See Figure 13.19.

$$B, C \in \alpha \Longrightarrow \alpha \subset \Pi,$$
$$A, C \in \beta \Longrightarrow \beta \subset \Pi,$$
$$A, B \in \gamma \Longrightarrow \gamma \subset \Pi,$$

i.e., the lines α, β, γ belong to the same plane Π, contradiction!

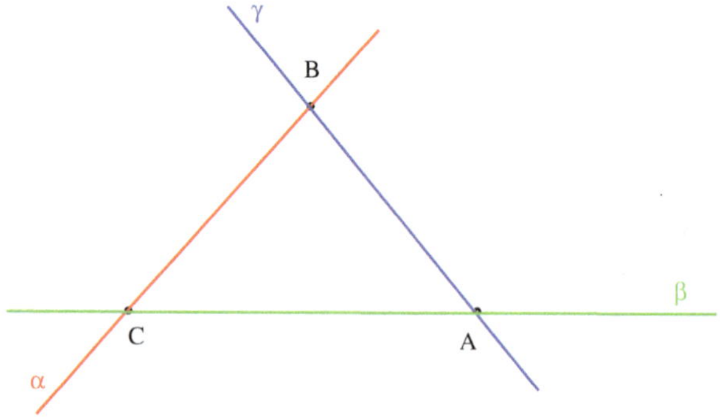

Figure 13.19: For Lemma 13.7.1.

Now, we assume $A \neq B = C$. From $B = C \in \gamma$, $A \in \gamma$, $B = C \in \alpha$, $A \in \alpha$, it follows that $\alpha = \gamma$ (they have two distinct points in common), contradiction!

In a similar way, the cases $B \neq C = A$ and $C \neq A = B$ lead to a contradiction.

Hence the only possible case is $A = B = C$, which means that the lines α, β, γ intersect at a point. ■

Proposition 13.7.2 *Let $ABCD$ be an orthogonal tetrahedron. Then the altitudes of the tetrahedron intersect at a point H, called the orthocenter of the tetrahedron.*

Proof: We shall refer to Figure 13.20. Let H_A, H_B be the orthocenters of triangles BCD and ACD, respectively. Since $ABCD$ is an orthogonal tetrahedron, from Vecten's theorem (Theorem 13.6.4) we have $AH_A \perp (BCD)$ and $BH_B \perp (ACD)$. Since H_A is the orthocenter of triangle BCD, we have $BB' \perp DC$, where $B' = BH_A \cap CD$. Using the three perpendicular lines theorem (Theorem 13.3.1), we obtain $AB' \perp CD$ and hence AB' is an altitude of triangle ACD and hence $H_B \in AB'$. Hence the points A, B, B', H_A, H_B are coplanar and AH_A, BH_B are altitudes in triangle ABB', because $AH_A \perp BB'$ and $BH_B \perp AB'$ (why?). Hence AH_A, BH_B intersect at a point H, which is the orthocenter of triangle ABB'.

Let H_C, H_D be the orthocenters of triangles ABD and ABC. In a similar way, any two altitudes among $\{AH_A, BH_B, CH_C, DH_D\}$ intersect

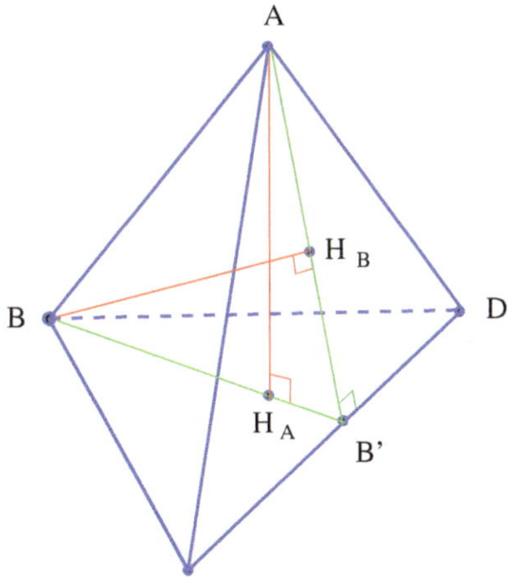

Figure 13.20: For Proposition 13.7.2.

at a point. Applying Lemma 13.7.1 it follows that all of the intersect at the point H. ∎

The converse statement is also true:

Proposition 13.7.3 *If the altitudes of the tetrahedron $ABCD$ intersect at a point, then the tetrahedron is orthogonal.*

Proof: Let AA_1, BB_1 be the altitudes from A and B of the tetrahedron $ABCD$, and let $\{H\} = AA_1 \cap BB_1$ be their intersecting point. See Figure 13.21. The vector $\overrightarrow{BA_1}$ can be written as a linear combination of the vectors $\overrightarrow{AA_1}$ and $\overrightarrow{BB_1}$ as $\overrightarrow{BA_1} = \overrightarrow{BH} + \overrightarrow{HA_1} = \beta\overrightarrow{BB_1} + \alpha\overrightarrow{AA_1}$, with α, $\beta > 0$. Then

$$\overrightarrow{CD} \cdot \overrightarrow{BA_1} = \beta\,\overrightarrow{CD} \cdot \overrightarrow{BB_1} + \alpha\,\overrightarrow{CD} \cdot \overrightarrow{AA_1} = 0,$$

since

$$\begin{aligned} BB_1 \perp (ACD), \quad CD \subset (ACD) \Longrightarrow BB_1 \perp CD, \\ AA_1 \perp (BCD), \quad CD \subset (ACD) \Longrightarrow AA_1 \perp CD. \end{aligned}$$

It follows that $BA_1 \perp CD$ *i.e.*, A_1 belongs to the altitude from B of the triangle BCD. In a similar way, A_1 belongs to the altitudes from D and C of triangle BCD, *i.e.*, A_1 is the orthocenter of triangle BCD. Applying Vecten's theorem (Theorem 13.6.4) yields that $ABCD$ is orthogonal. ∎

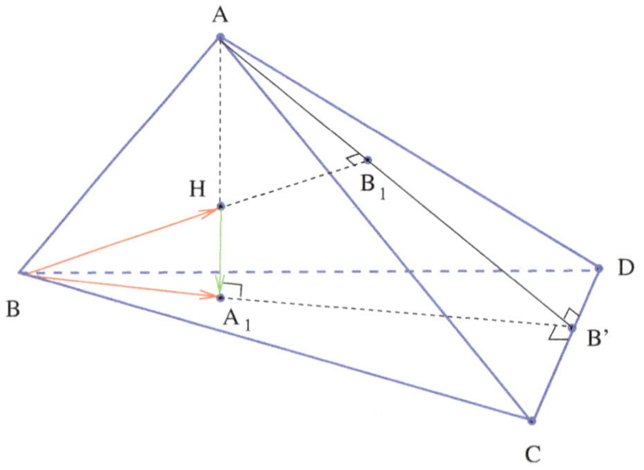

Figure 13.21: For Proposition 13.7.3.

In the following we shall present another convenient proof without using Vecten's theorem.

Proof: Since AA_1 and BB_1 intersect at H, we can write the vector $\overrightarrow{AB}$ as a linear combination of $\overrightarrow{AA_1}$ and $\overrightarrow{BB_1}$:

$$\overrightarrow{AB} = \overrightarrow{AH} + \overrightarrow{HB} = \lambda_1 \overrightarrow{AA_1} + \lambda_2 \overrightarrow{BB_1},$$

with $\lambda_1 > 0$, $\lambda_2 < 0$. Then

$$\overrightarrow{AB} \cdot \overrightarrow{CD} = \lambda_1 \underbrace{\overrightarrow{AA_1} \cdot \overrightarrow{CD}}_{=0} + \lambda_2 \underbrace{\overrightarrow{BB_1} \cdot \overrightarrow{CD}}_{=0} = 0.$$

Hence $AB \perp CD$. Similarly, we can prove that $AD \perp BC$, $BD \perp AC$, *i.e.*, the tetrahedron $ABCD$ is orthogonal. ■

Exercise 13.7.4 *Let ϕ be the angle between the edges AC and BD of the tetrahedron $ABCD$. Show that*

(i) $|\cos\phi| = \dfrac{\left||AD|^2 + |BC|^2 - |AB|^2 - |CD|^2\right|}{2|AC| \cdot |BD|}.$

(ii) The tetrahedron is orthogonal if and only if the sum of the squares of the opposite edges is the same.

Exercise 13.7.5 *Let H be the orthocenter of the orthogonal tetrahedron $ABCD$. Show that A is the orthocenter of the tetrahedron $BCDH$.*

Exercise 13.7.6 *Let $ABCD$ be an orthogonal tetrahedron and let H be its orthocenter. Show that $BCDH, ABCH, ADCH, ABDH$ are orthogonal tetrahedrons.*

Exercise 13.7.7 (Euler's line of a tetrahedron) *Let $ABCD$ be an orthogonal tetrahedron inscribed in the sphere $\mathcal{S}(O,R)$.*

(i) Show that $\overrightarrow{OA} + \overrightarrow{OB} + \overrightarrow{OC} + \overrightarrow{OD} = 2\overrightarrow{OH}$;

(ii) Show that the points O, H, G are collinear, where H and G are the orthocenter and the centroid of $ABCD$;

(iii) Prove that $2\overrightarrow{OG} = \overrightarrow{OH}$, i.e., G is the midpoint of the segment $[OH]$; OH is called the Euler line of the tetrahedron $ABCD$.

13.8 Equifacial tetrahedrons

In science one tries to tell people, in such a way as to be understood by everyone, something that no one ever knew before. But in poetry, it's the exact opposite.

Paul Dirac (1902–1984)

This section presents a type of tetrahedrons, which have all the faces of the same area.

Let $ABCD$ be a tetrahedron and let M, N, P, Q, R, S be the midpoints of the edges $[AB]$, $[CD]$, $[AC]$, $[BD]$, $[AD]$, $[BC]$, respectively. Consider the following notations $\overrightarrow{m} = \overrightarrow{AB}$, $\overrightarrow{n} = \overrightarrow{CD}$, $\overrightarrow{p} = \overrightarrow{AC}$, $\overrightarrow{q} = \overrightarrow{BD}$, $\overrightarrow{r} = \overrightarrow{AD}$, $\overrightarrow{s} = \overrightarrow{BC}$, see Figure 13.22.

Exercise 13.8.1 *Show that* $\overrightarrow{MN} = \frac{1}{2}(\overrightarrow{q} + \overrightarrow{p})$, $\overrightarrow{PQ} = \frac{1}{2}(\overrightarrow{m} + \overrightarrow{n})$, *and* $\overrightarrow{RS} = \frac{1}{2}(\overrightarrow{m} - \overrightarrow{n})$.

Lemma 13.8.2 *If the bimedians MN, PQ are perpendicular on the sides (AB, DC), (AC, BD), respectively, then the following conditions hold*

$$\overrightarrow{m} \cdot \overrightarrow{p} = \overrightarrow{q} \cdot \overrightarrow{n}, \qquad \overrightarrow{p} \cdot \overrightarrow{n} = \overrightarrow{m} \cdot \overrightarrow{q}, \qquad \overrightarrow{q} \cdot \overrightarrow{m} = \overrightarrow{n} \cdot \overrightarrow{p}.$$

Proof: Using Exercise 13.8.1, the medians are perpendicular on the

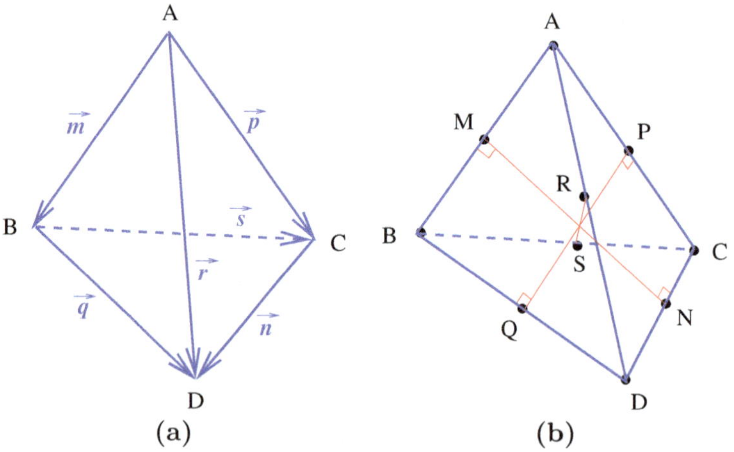

Figure 13.22: For Theorem 13.6.4.

opposite sides if and only if

$$(\overrightarrow{p}+\overrightarrow{q})\cdot\overrightarrow{n}=0, \qquad (\overrightarrow{p}+\overrightarrow{q})\cdot\overrightarrow{m}=0,$$
$$(\overrightarrow{m}+\overrightarrow{n})\cdot\overrightarrow{p}=0, \qquad (\overrightarrow{m}+\overrightarrow{n})\cdot\overrightarrow{q}=0,$$

which can be written as

$$\overrightarrow{p}\cdot\overrightarrow{n}+\overrightarrow{q}\cdot\overrightarrow{n}=0, \qquad \overrightarrow{p}\cdot\overrightarrow{m}+\overrightarrow{q}\cdot\overrightarrow{m}=0,$$
$$\overrightarrow{m}\cdot\overrightarrow{p}+\overrightarrow{n}\cdot\overrightarrow{p}=0, \qquad \overrightarrow{m}\cdot\overrightarrow{q}+\overrightarrow{n}\cdot\overrightarrow{q}=0.$$

Subtracting the above equations, yields

$$\overrightarrow{p}\cdot\overrightarrow{m}=\overrightarrow{n}\cdot\overrightarrow{q}, \qquad \overrightarrow{p}\cdot\overrightarrow{n}=\overrightarrow{m}\cdot\overrightarrow{q}, \qquad \overrightarrow{q}\cdot\overrightarrow{m}=\overrightarrow{n}\cdot\overrightarrow{p}.$$

■

Exercise 13.8.3 *Let $\overrightarrow{a}$, $\overrightarrow{b}$ be two vectors. Show that*

$$\overrightarrow{a}\cdot\overrightarrow{b}=\frac{1}{2}\left[|\overrightarrow{a}|^2+|\overrightarrow{b}|^2-|\overrightarrow{a}-\overrightarrow{b}|^2\right].$$

Proposition 13.8.4 *If the bimedians MN, PQ are perpendicular to the pairs of opposite edges (AB, DC) and (AC, BD), then $|AB| = |CD|$ and $|AC| = |BD|$.*

Proof: Using Exercise 13.8.3, Lemma 13.8.2 yields

$$\begin{aligned}
\overrightarrow{m}\cdot\overrightarrow{p} &= \overrightarrow{q}\cdot\overrightarrow{n} \iff \\
|\overrightarrow{m}|^2+|\overrightarrow{p}|^2-|\overrightarrow{m}-\overrightarrow{p}|^2 &= |\overrightarrow{q}|^2+|\overrightarrow{n}|^2-|\overrightarrow{q}-\overrightarrow{n}|^2 \iff \\
|\overrightarrow{m}|^2+|\overrightarrow{p}|^2-|\overrightarrow{s}|^2 &= |\overrightarrow{q}|^2+|\overrightarrow{n}|^2-|\overrightarrow{s}|^2 \iff \\
|\overrightarrow{m}|^2+|\overrightarrow{p}|^2 &= |\overrightarrow{q}|^2+|\overrightarrow{n}|^2. \qquad (13.8.6)
\end{aligned}$$

In a similar way, starting from $\overrightarrow{p}\cdot\overrightarrow{n}=\overrightarrow{m}\cdot\overrightarrow{q}$, we arrive at

$$|\overrightarrow{p}|^2+|\overrightarrow{n}|^2=|\overrightarrow{m}|^2+|\overrightarrow{q}|^2. \qquad (13.8.7)$$

Subtracting (13.8.6) and (13.8.7) yields

$$\begin{aligned}
|\overrightarrow{m}|^2-|\overrightarrow{n}|^2 &= |\overrightarrow{n}|^2-|\overrightarrow{m}|^2 \iff 2|\overrightarrow{m}|^2=2|\overrightarrow{n}|^2 \iff \\
|\overrightarrow{m}| &= |\overrightarrow{n}| \iff |AB|=|CD|.
\end{aligned}$$

From $\overrightarrow{q}\cdot\overrightarrow{m}=\overrightarrow{n}\cdot\overrightarrow{p}$, using the same idea of computations as above, we arrive at

$$|\overrightarrow{q}|^2+|\overrightarrow{m}|^2=|\overrightarrow{n}|^2+|\overrightarrow{p}|^2. \qquad (13.8.8)$$

Subtracting (13.8.6) and (13.8.8) yields

$$|\overrightarrow{p}|^2-|\overrightarrow{q}|^2=|\overrightarrow{q}|^2-|\overrightarrow{p}|^2,$$

which leads to $|\overrightarrow{p}|=|\overrightarrow{q}|$, or $|AC|=|BD|$. ■

Exercise 13.8.5 (The circumsphere) *Show that there is a sphere passing through any given tetrahedron's vertices.*

Exercise 13.8.6 *If each bimedian of a tetrahedron is perpendicular on the edges they bisect, then the faces of the tetrahedron are congruent triangles.*

Definition 13.8.7 *A tetrahedron with the faces of the same area is called an equifacial tetrahedron.*

Exercise 13.8.8 *Show that a tetrahedron $ABCD$ is equifacial if and only if*

$$dist(A,(BCD)) = dist(B,(ACD)) = dist(C,(ABD)) = dist(D,(ABC)).$$

Exercise 13.8.9 *A tetrahedron is equifacial if and only if each bimedian is perpendicular on the opposite edges they bisect.*

Exercise 13.8.10 *Let $ABCD$ be an equifacial tetrahedron. Show that for any point $M \in Int\, ABCD$ the expression*

$$dist(M,(BCD)) + dist(M,(ACD)) + dist(M,(ABC)) + dist(M,(ABD))$$

is constant. Is the converse true?

13.9 The medians of a tetrahedron

I will not add another word.
Horace

Let $ABCD$ be a tetrahedron and denote by G_A, G_B, G_C, G_D the centroids of the faces BCD, ACD, ABD, ABC, respectively. The line segments $[AG_A]$, $[BG_B]$, $[CG_C]$, $[DG_D]$ are called the *medians* of the tetrahedron $ABCD$. This section deals with the metric properties of the medians of a tetrahedron.

Theorem 13.9.1 (The median theorem) *The length of the median from the vertex A satisfies the relation*

$$|AG_A|^2 = \frac{1}{3}(|AB|^2 + |AC|^2 + |AD|^2) - \frac{1}{9}(|BC|^2 + |BD|^2 + |CD|^2).$$

Proof: Consider the Figure 13.23. Since G_A is the centroid of the triangle BCD, we have

$$3\overrightarrow{AG_A} = \overrightarrow{AB} + \overrightarrow{AC} + \overrightarrow{AD}.$$

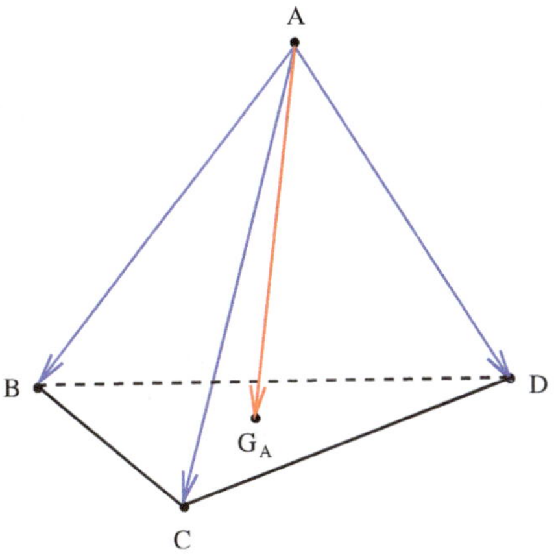

Figure 13.23: A median in a tetrahedron.

Taking the inner products and using the cosine theorem, we have

$$\begin{aligned}
9|AG_A|^2 &= |AB|^2 + |AC|^2 + |AD|^2 + 2\overrightarrow{AB} \cdot \overrightarrow{AC} + 2\overrightarrow{AB} \cdot \overrightarrow{AD} + 2\overrightarrow{AC} \cdot \overrightarrow{AD} \\
&= |AB|^2 + |AC|^2 + |AD|^2 + 2|AB|\,|AC| \cos \widehat{BAC} \\
&\quad + 2|AB|\,|AD| \cos \widehat{BAD} + 2|AD|\,|AC| \cos \widehat{DAC} \\
&= |AB|^2 + |AC|^2 + |AD|^2 + (|AB|^2 + |AC|^2 - |BC|^2) \\
&\quad (|AB|^2 + |AD|^2 - |BD|^2) + (|AC|^2 + |AD|^2 - |CD|^2) \\
&= 3(|AB|^2 + |AC|^2 + |AD|^2) - (|BC|^2 + |BD|^2 + |CD|^2).
\end{aligned}$$

Dividing by 9 leads to the desired result. ■

Exercise 13.9.2 *Using the median theorem prove that*

$$\begin{aligned}
|AG_A|^2 + |BG_B|^2 + |CG_C|^2 + |DG_D|^2 = \frac{4}{9}\Big[&|AB|^2 + |AC|^2 + |AD|^2 \\
&+ |BD|^2 + |BC|^2 + |CD|^2\Big].
\end{aligned}$$

13.10 Properties of the centroid of a tetrahedron

In this section we deal with some incidence properties of the centroid of a tetrahedron. To this end we shall find certain lines which intersect at the centroid.

We can introduce the tetrahedron centroid vectorially as follows.

Definition 13.10.1 *The centroid of a tetrahedron ABCD is the point G defined by*

$$\overrightarrow{MG} = \frac{1}{4}(\overrightarrow{MA} + \overrightarrow{MB} + \overrightarrow{MC} + \overrightarrow{MD}),$$

for any point M.

Remark 13.10.2 We note that the definition of the point G does not depend on the choice of the point M. By contradiction, we assume that starting from another point M' we arrive at a point G', with $G' \neq G$, namely we have

$$\overrightarrow{M'G'} = \frac{1}{4}(\overrightarrow{M'A} + \overrightarrow{M'B} + \overrightarrow{M'C} + \overrightarrow{M'D}).$$

Using vector addition and subtraction, we have

$$\begin{aligned}
\overrightarrow{M'M} &= \overrightarrow{M'G'} + \overrightarrow{G'G} + \overrightarrow{GM} = \overrightarrow{M'G'} - \overrightarrow{MG} + \overrightarrow{G'G} \\
&= \frac{1}{4}(\overrightarrow{M'A} + \overrightarrow{M'B} + \overrightarrow{M'C} + \overrightarrow{M'D}) - \frac{1}{4}(\overrightarrow{MA} + \overrightarrow{MB} + \overrightarrow{MC} + \overrightarrow{MD}) \\
&\quad + \overrightarrow{G'G} \\
&= \overrightarrow{MM'} + + \overrightarrow{G'G}.
\end{aligned}$$

It follows that $\overrightarrow{G'G} = 0$, so that $G = G'$, which is a contradiction!

We also make the remark that in the case $M = G$, then we obtain $\overrightarrow{GA} + \overrightarrow{GB} + \overrightarrow{GC} + \overrightarrow{GD} = 0$, which is another relation which defines the centroid G. The uniqueness of the centroid can be shown by a contradiction argument. If we assume there are two distinct points, G and G' satisfying the aforementioned property, then subtracting the sums we obtain $3\overrightarrow{GG'} = 0$, which implies $G = G'$.

Theorem 13.10.3 *The medians AG_A, BG_B, CG_C, DG_D of the tetrahedron ABCD intersect at a point G, which is the centroid of the tetrahedron. Moreover, the following relations hold*

$$\overrightarrow{AG} = \frac{3}{4}\overrightarrow{AG_A}, \qquad \overrightarrow{BG} = \frac{3}{4}\overrightarrow{BG_B}, \qquad \overrightarrow{CG} = \frac{3}{4}\overrightarrow{CG_C}.$$

Proof: Consider Figure 13.24. Using the definition of the centroid, we have

$$4\overrightarrow{MG} = \overrightarrow{MA} + \overrightarrow{MB} + \overrightarrow{MC} + \overrightarrow{MD} = \overrightarrow{MA} + 3\overrightarrow{MG_A},$$

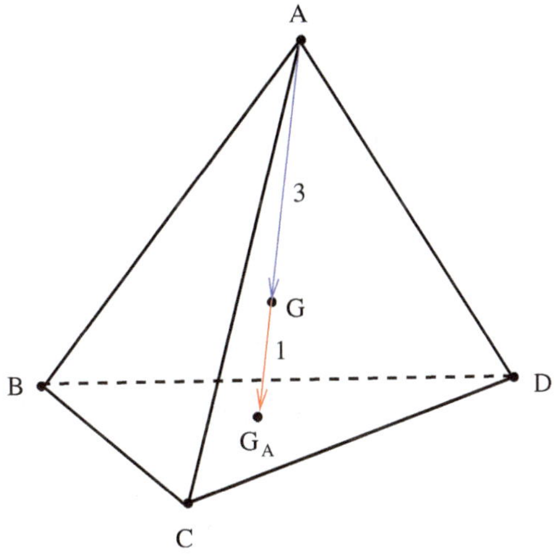

Figure 13.24: The centroid of a tetrahedron.

and dividing by 4, we obtain the following convex combination

$$\overrightarrow{MG} = \frac{1}{4}\overrightarrow{MA} + \frac{3}{4}\overrightarrow{MG_A},$$

which means that $G \in [AG_A]$ such that $\overrightarrow{AG} = 3\overrightarrow{GG_A}$, or $\overrightarrow{AG} = \frac{3}{4}\overrightarrow{AG_A}$.

In a similar way, we can show the following convex combinations take place:

$$\overrightarrow{MG} = \frac{1}{4}\overrightarrow{MB} + \frac{3}{4}\overrightarrow{MG_B},$$

$$\overrightarrow{MG} = \frac{1}{4}\overrightarrow{MC} + \frac{3}{4}\overrightarrow{MG_C},$$

$$\overrightarrow{MG} = \frac{1}{4}\overrightarrow{MD} + \frac{3}{4}\overrightarrow{MG_D},$$

which means

$$G \in [BG_B], \qquad G \in [CG_C], \qquad G \in [DG_D],$$

i.e., G is the intersection point of the medians. The centroid G lies at $1/4$ from each face and $3/4$ from the vertices. ■

Exercise 13.10.4 *Let G be the centroid of the tetrahedron $ABCD$.*

(*i*) *Show that*

$$16|AG|^2 = 3(|AB|^2 + |AC|^2 + |AD|^2) - (|BC|^2 + |BD|^2 + |CD|^2).$$

(*ii*) *Show that*

$$|AG|^2 + |BG|^2 + |CG|^2 + |DG|^2 = \frac{1}{4}(|AB|^2 + |AC|^2 + |AD|^2 + |BC|^2 + |BD|^2 + |DA|^2).$$

Solved Exercise 13.10.5 *Let $ABCD$ be a tetrahedron such that $\widehat{BAC} = \widehat{CAD} = \widehat{BAD} = 90°$. Find the length of the medians of $ABCD$.*

Proof: Consider the Figure 13.25. We shall compute $|AG_A|^2$. The Pythagorean theorem applied in triangles ΔBAC, ΔBAD, ΔACD yields

$$|BC|^2 = |AB|^2+|AC|^2, \quad |BD|^2 = |AB|^2+|AD|^2, \quad |CD|^2 = |AC|^2+|AD|^2.$$

Substituting into the median theorem we obtain

$$\begin{aligned} |AG_A|^2 &= \frac{1}{3}(|AB|^2 + |AD|^2 + |AC|^2) - \frac{1}{9}(|BC|^2 + |BD|^2 + |CD|^2) \\ &= \frac{1}{3}(|AB|^2 + |AD|^2 + |AC|^2) - \frac{1}{9}(|AB|^2 + |AC|^2 + |AD|^2) \\ &= \frac{2}{9}(|AB|^2 + |AD|^2 + |AC|^2). \end{aligned}$$

The length of the other three medians are left as an exercise. ∎

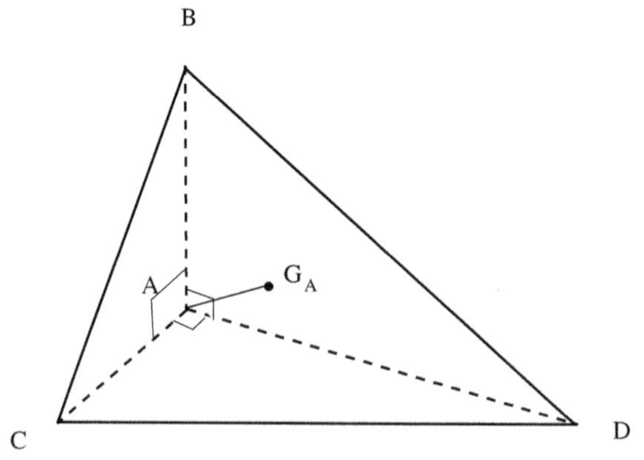

Figure 13.25: For Exercise 13.10.5.

Exercise 13.10.6 *The medians of a tetrahedron have the same length if and only if the opposite edges have the same length, namely, we have*

$$|AG_A| = |BG_B| = |CG_C| = |DG_D|$$

if and only if

$$|AB| = |CD|, \qquad |AD| = |BC|, \qquad |BD| = |AC|.$$

In 1773 the French mathematician Lagrange generalized to a tetrahedron the property proved by Leibnitz for a triangle.

Theorem 13.10.7 (Lagrange) *Let $ABCD$ be a tetrahedron and let G be its centroid. For any point M in the space we have*

$$|MA|^2+|MB|^2+|MC|^2+|MD|^2 = 4|MG|^2+|GA|^2+|GB|^2+|GC|^2+|GD|^2.$$

Proof: We have

$$\begin{aligned}
4\overrightarrow{MG} &= \overrightarrow{MA}+\overrightarrow{MB}+\overrightarrow{MC}+\overrightarrow{MD} \Longrightarrow \\
16|MG|^2 &= |MA|^2+|MB|^2+|MC|^2+|MD|^2+2\overrightarrow{MA}\cdot\overrightarrow{MB} \\
&\quad +2\overrightarrow{MA}\cdot\overrightarrow{MC}+2\overrightarrow{MA}\cdot\overrightarrow{MD}+2\overrightarrow{MB}\cdot\overrightarrow{MC}+2\overrightarrow{MB}\cdot\overrightarrow{MD} \\
&\quad +2\overrightarrow{MC}\cdot\overrightarrow{MD} \\
&= |MA|^2+|MB|^2+|MC|^2+|MD|^2+|MA|^2+|MB|^2-|AB|^2 \\
&\quad +|MA|^2+|MC|^2-|AC|^2+|MA|^2+|MD|^2-|AD|^2 \\
&\quad +|MB|^2+|MC|^2-|BC|^2+|MB|^2+|MD|^2 \\
&\quad -|BD|^2+|MC|^2+|MD|^2-|CD|^2 \\
&= 4(|MA|^2+|MB|^2+|MC|^2+|MD|^2) \\
&\quad -(|AB|^2+|AC|^2+|AD|^2+|BD|^2+|CD|^2+|BC|^2) \\
&= 4(|MA|^2+|MB|^2+|MC|^2+|MD|^2) \\
&\quad -4(|AB|^2+|AC|^2+|AD|^2+|BD|^2+|CD|^2+|BC|^2),
\end{aligned}$$

where we have used Exercise 13.10.4. Dividing by 4 leads to the desired result. ∎

Corollary 13.10.8 *Let $ABCD$ be a tetrahedron. Then the sum*

$$\mathcal{S}(M) = |MA|^2+|MB|^2+|MC|^2+|MD|^2$$

has a minimum for $M = G$, the centroid of the tetrahedron.

Exercise 13.10.9 *Prove Corollary 13.10.8.*

Exercise 13.10.10 *Let H be the orthocenter of the orthogonal tetrahedron $ABCD$. Show that*

$$\begin{aligned} 4|HG|^2 &= |HA|^2 + |HB|^2 + |HC|^2 + |HD|^2 \\ &\quad -\frac{1}{4}(|AB|^2 + |AC|^2 + |AD|^2 + |BC|^2 + |BD|^2 + |CD|^2). \end{aligned}$$

Exercise 13.10.11 *Let $ABCD$ be a tetrahedron inscribed in the sphere $\mathcal{S}(O, R)$. Find the distance $|OG|$ in terms of the circumradius R and the length of the edges of the tetrahedron. (G denotes the centroid.)*

13.11 Bimedians in a tetrahedron

The segments which join the midpoint of the opposite edges in a tetrahedron are called *bimedians*. A tetrahedron has three bimedians. In this section we shall deal with incidence and metric properties of the bimedians of a tetrahedron.

Theorem 13.11.1 *The bimedians of a tetrahedron intersect at a point, which is the midpoint of each bimedian.*

Proof: Let M, N, P, Q, R, S be the midpoints of the edges $[AB]$, $[CD]$, $[AD]$, $[BC]$, $[AC]$, $[BD]$, respectively, see Figure 13.26. The midpoint connector theorem yields $\overrightarrow{MQ} = \dfrac{1}{2}\overrightarrow{AC}$ and $\overrightarrow{PN} = \dfrac{1}{2}\overrightarrow{AC}$. Hence $\overrightarrow{MQ} = \overrightarrow{PN}$

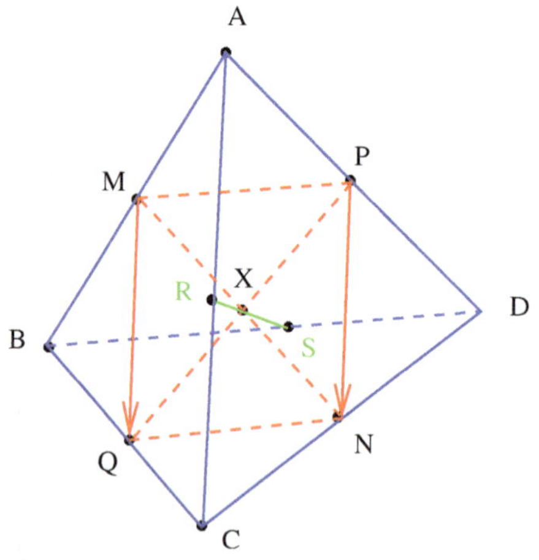

Figure 13.26: For Theorem 13.11.1.

and then $MPNQ$ is a parallelogram. Hence PQ and MN intersect at a point O, which is the midpoint of $[PQ]$ and $[MN]$. The same can be shown for the bimedians (MN, RS), (PQ, RS). Using Lemma 13.7.1 it follows that the bimedians intersect at the same point O. ■

Theorem 13.11.2 *The intersection point of the bimedians of a tetrahedron is the centroid of the tetrahedron.*

Proof: Let X be the point where the bimedians intersect. With the previous notations, we have $\overrightarrow{XM} + \overrightarrow{XN} = 0$, since X is the midpoint. Using Lemma 6.1.4 several times, we have

$$\begin{aligned} 0 &= \overrightarrow{XM} + \overrightarrow{XN} = \frac{1}{2}(\overrightarrow{XA} + \overrightarrow{XB}) + \frac{1}{2}(\overrightarrow{XC} + \overrightarrow{XD}) \\ &= \frac{1}{2}(\overrightarrow{XA} + \overrightarrow{XB} + \overrightarrow{XC} + \overrightarrow{XD}) \\ &= \frac{1}{2} \cdot 4\overrightarrow{XG} = 2\overrightarrow{XG}, \end{aligned}$$

where G denotes the centroid of the tetrahedron $ABCD$. Then $\overrightarrow{XG} = 0$ and hence $X = G$. ■

It is known that the bimedians are perpendicular to the edges they bisect, if and only if the opposite edges of the tetrahedron have the same length, see Proposition 13.8.4. Using this result, solve the following exercise.

Exercise 13.11.3 *Let M, N, E, F be the midpoints of the sides $[AD]$, $[BC]$, $[AB]$, $[DC]$ of the tetrahedron $ABCD$, see Figure 13.27.*

(i) Show that $\overrightarrow{ME} = \overrightarrow{FN}$, i.e., $MENF$ is a parallelogram.

(ii) Show that $MN \perp EF$ if and only if $|BD| = |AC|$.

Exercise 13.11.4 *Let $ABCD$ be an equifacial tetrahedron. Show that there is a sphere passing through the vertices A, B, C, D centered at the centroid G of the tetrahedron.*

The following results deal with the metric properties of the tetrahedron's bimedians.

Theorem 13.11.5 *With the previous notations, we have*

$$4|MN|^2 = |AC|^2 + |AD|^2 + |BC|^2 + |BD|^2 - |AB|^2 - |DC|^2.$$

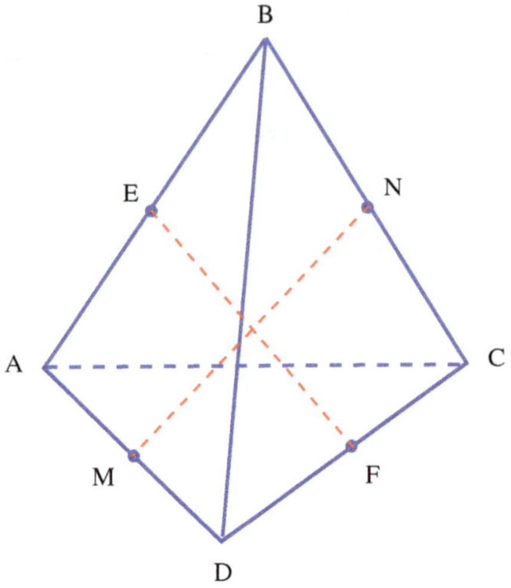

Figure 13.27: For Exercise 13.11.3.

Proof: Using the polygon rule, we can write

$$\begin{aligned}\overrightarrow{MN} &= \overrightarrow{MA}+\overrightarrow{AD}+\overrightarrow{DN}=\frac{1}{2}\overrightarrow{BA}+\overrightarrow{AD}+\frac{1}{2}\overrightarrow{DC}\\ &= \frac{1}{2}(\overrightarrow{BA}+\overrightarrow{AD}+\overrightarrow{DC})+\frac{1}{2}\overrightarrow{AD}=\frac{1}{2}(\overrightarrow{BC}+\overrightarrow{AD}).\end{aligned}$$

Taking the inner product by itself, and using Exercise 13.7.4, yields

$$\begin{aligned}4|MN|^2 &= (\overrightarrow{BC}+\overrightarrow{AD})\cdot(\overrightarrow{BC}+\overrightarrow{AD}) = |BC|^2+|AD|^2+2\overrightarrow{BC}\cdot\overrightarrow{AD}\\ &= |BC|^2+|AD|^2+2|BC|\cdot|AD|\cos(\widehat{\overrightarrow{BC},\overrightarrow{AD}})\\ &= |BC|^2+|AD|^2+2|BC|\cdot|AD|\frac{|BD|^2+|AC|^2-|AB|^2-|DC|^2}{2|BC|\cdot|AD|}\\ &= |BC|^2+|AD|^2+|BD|^2+|AC|^2-|AB|^2-|DC|^2.\end{aligned}$$

■

Corollary 13.11.6 *Let MN, PQ, RS be the bimedians of the tetrahedron $ABCD$. Then*

$$4(|MN|^2+|PQ|^2+|RS|^2) = |AB|^2+|AC|^2+|AD|^2+|BC|^2+|BD|^2+|CD|^2.$$

Proof: Using Theorem 13.11.5 for each bimedian and adding yields the desired relation. ■

Exercise 13.11.7 *With the previous notations, show that*
(*i*) $|MN|^2 + |PQ|^2 + |RS|^2 = |AG|^2 + |BG|^2 + |CG|^2 + |DG|^2$;
(*ii*) $\frac{16}{9}\left(|MN|^2+|PQ|^2+|RS|^2\right) = \left(|AG_A|^2+|BG_B|^2+|CG_C|^2+|DG_D|^2\right)$.

Hints and Solutions

This section contains answers, hints or brief solutions for selected exercises.

Chapter 1

Exercise 1.4.1. Proof by contradiction. Assume the lines intersect at two distinct points. Contradicts axiom I1.

Exercise 1.4.2. Proof by contradiction. Assume the line intersects the plane at two points, at least. Then by I2 the lines is contained in the plane, contradiction.

Exercise 1.4.3. Let A be the intersection point and let B and C be two other points, one on each of the lines. By axiom I3, the points A, B, C determine a plane which contains the lines.

Exercise 1.4.4. Consider two opposite sides of a tetrahedron.

Exercise 1.4.5. Consider the planes $\Pi_1 = (A_1, A_2, A_3)$ and $\Pi_2 = (B_1, B_2, B_3)$. Then the points U, $V.W$ belong to the intersection of Π_1 and Π_2, which is a line by axiom I4.

Exercise 1.4.7. Let $d_{12} = A_1A_2$ be the line determined by A_1 and A_2. Consider the line d_k determined by the points A_1, A_2, A_k. Since the lines d_{12} and d_k have two points in common, by I1 they coincide.

Exercise 1.4.8. Let $\mathcal{P}$ be the plane determined by the lines d_1 and d_2. Let Π_k be the plane determined by the lines d_1, d_2, d_k. Then $\mathcal{P} = \Pi_k$ since they have 3 points in common, see axiom I3.

Exercise 1.4.9. Let P_{123} be the point where the lines d_1, d_2, d_3 intersect. Then $P_{123} = P_{12k} = d_1 \cap d_2$. Also use the symmetries $P_{123} = P_{231} = P_{312} = P_{213} = P_{132} = P_{321}$.

Exercise 1.4.10. Proof by contradiction. Use Exercise 1.4.7.

Exercise 1.4.11. Proof by induction. $n(n-1)/2$.

Exercise 1.4.12. Proof by contradiction. Assume d_1 and d_2 not parallel and let P be their intersection point. Since $P \in \alpha$ and $P \in \gamma$, then $P \in d_3$, so the lines d_1, d_2, d_3 intersect at P, contradiction (the planes do not have a point in common).

Exercise 1.4.13. Subtract from the number of all segments joining two vertices, $N(N-1)/2$, the number of sides, N, and obtain $N(N-3)/2$.
Exercise 1.5.4. Consider a system of coordinates on the line AB. The midpoint M will have the unique coordinate $f(M) = \frac{1}{2}(f(A)+f(B))$, and so it is unique.
Exercise 1.5.5. The axioms $D1$–$D3$ are obviously satisfied. The property $D4$ can be proved in the following way. Denote $a = d(B,C)$, $b = d(A,C)$, $c = d(A,B)$. We need to show

$$\frac{a}{1+a} + \frac{c}{1+c} \geq \frac{b}{1+b}.$$

Consider the function $f(x) = \dfrac{x}{1+x}$, $x \in [0,\infty)$. The function f is increasing, so from the triangle inequality $a + c \geq b$ we get

$$f(a+c) \geq f(b). \tag{14.11.1}$$

The function f also satisfies the inequality

$$f(x) + f(y) \geq f(x+y).$$

Since $\dfrac{f(x)}{x} = \dfrac{1}{x+1}$ is decreasing for $x > 0$,

$$\frac{f(x+y)}{x+y} < \frac{f(x)}{x}, \qquad \frac{f(x+y)}{x+y} < \frac{f(y)}{y}.$$

Then

$$f(x+y) = \frac{f(x+y)}{x+y}(x+y) < \frac{f(x)}{x}x + \frac{f(y)}{y}y = f(x) + f(y).$$

From the above inequality and (14.11.1) yields

$$f(a) + f(c) \geq f(a+c) \geq f(b),$$

which is equivalent to $D4$.
Exercise 1.5.6. The axioms $D1$–$D3$ are obviously satisfied. The property $D4$ is equivalent to Cauchy's inequality

$$\sqrt{a_1^2 + b_1^2} + \sqrt{a_2^2 + b_2^2} \geq \sqrt{(a_1+a_2)^2 + (b_1+b_2)^2},$$

which can be shown by taking squares in both sides.
Exercise 1.5.7. (i) The equation $\dfrac{|OM|}{1+|OM|} = 1$ is never satisfied, so the circle of radius 1 is the empty set. (ii) The circle exists only in the case $r \in (0,1)$, case in which it is the usual Euclidean circle.

Exercise 1.5.8. It follows from the properties of the absolute value.
Exercise 1.5.9. It easily follows from the inequality $\sqrt{a+b} \le \sqrt{a} + \sqrt{b}$, for $a, b > 0$. The identity holds when either $a = 0$ or $b = 0$.
Exercise 1.5.10. (i) Use the definition. (ii) See Figure 1.4.
Exercise 1.5.11. Use the definition.
Exercise 1.5.12. $D4$ is an obvious computation using triangle inequality. The other properties are obviously satisfied.
Exercise 1.5.13. (i) It comes from the triangle inequality.
Exercise 1.5.15. Just P.
Exercise 1.5.16. Straightforward computation using the definition.
Exercise 1.6.4. If $u = 180 - x$ and $v = 180 - x$ then $u = v$.
Exercise 1.6.5. The angles $\widehat{BAC}$ and $\widehat{B'AC}$ have the same supplement $\widehat{BAC'}$. Apply Exercise 1.6.4.
Exercise 1.7.3. (i) If two points X and Y belong to $A \cap B$, then $X, Y \in A$ and $X, Y \in B$. Since A and B are convex, $[XY] \subset A$ and $[XY] \subset B$, and hence $[XY] \subset A \cap B$, i.e., $A \cap B$ is convex. The union $A \cup B$ is not always convex. Find a counterexample!
(ii) The same proof as (i).
(iii) Construct an infinite sequence of segments (convex sets) with empty intersection.
Exercise 1.7.6. $Int\Delta ABC = Int\widehat{BAC} \cap Int\widehat{ACB} \cap Int\widehat{ABC}$ is convex as an intersection of convex sets.
Exercise 1.7.7. Let E' be a point on $(AD$, $E \neq D$. Applying Pasch theorem in triangle $\Delta ACE'$ with transversal BD we obtain that there is a point $F = BD \cap AC$ between A and C. Applying Pasch again for triangle ΔBFC and transversal AD we obtain that AD intersects BC at an interior point E.
Exercise 1.7.9. We have $f(1) = 2$, $f(2) = 4 = f(1) + 2$, $f(3) = 7 = f(2) + 3$. In general $f(n) = f(n-1) + n$. By induction or by a direct computation we obtain $f(n) = 1 + \frac{n(n+1)}{2}$.
Variant: A point M in the plane of triangle ΔABC is unique characterized by the distances to the sides (normal coordinates). These distances are considered to be either positive or negative, if the point is above or below the support line of the side. The interior of the triangle is the region $(+++)$. Other regions are $(++-)$, $(+-+)$, $(-++)$, $(+--)$, $(-+-)$, $(--+)$. There is no region with coordinates $(---)$ (Why?). Hence there are only seven regions.
Exercise 1.7.10. Using normal coordinates, we have the following regions determined by the sides of a tetrahedron: $(++++)$ (the interior), $(+++-)$, $(++-+)$, $(+-++)$, $(-+++)$, $(++--)$, $(+-+-)$, $(+--+)$, $(-++-)$, $(-+-+)$, $(--++)$, $(---+)$, $(--+-)$, $(-+--)$, $(+---)$. There is no region described by $(----)$ (Why?). Hence there are 15 regions.

Exercise 1.7.11. Consider the analogy with a $3 \times 3 \times 3$ Rubik's cube. The inner cublet is the initial cube. There are 27 regions.
Exercise 1.7.14. (i) Divide the square using a median vertical line and a median horizontal line into four equal squares with side $1/2$. By Dirichlet's principle two of the five points will belong to the same square, so the distance between them is less than the diagonal of a small square, which is $\sqrt{2}/2$. (ii) In a similar way, divide the cube into 8 cublets of side $1/3$. By Dirichlet's principle, two of the points will belong to the same cublet, which has diagonal $\sqrt{3}/2$.
Exercise 1.7.15. Divide the square into n^2 equal squares, each having the diagonal equal to $\sqrt{2}/n$. Then apply Dirichlet's principle.
Exercise 1.7.16. (i) Divide the triangle into four equal triangles by joining the middle of the sides. By Dirichlet's principle two of the points will belong to the same triangle. Then apply that the distance between two points inside of an equilateral triangle is less than or equal to the side of the triangle. (ii) Iterating the procedure, we divide the triangle into 16 equal equilateral triangles, each having the side equal to $1/4$ and proceed as in part (i). (iii) Dividing the triangle into 4^n equilateral triangles, each of side $1/2^n$, then choosing $4^n + 1$ points inside an equilateral triangle with side equal to 1, we always get two points closer than $1/2^n$.
Exercise 1.8.6. We have $\Delta ABC \equiv \Delta ACB$ by (ASA) congruence case.
Exercise 1.8.7. Proof by construction. Let ΔABC and $\Delta A'B'C'$ be two triangles with $|AB| = |A'B'|$, $|BC| = |B'C'|$ and $|CA| = |C'A'|$. Overlap the triangles such that $A = A'$, $C = C'$ and we need to show $B = B'$. Assume $B' \in Int\,\Delta ABC$. Triangles $\Delta ABB'$ and $\Delta BCB'$ are isosceles, so $\widehat{ABB'} = \widehat{AB'B}$ and $\widehat{B'BC} = \widehat{BB'C}$. Adding we get $\widehat{ABC} = 360° - \widehat{AB'C}$, which contradicts that both angles $\widehat{ABC}$ and $\widehat{AB'C}$ have measures between) and 180°.
Exercise 1.8.8. $\Delta ABM \equiv \Delta ACM$ by (SSS) congruence case. Then $\widehat{BAM} = \widehat{MAC}$ and $\widehat{AMB} = \widehat{AMC}$. Since $\widehat{AMB} + \widehat{AMC} = 180°$, it follows that $\widehat{AMB} = 90°$.
Exercise 1.9.7. If there is a secant which forms equal alternate interior angles, then the lines are parallel. Hence any secant line will make equal alternate interior angles, see Theorem 1.9.5.
Exercise 1.10.1. The identities follow from congruent triangles. (i) Let X be the point where CD intersects AA'. Applying the midpoint connector theorem in $\Delta ABA'$ it follows that $|AX| = |XA'|$. Then $\Delta AXD \equiv \Delta A'XD$ by (SAS) congruence case (right triangles with common side XD). Hence $\widehat{ADC} = \widehat{CDA'}$. Similarly proof for the other identities.
Exercise 1.10.2. Let A' be the symmetric of A with respect to the midpoint M of the side BC. We have $\Delta ABM \equiv \Delta A'CM$ by (SAS) congruence

case. Then $|AB| = |CA'|$. Triangle inequality in $\Delta ACA'$ yields

$$2|AM| = |AA'| < |AC| + |CA'| = |AC| + |AB|,$$

then divide by 2.

Exercise 1.11.3. Let B on the circle, in the exterior of angle $\widehat{AOB}$. Let C' be the point where OC intersects the circle the second time. We have

$$\widehat{ABC} = 90^\circ - \widehat{ABC'} = 90^\circ - \frac{1}{2}\widehat{AOC'} = \frac{1}{2}(180^\circ - \widehat{AOC'}) = \frac{1}{2}\widehat{AOC}.$$

Exercise 1.11.5. By exterior angle identity we have

$$\widehat{BAD} = \widehat{ABE} + \widehat{AEB} = \frac{1}{2}arc(EC) + \frac{1}{2}arc(BD).$$

Exercise 1.11.6.

$$\begin{aligned}
\widehat{EDF} &= \widehat{EDA} + \widehat{ADC} - \widehat{FDC} \\
&= \frac{1}{2}(180^\circ - \widehat{EAD}) + \widehat{ADC} - \frac{1}{2}(180^\circ - \widehat{DCF}) \\
&= 90^\circ - \frac{1}{2}(360^\circ - \widehat{BAD} - \widehat{BAE}) + \widehat{ADC} - 90^\circ \\
&\quad + \frac{1}{2}(\widehat{DCB} + \widehat{BCF}) \\
&= -180^\circ + \frac{1}{2}\widehat{BAD} + \frac{1}{2}\widehat{BAE} + \widehat{ADC} + \frac{1}{2}\widehat{DCB} + \frac{1}{2}\widehat{BCF} \\
&= -180^\circ + \frac{1}{2}(\widehat{BAD} + \widehat{DCB}) + \frac{1}{2}\widehat{ABC} + \widehat{ADC} + \frac{1}{2}\widehat{ABC} \\
&= -180^\circ + \frac{1}{2}180 + (\widehat{ABC} + \widehat{ADC}) = -90^\circ + 180^\circ = 90^\circ.
\end{aligned}$$

Exercise 1.11.8. By the plane separation axiom, the line ℓ divides the plane in the half-planes H_1 and H_2. Show that one arc of the circle belongs to H_1 while the second belongs to H_2.

Exercise 1.11.9. Proof by contradiction. Assume the line ℓ intersects the circle at the second point B. Then the triangle ΔOAB is isosceles with the base angles 90°, contradiction.

Exercise 1.11.10. Proof by contradiction. Assume the line ℓ is not perpendicular on the radius OA. Then we can construct another line ℓ' through A such that $OA \perp \ell'$. By Exercise 1.11.9 we get ℓ' tangent line to the circle at A, so $\ell = \ell'$.

Exercise 1.11.12. By Exercise 1.11.10 $\widehat{AT_1O} = \widehat{AT_2O} = 90^\circ$. Since $\widehat{OT_1T_2} = \widehat{OT_2T_1}$ we get $\widehat{AT_1T_2} = \widehat{AT_2T_1}$ and triangle ΔAT_1T_2 is isosceles.

Exercise 1.11.13. By (SSS) congruence case we have $\Delta OAM \equiv \Delta OBM$. Hence $\widehat{OMA} = \widehat{OMB}$.

Exercise 1.11.14. Let $N \in AB$ such that $ON \perp AB$. Assume N is not the midpoint of AB. Let M be the midpoint of AB. By Exercise 1.11.13 we get $OM \perp AB$. Hence triangle ΔOMN has two right angles, contradiction.

Exercise 1.11.15. $|OO'| = R + R'$ and by the Pythagorean theorem $|TT'|^2 = |OO'|^2 - (R - R')^2$.

Exercise 1.11.16. The central angles $\widehat{AOB}$ and $\widehat{A'OB'}$ cut equal arcs on the circle, so they have equal measures. Then $\Delta OAB \equiv \Delta OA'B'$ by (SAS) congruence case. Then $|AB| = |A'B'|$.

Exercise 1.11.17. Show $\Delta AMB \equiv \Delta AMC$.

Exercise 1.11.19. (i) Since $\Delta ABD \equiv \Delta BCE$ by (SAS) congruence case, we get $\widehat{ECB} = \widehat{ABD}$. Then using the exterior angle identity we have

$$\widehat{DFC} = \widehat{FBC} + \widehat{FCB} = \widehat{DBC} + \widehat{ABD} = 60°.$$

Exercise 1.11.20. We have

$$\widehat{A} + \widehat{B} + \widehat{C} = \frac{1}{2}arc(BC) + \frac{1}{2}arc(CA) + \frac{1}{2}arc(AB) = \frac{360°}{2} = 180°.$$

Exercise 1.12.6. Apply Jensen's inequality for the concave function $f(x) = \sin^2 x$ on $(0°, 180°)$.

Exercise 1.12.7. Apply Jensen's inequality for the concave function $f(x) = \cos x$ on $(0°, 90°)$.

Exercise 1.12.8. Assume first that ΔABC is acute-angled and apply Jensen's inequality for the concave function $f(x) = \ln \cos x$ on $(0°, 90°)$. If ΔABC is not acute-angled, then it has one right or obtuse angle and two acute angles. In this case the product $\cos \widehat{A} \cos \widehat{B} \cos \widehat{C}$ is negative, and hence it is smaller that $1/8$.

Exercise 1.12.10. We apply the inequality of means for $x_1 = \frac{x}{y}$, $x_2 = \frac{y}{z}$, $x_3 = \frac{z}{x}$.

Chapter 2

Exercise 2.1.2. $1 + \tan^2 x = 1 + \frac{\sin^2 x}{\cos^2 x} = \frac{\sin^2 x + \cos^2 x}{\cos^2 x} = \frac{1}{\cos^2 x}$.

Exercise 2.1.8. Use formulas $\cos 2x = 1 - 2\sin^2 x = 2\cos^2 x - 1$.

Exercise 2.1.9. Let M be the midpoint of AB. Use $\sin \widehat{BOM} = \frac{|AB|}{2R}$ and $\widehat{BOM} = \frac{1}{2}\widehat{AOB}$.

Exercise 2.1.10. We have

$$\begin{aligned}\sin(x+y+z) &= \sin(z'+z) = \sin z'\cos z + \cos z'\sin z\\ &= \sin(x+y)\cos z + \cos(x+y)\sin z\\ &= \sin x\cos y\cos z + \cos x\sin y\cos z + \cos x\cos y\sin z\\ &\quad - \sin x\sin y\sin z.\end{aligned}$$

$$\begin{aligned}\cos(x+y+z) &= \cos(z'+z) = \cos z'\cos z - \sin z'\sin z\\ &= \cos(x+y)\cos z - \sin(x+y)\sin z\\ &= \cos x\cos y\cos z - \sin x\sin y\cos z - \sin x\cos y\sin z\\ &\quad - \cos x\sin y\sin z.\end{aligned}$$

Exercise 2.1.11.

$$\tan(x+y+z) = \tan(z'+z) = \frac{\tan z' + \tan z}{1-\tan z'\tan z} = \frac{\frac{\tan x+\tan y}{1-\tan x\tan y} + \tan z}{1 - \frac{\tan x+\tan y}{1-\tan x\tan y}\tan z}$$

$$= \frac{\tan x + \tan y + \tan z - \tan x\tan y\tan z}{1 - \tan x\tan y - \tan y\tan z - \tan z\tan x}.$$

Exercise 2.1.13. Let $x = \tan^{-1} 1$, $y = \tan^{-1} 2$ and $z = \tan^{-1} 3$ and note that $\tan x + \tan y + \tan z - \tan x\tan y\tan z = 1+2+3-1\cdot 2\cdot 3 = 0$. Then apply Exercise 2.1.11.

Exercise 2.2.2. Consider a circle of diameter BC passing through A. As a radius which is half the diameter, we have $|AM| = 5/2$.

Exercise 2.2.3. Use that any chord is shorter than the diameter of the circumcircle.

Exercise 2.2.4. Use Exercise 2.2.3 and add the inequalities. Since the identity can hold for at most one side, the sum will have a strict inequality.

Exercise 2.2.5. Use the law of sines for each angle and add the identities.

Exercise 2.2.6. Using the law of sines the relation $2a = b + c$ becomes

$$\begin{aligned}2\sin\widehat{A} &= \sin\widehat{B} + \sin\widehat{C} \iff\\ 2\sin\frac{\widehat{A}}{2}\cos\frac{\widehat{A}}{2} &= 2\sin\frac{\widehat{B}+\widehat{C}}{2}\cos\frac{\widehat{B}-\widehat{C}}{2} \iff\\ 2\sin\frac{\widehat{A}}{2}\cos\frac{\widehat{A}}{2} &= 2\sin(90-\widehat{A})\cos\frac{\widehat{\widehat{B}}-\widehat{C}}{2} \iff\\ \sin\frac{\widehat{A}}{2} &= \cos\frac{\widehat{B}-\widehat{C}}{2}.\end{aligned}$$

Exercise 2.3.5. Assume the altitudes are parallel and obtain a contradiction with the sum of the angles in triangle ΔABH equal to $180°$.

Exercise 2.3.6. Let $A' = AM \cap BC$. Then AA' splits triangle ΔABC into two triangles, so $\mathcal{A}_{ABC} = \mathcal{A}_{AA'B} + \mathcal{A}_{AA'C}$. Also, MA' splits triangle ΔBMC into two triangles, so $\mathcal{A}_{BMC} = \mathcal{A}_{BMA'} + \mathcal{A}_{MCA'}$. Subtracting yields

$$\mathcal{A}_{ABC} - \mathcal{A}_{BMC} = \mathcal{A}_{ABM} + \mathcal{A}_{ACM},$$

which is equivalent to the desired result.

Exercise 2.3.9. (i) If $N = 0$, the minimum value of k is 3, so $N + \frac{k}{2} = 1.5 > 1$.

(ii) Denote by ℓ the number of dots on the common sides, counting the endpoints. Then

$$\begin{aligned} N &= N_1 + N_2 + (\ell - 2) \\ k &= (k_1 - \ell) + (k_2 - \ell) + 2. \end{aligned}$$

Then we have

$$\begin{aligned} \left(N_1 + \frac{k_1}{2} - 1\right) + \left(N_2 + \frac{k_2}{2} - 1\right) &= (N_1 + N_2 + \ell - 2) \\ &\quad + \frac{k_1 + k_2 - 2\ell + 2}{2} - 1 \\ &= N + \frac{k}{2} - 1. \end{aligned}$$

(iii) Consider in the beginning a parallelogram with no interior points, $N = 0$, and show that its area is equal to 1. We can also verify the obvious identity

$$N + \frac{k}{2} - 1 = 1.$$

Then divide the polygon in a finite number of parallelograms and use case (ii) iteratively to find the area of the entire polygon.

Exercise 2.5.6. Express the triangle area in terms of r and R. We obtain $r = 1$ and $R = 5/2$.

Exercise 2.5.7. Find the area of the triangle and use formulas $\mathcal{A} = pr = abc/(4R)$.

Exercise 2.5.8. Use the formula $1/|AI| = (\sin \frac{A}{2})/r$ and two analog formulas and add them.

Exercise 2.5.9. Use $\sin^2 \frac{B}{2} = r^2/|BI|^2$ and the Pythagorean relation $|BI|^2 = r^2 + (p - b)^2$.

Exercise 2.5.10. $p = 10.5$, $p - 5 = 5.5$, $p - 7 = 3.5$ and $p - 9 = 1.5$.

Exercise 2.5.12. From $\mathcal{A} = ah_a/2$ we get $\frac{1}{h_a} = \frac{a}{2\mathcal{A}}$. Adding this relation with two similar relations and using $\mathcal{A} = pr$ yields the desired relation.

Exercise 2.5.13. Apply the means inequality $(xyz)^{1/3} \le \frac{1}{3}(x + y + z)$ for $x = h_a^{-1}$, $y = h_b^{-1}$, $z = h_c^{-1}$, and then apply Exercise 2.5.12.

Exercise 2.5.14. Decompose the area of ΔABC into the areas of triangles ΔAMB, ΔAMC and ΔBCM. The sum of the distances is equal to the altitude of triangle ΔABC.
Exercise 2.5.15. Show that triangles ΔADC and ΔBDC have the same area. Then subtract the area of ΔDOC.
Exercise 2.6.1. Use the decomposition $\mathcal{A}_{ABC} = \mathcal{A}_{ABI_a} + \mathcal{A}_{ACI_a} - \mathcal{A}_{BCI_a}$ and use $\mathcal{A}_{ABI_a} = |AB|r_a$ and the other similar relations.
Exercise 2.6.2. Using $\mathcal{A} = (p-a)r_a$ we get $\frac{1}{r_a} = \frac{p-a}{\mathcal{A}}$. Similar relations can be obtained for $\frac{1}{r_b}$ and $\frac{1}{r_c}$. Add the relations and use $\mathcal{A} = pr$.
Exercise 2.6.3. $\sqrt{rr_ar_br_c} = \sqrt{\frac{\mathcal{A}}{p}\frac{\mathcal{A}}{p-a}\frac{\mathcal{A}}{p-b}\frac{\mathcal{A}}{p-c}} = \mathcal{A}$ after using Heron's formula.
Exercise 2.6.4. (*i*) Apply the inequality between the geometric and arithmetic mean and Exercise 2.6.2. (*ii*) Apply part (*i*) and Exercise 2.6.3.
Exercise 2.6.5. (*i*) Consider the polynomial function

$$\begin{aligned} f(x) &= \left(x - \frac{1}{h_a}\right)\left(x - \frac{1}{h_b}\right)\left(x - \frac{1}{h_c}\right) \\ &= x^3 - \left(\frac{1}{h_a} + \frac{1}{h_b} + \frac{1}{h_c}\right)x^2 + \left(\frac{1}{r_ar_b} + \frac{1}{r_br_c} + \frac{1}{r_cr_a}\right)x - \frac{1}{h_ah_bh_c} \\ &= x^3 - \frac{1}{r}x^2 + \left(\frac{1}{r_ar_b} + \frac{1}{r_br_c} + \frac{1}{r_cr_a}\right)x - \frac{1}{h_ah_bh_c}, \end{aligned}$$

which has three real roots, $x_1 = 1/h_a$, $x_2 = 1/h_b$ and $x_3 = 1/h_c$. Then its derivative

$$f'(x) = 3x^2 - \frac{2}{r}x + \left(\frac{1}{r_ar_b} + \frac{1}{r_br_c} + \frac{1}{r_cr_a}\right)$$

vanishes at two points, so its discriminant is positive, $b^2 - 4ac \geq 0$. This produces the desired inequality. (*ii*) It is similar to part (*i*).
Exercise 2.6.6.

$$\frac{r_a}{r_b} = \frac{\mathcal{A}/(p-a)}{\mathcal{A}/(p-b)} = \frac{p-b}{p-a} = \frac{a-b+c}{b+c-a}.$$

Exercise 2.6.7. Use $r_a = \frac{\mathcal{A}}{p-a}$ and two similar relations.
Exercise 2.6.8. Since $\mathcal{A} = r_a(p-a)$ and $\mathcal{A} = rp$ we get $r_a(p-a) = rp$. If $r_a = 3r$ then $3r(p-a) = rp$, or $3(p-a) = p$ or $2a = b + c$.
Exercise 2.7.5. Let $\widehat{B} > \widehat{C}$.

$$\begin{aligned} \widehat{IAO} &= \widehat{BAO} - \widehat{BAI} = (90^\circ - \widehat{C}) - \frac{\widehat{A}}{2} \\ &= \frac{180^\circ - 2\widehat{C} - \widehat{A}}{2} = \frac{\widehat{B} - \widehat{C}}{2}. \end{aligned}$$

Exercise 2.7.6. Find the area and then use formula $R = \frac{abc}{4\mathcal{A}}$.
Exercise 2.7.7. Denote the cities by A, B and C and the Osiris temple by O, the circumcenter. The distances from the temple to each city is the circumradius

$$|OA| = |OB| = |OC| = R.$$

The distances between the cities are $|AB| = c$, $|AC| = b$ and $|BC| = a$. Assume $a \le b \le c$. The trip $OAOBOCO$ has the length $6R$. The trip $OBCA$ has the length $2R + a + b$. Since $a \le 2R$ and $b \le 2R$, the second itinerary is shorter.
Exercise 2.7.8. (*i*) Using Exercise 2.6.1 and Heron's formula we have

$$\begin{aligned} r_a + r_b + r_c - r &= \frac{\mathcal{A}}{p-a} + \frac{\mathcal{A}}{p-b} + \frac{\mathcal{A}}{p-c} - \frac{\mathcal{A}}{p} \\ &= \frac{\mathcal{A}}{\mathcal{A}^2}[(p-b)(p-c)p \\ &+ (p-a)(p-c)p + (p-a)(p-b)p - (p-a)(p-b)(p-c)] \\ &= \frac{1}{\mathcal{A}}abc = 4R, \end{aligned}$$

where we used the computation

$$\begin{aligned} &(p-b)(p-c)p + (p-a)(p-c)p + (p-a)(p-b)p - (p-a)(p-b)(p-c) \\ &= p^3 - (b+c)p^2 + bcp + p^3 - (a+c)p^2 + acp + p^3 - (a+b)p^2 + abp \\ &- [p^3 - (a+b+c)p^2 + (ab+bc+ca)p - abc] \\ &= 2p^3 - p^2(b+c+a+c+a+b-a-b-c) + abc \\ &= 2p^3 - p^2(a+b+c) + abc = abc. \end{aligned}$$

(*ii*) It follows from part (*i*) and from Euler's inequality, $R \ge 2r$.
Exercise 2.7.14. Let M be the midpoint of BC. The diameter through M perpendicular on BC cuts the circle at two points A_1 and A_2. If $|MA_1| < |MA_2|$, then choose $A = A_2$. The point A is constructed such that maximizes the distance from the circle to the chord BC. Then $\mathcal{A}_{ABC} = \frac{1}{2}|BC|dist(A, BC)$ is maximum. In order to show that $|AB| = |AC|$, we prove first that $\Delta AMB \equiv \Delta AMC$.
Exercise 2.7.15. (*i*) It follows from a repeated application of the Exercise 2.7.14. (*ii*) The area $\mathcal{A}_{ABC} = 2R^2 \sin\widehat{A} \sin\widehat{B} \sin\widehat{C}$ has the radius R constant, so its maximum occurs simultaneous with the maximum of $\sin\widehat{A} \sin\widehat{B} \sin\widehat{C}$. Then apply formula (1.12.3) from page 47.
Exercise 2.11.3. (*i*) Assume C' is not the midpoint. Let C'' be the midpoint of AC, $C' \ne C''$. By the MCT we have $B'C''||BC$. This contradicts the parallels postulate, since there are two parallel lines $B'C'$ and $B'C''$

through B' parallel to BC. It follows that $C'' = C'$. (*ii*) It follows from the MCT.

Exercise 2.12.3. (*i*) Use the similarity $\Delta ABD \sim \Delta CAD$. (*ii*) It follows from the means inequality. $\sqrt{|BD| \cdot |DC|} \leq \frac{|BD|+|DC|}{2} = \frac{|BC|}{2}$. (*iii*) Use triangle inequality $|AB| + |AC| > |BC|$. (*iv*) Writing the area in two ways $|AD| \cdot |BC| = \dfrac{|AB| \cdot |AC|}{|BC|}$ and solving for $|AD|$ yields $\dfrac{1}{|AD|^2} = \dfrac{|AB|^2 + |AC|^2}{|AB|^2|AC|^2} = \dfrac{1}{|AC|^2} + \dfrac{1}{|AB|^2}$.

Exercise 2.13.9. Using Carnot's relation, the inequality becomes $3R \geq 2(R + r)$, which is equivalent to Euler's inequality, $R \geq 2r$.

Exercise 2.13.10. (*i*) and (*ii*) follow from the congruence $\Delta ABD \equiv \Delta CDB$. (*iii*) Let O be the intersection of the diagonals. Show $\Delta ABO \equiv \Delta CDO$.

Exercise 2.13.11. It follows from congruence of triangles.

Exercise 2.13.12. Use Exercise 2.13.11.

Exercise 2.13.13. From $\Delta ABO \equiv \Delta ADO$ it follows that $\widehat{AOB} = \widehat{AOD}$. Then use $\widehat{AOB} + \widehat{AOD} = 180°$.

Exercise 2.13.14. Let $ABCD$ be a parallelogram and denote by S the intersection point of the angle bisectors of angles $\widehat{DAB}$ and $\widehat{ADC}$. Since $AB \| DC$, we have $\widehat{DAB} + \widehat{ADC} = 180°$. Then $\widehat{DAS} + \widehat{ADS} = \frac{1}{2}\widehat{DAB} + \frac{1}{2}\widehat{ADC} = 90°$, so $AS \perp DS$. Similarly, we can show that any two angle bisectors are perpendicular, so they form a rectangle.

Exercise 2.13.15. (*i*) and (*ii*) By Carnot's theorem the circumradii of triangles ΔBHC, ΔAHC and ΔAHB are equal to R, the circumradius of ΔABC. Since the circumcircles of triangles ΔBHC, ΔAHC and ΔAHB are passing through H, then we have the radii identity $|HO_A| = |HO_B| = |HO_C| = R$. This implies that the circumcircle of triangle $\Delta O_AO_BO_C$ is centered at H and its radius is R. (*iii*) By symmetry reasons $O_AO_B \perp CH$ (center axis perpendicular to the common chord). Since $CH \perp AB$, then $AB \| O_AO_B$. Similarly, we obtain $BC \| O_BO_C$ and $CA \| O_AO_C$. Hence $\Delta ABC \sim \Delta O_AO_BO_C$, and let ρ denote their similarity ratio. Since the triangles ΔABC and $\Delta O_AO_BO_C$ have equal circumradii, it follows that $\rho = 1$, so $\Delta O_AO_BO_C \equiv \Delta ABC$. (*iv*) We apply the law of cosines in

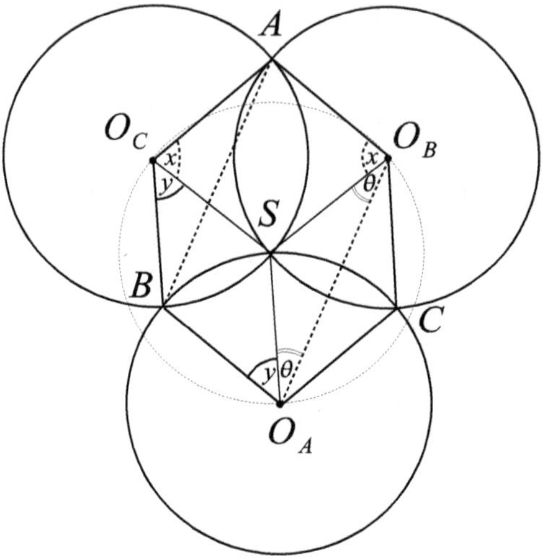

Figure 14.1: For Exercise 2.13.16.

triangle ΔHO_AO_B as

$$|O_AO_B|^2 = |HO_A|^2 + |HO_B|^2 - 2|HO_A|\,|HO_B| \cos \widehat{O_AHO_C}$$
$$= 2R^2 - 2R^2 \cos \widehat{O_AHO_C} = 4R^2 \frac{1 - \cos \widehat{O_AHO_C}}{2}$$
$$= 4R^2 \sin^2 \frac{\widehat{O_AHO_C}}{2}.$$

On another side, using the law of sines we have $|O_AO_B|^2 = c^2 = 4R^2 \sin^2 \widehat{C}$. Equating the last two relations yields $\widehat{C} = \frac{\widehat{O_AHO_C}}{2}$.

Exercise 2.13.16. Let O_A, O_B and O_C be the centers of the circles and S their intersection point. The following segments are equal as being radii of equal circles

$$|SO_A| = |SO_B| = |SO_C| = |O_BA| = |O_BC|$$
$$= |O_AB| = |O_AC| = |O_CB| = |O_CA|.$$

Therefore, AO_BSO_C, CO_BSO_A, BO_ASO_C are parallelograms as having equal opposite sides. First we note that the circle passing through O_A, O_B, O_C is equal to the given circles as having the same radius. The conclusion follows from the congruence $\Delta ABC \equiv \Delta O_AO_BO_C$. This follows from the (SSS) case as follows. Consider the equal angles denoted by x, y and θ as in Figure 14.1. Since the opposite sides in a parallelogram are parallel, we obtain $AO_B \| BO_A$. Since the secant O_AO_B forms supplementary angles, $\widehat{AO_BO_A} + \widehat{BO_AO_B} = 180°$, or $x+\theta+y+\theta = 180°$, so $x+y = 180° - 2\theta$. In ΔSO_AO_B the sum of angles is 180°, so $\widehat{O_ASO_B} = x + y$. Since $\widehat{BO_CA} = x + y$, then $\widehat{O_ASO_B} = \widehat{BO_CA}$. Therefore, $\Delta SO_AO_B \equiv \Delta O_CBA$ by (SAS) congruence case. It follows that $|AB| = |O_AO_B|$. Similarly we can show $|BC| = |O_BO_C|$ and $|CA| = |O_CO_A|$.

Exercise 2.14.4. Let $A'(-1,4)$ and $B'(-5,1)$ be the symmetric points of A and B with respect to the y-axis. Find the point $M(0,y)$ which is the intersection of the line $A'B$ with the y-axis.

Exercise 2.14.5. Let $M(0,y)$ variable on the y-axis and form the function $f(y) = |AM|^2 + |BM|^2 = (1-0)^2 + (4-y)^2 + (5-0)^2 + (1-y)^2$, and minimize $f(y)$ using the derivative test.

Exercise 2.14.6. If both A and B are outside of $\mathcal{C}$, then consider the point $M \in \mathcal{C}$ such that AM and BM make equal angles with the tangent line at M. For this position of M we have $|AM| + |BM|$ minimum. This phenomenon occurs whenever two points see each other in a spherical convex mirror. If A and B are both inside the circle, the solution is similar. If A is outside and B is inside, then take $M = AB \cap \mathcal{C}$, case in which $|AM| + |BM| = |AB|$.

Exercise 2.14.9. In the long run the ball trajectory will align to the focal axis F_1F_2.

Exercise 2.14.10. Since both balls will align to the focal axis, they will have collinear trajectories and will struck each other.

Exercise 2.15.7. Let $\widehat{A} = 90°$. Then $|AH| = 2R\cos\widehat{A} = 0$.

Exercise 2.15.9. From $\widehat{A} = 180° - 2\widehat{A}$ follows $\widehat{A} = 60°$.

Exercise 2.15.11. It suffices to show that $\widehat{C_1B_1H} = \widehat{HB_1A_1}$.

Exercise 2.15.12. They are similar by (AA).

Exercise 2.15.17. It follows from the fact that $\mathcal{A}_{med} = \frac{1}{4}\mathcal{A}_{ABC}$ and the inequality $\mathcal{A}_0 \le \frac{1}{4}\mathcal{A}_{ABC}$.

Exercise 2.16.3. Applying the law of cosines we have

$$\begin{aligned} bc\cos\widehat{A} + ca\cos\widehat{B} + ab\cos\widehat{C} &= \frac{1}{2}(b^2+c^2-a^2) + \frac{1}{2}(a^2+c^2-b^2) \\ &+ \frac{1}{2}(a^2+b^2-c^2) = \frac{1}{2}(a^2+b^2+c^2). \end{aligned}$$

Exercise 2.16.5. Use $\sin\frac{\widehat{A}}{2} = \sqrt{\frac{1-\cos\widehat{A}}{2}}$ and use a similar proof.

Exercise 2.17.2. Using Theorem 2.16.4, part 3) and Heron's formula, we have

$$(p-a)\tan\frac{\widehat{A}}{2} = (p-a)\frac{\sqrt{(p-b)(p-c)}}{\sqrt{p(p-a)}} = \sqrt{\frac{(p-a)(p-b)(p-c)}{p}}$$
$$= \frac{\sqrt{p(p-a)(p-b)(p-c)}}{p} = \frac{\mathcal{A}}{p} = \frac{pr}{p} = r.$$

Exercise 2.17.3.

$$r_a = \frac{\mathcal{A}}{p-a} = \frac{\sqrt{p(p-a)(p-b)(p-c)}}{p-a} = \sqrt{p}\frac{\sqrt{(p-b)(p-c)}}{\sqrt{(p-a)}}$$
$$= p\frac{\sqrt{(p-b)(p-c)}}{\sqrt{p(p-a)}} = p\tan\frac{\widehat{A}}{2}.$$

Exercise 2.17.4. (i) We have

$$\cos\frac{\widehat{A}}{2}\cos\frac{\widehat{B}}{2}\cos\frac{\widehat{C}}{2} = \sqrt{\frac{p(p-a)}{bc}}\sqrt{\frac{p(p-b)}{ac}}\sqrt{\frac{p(p-c)}{ab}}$$
$$= \frac{p\sqrt{p(p-a)(p-b)(p-c)}}{abc} = \frac{p}{abc}\mathcal{A} = \frac{p}{abc}\frac{abc}{4R} = \frac{p}{4R}.$$

(ii) It follows from the law of sines and part (i)

$$\sin\widehat{A} + \sin\widehat{B} + \sin\widehat{C} = \frac{a}{2R} + \frac{b}{2R} + \frac{c}{2R} = \frac{p}{R}$$
$$= 4\frac{p}{4R} = 4\cos\frac{\widehat{A}}{2}\cos\frac{\widehat{B}}{2}\cos\frac{\widehat{C}}{2}.$$

Exercise 2.18.2. Consider the cubic polynomial

$$f(x) = (x-a)(x-b)(x-c) = x^3 - (a+b+c)x^2 + (ab+bc+ca)x - abc,$$

which has three real roots, $x_1 = a$, $x_2 = b$ and $x_3 = c$. Then by Rolle's theorem its derivative, $f'(x) = 3x^2 - 4px + (ab+bc+ca)$, has two real roots, so $\Delta = b^2 - 4ac = 16p^2 - 12(ab+bc+ca) \geq 0$, or, $4p^2 \geq 3(ab+bc+ca)$. Using Proposition 2.18.1 we have $4p^2 \geq 3p^2 + 3r^2 + 12Rr$, which is equivalent to the desired inequality.

Exercise 2.18.3. It follows from Exercise 2.18.2 and Exercise 2.7.8.

Exercise 2.19.2. From $p_0 = h_a \sin\widehat{A} = h_b \sin\widehat{B} = h_c \sin\widehat{C}$ we have $\frac{1}{h_a} + \frac{1}{h_b} + \frac{1}{h_c} = \frac{\sin\widehat{A}+\sin\widehat{B}+\sin\widehat{C}}{p_0}$. Then use the relation $p_0 R = pr$ and the fact that $\sin\widehat{A} + \sin\widehat{B} + \sin\widehat{C} = \frac{p}{R}$.

Exercise 2.19.3. (i) $2R\sin\widehat{A}\sin\widehat{B}\sin\widehat{C} = p_0 \leq \frac{1}{2}p$, so $\sin\widehat{A}\sin\widehat{B}\sin\widehat{C} \leq \frac{1}{4}\frac{p}{R} = \frac{1}{4}(\sin\widehat{A}+\sin\widehat{B}+\sin\widehat{C})$.
(ii) Use $p_0 = \frac{1}{2}(a_0+b_0+c_0) = \frac{1}{2}(a\cos\widehat{A}+b\cos\widehat{B}+c\cos\widehat{C})$ and $p_0 \leq \frac{1}{2}p$.
Exercise 2.19.4. If $\widehat{A} = 90°$ the orthic triangle degenerates to the altitude AD, with $AD \perp BC$. Its perimeter is equal to $|AD|$. If $\widehat{A} > 90°$, Fagnano's theorem does not hold, since in this case the orthic triangle has two vertices in the exterior of the triangle sides and there are inscribed triangles with a smaller perimeter than the orthic triangle.
Exercise 2.20.2. It follow from (2.20.26) $|E_1E_2| = |AD|\cos 90°\,(\tan x + \tan y) = 0$.
Exercise 2.20.3. Substituting $\widehat{A} = \widehat{B} = 60°$ in relation (2.20.28) we obtain that y is the unique solution of the equation

$$\frac{2\sin(60° - 2y)}{\cos 60° + \cos(60° - 2y)} + \cot(60° + y) = 0.$$

We verify directly that for $y = 30°$ we have a solution. Then $x = \widehat{A} - y = 60° - 30° = 30°$, so AD is the angle bisector, which is also a median in ΔABC. Therefore, D is the midpoint of BC. In this case E_1E_2 is a midpoint connector, and $|E_1E_2| = a/2$.
Exercise 2.20.4. (i) $\mathcal{A}_{AFD} = \frac{1}{2}|AF|\,|AD|\sin\widehat{FAD} = \frac{1}{2}|AD|^2\cos\widehat{A}|\sin(y - x)|$. (ii) The minimum is equal to zero and it is realized when $x = y$, i.e., when AD is an angle bisector.
Exercise 2.20.5. Let $AD \perp BC$. Then $x = 90° - \widehat{B}$, $y = 90° - \widehat{C}$. Substituting in formula (2.20.26) yields

$$\begin{aligned}|E_1E_2| &= |AD|\cos\widehat{A}\,(\tan x + \tan y) = |AD|\cos\widehat{A}\,\big(\cot(\widehat{B}) + \cot(\widehat{C})\big)\\ &= \Big(|AD|\cot(\widehat{B}) + |AD|\cot(\widehat{C})\Big)\cos\widehat{A}\\ &= (|BD| + |DC|)\cos\widehat{A} = |BC|\cos\widehat{A} = a\cos\widehat{A} = a_0,\end{aligned}$$

which is a side of the orthic triangle.
Exercise 2.20.6. Using formula (2.20.26) as well as the cosines theorem and the expression of the angle bisector in terms of sides, see page 136, we have

$$\begin{aligned}|E_1E_2| &= 2|AD|\cos\widehat{A}\tan\frac{\widehat{A}}{2} = \frac{4bc}{b+c}\cos\frac{\widehat{A}}{2}\cos\widehat{A}\,\frac{\sin\frac{\widehat{A}}{2}}{\cos\frac{\widehat{A}}{2}}\\ &= \frac{4bc}{b+c}\cos\widehat{A}\sin\frac{\widehat{A}}{2} = \frac{4bc}{b+c}\,\frac{b^2+c^2-a^2}{2bc}\sqrt{\frac{(p-b)(p-c)}{bc}}\\ &= \frac{2(b^2+c^2-a^2)}{b+c}\sqrt{\frac{(p-b)(p-c)}{bc}}.\end{aligned}$$

Exercise 2.21.4. It follows from the law of cosines and the median theorem

$$\frac{a^2}{4} + bc\cos\widehat{A} = \frac{a^2}{4} + bc\frac{b^2+c^2-a^2}{2bc} = \frac{b^2+c^2}{2} - \frac{a^2}{4} = m_a^2.$$

Exercise 2.21.5. Add the formulas for the medians m_a^2, m_b^2, m_c^2.

Exercise 2.21.6. If $\widehat{A} = 90°$ substitute the Pythagorean theorem into the formula of m_a^2. If $m_a^2 = a^2/4$ then the median theorem yields the Pythagorean theorem.

Exercise 2.21.8. Apply Stewart's relation with $\delta = |AA'|$, $x = p - b$ and $y = p - c$ and obtain

$$|AA'|^2 = c^2(p-c) + b^2(p-b) - (p-b)(p-c).$$

Exercise 2.22.6. If $|AD| = |AE|$, then ΔADE is isosceles, so $\widehat{ADE} = \widehat{AED}$. Their supplement angles are also equal, $\widehat{ADB} = \widehat{AEC}$. Then using (SAS) congruence case we have $\Delta ABD \equiv \Delta ACE$. It follows that $|AB| = |AC|$.

Exercise 2.22.9. Since $(b-c)^2 \geq 0$ we get $2bc \leq b^2 + c^2$, with identity for $b = c$. Then $s_a = \frac{2bc}{b^2+c^2} m_a \leq m_a$.

Exercise 2.22.10. It follows from $\frac{s_a}{m_a} \leq 1$, see Exercise 2.22.9.

Exercise 2.22.11. Apply that $s_a = \frac{2bc}{b^2+c^2} m_a$ and use the median theorem.

Exercise 2.22.12. The inequality is equivalent to $m_a^2 > 0$ via the median theorem.

Exercise 2.22.17. The rank is $2-k$. If AD_1 and AD_2 are isogonal cevians, with $\frac{|BD_1|}{|D_1C|} = (\frac{c}{b})^k$, by Steiner's relation we have $\dfrac{|BD_1|}{|D_1C|}\dfrac{|BD_2|}{|D_2C|} = \left(\frac{c}{b}\right)^2$ and hence $\dfrac{|BD_2|}{|D_2C|} = \left(\frac{c}{b}\right)^{2-k}$.

Exercise 2.22.18. Let $\frac{x}{y} = \left(\frac{c}{b}\right)^k$ with $x + y = a$. Solving yields $x = \frac{a}{1+\left(\frac{b}{c}\right)^k}$ and $y = \frac{a}{1+\left(\frac{c}{b}\right)^k}$. Then apply Stewart's relation.

Exercise 2.23.1. Apply the law of sines in triangles ΔABD_1 and ΔAD_2C and obtain $|AD_1| = \frac{c\sin\widehat{B}}{\sin\left(\widehat{B}+\frac{\widehat{A}}{3}\right)}$ and $|AD_2| = \frac{b\sin\widehat{C}}{\sin\left(\widehat{C}+\frac{\widehat{A}}{3}\right)}$.

Exercise 2.24.3. Using the inequality between the harmonic and geometric mean yields

$$i_a = \frac{2bc}{b+c}\cos\frac{\widehat{A}}{2} \leq \frac{2bc}{b+c} \leq \sqrt{bc}.$$

The first inequality becomes identity when $\cos\frac{\widehat{A}}{2} = 1$, i.e., $\widehat{A} = 180°$, which never hold in a triangle.

Exercise 2.24.4. We have

$$\begin{aligned}
m_a^2 &= \frac{b^2+c^2}{2} - \frac{a^2}{4} = \frac{1}{4}\left(2(b^2+c^2) - a^2\right) \\
&= \frac{1}{4}\left((b^2+c^2+2bc) + (b^2+c^2-2bc) - a^2\right) \\
&= \frac{1}{4}\left((b+c)^2 + (b-c)^2 - a^2\right) \\
&= \frac{(b+c)^2}{4}\left(1 + \left(\frac{b-c}{b+c}\right)^2 - \frac{a^2}{(b+c)^2}\right) \geq \frac{(b+c)^2}{4}\left(1 - \frac{a^2}{(b+c)^2}\right) \\
&\geq bc\left(1 - \frac{a^2}{(b+c)^2}\right) = \frac{4bc}{(b+c)^2}\frac{(b+c)^2 - a^2}{4} \\
&= \frac{4}{(b+c)^2} bcp(p-a) = i_a^2,
\end{aligned}$$

where we used $4bc \leq (b+c)^2$ and the triangle inequality $a/(b+c) < 1$.

Exercise 2.24.5. Use $\cos 45^\circ = \frac{\sqrt{2}}{2}$.

Exercise 2.24.6. Use Exercise 2.24.5.

Exercise 2.24.7. Use $a = 5$, $b = 4$, $c = 3$, $m_a = 5/2$ and $s_a = \frac{2bc}{b^2+c^2} m_a$.

Exercise 2.24.8. (i) Assume $b > c$. Then $\widehat{HAI} = \widehat{BAI} - \widehat{BAH} = \frac{\widehat{A}}{2} - (90^\circ - \widehat{B}) = \frac{\widehat{B}-\widehat{C}}{2}$, where we used $\widehat{A} + \widehat{B} + \widehat{C} = 180^\circ$. (ii) Considering the right triangle made by the altitude and angle bisector from A we have $\frac{h_a}{i_a} = \cos \widehat{HAI} = \cos \frac{\widehat{B}-\widehat{C}}{2}$. (iii) If $h_a = i_a$ using part (ii) yields $\cos \frac{\widehat{B}-\widehat{C}}{2} = 1$, so $\widehat{B} = \widehat{C}$, and hence the triangle is isosceles.

Exercise 2.24.12. Just by direct computation.

Exercise 2.24.13. Use a direct computation.

Exercise 2.24.15. $h_a = h_b \iff \frac{\mathcal{A}}{a} = \frac{\mathcal{A}}{b} \iff a = b$.

Exercise 2.24.16. $m_a = m_b \iff m_a^2 = m_b^2 \iff \frac{b^2+c^2}{2} - \frac{a^2}{4} = \frac{a^2+c^2}{2} - \frac{b^2}{4} \iff b^2 = a^2 \iff a = b$.

Exercise 2.24.17. Apply the law of sines and reduce the inequality to Euler's inequality as follows

$$\begin{aligned}
\frac{\sin\widehat{A}\sin\widehat{B}\sin\widehat{C}}{\sin\widehat{A}+\sin\widehat{B}+\sin\widehat{C}} &= \frac{\frac{a}{2R}\cdot\frac{b}{2R}\cdot\frac{c}{2R}}{\frac{a+b+c}{2R}} = \frac{abc}{4R^2(a+b+c)} \\
&= \frac{abc}{4R}\frac{1}{2pR} = \frac{\mathcal{A}}{2pR} = \frac{pr}{2pR} = \frac{r}{2R} \leq \frac{1}{4},
\end{aligned}$$

since $R \geq 2r$.

Exercise 2.25.1. Using Proposition 2.18.1 we have

$$h_a + h_b + h_c = \frac{bc + ca + ab}{2R} = \frac{p^2 + r^2 + 4Rr}{2R} = \frac{p^2 + r^2}{2R} + 2r.$$

Exercise 2.25.2. (i) Use $\mathcal{A} = \frac{1}{2}h_a a$. (ii) Using the triangle inequality, $b + c > a$, we have $\frac{2p}{a} = \frac{a+b+c}{a} = 1 + \frac{b+c}{a} > 2$. Then by (i) yields $h_a > 2r$. Summing we obtain $h_a + h_b + h_c > 6r$.

Exercise 2.25.3. We substitute the altitude formula (2.25.40) into the identity $\frac{1}{h_a} + \frac{1}{h_b} + \frac{1}{h_c} = \frac{1}{r}$ and then use the Euler's inequality, $R \geq 2r$.

Exercise 2.26.3. Since a median bisects the area of a triangle, we have $\mathcal{A}_{ABA'} = \mathcal{A}_{ACA'}$ and $\mathcal{A}_{GBA'} = \mathcal{A}_{GCA'}$. Subtracting yields $\mathcal{A}_{ABG} = \mathcal{A}_{ACG}$. Similarly we can find two more identities.

Exercise 2.27.8. Writing the power of the points A, B and C with respect to the circumcircle we obtain

$$\begin{aligned} x(b+y) &= t(c+z) \Longrightarrow xb + xy = tc + tz \\ z(c+t) &= v(a+u) \Longrightarrow zc + zt = va + vu \\ u(a+v) &= y(b+x) \Longrightarrow au + uv = by + yx. \end{aligned}$$

Adding yields $xb+zc+au = tc+va+by$, or $a(u-v)+b(x-y)+c(z-t) = 0$.

Exercise 2.27.9. Apply Stewart's relation to find first $|AA'|$, see Exercise 2.21.8. Then evaluate the power of A with respect to the inner circle, $Power(A) = |AA''| \cdot |AA'| = |AI|^2 - R^2$, and use that $|AI| = r/\sin\frac{\widehat{A}}{2}$. Then solve for $|AA''|$.

Exercise 2.28.10. Using the area formulas $\mathcal{A} = \frac{abc}{4R} = pr$ we have

$$\begin{aligned} Power(I) = |AI| \cdot |ID| &= \frac{r}{\sin\frac{\widehat{A}}{2}} 2R \sin\frac{\widehat{A}}{2} = 2Rr = \frac{abc}{2\mathcal{A}} \cdot r = \frac{abcr}{2pr} \\ &= \frac{abc}{a+b+c}. \end{aligned}$$

Exercise 2.28.13. Using Corollary 2.28.4 and the fact that $|OG| = \frac{1}{3}|OH|$ we have

$$\begin{aligned} Power(G) = R^2 - |OG|^2 = R^2 - \frac{1}{9}|OH|^2 &= R^2 - \frac{1}{9}(9R^2 - (a^2 + b^2 + c^2)) \\ &= \frac{1}{3}(a^2 + b^2 + c^2). \end{aligned}$$

Exercise 2.28.14. Each term equals half the power of H. For instance $|HA| \cdot |HA_1| = \frac{1}{2}|HA| \cdot |HA_2| = \frac{1}{2}Power(H)$.

Exercise 2.28.16. By the law of cosines applied to ΔPOQ we have $|PQ|^2 = |OP|^2 + |OQ|^2 - 2|OP|\,|OQ|\cos\alpha$ and use that $Power(P) = R^2 - |OP|^2$ and $Power(Q) = R^2 - |OQ|^2$.

Exercise 2.31.4. Since $AB \| DC$, the secants AD and BC form two pairs of supplementary angles, $\widehat{DAB} + \widehat{ADC} = 180^\circ$ and $\widehat{ABC} + \widehat{BCD} = 180^\circ$. Then $\widehat{ADC} = \widehat{BCD}$ if and only if $\widehat{DAB} = \widehat{ABC}$.

Exercise 2.31.10. We shall extend the trapezoid to a parallelogram, by choosing $A' \in AB$, $D' \in DC$ such that $AA'D'D$ is a parallelogram. Let M' be the midpoint of $A'D'$. Then MM' is parallel to sides AA' and BB' and intersects BC at its midpoint, N. Therefore, $MN \| AB$. Since $|MN| = |NM'|$, and $|MM'| = |AA'| = |DD'|$ we obtain $|MN| = \frac{1}{2}|MM'|$.

Exercise 2.31.11. Let N be the midpoint of side BC. By contradiction, we assume that the parallel line through M to AB intersects BC at N', with $N' \neq N$. By Exercise 2.31.10 we have that $MN \| AB$. Since we also have $MN' \| AB$, we obtain two distinct parallel lines through M to AB, which contradicts the parallel postulate. Hence $N' = N$.

Exercise 2.31.12. Let AA' and AA_1 be the median and the altitude from vertex A of ΔABC. Then $OA'A_1H$ is a trapezoid with $HA_1 \| OA'$. Since A_1A' is a chord in Euler's circle, the perpendicular line through its midpoint intersects OH at its midpoint, see Exercise 2.31.11. The same argument can be applied for the other triangle sides. We obtain that the midpoint of OH belongs to all three perpendiculars, and hence it is the center of the Euler circle.

Exercise 2.32.2. Let A' and B' be the midpoints of sides BC and AC and denote by G the intersection of medians AA' and BB'. Applying Menelaus theorem for triangle $\Delta AA'C$ with transversal BB' we have $\frac{1}{2} \cdot \frac{1}{1} \cdot \frac{|AG|}{|GA'|} = 1$, from where $|AG| = 2|GA'|$. The same can be shown for each median.

Exercise 2.32.3. Apply Menelaus theorem.

Exercise 2.32.6. If $M = G$, then $\frac{|AG|}{|GA'|} = \frac{|BG|}{|GB'|} = \frac{|CG|}{|GC'|} = 2$, and hence their sum is equal to 6. Conversely, if the sum is equal to 6, then the inequalities in the proof of Corollary 2.32.5 becomes identities, i.e., A', B' and C' are midpoints of the sides. In this case $M = G$, since AA', BB' and CC' intersect at G.

Exercise 2.32.9. Apply Corollary 2.32.5 for $M = H$, the orthocenter.

Exercise 2.32.9. $\frac{|AG|}{|GA'|} = \frac{|AC'|}{|C'B|} + \frac{|AB'|}{|B'C|} = 2$.

Exercise 2.32.11. Let AA' be the symmedian from A. Since symmedians are cevians of rank 2, applying Van Aubel's theorem yields $\frac{|AK|}{|KA'|} = \frac{c^2}{a^2} + \frac{b^2}{a^2} = 1$, by Pythagorean theorem, so K is the midpoint of AA'. Note that reciprocal holds true, namely, if K is the midpoint of AA' then $\widehat{A} = 90^\circ$.

Exercise 2.32.15. In the case when the lines AD, BE and CF are parallel, then the points X, Y, and Z are still collinear (provided they exist).
Exercise 2.32.16. Apply Desargues' theorem for triangles ΔABC and $\Delta A_1B_1C_1$, whose omology center is the orthocenter H of ΔABC (since the altitudes AA_1, BB_1 and CC_1 intersect at H). Then the points A'', B'' and C'' belong to their omology axis, and hence they are collinear.
Exercise 2.32.17. Cevians AA', BB' and CC' intersect at the Gergonne's point, Γ, see Theorem 2.32.34, page 189. Then Γ is the omology center of triangles ΔABC and $\Delta A'B'C'$. Applying Desargues' theorem leads to the desired result.
Exercise 2.32.22. It is a limit case for Theorem 2.32.18, when the chords become tangents to the circle. The result is obtained by applying Pascal's theorem to the hexagram $ABBCDD$.
Exercise 2.32.23. The result is obtained by applying Pascal's theorem to the hexagram $AABCCD$.
Exercise 2.32.32. If AA', BB' and CC' are the medians, then $\frac{|BA'|}{|A'C|} = 1$, $\frac{|CB'|}{|B'A|} = 1$, $\frac{|AC'|}{|C'B|} = 1$. An application of Ceva's theorem provides that the medians AA', BB' and CC' are concurrent.
Exercise 2.32.33. Let AA', BB' and CC' be the angle bisectors of triangle ΔABC. Applying the angle bisector theorem we have $\frac{|BA'|}{|A'C|} = \frac{|AB|}{|AC|}$, $\frac{|CB'|}{|B'A|} = \frac{|BC|}{|BA|}$, $\frac{|AC'|}{|C'B|} = \frac{|AC|}{|BC|}$. Then after cancelations we obtain

$$\frac{|BA'|}{|A'C|}\frac{|CB'|}{|B'A|}\frac{|AC'|}{|C'B|} = 1$$

and an application of Ceva's theorem provides that the angle bisectors AA', BB' and CC' are concurrent.
Exercise 2.32.35. First we shall show that $|BA''| = p - c$, $|A''C| = p - b$, $|B''C| = p - a$, $|B''A| = p - c$, $|AC''| = p - b$ and $|BC''| = p - a$. To this end, let $x = |BA''|$ and $y = |A''C|$. We have $x + y = a$. Since the tangents from A to the excircle $\mathcal{C}(I_a)$ are equal, we have $c + x = b + y$. Solving the system yields $x = (a + b - c)/2 = p - c$ and $y = a - x = p - b$. Then after simplifications we obtain

$$\frac{|BA''|}{|A''C|}\frac{|CB''|}{|B''A|}\frac{|AC''|}{|C''B|} = 1.$$

Applying Ceva's theorem we obtain that AA'', BB'' and CC'' are concurrent.
Exercise 2.32.40. (*i*) Use that the isogonality of cevians is symmetric. (*ii*) If a cevian is inside of an angle, then its isogonal is also inside of the

same angle. (iii) Use that (AO, AH), (BO, BH) and (CO, CH) are pairs of isogonal cevians. (iv) The angle bisector is its own isogonal.

Exercise 2.32.41. To prove injectivity, we assume that $\mathcal{J}(M) = \mathcal{J}(M')$. Applying $\mathcal{J}$ yields $\mathcal{J}^2(M) = \mathcal{J}^2(M')$, and using the involutivity property we have $M = M'$. Now, if we choose $N \in Ext(ABC)$, then there are no points $M \in Int(ABC)$ such that $\mathcal{J}(M) = N$.

Exercise 2.32.44. The symmetric points of H with respect to the sides of triangle ΔABC belong to the circumcircle. Then apply Theorem 2.32.43.

Exercise 2.32.45. (i) By Theorem 2.32.43 the circumcenter of $\Delta O_A O_B O_C$ is the isogonal conjugate of O, which is the orthocenter H. (ii) Similarly, the circumcenter of $\Delta G_A G_B G_C$ is the Lemoine point, K.

Exercise 2.32.46. (i) Let AA' and AA_1 be the median and the altitude from A in triangle ΔABC. Since the centroid G is situated at $1/3$ from A' and $2/3$ from A, we obtain by similar triangles $\frac{|AA_1|}{d(G,BC)} = \frac{|AA'|}{|GA'|} = 3$, so $d(G, BC) = \frac{1}{3}h_a$. Hence $G(\frac{1}{3}h_a, \frac{1}{3}h_b, \frac{1}{3}h_c)$. (ii) Since O_9 is the midpoint of OG, we have

$$d(O_9, BC) = \frac{1}{2}\left(d(G, BC) + d(O, BC)\right) = \frac{1}{2}\left(\frac{1}{3}h_a + R\cos\widehat{A}\right).$$

Exercise 2.32.47. Using the additivity of areas, $ax_a + bx_b + cx_c = 2\mathcal{A}$. Since $a = b = c$ we obtain $x_a + x_b + x_c = \frac{2\mathcal{A}}{a} = \frac{2pr}{\frac{2}{3}p} = 3r$, where r is the inradius.

Exercise 2.32.50. The normal coordinates of G are given by Exercise 2.32.46, part (i). To find the normal coordinates of K use formulas (2.32.82) and (2.32.83).

Exercise 2.32.53. Denote $x_a = |BA'| = |A''C|$, $x_b = |CB'| = |B''A|$, $x_c = |AC'| = |C''B|$. By Ceva's theorem the cevians AA', BB', CC' intersect if and only if

$$\frac{x_a}{a - x_a}\frac{x_b}{b - x_b}\frac{x_c}{c - x_c} = 1.$$

Inverting, this becomes $\dfrac{a - x_a}{x_a}\dfrac{b - x_b}{x_b}\dfrac{c - x_c}{x_c} = 1$, which, by Ceva's theorem implies that the cevians AA'', BB'', CC'' intersect.

Exercise 2.32.57. With notations $A_1 = O$, $A_2 = B$, $A_3 = D$ and $A_4 = C$, relation (2.32.87) writes as $\frac{|A_1A_2|}{|A_1A_4|} = \frac{|A_2A_3|}{|A_3A_4|}$. This is equivalent to $\frac{|A_1A_2|}{|A_1A_4|} \cdot \frac{|A_3A_4|}{|A_2A_3|} = 1$, which means that A_1 is the harmonic conjugate of A_3 with respect to A_2A_4.

Exercise 2.32.58. Using that O_9 is the midpoint of HO and that G is situated at $1/3$ from O and $2/3$ from H we have the quotients $\frac{|HO_9|}{|HO|} = \frac{1}{2}$,

$\frac{|O_9G|}{|GO|} = \frac{1}{2}$, which shows that H is the harmonic conjugate of G with respect to O_9O.

Exercise 2.32.59. First we notice that the angle bisectors are perpendicular, $\widehat{A''AA'} = 90°$. By the angle bisector theorem, $\frac{|A'C|}{|BA'|} = \frac{|AC|}{|AB|}$. On the other side, using areas, we have

$$\frac{|A''B|}{|A''C|} = \frac{\mathcal{A}_{AA''B}}{\mathcal{A}_{AA''C}} = \frac{|AA''|\,|AB|\sin(90° - \frac{\widehat{A}}{2})}{|AA''|\,|AC|\sin(90° + \frac{\widehat{A}}{2})} = \frac{|AB|}{|AC|}.$$

Therefore, $\frac{|A''B|}{|A''C|} \cdot \frac{|A'C|}{|BA'|} = 1$, namely A'' is the harmonic conjugate of A' with respect to BC.

Exercise 2.32.60. Let A_3 be the midpoint of A_2A_4, so $\frac{|A_2A_3|}{|A_3A_4|} = 1$. We assume A_1 is the harmonic conjugate to A_3 with respect to A_2A_4, so $\frac{|A_1A_2|}{|A_1A_4|} = 1$. This implies $|A_1A_2| = |A_1A_4|$, so $A_2 = A_4$, contradiction.

Exercise 2.32.68. The orthopol in this case is the orthocenter of the triangle.

Exercise 2.33.4. (*ii*) Using formula (2.33.94) we obtain $|TA| + |TB| + |TC| = (b^2 + c^2 + bc\sqrt{3})^{1/2}$.

Exercise 2.37.2. In this case M is the point of intersection of diagonals. Since these are perpendicular, the desired angle is a right angle.

Chapter 3

Exercise 3.1.5. (*i*) Let $ABCD$ be a parallelogram. Since the opposite angles in a parallelogram are equal, $\widehat{A} = \widehat{C}$ and $\widehat{B} = \widehat{D}$. Assuming that $ABCD$ is cyclic, by Theorem 3.1.3 we have $\widehat{A}+\widehat{C} = 180°$ and $\widehat{B}+\widehat{D} = 180°$. It follows that $\widehat{A} = \widehat{C} = 90°$ and $\widehat{B} = \widehat{D} = 90°$, which means that $ABCD$ is a rectangle. (*ii*) If $ABCD$ is a rhombus, then, in particular, it is a parallelogram. Applying part (*i*) we get that $ABCD$ is a rectangle. Since all the sides are equal, then $ABCD$ becomes a square.

Exercise 3.1.8. See Proposition 2.31.5, page 164.

Exercise 3.2.4. Follow the lines of the proof of Theorem 3.2.1.

Exercise 3.2.14. (*i*) *Proof by construction:* Let M be such that $\widehat{ABM} = \widehat{DBC}$ and $\widehat{BAM} = \widehat{BDC}$. Then $\Delta BAM \sim \Delta BDC$, which implies $|BA| \cdot |DC| = |AM| \cdot |DB|$. Then construct $E \in BM$ such that $\widehat{BCE} = \widehat{ADB}$. Then $\Delta ABD \sim \Delta EBC$, so $|AD| \cdot |BC| = |BD| \cdot |EC|$. Let M' and E' be the orthogonal projections of M and E on AC. Summing, and using that M and E are on opposite sides of AC, yields

$$\begin{aligned}|BA| \cdot |DC| + |AD| \cdot |BC| &= (|AM| + |EC|)|DB| \geq (|AM'| + |E'C|)|DB| \\ &\geq |AC| \cdot |DB|.\end{aligned}$$

The fact that M and E are on opposite sides of AC can be proved by an argument involving contradiction as follows. Let's assume that M and E are on the same side of AC as B (a similar argument can be carried out if they are on the same side as the vertex D). Then from the aforementioned angle constructions, we have $\widehat{BAC} \geq \widehat{BDC}$ and $\widehat{BCA} \geq \widehat{BDA}$. Summing, we obtain $\widehat{D} = \widehat{BDC} + \widehat{BDA} \leq \widehat{BAC} + \widehat{BCA}$. Passing to supplements in triangles ΔABC and ΔADC, we obtain $\widehat{B} \leq \widehat{DAC} + \widehat{DCA}$. Summing yields the inequality $\widehat{B} + \widehat{D} \leq \widehat{A} + \widehat{C}$, which is a contradiction with the fact that the argument should hold for any convex quadrilateral. (*ii*) The inequality becomes identity in the case when both points M and E belong to AC and coincide. In this case the angles identity $\widehat{BAC} = \widehat{BDC}$, which implies that $ABCD$ is cyclic.
Proof by computation: We need to show the inequality $ac + bd \geq \delta_1\delta_2$, or equivalently, $\delta_1^2\delta_2^2 \leq a^2c^2 + d^2b^2 + 2abcd$. This inequality follows now from the Bretschneider identity, see page 271. Another proof using complex numbers can be found on page 334.
Exercise 3.2.15. Since $ADCB$ is a cyclic quadrilateral, using Ptolemy's first theorem we obtain $|AB|\cdot|DC|+|AD|\cdot|BC| = |BD|\cdot|AC|$. Simplifying by triangle side we obtain $|DA| + |DC| = |DB|$.
If the point D is arbitrary in the plane, the use of the Ptolemy's inequality, Exercise 3.2.14, yields $|AB| \cdot |DC| + |AD| \cdot |BC| \geq |BD| \cdot |AC|$, which after simplification recovers Pompeiu's inequality $|DA| + |DC| \geq |DB|$.
Exercise 3.2.18. By Corollary 3.2.17 we have $|AD| \cdot |DC| = |BD|^2 - \ell^2$. Then $|AD| \cdot |DC|$ is maximum when $|BD|$ is maximum, namely, when D is the midpoint of the arc AC.
Exercise 3.3.4. Let M and N be the midpoints of AC and BD, respectively. Then MN is a midpoint connector in a trapezoid, and hence $MN \| AB$. We obtain two trapezoids, $ABNM$ and $MNDC$. Denoting $h = dist(AB, MN) = dist(MN, CD)$, we obtain $\mathcal{A}_{ABNM} = \frac{h}{2}(|AB| + |MN|) < \frac{h}{2}(|CD| + |MN|) = \mathcal{A}_{NMCD}$.
Exercise 3.3.6. We express NF using the median theorem from triangle ΔAFD and then express AF and FD using the median theorem from triangles ΔABC and ΔBDC, respectively, as follows

$$\begin{aligned}
|NF|^2 &= \frac{1}{2}(|AF|^2 + |FD|^2) - \frac{1}{4}|AD|^2 \\
&= \frac{1}{2}\left(\frac{1}{2}(|AB|^2 + |AC|^2) - \frac{1}{4}|BC|^2 + \frac{1}{2}(|BD|^2 + |CD|^2) - \frac{1}{4}|AD|^2\right) \\
&\quad - \frac{1}{4}|AD|^2 \\
&= \frac{1}{4}\left(|AB|^2 + |CD|^2 + |AD|^2 + |BD|^2 + |AC|^2 - |BC|^2 - |AD|^2\right).
\end{aligned}$$

Exercise 3.3.7. The first relation follows from adding the expressions of $|EM|^2$ and $|NF|^2$ given by Exercise 3.3.6. The second relation follows by adding the expressions of $|EM|^2$, $|NF|^2$, and $|XY|^2$ given by Exercise 3.3.6 and Theorem 3.3.5.

Exercise 3.3.8. It follows from relation (3.3.10) using that $|XY|^2 \geq 0$. The identity occurs for the case $X = Y$, namely when diagonals bisect. Hence the quadrilateral is a parallelogram.

Exercise 3.3.9. Assume $|AB| < |DC|$ and that $\widehat{ABC} > 90°$ (at least one angle of a convex quadrilateral is obtuse). Let $B' \in (BC)$ such that $AB' \| MN$. Then $AB'CD$ is a trapezoid and we have $|MN| = (|AB'| + |DC|)/2 \leq (|AB| + |DC|)/2$, since $|AB| \leq |AB'|$ (in $\Delta ABB'$ the larger side corresponds to the larger angle).

Exercise 3.3.10. (i) Using Exercise 3.3.9 we have

$$\frac{1}{2}(|AB| + |CD|) \geq |MN| \Longleftrightarrow (|AB| + |CD|)^2 \geq 4|MN|^2.$$

Then using Exercise 3.3.6 this becomes

$$\begin{aligned}(|AB| + |CD|)^2 \geq 4|MN|^2 \Longleftrightarrow \\ |AB|^2 + |CD|^2 + 2|AB| \cdot |CD| \geq |AC|^2 + |BD|^2 + |AB|^2 + |CD|^2 \\ - |AD|^2 - |BC|^2 \Longleftrightarrow \\ 2|AB| \cdot |CD| + |AD|^2 + |BC|^2 \geq |AC|^2 + |BD|^2.\end{aligned}$$

(ii) The identity holds if we have identity in Exercise 3.3.9.

Exercise 3.4.2. (i) Use the proof idea of Theorem 3.4.1 by forming a parallelogram and using that its diagonals bisect. (ii) The orthocenter of triangle ΔXYZ is located at the intersection of the perpendicular from X to BD and the perpendicular from Y to AC. By part (i), this point is the anticenter of $ABCD$.

Exercise 3.5.1. Since $ABCD$ is a parallelogram, $\widehat{A} = \widehat{C}$. Since $ABCD$ is cyclic, $\widehat{A} + \widehat{C} = 180°$. Hence, $\widehat{A} = \widehat{C} = 90°$. Similarly, $\widehat{B} = \widehat{D} = 90°$. Therefore, $ABCD$ is a rectangle.

Exercise 3.6.8. Let O be the point of intersection of diagonals and denote by ϕ the acute angle made by them. We denote by $h_1 = dist(D, AC)$ and $h_2 = dist(B, AC)$. Then $h_1 = |DO| \sin\phi$ and $h_2 = |BO| \sin\phi$. Then $\mathcal{A}_{AED} + \mathcal{A}_{BEC} = \frac{1}{2}h_1|AE| + \frac{1}{2}h_2|EC| = \frac{1}{4}(h_1 + h_2)|AC| = \frac{1}{4}(|DO| + |BO|) \sin\phi|AC| = \frac{1}{4}|AC| \cdot |BD| \sin\phi = \frac{1}{2}\mathcal{A}_{ABDC}$.

Exercise 3.6.9. We first notice that the relation $\mathcal{A}_{AMB} + \mathcal{A}_{CMD} = \mathcal{A}_{BMC} + \mathcal{A}_{AMD}$ means $\mathcal{A}_{AMB} + \mathcal{A}_{CMD} = \frac{1}{2}\mathcal{A}_{ABCD}$. Now we use the fact that the set of points M such that $\mathcal{A}_{AMB} + \mathcal{A}_{CMD} =$ constant is a line. This is proved using vectors on page 440 in Solved Exercise 8.3.8. Another way to show this is as follows. Let $a = |AB|$, $b = |DC|$, $d_1 = dist(M, AB)$, and $d_2 = dist(M, CD)$. Then we need to show that $d_1a + d_2b =$ constant is equivalent to the fact that M describes a line. Let the lines AB and CD be given by the coordinate formulas $\alpha_1 x + \beta_1 y + \gamma_1 = 0$ and $\alpha_2 x + \beta_2 y + \gamma_2 = 0$. The distances from the point $M(x, y)$ to the aforementioned lines are $d_i = \frac{|\alpha_i x+\beta_i y+\gamma_i|}{\sqrt{\alpha_i^2+\beta_i^2}}$. Then the relation $d_1a + d_2b =$ constant becomes $Ax+By+C = 0$, for some constants A, B and C (depending on α_i, β_i, γ_i). Since this represents the equation of a line, then M describes a line. The last step now is to show that this line passes through the midpoints of diagonals, E and F. This follows from Exercise 3.6.8.

Exercise 3.6.10. Let I be the center of the circle. By Exercise 3.6.9, to show that I belongs to Newton's line it suffices to prove relation $\mathcal{A}_{AIB} + \mathcal{A}_{CID} = \mathcal{A}_{BIC} + \mathcal{A}_{AID}$. Using that all these triangles have the altitudes equal to the inradius, r, the relation is equivalent to $r(a + c) = r(b + d)$, which holds true via Pithot's theorem applied to $ABCD$.

Exercise 3.8.5. Using Proposition 3.6.4 and Heron's formula, Theorem 3.8.4, we have $r = \frac{\mathcal{A}}{p} = \frac{\sqrt{(p-a)(p-b)(p-c)(p-d)}}{p}$.

Exercise 3.9.9. (i) By Pithot's theorem, $a+c = d+b$, which is equivalent to $a - b = d - c$. Taking the square yields $a^2 + b^2 - 2ab = d^2 + c^2 - 2cd$. Using the cosine theorem, the previous relation becomes

$$|AC|^2 + 2ab\cos\widehat{B} - 2ab = |AC|^2 + 2bc\cos\widehat{D} - 2dc.$$

This is equivalent to $ab(1 - \cos\widehat{B})/2 = dc(\cos\widehat{D} - 1)/2$, or $ab\sin^2\frac{\widehat{B}}{2} = cd\sin^2\frac{\widehat{D}}{2}$.

Exercise 3.9.10. Using the cosine formula given by Lemma 3.8.2 and the area formula of a quadrilateral in terms of its diagonals, we have

$$\begin{aligned}\mathcal{A} &= \frac{1}{2}\delta_1\delta_2\sin\varphi = \frac{1}{2}\delta_1\delta_2\cos\varphi\tan\varphi\\ &= \frac{1}{2}\delta_1\delta_2\frac{|a^2 - b^2 + c^2 - d^2|}{2\delta_1\delta_2} = \frac{1}{4}|a^2 - b^2 + c^2 - d^2|\tan\varphi.\end{aligned}$$

Exercise 3.9.11. (i) The point P is the intersection of diagonals. Let E be another point in the plane. Then the triangle inequality provides $|AE| + |EC| \geq |AC|$ and $|BE| + |DE| \geq |DB|$. Therefore

$$|PA| + |PB| + |PC| + |PD| = |AC| + |DB| \leq |AE| + |EC| + |BE| + |DE|,$$

with identity for $E = P$. (ii) If $A_1A_2A_3A_4A_5A_6$ is a hexagon circumscribed to a circle, then diagonals A_1A_4, A_2A_5, A_3A_6 intersect at a point P (this is called Brianchon's theorem). Then the point P has the minimality property we are searching for. This follows from the triangle inequality as follows. For any arbitrary point E in the plane, we have

$$\sum_{i=1}^{6} |EA_i| = (|EA_1| + |EA_4|) + (|EA_2| + |EA_5|) + (|EA_3| + |EA_6|)$$

$$\geq |A_1A_4| + |A_2A_5| + |A_3A_6| = \sum_{i=1}^{6} |PA_i|.$$

Exercise 3.10.3. Use that the centroid G belongs to each symmetry axis; then their intersection is G.
Exercise 3.10.4. The centroid G belongs to both symmetry axis, so $\ell_1 \cap \ell_2 = \{G\}$. Since ℓ_1 and ℓ_2 are parallel, they must coincide.

Chapter 4

Exercise 4.1.2. It comes from (v) of Proposition 4.1.1.
Exercise 4.1.3. (i) $\frac{6}{5} - \frac{2i}{5}$, (ii) $2(i-1)$, (iii) $-1-i$, (iv) $-4(1+i)$.
Exercise 4.1.4. Use the algebraic form of complex numbers and do the algebra.
Exercise 4.1.5. (i) Find a counterexample! (ii) proportional complex numbers.
Exercise 4.1.6. (i) $z \in \mathbb{R}$, (ii) $z = 0$, (iii) $z \in \{0, 1, -\frac{1}{2} \pm \frac{\sqrt{3}}{2}i\}$, (iv) $z \in \{0, \pm 1, \pm i\}$.
Exercise 4.2.1. Taking the modulus yields $|z_1| = |z_2| \implies \rho_1 = \rho_2$. Equating $\sin\phi_1 = \sin\phi_2$ and $\cos\phi_1 = \cos\phi_2$ yields $\tan\phi_1 = \tan\phi_2$, and hence $\phi_1 = \phi_2$.
Exercise 4.2.2. (i) $z^{-1} = \bar{z}$, (ii) $z = \frac{\sqrt{2}}{2}(1+i)$.
Exercise 4.2.4. $\{1, -\frac{1}{2} \pm \frac{\sqrt{3}}{2}i\}$.
Exercise 4.2.5. $\{\pm 1, \pm i\}$.
Exercise 4.2.6. $\omega_0 + \omega_1 + \cdots + \omega_{n-1} = 1 + \omega_1 + \omega_1^2 + \cdots + \omega_1^{n-1} = \frac{1-\omega_1^n}{1-\omega_1} = 0$.
Exercise 4.2.7. Either by direct computation or by using Exercise 4.2.6.
Exercise 4.2.8. Since $z_1 \neq 0$, multiplying by z_1^{-1} we may assume $z_1 = 1$. Then show $z_2 = \bar{z_3}$, $Re(z_2) = Re(z_3) = -1/2$. It follows that $\{z_1, z_2, z_3\}$ are the 3rd roots of unity.
Exercise 4.2.9. Denote the sums by S_1 and S_2. Then evaluate $S_1 + iS_2$ using a geometric progression.
Exercise 4.3.1. Diagonals bisect.
Exercise 4.3.2. Use Exercise 4.3.1.
Exercise 4.3.3. The average of the affixes of the points $\{Q, N\}$, $\{P, M\}$, $\{S, L\}$ are equal, so the segments have a common midpoint.

Exercise 4.3.12. Write the affixes $z_{A'}$, $z_{B'}$, $z_{C'}$, $z_{B''}$, $z_{C''}$ in terms of z_A, z_B and z_C. Then find $t \in [0,1]$ such that $z_A = tz_{C''} + (1-t)z_{B''}$.
Exercise 4.3.13. It follows from the modulus inequality

$$\begin{aligned} |z_N - z_M| &= \frac{1}{2}|z_B + z_C - z_A - z_D| = \frac{1}{2}|(z_B - z_A) + (z_C - z_B)| \\ &\leq \frac{1}{2}|z_B - z_A| + \frac{1}{2}|z_C - z_B| = \frac{1}{2}(|AB| + |BC|). \end{aligned}$$

Exercise 4.4.2. It's an obvious consequence of the definition.
Exercise 4.4.3. It follows from Exercise 4.4.2.
Exercise 4.4.7. (*i*) Let $T(z) = 1/z$ and $U(z) = 1-z$. Then $T(\{-1, 1/2, 2\}) = \{-1, 2, 1/2\}$, and $U(\{-1, 1/2, 2\}) = \{2, 1/2, -1\}$. Yes. (*ii*) We have $1/z_1 = z_2$ and $1 - z_1 = z_2$. Solve the system!
Exercise 4.4.8. Use the definition, then solve for z_A in terms of z_B and z_C. The second relation can be shown in a similar way.
Exercise 4.5.7. Let M, N, S, T be the midpoints of the sides AB, DC, AD, and BC. Then

$$z_G = \frac{1}{4}(z_A + z_B + z_C + z_D) = \frac{1}{2}(z_M + z_N) = \frac{1}{2}(z_S + z_T).$$

Exercise 4.5.8. (*i*) Making a rotation we may assume that ℓ is the x-axis. Then $z_{A_j}, \bar{z}_{A_j} \in \{z_{A_1}, \ldots, z_{A_n}\}$ and $z_{A_j} + \bar{z}_{A_j} \in \mathbb{R}$. Hence $nz_G = \sum_j z_{A_j} \in \mathbb{R}$, i.e., $G \in \ell$. (*ii*) Applying (*i*) we get $G \in \ell_1$, $G \in \ell_2$ and hence $G \in \ell_1 \cap \ell_2$.
Exercise 4.5.14. Express z_{O_1}, z_{O_2}, z_{O_3} in terms of z_A, z_B, z_C and show that $z_{O_1} + z_{O_2} + z_{O_3} = z_A + z_B + z_C$.
Exercise 4.5.15. Express the affixes of G_a, G_b, G_c, G_d in terms of affixes of A, B, C, D and verify the relation $z_{G_a} + z_{G_b} + z_{G_c} + z_{G_d} = z_A + z_B + z_C + z_D$.
Exercise 4.5.17. Write z_{O_1}, z_{O_2}, z_{O_3}, z_{O_4} in terms of z_A, z_B, z_C, z_D and use a similar method as in Exercise 4.5.16.
Exercise 4.5.18. See Exercises 4.5.16 and 4.5.17.
Exercise 4.7.3. Use a similar method as the one described in Proposition 4.7.1.
Exercise 4.7.5. (*i*) Assume $z_O = 0$ and express all the affixes in terms of z_A, z_B, z_C. (*ii*) The intersection point has the affix proportional with $z_A + z_B + z_C$.
Exercise 4.8.6. $|(A, B; C, D)| = \frac{\rho_2}{\rho_1} = \frac{|AC| \cdot |BD|}{|BC| \cdot |AD|} > 1$.
Exercise 4.8.7. Write $z_{A_k} = e^{i\theta_k}$ and make use of trigonometry to transform the differences into products. Use also Exercise 4.3.8. Be patient!
Exercise 4.8.12. The midpoint of the arc AB.
Exercise 4.9.2. $\sqrt{2}/2$.

Chapter 5

Exercise 5.1.1. (*i*) The empty set, one point, a line; (*ii*) the empty set, one point, a line; a plane, a plane and a line.

Exercise 5.1.2. Use the method of double inclusion, namely, take an element from the set on the left side and show that it belongs to the set from the right side; then, repeat the procedure the other way. Most of the identities are obvious.

Exercise 5.1.3. (*i*) Let $x \in \overline{B}$, so $x \notin B$. Since $A \subset B$, then $x \notin A$, so $x \in \overline{A}$. (*ii*) Let $x \in \overline{\overline{A}}$, so $x \notin \overline{A}$, and then $x \in A$. Then $\overline{\overline{A}} \subset A$. The converse inclusion is similar. (*iii*) Let $x \in \overline{A \cap B}$, so $x \notin A \cap B$. This implies either $x \notin A$ or $x \notin B$, i.e., $x \in \overline{A}$ or $x \in \overline{B}$, which means $x \in \overline{A} \cup \overline{B}$. Hence $\overline{A \cap B} \subset \overline{A} \cup \overline{B}$. The converse inclusion is similar. (*iv*) See the argument in part (*iii*).

Exercise 5.2.3. Reflexivity follows from $X\mathcal{R}X$ because $[XX] = \{X\} \notin d$; the symmetry is obvious since the segments $[XY]$ and $[YX]$ are the same; the transitivity is based on the fact that $\Pi \backslash d$ is the reunion of two half-planes. Then if $X\mathcal{R}Y$ and $Y\mathcal{R}Z$, then X and Y as well as Y and Z belong to the same half-plane. Then X and Z are in relation $\mathcal{R}$ as they belong to the same half-plane.

Exercise 5.2.6. Use a proof by contradiction using cases. Assume that C is not between A and B, say that C belongs to the ray starting at B that does not contain A. This makes B to be between A and C.

Exercise 5.2.7. Take C to be the midpoint of AB.

Exercise 5.2.8. Since $d_-, d_+ \subset d$, then $d_+ \cup d_- \subset d$. To show the converse implication, let $M \in d \backslash \{O\}$. Since $\{\omega_-, \omega_+\}$ are total opposite relations, then M is either on the left or on the right side of O. Then $O\omega_+ M$ or $O\omega_- M$, i.e., $P \in d_+$ or $P \in d_-$. Hence $d \subset d_+ \cup d_-$.

Exercise 5.3.1. Reflexivity: $\ell || \ell$, because $\ell = \ell$; Symmetry: $\ell || \ell' \Leftrightarrow \ell' || \ell$ since $\ell \cap \ell' = \ell' \cap \ell = \emptyset$; Transitivity: Let ℓ, ℓ' and ℓ'' be three distinct lines, such that $\ell || \ell'$ and $\ell' || \ell''$. We assume by contradiction that ℓ and ℓ'' are not parallel, so let $P \in \ell \cap \ell''$. If $P \notin \ell'$, then through P we can construct two distinct parallel lines to ℓ', contradiction with the parallel postulate. Hence, $P \in \ell'$, which is in contradiction with the fact that the lines ℓ and ℓ' are parallel.

Exercise 5.3.2. It is similar to Exercise 5.3.1. The factor set is the set of directions in the plane.

Exercise 5.3.3. Reflexibility and symmetry are obvious. To show the transitivity, we consider three parallel oriented segments $\overline{AB}$, $\overline{CD}$ and $\overline{EF}$ and consider the two opposite orientations $\{\omega_-, \omega_+\}$. We assume $\overline{AB}\,\Omega\,\overline{CD}$, for instance $\overline{AB}, \overline{CD} \in \omega_+$. Let $\overline{CD}\,\Omega\,\overline{EF}$, i.e., $\overline{EF}, \overline{CD} \in \omega_+$. Then $\overline{AB}$, $\overline{EF} \in \omega_+$, namely $\overline{AB}\,\Omega\,\overline{EF}$, which proves transitivity.

Exercise 5.3.4. It follows from the fact that the equality relation of two real numbers is reflexive, symmetric and transitive.
Exercise 5.3.5. There are two cases: (*i*) $\overline{AB}$ and $\overline{CD}$ do not overlap. Then $|AC| = |AB| + |BC|$ and $|BD| = |BC| + |CD|$, from where $|AC| = |BD|$. (*ii*) $\overline{AB}$ and $\overline{CD}$ overlap. Let C be between A and B and B be between C and D. Then $|AC| = |AB| - |BC| = |CD| - |BC| = |BD|$.
Exercise 5.4.1. If the segments are noncollinear, then relations $\overline{AB}\,\mathcal{E}\,\overline{CD}$ and $\overline{BD}\,\mathcal{E}\,\overline{AC}$ are equivalent to the fact that $ABDC$ is a parallelogram. If the segments are collinear, then use Exercise 5.3.5 to show that they have the same length; their direction is the same as they belong to the same line.
Exercise 5.4.2. We have parallelograms $ABB'A'$ and $BB'C'C$. Then ΔABC and $A'B'C'$ are similar triangles by (AA) case, so $\overline{AC}||\overline{A'C'}$. We can show that these triangles are also congruent, so $|AC| = |A'C'|$. Hence, $\overline{AC}\,\mathcal{E}\,\overline{A'C'}$.
Exercise 5.4.3. Use that in a parallelogram diagonals bisect.
Exercise 5.4.5. Assume by contradiction that $M \neq M'$. Then $\overrightarrow{OM} \neq \overrightarrow{OM'}$ (since they have different directions), i.e., $p(M) \neq p(M')$, contradiction.
Exercise 5.4.6. Given $\overrightarrow{v}$, we shall construct a point M such that $\overrightarrow{OM} = \overrightarrow{v}$ as follows: let ℓ be a line through O such that ℓ is parallel with the support line of $\overrightarrow{v}$. Then choose $M \in \ell$ such that $|OM| = |\overrightarrow{v}|$ and $\overrightarrow{OM}$ has the same orientation as $\overrightarrow{v}$.
Exercise 5.5.1. (*i*) Use that $\overrightarrow{AB} + \overrightarrow{BC} = \overrightarrow{AC}$ and $\overrightarrow{AC} = -\overrightarrow{CA}$. (*ii*) It follows from the triangle inequality of real numbers as follows:

$$|\overrightarrow{AB} + \overrightarrow{BC}| = |\overrightarrow{AC}| = |AC| \leq |AB| + |BC| = |\overrightarrow{AB}| + |\overrightarrow{BC}|.$$

(*iii*) Let D be a point such that $BADC$ is a parallelogram. Using the parallelogram rule we have

$$|\overrightarrow{AB} - \overrightarrow{BC}| = |-\overrightarrow{BA} - \overrightarrow{BC}| = |\overrightarrow{BA} + \overrightarrow{BC}| = |\overrightarrow{BD}| = |BD|.$$

Since in triangle ΔABD we have $|BD| > \Big||AB| - |DA|\Big|$, the previous relation becomes $|\overrightarrow{AB} - \overrightarrow{BC}| \geq \Big||\overrightarrow{AB}| - |\overrightarrow{BC}|\Big|$.
Exercise 5.5.3. We can represent $\overrightarrow{v} = \overrightarrow{OM}$ and then $-\overrightarrow{v} = \overrightarrow{MO}$, so $\overrightarrow{a} - \overrightarrow{a} = \overrightarrow{a} + (-\overrightarrow{a}) = \overrightarrow{OM} + \overrightarrow{MO} = \overrightarrow{OO} = \overrightarrow{0}$, as a vector of length zero.
Exercise 5.5.9. Let M' be the midpoint of the segment $[AD]$. By Exercise 5.5.8 M' satisfies $\overrightarrow{M'D} + \overrightarrow{M'C} + \overrightarrow{M'B} = \frac{1}{2}\overrightarrow{AD} + 2\overrightarrow{AB}$. Subtracting from the hypothesis relation $\overrightarrow{MD} + \overrightarrow{MC} + \overrightarrow{MB} = \frac{1}{2}\overrightarrow{AD} + 2\overrightarrow{AB}$, yields $3\overrightarrow{MM'} = 0$, which implies $M = M'$.

Exercise 5.5.10. (*i*) Let O be the center of the regular hexagon $A_1\ A_2\ A_3\ A_4\ A_5\ A_6$. Then the triangles made by two consecutive vertices and the center O are equilateral triangles. Then $|A_1A_4| = |A_1O|+|OA_4| = 2|A_2A_3|$. Since $A_1A_2A_3A_4$ is an isosceles trapezoid, then $A_2A_3||A_1A_4$. Therefore, $\overrightarrow{A_1A_4} = 2\overrightarrow{A_2A_3} = 2(\overrightarrow{A_1A_3} - \overrightarrow{A_1A_2})$. (*iii*) The relation is equivalent to $\overrightarrow{A_1A_6}+2\overrightarrow{A_1A_2} = \overrightarrow{A_1A_3}$. This relation can be shown using the parallelogram rule as follows:

$$\overrightarrow{A_1A_6} + 2\overrightarrow{A_1A_2} = (\overrightarrow{A_1A_6} + \overrightarrow{A_1A_2}) + \overrightarrow{A_1A_2} = \overrightarrow{A_1O} + \overrightarrow{A_1A_2} = \overrightarrow{A_1A_3}.$$

Exercise 5.5.11. Using the reduction method, we obtain $\overrightarrow{x} = \frac{1}{3}(\overrightarrow{a} + \overrightarrow{b})$ and $\overrightarrow{y} = \frac{1}{3}(2\overrightarrow{b} - \overrightarrow{a})$.
Exercise 5.5.12. (*i*) Let O be the midpoint of AB and choose M to be the symmetric of C with respect to O, namely the points $\{M, A, B, C\}$ are the vertices of a parallelogram. Then using the parallelogram rule, we have $\overrightarrow{DA}+\overrightarrow{DB} = \overrightarrow{DC}$. (*ii*) We assume there is another point, D', with the same property. Subtracting relations $\overrightarrow{DA} + \overrightarrow{DB} = \overrightarrow{DC}$ and $\overrightarrow{D'A} + \overrightarrow{D'B} = \overrightarrow{D'C}$ yields $\overrightarrow{DD'} = 0$, so $D = D'$. (*iii*) It suffices to show that $\overrightarrow{MA} + \overrightarrow{MB} = \overrightarrow{MC} + \overrightarrow{MD}$. Each side can be evaluated using the parallelogram rule to obtain $\overrightarrow{MA} + \overrightarrow{MB} = \overrightarrow{MO}$ and $\overrightarrow{MC} + \overrightarrow{MD} = 2\overrightarrow{MO}$, and hence both sides are equal. (*iv*) Using part (*iii*) we have

$$\overrightarrow{MA} + \overrightarrow{MB} - 2\overrightarrow{MC} = (\overrightarrow{MA} + \overrightarrow{MB} - \overrightarrow{MC}) - \overrightarrow{MC} = \overrightarrow{MD} - \overrightarrow{MC} = \overrightarrow{CD}.$$

(*v*) By part (*iii*) it suffices to find a point M such that $\overrightarrow{MD} = \overrightarrow{AB}$.
Exercise 5.6.2. Apply a proof by contradiction.
Exercise 5.6.3. Since the vectors $\overrightarrow{v_1}, \overrightarrow{v_2}, \dots, \overrightarrow{v_n}$ are linear dependent, there is a vanishing linear combination $\sum_{i=1}^{n} \alpha_i v_i = 0$. Choosing $v_{n+1} = \cdots = v_{n+p} = 0$, we obtain $\sum_{i=1}^{n+p} \alpha_i v_i = 0$, namely the vectors $\overrightarrow{v_1}, \overrightarrow{v_2}, \dots, \overrightarrow{v_n}, \overrightarrow{v_{n+1}}, \dots, \overrightarrow{v_{n+p}}$ are also linearly dependent.
Exercise 5.6.4. Consider the vanishing linear combination $\sum_{i=1}^{p} \alpha_i v_i = 0$. Choosing $\alpha_{p+1} = \cdots = \alpha_n = 0$, we have the vanishing linear combination $\sum_{i=1}^{n} \alpha_i v_i = 0$. Using the linear independence of $\overrightarrow{v_1}, \overrightarrow{v_2}, \dots, \overrightarrow{v_n}$ it follows that all coefficients vanish, in particular, $\alpha_1 = \cdots = \alpha_p = 0$, so the vectors $\overrightarrow{v_1}, \overrightarrow{v_2}, \dots, \overrightarrow{v_p}$ are linearly independent.
Exercise 5.6.5. Let $\sum_{i=1}^{n} \alpha_i v_i = 0$. Choose a vector v_p with $\alpha_p \neq 0$. Divide by α_p in the previous linear combination and then solve for v_p in terms of the other vectors.
Exercise 5.6.13. (*i*) and (*ii*) are obvious. (*iii*) If $\overrightarrow{w_1}$ and $\overrightarrow{w_2}$ are collinear, there is $\lambda \neq 0$ such that $\overrightarrow{w_1} = \lambda\overrightarrow{w_2}$. Equating the coefficients of the basis we obtain $a_1 = \lambda a_2$ and $b_1 = \lambda b_2$. Eliminating λ yields $a_1b_2 = a_2b_1$.

Exercise 5.6.14. $\overrightarrow{w} = \alpha'\overrightarrow{u'} + \beta'\overrightarrow{v'}$, where $(\alpha', \beta')^T = (a_{ij})^{-1}(\alpha, \beta)^T$.

Exercise 5.6.19. Let $f(M) = \alpha\overrightarrow{MA} + \beta\overrightarrow{MB} + \gamma\overrightarrow{MC}$. Then $f(M) - f(M') = (\alpha + \beta + \gamma)\overrightarrow{MM'}$. Therefore, $f(M)$ does not depend on M iff $f(M) = f(M')$ for any M and M'. This is equivalent to $\alpha + \beta + \gamma = 0$.

Exercise 5.6.20. Let $f(M) = \sum_{i=1}^{n} a_i\overrightarrow{MA_i}$. Then $f(M) - f(M') = (\sum_{i=1}^{n} a_i)\overrightarrow{MM'}$ and repeat the argument from Exercise 5.6.19.

Chapter 6

Exercise 6.1.2. Subtracting yields $0 = (\overrightarrow{MA} + \overrightarrow{MB}) - (\overrightarrow{M'A} + \overrightarrow{M'B}) = 2\overrightarrow{MM'}$, which implies $M = M'$.

Exercise 6.3.3. We show that each expression individually, is equal to $3\overrightarrow{MG}$. For instance,

$$\begin{aligned} 3\overrightarrow{MA} + 2\overrightarrow{AA'} &= 3\overrightarrow{MA} + \overrightarrow{AB} + \overrightarrow{AC} = \overrightarrow{MA} + (\overrightarrow{MA} + \overrightarrow{AB}) + (\overrightarrow{MA} + \overrightarrow{AC}) \\ &= \overrightarrow{MA} + \overrightarrow{MB} + \overrightarrow{MC} = 3\overrightarrow{MG}. \end{aligned}$$

Exercise 6.3.11. Choose $M = O$ (the circumcenter) and use that $R = |MA| = |MB| = |MC|$ in part (*iii*) of Exercise 6.3.10.

Exercise 6.4.9. Using that the centroid G is defined by $\overrightarrow{GA} + \overrightarrow{GB} + \overrightarrow{GC} + \overrightarrow{GD} = 0$, we have

$$\begin{aligned} \overrightarrow{YA} + \overrightarrow{YB} + \overrightarrow{YC} + \overrightarrow{YD} &= (\overrightarrow{YG} + \overrightarrow{GA}) + (\overrightarrow{YG} + \overrightarrow{GB}) + (\overrightarrow{YG} + \overrightarrow{GC}) \\ &\quad + (\overrightarrow{YG} + \overrightarrow{GD}) \\ &= 4\overrightarrow{YG} + (\overrightarrow{GA} + \overrightarrow{GB} + \overrightarrow{GC} + \overrightarrow{GD}) = 4\overrightarrow{YG}. \end{aligned}$$

Exercise 6.4.10. Let O_1 and O_2 be the midpoints of diagonals AC and BD. We shall show that G is the midpoint of O_1O_2 as follows

$$\overrightarrow{GO_1} + \overrightarrow{GO_2} = \frac{1}{2}(\overrightarrow{GA} + \overrightarrow{GC}) + \frac{1}{2}(\overrightarrow{GB} + \overrightarrow{GD}) = 0,$$

and hence, G belongs to the segment $[O_1O_2]$.

Chapter 7

Exercise 7.2.2. Let θ be the angle made by $\overrightarrow{OA}$ and $\overrightarrow{OC}$. Then $\overrightarrow{OA} \cdot \overrightarrow{OC} = |OA| \cdot |OC| \cos\theta = |OC| \cdot |OB|$.

Exercise 7.2.4. We assume the points are ordered from left to right as A,

B, C, and D. Then

$$\begin{aligned}
&\overrightarrow{AB}\cdot\overrightarrow{CD}+\overrightarrow{BC}\cdot\overrightarrow{AD}+\overrightarrow{CA}\cdot\overrightarrow{BD}\\
&=|AB|\cdot|CD|+|BC|\cdot|AD|-|CA|\cdot|BD|\\
&=(|AC|-|BC|)|CD|+|BC|(|AC|+|CD|)-|CA|(|BC|+|CD|)\\
&=|AC|\,|CD|-|BC|\,|CD|+|BC|\,|AC|+|BC|\,|CD|\\
&\quad-|CA|\,|BC|-|CA|\,|CD|=0.
\end{aligned}$$

The other orderers can be treated similarly. A proof that is independent of order can be done using solved Exercise 7.2.5.

Exercise 7.2.8. By the parallelogram definition we have $\overrightarrow{AB}=\overrightarrow{DC}$, which implies $|AB|=|DC|$ and $AB\|DC$. Using triangle rule we have

$$\overrightarrow{BC}=\overrightarrow{BA}+\overrightarrow{AD}+\overrightarrow{DC}=(-\overrightarrow{AB}+\overrightarrow{DC})+\overrightarrow{AD}=\overrightarrow{AD}.$$

From $\overrightarrow{BC}=\overrightarrow{AD}$ we get $|AD|=|BC|$ and $BC\|BC$.

Exercise 7.3.1. It follows from the property of subtracting angles

$$\widehat{R_\phi\overrightarrow{u},R_\phi\overrightarrow{v}}=\widehat{\overrightarrow{v},R_\phi\overrightarrow{v}}-\theta=\widehat{u,R_\phi\overrightarrow{u}}-\theta=\widehat{\overrightarrow{u},\overrightarrow{v}},$$

where $\theta=\widehat{v,R_\phi\overrightarrow{u}}$.

Exercise 7.3.3. Let M and N be points in the plane such that $\overrightarrow{u}=\overrightarrow{OM}$ and $\overrightarrow{v}=\overrightarrow{ON}$. From $|\overrightarrow{u}|=|\overrightarrow{v}|$ it follows that M and N belong to the circle centered at O of radius $r=|\overrightarrow{u}|$. Let ϕ be the angle $\widehat{NOM}$. Then $\overrightarrow{v}=R_\phi\overrightarrow{u}$.

Exercise 7.3.4. These are straightforward verifications.

Exercise 7.4.3. Using that a rotation preserves angles between vectors and vectors magnitude, we have $|e_i'|=|R_\phi e_i|=|e_i|$ and

$$e_i'\cdot e_j'=|e_i'|\,|e_j'|\cos\widehat{e_i',e_j'}=|e_i|\,|e_j|\cos\widehat{e_i,e_j}=e_i\cdot e_j=\delta_{ij}.$$

Exercise 7.7.2. Use that for collinear vectors with the same orientation $\widehat{\overrightarrow{v},\overrightarrow{w}}=0$.

Exercise 7.7.3. Using the distributivity and commutativity properties we have

$$(\overrightarrow{u}-\overrightarrow{v})\cdot(\overrightarrow{u}+\overrightarrow{v})=\overrightarrow{u}\cdot\overrightarrow{u}+\overrightarrow{u}\cdot\overrightarrow{v}-\overrightarrow{v}\cdot\overrightarrow{u}-\overrightarrow{v}\cdot\overrightarrow{v}=|\overrightarrow{u}|^2-|\overrightarrow{v}|^2.$$

Exercise 7.7.4. (*i*) It follows from the law of cosines. (*ii*) Apply three versions of (*i*) and add. (*iii*) It follows from part (*ii*) and definition of the inner product.

Exercise 7.9.2. $m_a = a/2$, $m_b = \sqrt{4c^2+b^2}/2$, $m_b = \sqrt{4b^2+c^2}/2$.

Exercise 7.9.3. $m_a^2 = \frac{b^2+c^2}{2} - \frac{a^2}{4} = \frac{\ell^2+\ell^2}{2} - \frac{\ell^2}{4} = \frac{3}{4}\ell^2$, where ℓ is the side of the triangle.

Exercise 7.9.6. Let B' and C' be the midpoints of AC and AB, respectively, and $\{G\} = BB' \cap CC'$ the centroid. Then BB' and CC' are perpendicular iff the triangle $\Delta GB'C'$ has a right angle at G. This fact is equivalent to the Pythagorean relation $|GC'|^2 + |GB'|^2 = |B'C'|^2$. Substituting $|GB'|^2 = (\frac{m_b}{3})^2$, $|GC'|^2 = (\frac{m_c}{3})^2$ and $|B'C'|^2 = (\frac{a}{2})^2$ and using the median theorem to express m_b and m_c we obtain the equivalent relation $5a^2 = b^2+c^2$. Therefore, the relation $5a^2 = b^2+c^2$ implies that the medians BB' and CC' are perpendicular.

Exercise 7.9.7. Let $k = \frac{|BF|}{|FC|}$. Since $\overrightarrow{BF} = k\overrightarrow{FC}$ and $\overrightarrow{AE} = 2\overrightarrow{EB}$, by Lemma 6.1.9, page 371, we have

$$\overrightarrow{AF} = \frac{1}{k+1}\overrightarrow{AB} + \frac{k}{k+1}\overrightarrow{AC}, \qquad \overrightarrow{CE} = \frac{1}{3}\overrightarrow{CA} + \frac{2}{3}\overrightarrow{CB}.$$

Since $AF \perp CE$ we have $\overrightarrow{CE} \cdot \overrightarrow{AF} = 0$. Using distributivity we have

$$\begin{aligned}
\overrightarrow{CE} \cdot \overrightarrow{AF} &= \frac{1}{3(k+1)}\left(\overrightarrow{CA} \cdot \overrightarrow{AB} + k\overrightarrow{CA} \cdot \overrightarrow{AC} + 2\overrightarrow{CB} \cdot \overrightarrow{AB} + 2k\overrightarrow{CB} \cdot \overrightarrow{AC}\right)\\
&= \frac{1}{3(k+1)}\left(-\ell^2 \cos 60^\circ - k\ell^2 + 2\ell^2 \cos 60^\circ - 2k\ell^2 \cos 60^\circ\right)\\
&= \frac{-\ell^2}{3(k+1)}\left(-\frac{1}{2} - k + 1 - k\right) = 0,
\end{aligned}$$

so $k = 1/4$.

Exercise 7.10.5. Let D be the midpoint of $[AB]$. Applying the median theorem in triangle ΔMAB yields $|DM|^2 = \frac{1}{2}(|MA|^2 + |MB|^2) - \frac{1}{4}|AB|^2$. Then $|MA|^2+|MB|^2$ is minimum when $|DM|^2$ is minimum, i.e., $|DM| = 0$, or $M = D$.

Exercise 7.10.6. It follows from Leibnitz' relation

$$k = |MA|^2 + |MB|^2 + |MC|^2 = 3|MG|^2 + \frac{1}{3}(a^2+b^2+c^2),$$

which implies that $|MG|$ is constant.

Exercise 7.10.7. Let X and Y be the midpoints of AB and DC, respectively and denote by ℓ the side length of the square. The median theorem applied in triangles ΔABM and ΔDCM yields

$$|MX|^2 = \frac{1}{2}(|MA|^2+|MB|^2) - \frac{1}{4}\ell^2, \qquad |MY|^2 = \frac{1}{2}(|MD|^2+|MC|^2) - \frac{1}{4}\ell^2.$$

These imply

$$\frac{1}{2}(|MA|^2 + |MB|^2 + |MC|^2 + |MD|^2) = |MX|^2 + |MY|^2 + \ell^2/2.$$

Then $|MA|^2 + |MB|^2 + |MC|^2 + |MD|^2$ is minimum whenever $|MX|^2 + |MY|^2$ is minimum. By Exercise 7.10.5 this is minimum for $M = O$, the midpoint of XY. This point is also the center of the circle.

Exercise 7.11.1. To show that formulas (7.11.16) and (7.11.16) are equal, it suffices to prove that the two expressions for the sum $\sum |AI|^2$ are equal, namely, $p^2+r^2-8Rr = 3r^2-p^2+\sum a^2$. This is equivalent to $a^2+b^2+c^2 = 2p^2-2r^2-8Rr$, which is a formula proved in Proposition 2.18.1, page 117.

Exercise 7.13.3. AM is angle bisector iff $\widehat{BAM} = \widehat{MAC}$ iff $\frac{\overrightarrow{AB}\cdot\overrightarrow{AM}}{|AB|\,|AM|} = \frac{\overrightarrow{AC}\cdot\overrightarrow{AM}}{|AC|\,|AM|}$. Let $\frac{|BM|}{|MC|} = k$. Then by Lemma 6.1.9 $\overrightarrow{AM} = \frac{1}{k+1}\overrightarrow{AB} + \frac{k}{k+1}\overrightarrow{AC}$. Denoting $i_a = |AM|$ the previous identity becomes

$$\begin{aligned}
\frac{c}{i_a}\left[\frac{1}{k+1}c^2 + \frac{k}{k+1}\overrightarrow{AC}\cdot\overrightarrow{AB}\right] &= \frac{1}{bi_a}\left[\frac{1}{k+1}\overrightarrow{AB}\cdot\overrightarrow{AC} + \frac{k}{k+1}b^2\right] \iff \\
b\left[\frac{1}{k+1}c^2 + \frac{k}{k+1}bc\cos A\right] &= c\left[\frac{1}{k+1}cb\cos A + \frac{k}{k+1}b^2\right] \iff \\
\frac{bc}{k+1}[c-kb] &= \frac{bc\cos A}{k+1}[c-kb] \iff \\
[c-kb](1-\cos A) &= 0 \Longrightarrow k = c/b,
\end{aligned}$$

since $\cos A < 1$.

Exercise 7.13.4. Use formula (7.13.18) from page 415 with $k = \left(\frac{c}{b}\right)^2$.

Exercise 7.14.5. Expressing the bimedian MN by Proposition 6.4.2 we have $\overrightarrow{MN} = \frac{1}{2}(\overrightarrow{AD}+\overrightarrow{BC})$. Then $|MN|^2 = \overrightarrow{MN}\cdot\overrightarrow{MN} = \frac{1}{4}\Big(|AD|^2+|BC|^2+2\overrightarrow{AD}\cdot\overrightarrow{BC}\Big)$.

Exercise 7.15.4. It is a particular version of Bretschneider's relation.

Exercise 7.15.5. Use that in a rhombus diagonals are perpendicular.

Exercise 7.15.6. Let U and V be the midpoints of AB and CD, respectively. Applying the median theorem in triangles ΔMAB and ΔMSC and using that $|AB| = |CD|$ yields

$$\begin{aligned}
|MU|^2 &= \frac{1}{2}(|MA|^2 + |MB|^2) - \frac{1}{2}|AB|^2 \\
&= \frac{1}{2}(|MC|^2 + |MD|^2) - \frac{1}{2}|CD|^2 = |MV|^2,
\end{aligned}$$

and hence $|MU| = |MV|$. Then the locus of M is the perpendicular bisector of the segment UV, which is the bimedian corresponding to sides AB and

CD. This can be shown using vectors as follows. Let Q be the orthogonal projection of M on UV. Then $|MU|^2 = |MV|^2 \Leftrightarrow \overrightarrow{MU} \cdot \overrightarrow{MU} = \overrightarrow{MV} \cdot \overrightarrow{MV}$ $\Leftrightarrow (\overrightarrow{MQ}+\overrightarrow{QU})\cdot(\overrightarrow{MQ}+\overrightarrow{QU}) = (\overrightarrow{MQ}+\overrightarrow{QV})\cdot(\overrightarrow{MQ}+\overrightarrow{QV}) \Leftrightarrow |MQ|^2+|QU|^2 = |MQ|^2 + |QV|^2 \Leftrightarrow |QU| = |QV|$, i.e., Q is the midpoint of UV and hence M belongs to the perpendicular on UV passing through Q.

Exercise 7.15.7. We assume $AB||DC$. We shall show $|AC|^2 + |BD|^2 = |AD|^2 + |BC|^2 + 2|AB| \cdot |DC| = 0$. Using properties of scalar product, we have

$$\begin{aligned}
&|AC|^2 + |BD|^2 - |AD|^2 - |BC|^2 - 2|AB| \cdot |DC| \\
&= (\overrightarrow{AD} + \overrightarrow{DC})(\overrightarrow{AD} + \overrightarrow{DC}) + (\overrightarrow{BC} + \overrightarrow{CD})(\overrightarrow{BC} + \overrightarrow{CD}) \\
&\quad - |AD|^2 - |BC|^2 - 2|AB|\,|DC| \\
&= |AD|^2 + |DC|^2 + 2\overrightarrow{AD} \cdot \overrightarrow{DC} + |BC|^2 + |CD|^2 + 2\overrightarrow{BC} \cdot \overrightarrow{CD} \\
&\quad - |AD|^2 - |BC|^2 - 2|AB|\,|DC| \\
&= 2(|DC|^2 + \overrightarrow{AD} \cdot \overrightarrow{DC} + \overrightarrow{BC} \cdot \overrightarrow{CD}) - 2|AB|\,|DC| \\
&= 2\overrightarrow{DC} \cdot (\overrightarrow{DC} + \overrightarrow{AD} - \overrightarrow{BC} - \overrightarrow{AB}) = 2\overrightarrow{DC} \cdot (\overrightarrow{DC} + \overrightarrow{CB} + \overrightarrow{BA} + \overrightarrow{AD}) = 0.
\end{aligned}$$

Exercise 7.15.8. We consider a Cartesian system of coordinates centered at $D(0,0)$, with $C(c,0)$, $B(c,b)$ and $A(0,b)$. Then $|MB|^2 = (x-c)^2 + (y-b)^2$, $|AD|^2 = b^2$, $|MD|^2 = x^2 + y^2$, $|AB|^2 = c^2$, $|MA|^2 = x^2 + (y-b)^2$, $|BD|^2 = c^2 + b^2$. Substituting we show that relation (i) is verified by algebraic computation. A similar approach for (ii). Part (iii) follows by adding relations (i) and (ii) as well as an application of the Pythagorean theorem.

Exercise 7.15.9. A repetitive application of the median theorem yields

$$|OM|^2 = \frac{1}{2}(|AO|^2 + |OB|^2) - \frac{1}{4}|AB|^2, \; |ON|^2 = \frac{1}{2}(|OC|^2 + |OB|^2) - \frac{1}{4}|BC|^2$$
$$|OP|^2 = \frac{1}{2}(|OC|^2 + |OD|^2) - \frac{1}{4}|DC|^2, \; |OQ|^2 = \frac{1}{2}(|AO|^2 + |OD|^2) - \frac{1}{4}|AD|^2.$$

Summing, yields $\sum |OM|^2 = \sum |OA|^2 - \ell^2$, where ℓ is the side of the square.

Chapter 8

Exercise 8.1.1. Changing the sign of the entries of a row in a determinant changes the sign of the determinant.

Exercise 8.1.2. Form determinants having the vector components as rows and check their signs. (i) Positive oriented basis because $\begin{vmatrix} 1 & 3 & 0 \\ 0 & 1 & 4 \\ 3 & 0 & 2 \end{vmatrix} =$

$2 + 4 \cdot 4 \cdot 3 > 0$; (ii) Negative oriented; (iii) Not a basis, the determinant is zero.

Exercise 8.1.3. Let $P(x) = \begin{vmatrix} x & 0 & 1 \\ 1 & 0 & x \\ 1 & x & 0 \end{vmatrix} = -x(x-1)(x+1)$. We have $P(x) > 0$ for $x \in (-\infty, 0) \cup (0, 1)$.

Exercise 8.2.3. Using the definition of the vectorial product it is not hard to show $e_1 \times e_2 = -e_2 \times e_1 = e_3$. Thus, the first relation becomes $e_1 \times (u^1 e_1 + u^2 e_2) = u^2 e_3$. To show this, we verify the definition as follows: $e_3 \perp e_1$ and $e_3 \perp u^1 e_1 + u^2 e_2$; $\{e_1, u^1 e_1 + u^2 e_2, u^2 e_3\}$ is a positive definite basis since $\begin{vmatrix} 1 & 0 & \\ u^1 & u^2 & 0 \\ 0 & 0 & u^2 \end{vmatrix} = (u^2)^2 > 0$. We need to show now $|e_1||u^1 e_1 + u^2 e_2| \sin\phi = |u^2||e_3|$, or $\sin\phi = \frac{|u^2|}{\sqrt{(u^1)^2 +)u^2)^2}}$, or $\cos\phi = \frac{|u^1|}{\sqrt{(u^1)^2 +)u^2)^2}}$, which is equivalent to $\frac{e_1 \cdot (u^1 e_1 + u^2 e_2)}{|e_1||u^1 e_1 + u^2 e_2|} = \frac{|u^1|}{\sqrt{(u^1)^2 +)u^2)^2}}$, which holds true.

Exercise 8.2.5. Verify the definition properties of the vectorial product. For instance, to show that $\overrightarrow{u} \times \overrightarrow{v} \perp \overrightarrow{u}$ it suffices to show $(\overrightarrow{u} \times \overrightarrow{v}) \cdot \overrightarrow{u} = 0$, which can be verified algebraically as $u^1(u^2 v^3 - v^2 u^3) - u^2(u^1 v^3 - v^1 u^3) + u^3(u^1 v^2 - v^1 u^2) = 0$. The other properties can be shows as in Exercise 8.2.3.

Exercise 8.3.2. Let A', B' and C' be the midpoints of sides BC, AC and AB, respectively. Then use that

$$\overrightarrow{A'B'} \times \overrightarrow{A'C'} = \frac{1}{2}\overrightarrow{BA} \times \frac{1}{2}\overrightarrow{CA} = \frac{1}{4}\overrightarrow{BA} \times \overrightarrow{CA} = \frac{1}{4}\overrightarrow{AB} \times \overrightarrow{AC},$$

so $\mathcal{A}_{ABC} = \frac{1}{2}|\overrightarrow{A'B'} \times \overrightarrow{A'C'}| = \frac{1}{2}\frac{1}{4}|\overrightarrow{AB} \times \overrightarrow{AC}| = \frac{1}{4}\mathcal{A}_{A'B'C'}$.

Exercise 8.3.9. Let $\lambda \in \mathbb{R}$ be such that $\overrightarrow{CD} = \lambda\overrightarrow{AB}$ and define the point P such that $\overrightarrow{MP} = -(\overrightarrow{MA} + \overrightarrow{MC})$. Then $\overrightarrow{AB} \times \overrightarrow{AM} + \overrightarrow{CM} \times \overrightarrow{CD} = \overrightarrow{AB} \times (\overrightarrow{AM} - \lambda\overrightarrow{CM}) = \overrightarrow{AB} \times \overrightarrow{MP}$. The locus of M is a line parallel to ℓ_1 and ℓ_2.

Exercise 8.4.3. It follows from $\overrightarrow{D_2B_1} = \frac{2\overrightarrow{AB} + \overrightarrow{DC}}{3}$ by taking the scalar product with itself.

Exercise 8.4.5. $\Rightarrow$ Taking the square in relation $|\overrightarrow{u} + \overrightarrow{v}| = |\overrightarrow{u}| + |\overrightarrow{v}|$ we obtain $(\overrightarrow{u} + \overrightarrow{v}) \cdot (\overrightarrow{u} + \overrightarrow{v}) = |\overrightarrow{u}|^2 + 2|\overrightarrow{u}||\overrightarrow{v}| + |\overrightarrow{v}|^2$, or $|\overrightarrow{u}|^2 + 2\overrightarrow{u} \cdot \overrightarrow{v} + |\overrightarrow{v}|^2 = |\overrightarrow{u}|^2 + 2|\overrightarrow{u}||\overrightarrow{v}| + |\overrightarrow{v}|^2$. After cancellation we obtain $\overrightarrow{u} \cdot \overrightarrow{v} = |\overrightarrow{u}||\overrightarrow{v}|$ so $\cos\theta = 0$, or $\theta = 0$, i.e., $\overrightarrow{u}$ and $\overrightarrow{v}$ are collinear with the vectors oriented in the same sense. $\Leftarrow$ Conversely, let $\overrightarrow{u} = \lambda\overrightarrow{v}$, with $\lambda > 0$. Then we can verify that

$$|\overrightarrow{u} + \overrightarrow{v}| = |\lambda\overrightarrow{v} + \overrightarrow{v}| = \lambda|\overrightarrow{v}| + |\overrightarrow{v}| = |\overrightarrow{u}| + |\overrightarrow{v}|.$$

Exercise 8.4.7. $\mathcal{A}_{ABCD} = \frac{1}{2}\delta_1\delta_2 \sin\phi = \frac{1}{2}|\overrightarrow{AC} \times \overrightarrow{BD}|$, where ϕ is the acute angle between diagonals.

Exercise 8.4.8. Using Exercise 8.4.7 and the midpoint connector relations, we have

$$\mathcal{A}_{MNP} = \frac{1}{2}|\overrightarrow{MN} \times \overrightarrow{NP}| = \frac{1}{2}\frac{1}{4}|\overrightarrow{AC} \times \overrightarrow{BD}| = \frac{1}{4}\mathcal{A}_{ABCD}.$$

Similarly, $\mathcal{A}_{MQP} = \frac{1}{4}\mathcal{A}_{ABCD}$. Adding, $\mathcal{A}_{MNPQ} = \mathcal{A}_{MNP} + \mathcal{A}_{MQP} = \frac{1}{2}\mathcal{A}_{ABCD}$.

Exercise 8.4.9. We have $\overrightarrow{QM} = \overrightarrow{QA} + \overrightarrow{AM} = (1-\lambda)\overrightarrow{DA} + \lambda\overrightarrow{AB}$ and $\overrightarrow{MN} = \overrightarrow{MB} + \overrightarrow{BN} = (1-\lambda)\overrightarrow{AB} + \lambda\overrightarrow{BC}$. Adding yields

$$\begin{aligned}\overrightarrow{QN} &= \overrightarrow{QM} + \overrightarrow{MN} \\ &= (1-\lambda)(\overrightarrow{DA} + \overrightarrow{AB}) + \lambda(\overrightarrow{AB} + \overrightarrow{BC}) = (1-\lambda)\overrightarrow{DB} + \lambda\overrightarrow{AC}.\end{aligned}$$

Similarly, $\overrightarrow{MP} = (1-\lambda)\overrightarrow{AC} + \lambda\overrightarrow{BD}$. Taking the cross product

$$\begin{aligned}\overrightarrow{QN} \times \overrightarrow{MP} &= ((1-\lambda)\overrightarrow{DB} + \lambda\overrightarrow{AC}) \times ((1-\lambda)\overrightarrow{AC} + \lambda\overrightarrow{BD}) \\ &= (1-\lambda)^2\overrightarrow{DB} \times \overrightarrow{AC} + \lambda^2\overrightarrow{AC} \times \overrightarrow{BD} \\ &= \left(\lambda^2 + (1-\lambda)^2\right)\overrightarrow{DB} \times \overrightarrow{AC}.\end{aligned}$$

Therefore, $\mathcal{A}_{MNPQ} = \left(\lambda^2 + (1-\lambda)^2\right)\mathcal{A}_{ABCD}$. Then use that the quadratic polynomial $P(\lambda) = \lambda^2 + (1-\lambda)^2$ reaches its minimum for $\lambda = 1/2$.

Exercise 8.4.10. We express $\overrightarrow{NM}$ and $\overrightarrow{NP}$ as

$$\overrightarrow{NM} = \overrightarrow{NA} + \overrightarrow{AM} = \frac{1}{2}\overrightarrow{BA} + \frac{2}{3}\overrightarrow{AD}, \quad \overrightarrow{NP} = \overrightarrow{NB} + \overrightarrow{BP} = \frac{1}{2}\overrightarrow{AB} + \frac{1}{3}\overrightarrow{BC}.$$

and then take their cross product

$$\begin{aligned}\overrightarrow{NM} \times \overrightarrow{NP} &= \left(\frac{1}{2}\overrightarrow{BA} + \frac{2}{3}\overrightarrow{AD}\right) \times \left(\frac{1}{2}\overrightarrow{AB} + \frac{1}{3}\overrightarrow{BC}\right) \\ &= \frac{1}{3}\left(\frac{1}{2}\overrightarrow{BA} \times \overrightarrow{BC} + \overrightarrow{AD} \times \overrightarrow{AB}\right) = \frac{1}{2}\overrightarrow{BA} \times \overrightarrow{BC}.\end{aligned}$$

Then $\mathcal{A}_{MNP} = \frac{1}{2}|\overrightarrow{NM} \times \overrightarrow{NP}| = \frac{1}{4}|\overrightarrow{BA} \times \overrightarrow{BC}| = \frac{a^2}{4}$.

Chapter 9

Exercise 9.1.3. The relations follow from the determinant properties: (*i*) a cyclic permutation of the rows involves an even number of row transpositions; (*ii*) a determinant with two identical rows is equal to zero; (*iii*) the determinant is a multiplicative function of each row.

Exercise 9.1.4. *(i)* $\Rightarrow$ *(ii)* If $\overrightarrow{u}$, $\overrightarrow{v}$, $\overrightarrow{w}$ are linear dependent we have a, b, $c \in \mathbb{R}$ such that $a\overrightarrow{u} + b\overrightarrow{v} + c\overrightarrow{w} = 0$. If $c \neq 0$, take $\alpha = -a/c$ and $\beta = -b/c$. If $c = 0$, take $a = b = 0$. *(ii)* $\Rightarrow$ *(iv)* If $\overrightarrow{w} = \alpha\overrightarrow{u} + \beta\overrightarrow{v}$ then on components we have $w^j = \alpha u^j + \beta v^j$. Since the last row of the determinant is a linear combination of the first two rows, then the determinant is equal to zero. *(iv)* $\Rightarrow$ *(iii)* Apply the determinant formula of the mixed product. *(iv)* $\Rightarrow$ *(i)* We need to show the existence of a, b, $c \in \mathbb{R}$ such that $a\overrightarrow{u} + b\overrightarrow{v} + c\overrightarrow{w} = 0$, which on components becomes the linear system $au^i + bv^i + cw^i = 0$. Since (a, b, c) is a nonzero solution, the determinant of coefficients must be nonzero.

Exercise 9.1.5. It follows from the additivity property of determinants.

Exercise 9.2.2. *(i)* A simple computation implies

$$\begin{aligned}\overrightarrow{a} \times (\overrightarrow{b} \times \overrightarrow{c}) &= \overrightarrow{a} \times (\overrightarrow{b} \times (\overrightarrow{c} + s\,\overrightarrow{b})) = \overrightarrow{a} \times (\overrightarrow{b} \times \overrightarrow{c}) + s\,\overrightarrow{a} \times \underbrace{(\overrightarrow{b} \times \overrightarrow{b})}_{=0} \\ &= \overrightarrow{a} \times (\overrightarrow{b} \times \overrightarrow{c}).\end{aligned}$$

(ii) Using distributivity property yields

$$\begin{aligned}(\overrightarrow{a} \cdot \overrightarrow{c})\,\overrightarrow{b} - (\overrightarrow{a} \cdot \overrightarrow{b})\,\overrightarrow{c} &= (\overrightarrow{a} \cdot (\overrightarrow{c} + s\,\overrightarrow{b}))\,\overrightarrow{b} - (\overrightarrow{a} \cdot \overrightarrow{b})(\overrightarrow{c} + s\,\overrightarrow{b}) \\ &= (\overrightarrow{a} \cdot \overrightarrow{c})\,\overrightarrow{b} + s(\overrightarrow{a} \cdot \overrightarrow{b})\,\overrightarrow{b} - (\overrightarrow{a} \cdot \overrightarrow{b})\,\overrightarrow{c} \\ &- s(\overrightarrow{a} \cdot \overrightarrow{b})\,\overrightarrow{b} \\ &= (\overrightarrow{a} \cdot \overrightarrow{c})\,\overrightarrow{b} - (\overrightarrow{a} \cdot \overrightarrow{b})\,\overrightarrow{c}.\end{aligned}$$

Exercise 9.2.3. Both relations are homogenous in the vector c. This follows from the definition of the cross and inner products.

Exercise 9.2.6. *(i)* Use that $\overrightarrow{b} \times \overrightarrow{c} = (\lambda\overrightarrow{c}) \times \overrightarrow{c} = \lambda(\overrightarrow{c} \times \overrightarrow{c}) = 0$. *(ii)* A similar computation.

Exercise 9.3.3. *(i)* Let ϕ be the angle between vectors $\overrightarrow{a}$ and $\overrightarrow{b}$. Then we have

$$\begin{aligned}|\overrightarrow{a} \times \overrightarrow{b}|^2 &= |\overrightarrow{a}|^2 |\overrightarrow{b}|^2 \sin^2\phi = |\overrightarrow{a}|^2 |\overrightarrow{b}|^2 - |\overrightarrow{a}|^2 |\overrightarrow{b}|^2 \cos^2\phi \\ &= |\overrightarrow{a}|^2 \cdot |\overrightarrow{b}|^2 - (\overrightarrow{a} \cdot \overrightarrow{b})^2.\end{aligned}$$

(ii) $|\overrightarrow{a} \cdot \overrightarrow{b}| = |\overrightarrow{a}||\overrightarrow{b}||\cos\phi| \leq |\overrightarrow{a}||\overrightarrow{b}|$.

Exercise 9.3.5. It follows from Solved Exercise 9.3.4.

Exercise 9.3.6. It follows from Solved Exercise 9.3.4 and the skew-symmetry $(\overrightarrow{a} \times \overrightarrow{b}) \times (\overrightarrow{c} \times \overrightarrow{d}) = -(\overrightarrow{c} \times \overrightarrow{d}) \times (\overrightarrow{a} \times \overrightarrow{b})$.

Exercise 9.3.7. See Exercise 9.3.6.

Exercise 9.3.8. (*i*) Let $a_i = (a_i^k)_k$, $b_j = (b_j^k)_k$, and $c_i = (c_i^k)_k$, where $c_i^k = \sum_k a_i^k b_i^k$. Consider matrices $A = (a_i^k)$, $B = (b_j^k)$ and $C = (c_i^k)$. Then $C = AB^T$. Using $\det C = \det A \det B$ we obtain $\det(c_i^k) = \det(a_i^k)\,\det(b_j^k)$. Then use the determinant representation of the mixed product $[\overrightarrow{a_1}, \overrightarrow{a_2}, \overrightarrow{a_3}] = \det(a_i^k)$ and $[\overrightarrow{b_1}, \overrightarrow{b_2}, \overrightarrow{b_3}] = (b_j^k)$.

Exercise 9.3.11 We have

$$\overrightarrow{a} \cdot \overrightarrow{a'} = \overrightarrow{a} \cdot (\overrightarrow{b} \times \overrightarrow{c})/[\overrightarrow{a}, \overrightarrow{b}, \overrightarrow{c}] = [\overrightarrow{a}, \overrightarrow{b}, \overrightarrow{c}]/[\overrightarrow{a}, \overrightarrow{b}, \overrightarrow{c}] = 1.$$

$$\overrightarrow{a} \cdot \overrightarrow{b'} = \overrightarrow{a} \cdot (\overrightarrow{c} \times \overrightarrow{a})/[\overrightarrow{a}, \overrightarrow{b}, \overrightarrow{c}] = [\overrightarrow{a}, \overrightarrow{c}, \overrightarrow{a}]/[\overrightarrow{a}, \overrightarrow{b}, \overrightarrow{c}] = 0.$$

The other relations are similar.

Exercise 9.3.13. Since $\overrightarrow{a}$, $\overrightarrow{b}$ and $\overrightarrow{c}$ are linearly independent, we can write $\overrightarrow{v} = \alpha\overrightarrow{a} + \beta\overrightarrow{b} + \gamma\overrightarrow{c}$. To determine the coefficients α, β and γ we take inner products of $\overrightarrow{v}$ with $\overrightarrow{a'}$, $\overrightarrow{b'}$ and $\overrightarrow{c'}$ and use Exercise 9.3.11 to obtain

$$\begin{aligned}
\overrightarrow{v} \cdot \overrightarrow{a'} &= \alpha\overrightarrow{a} \cdot \overrightarrow{a'} + \beta\overrightarrow{b} \cdot \overrightarrow{a'} + \gamma\overrightarrow{c} \cdot \overrightarrow{a'} = \alpha \\
\overrightarrow{v} \cdot \overrightarrow{b'} &= \alpha\overrightarrow{a} \cdot \overrightarrow{b'} + \beta\overrightarrow{b} \cdot \overrightarrow{b'} + \gamma\overrightarrow{c} \cdot \overrightarrow{b'} = \beta \\
\overrightarrow{v} \cdot \overrightarrow{c'} &= \alpha\overrightarrow{a} \cdot \overrightarrow{c'} + \beta\overrightarrow{b} \cdot \overrightarrow{c'} + \gamma\overrightarrow{c} \cdot \overrightarrow{c'} = \gamma.
\end{aligned}$$

Substituting α, β and γ we obtain the desired relation.

Exercise 9.3.14. $\mathcal{V} = [\overrightarrow{u}, \overrightarrow{v}, \overrightarrow{w}] = \det\begin{pmatrix} 1 & 0 & 3 \\ 3 & 1 & 2 \\ 0 & 1 & 5 \end{pmatrix} = 18.$

Exercise 9.3.15. Using that the mixed product is zero for two equal arguments, we can simplify the relation to

$$\begin{aligned}
[2\overrightarrow{u} + \overrightarrow{v}, \overrightarrow{v} + \overrightarrow{w}, 3\overrightarrow{w} + \overrightarrow{u}] &= 6[\overrightarrow{u}, \overrightarrow{v}, \overrightarrow{w}] + [\overrightarrow{v}, \overrightarrow{w}, \overrightarrow{u}] = 7[\overrightarrow{u}, \overrightarrow{v}, \overrightarrow{w}] \\
&= 7\det\begin{pmatrix} 1 & 0 & 3 \\ 2 & 1 & 0 \\ 0 & 4 & 6 \end{pmatrix} = 210.
\end{aligned}$$

Exercise 9.3.18. Use Exercise 9.3.3, part (*i*).

Chapter 10

Exercise 10.1.14. (*i*) Let $a = |AB| = |DC|$, $b = |BC| = |AD|$ and $\theta = \widehat{BCD}$. Then

$$\begin{aligned}
\overrightarrow{C^*B^*} \cdot \overrightarrow{C^*D^*} &= (\overrightarrow{C^*C} + \overrightarrow{CB^*}) \cdot (\overrightarrow{C^*D} + \overrightarrow{DD^*}) \\
&= \underbrace{\overrightarrow{C^*C} \cdot \overrightarrow{C^*D}}_{=0} + \overrightarrow{C^*C} \cdot \overrightarrow{DD^*} + \overrightarrow{CB^*} \cdot \overrightarrow{C^*D} + \underbrace{\overrightarrow{CB^*} \cdot \overrightarrow{DD^*}}_{=0} \\
&= \frac{a}{\sqrt{2}}\frac{b}{\sqrt{2}}\cos(45^\circ) - \frac{a}{\sqrt{2}}\frac{b}{\sqrt{2}}\cos(45^\circ) = 0,
\end{aligned}$$

so $C^*B^* \perp C^*D^*$. Similarly, we can show that the other sides are perpendicular. We also have

$$\begin{aligned}
|C^*B^*|^2 &= \overrightarrow{C^*B^*} \cdot \overrightarrow{C^*B^*} = (\overrightarrow{C^*C} + \overrightarrow{CB^*}) \cdot (\overrightarrow{C^*C} + \overrightarrow{CB^*}) \\
&= |C^*C|^2 + |CB^*|^2 + 2\overrightarrow{C^*C} \cdot \overrightarrow{CB^*} \\
&= \frac{a^2}{2} + \frac{b^2}{2} + ab\cos(\theta + 90^\circ).
\end{aligned}$$

$$\begin{aligned}
|C^*D^*|^2 &= \overrightarrow{C^*D^*} \cdot \overrightarrow{C^*D^*} = (\overrightarrow{C^*D} + \overrightarrow{DD^*}) \cdot (\overrightarrow{C^*D} + \overrightarrow{DD^*}) \\
&= |C^*D|^2 + |DD^*|^2 + 2\overrightarrow{C^*D} \cdot \overrightarrow{DD^*} \\
&= \frac{a^2}{2} + \frac{b^2}{2} + ab\cos(\theta + 90^\circ).
\end{aligned}$$

Hence $|C^*B^*| = |C^*D^*|$. It follows that $A^*B^*C^*D^*$ is a square.
(*ii*) The lines intersect at a point, which is their common midpoint. This follows from the fact that diagonals in a parallelogram bisect each other.
(*iii*) Use, for instance, $\Delta A''AA' \equiv \Delta C''CC'$.
(*iv*) It suffices to show $\overrightarrow{A'A''} \cdot \overrightarrow{AC} = 0$.

$$\begin{aligned}
\overrightarrow{A'A''} \cdot \overrightarrow{AC} &= (\overrightarrow{A'A} + \overrightarrow{AA''}) \cdot (\overrightarrow{AD} + \overrightarrow{AB}) \\
&= \underbrace{\overrightarrow{A'A} \cdot \overrightarrow{AD}}_{=0} + \overrightarrow{A'A} \cdot \overrightarrow{AB} + \overrightarrow{AA''} \cdot \overrightarrow{AD} + \underbrace{\overrightarrow{AA''} \cdot \overrightarrow{AB}}_{=0} \\
&= -\overrightarrow{AA'} \cdot \overrightarrow{AB} + \overrightarrow{AA''} \cdot \overrightarrow{AD} \\
&= -\frac{ab}{2}\cos(\theta + 90^\circ) + \frac{ab}{2}\cos(\theta + 90^\circ) = 0.
\end{aligned}$$

Exercise 10.1.15. (*i*) We have

$$\begin{aligned}
2\overrightarrow{AA'} \cdot \overrightarrow{EI} &= (\overrightarrow{AB} + \overrightarrow{AC}) \cdot (\overrightarrow{EA} + \overrightarrow{AI}) \\
&= \overrightarrow{AB} \cdot \overrightarrow{EA} + \overrightarrow{AB} \cdot \overrightarrow{AI} + \overrightarrow{AC} \cdot \overrightarrow{EA} + \overrightarrow{AC} \cdot \overrightarrow{AI} \\
&= 0 + cb\cos(\widehat{A} + 90^\circ) - cb\cos(\widehat{A} + 90^\circ) + 0 = 0.
\end{aligned}$$

$$\begin{aligned}
|EI|^2 &= \overrightarrow{EI} \cdot \overrightarrow{EI} = |EA|^2 + |AI|^2 - 2|EA|\,|AI|\cos(180^\circ - \widehat{A}) \\
&= c^2 + b^2 + 2bc\cos A = 2(b^2 + c^2) - a^2 \\
&= 4\left(\frac{b^2 + c^2}{2} - \frac{a^2}{4}\right) = 4|AA'|^2.
\end{aligned}$$

(iv) To show that $AX \perp YZ$ we evaluate the inner product of $\overrightarrow{AX} = \overrightarrow{AB} + \overrightarrow{BX}$ and $\overrightarrow{ZY} = \overrightarrow{AY} - \overrightarrow{AZ}$

$$\begin{aligned}
\overrightarrow{AX} \cdot \overrightarrow{ZY} &= (\overrightarrow{AB} + \overrightarrow{BX}) \cdot (\overrightarrow{AY} - \overrightarrow{AZ}) \\
&= \overrightarrow{AB} \cdot \overrightarrow{AY} - \overrightarrow{AB} \cdot \overrightarrow{AZ} + \overrightarrow{BX} \cdot \overrightarrow{AY} - \overrightarrow{BX} \cdot \overrightarrow{AZ} \\
&= \frac{cb}{\sqrt{2}} \cos(\widehat{A} + 45^\circ) - \frac{c^2}{\sqrt{2}} \cos 45^\circ + \frac{ab}{2} \cos(90^\circ - \widehat{C}) + \frac{ac}{2} \cos \widehat{B} \\
&= \frac{cb}{\sqrt{2}} \frac{1}{\sqrt{2}} (\cos \widehat{A} - \sin \widehat{A}) - \frac{c^2}{2} + \frac{ab}{2} \sin \widehat{C} + \frac{ac}{2} \frac{a^2 + c^2 - b^2}{2ac} \\
&= \frac{bc}{2} \frac{b^2 + c^2 - a^2}{2bc} - \frac{abc}{4R} - \frac{c^2}{2} + \frac{abc}{4R} + \frac{1}{4}a^2 + \frac{1}{4}c^2 - \frac{1}{4}b^2 = 0.
\end{aligned}$$

To show that $|AX| = |ZY|$ we compute each of them individually as

$$\begin{aligned}
|AX|^2 &= c^2 + \frac{a^2}{2} - \frac{2ca}{\sqrt{2}} \cos(\widehat{B} + 45^\circ) \\
&= c^2 + \frac{a^2}{2} - ac(\cos \widehat{B} - \sin \widehat{B}) = \frac{c^2}{2} + \frac{b^2}{2} + \frac{abc}{2R}.
\end{aligned}$$

$$\begin{aligned}
|ZY|^2 &= \frac{c^2}{2} + \frac{b^2}{2} - 2\frac{bc}{2} \cos(\widehat{A} + 90^\circ) = \frac{b^2 + c^2}{2} + bc \sin \widehat{A} \\
&= \frac{c^2}{2} + \frac{b^2}{2} + \frac{abc}{2R}.
\end{aligned}$$

Exercise 10.2.2. It follows from Menelaus theorem.

Exercise 10.3.2. Apply Proposition 2.7.9 from page 73 and Proposition 2.13.6 from page 94.

Exercise 10.4.3. Let A', B' and C' be the midpoints of sides BC, AC and AB of triangle ΔABC. The perpendicular through A' on BC is the perpendicular bisector of BC and it passes through the circumcenter, O, of ΔABC. Similarly, the perpendiculars through B' and C' on AC and AB pass through O. The relation is symmetric in the following sense. The perpendicular through A on $B'C'$ is the same as the altitude from A of ΔABC. Similarly, the perpendiculars from B and C on $A'C'$ and $A'B'$ are altitudes, and hence they intersect at the orthocenter, H.

Exercise 10.5.4. Since

$$\overrightarrow{BA_1} = \overrightarrow{BM} + \overrightarrow{MA_1}, \quad \overrightarrow{CB_1} = \overrightarrow{CM} + \overrightarrow{MB_1}, \quad \overrightarrow{AC_1} = \overrightarrow{AM} + \overrightarrow{MC_1}$$

we obtain by taking the inner products

$$|BA_1|^2 = |BM|^2 + |MA_1|^2 + 2\overrightarrow{BM} \cdot \overrightarrow{MA_1}$$
$$|CB_1|^2 = |CM|^2 + |MB_1|^2 + 2\overrightarrow{CM} \cdot \overrightarrow{MB_1}$$
$$|AC_1|^2 = |AM|^2 + |MC_1|^2 + 2\overrightarrow{AM} \cdot \overrightarrow{MC_1}.$$

Also, using

$$\overrightarrow{CA_1} = \overrightarrow{CM} + \overrightarrow{MA_1}, \quad \overrightarrow{AB_1} = \overrightarrow{AM} + \overrightarrow{MB_1}, \quad \overrightarrow{BC_1} = \overrightarrow{BM} + \overrightarrow{MC_1}$$

we obtain

$$|CA_1|^2 = |CM|^2 + |MA_1|^2 + 2\overrightarrow{CM} \cdot \overrightarrow{MA_1}$$
$$|AB_1|^2 = |AM|^2 + |MB_1|^2 + 2\overrightarrow{AM} \cdot \overrightarrow{MB_1}$$
$$|BC_1|^2 = |BM|^2 + |MC_1|^2 + 2\overrightarrow{BM} \cdot \overrightarrow{MC_1}.$$

Summing, we obtain that relation

$$|BA_1|^2 + |CB_1|^2 + |AC_1|^2 = |A_1C|^2 + |B_1A|^2 + |C_1B|^2$$

is equivalent to

$$\overrightarrow{BM}\cdot\overrightarrow{MA_1}+\overrightarrow{CM}\cdot\overrightarrow{MB_1}+\overrightarrow{AM}\cdot\overrightarrow{MC_1} = \overrightarrow{CM}\cdot\overrightarrow{MA_1}+\overrightarrow{AM}\cdot\overrightarrow{MB_1}+\overrightarrow{BM}\cdot\overrightarrow{MC_1}.$$

This relation is equivalent to

$$(\overrightarrow{BM} - \overrightarrow{CM}) \cdot \overrightarrow{MA_1} + (\overrightarrow{CM} - \overrightarrow{AM}) \cdot \overrightarrow{MB_1} + (\overrightarrow{AM} - \overrightarrow{BM}) \cdot \overrightarrow{MC_1} = 0,$$

or to

$$\overrightarrow{BC} \cdot \overrightarrow{MA_1} + \overrightarrow{CA} \cdot \overrightarrow{MB_1} + \overrightarrow{AB} \cdot \overrightarrow{MC_1} = 0,$$

which holds true since $\overrightarrow{BC} \perp \overrightarrow{MA_1}$, $\overrightarrow{CA} \perp \overrightarrow{MB_1}$ and $\overrightarrow{AB} \perp \overrightarrow{MC_1}$. The reciprocal statement can be proved by contradiction.

Exercise 10.7.2. It suffices to show $(\overrightarrow{AB} \cdot \overrightarrow{BD})\overrightarrow{BC} = (\overrightarrow{AB} \cdot \overrightarrow{BC})\overrightarrow{BD}$. Both sides have the same direction, sense and length, equal to $|AB|\,|BD|$ $\cos\widehat{B}|BC|$. Hence, both sides are equal.

Exercise 10.7.3. Both vectors have the same direction, sense and length, so they are equal.

Chapter 11

Exercise 11.1.3. We just verify that α, β and γ satisfy Definition 11.1.1.

Exercise 11.1.4. The centroid G satisfies $\overrightarrow{GA}+\overrightarrow{GB}+\overrightarrow{GC}=0$. Taking the cross product to $\overrightarrow{GB}$ yields $\overrightarrow{GA}\times\overrightarrow{GB}+\underbrace{\overrightarrow{GB}\times\overrightarrow{GB}}_{=0}+\overrightarrow{GC}\times\overrightarrow{GB}=0$, which is equivalent to $\overrightarrow{GA}\times\overrightarrow{GB}=\overrightarrow{GB}\times\overrightarrow{GC}$. Taking the cross product to $\overrightarrow{GC}$ yields the other identify.

Exercise 11.1.7. The baricentric coordinates of A corresponding to the vanishing areas are equal to zero, so $\beta=\gamma=0$. Since their sum is equal to 1, we have $\alpha=1$.

Exercise 1.11.8. Let A' be the midpoint of $[BC]$. Then $\mathcal{A}_{ABA'}=\mathcal{A}_{AA'C}=\frac{1}{2}\mathcal{A}_{ABC}$. Then by Exercise 11.1.3 we have $\beta=\frac{\mathcal{A}_{AA'C}}{\mathcal{A}_{ABC}}=\frac{1}{2}$ and $\gamma=\frac{\mathcal{A}_{ABA'}}{\mathcal{A}_{ABC}}=\frac{1}{2}$, so $\alpha=1-\beta-\gamma=0$.

Exercise 11.1.11. Use that $\widehat{BOC}=2(180^\circ-\widehat{A})=360^\circ-2\widehat{A})$.

Exercise 11.2.8. Consider the vertex $A(1,0,0)$. Then the line OI passes through A iff

$$\begin{aligned}
0&=\begin{vmatrix} 1 & 0 & 0\\ \frac{\frac{a}{2p}}{\frac{\cos A}{2\sin B\sin C}} & \frac{\frac{b}{2p}}{\frac{\cos B}{2\sin A\sin C}} & \frac{\frac{c}{2p}}{\frac{\cos C}{2\sin A\sin B}} \end{vmatrix} = \begin{vmatrix} \frac{\frac{b}{2p}}{\frac{\cos B}{2\sin A\sin C}} & \frac{\frac{c}{2p}}{\frac{\cos C}{2\sin A\sin B}} \end{vmatrix}\\
&=\frac{1}{4p\sin A}\begin{vmatrix} b & c\\ \frac{\cos B}{\sin C} & \frac{\cos C}{\sin B}\end{vmatrix}=\frac{2R}{4p\sin A}\begin{vmatrix} b & c\\ \frac{\cos B}{c} & \frac{\cos C}{b}\end{vmatrix}\\
&=\frac{R}{2p\sin A}(\cos C-\cos B).
\end{aligned}$$

Then $\cos C=\cos B$, which implies $\widehat{C}=\widehat{B}$, so the triangle is isosceles with $c=b$.

Exercise 11.4.4. We recall first the following formulas from pages 55 and 75

$$\tan\widehat{A}+\tan\widehat{B}+\tan\widehat{C}=\tan\widehat{A}\tan\widehat{B}\tan\widehat{C},$$

$$\sin 2\widehat{A}+\sin 2\widehat{B}+\sin 2\widehat{C}=4\sin\widehat{A}\sin\widehat{B}\sin\widehat{C}.$$

Then applying Theorem 11.4.1 from page 510 and the law of sines and the previous formulas, we have

$$\begin{aligned}|OH|^2 &= \alpha|OA|^2 + \beta|OB|^2 + \gamma|OC|^2 - \alpha\beta c^2 - \alpha\gamma b^2 - \beta\gamma a^2\\ &= R^2\cot B\cot C + R^2\cot A\cot C + R^2\cot A\cot B\\ &\quad - \cot B\cot^2 C\cot A\, c^2 - \cot^2 B\cot C\cot A\, b^2\\ &\quad - \cot B\cot C\cot^2 A\, a^2\\ &= \cot A\cot B\cot C\Big(R^2(\tan A + \tan B + \tan C)\\ &\quad - 4R^2\sin^2 C\cot C - 4R^2\sin^2 B\cot B - 4R^2\sin^2 A\cot A\Big)\\ &= R^2 - 4R^2\cot A\cot B\cot C\Big(\sin A\cos A + \sin B\cos B\\ &\quad + \sin C\cos C\Big)\\ &= R^2 - 2R^2\cot A\cot B\cot C\Big(\sin 2A + \sin 2B + \sin 2C\Big)\\ &= R^2 - 2R^2\cot A\cot B\cot C\, 4\sin A\sin B\sin C\\ &= R^2(1 - 8\cos A\cos B\cos C).\end{aligned}$$

One can also use Theorem 11.4.6.

Exercise 11.4.5. (i) Let $I_a(\alpha, \beta, \gamma)$. Since I_a is outside ΔABC but still inside the angle $\widehat{BAC}$, we have $\alpha < 0$ and $\beta > 0$, $\gamma > 0$. Using the area formula $\mathcal{A} = (p-a)r_a = (p-b)r_b = (p-c)r_c$, see page 70, we have

$$\alpha = -\frac{\mathcal{A}_{BI_aC}}{\mathcal{A}_{ABC}} - \frac{\frac{1}{2}ar_a}{(p-a)r_a} = \frac{-a}{2(p-a)}.$$

The other baricentric coordinates can be shown in similar way. (ii) Use Theorem 11.4.1 from page 510.

Exercise 11.4.7. Using Theorem 11.4.1 applied to $G(\frac{1}{3}, \frac{1}{3}, \frac{1}{3})$ and $|HA| = 2R\cos A$, we obtain

$$\begin{aligned}|HG|^2 &= \frac{1}{3}|HA|^2 + \frac{1}{3}|HB|^2 + \frac{1}{3}|HC|^2 - \frac{1}{9}(a^2+b^2+c^2)\\ &= \frac{4R^2}{3}(\cos^2 A + \cos^2 B + \cos^2 C) - \frac{1}{9}(a^2+b^2+c^2)\\ &= \frac{4R^2}{3}(1 - \cos A\cos B\cos C) - \frac{1}{9}(a^2+b^2+c^2).\end{aligned}$$

Applying the law of cosines yields

$$\begin{aligned}|HG|^2 &= \frac{4R^2}{3} - \frac{R^2}{3}\frac{(a^2+b^2-c^2)(a^2-b^2+c^2)(-a^2+b^2+c^2)}{(abc)^2}\\ &\quad - \frac{1}{9}(a^2+b^2+c^2).\end{aligned}$$

Exercise 11.4.8. Use Theorem 11.4.1.
Exercise 11.4.11. Let $M(x,y,z)$ be a point on the circumcircle. Using $\mathcal{C}(O,R) = \{M; |OM| = R\}$, then Theorem 11.4.1 yields

$$|OM|^2 = xR^2 + yR^2 + zR^2 - xyc^2 - yza^2 - zxb^2.$$

Replacing $|OM|^2$ by R^2 and using that $x+y+z = 1$, we obtain the following characterization of the circumcircle in baricentric coordinates

$$\mathcal{C}(O,R) = \{M(x,y,z); xyc^2 + yza^2 + zxb^2 = 0\}.$$

Exercise 11.4.12. The point $M(x,y,z)$ belongs to the incircle iff $|IM| = r$. By Theorem 11.4.1 we have $|IM|^2 = x|AI|^2 + y|BI|^2 + z|CI|^2 - xyc^2 - yza^2 - zxb^2$. Using $|AI| = r/\sin\frac{A}{2}$, we obtain

$$r^2 = x\frac{r^2}{\sin^2\frac{A}{2}} + y\frac{r^2}{\sin^2\frac{B}{2}} + z\frac{r^2}{\sin^2\frac{C}{2}} - xyc^2 - yza^2 - zxb^2.$$

Dividing by r^2, using that $1 = x + y + z$ and $1 - \frac{1}{\sin^2 x} = \cot^2 x$, we obtain

$$r^2\left(x\cot^2\frac{A}{2} + y\cot^2\frac{B}{2} + z\cot^2\frac{C}{2}\right) = xyc^2 + yza^2 + zxb^2.$$

For the second part, use $|I_aM| = r_a$ and $|AI_a| = r_a/\sin\frac{A}{2}$ and obtain a similar relation.
Exercise 11.4.13. It is a consequence of the midpoint connector theorem. Let $d_M = dist(M,BC)$, $d_{M'} = dist(M',BC)$ and $d_X = dist(X,BC)$. Then by the midpoint connector theorem in a trapezoid we have $\frac{1}{2}(d_M + d_{M'}) = d_X$. Multiplying by $|BC|/2$ yields $\dfrac{1}{2}(\mathcal{A}_{MBC} + \mathcal{A}_{M'BC}) = \mathcal{A}_{XBC}$. Dividing by $\mathcal{A}_{ABC}$ leads to $x = \frac{\alpha+\alpha'}{2}$.
Exercise 11.4.14. We have $(x,y,z) = \frac{1}{1+\lambda}(\alpha,\beta,\gamma) + \frac{\lambda}{1+\lambda}(\alpha',\beta',\gamma')$.
Exercise 11.4.15. (i) Since O_9 is the midpoint of OH, by Exercise 11.4.13 we have $\alpha_{O_9} = \frac{1}{2}(\alpha_O + \alpha_H)$, $\beta_{O_9} = \frac{1}{2}(\beta_O + \beta_H)$, $\gamma_{O_9} = \frac{1}{2}(\gamma_O + \gamma_H)$. From the proof of Theorem 11.2.5 we have $\alpha_O = \frac{1}{2}\alpha_H + \frac{1}{2}$ and other two similar relations. Then $\alpha_{O_9} = \frac{1}{2}\left(\frac{1}{2}\alpha_H + \frac{1}{2} + \alpha_H\right) = \frac{3}{4}\alpha_H + \frac{1}{4} = \frac{3}{4}\cot\widehat{B}\cot\widehat{C} + \frac{1}{4}$.
(ii) Using that the radius of the Euler circle is $R/2$, Theorem 11.4.1 yields

$$\frac{R^2}{4} = x|AO_9|^2 + y|BO_9|^2 + z|CO_9|^2 - xyc^2 - yza^2 - xzb^2,$$

where $M(x,y,z)$ is a point on the circle. By the median theorem $|AO_9|^2 = \frac{1}{2}(|AH|^2 + |AO|^2) - \frac{1}{4}|OH|^2$ and two more similar relations. Substituting in the previous relation and using that $|AO| = R$ and $x + y + z = 1$, yields

$$\frac{R^2}{4} = \frac{1}{2}(x|AH|^2 + y|BH|^2 + z|CH|^2) + \frac{R^2}{2} - \frac{1}{4}|OH|^2 - xyc^2 - yza^2 - xzb^2.$$

Using $|AH| = 2R\cos A$, $|BH| = 2R\cos B$, $|CH| = 2R\cos C$ and $|OH|^2 = 9R^2 - (a^2+b^2+c^2)$, we obtain

$$2R^2 = 2R^2(x\cos^2 A + y\cos^2 B + z\cos^2 C) + \frac{1}{4}(a^2+b^2+c^2) - xyc^2 - yza^2 - xzb^2.$$

Using $x+y+z=1$ and $1-\cos^2 A = \sin^2 A$, we obtain

$$2R^2(x\sin^2 A + y\sin^2 B + z\sin^2 C) = \frac{1}{4}(a^2+b^2+c^2) - xyc^2 - yza^2 - xzb^2.$$

Using $a^2 = 4R^2\sin^2 A$ and two more similar relations, yields

$$\frac{1}{2}(xa^2+yb^2+zc^2) = \frac{1}{4}(a^2+b^2+c^2) - xyc^2 - yza^2 - xzb^2,$$

which is the equation of the Euler circle in baricentric coordinates. (*iii*) Use Theorem 11.4.1 to show $|IO_9|^2 = \left(\frac{R}{2} - r\right)^2$, i.e., the distance between the centers is the difference between the radii.

Chapter 12

Exercise 12.1.10. Since

$$\begin{aligned} L(O,\overrightarrow{v_j}) &= \overrightarrow{OA_j} \times \overrightarrow{v_j} = (\overrightarrow{OO'} + \overrightarrow{O'A_j}) \times \overrightarrow{v_j} \\ &= \overrightarrow{OO'} \times \overrightarrow{v_j} + L(O',\overrightarrow{v_j}), \end{aligned}$$

summing over j and using the distributivity of the cross product, we obtain the desired result.

Exercise 12.1.11. It follows from Exercise 12.1.10 by taking $O=G$, $O'=M$, $\overrightarrow{v_j} = \overrightarrow{A_jG}$ and using $\sum_{j=1}^{n-1}\overrightarrow{A_jG} = 0$ and $L(G,\overrightarrow{A_jG}) = \overrightarrow{GA_j} \times \overrightarrow{A_jG} = 0$.

Exercise 12.4.5. Let A_1, B_1 and C_1 be the orthogonal projections of vertices A, B and C onto the line. Using the similar triangles $\Delta AGN \sim \Delta BB_1N$ and $\Delta AGM \sim \Delta CC_1M$, we have

$$\frac{|BN|}{|NA|} = \frac{|BB_1|}{|AA_1|}, \qquad \frac{|CM|}{|MA|} = \frac{|CC_1|}{|AA_1|}.$$

Adding yields

$$\frac{|BN|}{|NA|} + \frac{|CM|}{|MA|} = \frac{|BB_1|}{|AA_1|} + \frac{|CC_1|}{|AA_1|} = \frac{|BB_1| + |CC_1|}{|AA_1|} = 1,$$

where we used Remark 12.4.4.

Chapter 13

Exercise 13.1.3. A line is determined by a point and a direction. In this case the point is A and the direction is given by the vector $\overrightarrow{v_1} \times \overrightarrow{v_2}$, where $\{\overrightarrow{v_1}, \overrightarrow{v_2}\}$ is a system of two linearly independent vectors in the plane. To show the uniqueness part, we notice that being given two bases, $\{\overrightarrow{v_1}, \overrightarrow{v_2}\}$ and $\{\overrightarrow{w_1}, \overrightarrow{w_2}\}$ in the plane, the vectors $\overrightarrow{v_1} \times \overrightarrow{v_2}$ and $\overrightarrow{w_1} \times \overrightarrow{w_2}$ are parallel. To this end, we represent one basis in terms of the other, $\overrightarrow{w_1} = a_1\overrightarrow{v_1} + a_2\overrightarrow{v_2}$ and $\overrightarrow{w_2} = b_1\overrightarrow{v_1} + b_2\overrightarrow{v_2}$ and show by a simple computation that $(\overrightarrow{w_1} \times \overrightarrow{w_2}) \times (\overrightarrow{v_1} \times \overrightarrow{v_2}) = 0$.
Exercise 13.1.5. Let $\Pi \perp \mathcal{P}$, so there is a line $d \subset \Pi$ such that $d \perp \mathcal{P}$. We note first that the planes Π and $\mathcal{P}$ are not parallel (if they were the line d would be parallel to $\mathcal{P}$, contradiction). Let $\ell = \Pi \cap \mathcal{P}$, so $d \perp \ell$. Let $\{A\} = d \cap \ell$ and consider the perpendicular to ℓ at A contained in the plane $\mathcal{P}$, and denote it by m. Then $m \perp d$ and $m \perp \ell$, so $m \perp \Pi$. Since $d \subset \mathcal{P}$, then $\mathcal{P} \perp \Pi$.
Exercise 13.1.6. Consider three planes, Π_1, Π_2 and Π_3, such that $\Pi_1 || \Pi_3$ and take Π_2 perpendicular on both Π_1 and Π_3. Then $\Pi_1 \perp \Pi_2$ and $\Pi_2 \perp \Pi_3$, but Π_1 not perpendicular on Π_3, because they are parallel. The second part follows by a contradiction argument. Let $\Pi_1 \neq \Pi_3$ and assume there is a point $P \in \Pi_1 \cap \Pi_3$. Then there are two perpendicular lines from P to Π_2: one inside Π_1 and the other inside Π_3. This contradicts the uniqueness of the perpendicular line from a point to a plane, see Exercise 13.1.3.
Exercise 13.1.7. We adopt a proof by contradiction. Let B be the point where the perpendicular line from A intersects $\mathcal{P}$. Then $B \notin \Pi$, because otherwise, $AB \subset \Pi$, contradiction. Let $\ell = \Pi \cap \mathcal{P}$ and $C \in \ell$ such that $AC \perp \ell$. Since $\Pi \perp \mathcal{P}$, then $AC \perp \mathcal{P}$ and hence $AC \perp CB$. We obtain that ΔABC has two right angles, at C and B, which is a contradiction.
Exercise 13.3.2. (i) Since $MB \perp BC$ we have $\overrightarrow{MB} \cdot \overrightarrow{BC} = 0$. Also, since $MA \perp \mathcal{P}$ and $BC \subset \mathcal{P}$ we have $\overrightarrow{MA} \cdot \overrightarrow{BC} = 0$. By distributivity we have $\overrightarrow{AB} \cdot \overrightarrow{BC} = (\overrightarrow{AM} + \overrightarrow{MB}) \cdot \overrightarrow{BC} = -\overrightarrow{MA} \times \overrightarrow{BC} + \overrightarrow{MB} \cdot \overrightarrow{BC} = 0$. Hence, $AB \perp BC$. (ii) Similarly, we have $\overrightarrow{MA} \cdot \overrightarrow{BC} = (\overrightarrow{MB} + \overrightarrow{BA}) \cdot \overrightarrow{BC} = \overrightarrow{MB} \cdot \overrightarrow{BC} + \overrightarrow{BA} \cdot \overrightarrow{BC} = 0$.
Exercise 13.5.5. $V = \frac{1}{6}\left| \begin{vmatrix} 1 & 0 & 3 \\ 0 & 3 & 2 \\ 2 & 6 & 1 \end{vmatrix} \right| = \frac{9}{2}$.
Exercise 13.5.6. Use the (SSS) congruence case and the fact that in any parallelogram the opposite sides are equal.
Exercise 13.5.10. The parallelepiped is obtained by gluing together the initial triangular prismatoid and another triangular prismatoid obtained by a symmetry of the first one with respect to one of its sides.
Exercise 13.5.14. It follows from Proposition 13.5.13 by choosing $\sin\phi = 1$.

Exercise 13.5.15. Use Proposition 13.5.13 and Proposition 13.8.4
Exercise 13.5.19. It follows using algebraic manipulations from Proposition 13.5.18.
Exercise 13.5.20. We reduce the 5×5 determinant given in the exercise to the 3×3 determinant given in Proposition 13.5.18 as follows. We denote by R_1, R_2, R_3, R_4, R_5 the rows and by C_1, C_2, C_3, C_4, C_5 the columns of the determinant. We perform the following elementary transformations of the determinant, which does not change its value: Subtract R_2 from R_3, R_4 and R_5. Subtract C_2 from C_3, C_4 and C_5. Then expand the determinant over C_1, which has only one non-zero entry on position $(2, 1)$, yielding a 4×4 determinant. Then expand the new determinant over its first row and obtain exactly the determinant given in Proposition 13.5.18.
Exercise 13.5.21. We use that four points are collinear iff the volume of the tetrahedron with vertices at that points is equal to zero. Then use Exercise 13.5.19.
Exercise 13.6.7. It follows from Vecten's theorem.
Exercise 13.6.8. We assume the tetrahedron has all faces of equal area. Let M be the midpoint of the side AB. Let U, N, V be the orthogonal projections of B, M and A on the side CD, respectively. Since M is a midpoint, then N is the midpoint of UV. From $\mathcal{A}_{ACD} = \mathcal{A}_{BCD}$ we obtain $|AV| = |BU|$. Then $\Delta AVN \equiv \Delta BUN$ as two right triangles with two equal sides. Then $|AN| = |BN|$ and hence ΔNAB is an isosceles triangle. Therefore, the median NM becomes altitude, so $NM \perp AB$. Then MN is the common perpendicular of AB and CD. To show that N is the midpoint of CD, we proceed by contradiction. Let N' be the midpoint of CD and M' be the orthogonal projection on AB. By a similar procedure we can show that $M'N' \perp CD$, so $M'N'$ is perpendicular to both AB and CD. From the uniqueness of the common perpendicular, we obtain $N = N'$.
Exercise 13.6.9. "$\Longleftarrow$" First, we notice that if the opposite edges are equal, then the faces of the tetrahedron are congruent triangles, by (SSS) congruence case, and hence they have equal areas. For instance, $\Delta ACD \equiv \Delta ACB$, since AC is a common side and $|AD| = |BC|$, $|AB| = |CD|$ as opposite sides. Writing the volume as the product between the area of each face and the length of the corresponding altitude, we obtain that all altitudes have equal lengths.
"$\Longrightarrow$" Assuming that all altitudes have equal length, using the volume formula yields that all faces have equal areas. Then by Exercise 13.6.8 the bimedian MN is perpendicular on both AB and CD. By similar reasons as in Exercise 13.6.8 we have $|CU| = |VD|$. Since $|BU| = |AV|$ (because they are altitudes of triangles of equal areas), we obtain $\Delta BCU \equiv \Delta ADV$, as right triangles with two equal sides. Therefore, $|BC| = |AD|$. A similar argument can be applied for the other pairs of opposite sides.

Exercise 13.7.4. (*i*) It follows by computing the inner product $\overrightarrow{AC} \cdot \overrightarrow{BD}$ in two ways and then equating the results. On one side, we have

$$\begin{aligned}\overrightarrow{AC} \cdot \overrightarrow{BD} &= (\overrightarrow{AD} + \overrightarrow{DC}) \cdot (\overrightarrow{BA} + \overrightarrow{AD}) = \overrightarrow{AD} \cdot \overrightarrow{BA} + |AD|^2 \\ &\quad + \overrightarrow{DC} \cdot \overrightarrow{BA} + \overrightarrow{DC} \cdot \overrightarrow{AD} \\ &= (\overrightarrow{AB} + \overrightarrow{BD}) \cdot \overrightarrow{BA} + |AD|^2 + \overrightarrow{DC} \cdot \overrightarrow{BA} + \overrightarrow{DC} \cdot \overrightarrow{AD} \\ &= -|AB|^2 + \overrightarrow{BD} \cdot \overrightarrow{BA} + |AD|^2 + \overrightarrow{DC} \cdot \overrightarrow{BA} + \overrightarrow{DC} \cdot \overrightarrow{AD} \\ &= -|AB|^2 + \overrightarrow{BA} \cdot (\overrightarrow{BD} + \overrightarrow{DC}) + |AD|^2 + \overrightarrow{DC} \cdot \overrightarrow{AD} \\ &= -|AB|^2 + |AD|^2 + \overrightarrow{BA} \cdot \overrightarrow{BC} + \overrightarrow{DC} \cdot \overrightarrow{AD}.\end{aligned}$$

We evaluate the expression $\mathcal{E} = \overrightarrow{BA} \cdot \overrightarrow{BC} + \overrightarrow{DC} \cdot \overrightarrow{AD}$ as follows:

$$\begin{aligned}\mathcal{E} &= (\overrightarrow{BC} + \overrightarrow{CA}) \cdot \overrightarrow{BC} - (\overrightarrow{DC} + \overrightarrow{CA}) \cdot \overrightarrow{DC} \\ &= |BC|^2 + \overrightarrow{CA} \cdot \overrightarrow{BC} - |DC|^2 - \overrightarrow{CA} \cdot \overrightarrow{DC} \\ &= |BC|^2 - |DC|^2 + \overrightarrow{CA} \cdot (\overrightarrow{BC} + \overrightarrow{CD}) \\ &= |BC|^2 - |DC|^2 - \overrightarrow{AC} \cdot \overrightarrow{BD}.\end{aligned}$$

Substituting in the previous expression and solving for the inner product we get $2\overrightarrow{AC} \cdot \overrightarrow{BD} = |AD|^2 + |BC|^2 - |AB|^2 - |DC|^2$. Using $\overrightarrow{AC} \cdot \overrightarrow{BD} = |AC|\,|BD| \cos\phi$, we obtain the desired result by equating the expressions. (*ii*) It follows from part (*i*) by taking $\cos\phi = 0$.

Exercise 13.7.5. Let $ABCD$ be an orthocentric tetrahedron with orthocenter H. To show that $BCDH$ is an orthocentric tetrahedron it suffice to prove that the opposite sides are perpendicular. To this end, from $DH \perp (ABC)$ and $BC \subset (ABC)$ it follows that $DH \perp BC$. Similarly we arrive to $BH \perp DC$ and $CH \perp BD$. Since $BCDH$ is an orthocentric, it has an orthocenter. We shall show that this point is A. We need to prove that $HA \perp (BCD)$, $BA \perp (HCD)$, $DA \perp BC$, $CA \perp (BHD)$. The first perpendicularity relation, $HA \perp (BCD)$, follows from the fact that AH is an altitude. Next, we show the second relation, $BA \perp (HCD)$. We have $CD, HD \subset (HCD)$ and $BA \perp CD$ (the opposite sides are perpendicular). Also, from $DH \perp (ABC)$ and $AB \subset (ABC)$ we get $HD \perp BA$, so BA is perpendicular to both HD and CD, and hence to (HCD). The other relations can be shown similarly.

Exercise 13.7.6. See first part of the proof of Exercise 13.7.5.

Exercise 13.7.7. (*i*) and (*ii*) Let H_A and O_A be the orthogonal projections of H and O on the face (BCD). Then H_A is the orthocenter and O_A is the circumcenter of triangle ΔBCD. Then O_AH_A is the Euler line of

triangle ΔBCD. On this line we choose the points G_A and O_9 to be the centroid and the center of the Euler circle of triangle ΔBCD. We know that $\overrightarrow{O_AG_A} = \frac{1}{3}\overrightarrow{O_AH_A}$ and $\overrightarrow{O_AO_9} = \overrightarrow{O_9H_A}$. Since $AH_A \perp (BCD)$ and $OO_A \perp (BCD)$, then $AH_A || OO_A$, which implies that the points A, H_A, O and A_A are coplanar on a plane denoted by Π. We note that $\Pi \perp (BCD)$, since $AH_A \subset \Pi$. We also have O_A, G_A, O_9, H_A, O, H, $A \in \Pi$. Let $G \in AG_A$ such that $\overrightarrow{GG_A} = \frac{1}{4}\overrightarrow{AG_A}$ be the centroid of $ABCD$. Since $\frac{|G_AO_9|}{|G_AH_A|} = \frac{1}{4}$ and $\frac{|G_AG|}{G_AA} = \frac{1}{4}$ by Thales' theorem we get $GO_9 || AH_A$, so $GO_9 \perp (BCD)$. It follows that the perpendicular line on the center of Euler's circle, O_9, passes through the tetrahedron centroid, G. This property applies to all tetrahedron faces. Now, in the trapezoid OO_AH_AH (with $OO_A || HH_A$) O_9 is the midpoint of O_AH_A. By the midpoint connector theorem, the perpendicular on O_AH_A at O_9 intersects OH at its midpoint. This applies again for all tetrahedron faces. It follows that G is the midpoint of OH. Consequently, O, G and H are collinear points.
(*iii*) Using part (*ii*) we have $\overrightarrow{OA} + \overrightarrow{OB} + \overrightarrow{OC} + \overrightarrow{OD} = 4\overrightarrow{OG} = 2\overrightarrow{OH}$.
Exercise 13.8.1. Using the vectorial version of the median theorem, we have

$$\begin{aligned}\overrightarrow{MN} &= \frac{1}{2}(\overrightarrow{MC} + \overrightarrow{MD}) = \frac{1}{2}(\overrightarrow{MA} + \overrightarrow{AC} + \overrightarrow{MB} + \overrightarrow{BD}) \\ &= \frac{1}{2}(\overrightarrow{AC} + \overrightarrow{BD}) = \frac{1}{2}(\overrightarrow{p} + \overrightarrow{q}).\end{aligned}$$

The other relations can be shown similarly.
Exercise 13.8.3. From $|\overrightarrow{a} - \overrightarrow{b}|^2 = (\overrightarrow{a} - \overrightarrow{b}) \cdot (\overrightarrow{a} - \overrightarrow{b}) = |\overrightarrow{a}|^2 + |\overrightarrow{b}|^2 - 2\overrightarrow{a} \cdot \overrightarrow{b}$ solve for $\overrightarrow{a} \cdot \overrightarrow{b}$.
Exercise 13.8.5. Let O_A, O_B, O_C and O_D be the circumcenters of triangles ΔBCD, ΔACD, ΔABD and ΔABC, respectively. We also denote by ℓ_A, ℓ_B, ℓ_C, and ℓ_D the perpendicular lines through O_A, O_B, O_C and O_D, respectively, on the tetrahedron's faces. Any point $P \in \ell_A$ satisfies $|PB| = |PC| = |PD|$. Similar relations hold for the other perpendicular lines. Now, in order to show the existence of a sphere passing through the tetrahedron's vertices it suffices to show that the lines ℓ_A, ℓ_B, ℓ_C and ℓ_D intersect at the same point, O. This point, if exists, satisfies the property $|OA| = |OB| = OC| = |OD|$. To this end, we shall denote by μ_{BC} the plane perpendicular to BC, which passes through the midpoint of BC, and note that any point $P \in \mu_{BC}$ satisfies $|PB| = |PC|$. We employ the similar notations μ_{CD}, μ_{AD}, μ_{AB} and μ_{BD}. We note that $\ell_A = \mu_{BC} \cap \mu_{CD} \cap \mu_{BD}$ and $\ell_B = \mu_{CD} \cap \mu_{AB} \cap \mu_{AC}$ and then $\ell_A \subset \mu_{CD}$ and $\ell_B \subset \mu_{CD}$. Since ℓ_A is not parallel to ℓ_B, let $\{O\} = \ell_A \cap \ell_B$. Then $|OB| = |OC| = |OD| = |OA|$. We arrive to the same point O if start from any other two perpendicular lines.

Exercise 13.8.6. If each bimedian of a tetrahedron is perpendicular on the edges they bisect, by Proposition 13.8.4 the opposite sides are equal. Then use (SSS) congruence case to show that the faces of the tetrahedron are congruent triangles.
Exercise 13.8.8. " $\Longleftarrow$ " It follows from expressing the tetrahedron's volume as products between faces areas and distances between faces and opposite vertices. " $\Longrightarrow$ " If the tetrahedron is equifacial, then by Exercise 13.6.8 the bimedians are perpendicular on the opposite sides. Then by Proposition 13.8.4 the opposite sides are equal. Then apply Exercise 13.6.9.
Exercise 13.8.9. Use Exercise 13.6.8 and Proposition 13.8.4 to find that the opposite sides are equal. Then use (SSS) congruence case to show that all faces are congruent triangles, and hence, they have the same area.
Exercise 13.8.10. It follows from the addition property of volumes

$$\mathcal{V}_{ABCD} = \mathcal{V}_{MBCD} + \mathcal{V}_{MACD} + \mathcal{V}_{MABC} + \mathcal{V}_{MABD}$$

and the formula of the tetrahedron volume involving the area of the base and the distance to the opposite vertex. The converse holds true. Let $f(M) = dist(M,(BCD)) + dist(M,(ACD)) + dist(M,(ABC)) + dist(M,(ABD))$. Then $f(A) = f(B) = f(C) = f(D)$, which implies

$$dist(A,(BCD)) = dist(B,(ACD)) = dist(D,(ABC)) = dist(C,(ABD)).$$

Applying Exercise 13.8.8 we obtain an equifacial tetrahedron.
Exercise 13.9.2. We denote $S = |AB|^2+|AC|^2+|AD|^2+|BD|^2+|BC|^2+|CD|^2$. Applying the median theorem for each median and taking the sum, we obtain

$$\sum |AG_A|^2 = \frac{2}{3}S - \frac{2}{9}S = \frac{4}{9}S.$$

Exercise 13.10.4. (i) Use $|AG|^2 = \frac{9}{16}|AG_A|^2$ and the expression of $|AG_A|^2$ given by the median theorem.
(ii) Use $\overrightarrow{MA} + \overrightarrow{MB} + \overrightarrow{MC} + \overrightarrow{MD} = 4\overrightarrow{MG}$ and choose $M = A$.
Exercise 13.10.6. " $\Longrightarrow$ " Assume $|AG_A| = |BG_B| = |CG_C| = |DG_D|$. From the median theorem we have

$$9|AG_A|^2 = 2(|AB|^2 + |AC|^2 + |AD|^2) - (|BC|^2 + |CD|^2 + |BD|^2)$$
$$9|BG_B|^2 = 2(|AB|^2 + |BD|^2 + |BC|^2) - (|AC|^2 + |AD|^2 + |CD|^2).$$

Subtracting and using $|AG_A|^2 = |BG_B|^2$ yields $|AC|^2 + |AD|^2 = |BD|^2 + |BC|^2$. This is equivalent to

$$\frac{1}{2}(|AC|^2 + |AD|^2) - \frac{1}{4}|DC|^2 = \frac{1}{2}(|BD|^2 + |BC|^2) - \frac{1}{4}|DC|^2,$$

which means $|AM|^2 = |BM|^2$, where M denotes the midpoint of $[CD]$. Hence, $|AM| = |BM|$ and triangle ΔAMB is isosceles. Therefore, its median and altitude from M coincide. So, if N is the midpoint of $[AB]$, then MN is perpendicular to AB. Repeating the argument we obtain that the bimedians are perpendicular on the sides they bisect. Then apply an argument as in Exercise 13.8.6 to show that the opposite sides are equal.
" $\Longleftarrow$ " By direct computation and using the median theorem.
Exercise 13.10.9. By Lagrange's theorem, page 575, we have

$$\mathcal{S}(M) = |MA|^2+|MB|^2+|MC|^2+|MD|^2 \geq |GA|^2+|GB|^2+|GC|^2+|GD|^2,$$

with identity reached for $M = G$.
Exercise 13.10.10. From Lagrange's relation, with $M = H$, we have

$$|HA|^2+|HB|^2+|HC|^2+|HD|^2 = 4|HG|^2+|GA|^2+|GB|^2+|GC|^2+|GD|^2$$
$$= 4|HG|^2 + \frac{1}{4}(|AB|^2 + |AC|^2 + |AD|^2 + |BC|^2 + |BD|^2 + |CD|^2),$$

by Exercise 13.10.4, part (ii). Solving for $4|HG|^2$ yields the desired relation.
Exercise 13.10.11. We apply Lagrange relation with $M = O$ and use that $|AO| = |BO| = |CO| = |DO| = R$ and

$$|GA|^2 + |GB|^2 + |GC|^2 + |GD|^2 = \frac{1}{4}(|AB|^2 + |AC|^2 + |AD|^2 + |BC|^2 + |BD|^2 + |CD|^2)$$

to obtain

$$|OG|^2 = R^2 - \frac{1}{16}(|AB|^2 + |AC|^2 + |AD|^2 + |BC|^2 + |BD|^2 + |CD|^2).$$

We make the remark that in any tetrahedron we have the inequality

$$16R^2 \geq |AB|^2 + |AC|^2 + |AD|^2 + |BC|^2 + |BD|^2 + |CD|^2.$$

The identity is reached in the case $O = G$. Since G is the midpoint of OH, see Exercise 13.7.7, then $O = H$. The projection of these points on the face (BCD) is $O_A = H_A$, which means that ΔBCD is equilateral. Similarly, all faces are equilateral triangles.
Exercise 13.11.3. (i) By the midpoint connector theorem, $\overrightarrow{ME} = \frac{1}{2}\overrightarrow{DB} = \overrightarrow{FN}$, so $MENF$ is a parallelogram. (ii) If $MN \perp EF$ we have a parallelogram whose diagonals are perpendicular, so $MENF$ is a square. Then $|BD| = 2|EM| = 2|MF| = |AC|$. The argument can be easily inverted.
Exercise 13.11.4. We employ the notations from Exercise 13.11.3. Since $MENF$ is a parallelogram, its diagonals bisect. Let $L = MN \cap EF$. It

suffices to show that $|LA| = |LB| = |LC| = |LD|$. Since the tetrahedron is equifacial, its bimedians are perpendicular on the sides they bisect, so $EF \perp AB$. Then ΔLAB is isosceles, since LE is both median and altitude. Therefore, $|LA| = |LB|$. Considering the other cases we obtain the desired identity. Therefore, the sphere is centered at L. We still need to show that L is the centroid, $L = G$. For this it suffices to show

$$\overrightarrow{LA} + \overrightarrow{LB} + \overrightarrow{LC} + \overrightarrow{LD} = 0.$$

We have $\overrightarrow{LA} + \overrightarrow{LB} = 2\overrightarrow{LE}$, since E is midpoint. Also, $\overrightarrow{LC} + \overrightarrow{LD} = 2\overrightarrow{LF}$, since F is midpoint. Using $\overrightarrow{LE} + \overrightarrow{LF} = 0$ (since L is the midpoint of EF), we obtain the desired vectorial identity.

Exercise 13.11.7. Apply Corollary 13.11.6 and Exercise 13.10.4.

Bibliographical Remarks

The preparation of this book involved extensive use of materials written in Romanian, many of which have never been translated into English. These Romanian sources provided a unique perspective, contributing to the originality of the work. Additionally, this book was influenced by French literature, a significant source for modern classical geometry, as well as Russian literature. Although a relatively small number of English books were consulted, the book presents original results derived from processing information from these diverse sources. The author's contribution also includes the exposition and presentation of traditional classical ideas.

For the section on the foundations of geometry, key references include works such as [35], [37], [86], [73], [80], [76], [41], [25], [58], [31], [2], [3], and [10]. The trigonometric properties of angles were referenced from works by [42], [83] and [22]. For the section on triangle geometry, the book consulted works by [48], [4], [21], [55], [83], [69], [12], [79], and [7]. The quadrilateral geometry part was based on the works of [14], [79], [83], and [23].

The chapter on complex numbers in geometry utilized resources from [30], [54], [83], [11], [51], and [60]. The discussion on vectorial geometry was derived from [34], [47], [28], [82], [57], [40], [26], [61], [6], and [16]. The section on the solid geometry of tetrahedrons was inspired by [13], [8], [18], and [63], while the baricentric coordinates were inspired by [17].

Further recommended readings may refer to works by [37], [38], [67], [68], [27], [19], [24], [59], [33], [70], [45], [2], and [15].

The role of computer exploration in geometry is acknowledged, with references to works by [1], [39], and [72], GEOGEBRA software was used for the verification of geometric properties.

This work has been indirectly influenced by the French school of geometry, as seen in works by [75], [74], [29], [66], and [49].

Additionally, the Romanian school's influence is evident through works by [85], [84], [64], [62], [32], and [5].

Books dedicated to beginners, such as those by [81], [77], [44], [71], [43], [9], [52], [53], [36], [78], [20], [56], [46], and [50], were also consulted.

Bibliography

[1] H. Abelson and A. diSessa. *Turtle Geometry.* MIT Press, Cambridge, 1986.

[2] C. F. Adler. *Modern Geometry An Integrated First Course.* McGraw-Hill, New York, 1958.

[3] C. F. Adler. *A New Look at Geometry.* John Day, New York, 1966.

[4] N. Altshiller-Court. *College Geometry, an Introduction to the Modern Geometry of the Triangle and the Circle.* Barnes and Noble Inc., New York, 1952.

[5] C. Avadanei, C. Bors, N. Avadanei, and C. Ciurea. *De la matematica elementara spre matematica superioara.* Ed. Academiei RSR, Bucuresti, 1987.

[6] T. Banckoff and J. Wermer. *Linear Algebra through Geometry.* Springer-Verlag, New York, 1983.

[7] A. E. Beju and I. Beju. *Compendiu de matematica.* Editura stiintifica si enciclopedica, 1983.

[8] G. Biggiogero. *La geometria del tetraedro.* in L. Benedetti, Fondamenti di geometria, Hoepli, Milano, Vol. 2, 1942, pp. 215–253, 1964.

[9] G. D. Birkhoff and R. Beatley. *Basic Geometry.* Chelsea Publishing Company, New York, 1959.

[10] L. M. Blumenthal. *A Modern View of Geometry.* Freeman and Comp., San Francisco, 1961.

[11] C. Bors. *Numere complexe.* Editura tehnica, Bucuresti, 1962.

[12] D. Branzei. *Geometrie circumstantiala.* Junimea, Iasi, 1983.

[13] D. Branzei, S. Anita, and C. Cocea. *Planul si spatiul euclidian.* Ed. Academiei RSR, Bucuresti, 1986.

[14] D. Branzei, S. Anita, E. Onofras, and G. Isvoranu. *Bazele rationamentului geometric.* Ed. Academiei RSR, Bucuresti, 1983.

[15] L. Carlitz. *Some inequalities for the traingle.* In: Publ. Elektr. Fak. Univ. Beog. Ser. Mat-Fiz. nr. 357–380, pp. 7–10, 2007.

[16] J. S. R. Chisholm. *Vectors in three dimensional space.* Cambridge University Press, 1978.

[17] C. Cosnita. *Coordonate baricentrice.* In Numerus, Vol. 7, pp. 202–210, 1941.

[18] P. Couderc and A. Balliccioni. *Premier livre du tetraedre.* Gauthier-Villars, Paris, 1935.

[19] R. Courant and H. Robbins. *What is Mathematics?* Oxford University Press, New York, 1941.

[20] H. Coxeter. *Introduction to Geometry.* Wiley and Sons, New York-London, 1961.

[21] D. R. Davis. *Modern College Geometry.* Addison-Wesley, Cambridge, 1949.

[22] I. C. Draghicescu and V. Masgras. *Probleme de geometrie.* Editura tehnica, 1987.

[23] L. Duican and I. Duican. *Transformari geometrice.* Editura stiintifica si enciclopedica, Bucuresti, 1987.

[24] H. Eves. *A Survey of Geometry.* Allyn and Bacon, Inc., Boston, 1972.

[25] H. G. Forder. *Fundamentele geometriei euclidiene.* Editura stiintifica, Bucuresti, 1970.

[26] G. Girard and C. Thierce. *Alef zero Geometrie, Vol. I–IV.* Editura didactica si pedagogica, Bucuresti, 1973.

[27] M. J. Greenberg. *Euclidean and Non-Euclidean Geometries, 3rd ed.* Freeman and Comp., New York, 1970.

[28] M. Gutmann. *Probleme de calcul vectorial.* Editura tehnica, Bucuresti, 1961.

[29] J. Hadamard. *Lecons de geometrie elementaire.* A. Colin, 1931–1932.

[30] L-S. Hahn. *Complex Numbers and Geometry.* AMS/MAA Textbooks Vol. 52, 1994.

[31] A. Haimovici. *Elemente de geometrie a planului.* Editura didactica si pedagogica, Bucuresti, 1968.

[32] A. Haimovici. *Lectii de geometrie elementara.* Univ. A.I. Cuza, Iasi, 1975.

[33] D. W. Hall and S. Szabo. *Plane Geometry An Approach through Isometries.* Pretice-Hall, Inc., Englewood Cliffs, 1971.

[34] M. Hausner. *A Vector Space Approach to Geometry.* Dover, New York, 1965.

[35] Sir T. L. Heath. *The thirteen books of Euclid's Elements, Vols. 1, 2, 3.* Dover, 1956.

[36] D. W. Henderson. *Experiencing Geometry on plane and sphere.* Prentice Hall, 1996.

[37] D. Hilbert. *Grundlagen der Geometrie.* Teubner, Leipzig, 1899, 1909.

[38] D. Hilbert and S. Cohn-Vossen. *Geometry and the Imagination.* Chelsea Publishing Company, New York, 1952.

[39] M. Hvidsten. *Geometry with Geometry Explorer.* McGraw Hill, 2006.

[40] P. P. Teodorescu, I. Beju, and E. Soos. *Tehnici de calcul vectorial cu aplicatii.* Editura tehnica, Bucuresti, 1976.

[41] A. M. Iaglom and I. M. Iaglom. *Probleme neelementare tratate elementar.* Editura tehnica, Bucuresti, 1962.

[42] S. Ianus, N. Soare, L. Niculescu, S. Dragomir, and M. Tenea. *Probleme de geometrie si de trigonometrie.* Editura didactica si pedagogica, Bucuresti, 1983.

[43] D. C. Kay. *College Geometry.* Holt, Rinehart and Winston, Inc., New York, 1969.

[44] M. L. Keedy and Ch. W. Nelson. *Geometry A Modern Introduction.* Addison Wesley, Massachusetts, 1965.

[45] F. Klein. *Famous Problems of Elementary Geometry.* Dover, New York, 1956.

[46] A. Koeberlein. *Elementary Geometry for College Students, 4th ed.* Houghton Mifflin Company, Boton, 2007.

[47] I. E. Ladon. *Bazele calculului vectorial.* Editura de stat, Bucuresti, 1950.

[48] Tr. Lalescu. *La géométrie du triangle.* Paris., Vuibert, 1937.

[49] E. Lemoine. *Les lieux geometriques en mathematiques speciales.* University of Michigan Library, 2010.

[50] M. L. Lial, B. A. Brown, A. R. Steffensen, and L. M. Johnson. *Essentials of Geometry for College Students, 2nd ed.* Pearson Addison-Wesley, 2004.

[51] A. I. Markusevici. *Numere complexe si reprezentari conforme.* Editura tehnica, Bucuresti, 1961.

[52] N. N. Mihaileanu. *Elemente de geometrie neeuclideana.* Editura Academiei, Bucuresti, 1954.

[53] N. N. Mihaileanu. *Complemente de geometrie sintetica.* Editura didactica si pedagogica, Bucuresti, 1965.

[54] N. N. Mihaileanu. *Utilizarea numerelor complexe in geometrie.* Editura tehnica, Bucuresti, 1968.

[55] C. Mihalescu. *Geometria elementelor remarcabile.* Editura tehnica, Bucuresti, 1957.

[56] R. Miron. *Geometrie elementara.* Editura didactica si pedagogica, Bucuresti, 1968.

[57] R. Miron. *Introducere vectoriala in geometria analitica plana.* Editura didactica si pedagogica, Bucuresti, 1970.

[58] R. Miron and D. Branzei. *Fundamentele arithmeticii si geometriei.* Ed. Academiei RSR, Bucuresti, 1983.

[59] E. Moise. *Elementary Geometry from an Advanced Standpoint.* Addison-Wesley, Inc., Massachussetts, 1963.

[60] M. O. Moscovici. *O interpretare geometrica a numerelor complexe.* Editura tehnica, Bucuresti, 1956.

[61] J. A. Murtha and E. R. Willard. *Linear Algebra and Geometry.* Reinhart and Winston Inc., New York, 1966.

[62] A. Myller. *Geometrie analitica.* Editura didactica si pedagogica, Bucuresti, 1972.

[63] B. Niewenglowski and L. Gerard. *Cours de geometrie elementaire (2 vols.).* G. Carre, Paris, 1898–1899.

[64] V. Oproiu. *Curs de geometrie.* Univ. A. I. Cuza, Iasi, 1981.

[65] L. Panaitopol. *article in.* Gazeta Matematica, no. 11, 1980.

[66] G. Papelier. *Exercices de géometrie moderne.* Paris, Vuibert, 1927.

[67] D. Pedoe. *A course of Geometry.* Univ. Press, Cambridge, 1970.

[68] D. Pedoe. *Geometry A Comprehensive Course.* Dover Publications, New York, 1970.

[69] D. Pompeiu. *Opera matematica.* Editura Academiei, Bucuresti, 1959.

[70] A. S. Posamentier and Ch. T. Salkind. *Challenging Problems in Geometry.* MacMillan Company, London, 1970.

[71] W. Prenowitz and M. Jordan. *Basic Concepts of Geometry.* Ardsley House Publishers, 1989.

[72] B. E. Reynolds and W. E. Fenton. *College Geometry using the Geometer's Sketchpad.* Key College Publishing, 2006.

[73] G. B. Robinson. *The Foundations of Geometry.* Toronto Press, Toronto, 1946.

[74] E. Rouché and Ch. de Comberousse. *Elements de géometrie (2 vols.).* Paris, Gauthier-Villars, 1912.

[75] E. Rouché and Ch. de Comberousse. *Traité de géometrie (2 vols.).* Paris, Gauthier-Villars, 1922.

[76] E. Rusu. *Geometrie elementara.* Editura didactica si pedagogica, Bucuresti, 1964.

[77] J. Smart. *Introductory Geometry: An Informal Approach, 2nd ed.* Brooks/Cole Publishing Company, Monterey, 1972.

[78] D. A. Thomas. *Modern Geometry.* Thomson Learning, 2002.

[79] G. Titeica. *Probleme de geometrie.* Editura tehnica, Bucuresti, 1981.

[80] A. Tuller. *A Modern Introduction to Geometries.* D. Van Nostrand Company, Inc., Princeton, 1967.

[81] A. S. Tussy and R. D. Gustafson. *Basic Geometry for College Students, 2nd ed.* Brooks/Cole Cengage Learning, 2010.

[82] C. Udriste. *Probleme de algebra liniara, geometrie analitica si differentiala.* Editura didactica si pedagogica, Bucuresti, 1981.

[83] V. Gh. Voda. *Vraja geometriei demodate.* Editura Albatros, 1983.

[84] Gh. Vranceanu. *Geometrie elementara din punct de vedere modern.* Editura tehnica, Bucuresti, 1967.

[85] Gh. Vranceanu. *Geometrie analitica, proiectiva si differentiala.* Editura didactica si pedagogica, Bucuresti, 1967, 1974.

[86] R. L. Wilder. *Introduction to the Foundations of Mathematics.* John Wiley and Sons, 1952.

Index

www.ingramcontent.com/pod-product-compliance
Lightning Source LLC
Chambersburg PA
CBHW050446051126
54379CB00004B/115
* 9 7 8 9 8 1 9 8 0 6 1 6 4 *